COMPLETE SOLUTIONS FOR

CHEMICAL PRINCIPLES
Second Edition

Kenneth C. Brooks
New Mexico State University, Las Cruces

Thomas J. Hummel
University of Illinois, Urbana-Champaign

Steven S. Zumdahl
University of Illinois, Urbana-Champaign

D.C. Heath and Company
Lexington, Massachusetts   Toronto

Address editorial correspondence to:

D. C. Heath and Company
125 Spring Street
Lexington, MA 02173

Published simultaneously in Canada.

Printed in the United States of America.

International Standard Book Number: 0-669-39323-1

10 9 8 7 6 5 4 3 2 1

TABLE OF CONTENTS

Page

CHAPTER TWO

ATOMS, MOLECULES AND IONS

Development of the Atomic Theory

1. $\dfrac{1.188}{1.188} = 1.000; \quad \dfrac{2.375}{1.188} = 1.999; \quad \dfrac{3.563}{1.188} = 2.999$

   The masses of fluorine are simple ratios of whole numbers to each other, 1:2:3.

2. From Avogadro's hypothesis, volume ratios are equal to molecule ratios at constant temperature and pressure. Therefore, $Cl_2 + 3 F_2 \rightarrow 2 X$. Two molecules of X contain 6 atoms of F and two atoms of Cl. Therefore, the formula of X is ClF_3.

3. a. The composition of a substance depends on the numbers of atoms of each element making up the compound (i.e. the formula of the compound) and not on the composition of the mixture from which it was formed.

   b. $H_2 + Cl_2 \rightarrow 2$ HCl. The volume of HCl produced is twice the volume of H_2 (or Cl_2) used.

4. To get the atomic mass of H to be 1.00, we divide the mass that reacts with 1.00 g of oxygen by 0.126. $\dfrac{0.126}{0.126} = 1.00$

   To get Na, Mg, and O on the same scale, we do the same division.

   Na: $\dfrac{2.875}{0.126} = 22.8; \quad$ Mg: $\dfrac{1.500}{0.126} = 11.9; \quad$ O: $\dfrac{1.00}{0.1260} = 7.94$

	H	O	Na	Mg
Scale	1.00	7.94	22.8	11.9
Accepted Value	1.01 (1.0079)	15.999	22.99	24.31

   The atomic masses of O and Mg are incorrect. The atomic masses of H and Na are close. Something must be wrong about the assumed formulas of the compounds. It turns out the correct formulas are H_2O, Na_2O, and MgO. The smaller discrepancies result from the error in the atomic mass of H.

1

5.   a.   Atoms have mass and are neither destroyed nor created by chemical reactions. Therefore, mass is neither created nor destroyed by chemical reactions. Mass is conserved.

      b.   The composition of a substance depends on the number and kinds of atoms that form it.

      c.   Compounds of the same elements differ only in numbers of atoms of the elements forming them, i.e., NO, N_2O, NO_2.

6.   Some elements exist as molecular substances. That is, hydrogen normally exists as H_2 molecules, not single hydrogen atoms. The same is true for N_2, O_2, F_2, Cl_2, etc.

7.   Yes, many questions can be raised from Dalton's theory. For example: What are the masses of atoms? Are the atoms really structureless? What forces hold atoms together in compounds?, etc.

8.   We now know that some atoms of the same element have different masses. We have had to include the existence of isotopes in our models.

The Nature of the Atom

9.   Deflection of cathode rays by magnetic and electric fields led to the conclusion that they were negatively charged. The ray was produced at the negative electrode and repelled by the negative pole of the applied electric field.

10. β particles are electrons. A cathode ray is a stream of electrons (β particles).

11.   Density of hydrogen nucleus:

$$V_{nucleus} = \frac{4}{3}\pi r^3 = \frac{4}{3}(3.14)(5 \times 10^{-14} \text{ cm})^3 = 5 \times 10^{-40} \text{ cm}^3$$

$$d = \frac{1.67 \times 10^{-24} \text{ g}}{5 \times 10^{-40} \text{ cm}^3} = 3 \times 10^{15} \text{ g/cm}^3$$

Density of H-atom:

$$V_{atom} = \frac{4}{3}(3.14)(1 \times 10^{-8} \text{ cm})^3 = 4 \times 10^{-24} \text{ cm}^3$$

$$d = \frac{1.67 \times 10^{-24} + 9 \times 10^{-28} \text{ g}}{4 \times 10^{-24} \text{ cm}^3} = 0.4 \text{ g/cm}^3$$

12.   Divide all charges by 6.40×10^{-13}.

$$\frac{2.56 \times 10^{-12}}{6.40 \times 10^{-13}} = 4.00 \qquad \frac{7.68}{0.640} = 12.00 \qquad \frac{3.84}{0.640} = 6.00$$

Since all charges are whole number multiples of 6.40×10^{-13} zirkombs then the charge on one electron could be 6.40×10^{-13} zirkombs. However, 6.40×10^{-13} zirkombs could be the charge of two electrons (or three electrons, etc.). All one can conclude is that the charge of an electron is 6.40×10^{-13} zirkombs or an integer fraction of 6.40×10^{-13}.

13.   J. J. Thomson discovered electrons. Antoine Henri Becquerel discovered radioactivity. Lord Rutherford proposed the nuclear model of the atom. Dalton's original model proposed that atoms were indivisible particles (that is, atoms had no internal structure). Thomson and Becquerel discovered sub-atomic particles and Rutherford's model attempted to describe the internal structure of the atom.

Elements and the Periodic Table

14.   The atomic number of an element is equal to the number of protons in the nucleus of an atom of that element. The mass number is the sum of the number of protons plus neutrons in the nucleus. The atomic mass is the actual mass of a particular isotope (including electrons). As we will see in chapter three, the average mass of an atom is taken from a measurement made on a large number of atoms. The average atomic mass value is listed on the periodic table.

15.   promethium (Pm) and technetium (Tc) Elements with atomic masses in parentheses have no stable isotopes.

16.   a. Eight, Li to Ne   b. Eight, Na to Ar

       c. Eighteen, K to Kr   d. Four, Fe, Ru, Os, and Uno (#108)

       e. Five, O, S, Se, Te, Po   f. Three, Ni, Pd, Pt

17.   a. P   b. I   c. K   d. Yb

18.   a. $^{24}_{12}$Mg: 12 protons, 12 neutrons, 12 electrons

       b. $^{24}_{12}$Mg^{2+}: 12 p, 12 n, 10 e   c. $^{59}_{27}$Co^{2+}: 27 p, 32 n, 25 e

       d. $^{59}_{27}$Co^{3+}: 27 p, 32 n, 24 e   e. $^{59}_{27}$Co: 27 p, 32 n, 27 e

       f. $^{79}_{34}$Se: 34 p, 45 n, 34 e   g. $^{79}_{34}$Se^{2-}: 34 p, 45 n, 36 e

       h. $^{63}_{28}$Ni: 28 p, 35 n, 28 e   i. $^{59}_{28}$Ni^{2+}: 28 p, 31 n, 26 e

19.

Symbol	Number of protons	Number of neutrons	Number of electrons	Net charge
$^{75}_{33}As^{3+}$	33	42	30	3+
$^{32}_{16}S^{2-}$	16	16	18	2-
$^{204}_{81}Tl^{+}$	81	123	80	1+
$^{197}_{79}Au$	79	118	79	0
$^{197}_{79}Au^{3+}$	79	118	76	3+

20. a. Lose one e^- to form Na^+.

b. Lose two e^- to form Sr^{2+}.

c. Lose two e^- to form Ba^{2+}.

d. Gain one e^- to form I^-.

e. Lose three e^- to form Al^{3+}.

f. Gain two e^- to form S^{2-}.

g. Lose three e^- to form B^{3+}.

h. Lose one e^- to form Cs^+.

i. Gain two e^- to form Se^{2-}.

21.   Carbon is a nonmetal. Silicon and germanium are metalloids. Tin and lead are metals. Thus, metallic character increases as one goes down a family in the periodic table.

22.   The metallic character decreases from left to right.

Nomenclature

23. a. sodium perchlorate

b. magnesium phosphate

c. aluminum sulfate

d. sulfur difluoride

e. sulfur hexafluoride

f. sodium hydrogen phosphate

g. sodium dihydrogen phosphate   h. lithium nitride

i. sodium hydroxide   j. magnesium hydroxide

k. aluminum hydroxide   l. germanium(IV) oxide

24.   a. sodium bromide   b. barium bromide

c. rubidium chloride   d. cesium chloride

e. aluminum fluoride

f. hydrogen bromide (or hydrobromic acid if dissolved in water)

g. nitrogen monoxide or nitric oxide (common name)

h. nitrogen dioxide   i. dinitrogen tetroxide

j. nitrogen trifluoride   k. dinitrogen tetrafluoride

l.   iron(II) sulfate

25.   a. nitric acid   b. nitrous acid

c. phosphoric acid   d. phosphorous acid

e. sodium hydrogen sulfate or sodium bisulfate (common name)

f. calcium hydrogen sulfite or calcium bisulfite (common name)

g. sodium bromate   h. iron(III) periodate

i. ruthenium(III) nitrate   j. vanadium(V) oxide

k.   platinum(IV) chloride   l. platinum(II) chloride

26.   a. SO_2 b. SO_3 c. Na_2SO_3 d. $KHSO_3$

e. Li_3N f. $Cr_2(CO_3)_3$ g. $Cr(C_2H_3O_2)_2$ h. SnF_4

i. NH_4HSO_4: Composed of NH_4^+ and HSO_4^- ions.

j. $(NH_4)_2HPO_4$ k. $KClO_4$ l. NaH

27.   a. Na_2O b. Na_2O_2 c. KCN d. $Cu(NO_3)_2$

e. $SiCl_4$ f. PbO g. PbO_2 h. $CuCl$

i. GaAs: We would predict the inert stable ions to be Ga^{3+} and As^{3-}.

j. CdS ℓ   k. ZnS   l. Hg_2Cl_2: Mercury(I) exists as Hg_2^{2+}.

28.   a. $Pb(C_2H_3O_2)_2$: lead(II) acetate   b. $CuSO_4$: copper(II) sulfate

c. CaO: calcium oxide   d. $MgSO_4$: magnesium sulfate

e. $Mg(OH)_2$: magnesium hydroxide   f. $CaSO_4$: calcium sulfate

g. N_2O: dinitrogen monoxide or nitrous oxide

Additional Exercises

29.   There should be no difference. The composition of insulin from both sources will be the same and therefore, it will have the same activity regardless of the source. As a practical note, trace contaminants in the two types of insulin may be different. These trace components may be important.

30. ^{98}Tc: 43 protons and 55 neutrons; ^{99}Tc: 43 protons and 56 neutrons

Tc is in the same family as Mn. We would expect it to have similar properties.

permanganate: MnO_4^-; pertechnetate: TcO_4^-; ammonium pertechnetate: NH_4TcO_4

31.   The solid residue must have come from the flask.

32.   In the case of sulfur, SO_4^{2-} is sulfate and SO_3^{2-} is sulfite. By analogy:

SeO_4^{2-}: selenate; SeO_3^{2-}: selenite; TeO_4^{2-} tellurate; TeO_3^{2-}: tellurite

33.   The PO_4^{3-} ion is phosphate and PO_3^{3-} is phosphite. By analogy:

$Mg_3(AsO_4)_2$: magnesium arsenate; H_3AsO_4: arsenic acid; Na_3SbO_4: sodium antimonate;

Na_3AsO_3: sodium asenite; Na_2HAsO_4: sodium hydrogen arsenate

34.   If the formula was Be_2O_3, then 2 times the atomic mass of Be would combine with three times the atomic mass of oxygen, or:

$$\frac{2\,A}{3(15.999)} = \frac{0.5633}{1.000}$$ Solving, A = atomic mass Be = 13.52.

35.   a. Ba^{2+} and O^{2-}: BaO, barium oxide

b. Li^+ and H^-: LiH, lithium hydride

c. In^{3+} and F^-: InF_3, indium(III) fluoride

d. Ca^{2+} and N^{3-}: Ca_3N_2, calcium nitride

e. B^{3+} and O^{2-}; B_2O_3, diboron trioxide or boron oxide

f. OF_2, oxygen difluoride

g. Al^{3+} and H^-; AlH_3, aluminum hydride

h. $In^{3+}$and P^{3-}; InP, indium(III) phosphide

i. Mg^{2+} and F^-; MgF_2, magnesium fluoride

36.   hydrazine: 1.44×10^{-1} g H/g N

ammonia: 2.16×10^{-1} g H/g N

hydrogen azide: 2.40×10^{-2} g H/g N

Let's try all of the ratios:

$$\frac{0.216}{0.144} = 1.50 = \frac{3}{2}; \ \frac{0.144}{0.0240} = 6.00; \ \frac{0.216}{0.0240} = 9.00$$

They can all be expressed as simple whole number ratios. The g H/g N in hydrazine, ammonia, and hydrogen azide are in the ratios 6:9:1.

37.   a.   Atomic number is 36. Kr   b.   Atomic number is 52. Te

c.   Atomic number is 20. Ca   d.   Atomic number is 47. Ag

e.   Atomic number is 94. Pu

38.   Yes, 1.0 g H would react with 37.0 g ^{37}Cl and 1.0 g H would react with 35.0 g ^{35}Cl.

No, the ratio of H/Cl would always be 1 g H/37 g Cl for ^{37}Cl and 1 g H/35 g Cl for ^{35}Cl. As long as we had pure ^{35}Cl or pure ^{37}Cl, the above ratios will always hold. If we have a mixture (such as the natural abundance of chlorine), the ratio will also be constant as long as the composition of the mixture of the two isotopes does not change.

CHAPTER THREE

STOICHIOMETRY

Atomic Masses and the Mass Spectrometer

1. A = atomic mass = 0.7899(23.9850 amu) + 0.1000(24.9858 amu) + 0.1101(25.9826 amu)

    A = 18.95 amu + 2.499 amu + 2.861 amu = 24.31 amu

2. Let x = % of ^{151}Eu and y = % of ^{153}Eu, then $x + y = 100$ and $y = 100 - x$.

    $$151.96 = \frac{x(150.9196) + (100 - x)(152.9209)}{100}$$

    $15196 = 150.9196x + 15292.09 - 152.9209x$, $\quad -96 = -2.0013x$

    $x = 48\%$; 48% ^{151}Eu and 100 - 48 = 52% ^{153}Eu

3. $186.207 = 0.6260 (186.956) + 0.3740(A)$, $\quad 186.207 - 117.0 = 0.3740(A)$

    $$A = \frac{69.2}{0.3740} = 185 \text{ amu}$$

4. $A = 0.2100 (283.9 \text{ amu}) + 0.3154 (284.8 \text{ amu}) + 0.4746 (287.8 \text{ amu}) = 286.0 \text{ amu}$

5. $A = 0.0140(203.973) + 0.2410(205.9745) + 0.2210(206.9759) + 0.5240(207.9766)$

    $A = 2.86 + 49.64 + 45.74 + 109.0 = 207.2 \text{ amu}$; The element is Pb.

6. There are three peaks in the mass spectrum, each 2 mass units apart. This is consistent with two isotopes, differing in mass by two mass units. The peak at 157.84 corresponds to a Br_2 molecule composed of two atoms of the lighter isotope. This isotope has mass equal to 157.84/2 or 78.92. This corresponds to ^{79}Br. The second isotope is ^{81}Br with mass equal to 161.84/2 = 80.92. The peaks in the mass spectrum correspond to $^{79}Br_2$, $^{79}Br^{81}Br$ and $^{81}Br_2$ in order of increasing mass. The intensities of the highest and lowest mass tell us the two isotopes are present at about equal abundance. The actual abundance is 50.69% ^{79}Br and 49.31% ^{81}Br. The calculation of the abundance from the mass spectrum is beyond the scope of this text.

7.   GaAs can be either ^{69}GaAs or ^{71}GaAs. The mass spectrum for GaAs will have 2 peaks at 144
        (69 + 75) and 146 (71 + 75) with intensities in the ratio of 60:40 or 3:2.

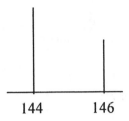

$$144 \qquad 146$$

Ga$_2$As$_2$ can be ^{69}Ga$_2$As$_2$, ^{69}Ga^{71}GaAs$_2$ or ^{71}Ga$_2$As$_2$. The mass spectrum will have 3 peaks at 288,
290, and 292 with intensities in the ratio of 36:48:16 or 9:12:4. We get this ratio from the
following probability table:

	^{69}Ga (0.6)	^{71}Ga (0.4)
^{69}Ga (0.6)	0.36	0.24
^{71}Ga (0.4)	0.24	0.16

$$288 \qquad 290 \qquad 292$$

Moles and Molar Masses

8.   a. 3.00×10^{20} molecules HF $\times \dfrac{1 \text{ mol HF}}{6.022 \times 10^{23} \text{ molecules}} \times \dfrac{20.01 \text{ g HF}}{\text{mol HF}} = 9.97 \times 10^{-3}$ g HF

     b. 3.00×10^{-3} mol HF $\times \dfrac{20.01 \text{ g HF}}{\text{mol HF}} = 6.00 \times 10^{-2}$ g HF

     c. 1.5×10^{2} mol HF $\times \dfrac{20.01 \text{ g HF}}{\text{mol HF}} = 3.0 \times 10^{3}$ g HF

     d.   1 molecule HF $\times \dfrac{1 \text{ mol HF}}{6.022 \times 10^{23} \text{ molecules HF}} \times \dfrac{20.01 \text{ g HF}}{\text{mol HF}} = 3.323 \times 10^{-23}$ g HF

     e. 2.00×10^{-15} mol HF $\times \dfrac{20.01 \text{ g HF}}{\text{mol HF}} = 4.00 \times 10^{-14}$ g HF $= 40.0$ fg HF

     f.   18.0 pmol HF $\times \dfrac{1 \text{ mol HF}}{10^{12} \text{ pmol}} \times \dfrac{20.01 \text{ g HF}}{\text{mol HF}} = 3.60 \times 10^{-10}$ g HF $= 360.$ pg HF

g. $5.0 \text{ nmol HF} \times \dfrac{1 \text{ mol HF}}{10^9 \text{ nmol}} \times \dfrac{20.01 \text{ g HF}}{\text{mol HF}} = 1.0 \times 10^{-7} \text{ g HF} = 1.0 \times 10^2 \text{ ng}$

9.   a. $100 \text{ molecules H}_2\text{O} \times \dfrac{1 \text{ mol H}_2\text{O}}{6.022 \times 10^{23} \text{ molecules H}_2\text{O}} = 1.661 \times 10^{-22} \text{ mol H}_2\text{O}$

b. $100.0 \text{ g H}_2\text{O} \times \dfrac{1 \text{ mol H}_2\text{O}}{18.015 \text{ g H}_2\text{O}} = 5.551 \text{ mol H}_2\text{O}$

c. $500 \text{ atoms Fe} \times \dfrac{1 \text{ mol Fe}}{6.022 \times 10^{23} \text{ atoms}} = 8.303 \times 10^{-22} \text{ mol Fe}$

d. $500.0 \text{ g Fe} \times \dfrac{1 \text{ mol Fe}}{55.85 \text{ g Fe}} = 8.953 \text{ mol Fe}$

e. $150 \text{ molecules N}_2 \times \dfrac{1 \text{ mol N}_2}{6.022 \times 10^{23} \text{ molecules N}_2} = 2.491 \times 10^{-22} \text{ mol N}_2$

f. $150.0 \text{ g Fe}_2\text{O}_3 \times \dfrac{1 \text{ mol}}{159.70 \text{ g}} = 0.9393 \text{ mol Fe}_2\text{O}_3$

g. $10.0 \text{ mg NO}_2 \times \dfrac{1 \text{ g}}{10^3 \text{ mg}} \times \dfrac{1 \text{ mol}}{46.01 \text{ g}} = 2.17 \times 10^{-4} \text{ mol NO}_2$

h. $1.0 \text{ fmol NO}_2 \times \dfrac{1 \text{ mol}}{10^{15} \text{ fmol}} = 1.0 \times 10^{-15} \text{ mol NO}_2$

i. $1.5 \times 10^{16} \text{ molecules BF}_3 \times \dfrac{1 \text{ mol}}{6.02 \times 10^{23} \text{ molecules}} = 2.5 \times 10^{-8} \text{ mol BF}_3$

j. $2.6 \text{ mg BF}_3 \times \dfrac{1 \text{ g}}{10^3 \text{ mg}} \times \dfrac{1 \text{ mol}}{67.8 \text{ g}} = 3.8 \times 10^{-5} \text{ mol BF}_3$

10.   a. $14 \text{ mol C} \left(\dfrac{12.011 \text{ g}}{\text{mol C}} \right) + 18 \text{ mol H} \left(\dfrac{1.0079 \text{ g}}{\text{mol H}} \right) + 2 \text{ mol N} \left(\dfrac{14.007 \text{ g}}{\text{mol N}} \right)$

$+ 5 \text{ mol O} \left(\dfrac{15.999 \text{ g}}{\text{mol O}} \right) = 294.305 \text{ g/mol}$

b. $10.0 \text{ g aspartame} \times \dfrac{1 \text{ mol}}{294.3 \text{ g}} = 3.40 \times 10^{-2} \text{ mol}$

c. $1.56 \text{ mol} \times \dfrac{294.3 \text{ g}}{\text{mol}} = 459 \text{ g}$

d. $5.0 \text{ mg} \times \dfrac{1 \text{ g}}{1000 \text{ mg}} \times \dfrac{1 \text{ mol}}{294.3 \text{ g}} \times \dfrac{6.02 \times 10^{23} \text{ molecules}}{\text{mol}} = 1.0 \times 10^{19} \text{ molecules}$

e. $1.2 \text{ g aspartame} \times \dfrac{1 \text{ mol aspartame}}{294.3 \text{ g aspartame}} \times \dfrac{2 \text{ mol N}}{\text{mol aspartame}} \times \dfrac{6.02 \times 10^{23} \text{ atoms N}}{\text{mol N}}$

$$= 4.9 \times 10^{21} \text{ atoms of nitrogen}$$

f. $1.0 \times 10^9 \text{ molecules} \times \dfrac{1 \text{ mol}}{6.02 \times 10^{23} \text{ molecules}} \times \dfrac{294.3 \text{ g}}{\text{mol}} = 4.9 \times 10^{-13} \text{ g or 490 fg}$

g. $1 \text{ molecule aspartame} \times \dfrac{1 \text{ mol}}{6.022 \times 10^{23} \text{ molecules}} \times \dfrac{294.305 \text{ g}}{\text{mol}} = 4.887 \times 10^{-22} \text{ g}$

11. a. Molar mass $= 21(12.011) + 30(1.0079) + 5(15.999) = 362.463 \text{ g/mol}$

b. $275 \text{ mg} \times \dfrac{1 \text{ g}}{1000 \text{ mg}} \times \dfrac{1 \text{ mol}}{362.46 \text{ g}} = 7.59 \times 10^{-4} \text{ mol}$

c. $0.600 \text{ mol} \times \dfrac{362.46 \text{ g}}{\text{mol}} = 217 \text{ g}$

d. $1.00 \times 10^{-12} \text{ g} \times \dfrac{1 \text{ mol C}_{21}\text{H}_{30}\text{O}_5}{362.46 \text{ g}} \times \dfrac{30 \text{ mol H}}{\text{mol C}_{21}\text{H}_{30}\text{O}_5} \times \dfrac{6.022 \times 10^{23} \text{ atoms H}}{\text{mol H}}$

$$= 4.98 \times 10^{10} \text{ atoms H}$$

e. $1.00 \times 10^9 \text{ molecules} \times \dfrac{1 \text{ mol}}{6.022 \times 10^{23} \text{ molecules}} \times \dfrac{362.46 \text{ g}}{\text{mol}} = 6.02 \times 10^{-13} \text{ g}$

f. $1 \text{ molecule} \times \dfrac{1 \text{ mol}}{6.022 \times 10^{23} \text{ molecules}} \times \dfrac{362.46 \text{ g}}{\text{mol}} = 6.019 \times 10^{-22} \text{ g}$

12. $1.0 \text{ lb flour} \times \dfrac{454 \text{ g flour}}{\text{lb flour}} \times \dfrac{30.0 \times 10^{-9} \text{ g EDB}}{\text{g flour}} \times \dfrac{1 \text{ mol EDB}}{187.9 \text{ g}} \times \dfrac{6.02 \times 10^{23} \text{ molecules}}{\text{mol EDB}}$

$$= 4.4 \times 10^{16} \text{ molecules of EDB}$$

Percent Composition

13. In 1 mole of $YBa_2Cu_3O_7$ there are 1 mole of Y, 2 moles of Ba, 3 moles of Cu, and 7 moles of O.

$$\text{Molar mass} = 1 \text{ mol Y} \left(\dfrac{88.91 \text{ g Y}}{\text{mol Y}} \right) + 2 \text{ mol Ba} \left(\dfrac{137.3 \text{ g Ba}}{\text{mol Ba}} \right)$$

$$+ 3 \text{ mol Cu} \left(\dfrac{63.55 \text{ g Cu}}{\text{mol Cu}} \right) + 7 \text{ mol O} \left(\dfrac{16.00 \text{ g O}}{\text{mol O}} \right)$$

Molar mass = 88.91 + 274.6 + 190.65 + 112.00 = 666.2 g/mol

$$\% \text{ Y} = \frac{88.91 \text{ g}}{666.2 \text{ g}} \times 100 = 13.35\% \text{ Y}; \quad \% \text{ Ba} = \frac{274.6 \text{ g}}{666.2 \text{ g}} \times 100 = 41.22\% \text{ Ba}$$

$$\% \text{ Cu} = \frac{190.65 \text{ g}}{666.2 \text{ g}} \times 100 = 28.62\% \text{ Cu}; \quad \% \text{ O} = \frac{112.0 \text{ g}}{666.2 \text{ g}} \times 100 = 16.81\% \text{ O}$$

or % O = 100.00 - (13.35 + 41.22 + 28.62) = 100.00 - (83.19) = 16.81% O

14.   a. PF_3: molar mass = 31.0 + 3(19.0) = 88.0 g/mol

$$\text{mass } \% \text{ P} = \frac{31.0 \text{ g}}{88.0 \text{ g}} \times 100 = 35.2\%$$

b. P_4O_{10}: molar mass = 4(31.0) + 10(16.0) = 284.0 g/mol

$$\text{mass } \% \text{ P} = \frac{124. \text{ g}}{284. \text{ g}} \times 100 = 43.7\%$$

c. $(NPCl_2)_3$: molar mass = 3(14.0) + 3(31.0) + 6(35.45) = 347.7 g/mol

$$\text{mass } \% \text{ P} = \frac{93.0 \text{ g}}{347.7 \text{ g}} \times 100 = 26.7\%$$

d.   InP: molar mass = 114.8 + 31.0 = 145.8 g/mol

$$\text{mass } \% \text{ P} = \frac{31.0 \text{ g}}{145.8 \text{ g}} \times 100 = 21.3\%$$

The order from lowest to highest percentage of phosphorus is:

$$\text{InP} < (NPCl_2)_3 < PF_3 < P_4O_{10}$$

15.   a.   NO; $\% \text{ N} = \dfrac{14.007 \text{ g N}}{30.006 \text{ g NO}} \times 100 = 46.681\% \text{ N}$

b. NO_2; $\% \text{ N} = \dfrac{14.007 \text{ g N}}{46.005 \text{ g NO}_2} \times 100 = 30.447\% \text{ N}$

c. N_2O_4; $\% \text{ N} = \dfrac{28.014 \text{ g N}}{92.010 \text{ g N}_2O_4} \times 100 = 30.447 \% \text{ N}$

d. N_2O; $\% \text{ N} = \dfrac{28.014 \text{ g N}}{44.013 \text{ g N}_2O} \times 100 = 63.649\% \text{ N}$

16. $\% \text{ Co} = 4.34 = \dfrac{58.95 \text{ g Co}}{\text{molar mass}} \times 100$, molar mass = 1360 g/mol

17.   Assuming 100.0 g hemoglobin (Hb):

$$\frac{100.0 \text{ g Hb}}{0.342 \text{ g Fe}} \times \frac{55.85 \text{ g Fe}}{\text{mol Fe}} \times \frac{4 \text{ mol Fe}}{\text{mol Hb}} = \frac{65,300 \text{ g Hb}}{\text{mol Hb}}$$

18.   If we have 100.0 g of Portland cement, we have 50. g Ca_3SiO_5, 25 g Ca_2SiO_4, 12 g $Ca_3Al_2O_6$, 8.0 g Ca_2AlFeO_5, and 3.5 g $CaSO_4 \cdot 2H_2O$.

Percent Ca:

$$50. \text{ g Ca}_3\text{SiO}_5 \times \frac{1 \text{ mol Ca}_3\text{SiO}_5}{228.33 \text{ g Ca}_3\text{SiO}_5} \times \frac{3 \text{ mol Ca}}{1 \text{ mol Ca}_3\text{SiO}_5} \times \frac{40.08 \text{ g Ca}}{1 \text{ mol Ca}} = 26 \text{ g Ca}$$

$$25 \text{ g Ca}_2\text{SiO}_4 \times \frac{80.16 \text{ g Ca}}{172.25 \text{ g Ca}_2\text{SiO}_4} = 12 \text{ g Ca}$$

$$12 \text{ g Ca}_3\text{Al}_2\text{O}_6 \times \frac{120.24 \text{ g Ca}}{270.20 \text{ g Ca}_3\text{Al}_2\text{O}_6} = 5.3 \text{ g Ca}$$

$$8.0 \text{ g Ca}_2\text{AlFeO}_5 \times \frac{80.16 \text{ g Ca}}{242.99 \text{ g Ca}_2\text{AlFeO}_5} = 2.6 \text{ g Ca}$$

$$3.5 \text{ g CaSO}_4 \cdot 2\text{H}_2\text{O} \times \frac{40.08 \text{ g Ca}}{172.18 \text{ g CaSO}_4 \cdot 2\text{H}_2\text{O}} = 0.81 \text{ g Ca}$$

Mass of Ca = 26 + 12 + 5.3 + 2.6 + 0.81 = 47 g

$$\% \text{ Ca} = \frac{47 \text{ g Ca}}{100.0 \text{ g cement}} \times 100 = 47\% \text{ Ca}$$

Percent Al:

$$12 \text{ g Ca}_3\text{Al}_2\text{O}_6 \times \frac{53.96 \text{ g Al}}{270.20 \text{ g Ca}_3\text{Al}_2\text{O}_6} = 2.4 \text{ g Al}$$

$$8.0 \text{ g Ca}_2\text{AlFeO}_5 \times \frac{26.98 \text{ g Al}}{242.99 \text{ g Ca}_2\text{AlFeO}_5} = 0.89 \text{ g Al}$$

$$\% \text{ Al} = \frac{2.4 + 0.89}{100.0} \times 100 = 3.3\% \text{ Al}$$

Percent Fe:

$$8.0 \text{ g Ca}_2\text{AlFeO}_5 \times \frac{55.85 \text{ g Fe}}{242.99 \text{ g Ca}_2\text{AlFeO}_5} = 1.8 \text{ g Fe}$$

$$\% \text{ Fe} = \frac{1.8}{100.0} \times 100 = 1.8\% \text{ Fe}$$

Empirical and Molecular Formulas

19.   Out of 100.00 g of compound, there are:

$$48.64 \text{ g C} \times \frac{1 \text{ mol C}}{12.011 \text{ g C}} = 4.050 \text{ mol C}$$

$$8.16 \text{ g H} \times \frac{1 \text{ mol H}}{1.008 \text{ g H}} = 8.10 \text{ mol H}$$

% O = 100.00 - 48.64 - 8.16 = 43.20%

$$43.20 \text{ g O} \times \frac{1 \text{ mol O}}{15.999 \text{ g O}} = 2.700 \text{ mol O}$$

Dividing each mole value by the smallest number:

$$\frac{4.050}{2.700} = 1.500 \qquad \frac{8.10}{2.700} = 3.00 \qquad \frac{2.700}{2.700} = 1.000$$

Since whole number ratio is required, C:H:O ratio is 1.5:3:1 or 3:6:2.

So, empirical formula is $C_3H_6O_2$.

20.   First, get composition in mass percent. We assume all of the carbon in 0.213 g CO_2 came from the 0.157 g of the compound and that all of the hydrogen in the 0.0310 g H_2O came from the 0.157 g of the compound.

$$0.213 \text{ g CO}_2 \times \frac{12.01 \text{ g C}}{44.01 \text{ g CO}_2} = 0.0581 \text{ g C}$$

$$\% \text{ C} = \frac{0.0581 \text{ g C}}{0.157 \text{ g compound}} \times 100 = 37.0\% \text{ C}$$

$$0.0310 \text{ g H}_2\text{O} \times \frac{2.016 \text{ g H}}{18.02 \text{ g H}_2\text{O}} = 3.47 \times 10^{-3} \text{ g H}$$

$$\% \text{ H} = \frac{3.47 \times 10^{-3} \text{ g}}{0.157 \text{ g}} = 2.21\% \text{ H}$$

We get % N from the second experiment:

$$0.0230 \text{ g NH}_3 \times \frac{14.01 \text{ g N}}{17.03 \text{ g NH}_3} = 1.89 \times 10^{-2} \text{ g N}$$

$$\% \text{ N} = \frac{1.89 \times 10^{-2} \text{ g}}{0.103 \text{ g}} \times 100 = 18.3\% \text{ N}$$

The mass percent of oxygen is obtained by difference:

% O = 100.0 - (37.0 + 2.21 + 18.3) = 42.5%

So, out of 100.0 g of compound, there are:

$$37.0 \text{ g C} \times \frac{1 \text{ mol}}{12.01 \text{ g}} = 3.08 \text{ mol C}$$

$$2.21 \text{ g H} \times \frac{1 \text{ mol H}}{1.008 \text{ g H}} = 2.19 \text{ mol H}$$

$$18.3 \text{ g N} \times \frac{1 \text{ mol N}}{14.01 \text{ g N}} = 1.31 \text{ mol N}$$

$$42.5 \text{ g O} \times \frac{1 \text{ mol O}}{16.00 \text{ g O}} = 2.66 \text{ mol O}$$

Lastly, and often the hardest part, we need to find simple whole number ratios. We do this by trial and error.

$$\frac{2.19}{1.31} = 1.67 = 1\frac{2}{3} = \frac{5}{3}$$

So, we try $\dfrac{1.31}{3} = 0.437$ as lowest common denominator.

$$\frac{3.08}{0.437} \approx 7 \qquad \frac{2.19}{0.437} \approx 5 \qquad \frac{1.31}{0.437} = 3 \qquad \frac{2.66}{0.437} \approx 6$$

Empirical formula is $C_7H_5N_3O_6$.

21.   Let's first get the elemental composition of the confiscated substance:

$$150.0 \text{ mg CO}_2 \times \frac{12.011 \text{ mg C}}{44.009 \text{ mg CO}_2} = 40.94 \text{ mg C}$$

$$\% \text{ C} = \frac{40.94 \text{ mg}}{50.86 \text{ mg}} \times 100 = 80.50\% \text{ C}$$

$$46.05 \text{ mg H}_2\text{O} \times \frac{2.0158 \text{ mg H}}{18.015 \text{ mg H}_2\text{O}} = 5.153 \text{ mg H}$$

$$\% \text{ H} = \frac{5.153 \text{ mg}}{50.86 \text{ mg}} \times 100 = 10.13\% \text{ H}$$

% N = 9.39%

% O = 100.00 - (80.50 + 10.13 + 9.39) = 0.02%

Composition of cocaine: $C_{17}H_{21}NO_4$

molar mass of cocaine = $17(12.01) + 21(1.008) + 1(14.01) + 4(16.00) = 303.35$ g/mol

$$\% \text{ C} = \frac{204.17 \text{ g}}{303.35 \text{ g}} \times 100 = 67.31\% \qquad \% \text{ N} = \frac{14.01 \text{ g}}{303.35 \text{ g}} \times 100 = 4.618\%$$

$$\% \text{ H} = \frac{21.168}{303.35 \text{ g}} \times 100 = 6.978\% \qquad \% \text{ O} = \frac{64.00 \text{ g}}{303.35 \text{ g}} \times 100 = 21.10\%$$

Obviously, the composition by mass is not the same. The chemist can conclude the compound is not cocaine, assuming he analyzed a pure substance.

22. $0.979 \text{ g Na} \times \dfrac{1 \text{ mol Na}}{22.99 \text{ g Na}} = 4.26 \times 10^{-2} \text{ mol Na}$

$1.365 \text{ g S} \times \dfrac{1 \text{ mol S}}{32.07 \text{ g S}} = 4.256 \times 10^{-2} \text{ mol S}$

$1.021 \text{ g O} \times \dfrac{1 \text{ mol O}}{15.999 \text{ g O}} = 6.382 \times 10^{-2} \text{ mol O}$

$$\frac{6.382 \times 10^{-2} \text{ mol O}}{4.256 \times 10^{-2} \text{ mol S}} = 1.500 = \frac{1.5 \text{ mol O}}{\text{mol S}} = \frac{3 \text{ mol O}}{2 \text{ mol S}}, \frac{\text{mol Na}}{\text{mol S}} \approx 1$$

Empirical formula: $Na_2S_2O_3$

23.   Out of 100.0 g of the pigment, there are:

$$59.9 \text{ g Ti} \times \frac{1 \text{ mol Ti}}{47.88 \text{ g Ti}} = 1.25 \text{ mol Ti}$$

$$40.1 \text{ g O} \times \frac{1 \text{ mol O}}{16.00 \text{ g O}} = 2.51 \text{ mol O}$$

Empirical formula: TiO_2

24.   First, determine composition by mass percent:

$$16.01 \text{ mg CO}_2 \times \frac{12.011 \text{ mg C}}{44.009 \text{ mg CO}_2} = 4.369 \text{ mg C}$$

$$\% \text{ C} = \frac{4.369 \text{ mg C}}{10.68 \text{ mg compound}} \times 100 = 40.91\% \text{ C}$$

$$4.37 \text{ mg H}_2\text{O} \times \frac{2.016 \text{ mg H}}{18.02 \text{ mg H}_2\text{O}} = 0.489 \text{ mg H}$$

$$\% \text{ H} = \frac{0.489 \text{ mg}}{10.68 \text{ mg}} \times 100 = 4.58\% \text{ H}$$

$$\% \text{ O} = 100.00 - (40.91 + 4.58) = 54.51\% \text{ O}$$

So, if we have 100.00 g of the compound, we have:

$$40.91 \text{ g C} \times \frac{1 \text{ mol C}}{12.011 \text{ g C}} = 3.406 \text{ mol C}$$

$$4.58 \text{ g H} \times \frac{1 \text{ mol H}}{1.008 \text{ g H}} = 4.54 \text{ mol H}$$

$$54.51 \text{ g O} \times \frac{1 \text{ mol O}}{16.00 \text{ g O}} = 3.407 \text{ mol O}$$

Look for common ratios: $\dfrac{3.406}{4.54} = 0.750 = \dfrac{3}{4}$

Therefore, empirical formula is $C_3H_4O_3$.

The approximate mass of one $C_3H_4O_3$ unit is $3(12) + 4(1) + 3(16) = 88$ g.

Since $\dfrac{176.1}{88} = 2$, then the molecular formula is $C_6H_8O_6$.

25. $156.8 \text{ mg CO}_2 \times \dfrac{12.011 \text{ mg C}}{44.009 \text{ mg CO}_2} = 42.79 \text{ mg C}$

$42.8 \text{ mg H}_2\text{O} \times \dfrac{2.016 \text{ mg H}}{18.02 \text{ mg H}_2\text{O}} = 4.79 \text{ mg H}$

$\% \text{ C} = \dfrac{42.79 \text{ mg}}{47.6 \text{ mg}} \times 100 = 89.9\% \text{ C}; \ \% \text{ H} = 100.0 - 89.9 = 10.1\% \text{ H}$

Out of 100.0 g Cumene, we have:

$$89.9 \text{ g C} \times \frac{1 \text{ mol C}}{12.01 \text{ g C}} = 7.49 \text{ mol C}; \ 10.1 \text{ g H} \times \frac{1 \text{ mol H}}{1.008 \text{ g H}} = 10.0 \text{ mol H}$$

$\dfrac{7.49}{10.0} = 0.749 \approx 0.75$ which is a 3:4 ratio.   Empirical formula: C_3H_4

Mass of one empirical formula $\approx 3(12) + 4(1) = 40$

So molecular formula is $(C_3H_4)_3$ or C_9H_{12}.

26.   a.   Only acrylonitrile contains nitrogen. If we have 100.00 g of polymer:

$$8.80 \text{ g N} \times \frac{1 \text{ mol C}_3\text{H}_3\text{N}}{14.01 \text{ g N}} \times \frac{53.06 \text{ g C}_3\text{H}_3\text{N}}{1 \text{ mol C}_3\text{H}_3\text{N}} = 33.3 \text{ g C}_3\text{H}_3\text{N}$$

$$\% \text{ C}_3\text{H}_3\text{N} = \frac{33.3 \text{ g C}_3\text{H}_3\text{N}}{100.00 \text{ g polymer}} = 33.3\%$$

Only butadiene in polymer reacts with Br_2:

$$0.605 \text{ g Br}_2 \times \frac{1 \text{ mol Br}_2}{159.8 \text{ g Br}_2} \times \frac{1 \text{ mol C}_4\text{H}_6}{\text{mol Br}_2} \times \frac{54.09 \text{ g C}_4\text{H}_6}{\text{mol C}_4\text{H}_6} = 0.205 \text{ g C}_4\text{H}_6$$

$$\% \text{ C}_4\text{H}_6 = \frac{0.205 \text{ g}}{1.20 \text{ g}} \times 100 = 17.1\% \text{ C}_4\text{H}_6$$

b. If we have 100.0 g of polymer:

$$33.3 \text{ g C}_3\text{H}_3\text{N} \times \frac{1 \text{ mol C}_3\text{H}_3\text{N}}{53.06 \text{ g}} = 0.628 \text{ mol C}_3\text{H}_3\text{N}$$

$$17.1 \text{ g C}_4\text{H}_6 \times \frac{1 \text{ mol C}_4\text{H}_6}{54.09 \text{ g C}_4\text{H}_6} = 0.316 \text{ mol C}_4\text{H}_6$$

$$49.6 \text{ g C}_8\text{H}_8 \times \frac{1 \text{ mol C}_8\text{H}_8}{104.14 \text{ g C}_8\text{H}_8} = 0.476 \text{ mol C}_8\text{H}_8$$

Dividing by 0.316: $\frac{0.628}{0.316} = 1.99$, $\frac{0.316}{0.316} = 1.00$, $\frac{0.476}{0.316} = 1.51$

This is close to a ratio of 4:2:3.   Thus, there are 4 acrylonitrile to 2 butadiene to 3 styrene molecules in the polymer or $(A_4B_2S_3)_n$.

Balancing Chemical Equations

27.   Unbalanced equation:

$$CaF_2 \cdot 3Ca_3(PO_4)_2(s) + H_2SO_4(aq) \rightarrow H_3PO_4(aq) + HF(aq) + CaSO_4 \cdot 2H_2O(s)$$

Balancing Ca^{2+}, F^-, and PO_4^{3-}:

$$CaF_2 \cdot 3Ca_3(PO_4)_2(s) + H_2SO_4(aq) \rightarrow 6 \text{ H}_3PO_4(aq) + 2 \text{ HF}(aq) + 10 \text{ CaSO}_4 \cdot 2H_2O(s)$$

On the right hand side there are 20 hydrogen atoms, 10 sulfates, and 20 water molecules. We can balance the hydrogen and sulfate with 10 sulfuric acid molecules. The extra waters came from the water in the sulfuric acid solution. The balanced equation is:

$$CaF_2 \cdot 3Ca_3(PO_4)_2(s) + 10 \text{ H}_2SO_4(aq) + 20 \text{ H}_2O(l) \rightarrow 6 \text{ H}_3PO_4(aq) + 2 \text{ HF}(aq)$$

$$+ 10 \text{ CaSO}_4 \cdot 2H_2O(s)$$

28.   a. $Fe + O_2 \rightarrow Fe_2O_3$

$$2 \text{ Fe}(s) + 3/2 \text{ O}_2(g) \rightarrow Fe_2O_3(s)$$

Multiply all coefficients by 2 to convert the fraction coefficient of oxygen to a whole number.

$$4\ Fe(s) + 3\ O_2(g) \rightarrow 2\ Fe_2O_3(s)$$

b. $C_6H_{12}O_6 \rightarrow C_2H_5OH + CO_2$

Balance C-atoms first.   6-C on left, 3-C on right.

First, try multiplying both products by 2:

$$C_6H_{12}O_6 \rightarrow 2\ C_2H_5OH + 2\ CO_2$$

O and H are also balanced, and the balanced equation is:

$$C_6H_{12}O_6(aq) \rightarrow 2\ C_2H_5OH(aq) + 2\ CO_2(g)$$

c. $Ca(s) + 2\ H_2O(l) \rightarrow Ca(OH)_2(aq) + H_2(g)$

d. $Ba(OH)_2(aq) + H_2SO_4(aq) \rightarrow BaSO_4(s) + 2\ H_2O(l)$

29. $Pb(NO_3)_2(aq) + H_3AsO_4(aq) \rightarrow PbHAsO_4(s) + 2\ HNO_3(aq)$

Note: The insecticide used is $PbHAsO_4$ and is commonly called lead arsenate. This is not the correct name, however. Correctly, lead arsenate would be $Pb_3(AsO_4)_2$ and $PbHAsO_4$ should be named lead hydrogen arsenate.

30. $2\ NaCl(aq) + 2\ H_2O(l) \rightarrow Cl_2(g) + H_2(g) + 2\ NaOH(aq)$

Reaction Stoichiometry

31. $1.000\ kg\ Al \times \dfrac{1000\ g\ Al}{kg\ Al} \times \dfrac{1\ mol\ Al}{26.98\ g\ Al} \times \dfrac{3\ mol\ NH_4ClO_4}{3\ mol\ Al} \times \dfrac{117.49\ g\ NH_4ClO_4}{mol\ NH_4ClO_4} = 4355\ g$

32. $1.0 \times 10^6\ kg\ HNO_3 \times \dfrac{1000\ g\ HNO_3}{kg\ HNO_3} \times \dfrac{1\ mol\ HNO_3}{63.0\ g\ HNO_3} = 1.6 \times 10^7\ mol\ HNO_3$

We need to get the relationship between moles of HNO_3 and moles of NH_3. We have to use all 3 equations:

$$\frac{2\ mol\ HNO_3}{3\ mol\ NO_2} \times \frac{2\ mol\ NO_2}{2\ mol\ NO} \times \frac{4\ mol\ NO}{4\ mol\ NH_3} = \frac{8\ mol\ HNO_3}{12\ mol\ NH_3}$$

Thus, we can produce 8 mol HNO_3 for every 12 mol NH_3 that we begin with.

$$1.6 \times 10^7 \text{ mol HNO}_3 \times \frac{12 \text{ mol NH}_3}{8 \text{ mol HNO}_3} \times \frac{17.0 \text{ g NH}_3}{\text{mol NH}_3} = 4.1 \times 10^8 \text{ g or } 4.1 \times 10^5 \text{ kg}$$

This is an oversimplified answer. In practice the NO produced in the 3rd step is recycled back into the process in the second step.

33. $$1.0 \times 10^2 \text{ g Ca}_3(\text{PO}_4)_2 \times \frac{1 \text{ mol Ca}_3(\text{PO}_4)_2}{310.2 \text{ g Ca}_3(\text{PO}_4)_2} \times \frac{3 \text{ mol H}_2\text{SO}_4}{\text{mol Ca}_3(\text{PO}_4)_2}$$

$$\times \frac{98.09 \text{ g H}_2\text{SO}_4}{\text{mol H}_2\text{SO}_4} = 95 \text{ g H}_2\text{SO}_4 \text{ are needed}$$

$$95 \text{ g H}_2\text{SO}_4 \times \frac{100 \text{ g concentrated reagent}}{98 \text{ g H}_2\text{SO}_4} = 97 \text{ g of concentrated sulfuric acid}$$

34.   a. $$1.0 \times 10^2 \text{ mg NaHCO}_3 \times \frac{1 \text{ g}}{1000 \text{ mg}} \times \frac{1 \text{ mol NaHCO}_3}{84.0 \text{ g NaHCO}_3} \times \frac{1 \text{ mol C}_6\text{H}_8\text{O}_7}{3 \text{ mol NaHCO}_3}$$

$$\times \frac{192.1 \text{ g C}_6\text{H}_8\text{O}_7}{\text{mol C}_6\text{H}_8\text{O}_7} = 0.076 \text{ g or } 76 \text{ mg C}_6\text{H}_8\text{O}_7$$

        b. $$0.10 \text{ g NaHCO}_3 \times \frac{1 \text{ mol NaHCO}_3}{84.0 \text{ g NaHCO}_3} \times \frac{3 \text{ mol CO}_2}{3 \text{ mol NaHCO}_3} \times \frac{44.0 \text{ g CO}_2}{\text{mol CO}_2} = 0.052 \text{ g or } 52 \text{ mg CO}_2$$

35.   a. $$\text{C}_8\text{H}_{18}(\text{l}) + \frac{25}{2} \text{ O}_2(\text{g}) \rightarrow 8 \text{ CO}_2(\text{g}) + 9 \text{ H}_2\text{O}(\text{g})$$

        or $$2 \text{ C}_8\text{H}_{18}(\text{l}) + 25 \text{ O}_2(\text{g}) \rightarrow 16 \text{ CO}_2(\text{g}) + 18 \text{ H}_2\text{O}(\text{g})$$

        b. $$1.2 \times 10^{10} \text{ gallon} \times \frac{4 \text{ qt}}{\text{gal}} \times \frac{946 \text{ mL}}{\text{qt}} \times \frac{0.692 \text{ g}}{\text{mL}} = 3.1 \times 10^{13} \text{ g of gasoline}$$

        $$3.1 \times 10^{13} \text{ g C}_8\text{H}_{18} \times \frac{1 \text{ mol C}_8\text{H}_{18}}{114.2 \text{ g C}_8\text{H}_{18}} \times \frac{16 \text{ mol CO}_2}{2 \text{ mol C}_8\text{H}_{18}} \times \frac{44.0 \text{ g CO}_2}{\text{mol CO}_2} = 9.6 \times 10^{13} \text{ g CO}_2$$

36.   Total mass of copper used:

        $$10,000 \text{ boards} \times \frac{(8.0 \text{ cm} \times 16.0 \text{ cm} \times 0.060 \text{ cm})}{\text{board}} \times \frac{8.96 \text{ g}}{\text{cm}^3} = 6.9 \times 10^5 \text{ g Cu}$$

        Amount of Cu to be recovered $= 0.80 \times 6.9 \times 10^5 \text{ g} = 5.5 \times 10^5 \text{ g Cu}$

        $$5.5 \times 10^5 \text{ g Cu} \times \frac{1 \text{ mol Cu}}{63.55 \text{ g Cu}} \times \frac{1 \text{ mol Cu(NH}_3)_4\text{Cl}_2}{\text{mol Cu}} \times \frac{202.6 \text{ g Cu(NH}_3)_4\text{Cl}_2}{\text{mol Cu(NH}_3)_4\text{Cl}_2}$$

$$= 1.8 \times 10^6 \text{ g Cu(NH}_3)_4\text{Cl}_2$$

$$5.5 \times 10^5 \text{ g Cu} \times \frac{1 \text{ mol Cu}}{63.55 \text{ g Cu}} \times \frac{4 \text{ mol NH}_3}{\text{mol Cu}} \times \frac{17.03 \text{ g NH}_3}{\text{mol NH}_3} = 5.9 \times 10^5 \text{ g NH}_3$$

Limiting Reactants and Percent Yield

37. $4 \text{ Al} + 3 \text{ O}_2 \rightarrow 2 \text{ Al}_2\text{O}_3$

a. $1.0 \text{ mol Al} \times \dfrac{3 \text{ mol O}_2}{4 \text{ mol Al}} = 0.75 \text{ mol O}_2$; Al is limiting.

b. $2.0 \text{ mol Al} \times \dfrac{3 \text{ mol O}_2}{4 \text{ mol Al}} = 1.5 \text{ mol O}_2$; Al is limiting.

c. $0.50 \text{ mol Al} \times \dfrac{3 \text{ mol O}_2}{4 \text{ mol Al}} = 0.38 \text{ mol O}_2$; Al is limiting.

d. $64.75 \text{ g Al} \times \dfrac{1 \text{ mol Al}}{26.98 \text{ g Al}} \times \dfrac{3 \text{ mol O}_2}{4 \text{ mol Al}} \times \dfrac{32.00 \text{ g O}_2}{\text{mol O}_2} = 57.60 \text{ g O}_2$; Al is limiting.

e. $75.89 \text{ g Al} \times \dfrac{1 \text{ mol Al}}{26.98 \text{ g Al}} \times \dfrac{3 \text{ mol O}_2}{4 \text{ mol Al}} \times \dfrac{32.00 \text{ g O}_2}{\text{mol O}_2} = 67.51 \text{ g O}_2$; Al is limiting.

f. $51.28 \text{ g Al} \times \dfrac{1 \text{ mol Al}}{26.98 \text{ g Al}} \times \dfrac{3 \text{ mol O}_2}{4 \text{ mol Al}} \times \dfrac{32.00 \text{ g O}_2}{\text{mol O}_2} = 45.62 \text{ g O}_2$; Al is limiting.

38. $2 \text{ Cu(s)} + \text{S(s)} \rightarrow \text{Cu}_2\text{S(s)}$ or $16 \text{ Cu(s)} + \text{S}_8\text{(s)} \rightarrow 8 \text{ Cu}_2\text{S(s)}$

$$2.00 \text{ g Cu} \times \frac{1 \text{ mol Cu}}{63.55 \text{ g Cu}} \times \frac{1 \text{ mol Cu}_2\text{S}}{2 \text{ mol Cu}} \times \frac{159.2 \text{ g Cu}_2\text{S}}{\text{mol Cu}_2\text{S}} = 2.51 \text{ g Cu}_2\text{S is theoretical yield.}$$

$$\% \text{ yield} = \frac{\text{Actual yield}}{\text{Theoretical yield}} \times 100 = \frac{2.31 \text{ g}}{2.51 \text{ g}} \times 100 = 92.0\%$$

39. $10.0 \text{ g Al} \times \dfrac{1 \text{ mol Al}}{26.98 \text{ g Al}} \times \dfrac{2 \text{ mol AlBr}_3}{2 \text{ mol Al}} \times \dfrac{266.7 \text{ g AlBr}_3}{\text{mol AlBr}_3} = 98.9 \text{ g AlBr}_3$

$$\% \text{ yield} = \frac{79.8 \text{ g}}{98.9 \text{ g}} \times 100 = 80.7\%$$

40.   a. $1.00 \text{ g Ag} \times \dfrac{1 \text{ mol Ag}}{107.9 \text{ g Ag}} \times \dfrac{8 \text{ mol Ag}_2\text{S}}{16 \text{ mol Ag}} = 4.63 \times 10^{-3} \text{ mol Ag}_2\text{S}$

$$2.00 \text{ g S}_8 \times \frac{1 \text{ mol S}_8}{256.6 \text{ g S}_8} \times \frac{8 \text{ mol Ag}_2\text{S}}{\text{mol S}_8} = 6.24 \times 10^{-2} \text{ mol Ag}_2\text{S}$$

When 4.63×10^{-3} mol Ag_2S is formed, all of the Ag will be consumed.

Ag is limiting. 4.63×10^{-3} mol $Ag_2S \times 247.9$ g/mol = 1.15 g Ag_2S

b. S_8 is left unreacted. 0.15 g sulfur (1.15 g Ag_2S - 1.00 g Ag) is required to make 1.15 g of Ag_2S. Therefore, 1.85 g S_8 remains.

41. $$5.00 \times 10^6 \text{ g NH}_3 \times \frac{1 \text{ mol NH}_3}{17.03 \text{ g NH}_3} \times \frac{2 \text{ mol HCN}}{2 \text{ mol NH}_3} = 2.94 \times 10^5 \text{ mol HCN}$$

$$5.00 \times 10^6 \text{ g O}_2 \times \frac{1 \text{ mol O}_2}{32.00 \text{ g O}_2} \times \frac{2 \text{ mol HCN}}{3 \text{ mol O}_2} = 1.04 \times 10^5 \text{ mol HCN}$$

$$5.00 \times 10^6 \text{ g CH}_4 \times \frac{1 \text{ mol CH}_4}{16.04 \text{ g CH}_4} \times \frac{2 \text{ mol HCN}}{2 \text{ mol CH}_4} = 3.12 \times 10^5 \text{ mol HCN}$$

O_2 is limiting. Therefore:

$$1.04 \times 10^5 \text{ mol HCN} \times \frac{27.03 \text{ g HCN}}{\text{mol HCN}} = 2.81 \times 10^6 \text{ g HCN}$$

$$1.04 \times 10^5 \text{ mol HCN} \times \frac{6 \text{ mol H}_2\text{O}}{2 \text{ mol HCN}} \times \frac{18.02 \text{ g H}_2\text{O}}{\text{mol H}_2\text{O}} = 5.62 \times 10^6 \text{ g H}_2\text{O}$$

42. $2 C_3H_6 + 2 NH_3 + 3 O_2 \rightarrow 2 C_3H_3N + 6 H_2O$

a. $$1.00 \times 10^3 \text{ g C}_3\text{H}_6 \times \frac{1 \text{ mol C}_3\text{H}_6}{42.08 \text{ g C}_3\text{H}_6} \times \frac{2 \text{ mol C}_3\text{H}_3\text{N}}{2 \text{ mol C}_3\text{H}_6} = 23.8 \text{ mol C}_3\text{H}_3\text{N}$$

$$1.50 \times 10^3 \text{ g NH}_3 \times \frac{1 \text{ mol NH}_3}{17.03 \text{ g NH}_3} \times \frac{2 \text{ mol C}_3\text{H}_3\text{N}}{2 \text{ mol NH}_3} = 88.1 \text{ mol C}_3\text{H}_3\text{N}$$

$$2.00 \times 10^3 \text{ g O}_2 \times \frac{1 \text{ mol O}_2}{32.00 \text{ g O}_2} \times \frac{2 \text{ mol C}_3\text{H}_3\text{N}}{3 \text{ mol O}_2} = 41.7 \text{ mol C}_3\text{H}_3\text{N}$$

Therefore, C_3H_6 is limiting and the mass of acrylonitrile produced is:

$$23.8 \text{ mol} \times \frac{53.06 \text{ g C}_3\text{H}_3\text{N}}{\text{mol}} = 1.26 \times 10^3 \text{ g}$$

b. $$23.8 \text{ mol C}_3\text{H}_3\text{N} \times \frac{6 \text{ mol H}_2\text{O}}{2 \text{ mol C}_3\text{H}_3\text{N}} \times \frac{18.02 \text{ g H}_2\text{O}}{\text{mol H}_2\text{O}} = 1.29 \times 10^3 \text{ g H}_2\text{O}$$

Amount NH_3 needed:

$$23.8 \text{ mol } C_3H_3N \times \frac{2 \text{ mol } NH_3}{2 \text{ mol } C_3H_3N} \times \frac{17.03 \text{ g } NH_3}{\text{mol } NH_3} = 405 \text{ g } NH_3$$

Amount NH_3 left = 1.50×10^3 g - 405 g = 1.10×10^3 g

Amount O_2 needed:

$$23.8 \text{ mol } C_3H_3N \times \frac{3 \text{ mol } O_2}{2 \text{ mol } C_3H_3N} \times \frac{32.00 \text{ g } O_2}{\text{mol } O_2} = 1.14 \times 10^3 \text{ g } O_2$$

Amount O_2 left = 2.00×10^3 g - 1.14×10^3 g = 860 g

1.10×10^3 g NH_3 and 860 g O_2 left unreacted.

43. $C_6H_{10}O_4 + 2 NH_3 + 4 H_2 \rightarrow C_6H_{16}N_2 + 4 H_2O$

Adip (Adipic acid)   HMD

a. 1.00×10^3 g Adip $\times \dfrac{1 \text{ mol Adip}}{146.1 \text{ g Adip}} \times \dfrac{1 \text{ mol HMD}}{\text{mol Adip}} \times \dfrac{116.2 \text{ g HMD}}{\text{mol HMD}} = 795$ g HMD

b.   % Yield = $\dfrac{765 \text{ g}}{795 \text{ g}} \times 100 = 96.2\%$

44.   a.   From the reaction stoichiometry we would expect 1 mole of acetaminophen for every mole of $C_6H_5O_3N$ reacted. The actual yield is 3 moles of acetaminophen compared to a theoretical yield of 4 moles of acetaminophen. By mass (where M = molar mass acetaminophen):

$$\% \text{ Yield} = \frac{3 \text{ mol} \times M}{4 \text{ mol} \times M} \times 100 = 75\%$$

b.   The product of the percent yields of the individual steps must equal the overall yield, 75%.

$(0.87)(0.98)(x) = 0.75, \quad x = 0.88$

Step III has a % yield = 88%.

Additional Exercises

45. $\dfrac{9.123 \times 10^{-23} \text{ g}}{\text{atom}} \times \dfrac{6.022 \times 10^{23} \text{ atom}}{\text{mol}} = \dfrac{54.94 \text{ g}}{\text{mol}}$

The atomic mass is 54.94. The element is manganese (Mn).

46.   Mass of repeating unit is approximately:

$$8 \text{ mol C} \left(\frac{12 \text{ g}}{\text{mol C}} \right) + 14 \text{ mol H} \left(\frac{1 \text{ g}}{\text{mol H}} \right) + 2 \text{ mol O} \left(\frac{16 \text{ g}}{\text{mol O}} \right) \approx 142 \text{ g}$$

$$\frac{100{,}000}{142} = 704 \text{ or about 700 monomer units}$$

47.   a. $M = 195.1 + 2(14.01) + 6(1.008) + 2(35.45) = 300.1 \text{ g/mol}$

$$\% \text{ Pt} = \frac{195.1 \text{ g}}{300.1 \text{ g}} \times 100 = 65.01\% \text{ Pt}; \quad \% \text{ N} = \frac{28.02 \text{ g}}{300.1 \text{ g}} \times 100 = 9.335\% \text{ N};$$

$$\% \text{ H} = \frac{6.048 \text{ g}}{300.1 \text{ g}} \times 100 = 2.015\% \text{ H}; \quad \% \text{ Cl} = \frac{70.90 \text{ g}}{300.1 \text{ g}} \times 100 = 23.63\% \text{ Cl}$$

65.01% Pt; 9.335% N; 2.015% H; 23.63% Cl

   b. $100. \text{ g K}_2\text{PtCl}_4 \times \dfrac{1 \text{ mol K}_2\text{PtCl}_4}{415.1 \text{ g K}_2\text{PtCl}_4} \times \dfrac{1 \text{ mol cisplatin}}{\text{mol K}_2\text{PtCl}_4} \times \dfrac{300.1 \text{ g cisplatin}}{\text{mol cisplatin}} = 72.3 \text{ g cisplatin}$

   $100. \text{ g K}_2\text{PtCl}_4 \times \dfrac{1 \text{ mol K}_2\text{PtCl}_4}{415.1 \text{ g K}_2\text{PtCl}_4} \times \dfrac{2 \text{ mol KCl}}{\text{mol K}_2\text{PtCl}_4} \times \dfrac{74.55 \text{ g KCl}}{\text{mol KCl}} = 35.9 \text{ g KCl}$

48.   Since 4.784 g In combines with 1.000 g O, then out of 5.784 g of the oxide there will be 4.784 g In and 1.000 g O.

$$\% \text{ In} = \frac{4.784 \text{ g}}{5.784 \text{ g}} \times 100 = 82.71\% \text{ In}; \quad \% \text{ O} = 100.00 - 82.71 = 17.29\% \text{ O}$$

Assume In_2O_3 is the formula and that we know that the atomic mass of oxygen is 16.00. This was known at that time. Out of 100.00 g compound:

$$17.29 \text{ g O} \times \frac{1 \text{ mol O}}{16.00 \text{ g O}} \times \frac{2 \text{ mol In}}{3 \text{ mol O}} = 0.7204 \text{ mol In}$$

So, 0.7204 mol In has a mass of 82.71 g, or the atomic mass is:

$$\frac{82.71 \text{ g}}{0.7204 \text{ mol}} = 114.8 \text{ g/mol}$$

This is in good agreement with the modern value.

If the formula is InO:

$$17.29 \text{ g O} \times \frac{1 \text{ mol O}}{16.00 \text{ g O}} \times \frac{1 \text{ mol In}}{\text{mol O}} = 1.081 \text{ mol In}$$

Thus, 1.081 mol of In has a mass of 82.71 g or an atomic mass of:

$$\frac{82.71 \text{ g}}{1.081 \text{ mol}} = 76.51 \text{ g/mol}$$

Obviously, Mendeleev was correct.

49.   Out of 100.00 g of compound there are:

$$83.53 \text{ g Sb} \times \frac{1 \text{ mol Sb}}{121.8 \text{ g Sb}} = 0.6858 \text{ mol Sb}$$

$$16.47 \text{ g O} \times \frac{1 \text{ mol O}}{15.599 \text{ g O}} = 1.029 \text{ mol O}$$

$$\frac{0.6858}{1.029} = 0.6665 \approx \frac{2}{3} \qquad \text{Empirical formula: } Sb_2O_3$$

Mass of Sb_2O_3 unit is $2(121.8) + 3(16.0) = 291.6$

Mass of Sb_4O_6 is 583.2, which is in the correct range. Therefore the molecular formula is Sb_4O_6.

50.   The 1000. kg of wet cereal contains 580 kg H_2O and 420 kg of cereal. We want the final product to contain 20.% H_2O. Let x = mass of H_2O in final product.

$$\frac{x}{420 + x} = 0.20, \ x = 84 + 0.20 \ x, \ x = 105 \approx 110 \text{ kg } H_2O$$

The amount of water to be removed is 580 - 110 = 470 kg/hr.

51.   Consider the case of aluminum plus oxygen. Aluminum forms Al^{3+} ions; oxygen forms O^{2-} anions. The simplest compound of the two elements is Al_2O_3. Similarly we would expect the formula of any group VI element with Al to be Al_2X_3. Assuming this, out of 100.00 g of compound there are 18.56 g Al and 81.44 g of the unknown element.

$$18.56 \text{ g Al} \times \frac{1 \text{ mol Al}}{26.98 \text{ g Al}} \times \frac{3 \text{ mol X}}{2 \text{ mol Al}} = 1.032 \text{ mol X}$$

100.00 g of the compound must contain 1.032 mol of X, if the formula is Al_2X_3. Therefore:

$$\text{Atomic mass X} = \frac{81.44 \text{ g X}}{1.032 \text{ mol X}} = 78.91 \text{ g X/mol}$$

The unknown element is selenium, Se, and the formula is Al_2Se_3.

52.   The reaction is: $BaX_2(aq) + H_2SO_4(aq) \rightarrow BaSO_4(s) + 2HX(aq)$

$$0.124 \text{ g } BaSO_4 \times \frac{137.3 \text{ g Ba}}{233.4 \text{ g } BaSO_4} = 0.0729 \text{ g Ba}$$

$$\% \ Ba = \frac{0.0729 \ g \ Ba}{0.158 \ g \ BaX_2} \times 100 = 46.1\% \ Ba$$

The formula is BaX_2 (from positions of the elements in the periodic table) and 100.0 g of compound contains 46.1 g Ba and 53.9 g of the unknown halogen.   There must also be:

$$46.1 \ g \ Ba \times \frac{1 \ mol \ Ba}{137.3 \ g \ Ba} \times \frac{2 \ mol \ X}{mol \ Ba} = 0.672 \ mol \ of \ the \ halogen \ in \ 100.0 \ g \ of \ BaX_2$$

$$Atomic \ mass \ of \ the \ halogen = \frac{53.9 \ g}{0.672 \ mol} = 80.2 \ g/mol$$

This atomic mass is close to that of bromine. The formula of the compound is $BaBr_2$.

53. $41.98 \ mg \ CO_2 \times \frac{12.01 \ mg \ C}{44.01 \ mg \ CO_2} = 11.46 \ mg \ C; \ \% \ C = \frac{11.46 \ mg}{19.81 \ mg} \times 100 = 57.85\% \ C$

$6.45 \ mg \ H_2O \times \frac{2.016 \ mg \ H}{18.02 \ mg \ H_2O} = 0.722 \ mg \ H; \ \% \ H = \frac{0.722 \ mg}{19.81 \ mg} \times 100 = 3.64\% \ H$

$\% \ O = 100.00 - (57.85 + 3.64) = 38.51\% \ O$

Out of 100.00 g terephthalic acid, there are:

$$57.85 \ g \ C \times \frac{1 \ mol \ C}{12.011 \ g \ C} = 4.816 \ mol \ C; \ 3.64 \ g \ H \times \frac{1 \ mol \ H}{1.008 \ g \ H} = 3.61 \ mol \ H;$$

$$38.51 \ g \ O \times \frac{1 \ mol \ O}{15.999 \ g \ O} = 2.407 \ mol \ O$$

$$\frac{4.816}{2.407} \approx 2, \ \frac{3.61}{2.407} = 1.5, \ \frac{2.407}{2.407} = 1$$

C:H:O ratio is 2:1.5:1 or 4:3:2. Empirical formula: $C_4H_3O_2$

Mass of $C_4H_3O_2 \approx 4(12) + 3(1) + 2(16) = 83$

$\frac{166}{83} = 2$; Molecular formula: $C_8H_6O_4$

54. $2.00 \times 10^6 \ g \ CaCO_3 \times \frac{1 \ mol \ CaCO_3}{100.1 \ g \ CaCO_3} \times \frac{1 \ mol \ CaO}{mol \ CaCO_3} \times \frac{56.08 \ g \ CaO}{mol \ CaO} = 1.12 \times 10^6 \ g \ CaO$

55. $Ca_3(PO_4)_2 + 3 \ H_2SO_4 \rightarrow 3 \ CaSO_4 + 2 \ H_3PO_4$

$1.0 \times 10^3 \ g \ Ca_3(PO_4)_2 \times \frac{1 \ mol \ Ca_3(PO_4)_2}{310.2 \ g \ Ca_3(PO_4)_2} \times \frac{3 \ mol \ CaSO_4}{mol \ Ca_3(PO_4)_2} = 9.7 \ mol \ CaSO_4$

$$1.0 \times 10^3 \text{ g con } H_2SO_4 \times \frac{98 \text{ g } H_2SO_4}{100 \text{ g con } H_2SO_4} = 980 \text{ g } H_2SO_4$$

$$980 \text{ g } H_2SO_4 \times \frac{1 \text{ mol } H_2SO_4}{98.1 \text{ g } H_2SO_4} \times \frac{3 \text{ mol } CaSO_4}{3 \text{ mol } H_2SO_4} = 10. \text{ mol } CaSO_4$$

The calcium phosphate is the limiting reagent.

$$9.7 \text{ mol } CaSO_4 \times \frac{136.2 \text{ g } CaSO_4}{\text{mol } CaSO_4} = 1300 \text{ g } CaSO_4$$

$$9.7 \text{ mol } CaSO_4 \times \frac{2 \text{ mol } H_3PO_4}{3 \text{ mol } CaSO_4} \times \frac{98.0 \text{ g } H_3PO_4}{\text{mol } H_3PO_4} = 630 \text{ g } H_3PO_4$$

56.   Mass of Ni in sample:

$$98.4 \text{ g Ni(CO)}_4 \times \frac{58.69 \text{ g Ni}}{170.7 \text{ g Ni(CO)}_4} = 33.8 \text{ g Ni}$$

$$\% \text{ Ni} = \frac{33.8 \text{ g}}{94.2 \text{ g}} \times 100 = 35.9\% \text{ Ni}$$

57. $Hg + Br_2 \rightarrow HgBr_2$

a. $10.0 \text{ g Hg} \times \dfrac{1 \text{ mol Hg}}{200.6 \text{ g Hg}} \times \dfrac{1 \text{ mol } HgBr_2}{\text{mol Hg}} = 4.99 \times 10^{-2} \text{ mol } HgBr_2$

$10.0 \text{ g } Br_2 \times \dfrac{1 \text{ mol } Br_2}{159.8 \ Br_2} \times \dfrac{1 \text{ mol } HgBr_2}{\text{mol } Br_2} = 6.26 \times 10^{-2} \text{ mol } HgBr_2$

Hg is limiting. $HgBr_2$ produced is:

$$4.99 \times 10^{-2} \text{ mol } HgBr_2 \times \frac{360.4 \text{ g } HgBr_2}{\text{mol } HgBr_2} = 17.98 \text{ g} \approx 18.0 \text{ g } HgBr_2$$

18.0 g $HgBr_2$ with 2.0 g Br_2 left unreacted.

b. $5.00 \text{ mL Hg} \times \dfrac{13.5 \text{ g Hg}}{\text{mL}} = 67.5 \text{ g Hg};\ \ 5.00 \text{ ml } Br_2 \times \dfrac{3.12 \text{ g } Br_2}{\text{mL } Br_2} = 15.6 \text{ g } Br_2$

$67.5 \text{ g Hg} \times \dfrac{1 \text{ mol Hg}}{200.6 \text{ g Hg}} \times \dfrac{1 \text{ mol } HgBr_2}{\text{mol Hg}} = 0.336 \text{ mol } HgBr_2$

$15.6 \text{ g } Br_2 \times \dfrac{1 \text{ mol } Br_2}{159.8 \text{ g } Br_2} \times \dfrac{1 \text{ mol } HgBr_2}{\text{mol } Br_2} = 0.0976 \text{ mol } HgBr_2$

Br_2 is limiting.

$$0.0976 \text{ mol HgBr}_2 \times \frac{360.4 \text{ g HgBr}_2}{\text{mol HgBr}_2} = 35.2 \text{ g HgBr}_2$$

58. $4.0 \text{ g H}_2 \times \dfrac{1 \text{ mol H}_2}{2.016 \text{ g H}_2} \times \dfrac{2 \text{ mol H}}{1 \text{ mol H}_2} \times \dfrac{6.02 \times 10^{23} \text{ atoms H}}{1 \text{ mol H}} = 2.4 \times 10^{24} \text{ atoms}$

$4.0 \text{ g He} \times \dfrac{1 \text{ mol He}}{4.003 \text{ g He}} \times \dfrac{6.02 \times 10^{23} \text{ atoms He}}{1 \text{ mol He}} = 6.0 \times 10^{23} \text{ atoms}$

$1.0 \text{ mol F}_2 \times \dfrac{2 \text{ mol F}}{1 \text{ mol F}_2} \times \dfrac{6.02 \times 10^{23} \text{ atoms F}}{1 \text{ mol F}} = 1.2 \times 10^{24} \text{ atoms}$

$44.0 \text{ g CO}_2 \times \dfrac{1 \text{ mol CO}_2}{44.01 \text{ g CO}_2} \times \dfrac{3 \text{ mol atom(1 C + 2 O)}}{1 \text{ mol CO}_2} \times \dfrac{6.022 \times 10^{23} \text{ atoms}}{1 \text{ mol atoms}} = 1.81 \times 10^{24} \text{ atoms}$

$146. \text{ g SF}_6 \times \dfrac{1 \text{ mol SF}_6}{146.07 \text{ g SF}_6} \times \dfrac{7 \text{ mol atoms (1 S + 6 F)}}{1 \text{ mol SF}_6} \times \dfrac{6.022 \times 10^{23} \text{ atoms}}{1 \text{ mol atoms}} = 4.21 \times 10^{24} \text{ atoms}$

$146 \text{ g SF}_6 > 4.0 \text{ g H}_2 > 44.0 \text{ g CO}_2 > 1.0 \text{ mol F}_2 > 4.0 \text{ g He}$

59. $^{12}C_2{}^1H_6$: $2(12.000000) + 6(1.007825) = 30.046950$ amu

$^1H_2{}^{16}O$: $1(12.000000) + 2(1.007825) + 1(15.994915) = 30.010565$ amu

$^{14}N^{16}O$: $1(14.003074) + 1(15.994915) = 29.997989$ amu

The peak results from $^{12}C^1H_2{}^{16}O$.

60. $\dfrac{^{85}Rb}{^{87}Rb} = 2.591$ If we had exactly 100 atoms: x = number of ^{85}Rb,
                                          100 - x = number of ^{87}Rb

$\dfrac{x}{100 - x} = 2.591,$ $x = 259.1 - 2.591 \, x,$ $x = \dfrac{259.1}{3.591} = 72.15\%$ ^{85}Rb

$0.7215 (84.9117) + 0.2785 (A) = 85.4678$

$A = \dfrac{85.4678 - 61.26}{0.2785} = 86.92$ amu

61.   Compound I: Out of 100.00 g sample:

$9.93 \text{ g C} \times \dfrac{1 \text{ mol C}}{12.01 \text{ g C}} = \dfrac{0.827 \text{ mol C}}{0.827} = 1$

$$58.6 \text{ g Cl} \times \frac{1 \text{ mol Cl}}{35.45 \text{ g Cl}} = \frac{1.65 \text{ mol Cl}}{0.827} = 2$$

$$31.4 \text{ g F} \times \frac{1 \text{ mol F}}{19.00 \text{ g F}} = \frac{1.65 \text{ mol F}}{0.827} = 2$$

Empirical formula: CCl_2F_2

Compound II: Out of 100.0 g sample:

$$11.5 \text{ g C} \times \frac{1 \text{ mol C}}{12.01 \text{ g C}} = \frac{0.958 \text{ C}}{0.956} = 1$$

$$33.9 \text{ g Cl} \times \frac{1 \text{ mol Cl}}{35.45 \text{ g Cl}} = \frac{0.956 \text{ Cl}}{0.956} = 1$$

$$54.6 \text{ g F} \times \frac{1 \text{ mol F}}{19.00 \text{ g F}} = \frac{2.87 \text{ F}}{0.956} = 3$$

Empirical formula: $CClF_3$

Let's consider masses of Cl and F that combine with 1.00 g C:

> Compound I: 1.00 g C   5.90 g Cl   3.16 g F
> Compound II: 1.00 g C   2.95 g Cl   4.75 g F

Cl ratio: $\dfrac{I}{II} = \dfrac{5.90}{2.95} = \dfrac{2}{1}$

> Small whole number ratios as predicted
> by law of multiple proportions.

F ratio: $\dfrac{I}{II} = \dfrac{3.16}{4.75} \approx \dfrac{2}{3}$

62.   The volume of a gas is proportional to the number of molecules of gas. Thus the formulas are:

> I: H_3N,   II: H_4N_2,   III: HN_3

The mass ratios are:

> I: $\dfrac{4.634 \text{ g N}}{\text{g H}}$,   II: $\dfrac{6.949 \text{ g N}}{\text{g H}}$, III: $\dfrac{41.7 \text{ g N}}{\text{g H}}$

If we set the atomic mass of H equal to 1.008, then the atomic mass for nitrogen is:

> I: 14.01,   II: 14.01,   III. 14.0

For example for Compound I: $\dfrac{A}{3(1.008)} = \dfrac{4.634}{1}$, A = 14.01

63.   Consider a 100.0 g sample:

$$93.31 \text{ g Fe} \times \frac{1 \text{ mol Fe}}{55.85 \text{ g Fe}} = \frac{1.671 \text{ mol Fe}}{0.557} = 3$$

$$6.69 \text{ g C} \times \frac{1 \text{ mol C}}{12.01 \text{ g C}} = \frac{0.557 \text{ mol C}}{0.557} = 1$$

Empirical Formula: Fe_3C

64. $PaO_2 + O_2 \rightarrow Pa_xO_y$ (unbalanced)

$$0.200 \text{ g PaO}_2 \times \frac{231 \text{ g Pa}}{263 \text{ g PaO}_2} = 0.1757 \text{ g Pa} \quad \text{(We will carry an extra S.F.)}$$

$0.2081 \text{ g Pa}_xO_y - 0.1757 \text{ g Pa} = 0.0324 \text{ g O}$

$$0.1757 \text{ g Pa} \times \frac{1 \text{ mol Pa}}{231 \text{ g Pa}} = 7.61 \times 10^{-4} \text{ mol Pa}$$

$$0.0324 \text{ g O} \times \frac{1 \text{ mol O}}{16.00 \text{ g O}} = 2.025 \times 10^{-3} \text{ mol O}$$

$$\frac{\text{mol O}}{\text{mol Pa}} = \frac{2.025 \times 10^{-3} \text{ mol O}}{7.61 \times 10^{-4} \text{ mol Pa}} = 2.66 \approx 2\frac{2}{3} = \frac{8 \text{ mol O}}{3 \text{ mol Pa}}$$

Empirical formula: Pa_3O_8

65. $10.00 \text{ g XCl}_2 \rightarrow 12.55 \text{ g XCl}_4$

XCl_4 contains 2.55 g Cl and 10.00 g XCl_2. Therefore, XCl_2 must contain 2.55 g Cl and 7.45 g X.

$$2.55 \text{ g Cl} \times \frac{1 \text{ mol Cl}}{35.45 \text{ g Cl}} \times \frac{1 \text{ mol XCl}_2}{2 \text{ mol Cl}} \times \frac{1 \text{ mol X}}{1 \text{ mol XCl}_2} = 3.60 \times 10^{-2} \text{ mol X}$$

$$\frac{7.45 \text{ g X}}{3.60 \times 10^{-2} \text{ mol X}} = \frac{207 \text{ g}}{\text{mol X}} \; ; \quad \text{X is Pb.}$$

66. $$1.375 \text{ g AgI} \times \frac{1 \text{ mol AgI}}{234.8 \text{ g AgI}} = 5.856 \times 10^{-3} \text{ mol AgI} = 5.856 \times 10^{-3} \text{ mol I}$$

$$1.375 \text{ g AgI} \times \frac{126.9 \text{ g I}}{234.8 \text{ g AgI}} = 0.7431 \text{ g I}$$

XI_2 contains 0.7431 g I and 0.257 g X.

$$5.856 \times 10^{-3} \text{ mol I} \times \frac{1 \text{ mol X}}{2 \text{ mol I}} = 2.928 \times 10^{-3} \text{ mol X}$$

$$\frac{0.257 \text{ g X}}{2.928 \times 10^{-3} \text{ mol X}} = \frac{87.8 \text{ g}}{\text{mol}} \qquad (\text{X is Sr.})$$

67.   For a gas, density and molar mass are proportional.

$$M = 2.393 \,(32.00) = \frac{76.58 \text{ g}}{\text{mol}}$$

$$0.803 \text{ g H}_2\text{O} \times \frac{2 \text{ mol H}}{18.02 \text{ g H}_2\text{O}} = 8.91 \times 10^{-2} \text{ mol H}$$

$$\frac{8.91 \times 10^{-2} \text{ mol H}}{2.23 \times 10^{-2} \text{ mol XH}_n} = \frac{4 \text{ mol H}}{\text{mol XH}_n}$$

$$A_x = 76.58 - 4(1.008 \text{ g}) = \frac{72.55 \text{ g}}{\text{mol}}; \quad \text{The element is Ge.}$$

68. $$\frac{6.05 \text{ g "Il"}}{1.00 \text{ g O}} \times \frac{16.00 \text{ g O}}{1 \text{ mol O}} = \frac{96.8 \text{ g "Il"}}{\text{mol O}}$$

If formula of the oxide is: IlO $A = 96.8$

$\qquad\qquad\qquad\qquad\qquad\qquad IlO_2$ $A = 193.6$

$\qquad\qquad\qquad\qquad\qquad\qquad Il_2O_3$ $A = 145.2$

Since $A \approx 150$, then Il_2O_3 with $A = 145.2$ is the best fit. "Illinium" is really Promethium, Pm.

69. $4.000 \text{ g M}_2\text{S}_3 \rightarrow 3.723 \text{ g MO}_2$

There are twice as many moles of MO_2 as of M_2S_3.

$$2\left(\frac{4.000 \text{ g}}{2 \text{ A} + 3(32.07)}\right) = \frac{3.723 \text{ g}}{\text{A} + 2(16.00)}, \quad \frac{8.000}{2 \text{ A} + 96.21} = \frac{3.723}{\text{A} + 32.00}$$

$8.000 \text{ A} + 256.0 = 7.446 \text{ A} + 358.2, \quad 0.554 \text{ A} = 102.2$

$$A = \frac{184 \text{ g}}{\text{mol}}; \qquad \text{The metal is W.}$$

70. X_2Z: 40.0% X and 60.0% Z by mass

$$\frac{\text{mol X}}{\text{mol Z}} = 2 = \frac{40.0/A_x}{60.0/A_z} = \frac{40.0\ A_z}{60.0\ A_x} \text{ or } A_z = 3\ A_x$$

For XZ_2, molar mass $= A_x + 2\ A_z = A_x + 2(3\ A_x) = 7\ A_x$

$$\% \text{ X} = \frac{A_x}{7A_x} \times 100 = 14.3\% \text{ X} \qquad \% \text{ Z} = 100.0 - \% \text{ X} = 85.7\% \text{ Z}$$

71.   The balanced equations are:

$$C + 1/2\ O_2 \rightarrow CO \text{ and } C + O_2 \rightarrow CO_2$$

If we have 100.0 mol of products, we have 72.0 mol CO_2, 16.0 mol CO, and 12.0 mol O_2. The initial mixture contained $72.0 + 16.0 = 88.0$ mol C and 72.0 (from CO_2) $+ \dfrac{16.0}{2}$ (from CO) + 12.0 (unreacted) = 92.0 mol O_2. Initial reaction mixture contained:

$$\frac{92.0 \text{ mol O}_2}{88.0 \text{ mol C}} = 1.05 \text{ mol O}_2/\text{mol C}$$

72. $10{,}000 \text{ kg waste} \times \dfrac{3.0 \text{ kg NH}_4^+}{100 \text{ kg waste}} \times \dfrac{1000 \text{ g}}{\text{kg}} \times \dfrac{1 \text{ mol NH}_4{+}}{18.04 \text{ g NH}_4^+} \times \dfrac{1 \text{ mol C}_5\text{H}_7\text{O}_2\text{N}}{55 \text{ mol NH}_4^+}$

$$\times \frac{113.12 \text{ g C}_5\text{H}_7\text{O}_2\text{N}}{\text{mol C}_5\text{H}_7\text{O}_2\text{N}} = 3.4 \times 10^4 \text{ g tissue if all NH}_4^+ \text{ converted}$$

Since only 95% of the NH_4^+ ions react:

mass of tissue $= (0.95)\ (3.4 \times 10^4 \text{ g}) = 3.2 \times 10^4 \text{ g or 32 kg}$

73.   a. $1500 \text{ kg Ca}_3(\text{PO}_4)_2 \times \dfrac{1 \text{ kmol Ca}_3(\text{PO}_4)_2}{310.18 \text{ kg Ca}_3(\text{PO}_4)_2} \times \dfrac{2 \text{ kmol P}}{\text{kmolCa}_3(\text{PO}_4)_2} \times \dfrac{30.97 \text{ kg P}}{\text{kmol P}} = 3.0 \times 10^2 \text{ kg P}$

$250 \text{ kg C} \times \dfrac{1 \text{ kmol C}}{12.01 \text{ kg C}} \times \dfrac{2 \text{ kmol P}}{5 \text{ kmol C}} \times \dfrac{30.97 \text{ kg P}}{\text{kmol P}} = 260 \text{ kg P}$

$1000 \text{ kg SiO}_2 \times \dfrac{1 \text{ kmol SiO}_2}{60.09 \text{ kg SiO}_2} \times \dfrac{2 \text{ kmol P}}{3 \text{ kmol SiO}_2} \times \dfrac{30.97 \text{ kg P}}{\text{kmol P}} = 340 \text{ kg P}$

C is the limiting reagent.

b.   From a, the theoretical yield is 260 kg P.

c. The total reactant mass we began with was 2750 kg (assuming extra S.F.). After the reaction, this solid mass will be decreased by the amount of $CO(g)$ and $P(l)$ formed. If we let x = kmol of $Ca_3(PO_4)_2$ that reacted, then the balanced equation is:

$$xCa_3(PO_4)_2(s) + 5xC(s) + 3xSiO_2(s) \rightarrow 3xCaSiO_3(s) + 5xCO(g) + 2xP(l)$$

We began with (assuming extra S.F.):

$$250 \text{ kg C} \times \frac{1 \text{ kmol C}}{12.01 \text{ kg C}} = 20.8 \text{ kmol C}$$

In the slag, % C = $\dfrac{\text{mass C}}{\text{mass slag}} \times 100$:

$$3.8 = \frac{(20.8 \text{ kmol C} - 5x) \, 12.01 \text{ kg/kmol}}{2750 \text{ kg} - 5x(28.01 \text{ kg/kmol}) - 2x(30.97 \text{ kg/kmol})} \times 100$$

$$0.038 \, [2750 - 5x(28.01) - 2x \, (30.97)] = (20.8 - 5x) \, 12.01$$

$$104.5 - 5.32x - 2.35x = 249.8 - 60.05x, \quad 52.38x = 145.3, \quad x = 2.77 \approx 2.8$$

$x = 2.8, \quad 5x = 14$ kmol of C reacted

$$14 \text{ kmol C} \times \frac{2 \text{ kmol P}}{5 \text{ kmol C}} \times \frac{30.97 \text{ kg P}}{\text{kmol P}} = 170 \text{ kg P actually produced}$$

$$\% \text{ yield} = \frac{170 \text{ kg P}}{260 \text{ kg P}} \times 100 = 65\%$$

Checking with % P calculation:

$$1500/310.18 = 4.84 \text{ kmol } Ca_3(PO_4)_2 \text{ initially}$$

$$5.8 = \frac{(4.84 - x) \, (2) \, (30.97) \text{ kg P}}{2750 - 5x \, (28.01) - 2x \, (30.97) \text{ kg slag}} \times 100$$

$$0.058 \, [2750 - 5x(28.01) - 2x(30.97)] = (4.84 - x) \, (2) \, (30.97)$$

$$159.5 - 8.12x - 3.59x = 299.8 - 61.94x, \quad 50.23x = 140.3, \quad x = 2.79 \approx 2.8$$

We actually produce $2x = 2(2.8)(30.97) = 170$ kg P. This agrees with our previous answer.

$$\% \text{ yield} = \frac{170}{260} \times 100 = 65\%$$

74. $LaH_{2.90}$ is the formula. If only La^{3+} is present, LaH_3 would be the formula. If only La^{2+} is present, LaH_2 would be the formula.

Let x = mol La^{2+} and y = mol La^{3+}:

$(La^{2+})_x(La^{3+})_yH_{(2x+3y)}$ where $x + y = 1.00$ and $2x + 3y = 2.90$

Solving by simultaneous equations:

$$
\begin{aligned}
2x + 3y &= 2.90 \\
-2x - 2y &= -2.00 \\
\hline
y &= 0.90 \text{ and } x = 0.10
\end{aligned}
$$

$LaH_{2.90}$ contains $\dfrac{1}{10}$ La^{2+} or 10.% La^{2+} and $\dfrac{9}{10}$ La^{3+} or 90.% La^{3+}.

75. $MCO_3(s) + 2 H^+(aq) \rightarrow M^{2+}(aq) + H_2O(l) + CO_2(g)$

$$0.421 \text{ g } CO_2 \times \frac{1 \text{ mol } CO_2}{44.01 \text{ g } CO_2} \times \frac{1 \text{ mol } MCO_2}{1 \text{ mol } CO_2} = 9.57 \times 10^{-3} \text{ mol of } MCO_3 \text{ present}$$

Let x = g $SrCO_3$ and y = g $BaCO_3$:

Mass balance: $x + y = 1.60$ g

Mole balance: $\dfrac{x}{147.6} + \dfrac{y}{197.3} = 9.57 \times 10^{-3}$ total mol or $1.337 x + y = 1.89$

Solving:

$$
\begin{aligned}
1.337 x + y &= 1.89 \\
-x - y &= -1.60 \\
\hline
0.337 x &= 0.29
\end{aligned}
$$

$x = 0.86$ g $SrCO_3$ and $y = 0.74$ g $BaCO_3$

% $SrCO_3 = \dfrac{0.86 \text{ g}}{1.60 \text{ g}} \times 100 = 54\%$ by mass; % $BaCO_3 = 46\%$

Note: with no rounding, % $SrCO_3$ = 53.4%; % $BaCO_3$ = 46.6%

76. $NaCl + Ag^+ \rightarrow AgCl(s)$; $KCl + Ag^+ \rightarrow AgCl(s)$

$$8.5904 \text{ g } AgCl \times \frac{1 \text{ mol } AgCl}{143.4 \text{ g } AgCl} \times \frac{1 \text{ mol } Cl^-}{1 \text{ mol } AgCl} = 5.991 \times 10^{-2} \text{ mol } Cl^-$$

Let x = g NaCl and y = g KCl:

$x + y = 4.000$ g and $\dfrac{x}{58.44} + \dfrac{y}{74.55} = 5.991 \times 10^{-2}$ total mol Cl^-

Solving using simultaneous equations:

$$
\begin{aligned}
1.276 x + y &= 4.466 \\
-x - y &= -4.000 \\
\hline
0.276 x &= 0.466
\end{aligned}
$$

x = 1.69 g NaCl and y = 2.31 g KCl

$$\% \text{ NaCl} = \frac{1.69}{4.000} \times 100 = 42.3\% \text{ NaCl}; \ \% \text{ KCl} = 57.7\%$$

77. $1.297 \text{ g Cu} \times \dfrac{1 \text{ mol Cu}}{63.55 \text{ g Cu}} = 2.041 \times 10^{-2} \text{ mol Cu}$

Let x = g Cu_2O and y = g CuO: x + y = 1.500 and

$$2\left(\frac{x}{143.1}\right) + \frac{y}{79.55} = 2.041 \times 10^{-2} \text{ total mol Cu or } 1.112 \text{ x} + \text{y} = 1.624$$

Solving:

$$\begin{array}{r} 1.112 \text{ x} + \text{y} = 1.624 \\ - \text{x} - \text{y} = -1.500 \\ \hline 0.112 \text{ x} \phantom{- \text{y} =} = 0.124 \end{array}$$

x = 1.11 g Cu_2O; 74.0% Cu_2O by mass and 26.0% CuO by mass

78.   a C_8H_{18} (l) + b O_2 (g) → c CO_2 (g) + d CO (g) + e CH_4 (g) + f H_2 (g) + g H_2O (g)

Volume ratios are proportional to mole ratios. Volume percent of O_2 reacted is:

$$\frac{21 \text{ mol } \% \text{ O}_2}{78 \text{ mol } \% \text{ N}_2} = \frac{x}{82.1 \text{ vol } \% \text{ N}_2}, \ x = 22 \text{ vol } \% \text{ O}_2 = \text{volume percent of O}_2 \text{ reacted}$$

Determining mole ratios in balanced equation:

$$\frac{b}{c} = \frac{22}{11.5} = 1.9, \ c = 0.52 \text{ b}; \ \frac{b}{d} = \frac{22}{4.4} = 5.0, \ d = 0.20 \text{ b};$$

$$\frac{b}{e} = \frac{22}{0.5} = 44 \approx 40, \ e = 0.02 \text{ b}; \ \frac{b}{f} = \frac{22}{1.5} = 15, \ f = 0.068 \text{ b}$$

Balancing the carbon atoms: 8a = c + d + e

Let's assume a = 1, so 1 mol of gasoline consumed:

8 = 0.52 b + 0.20 b + 0.02 b = 0.74 b, b = 11

Using b to determine c, d, e, and f:

c = 5.7; d = 2.2; e = 0.2; f = 0.75

Balancing the hydrogen atoms: $18\ a = 4\ e + 2\ f + 2\ g$, $18 = 4(0.2) + 2(0.75) + 2\ g$, $g = 7.9$

$$C_8H_{18} + 11\ O_2 \rightarrow 5.7\ CO_2 + 2.2\ CO + 0.2\ CH_4 + 0.75\ H_2 + 7.9\ H_2O$$

79. $A + B \rightarrow AB$

6.56 g AB was 76.2% yield, so theoretical yield is $\dfrac{6.56}{0.762} = 8.61$ g AB.

8.61 g AB contains 4.72 g A, g B = 8.61 - 4.72 = 3.89 g B

80.   We would see the peaks corresponding to:

$^{10}B\ ^{35}Cl_3(115\ amu)$, $^{10}B\ ^{35}Cl_2{}^{37}Cl(117)$, $^{10}B\ ^{35}Cl\ ^{37}Cl_2(119)$, $^{10}B\ ^{37}Cl_3(121)$,

$^{11}B\ ^{35}Cl_3(116)$, $^{11}B\ ^{35}Cl_2\ ^{37}Cl(118)$, $^{11}B\ ^{35}Cl\ ^{37}Cl_2(120)$, $^{11}B\ ^{37}Cl_3(122)$

We would see a total of 8 peaks at masses 115, 116, 117, 118, 119, 120, 121, and 122. The largest peak would have the two most abundant isotopes ($^{11}B^{35}Cl_3$) and occur at mass 116. The smallest peak would correspond to the least abundant isotopes ($^{10}B^{37}Cl_3$) and occur at mass 121.

CHAPTER FOUR

TYPES OF CHEMICAL REACTIONS AND SOLUTION STOICHIOMETRY

Aqueous Solutions: Strong and Weak Electrolytes

1.   "Slightly soluble" refers to substances that dissolve only to a small extent. A slightly soluble salt may still dissociate completely to ions and, hence, be a strong electrolyte. An example of such a substance is $Mg(OH)_2$. It is a strong electrolyte, but not very soluble. A weak electrolyte is a substance that doesn't dissociate completely to produce ions. A weak electrolyte may be very soluble in water, or it may not be very soluble.   Acetic acid is an example of a weak electrolyte that is very soluble in water.

2.   Measure the electrical conductivity of a solution and compare it to the conductivity of a solution of equal concentration of a strong electrolyte.

3. $MgSO_4(s) \rightarrow Mg^{2+}(aq) + SO_4^{2-}(aq)$

   $NH_4NO_3(s) \rightarrow NH_4^+(aq) + NO_3^-(aq)$

Solution Concentration: Molarity

4.   a. $1.00 \text{ L solution} \times \dfrac{0.50 \text{ mol } H_2SO_4}{L} = 0.50 \text{ mol } H_2SO_4$

   $0.50 \text{ mol } H_2SO_4 \times \dfrac{1 \text{ L}}{18 \text{ mol } H_2SO_4} = 2.8 \times 10^{-2} \text{ L conc. } H_2SO_4 \text{ or } 28 \text{ mL}$

   Dilute 28 mL of concentrated H_2SO_4 to a total volume of 1.00 L with water.

   b.   We will need 0.50 mol HCl.

   $0.50 \text{ mol HCl} \times \dfrac{1 \text{ L}}{12 \text{ mol HCl}} = 4.2 \times 10^{-2} \text{ L} = 42 \text{ mL}$

   Dilute 42 mL of concentrated HCl to a final volume of 1.00 L.

   c.   We need 0.50 mol $NiCl_2$.

$$0.50 \text{ mol NiCl}_2 \times \frac{1 \text{ mol NiCl}_2 \cdot 6H_2O}{\text{mol NiCl}_2} \times \frac{237.7 \text{ g NiCl}_2 \cdot 6H_2O}{\text{mol NiCl}_2 \cdot 6H_2O} = 120 \text{ g NiCl}_2 \cdot 6H_2O$$

Dissolve 120 g $NiCl_2 \cdot 6H_2O$ in water, and add water until the total volume of the solution is 1.00 L.

d. $$1.00 \text{ L} \times \frac{0.50 \text{ mol HNO}_3}{\text{L}} = 0.50 \text{ mol HNO}_3$$

$$0.50 \text{ mol HNO}_3 \times \frac{1 \text{ L}}{16 \text{ mol}} = 0.031 \text{ L} = 31 \text{ mL}$$

Dissolve 31 mL of concentrated reagent in water. Dilute to a total volume of 1.00 L.

e.   We need 0.50 mol Na_2CO_3.

$$0.50 \text{ mol Na}_2CO_3 \times \frac{106.0 \text{ g Na}_2CO_3}{\text{mol}} = 53 \text{ g Na}_2CO_3$$

Dissolve 53 g Na_2CO_3 in water, dilute to 1.00 L.

5.   a. $$4.592 \text{ g NaHCO}_3 \times \frac{1 \text{ mol NaHCO}_3}{84.01 \text{ g NaHCO}_3} = 5.466 \times 10^{-2} \text{ mol}$$

$$M = \frac{5.466 \times 10^{-2} \text{ mol}}{250.0 \text{ mL}} \times \frac{1000 \text{ mL}}{\text{L}} = 0.2186 \text{ mol/L}$$

b. $$0.2759 \text{ g K}_2Cr_2O_7 \times \frac{1 \text{ mol K}_2Cr_2O_7}{294.20 \text{ g K}_2Cr_2O_7} = 9.378 \times 10^{-4} \text{ mol}$$

$$M = \frac{9.378 \times 10^{-4} \text{ mol}}{500.0 \times 10^{-3} \text{ L}} = 1.876 \times 10^{-3} \text{ mol/L}$$

c. $$0.1025 \text{ g Cu} \times \frac{1 \text{ mol Cu}}{63.55 \text{ g Cu}} = 1.613 \times 10^{-3} \text{ mol Cu} = 1.613 \times 10^{-3} \text{ mol Cu}^{2+}$$

$$M = \frac{1.613 \times 10^{-3} \text{ mol Cu}^{2+}}{200.0 \text{ mL}} \times \frac{1000 \text{ mL}}{\text{L}} = 8.065 \times 10^{-3} \text{ mol/L}$$

6. $$\frac{50.0 \times 10^{-3} \text{ g Myocrisin}}{0.500 \times 10^{-3} \text{ L}} \times \frac{1 \text{ mol Myocrisin}}{390.1 \text{ g}} = 0.256 \ M$$

$$\frac{300.0 \times 10^{-6} \text{ g Au}}{100.0 \text{ mL}} \times \frac{1 \text{ mol Au}}{197.0 \text{ g Au}} \times \frac{1 \text{ mol Myocrisin}}{\text{mol Au}} \times \frac{1000 \text{ mL}}{\text{L}} = 1.523 \times 10^{-5} \ M$$

7. $$25.0 \text{ g (NH}_4)_2SO_4 \times \frac{1 \text{ mol}}{132.15 \text{ g}} = 1.89 \times 10^{-1} \text{ mol}$$

$$\text{Molarity} = \frac{1.89 \times 10^{-1} \text{ mol}}{100.0 \text{ mL}} \times \frac{1000 \text{ mL}}{\text{L}} \quad \frac{1.89 \text{ mol}}{\text{L}}$$

Moles of $(NH_4)_2SO_4$ in final solution $= 10.00 \times 10^{-3}$ L $\times \dfrac{1.89 \text{ mol}}{L} = 1.89 \times 10^{-2}$ mol

Molarity of final solution $= \dfrac{1.89 \times 10^{-2} \text{ mol}}{(10.00 + 50.00) \text{ mL}} \times \dfrac{1000 \text{ mL}}{L} = \dfrac{0.315 \text{ mol } (NH_4)_2SO_4}{L}$

$M_{NH_4^+} = 2(0.315) = 0.630 \ M; \ M_{SO_4^{2-}} = 0.315 \ M$

8. $75.0 \text{ mL} \times \dfrac{0.79 \text{ g}}{mL} \times \dfrac{1 \text{ mol}}{46.1 \text{ g}} = 1.3$ mol ethanol; molarity $= \dfrac{1.3 \text{ mol}}{0.250 \text{ L}} = \dfrac{5.2 \text{ mol}}{L}$

9.   a.   5.0 ppb Hg in water $= \dfrac{5.0 \text{ ng Hg}}{mL \ H_2O} = \dfrac{5.0 \times 10^{-9} \text{ g Hg}}{mL \ H_2O}$

$\dfrac{5.0 \times 10^{-9} \text{ g Hg}}{mL} \times \dfrac{1 \text{ mol Hg}}{200.6 \text{ g Hg}} \times \dfrac{1000 \text{ mL}}{L} = 2.5 \times 10^{-8} \ M$

b. $\dfrac{1.0 \times 10^{-9} \text{ g CHCl}_3}{mL} \times \dfrac{1 \text{ mol CHCl}_3}{119.4 \text{ g CHCl}_3} \times \dfrac{1000 \text{ mL}}{L} = 8.4 \times 10^{-9} \ M$

c.   10.0 ppm As $= \dfrac{10.0 \ \mu\text{g As}}{mL} = \dfrac{10.0 \times 10^{-6} \text{ g As}}{mL}$

$\dfrac{10.0 \times 10^{-6} \text{ g As}}{mL} \times \dfrac{1 \text{ mol As}}{74.92 \text{ g As}} \times \dfrac{1000 \text{ mL}}{L} = 1.33 \times 10^{-4} \ M$

d. $\dfrac{0.10 \times 10^{-6} \text{ g DDT}}{mL} \times \dfrac{1 \text{ mol DDT}}{354.5 \text{ g DDT}} \times \dfrac{1000 \text{ mL}}{L} = 2.8 \times 10^{-7} \ M$

10.   a.   ppm $= \dfrac{\mu\text{g Na}}{mL}$

$150 \times 10^{-3} \text{ g Na}_2CO_3 \times \dfrac{46.0 \text{ g Na}^+}{106.0 \text{ g Na}_2CO_3} = 6.5 \times 10^{-2} \text{ g Na}^+$

$\dfrac{6.5 \times 10^{-2} \text{ g Na}^+}{1.0 \text{ L}} \times \dfrac{1 \text{ L}}{1000 \text{ mL}} = \dfrac{6.5 \times 10^{-5} \text{ g Na}^+}{mL} = \dfrac{65 \times 10^{-6} \text{ g}}{mL} = 65$ ppm

b.   ppb $= \dfrac{\text{ng dioctylphthalate}}{mL}$

$\dfrac{2.5 \times 10^{-3} \text{ g}}{500.0 \text{ mL}} = \dfrac{5.0 \times 10^{-6} \text{ g}}{mL} = \dfrac{5000 \times 10^{-9} \text{ g}}{mL} = \dfrac{5.0 \times 10^{3} \text{ ng}}{mL} = 5.0 \times 10^{3}$ ppb

11.   Stock solution $= \dfrac{10.0 \text{ mg}}{500.0 \text{ mL}} = \dfrac{10.0 \times 10^{-3} \text{ g}}{500.0 \text{ mL}} = \dfrac{2.00 \times 10^{-5} \text{ g}}{mL}$

$100.0 \times 10^{-6} \text{ L stock} \times \dfrac{1000 \text{ mL}}{L} \times \dfrac{2.00 \times 10^{-5} \text{ g steroid}}{mL} = 2.00 \times 10^{-6} \text{ g steroid}$

This is diluted to a final volume of 100.0 mL.

$$\text{ppb steroid} = \frac{\text{g steroid}}{\text{g solution}} \times 10^9 = \frac{\text{ng steroid}}{\text{mL aqueous solution}}$$

$$\text{ppb steroid} = \frac{2.00 \times 10^{-6}\,\text{g}}{100.0\,\text{mL}} = \frac{2.00 \times 10^{-8}\,\text{g}}{\text{mL}} = \frac{20.0 \times 10^{-9}\,\text{g}}{\text{mL}} = 20.0\,\text{ppb}$$

$$\frac{20.0 \times 10^{-9}\,\text{g steroid}}{\text{mL}} \times \frac{1000\,\text{mL}}{\text{L}} \times \frac{1\,\text{mol steroid}}{336.4\,\text{g steroid}} = 5.95 \times 10^{-8}\,M$$

12.
$$\frac{30.\,\text{mg Mg}^{2+}}{\text{L}} \times \frac{1\,\text{g}}{1000\,\text{mg}} \times \frac{1\,\text{mol Mg}^{2+}}{24.3\,\text{g Mg}^{2+}} = \frac{1.2 \times 10^{-3}\,\text{mol}}{\text{L}}$$

$$\frac{75\,\text{mg Ca}^{2+}}{\text{L}} \times \frac{1\,\text{g}}{1000\,\text{mg}} \times \frac{1\,\text{mol Ca}^{2+}}{40.1\,\text{g Ca}^{2+}} = \frac{1.9 \times 10^{-3}\,\text{mol}}{\text{L}}$$

$$\frac{1250\,\text{mg Mg}^{2+}}{\text{L}} \times \frac{1\,\text{g}}{1000\,\text{mg}} \times \frac{1\,\text{mol Mg}^{2+}}{24.31\,\text{g}} = \frac{0.0514\,\text{mol}}{\text{L}}$$

$$\frac{200.\,\text{mg Ca}^{2+}}{\text{L}} \times \frac{1\,\text{g}}{1000\,\text{mg}} \times \frac{1\,\text{mol Ca}^{2+}}{40.08\,\text{g}} = \frac{4.99 \times 10^{-3}\,\text{mol}}{\text{L}}$$

13.
$$1\,\text{ppm} = \frac{1\,\mu\text{g}}{\text{mL}} = \frac{1\,\text{mg}}{\text{L}}$$

$$\frac{1 \times 10^{-3}\,\text{g F}^-}{\text{L}} \times \frac{1\,\text{mol F}^-}{19.0\,\text{g F}^-} = \frac{5 \times 10^{-5}\,\text{mol}}{\text{L}}$$

$$2\,\text{ppm F}^- = \frac{1 \times 10^{-4}\,\text{mol}}{\text{L}}; \quad 3\,\text{ppm F}^- = \frac{1.6 \times 10^{-4}\,\text{mol}}{\text{L}} = \frac{2 \times 10^{-4}\,\text{mol}}{\text{L}}$$

$$\frac{50. \times 10^{-3}\,\text{g}}{\text{L}} \times \frac{1\,\text{mol F}^-}{19.0\,\text{g F}^-} = \frac{2.6 \times 10^{-3}\,\text{mol}}{\text{L}}$$

14.   We want 100.0 mL of each standard. To make the 100. ppm standard:

$$\frac{100.\,\mu\text{g Cu}}{\text{mL}} \times 100.0\,\text{mL solution} = 1.00 \times 10^4\,\mu\text{g Cu needed}$$

$$1.00 \times 10^4\,\mu\text{g Cu} \times \frac{1\,\text{mL stock}}{1000.0\,\mu\text{g}} = 10.0\,\text{mL of stock solution}$$

Therefore, to make 100.0 mL of 100. ppm solution, transfer 10.0 mL of the 1000.0 ppm stock solution to a 100 mL volumetric flask and dilute to the mark.

Similarly for the:

75.0 ppm standard, dilute 7.50 mL of the 1000.0 ppm stock to 100.0 mL.

50.0 ppm standard, dilute 5.00 mL of the 1000.0 ppm stock to 100.0 mL.

25.0 ppm standard, dilute 2.50 mL of the 1000.0 ppm stock to 100.0 mL.

10.0 ppm standard, dilute 1.00 mL of the 1000.0 ppm stock to 100.0 mL.

15. Stock solution:

$$1.584 \text{ g Mn} \times \frac{1 \text{ mol Mn}}{54.94 \text{ g Mn}} = 2.883 \times 10^{-2} \text{ mol Mn}; \quad M = \frac{2.883 \times 10^{-2} \text{ mol}}{L}$$

Solution A contains:

$$50.00 \text{ mL} \times \frac{1 \text{ L}}{1000 \text{ mL}} \times \frac{2.883 \times 10^{-2} \text{ mol}}{L} = 1.442 \times 10^{-3} \text{ mol}$$

$$\text{molarity} = \frac{1.442 \times 10^{-3} \text{ mol}}{1000.0 \text{ mL}} \times \frac{1000 \text{ mL}}{L} = 1.442 \times 10^{-3} \ M$$

Solution B contains:

$$10.0 \text{ mL} \times \frac{1 \text{ L}}{1000 \text{ mL}} \times \frac{1.442 \times 10^{-3} \text{ mol}}{L} = 1.442 \times 10^{-5} \text{ mol}$$

$$\text{molarity} = \frac{1.442 \times 10^{-5} \text{ mol}}{0.250 \text{ L}} = 5.768 \times 10^{-5} \ M$$

Solution C contains:

$$10.00 \times 10^{-3} \text{ L} \times \frac{5.768 \times 10^{-5} \text{ mol}}{L} = 5.768 \times 10^{-7} \text{ mol}$$

$$\text{molarity} = \frac{5.768 \times 10^{-7} \text{ mol}}{0.500 \text{ L}} = 1.154 \times 10^{-6} \ M$$

16. a. Procedure i. The greatest error is on the measurement of the sample taken using the 10 mL pipet. The error is $\frac{0.01}{10.00} \times 100 = 0.1\%$. Procedure ii: The error in measuring the 0.050 mL sample is $\frac{1 \ \mu L}{0.050 \text{ mL} \times 1000 \ \mu L/\text{mL}} \times 100 = 2\%$. Three dilutions with a 0.1% error will be more accurate than a single dilution with 2% error.

b. $\dfrac{0.200 \text{ mol Fe}^{2+}}{L} \times \dfrac{55.85 \text{ g Fe}^{2+}}{\text{mol Fe}^{2+}} = 11.2 \text{ g/L} = 11.2 \text{ mg/mL}$

The conversions from g to mg and from L to mL cancel each other.

Precipitation Reactions

17.   a. $(NH_4)_2SO_4(aq) + Ba(NO_3)_2(aq) \rightarrow 2\ NH_4NO_3(aq) + BaSO_4(s)$

$Ba^{2+}(aq) + SO_4^{2-}(aq) \rightarrow BaSO_4(s)$

b. $Pb(NO_3)_2(aq) + 2\ NaCl(aq) \rightarrow PbCl_2(s) + 2\ NaNO_3(aq)$

$Pb^{2+}(aq) + 2\ Cl^-(aq) \rightarrow PbCl_2(s)$

c.   Potassium phosphate and sodium nitrate are both soluble in water. No reaction occurs.

d.   No reaction occurs.

e. $CuCl_2(aq) + 2\ NaOH(aq) \rightarrow Cu(OH)_2(s) + 2\ NaCl(aq)$

$Cu^{2+}(aq) + 2\ OH^-(aq) \rightarrow Cu(OH)_2(s)$

18.

a.

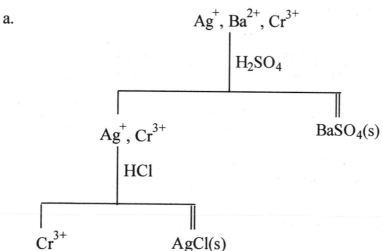

b.   c.

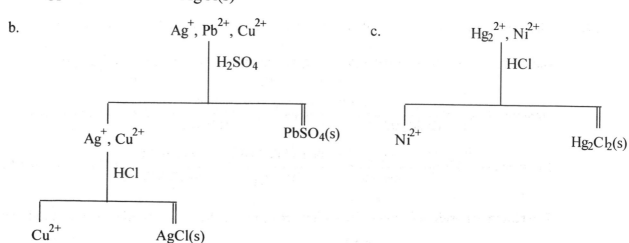

19.   Three possibilities are:

addition of H_2SO_4 solution to give a white ppt. of $PbSO_4$.

addition of HCl solution to give a white ppt. of $PbCl_2$.

addition of K_2CrO_4 solution to give a bright yellow ppt. of $PbCrO_4$.

20.   Since no precipitates formed upon addition of NaCl or Na_2SO_4, we can conclude that Hg_2^{2+} and Ba^{2+} are not present. The precipitate with NaOH is $Mn(OH)_2$. Therefore, Mn^{2+} was present.

21.   The reaction is: $Ni(NO_3)_2(aq) + 2\ NaOH(aq) \rightarrow Ni(OH)_2(s) + 2\ NaNO_3(aq)$

$$50.00 \text{ mL Ni(NO}_3)_2 \times \frac{1 \text{ L}}{1000 \text{ mL}} \times \frac{0.175 \text{ mol Ni(NO}_3)_2}{\text{L Ni(NO}_3)_2} \times \frac{2 \text{ mol NaOH}}{1 \text{ mol Ni(NO}_3)_2}$$

$$\times \frac{1 \text{ L NaOH}}{0.100 \text{ mol NaOH}} = 0.175 \text{ L or } 175 \text{ mL}$$

22.   The molar mass of $Al(C_9H_6NO)_3$ is:

$$1 \text{ mol Al} \left(\frac{26.98 \text{ g}}{\text{mol Al}} \right) + 27 \text{ mol C} \left(\frac{12.01 \text{ g}}{\text{mol C}} \right) + 3 \text{ mol N} \left(\frac{14.01 \text{ g}}{\text{mol N}} \right)$$

$$+ 3 \text{ mol O} \left(\frac{16.00 \text{ g}}{\text{mol O}} \right) + 18 \text{ mol H} \left(\frac{1.008 \text{ g}}{\text{mol H}} \right) = \frac{459.42 \text{ g}}{\text{mol}}$$

$$0.1248 \text{ g Al (C}_9H_6NO)_3 \times \frac{26.98 \text{ g Al}}{459.42 \text{ g Al(C}_9H_6NO)_3} = 7.329 \times 10^{-3} \text{ g Al}$$

$$\% \text{ Al} = \frac{7.329 \times 10^{-3} \text{ g}}{1.8571 \text{ g}} \times 100 = 0.3946\% \text{ Al}$$

23.   The unbalanced reaction is: $BaCl_2(aq) + Fe_2(SO_4)_3(aq) \rightarrow BaSO_4(s) + FeCl_3(aq)$

Balancing the equation: $3\ BaCl_2(aq) + Fe_2(SO_4)_3(aq) \rightarrow 3\ BaSO_4(s) + 2\ FeCl_3(aq)$

$$100.0 \text{ mL BaCl}_2 \times \frac{1 \text{ L}}{1000 \text{ mL}} \times \frac{0.100 \text{ mol BaCl}_2}{\text{L}} \times \frac{3 \text{ mol BaSO}_4}{3 \text{ mol BaCl}_2} = 1.00 \times 10^{-2} \text{ mol BaSO}_4$$

$$100.0 \text{ mL Fe}_2(SO_4)_3 \times \frac{1 \text{ L}}{1000 \text{ mL}} \times \frac{0.100 \text{ mol Fe}_2(SO_4)_3}{\text{L Fe}_2(SO_4)_3} \times \frac{3 \text{ mol BaSO}_4}{\text{mol Fe}_2(SO_4)_3} = 3.00 \times 10^{-2} \text{ mol BaSO}_4$$

The barium chloride solution is the limiting reagent and the mass of barium sulfate produced is:

$$1.00 \times 10^{-2} \text{ mol BaSO}_4 \times \frac{233.4 \text{ g}}{\text{mol}} = 2.33 \text{ g BaSO}_4$$

24. $1.00 \text{ L} \times \dfrac{0.200 \text{ mol Na}_2\text{S}_2\text{O}_3}{\text{L}} \times \dfrac{1 \text{ mol AgBr}}{2 \text{ mol Na}_2\text{S}_2\text{O}_3} \times \dfrac{187.8 \text{ g AgBr}}{\text{mol AgBr}} = 18.8 \text{ g AgBr}$

25. $0.1824 \text{ g TlI} \times \dfrac{204.4 \text{ g Tl}}{331.3 \text{ gTlI}} \times \dfrac{504.9 \text{ g Tl}_2\text{SO}_4}{408.8 \text{ g Tl}} = 0.1390 \text{ g Tl}_2\text{SO}_4$

$\dfrac{0.1390 \text{ g Tl}_2\text{SO}_4}{9.486 \text{ g pesticide}} \times 100 = 1.465\% \text{ Tl}_2\text{SO}_4$

26. $0.5032 \text{ g BaSO}_4 \times \dfrac{32.07 \text{ g S}}{233.4 \text{ g BaSO}_4} \times \dfrac{183.19 \text{ g saccharine}}{32.07 \text{ g S}} = 0.3949 \text{ g saccharin}$

$\dfrac{\text{Avg. Mass}}{\text{Tablet}} = \dfrac{0.3949 \text{ g}}{10 \text{ tablets}} = \dfrac{3.949 \times 10^{-2} \text{ g}}{\text{tablet}} = \dfrac{39.49 \text{ mg}}{\text{tablet}}$

Avg. Mass % $= \dfrac{0.3949 \text{ g saccharine}}{0.5894 \text{ g}} \times 100 = 67.00\% \text{ saccharin by mass}$

27. $37.20 \text{ ml} \times \dfrac{0.1000 \text{ mmol Ag}^+}{\text{mL}} \times \dfrac{1 \text{ mmol Cl}^-}{\text{mmol Ag}^+} \times \dfrac{1 \text{ mmol douglasite}}{4 \text{ mmol Cl}^-}$

$\times \dfrac{311.88 \text{ mg douglasite}}{\text{mmol}} = 290.0 \text{ mg douglasite}$

$\dfrac{290.0 \text{ mg}}{455.0 \text{ mg}} \times 100 = 63.74\%$

28. $100.0 \text{ mL} \times \dfrac{0.0426 \text{ mmol Ba}^{2+}}{\text{mL}} = 4.26 \text{ mmol Ba}^{2+}$

$50.0 \text{ mL} \times \dfrac{0.2000 \text{ mmol SO}_4^{2-}}{\text{mL}} = 10.0 \text{ mmol SO}_4^{2-}$

The Ba^{2+} is limiting. We form 4.26 mmol $BaSO_4$.

$4.26 \times 10^{-3} \text{ mol BaSO}_4 \times \dfrac{233.4 \text{ g BaSO}_4}{\text{mol}} = 0.994 \text{ g BaSO}_4$

Acid-Base Reactions

29.   a. $NH_3(aq) + HNO_3(aq) \rightarrow NH_4NO_3(aq)$ (molecular equation)

$NH_3(aq) + H^+(aq) + NO_3^-(aq) \rightarrow NH_4^+(aq) + NO_3^-(aq)$ (complete ionic equation)

$NH_3(aq) + H^+(aq) \rightarrow NH_4^+(aq)$ (net ionic equation)

b. $$Ba(OH)_2(aq) + 2\ HCl(aq) \rightarrow 2\ H_2O(l) + BaCl_2(aq)$$

$$Ba^{2+}(aq) + 2\ OH^-(aq) + 2\ H^+(aq) + 2\ Cl^-(aq) \rightarrow Ba^{2+}(aq) + 2\ Cl^-(aq) + 2\ H_2O(l)$$

$$OH^-(aq) + H^+(aq) \rightarrow H_2O(l)$$

c. $$3\ HClO_4(aq) + Fe(OH)_3(s) \rightarrow 3\ H_2O(l) + Fe(ClO_4)_3(aq)$$

$$3\ H^+(aq) + 3\ ClO_4^-(aq) + Fe(OH)_3(s) \rightarrow 3\ H_2O(l) + Fe^{3+}(aq) + 3\ ClO_4^-(aq)$$

$$3\ H^+(aq) + Fe(OH)_3(s) \rightarrow 3\ H_2O(l) + Fe^{3+}(aq)$$

d. $$AgOH(s) + HBr(aq) \rightarrow AgBr(s) + H_2O(l)$$

$$AgOH(s) + H^+(aq) + Br^-(aq) \rightarrow AgBr(s) + H_2O(l)$$

$$AgOH(s) + H^+(aq) + Br^-(aq) \rightarrow AgBr(s) + H_2O(l)$$

30. . a. $$MgO(s) + 2\ HCl(aq) \rightarrow MgCl_2(aq) + H_2O(l)$$

$$Mg(OH)_2(s) + 2\ HCl(aq) \rightarrow MgCl_2(aq) + 2\ H_2O(l)$$

$$Al(OH)_3(s) + 3\ HCl(aq) \rightarrow AlCl_3(aq) + 3\ H_2O(l)$$

b. Let's calculate the number of moles of HCl neutralized per gram. We can get these directly from the balanced equation.

$$\frac{2\ \text{mol HCl}}{\text{mol MgO}} \times \frac{1\ \text{mol MgO}}{40.3\ \text{g MgO}} = \frac{5.0 \times 10^{-2}\ \text{mol HCl}}{\text{g MgO}}$$

$$\frac{2\ \text{mol HCl}}{\text{mol Mg(OH)}_2} \times \frac{1\ \text{mol Mg(OH)}_2}{58.3\ \text{g Mg(OH)}_2} = \frac{3.4 \times 10^{-2}\ \text{mol HCl}}{\text{g Mg(OH)}_2}$$

$$\frac{3\ \text{mol HCl}}{\text{mol Al(OH)}_3} \times \frac{1\ \text{mol Al(OH)}_3}{78.0\ \text{g Al(OH)}_3} = \frac{3.8 \times 10^{-2}\ \text{mol HCl}}{\text{g Al(OH)}_3}$$

Therefore, one gram of magnesium oxide would neutralize the most 0.10 M HCl.

31. We get the empirical formula from the elemental analysis. Out of 100.00 g carminic acid there are:

$$53.66\ \text{g C} \times \frac{1\ \text{mol C}}{12.011\ \text{g C}} = 4.468\ \text{mol C}$$

$$4.09\ \text{g H} \times \frac{1\ \text{mol H}}{1.008\ \text{g H}} = 4.06\ \text{mol H}$$

$$42.25\ \text{g O} \times \frac{1\ \text{mol O}}{15.999\ \text{g O}} = 2.641\ \text{mol O}$$

Taking ratios in the usual way: $\dfrac{4.468}{4.06} = 1.10 = \dfrac{11}{10}$

So let's try $\dfrac{4.06}{10} = 0.406$ as a common factor.

$\dfrac{4.468}{0.406} = 11.0$ $\quad$ $\dfrac{4.06}{0.406} = 10.0$ $\quad$ $\dfrac{2.641}{0.406} = 6.50$

$C_{22}H_{20}O_{13}$ is the empirical formula, to make all units whole numbers.

We can get the molecular weight from the titration data:

$$18.02 \times 10^{-3} \text{ L soln} \times \frac{0.0406 \text{ mol NaOH}}{\text{L soln}} \times \frac{1 \text{ mol carminic acid}}{\text{mol NaOH}}$$

$$= 7.32 \times 10^{-4} \text{ mol carminic acid}$$

Molar mass $= \dfrac{0.3602 \text{ g}}{7.32 \times 10^{-4} \text{ mol}} = \dfrac{492 \text{ g}}{\text{mol}}$

The mass of $C_{22}H_{20}O_{13} \approx 22(12) + 20(1) + 13(16) = 492$ g. Therefore, the molecular formula of carminic acid is $C_{22}H_{20}O_{13}$.

32.   If we begin with 50.00 mL of 0.100 M NaOH, then:

$$50.00 \times 10^{-3} \text{ L} \times \frac{0.100 \text{ mol}}{\text{L}} = 5.00 \times 10^{-3} \text{ mol NaOH to be neutralized.}$$

a.   NaOH(aq) + HCl(aq) → NaCl(aq) + H_2O(l)

$$5.00 \times 10^{-3} \text{ mol NaOH} \times \frac{1 \text{ mol HCl}}{\text{mol NaOH}} \times \frac{1 \text{ L soln}}{0.100 \text{ mol}} = 5.00 \times 10^{-2} \text{ L or 50.0 mL}$$

b.   2 NaOH + H_2SO_3 → 2 H_2O + Na_2SO_3

$$5.00 \times 10^{-3} \text{ mol NaOH} \times \frac{1 \text{ mol } H_2SO_3}{2 \text{ mol NaOH}} \times \frac{1 \text{ L soln}}{0.100 \text{ mol } H_2SO_3} = 2.50 \times 10^{-2} \text{ L or 25.0 mL}$$

c.   3 NaOH + H_3PO_4 → Na_3PO_4 + 3 H_2O

$$5.00 \times 10^{-3} \text{ mol NaOH} \times \frac{1 \text{ mol } H_3PO_4}{3 \text{ mol NaOH}} \times \frac{1 \text{ L soln}}{0.200 \text{ mol } H_3PO_4} = 8.33 \times 10^{-3} \text{ L or 8.33 mL}$$

d. HNO_3(aq) + NaOH(aq) → H_2O(l) + $NaNO_3$(aq)

$$5.00 \times 10^{-3} \text{ mol NaOH} \times \frac{1 \text{ mol } HNO_3}{\text{mol NaOH}} \times \frac{1 \text{ L soln}}{0.150 \text{ mol } HNO_3} = 3.33 \times 10^{-2} \text{ L or 33.3 mL}$$

e. $HC_2H_3O_2(aq) + NaOH(aq) \rightarrow H_2O(l) + NaC_2H_3O_2(aq)$

$$5.00 \times 10^{-3} \text{ mol NaOH} \times \frac{1 \text{ mol } HC_2H_3O_2}{\text{mol NaOH}} \times \frac{1 \text{ L soln}}{0.200 \text{ mol } HC_2H_3O_2} = 2.50 \times 10^{-2} \text{ L or } 25.0 \text{ mL}$$

f. $H_2SO_4(aq) + 2 \text{ NaOH}(aq) \rightarrow 2 \text{ H}_2O(l) + Na_2SO_4(aq)$

$$5.00 \times 10^{-3} \text{ mol NaOH} \times \frac{1 \text{ mol } H_2SO_4}{2 \text{ mol NaOH}} \times \frac{1 \text{ L soln}}{0.300 \text{ mol } H_2SO_4} = 8.33 \times 10^{-3} \text{ L or } 8.33 \text{ mL}$$

33.   Since KHP is a monoprotic acid, the reaction is (HX is an abbreviation for KHP):

$$NaOH(aq) + HX(aq) \rightarrow NaX(aq) + H_2O(l)$$

$$0.1082 \text{ g KHP} \times \frac{1 \text{ mol KHP}}{204.22 \text{ g KHP}} \times \frac{1 \text{ mol NaOH}}{\text{mol KHP}} = 5.298 \times 10^{-4} \text{ mol NaOH}.$$

There is 5.298×10^{-4} mol of sodium hydroxide in 20.46 mL of solution. Therefore, the concentration of sodium hydroxide is:

$$\frac{5.298 \times 10^{-4} \text{ mol}}{20.46 \times 10^{-3} \text{ L}} = 2.589 \times 10^{-2} \text{ } M$$

34.   The pertinent reactions are:

$$2 \text{ NaOH}(aq) + H_2SO_4(aq) \rightarrow Na_2SO_4(aq) + 2 \text{ H}_2O(l)$$

$$HCl(aq) + NaOH(aq) \rightarrow NaCl(aq) + H_2O(l)$$

Amount of NaOH added $= 0.0500 \text{ L} \times \dfrac{0.213 \text{ mol}}{\text{L}} = 1.07 \times 10^{-2} \text{ mol NaOH}$

Amount of NaOH neutralized by HCl:

$$0.01321 \text{ L HCl} \times \frac{0.103 \text{ mol HCl}}{\text{L HCl}} \times \frac{1 \text{ mol NaOH}}{\text{mol HCl}} = 1.36 \times 10^{-3} \text{ mol NaOH}$$

The difference, 9.3×10^{-3} mol, is the amount of NaOH neutralized by the sulfuric acid.

$$9.3 \times 10^{-3} \text{ mol NaOH} \times \frac{1 \text{ mol } H_2SO_4}{2 \text{ mol NaOH}} = 4.7 \times 10^{-3} \text{ mol } H_2SO_4$$

Concentration of $H_2SO_4 = \dfrac{4.7 \times 10^{-3} \text{ mol}}{0.1000 \text{ L}} = 4.7 \times 10^{-2} \text{ } M$

35. $2.58 \text{ g Ca} \times \dfrac{1 \text{ mol Ca}}{40.08 \text{ g Ca}} \times \dfrac{1 \text{ mol Ca(OH)}_2}{\text{mol Ca}} \times \dfrac{2 \text{ mol OH}^-}{\text{mol Ca(OH)}_2} = 0.129 \text{ mol OH}^-$

Molarity $= \dfrac{0.129 \text{ mol}}{100.0 \times 10^{-3} \text{ L}} = 1.29 \ M$

36. $35.08 \text{ mL NaOH} \times \dfrac{1 \text{ L}}{1000 \text{ mL}} \times \dfrac{2.12 \text{ mol NaOH}}{\text{L soln}} \times \dfrac{1 \text{ mol H}_2\text{SO}_4}{2 \text{ mol NaOH}} = 3.72 \times 10^{-2} \text{ mol H}_2\text{SO}_4$

Molarity $= \dfrac{3.72 \times 10^{-2} \text{ mol}}{10.00 \text{ mL}} \times \dfrac{1000 \text{ mL}}{\text{L}} = 3.72 \ M$

37. $CH_3CO_2H(aq) + NaOH(aq) \rightarrow H_2O(l) + CH_3CO_2Na(aq)$

a. $16.58 \times 10^{-3} \text{ L soln} \times \dfrac{0.5062 \text{ mol NaOH}}{\text{L soln}} \times \dfrac{1 \text{ mol acetic acid}}{\text{mol NaOH}} = 8.393 \times 10^{-3} \text{ mol acetic acid}$

Concentration of acetic acid $= \dfrac{8.393 \times 10^{-3} \text{ mol}}{0.01000 \text{ L}} = 0.8393 \ M$

b. If we have 1.000 L of solution:

Total mass $= 1000. \text{ mL} \times \dfrac{1.006 \text{ g}}{\text{mL}} = 1006 \text{ g}$

Mass of acetic acid $= 0.8393 \text{ mol} \times \dfrac{60.052 \text{ g}}{\text{mol}} = 50.40 \text{ g}$

% acetic acid $= \dfrac{50.40 \text{ g}}{1006 \text{ g}} \times 100 = 5.010\%$

38. $HCl(aq) + NaOH(aq) \rightarrow H_2O(l) + NaCl(aq)$

$24.16 \times 10^{-3} \text{ L NaOH soln} \times \dfrac{0.106 \text{ mol NaOH}}{\text{L}} \times \dfrac{1 \text{ mol HCl}}{\text{mol NaOH}} = 2.56 \times 10^{-3} \text{ mol HCl}$

Molarity of HCl $= \dfrac{2.56 \times 10^{-3} \text{ mol}}{25.00 \times 10^{-3} \text{ L}} = \dfrac{0.102 \text{ mol}}{\text{L}}$

39. $15.0 \text{ g NaOH} \times \dfrac{1 \text{ mol NaOH}}{40.00 \text{ g}} = 0.375 \text{ mol NaOH}$

$0.1500 \text{ L} \times \dfrac{0.250 \text{ mol HNO}_3}{\text{L}} = 0.0375 \text{ mol HNO}_3$

The reaction is: $HNO_3(aq) + NaOH(aq) \rightarrow NaNO_3(aq) + H_2O(l)$

We have excess NaOH. The solution will be basic. In the final solution there will be 0.0375 mol NO_3^-, (0.375 - 0.0375) = 0.338 mol OH^-, and 0.375 mol Na^+.

$$C_{NO_3^-} = \frac{0.0375 \text{ mol}}{0.1500 \text{L}} = \frac{0.250 \text{ mol NO}_3^-}{L}; \ \ C_{Na^+} = \frac{0.375 \text{ mol}}{0.1500 \text{ L}} = \frac{2.50 \text{ mol Na}^+}{L}$$

$$C_{OH^-} = \frac{0.338 \text{ mol}}{0.1500 \text{ L}} = \frac{2.25 \text{ mol OH}^-}{L}$$

40. $39.47 \times 10^{-3} \text{ L HCl} \times \frac{0.0984 \text{ mol HCl}}{L} \times \frac{1 \text{ mol NH}_3}{\text{mol HCl}} = 3.88 \times 10^{-3} \text{ mol NH}_3$

Molarity of $NH_3 = \frac{3.88 \times 10^{-3} \text{ mol}}{50.00 \times 10^{-3} \text{ L}} = \frac{0.0776 \text{ mol}}{L}$

41. $Ba(OH)_2(aq) + 2 \text{ HCl}(aq) \rightarrow BaCl_2(aq) + 2 \text{ H}_2O(l); \ \ H^+(aq) + OH^-(aq) \rightarrow H_2O(l)$

$$75.0 \times 10^{-3} \text{ L} \times \frac{0.250 \text{ mol HCl}}{L} = 1.88 \times 10^{-2} \text{ mol HCl} = 1.88 \times 10^{-2} \text{ mol H}^+$$

$$225.0 \times 10^{-3} \text{ L} \times \frac{0.0550 \text{ mol Ba(OH)}_2}{L} \times \frac{2 \text{ mol OH}^-}{\text{mol Ba(OH)}_2} = 2.48 \times 10^{-2} \text{ mol OH}^-$$

We have an excess of OH^-. Since 1.88×10^{-2} mol OH^- will neutralize all of the H^+ present, we will have $(2.48 - 1.88) \times 10^{-2} = 0.60 \times 10^{-2}$ mol OH^- left unreacted.

$$C_{OH^-} = \frac{6.0 \times 10^{-3} \text{ mol}}{300.0 \text{ mL}} \times \frac{1000 \text{ mL}}{L} = \frac{2.0 \times 10^{-2} \text{ mol OH}^-}{L}$$

42. $H_2SO_4(aq) + 2 \text{ NaOH}(aq) \rightarrow Na_2SO_4(aq) + 2 \text{ H}_2O(l)$

$$28.44 \text{ mL} \times \frac{0.1000 \text{ mmol NaOH}}{\text{mL}} \times \frac{1 \text{ mmol H}_2\text{SO}_4}{2 \text{ mmol NaOH}} \times \frac{1 \text{ mmol SO}_2}{\text{mmol H}_2\text{SO}_4} \times \frac{32.07 \text{ mg S}}{\text{mmol SO}_2}$$

$$= 45.60 \text{ mg S} = 4.560 \times 10^{-2} \text{ g S}$$

$\% \text{ S} = \frac{0.04560 \text{ g}}{1.325 \text{ g}} \times 100 = 3.442\%$ by mass

Oxidation-Reduction Reactions

43.   a. K, +1: O, -2; Mn, +7   b. Ni, +4:, O, -2

c. $K_4Fe(CN)_6$; Fe, +2

K^+ ions and $Fe(CN)_6^{4-}$ anions; $Fe(CN)_6^{4-}$ composed of Fe^{2+} and CN^- ions.

d. $(NH_4)_2HPO_4$ is made of NH_4^+ cations and HPO_4^{2-} anions. Assign +1 as oxidation number of H and -2 as oxidation number of O. Then we get: N, -3: P, +5

e. P, +3: O, -2   f. O, -2: Fe, +8/3

g. O, -2: F, -1: Xe, +6   h. S, +4: F, -1

i. C, +2: O, -2   j. Na, +1: O, -2: C, +3

44. OCl^-: Oxidation number of oxygen is (-2).

       $-2 + x = -1$, $x = +1$: The oxidation number of Cl in OCl^- is +1.

       ClO_2^-: $2(-2) + x = -1$, $x = +3$ ClO_3^-: $3(-2) + x = -1$, $x = +5$

       ClO_4^-: $4(-2) + x = -1$, $x = +7$

45.   a. -3   b. -3   c. $2(x) + 4 = 0$, $x = -2$

       d. +2   e. +1   f. +4

       g. +3   h. +5   i. 0

46.   a. UO_2^{2+}: O, -2: U, $x + 2(-2) = +2$, $x = \underline{+6}$

       b. As_2O_3: O, -2: As, $2(x) + 3(-2) = 0$, $x = \underline{+3}$

       c. $NaBiO_3$: Na, +1: O, -2: Bi, $+1 + x + 3(-2) = 0$, $x = \underline{+5}$

       d. As_4: As, 0

       e. $HAsO_2$: assign H = +1 and O = -2: As, $+1 + x + 2(-2) = 0$; $x = \underline{+3}$

       f. $Mg_2P_2O_7$: Composed of Mg^{2+} ions and $P_2O_7^{4-}$ ions. Mg, +2: O, -2: P, +5

       g. $Na_2S_2O_3$: Composed of Na^+ ions and $S_2O_3^{2-}$ ions. Na, +1: O, -2: S, +2

       h. Hg_2Cl_2: Hg, +1: Cl, -1

       i. $Ca(NO_3)_2$: Composed of Ca^{2+} ions and NO_3^- ions. Ca, +2: O, -2: N, +5

47.

	Redox?	Oxidizing Agent	Reducing Agent	Substance Oxidized	Substance Reduced
a.	Yes	O_2	CH_4	CH_4 (C)	O_2 (O)
b.	Yes	HCl	Zn	Zn	HCl (H)
c.	No	-	-	-	-
d.	Yes	O_3	NO	NO (N)	O_3 (O)
e.	Yes	H_2O_2	H_2O_2	H_2O_2 (O)	H_2O_2 (O)
f.	Yes	CuCl	CuCl	CuCl (Cu)	CuCl (Cu)
g.	No	-	-	-	-
h.	No	-	-	-	-
i.	Yes	$SiCl_4$	Mg	Mg	$SiCl_4$ (Si)

In c, g, and h no oxidation numbers change from reactants to products.

48.   a.   The first step is to assign oxidation numbers to all atoms (see numbers above the atoms).

$$\overset{-3\ +1}{C_2H_6} + \overset{0}{O_2} \rightarrow \overset{+4\ -2}{CO_2} + \overset{+1\ -2}{H_2O}$$

Each carbon atom changes from -3 to +4, an increase of seven. Each oxygen atom changes from 0 to -2, a decrease of 2. We need 7/2 O-atoms for every C-atom.

$$C_2H_6 + 7/2\ O_2 \rightarrow CO_2 + H_2O$$

Balancing the remainder of the equation by inspection:

$$C_2H_6(g) + 7/2\ O_2(g) \rightarrow 2\ CO_2(g) + 3\ H_2O(g)$$

or

$$2\ C_2H_6(g) + 7\ O_2(g) \rightarrow 4\ CO_2(g) + 6\ H_2O(g)$$

   b.   The oxidation state of magnesium changes from 0 to +2, an increase of 2. The oxidation state of hydrogen changes from +1 to 0, a decrease of 1. We need 2 H-atoms for every Mg-atom. The balanced equation is:

$$Mg(s) + 2\ HCl(aq) \rightarrow Mg^{2+}(aq) + 2\ Cl^-(aq) + H_2(g)$$

   c.   The oxidation state of copper increases by 2 (0 to +2) and the oxidation state of silver decreases by 1 (+1 to 0). We need 2 Ag-atoms for every Cu-atom. The balanced equation is:

$$Cu(s) + 2\ Ag^+(aq) \rightarrow Cu^{2+}(aq) + 2\ Ag(s)$$

   d.   The oxidation state of copper increases by 2 (0 to +2) and the oxidation state of nitrogen decreases by 3 (+5 to +2 in NO). We need 3 Cu-atoms for every 2 N-atoms which are reduced.

$$3\ Cu + 2\ HNO_3 \rightarrow 3\ Cu(NO_3)_2 + 2\ NO$$

Balance the rest by inspection. Six NO_3^- ions must be added to the reactants to balance the six NO_3^- ions on the product side. Note: Some of the nitrate ions were not reduced.

$$3\ Cu + 8\ HNO_3 \rightarrow 3\ Cu(NO_3)_2 + 2\ NO$$

Left: 8-H and 24-O;   Right: 0-H and 20-O; Left has 8-H and 4-O extra, enough for 4 H_2O. The balanced equation is :

$$3\ Cu(s) + 8\ HNO_3(aq) \rightarrow 3\ Cu(NO_3)_2(aq) + 2\ NO(g) + 4\ H_2O(l)$$

e. The equation is balanced. Each hydrogen atom gains one electron $(+1 \rightarrow 0)$ and each zinc atom loses two electrons $(0 \rightarrow +2)$. We need 2 H-atoms for every Zn-atom. This is the ratio in the given equation:

$$Zn(s) + H_2SO_4(aq) \rightarrow ZnSO_4(aq) + H_2(g)$$

49. a. Review section 4.11 of the text for rules on balancing by the half-reaction method. The first step is to balance the two half-reactions.

$$(Cu \rightarrow Cu^{2+} + 2\ e^-) \times 3 \qquad\qquad HNO_3 \rightarrow NO + 2\ H_2O$$
$$(3\ e^- + 3\ H^+ + HNO_3 \rightarrow NO + 2\ H_2O) \times 2$$

Adding the two balanced half reactions so electrons cancel:

$$3\ Cu \rightarrow 3\ Cu^{2+} + 6\ e^-$$
$$6\ e^- + 6\ H^+ + 2\ HNO_3 \rightarrow 2\ NO + 4\ H_2O$$

$$\overline{\hspace{3cm}}$$

$$3\ Cu(s) + 6\ H^+(aq) + 2\ HNO_3(aq) \rightarrow 3\ Cu^{2+}(aq) + 2\ NO(g) + 4\ H_2O(l)$$

or $\qquad\qquad 3\ Cu(s) + 8\ HNO_3(aq) \rightarrow 3\ Cu(NO_3)_2(aq) + 2\ NO(g) + 4\ H_2O(l)$

b. $(2\ Cl^- \rightarrow Cl_2 + 2\ e^-) \times 3 \qquad\qquad Cr_2O_7^{2-} \rightarrow 2\ Cr^{3+} + 7\ H_2O$
$$6\ e^- + 14\ H^+ + Cr_2O_7^{2-} \rightarrow 2\ Cr^{3+} + 7\ H_2O$$

Adding the two balanced half reactions so electrons cancel:

$$6\ Cl^- \rightarrow 3\ Cl_2 + 6\ e^-$$
$$6\ e^- + 14\ H^+ + Cr_2O_7^{2-} \rightarrow 2\ Cr^{3+} + 7\ H_2O$$

$$\overline{\hspace{3cm}}$$

$$14\ H^+(aq) + Cr_2O_7^{2-}(aq) + 6\ Cl^-(aq) \rightarrow 3\ Cl_2(g) + 2\ Cr^{3+}(aq) + 7\ H_2O(l)$$

c. $\qquad\qquad Pb \rightarrow PbSO_4 \qquad\qquad\qquad\qquad PbO_2 \rightarrow PbSO_4$
$Pb + H_2SO_4 \rightarrow PbSO_4 + 2\ H^+ \qquad\quad PbO_2 + H_2SO_4 \rightarrow PbSO_4 + 2\ H^+$
$Pb + H_2SO_4 \rightarrow PbSO_4 + 2\ H^+ + 2\ e^- \qquad 2\ e^- + 2\ H^+ + PbO_2 + H_2SO_4 \rightarrow PbSO_4 + 2\ H_2O$

Add the two half reactions:

$$2\ e^- + 2\ H^+ + PbO_2 + H_2SO_4 \rightarrow PbSO_4 + 2\ H_2O$$
$$Pb + H_2SO_4 \rightarrow PbSO_4 + 2\ H^+ + 2\ e^-$$

$$\overline{\hspace{3cm}}$$

$$Pb(s) + 2\ H_2SO_4(aq) + PbO_2(s) \rightarrow 2\ PbSO_4(s) + 2\ H_2O(l)$$

This is the reaction that occurs in an automobile lead storage battery.

d. $Mn^{2+} \rightarrow MnO_4^-$
                    $(4 H_2O + Mn^{2+} \rightarrow MnO_4^- + 8 H^+ + 5 e^-) \times 2$

                                                    $NaBiO_3 \rightarrow Bi^{3+} + Na^+$
                                          $6 H^+ + NaBiO_3 \rightarrow Bi^{3+} + Na^+ + 3 H_2O$
                              $(2 e^- + 6 H^+ + NaBiO_3 \rightarrow Bi^{3+} + Na^+ + 3 H_2O) \times 5$

                      $8 H_2O + 2 Mn^{2+} \rightarrow 2 MnO_4^- + 16 H^+ + 10 e^-$
            $10 e^- + 30 H^+ + 5 NaBiO_3 \rightarrow 5 Bi^{3+} + 5 Na^+ + 15 H_2O$

_____

$8 H_2O + 30 H^+ + 2 Mn^{2+} + 5 NaBiO_3 \rightarrow 2 MnO_4^- + 5 Bi^{3+} + 5 Na^+ + 15 H_2O + 16 H^+$

Simplifying:

$14 H^+ (aq) + 2 Mn^{2+}(aq) + 5 NaBiO_3(s) \rightarrow 2 MnO_4^-(aq) + 5 Bi^{3+}(aq) + 5 Na^+(aq) + 7 H_2O(l)$

e. $H_3AsO_4 \rightarrow AsH_3$ $(Zn \rightarrow Zn^{2+} + 2 e^-) \times 4$
                        $H_3AsO_4 \rightarrow AsH_3 + 4 H_2O$
            $8 e^- + 8 H^+ + H_3AsO_4 \rightarrow AsH_3 + 4 H_2O$

                      $8 e^- + 8 H^+ + H_3AsO_4 \rightarrow AsH_3 + 4 H_2O$
                                    $4 Zn \rightarrow 4 Zn^{2+} + 8 e^-$

_____

        $8 H^+(aq) + H_3AsO_4(aq) + 4 Zn(s) \rightarrow 4 Zn^{2+}(aq) + AsH_3(g) + 4 H_2O(l)$

f. $As_2O_3 \rightarrow H_3AsO_4$ $NO_3^- \rightarrow NO + 2 H_2O$
      $As_2O_3 \rightarrow 2 H_3AsO_4$ $4 H^+ + NO_3^- \rightarrow NO + 2 H_2O$
                                                              $(3 e^- + 4 H^+ + NO_3^- \rightarrow NO + 2 H_2O) \times 4$

      left 3 - O, right 8 - O

Balance the oxygen atoms first, using H_2O, then balance H using H^+.

$(5 H_2O + As_2O_3 \rightarrow 2 H_3AsO_4 + 4 H^+ + 4 e^-) \times 3$

Common factor is a transfer of 12 e$^-$. Adding the balanced half reactions:

                  $12 e^- + 16 H^+ + 4 NO_3^- \rightarrow 4 NO + 8 H_2O$
                        $15 H_2O + 3 As_2O_3 \rightarrow 6 H_3AsO_4 + 12 H^+ + 12 e^-$

_____

$7 H_2O(l) + 4 H^+(aq) + 3 As_2O_3(s) + 4 NO_3^-(aq) \rightarrow 4 NO(g) + 6 H_3AsO_4(aq)$

g. $(2 Br^- \rightarrow Br_2 + 2 e^-) \times 5$ $MnO_4^- \rightarrow Mn^{2+} + 4 H_2O$
                                                              $(5 e^- + 8 H^+ + MnO_4^- \rightarrow Mn^{2+} + 4 H_2O) \times 2$

$$10\ Br^- \rightarrow 5\ Br_2 + 10\ e^-$$
$$10\ e^- + 16\ H^+ + 2\ MnO_4^- \rightarrow 2\ Mn^{2+} + 8\ H_2O$$

$$16\ H^+(aq) + 2\ MnO_4^-(aq) + 10\ Br^-(aq) \rightarrow 5\ Br_2(l) + 2\ Mn^{2+}(aq) + 8\ H_2O(l)$$

h. $CH_3OH \rightarrow CH_2O$ $\qquad\qquad\qquad\qquad\qquad\qquad Cr_2O_7^{2-} \rightarrow Cr^{3+}$
$\quad (CH_3OH \rightarrow CH_2O + 2\ H^+ + 2\ e^-) \times 3 \qquad\qquad 14\ H^+ + Cr_2O_7^{2-} \rightarrow 2\ Cr^{3+} + 7\ H_2O$
$\qquad\qquad\qquad\qquad\qquad\qquad\qquad\qquad\quad 6\ e^- + 14\ H^+ + Cr_2O_7^{2-} \rightarrow 2\ Cr^{3+} + 7\ H_2O$

$$3\ CH_3OH \rightarrow 3\ CH_2O + 6\ H^+ + 6\ e^-$$
$$6\ e^- + 14\ H^+ + Cr_2O_7^{2-} \rightarrow 2\ Cr^{3+} + 7\ H_2O$$

$$8\ H^+(aq) + 3\ CH_3OH(aq) + Cr_2O_7^{2-}(aq) \rightarrow 2\ Cr^{3+}(aq) + 3\ CH_2O(aq) + 7\ H_2O(l)$$

50. a. For basic solutions, use the same half reaction balancing procedure as with acidic solutions, with an extra step. The extra step is to convert the H^+ into H_2O by adding equal moles of OH^- to each side of reaction. This converts the reaction to a basic solution while keeping it balanced.

$\qquad\qquad\qquad Al \rightarrow Al(OH)_4^- \qquad\qquad\qquad\qquad\qquad\qquad MnO_4^- \rightarrow MnO_2$
$\qquad\qquad 4\ OH^- + Al \rightarrow Al(OH)_4^- \qquad\qquad\qquad 4\ OH^- + 4\ H^+ + MnO_4^- \rightarrow MnO_2 + 2\ H_2O + 4\ OH^-$
$\qquad\qquad 4\ OH^- + Al \rightarrow Al(OH)_4^- + 3\ e^- \qquad\qquad\quad 2\ H_2O + MnO_4^- \rightarrow MnO_2 + 4\ OH^-$
$\qquad\qquad\qquad\qquad\qquad\qquad\qquad\qquad\qquad\qquad 3e^- + 2\ H_2O + MnO_4^- \rightarrow MnO_2 + 4OH^-$

$$4\ OH^- + Al \rightarrow Al(OH)_4^- + 3\ e^-$$
$$3\ e^- + 2\ H_2O + MnO_4^- \rightarrow MnO_2 + 4\ OH^-$$

$$2\ H_2O(l) + Al(s) + MnO_4^-(aq) \rightarrow Al(OH)_4^-(aq) + MnO_2(s)$$

b. $\qquad\quad Cl_2 \rightarrow Cl^- \qquad\qquad\qquad\qquad\qquad\qquad\qquad Cl_2 \rightarrow ClO^-$
$\qquad 2\ e^- + Cl_2 \rightarrow 2\ Cl^- \qquad\qquad\qquad\qquad\quad 2\ H_2O + Cl_2 \rightarrow 2\ ClO^- + 4\ H^+$
$\qquad\qquad\qquad\qquad\qquad\qquad\qquad 4\ OH^- + 2\ H_2O + Cl_2 \rightarrow 2\ ClO^- + 4\ H^+ + 4\ OH^-$
$\qquad\qquad\qquad\qquad\qquad\qquad\qquad\qquad 4\ OH^- + Cl_2 \rightarrow 2\ ClO^- + 2\ H_2O + 2\ e^-$

$$2\ e^- + Cl_2 \rightarrow 2\ Cl^-$$
$$4\ OH^- + Cl_2 \rightarrow 2\ ClO^- + 2\ H_2O + 2\ e^-$$

$$4\ OH^- + 2\ Cl_2 \rightarrow 2\ Cl^- + 2\ ClO^- + 2\ H_2O$$

Reducing: $2\ OH^-(aq) + Cl_2(g) \rightarrow Cl^-(aq) + ClO^-(aq) + H_2O(l)$

c. $\qquad\qquad\qquad\qquad NO_2^- \rightarrow NH_3$
$\qquad\qquad\qquad 7\ H^+ + NO_2^- \rightarrow NH_3 + 2\ H_2O$
$\qquad\qquad 7\ OH^- + 7\ H^+ + NO_2^- \rightarrow NH_3 + 2\ H_2O + 7\ OH^-$
$\qquad\qquad\qquad 5\ H_2O + NO_2^- \rightarrow NH_3 + 7\ OH^-$
$\qquad\qquad 6\ e^- + NO_2^- + 5\ H_2O \rightarrow NH_3 + 7\ OH^-$

$$Al \rightarrow AlO_2^-$$
$$4\ OH^- + 2\ H_2O + Al \rightarrow AlO_2^- + 4\ H^+ + 4\ OH^-$$
$$4\ OH^- + Al \rightarrow AlO_2^- + 2\ H_2O$$
$$(4\ OH^- + Al \rightarrow AlO_2^- + 2\ H_2O + 3\ e^-) \times 2$$

$$6e^- + NO_2^- + 5\ H_2O \rightarrow NH_3 + 7\ OH^-$$
$$8\ OH^- + 2\ Al \rightarrow 2\ AlO_2^- + 4\ H_2O + 6\ e^-$$
_____
$$OH^-(aq) + H_2O(l) + NO_2^-(aq) + 2\ Al(s) \rightarrow NH_3(g) + 2\ AlO_2^-(aq)$$

d. $S^{2-} \rightarrow S$ $MnO_4^- \rightarrow MnS$
     $(S^{2-} \rightarrow S + 2\ e^-) \times 5$ $MnO_4^- + S^{2-} \rightarrow MnS$
                         $8\ OH^- + 8\ H^+ + MnO_4^- + S^{2-} \rightarrow MnS + 4\ H_2O + 8\ OH^-$
                          $4\ H_2O + MnO_4^- + S^{2-} \rightarrow MnS + 8\ OH^-$
                        $(5\ e^- + 4\ H_2O + MnO_4^- + S^{2-} \rightarrow MnS + 8\ OH^-) \times 2$

Common factor is a transfer of 10 e⁻. Adding the balanced half reactions:

$$5\ S^{2-} \rightarrow 5\ S + 10\ e^-$$
$$10\ e^- + 8\ H_2O + 2\ MnO_4^- + 2\ S^{2-} \rightarrow 2\ MnS + 16\ OH^-$$
_____
$$8\ H_2O(l) + 2\ MnO_4^-(aq) + 7\ S^{2-}(aq) \rightarrow 5\ S(s) + 2\ MnS(s) + 16\ OH^-(aq)$$

e. $CN^- \rightarrow CNO^-$
       $H_2O + CN^- \rightarrow CNO^- + 2\ H^+$
   $2\ OH^- + H_2O + CN^- \rightarrow CNO^- + 2\ H^+ + 2\ OH^-$
    $(2\ OH^- + CN^- \rightarrow CNO^- + H_2O + 2\ e^-) \times 3$

                         $MnO_4^- \rightarrow MnO_2$
                  $4\ H^+ + MnO_4^- \rightarrow MnO_2 + 2\ H_2O$
            $4\ H^+ + 4\ OH^- + MnO_4^- \rightarrow MnO_2 + 2\ H_2O + 4\ OH^-$
             $(3\ e^- + MnO_4^- + 2\ H_2O \rightarrow MnO_2 + 4\ OH^-) \times 2$

Common factor is a transfer of 6 electrons.

$$6\ OH^- + 3\ CN^- \rightarrow 3\ CNO^- + 3\ H_2O + 6\ e^-$$
$$6\ e^- + 2\ MnO_4^- + 4\ H_2O \rightarrow 2\ MnO_2 + 8\ OH^-$$
_____
$$3\ CN^-(aq) + 2\ MnO_4^-(aq) + H_2O(l) \rightarrow 3\ CNO^-(aq) + 2\ MnO_2(s) + 2\ OH^-(aq)$$

51.   a. HCl(aq) dissociates to $H^+(aq) + Cl^-(aq)$. For simplicity let's use H^+ and Cl^- separately.

             $H^+ \rightarrow H_2$ $Fe \rightarrow HFeCl_4$
      $(2\ H^+ + 2e^- \rightarrow H_2) \times 3$ $(H^+ + 4\ Cl^- + Fe \rightarrow HFeCl_4 + 3e^-) \times 2$

$$6 \, H^+ + 6 \, e^- \rightarrow 3 \, H_2$$
$$2 \, H^+ + 8 \, Cl^- + 2 \, Fe \rightarrow 2 \, HFeCl_4 + 6 \, e^-$$

$$8 \, H^+ + 8 \, Cl^- + 2 \, Fe \rightarrow 2 \, HFeCl_4 + 3 \, H_2$$

or $8 \, HCl(aq) + 2 \, Fe(s) \rightarrow 2 \, HFeCl_4(aq) + 3 \, H_2(g)$

b. $IO_3^- \rightarrow I_3^-$ $I^- \rightarrow I_3^-$
                          $3 \, IO_3^- \rightarrow I_3^-$ $(3 \, I^- \rightarrow I_3^- + 2e^-) \times 8$
                          $3 \, IO_3^- \rightarrow I_3^- + 9 \, H_2O$
              $16 \, e^- + 18 \, H^+ + 3 \, IO_3^- \rightarrow I_3^- + 9 \, H_2O$

$$16 \, e^- + 18 \, H^+ + 3 \, IO_3^- \rightarrow I_3^- + 9 \, H_2O$$
$$24 \, I^- \rightarrow 8 \, I_3^- + 16 \, e^-$$

$$18 \, H^+ + 24 \, I^- + 3 \, IO_3^- \rightarrow 9 \, I_3^- + 9 \, H_2O$$

Reducing: $6 \, H^+(aq) + 8 \, I^-(aq) + IO_3^-(aq) \rightarrow 3 \, I_3^-(aq) + 3 \, H_2O(l)$

c. $(Ce^{4+} + e^- \rightarrow Ce^{3+}) \times 97$

$$Cr(NCS)_6^{4-} \rightarrow Cr^{3+} + NO_3^- + CO_2 + SO_4^{2-}$$
$$54 \, H_2O + Cr(NCS)_6^{4-} \rightarrow Cr^{3+} + 6 \, NO_3^- + 6 \, CO_2 + 6 \, SO_4^{2-} + 108 \, H^+$$

Charge on left is -4. Charge on right is $(+3) + 6(-1) + 6(-2) + 108(+1) = +93$. Add 97 e^- to the right to balance the charge. Balancing the overall equation:

$$54 \, H_2O + Cr(NCS)_6^{4-} \rightarrow Cr^{3+} + 6 \, NO_3^- + 6 \, CO_2 + 6 \, SO_4^{2-} + 108 \, H^+ + 97 \, e^-$$
$$97 \, e^- + 97 \, Ce^{4+} \rightarrow 97 \, Ce^{3+}$$

$$97 \, Ce^{4+}(aq) + 54 \, H_2O(l) + Cr(NCS)_6^{4-}(aq) \rightarrow 97 \, Ce^{3+}(aq) + Cr^{3+}(aq) + 6 \, NO_3^-(aq) + 6 \, CO_2(g)$$

$$+ 6 \, SO_4^{2-}(aq) + 108 \, H^+(aq)$$

This is very complicated. A check of the net charge is a good check to see if the equation is balanced. Left: charge = $97(+4) - 4 = +384$. Right: charge = $97(+3) + 3 + 6(-1) + 6(-2) + 108(+1) = +384$.

d. $Cl_2 \rightarrow Cl^-$
                                                                      $(2 \, e^- + Cl_2 \rightarrow 2 \, Cl^-) \times 27$

$$CrI_3 \rightarrow CrO_4^{2-} + IO_4^-$$
$$32 \, OH^- + 16 \, H_2O + CrI_3 \rightarrow CrO_4^{2-} + 3 \, IO_4^- + 32 \, H^+ + 32 \, OH^-$$
$$32 \, OH^- + CrI_3 \rightarrow CrO_4^{2-} + 3 \, IO_4^- + 16 \, H_2O$$

Net charge on left is -32. Net charge on right is -5. Add 27 e$^-$ to the right.

$(32 \text{ OH}^- + \text{CrI}_3 \rightarrow \text{CrO}_4^{2-} + 3 \text{ IO}_4^- + 16 \text{ H}_2\text{O} + 27 \text{ e}^-) \times 2$

Common factor is a transfer of 54 e$^-$.

$$54 \text{ e}^- + 27 \text{ Cl}_2 \rightarrow 54 \text{ Cl}^-$$
$$64 \text{ OH}^- + 2 \text{ CrI}_3 \rightarrow 2 \text{ CrO}_4^{2-} + 6 \text{ IO}_4^- + 32 \text{ H}_2\text{O} + 54 \text{ e}^-$$

$$\overline{64 \text{ OH}^-(aq) + 2 \text{ CrI}_3(s) + 27 \text{ Cl}_2(g) \rightarrow 54 \text{ Cl}^-(aq) + 2 \text{ CrO}_4^{2-}(aq) + 6 \text{ IO}_4^-(aq) + 32 \text{ H}_2\text{O}(l)}$$

e.
$$\text{Ce}^{4+} \rightarrow \text{Ce(OH)}_3$$
$$(\text{e}^- + 3 \text{ OH}^- + \text{Ce}^{4+} \rightarrow \text{Ce(OH)}_3) \times 61$$

$$\text{Fe(CN)}_6^{4-} \rightarrow \text{Fe(OH)}_3 + \text{CO}_3^{2-} + \text{NO}_3^-$$
$$3 \text{ OH}^- + \text{Fe(CN)}_6^{4-} \rightarrow \text{Fe(OH)}_3 + 6 \text{ CO}_3^{2-} + 6 \text{ NO}_3^-$$

There are 36 extra O atoms on the right side. Add 36 H$_2$O to left, 72 H$^+$ to right, and neutralize with 72 OH$^-$ on each side.

$$72 \text{ OH}^- + 36 \text{ H}_2\text{O} + 3 \text{ OH}^- + \text{Fe(CN)}_6^{4-} \rightarrow \text{Fe(OH)}_3 + 6 \text{ CO}_3^{2-} + 6 \text{ NO}_3^- + 72 \text{ H}^+ + 72 \text{ OH}^-$$
$$75 \text{ OH}^- + \text{Fe(CN)}_6^{4-} \rightarrow \text{Fe(OH)}_3 + 6 \text{ CO}_3^{2-} + 6 \text{ NO}_3^- + 36 \text{ H}_2\text{O}$$

Net charge on left is -79. Net charge on right is -18. Add 61 e$^-$ to the right side to balance charge, then add the two balanced half reactions so electrons cancel.

$$75 \text{ OH}^- + \text{Fe(CN)}_6^{4-} \rightarrow \text{Fe(OH)}_3 + 6 \text{ CO}_3^{2-} + 6 \text{ NO}_3^- + 36 \text{ H}_2\text{O} + 61 \text{ e}^-$$
$$61 \text{ e}^- + 183 \text{ OH}^- + 61 \text{ Ce}^{4+} \rightarrow 61 \text{ Ce(OH)}_3$$

$$\overline{258 \text{ OH}^-(aq) + \text{Fe(CN)}_6^{4-}(aq) + 61 \text{ Ce}^{4+}(aq) \rightarrow \text{Fe(OH)}_3(s) + 61 \text{ Ce(OH)}_3(s) + 6 \text{ CO}_3^{2-}(aq)}$$

$$+ 6 \text{ NO}_3^-(aq) + 36 \text{ H}_2\text{O}(l)$$

f.
$$\text{Fe(OH)}_2 \rightarrow \text{Fe(OH)}_3 \qquad\qquad \text{H}_2\text{O}_2 \rightarrow \text{OH}^-$$
$$(\text{OH}^- + \text{Fe(OH)}_2 \rightarrow \text{Fe(OH)}_3 + \text{e}^-) \times 2 \qquad 2 \text{ e}^- + \text{H}_2\text{O}_2 \rightarrow 2 \text{ OH}^-$$

$$2 \text{ e}^- + \text{H}_2\text{O}_2 \rightarrow 2 \text{ OH}^-$$
$$2 \text{ OH}^- + 2 \text{ Fe(OH)}_2 \rightarrow 2 \text{ Fe(OH)}_3 + 2 \text{ e}^-$$

$$\overline{2 \text{ Fe(OH)}_2(s) + \text{H}_2\text{O}_2(aq) \rightarrow 2 \text{ Fe(OH)}_3(s)}$$

52. a. $4 \text{ NH}_3(g) + 5 \text{ O}_2(g) \rightarrow 4 \text{ NO}(g) + 6 \text{ H}_2\text{O}(g)$
          -3 +1   0   +2 -2   +1 -2   oxidation numbers

$2 \text{ NO}(g) + \text{O}_2(g) \rightarrow 2 \text{ NO}_2(g)$
+2 -2   0   +4 -2

$3 NO_2(g) + H_2O(l) \rightarrow 2 HNO_3(aq) + NO(g)$
  +4 -2   +1 -2   +1 +5 -2   +2 -2

All three reactions are oxidation-reduction reactions. All contain two species which show a change in oxidation numbers.

b. $4 NH_3 + 5 O_2 \rightarrow 4 NO + 6 H_2O$ O_2 is the oxidizing agent. NH_3 is the reducing agent.

$2 NO + O_2 \rightarrow 2 NO_2$ O_2 is the oxidizing agent. NO is the reducing agent.

$3 NO_2 + H_2O \rightarrow 2 HNO_3 + NO$ NO_2 is both the oxidizing and reducing agent.

c. $5.0 \times 10^6 \text{ g NH}_3 \times \dfrac{1 \text{ mol NH}_3}{17.0 \text{ g NH}_3} \times \dfrac{4 \text{ mol NO}}{4 \text{ mol NH}_3} = 2.9 \times 10^5 \text{ mol NO}$

$5.0 \times 10^7 \text{ g O}_2 \times \dfrac{1 \text{ mol O}_2}{32.0 \text{ g O}_2} \times \dfrac{4 \text{ mol NO}}{5 \text{ mol O}_2} = 1.3 \times 10^6 \text{ mol NO}$

The NH_3 will be consumed completely before the O_2 will. NH_3 is limiting.

$2.9 \times 10^5 \text{ mol NO} \times \dfrac{30.0 \text{ g NO}}{\text{mol NO}} = 8.7 \times 10^6 \text{ g NO}$

53. $Cl + H_2SO_4 + MnO_2 \rightarrow Na_2SO_4 + MnCl_2 + Cl_2 + H_2O$
        +1 -1   +1+6 -2   +4 -2   +1 +6 -2   +2 -1   0   +1 -2   oxidation numbers

Cl is oxidized (-1 to 0) and Mn is reduced (+4 $\rightarrow$ +2). Need two Cl for every one Mn. However, need 2 extra Cl⁻ to balance Cl⁻ in $MnCl_2$.

$4 NaCl + H_2SO_4 + MnO_2 \rightarrow Na_2SO_4 + MnCl_2 + Cl_2 + H_2O$

Balance Na^+ and SO_4^{2-}: $4 NaCl + 2 H_2SO_4 + MnO_2 \rightarrow 2 Na_2SO_4 + MnCl_2 + Cl_2 + H_2O$

Left: 4-H and 10-O; Right: 8-O not counting H_2O; Need 2 H_2O on right to balance oxygen and hydrogen.

$4 NaCl(aq) + 2 H_2SO_4(aq) + MnO_2(s) \rightarrow 2 Na_2SO_4(aq) + MnCl_2(aq) + Cl_2(g) + 2 H_2O(l)$

54. $Mn \rightarrow Mn^{2+} + 2 e^-$ $HNO_3 \rightarrow NO_2$
                                                        $HNO_3 \rightarrow NO_2 + H_2O$
                                            $(e^- + H^+ + HNO_3 \rightarrow NO_2 + H_2O) \times 2$

                            $Mn \rightarrow Mn^{2+} + 2 e^-$
                    $2 e^- + 2 H^+ + 2 HNO_3 \rightarrow 2 NO_2 + 2 H_2O$
_____

$2 H^+(aq) + Mn(s) + 2 HNO_3(aq) \rightarrow Mn^{2+}(aq) + 2 NO_2(g) + 2 H_2O(l)$

$$(4\ H_2O + Mn^{2+} \rightarrow MnO_4^- + 8\ H^+ + 5\ e^-) \times 2 \qquad (2\ e^- + 2\ H^+ + IO_4^- \rightarrow IO_3^- + H_2O) \times 5$$

$$8\ H_2O + 2\ Mn^{2+} \rightarrow 2\ MnO_4^- + 16\ H^+ + 10\ e^-$$
$$10\ e^- + 10\ H^+ + 5\ IO_4^- \rightarrow 5\ IO_3^- + 5\ H_2O$$

$$\overline{3\ H_2O(l) + 2\ Mn^{2+}(aq) + 5\ IO_4^-(aq) \rightarrow 2\ MnO_4^-(aq) + 5\ IO_3^-(aq) + 6\ H^+(aq)}$$

55. $Au + 4\ Cl^- \rightarrow AuCl_4^- + 3\ e^- \qquad\qquad 3\ e^- + 4\ H^+ + NO_3^- \rightarrow NO + 2\ H_2O$

Adding the two balanced half reactions:

$$Au(s) + 4\ Cl^-(aq) + 4\ H^+(aq) + NO_3^-(aq) \rightarrow AuCl_4^-(aq) + NO(g) + 2\ H_2O(l)$$

56. $(H_2C_2O_4 \rightarrow 2\ CO_2 + 2\ H^+ + 2\ e^-) \times 5 \qquad (5\ e^- + 8\ H^+ + MnO_4^- \rightarrow Mn^{2+} + 4\ H_2O) \times 2$

$$5\ H_2C_2O_4 \rightarrow 10\ CO_2 + 10\ H^+ + 10\ e^-$$
$$10\ e^- + 16\ H^+ + 2\ MnO_4^- \rightarrow 2\ Mn^{2+} + 8\ H_2O$$

$$\overline{6\ H^+ + 5\ H_2C_2O_4 + 2\ MnO_4^- \rightarrow 10\ CO_2 + 2\ Mn^{2+} + 8\ H_2O}$$

$$0.1058\ g\ \text{oxalic acid} \times \frac{1\ \text{mol oxalic acid}}{90.034\ g} \times \frac{2\ \text{mol}\ MnO_4^-}{5\ \text{mol oxalic acid}} = 4.700 \times 10^{-4}\ mol$$

$$\text{Molarity} = \frac{4.700 \times 10^{-4}\ mol}{28.97\ mL} \times \frac{1000\ mL}{L} = 1.622 \times 10^{-2}\ M$$

57.   a. $Fe^{2+} \rightarrow Fe^{3+} + e^-$

$$5\ e^- + 8\ H^+ + MnO_4^- \rightarrow Mn^{2+} + 4\ H_2O$$

Balanced equation is:

$$8\ H^+ + MnO_4^- + 5\ Fe^{2+} \rightarrow 5\ Fe^{3+} + Mn^{2+} + 4\ H_2O$$

$$20.62 \times 10^{-3}\ L\ \text{soln} \times \frac{0.0216\ mol\ MnO_4^-}{L\ \text{soln}} \times \frac{5\ mol\ Fe^{2+}}{mol\ MnO_4^-} = 2.23 \times 10^{-3}\ mol\ Fe^{2+}$$

$$\text{Molarity} = \frac{2.23 \times 10^{-3}\ mol\ Fe^{2+}}{50.00 \times 10^{-3}\ L} = 4.46 \times 10^{-2}\ M$$

b. $Fe^{2+} \rightarrow Fe^{3+} + e^-$

$$6\ e^- + 14\ H^+ + Cr_2O_7^{2-} \rightarrow 2\ Cr^{3+} + 7\ H_2O$$

The balanced equation is:

$$14\ H^+ + Cr_2O_7^{2-} + 6\ Fe^{2+} \rightarrow 6\ Fe^{3+} + 2\ Cr^{3+} + 7\ H_2O$$

$$50.00 \times 10^{-3}\ L \times \frac{4.46 \times 10^{-2}\ mol\ Fe^{2+}}{L} \times \frac{1\ mol\ Cr_2O_7^{2-}}{6\ mol\ Fe^{2+}} \times \frac{1\ L}{0.0150\ mol\ Cr_2O_7^{2-}}$$

$$= 2.48 \times 10^{-2}\ L\ or\ 24.8\ mL$$

58.   This is the same titration as in problem 4.57. The stoichiometry of the reaction depends on the reaction:

$$8\ H^+ + MnO_4^- + 5\ Fe^{2+} \rightarrow 5\ Fe^{3+} + Mn^{2+} + 4\ H_2O$$

From the titration data we can get the number of moles of Fe^{2+}. We then convert this to a mass of iron and calculate the mass percent of iron in the sample.

$$38.37 \times 10^{-3}\ L\ MnO_4^- \times \frac{0.0198\ mol\ MnO_4^-}{L} \times \frac{5\ mol\ Fe^{2+}}{mol\ MnO_4^-} = 3.80 \times 10^{-3}\ mol\ Fe^{2+}$$

$$3.80 \times 10^{-3}\ mol\ Fe \times \frac{55.85\ g\ Fe}{mol\ Fe} = 0.212\ g\ Fe$$

$$\%\ Fe = \frac{0.212\ g}{0.6128\ g} \times 100 = 34.6\%\ Fe$$

59.
$$(5\ e^- + 8\ H^+ + MnO_4^- \rightarrow Mn^{2+} + 4\ H_2O) \times 2$$
$$(C_2O_4^{2-} \rightarrow 2\ CO_2 + 2\ e^-) \times 5$$

$$\overline{2\ MnO_4^- + 16\ H^+ + 5\ C_2O_4^{2-} \rightarrow 10\ CO_2 + 2\ Mn^{2+} + 8\ H_2O}$$

$$0.250\ g\ Na_2C_2O_4 \times \frac{1\ mol\ Na_2C_2O_4}{134.0\ g\ Na_2C_2O_4} \times \frac{10\ mol\ CO_2}{5\ mol\ Na_2C_2O_4} = 3.73 \times 10^{-3}\ mol\ CO_2$$

$$50.00 \times 10^{-3}\ L\ KMnO_4\ soln \times \frac{0.0200\ mol\ MnO_4^-}{L} \times \frac{10\ mol\ CO_2}{2\ mol\ MnO_4^-} = 5.00 \times 10^{-3}\ mol\ CO_2$$

The $Na_2C_2O_4$ is the limiting reagent. Amount CO_2 produced is:

$$3.73 \times 10^{-3}\ mol\ CO_2 \times \frac{44.01\ g\ CO_2}{mol\ CO_2} = 0.164\ g\ CO_2$$

60. $1.06\ g\ Cu \times \dfrac{1\ mol\ Cu}{63.55\ g} \times \dfrac{2\ mol\ Ag^+}{mol\ Cu} = 3.34 \times 10^{-2}\ mol\ Ag^+$ required to dissolve all of the Cu.

We have: 250×10^{-3} L $\times \dfrac{0.20 \text{ mol Ag}^+}{\text{L}} = 5.0 \times 10^{-2}$ mol Ag$^+$

Yes, all of the copper will dissolve.

61.   a. $16 \text{ e}^- + 18 \text{ H}^+ + 3 \text{ IO}_3^- \rightarrow \text{I}_3^- + 9 \text{ H}_2\text{O}$ $(3 \text{ I}^- \rightarrow \text{I}_3^- + 2 \text{ e}^-) \times 8$

$$24 \text{ I}^- \rightarrow 8 \text{ I}_3^- + 16 \text{ e}^-$$
$$16 \text{ e}^- + 18 \text{ H}^+ + 3 \text{ IO}_3^- \rightarrow \text{I}_3^- + 9 \text{ H}_2\text{O}$$
$$\overline{\rule{0pt}{1em}\hspace{7cm}}$$
$$18 \text{ H}^+ + 24 \text{ I}^- + 3 \text{ IO}_3^- \rightarrow 9 \text{ I}_3^- + 9 \text{ H}_2\text{O}$$

   or $6 \text{ H}^+(aq) + 8 \text{ I}^-(aq) + \text{IO}_3^-(aq) \rightarrow 3 \text{ I}_3^-(aq) + 3 \text{ H}_2\text{O}(l)$

   b. 0.6013 g KIO$_3$ $\times \dfrac{1 \text{ mol KIO}_3}{214.0 \text{ g KIO}_3} = 2.810 \times 10^{-3}$ mol KIO$_3$

   2.810×10^{-3} mol KIO$_3$ $\times \dfrac{8 \text{ mol KI}}{\text{mol KIO}_3} \times \dfrac{166.0 \text{ g KI}}{\text{mol KI}} = 3.732$ g KI

   2.810×10^{-3} mol KIO$_3$ $\times \dfrac{6 \text{ mol HCl}}{\text{mol KIO}_3} \times \dfrac{1 \text{ L}}{3.00 \text{ mol HCl}} = 5.62 \times 10^{-3}$ L $= 5.62$ mL HCl

   c. $\text{I}_3^- + 2 \text{ e}^- \rightarrow 3 \text{ I}^-$

   $2 \text{ S}_2\text{O}_3^{2-} \rightarrow \text{S}_4\text{O}_6^{2-} + 2 \text{ e}^-$

   Adding the balanced half reactions gives:

   $2 \text{ S}_2\text{O}_3^{2-}(aq) + \text{I}_3^-(aq) \rightarrow 3 \text{ I}^-(aq) + \text{S}_4\text{O}_6^{2-}(aq)$

   d. 25.00×10^{-3} L IO$_3^-$ $\times \dfrac{0.0100 \text{ mol IO}_3^-}{\text{L}} \times \dfrac{3 \text{ mol I}_3^-}{\text{mol IO}_3^-} \times \dfrac{2 \text{ mol S}_2\text{O}_3^{2-}}{\text{mol I}_3^-} = 1.50 \times 10^{-3}$ mol S$_2$O$_3^{2-}$

   $M_{\text{S}_2\text{O}_3^{2-}} = \dfrac{1.50 \times 10^{-3} \text{ mol}}{32.04 \times 10^{-3} \text{ L}} = 0.0468 \; M$

   e. 0.5000 L $\times \dfrac{0.0100 \text{ mol KIO}_3}{\text{L}} \times \dfrac{214.0 \text{ g KIO}_3}{\text{mol KIO}_3} = 1.07$ g KIO$_3$

   Place 1.07 g KIO$_3$ in a 500 mL volumetric flask; add water to dissolve KIO$_3$; continue adding water to the mark.

Additional Exercises

62.   Zn$_2$P$_2$O$_7$: $2(65.38) + 2(30.97) + 7(16.00) = \dfrac{304.69 \text{ g}}{\text{mol}}$

$$0.4089 \text{ g Zn}_2\text{P}_2\text{O}_7 \times \frac{130.76 \text{ g Zn}}{304.69 \text{ g Zn}_2\text{P}_2\text{O}_7} = 0.1755 \text{ g Zn}$$

$$\% \text{ Zn} = \frac{0.1755 \text{ g}}{1.200 \text{ g}} \times 100 = 14.63\% \text{ Zn}$$

63. $\text{Na}_2\text{CrO}_4(aq) + \text{Pb(NO}_3)_2(aq) \rightarrow \text{PbCrO}_4(s) + 2 \text{ NaNO}_3(aq)$

$$0.1000 \text{ L} \times \frac{0.4100 \text{ mol Na}_2\text{CrO}_4}{\text{L}} \times \frac{1 \text{ mol chrome yellow}}{\text{mol Na}_2\text{CrO}_4} = 4.100 \times 10^{-2} \text{ mol}$$

$$0.1000 \text{ L} \times \frac{0.3200 \text{ mol Pb(NO}_3)_2}{\text{L}} \times \frac{1 \text{ mol chrome yellow}}{\text{mol Pb(NO}_3)_2} = 3.200 \times 10^{-2} \text{ mol}$$

$\text{Pb(NO}_3)_2$ is limiting. Therefore:

$$3.200 \times 10^{-2} \text{ mol PbCrO}_4 \times \frac{323.2 \text{ g PbCrO}_4}{\text{mol}} = 10.34 \text{ g PbCrO}_4$$

64.   a.   No, no element shows a change in oxidation number.

   b. $1.0 \text{ L oxalic acid} \times \dfrac{0.14 \text{ mol oxalic acid}}{\text{L}} \times \dfrac{1 \text{ mol Fe}_2\text{O}_3}{6 \text{ mol oxalic acid}} \times \dfrac{159.7 \text{ g Fe}_2\text{O}_3}{\text{mol Fe}_2\text{O}_3} = 3.7 \text{ g Fe}_2\text{O}_3$

65.   a. $\text{Fe}^{3+}(aq) + 3 \text{ OH}^-(aq) \rightarrow \text{Fe(OH)}_3(s)$

        Fe(OH)_3: $55.85 + 3(16.00) + 3(1.008) = 106.87 \text{ g/mol}$

$$0.107 \text{ g Fe(OH)}_3 \times \frac{55.85 \text{ g Fe}}{106.9 \text{ g Fe(OH)}_3} = 0.0559 \text{ g Fe}$$

   b. $\text{Fe(NO}_3)_3$: $55.85 + 3(14.01) + 9(16.00) = 241.86 \text{ g/mol}$

$$0.0559 \text{ g Fe} \times \frac{241.9 \text{ g Fe(NO}_3)_3}{55.85 \text{ g Fe}} = 0.242 \text{ g Fe(NO}_3)_3$$

   c. $\% \text{ Fe(NO}_3)_3 = \dfrac{0.242 \text{ g}}{0.456 \text{ g}} \times 100 = 53.1\%$

66.   Molar masses: KCl, $39.10 + 35.45 = 74.55 \text{ g/mol}$; KBr, $39.10 + 79.90 = 119.00 \text{ g/mol}$

        AgCl, $107.9 + 35.45 = 143.4 \text{ g/mol}$; AgBr, $107.9 + 79.90 = 187.8 \text{ g/mol}$

Let x = number of moles of KCl in mixture and y = number of moles of KBr in mixture.
Since $\text{Ag}^+ + \text{Cl}^- \rightarrow \text{AgCl}$ and $\text{Ag}^+ + \text{Br}^- \rightarrow \text{AgBr}$, then x = moles AgCl and y = moles AgBr.
Setting up simultaneous equations from the given information:

0.1024 g = 74.55 x + 119.0 y

0.1889 g = 143.4 x + 187.8 y

Multiply the first equation by $\dfrac{187.8}{119.0}$, and subtract from the second.

$$
\begin{aligned}
0.1889 &= 143.4\,x + 187.8\,y \\
-0.1616 &= -117.7\,x - 187.8\,y \\
\hline
0.0273 &= 25.7\,x, \qquad x = 1.06 \times 10^{-3} \text{ mol KCl}
\end{aligned}
$$

$$1.06 \times 10^{-3} \text{ mol KCl} \times \dfrac{74.55 \text{ g KCl}}{\text{mol KCl}} = 0.0790 \text{ g KCl}$$

% KCl = $\dfrac{0.0790 \text{ g}}{0.1024 \text{ g}}$ × 100 = 77.1%, % KBr = 100.0 - 77.1 = 22.9%

Alternatively, we can calculate the mass of AgX, X = Cl or Br, assuming the sample was pure KCl or KBr and set up a ratio.

If pure KCl: 0.1024 g KCl × $\dfrac{143.4 \text{ g AgCl}}{74.55 \text{ g KCl}}$ = 0.1970 g AgCl

If pure KBr: 0.1024 g KBr × $\dfrac{187.8 \text{ g AgBr}}{119.0 \text{ g KBr}}$ = 0.1616 g AgBr

So, 0% KCl will give 0.1616 g AgX and 100% KCl will give 0.1970 g AgX. The mass in excess of 0.1616 g will give us % KCl.

$\dfrac{0.1889 - 0.1616}{0.1970 - 0.1616}$ × 100 = % KCl = 77.1% and % KBr = 22.9%

67. $Fe(NO_3)_3 + 3 \text{ NaOH} \rightarrow Fe(OH)_3 + 3 \text{ NaNO}_3$

$$75.0 \times 10^{-3} \text{ L soln} \times \dfrac{0.105 \text{ mol Fe(NO}_3)_3}{\text{L soln}} \times \dfrac{1 \text{ mol Fe(OH)}_3}{\text{mol Fe(NO}_3)_3} = 7.88 \times 10^{-3} \text{ mol Fe(OH)}_3$$

$$125 \times 10^{-3} \text{ L soln} \times \dfrac{0.150 \text{ mol NaOH}}{\text{L soln}} \times \dfrac{1 \text{ mol Fe(OH)}_3}{3 \text{ mol NaOH}} = 6.25 \times 10^{-3} \text{ mol Fe(OH)}_3$$

The NaOH is limiting.

$$6.25 \times 10^{-3} \text{ mol Fe(OH)}_3 \times \dfrac{106.9 \text{ g Fe(OH)}_3}{\text{mol}} = 0.668 \text{ g or } 668 \text{ mg Fe(OH)}_3$$

68. $22.90 \text{ mL} \times 0.1000 \text{ mmol/mL} = 2.290 \text{ mmol Ag}^+ = 2.290 \times 10^{-3} \text{ mol Cl}^-$ total

mol NaCl (n_{NaCl}) + mol KCl (n_{KCl}) = 2.290×10^{-3} mol Cl$^-$, $n_{NaCl} = 2.290 \times 10^{-3} - n_{KCl}$

Since the molar mass of NaCl is 58.44 and the molar mass of KCl is 74.55, then:

$58.44 \ n_{NaCl} + 74.55 \ n_{KCl} = 0.1586$ g

$58.44 \ (2.290 \times 10^{-3} - n_{KCl}) + 74.55 \ n_{KCl} = 0.1586$, $16.11 \ n_{KCl} = 0.0248$, $n_{KCl} = 1.54 \times 10^{-3}$ mol

$\% \text{ KCl} = \dfrac{1.54 \times 10^{-3} \text{ mol} \times 74.55 \text{ g/mol}}{0.1586 \text{ g}} \times 100 = 72.4\%$ KCl

% NaCl = 100.0 - 72.4 = 27.6% NaCl

69. $50.00 \text{ mL} \times \dfrac{0.0565 \text{ mmol}}{\text{mL}} = 2.83 \text{ mmol Ag}^+$ total

$8.32 \text{ mL} \times \dfrac{0.0510 \text{ mmol KSCN}}{\text{mL}} \times \dfrac{1 \text{ mmol Ag}^+}{\text{mmol KSCN}} = 0.424 \text{ mmol Ag}^+$ unreacted

Thus, 2.83 - 0.424 = 2.41 mmol Ag$^+$ reacted with I$^-$. Since this is a 1:1 reaction, the solution contained 2.41 mmol I$^-$.

$C_{KI} = \dfrac{2.41 \text{ mmol}}{25.00 \text{ mL}} = 0.0964 \ M$

70.   a. $C_{12}H_{10-n}Cl_n + n \text{ Ag}^+ \rightarrow n \text{ AgCl}$; molar mass (AgCl) = 143.4 g/mol

molar mass (PCB) = 12(12.01) + (10-n) (1.008) + n(35.45) = 154.20 + 34.44 n

Since n mol AgCl are produced for every 1 mol PCB reacted, then n(143.4) grams of AgCl will be produced for every (154.20 + 34.44 n) grams of PCB reacted.

$\dfrac{\text{mass of AgCl}}{\text{mass of PCB}} = \dfrac{143.4 \text{ n}}{154.20 + 34.44 \text{ n}}$ or mass$_{AgCl}$ (154.20 + 34.44 n) = mass$_{PCB}$ (143.4 n)

b.   0.4971 (154.20 + 34.44 n) = 0.1947 (143.4 n), 76.65 + 17.12 n = 27.92 n

76.65 = 10.80 n, n = 7.097

71.   a.   Flow rate = 5.00×10^4 L/s + 3.50×10^3 L/s = 5.35×10^4 L/s

b. $C_{HCl} = \dfrac{3.50 \times 10^3 \ (65.0)}{5.35 \times 10^4} = 4.25 \text{ ppm HCl}$

c.   1 ppm = 1 mg/kg H_2O = 1 mg/L   (since density = 1.00 g/mL)

$$8.00 \text{ hr} \times \frac{60 \text{ min}}{\text{hr}} \times \frac{60 \text{ s}}{\text{min}} \times \frac{1.80 \times 10^4 \text{ L}}{\text{s}} \times \frac{4.25 \text{ mg HCl}}{\text{L}} \times \frac{1 \text{ g}}{1000 \text{ mg}} = 2.20 \times 10^6 \text{ g HCl}$$

$$2.20 \times 10^6 \text{ g HCl} \times \frac{1 \text{ mol HCl}}{36.46 \text{ g HCl}} \times \frac{1 \text{ mol CaO}}{2 \text{ mol HCl}} \times \frac{56.08 \text{ g CaO}}{\text{mol CaO}} = 1.69 \times 10^6 \text{ g CaO}$$

d.   The concentration of Ca^{2+} going into the second plant was:

$$\frac{5.00 \times 10^4 (10.2)}{5.35 \times 10^4} = 9.53 \text{ ppm}$$

The second plant used: 1.80×10^4 L/s $\times (8.00 \times 60 \times 60 \text{ sec}) = 5.18 \times 10^8$ L of water.

$$1.69 \times 10^6 \text{ g CaO} \times \frac{40.08 \text{ g Ca}^{2+}}{56.08 \text{ g CaO}} = 1.21 \times 10^6 \text{ g Ca}^{2+} \text{ was added to this water.}$$

$$C_{Ca^{2+}} \text{ (plant water)} = 9.53 + \frac{1.21 \times 10^9 \text{ mg}}{5.18 \times 10^8 \text{ L}} = 9.53 + 2.34 = 11.87 \text{ ppm}$$

Since 90.0% of this water is returned, $1.80 \times 10^4 \times 0.900 = 1.62 \times 10^4$ L/s of water with 11.87 ppm Ca^{2+} is mixed with $(5.35 - 1.80) \times 10^4 = 3.55 \times 10^4$ L/s of water containing 9.53 ppm Ca^{2+}.

$$C_{Ca^{2+}} \text{ (final)} = \frac{(1.62 \times 10^4 \text{ L/s})(11.87 \text{ ppm}) + (3.55 \times 10^4 \text{ L/s})(9.53 \text{ ppm})}{1.62 \times 10^4 \text{ L/s} + 3.55 \times 10^4 \text{ L/s}} = 10.3 \text{ ppm}$$

72.   a.
$$7 H_2O + 2 Cr^{3+} \rightarrow Cr_2O_7^{2-} + 14 H^+ + 6 e^-$$
$$(2 e^- + S_2O_8^{2-} \rightarrow 2 SO_4^{2-}) \times 3$$

$$7 H_2O(l) + 2 Cr^{3+}(aq) + 3 S_2O_8^{2-}(aq) \rightarrow Cr_2O_7^{2-}(aq) + 14 H^+(aq) + 6 SO_4^{2-}(aq)$$

$$(Fe^{2+} \rightarrow Fe^{3+} + e^-) \times 6$$
$$6 e^- + 14 H^+ + Cr_2O_7^{2-} \rightarrow 2 Cr^{3+} + 7 H_2O$$

$$14 H^+(aq) + 6 Fe^{2+}(aq) + Cr_2O_7^{2-}(aq) \rightarrow 2 Cr^{3+}(aq) + 6 Fe^{3+}(aq) + 7 H_2O(l)$$

b. $$8.58 \text{ mL} \times \frac{0.0520 \text{ mmol Cr}_2O_7^{2-}}{\text{mL}} \times \frac{6 \text{ mmol Fe}^{2+}}{\text{mmol Cr}_2O_7^{2-}} = 2.68 \text{ mmol of excess Fe}^{2+}$$

$$Fe^{2+} \text{ (total)} = 3.000 \text{ g Fe(NH}_4)_2 \text{ (SO}_4)_2 \cdot 6H_2O \times \frac{1 \text{ mol}}{392.17 \text{ g}} \times \frac{1000 \text{ mmol}}{1 \text{ mol}} = 7.650 \text{ mmol Fe}^{2+}$$

7.650 - 2.68 = 4.97 mmol Fe^{2+} reacted with $Cr_2O_7^{2-}$ generated from the Cr plating.

The Cr plating contained:

$$4.97 \text{ mmol Fe}^{2+} \times \frac{1 \text{ mmol Cr}_2O_7^{2-}}{6 \text{ mmol Fe}^{2+}} \times \frac{2 \text{ mmol Cr}^{3+}}{\text{mmol Cr}_2O_7^{2-}} = 1.66 \text{ mmol Cr}^{3+} = 1.66 \text{ mmol Cr}$$

$$1.66 \times 10^{-3} \text{ mol Cr} \times \frac{52.00 \text{ g Cr}}{\text{mol Cr}} = 8.63 \times 10^{-2} \text{ g Cr}$$

$$V_{Cr} = 8.63 \times 10^{-2} \text{ g} \times \frac{1 \text{ cm}^3}{7.19 \text{ g}} = 1.20 \times 10^{-2} \text{ cm}^3 = \text{area} \times \text{thickness}$$

$$\text{thickness of Cr plating} = = \frac{1.20 \times 10^{-2} \text{ cm}^3}{40.0 \text{ cm}^2} = 3.00 \times 10^{-4} \text{ cm} = 3.00 \ \mu\text{m}$$

73.   a. $YBa_2Cu_3O_{6.5}$:

$$+3 + 2(+2) + 3x + 6.5(-2) = 0$$

$$7 + 3x - 13 = 0, \ 3x = 6, \ x = +2 \qquad \text{Only Cu}^{2+} \text{ present.}$$

$YBa_2Cu_3O_7$:

$$+3 + 2(+2) + 3x + 7(-2) = 0, \ x = +2 \ 1/3 \text{ or } 2.33$$

This corresponds to two Cu^{2+} and one Cu^{3+}.

$YBa_2Cu_3O_8$:

$$+3 + 2(+2) + 3x + 8(-2) = 0, \ x = 3 \qquad \text{Only Cu}^{3+} \text{ present.}$$

b. $(e^- + Cu^{2+} + I^- \rightarrow CuI) \times 2 \qquad\qquad 2 \ e^- + Cu^{3+} + I^- \rightarrow CuI$
    $\qquad\quad 3I^- \rightarrow I_3^- + 2 \ e^- \qquad\qquad\qquad\qquad\qquad 3I^- \rightarrow I_3^- + 2 \ e^-$

$$\overline{\qquad\quad 2 \ Cu^{2+} + 5 \ I^- \rightarrow 2 \ CuI + I_3^- \qquad\qquad\qquad Cu^{3+} + 4 \ I^- \rightarrow CuI + I_3^- \qquad}$$

$$2 \ S_2O_3^{2-} \rightarrow S_4O_6^{2-} + 2 \ e^-$$
$$2 \ e^- + I_3^- \rightarrow 3 \ I^-$$

$$\overline{\qquad 2 \ S_2O_3^{2-} + I_3^- \rightarrow 3 \ I^- + S_4O_6^{2-} \qquad}$$

c.   Step II data: All Cu is converted to Cu^{2+}. Note: superconductor abbreviated as "123."

$$22.57 \text{ mL} \times \frac{0.1000 \text{ mmol S}_2O_3^{2-}}{\text{mL}} \times \frac{1 \text{ mmol I}_3^-}{2 \text{ mmol S}_2O_3^{2-}} \times \frac{2 \text{ mmol Cu}^{2+}}{\text{mmol I}_3^-} = 2.257 \text{ mmol Cu}^{2+}$$

$$2.257 \text{ mmol Cu} \times \frac{1 \text{ mmol "123"}}{3 \text{ mmol Cu}} = 0.7523 \text{ mmol "123"}$$

$$\text{Molar mass of } YBa_2Cu_3O_x = \frac{504.2 \text{ mg}}{0.7523 \text{ mmol}} = 670.2 \text{ g/mol}$$

$$670.2 = 88.91 + 2(137.3) + 3(63.55) + x(16.00), \quad 670.2 = 554.2 + x(16.00)$$

$x = 7.250$; Formula is $YBa_2Cu_3O_{7.25}$.

Check with Step I data: Both $Cu^{2+} + Cu^{3+}$ present.

$$37.77 \text{ mL} \times \frac{0.1000 \text{ mmol } S_2O_3^{2-}}{\text{mL}} \times \frac{1 \text{ mmol } I_3^-}{2 \text{ mmol } S_2O_3^{2-}} = 1.889 \text{ mmol } I_3^-$$

We get 1 mmol I_3^- per mmol Cu^{3+} and 1 mmol I_3^- per 2 mmol Cu^{2+}.

$$n_{Cu^{3+}} + \frac{n_{Cu^{2+}}}{2} = 1.889 \text{ mmol}$$

In addition: $\dfrac{562.5 \text{ g}}{670.2 \text{ g/mol}} = 0.8393 \text{ mmol "123"}$

This amount of "123" contains: $3(0.8393) = 2.518 \text{ mmol Cu} = n_{Cu^{3+}} + n_{Cu^{2+}}$

Solving by simultaneous equations:

$$n_{Cu^{3+}} + n_{Cu^{2+}} = \ \ 2.518$$

$$-n_{Cu^{3+}} - \frac{n_{Cu^{2+}}}{2} = -1.889$$

$$\rule{5cm}{0.4pt}$$

$$\frac{n_{Cu^{2+}}}{2} = 0.629$$

$n_{Cu^{2+}} = 1.26 \text{ mmol } Cu^{2+}$; $n_{Cu^{3+}} = 2.518 - 1.26 = 1.26 \text{ mmol } Cu^{3+}$

This sample of superconductor contains equal moles of Cu^{2+} and Cu^{3+}. Therefore, 1 mol of $YBa_2Cu_3O_x$ contains 1.50 mol Cu^{2+} and 1.50 mol Cu^{3+}. Solving for x using oxidation numbers:

$$+3 + 2(+2) + 1.50\,(+2) + 1.50(+3) + x(-2) = 0, \quad 14.50 = 2x, \quad x = 7.25$$

The two experiments give the same result, $x = 7.25$ with formula $YBa_2Cu_3O_{7.25}$.

Average oxidation state of Cu:

$$+3 + 2(+2) + 3(x) + 7.25(-2) = 0, \quad 3x = 7.50, \quad x = 2.50$$

As determined from step I data, this superconductor sample contains equal moles of Cu^{2+} and Cu^{3+}, giving an average oxidation state of 2.50.

74. Fe^{2+} will react with MnO_4^- (purple) producing Fe^{3+} and Mn^{2+}(almost colorless). There is no reaction between MnO_4^- and Fe^{3+}. Therefore, add a few drops of the potassium permanganate solution. If the purple color persists, the solution contains iron(III) sulfate. If the color disappears, iron(II) sulfate is present.

75. $0.298 \text{ g BaSO}_4 \times \dfrac{96.07 \text{ g SO}_4^{2-}}{233.4 \text{ g BaSO}_4} = 0.123 \text{ g sulfate}$

$\% \text{ sulfate} = \dfrac{0.123 \text{ g SO}_4^{2-}}{0.205 \text{ g}} = 60.0\%$

Assume we have 100.0 g of the mixture of Na_2SO_4 and K_2SO_4. There is:

$60.0 \text{ g SO}_4^{2-} \times \dfrac{1 \text{ mol}}{96.07 \text{ g}} = 0.625 \text{ mol SO}_4^{2-}$

There must be $2 \times 0.625 = 1.25$ mol of cations to balance charge.

Let x = number of moles of K^+ and y = number of moles of Na^+. Then, $x + y = 1.25$.

The total mass of Na^+ and K^+ must be 40.0 g, so:

$x \text{ mol K}^+ \left(\dfrac{39.10 \text{ g}}{\text{mol}} \right) + y \text{ mol Na}^+ \left(\dfrac{22.99 \text{ g}}{\text{mol}} \right) = 40.0 \text{ g}$

So we have two equations in two unknowns:

$x + y = 1.25 \text{ and } 39.10 \, x + 22.99 \, y = 40.0$

Solving: $x = 1.25 - y, \ 39.10(1.25 - y) + 22.99 \, y = 40.0$

$48.9 - 39.10 \, y + 22.99 \, y = 40.0, \ -16.11 \, y = -8.9$

$y = 0.55 \text{ mol Na}^+ \text{ and } x = 1.25 - 0.55 = 0.70 \text{ mol K}^+$

Therefore:

$0.70 \text{ mol K}^+ \times \dfrac{1 \text{ mol K}_2SO_4}{2 \text{ mol K}^+} = 0.35 \text{ mol K}_2SO_4$

$0.35 \text{ mol K}_2SO_4 \times \dfrac{174.3 \text{ g}}{\text{mol}} = 61 \text{ g K}_2SO_4$

Since we assumed 100.0 g, then the mixture is 61% K_2SO_4 and 39% Na_2SO_4.

76.   First we will calculate the molarity while ignoring the uncertainty.

$0.150 \text{ g} \times \dfrac{1 \text{ mol}}{58.44 \text{ g}} = 2.57 \times 10^{-3} \text{ moles}; \ \text{Molarity} = \dfrac{2.57 \times 10^{-3} \text{ mol}}{0.1000 \text{ L}} = \dfrac{2.57 \times 10^{-2} \text{ mol}}{\text{L}}$

The maximum value for the molarity is $= \dfrac{0.153 \text{ g} \times \dfrac{1 \text{ mol}}{58.44 \text{ g}}}{0.0995 \text{ L}} = \dfrac{2.63 \times 10^{-2} \text{ mol}}{\text{L}}$

The minimum value for the molarity is $= \dfrac{0.147 \text{ g} \times \dfrac{1 \text{ mol}}{58.44 \text{ g}}}{0.1005 \text{ L}} = \dfrac{2.50 \times 10^{-2} \text{ mol}}{\text{L}}$

The range of the molarity is $0.0250\ M$ to $0.0263\ M$ or we can express this range as $0.0257\ M \pm 0.0007\ M$.

77.   The amount of KHP used $= 0.4016 \text{ g} \times \dfrac{1 \text{ mol}}{204.22 \text{ g}} = 1.967 \times 10^{-3}$ mol KHP

Since one mole of NaOH reacts completely with one mole of KHP, the NaOH solution contains 1.967×10^{-3} mol NaOH.

Molarity of NaOH $= \dfrac{1.967 \times 10^{-3} \text{ mol}}{25.06 \times 10^{-3} \text{ L}} = \dfrac{7.849 \times 10^{-2} \text{ mol}}{\text{L}}$

Maximum molarity $= \dfrac{1.967 \times 10^{-3} \text{ mol}}{25.01 \times 10^{-3} \text{ L}} = \dfrac{7.865 \times 10^{-2} \text{ mol}}{\text{L}}$

Minimum molarity $= \dfrac{1.967 \times 10^{-3} \text{ mol}}{25.11 \times 10^{-3} \text{ L}} = \dfrac{7.834 \times 10^{-2} \text{ mol}}{\text{L}}$

We can express this as $0.07849\ M \pm 0.00016\ M$. An alternate is to express the molarity as $0.0785\ M \pm 0.0002\ M$. This second way shows the actual number of significant figures in the molarity.   The advantage of the first method is that it shows that we made all of our individual measurements to four significant figures.

78.   Desired uncertainty is 1% of 0.02 or ± 0.0002. So we want the solution to be $0.0200 \pm 0.0002\ M$ or the concentration should be between 0.0198 and 0.0202 M. We should use a 1-L volumetric flask to make the solution. They are good to $\pm 0.1\%$. We want to weigh out between 0.0198 mol and 0.0202 mol of KIO_3.

Molar mass of $KIO_3 = 39.10 + 126.9 + 3(16.00) = \dfrac{214.0 \text{ g}}{\text{mol}}$

0.0198 mol $\times \dfrac{214.0 \text{ g}}{\text{mol}} = 4.2372$ g   (For now we will carry significant figures.)

0.0202 mol $\times \dfrac{214.0 \text{ g}}{\text{mol}} = 4.3228$ g

We should weigh out between 4.24 and 4.32 g of KIO_3. We should weigh it to the nearest mg or 0.1 mg. Dissolve the KIO_3 in water and dilute to the mark in a one liter volumetric flask. This will produce a solution whose concentration is within the limits and is known to at least the fourth decimal place.

CHAPTER FIVE

GASES

Pressure

1. $4.75 \text{ cm} \times \dfrac{10 \text{ mm}}{\text{cm}} = 47.5 \text{ mm Hg or } 47.5 \text{ torr}$

    $47.5 \text{ torr} \times \dfrac{1 \text{ atm}}{760 \text{ torr}} = 6.25 \times 10^{-2} \text{ atm}$

    $6.25 \times 10^{-2} \text{ atm} \times \dfrac{1.013 \times 10^5 \text{ Pa}}{\text{atm}} = 6.33 \times 10^3 \text{ Pa}$

2. If the levels of Hg in each arm of the manometer are equal, then the pressure in the flask is equal to atmospheric pressure. When they are unequal, the difference in height in mm will be equal to the difference in pressure in mm Hg between the flask and the atmosphere. Which level is higher will tell us whether the pressure in the flask is less than or greater than atmospheric.

    a. $P_{flask} < P_{atm}$; $P_{flask} = 760. - 118 = 642 \text{ torr}$

    $642 \text{ torr} \times \dfrac{1 \text{ atm}}{760 \text{ torr}} = 0.845 \text{ atm}$

    $0.845 \text{ atm} \times \dfrac{1.013 \times 10^5 \text{ Pa}}{\text{atm}} = 8.56 \times 10^4 \text{ Pa}$

    b. $P_{flask} > P_{atm}$; $P_{flask} = 760. \text{ torr} + 215 \text{ torr} = 975 \text{ torr}$

    $975 \text{ torr} \times \dfrac{1 \text{ atm}}{760 \text{ torr}} = 1.28 \text{ atm}$

    $1.28 \text{ atm} \times \dfrac{1.013 \times 10^5 \text{ Pa}}{\text{atm}} = 1.30 \times 10^5 \text{ Pa}$

    c. $P_{flask} = 635 - 118 = 517 \text{ torr}$; $P_{flask} = 635 + 215 = 850. \text{ torr}$

3. Suppose we have a column of Hg $1.00 \text{ cm} \times 1.00 \text{ cm} \times 76.0 \text{ cm} = V = 76.0 \text{ cm}^3$:

    $\text{mass} = 76.0 \text{ cm}^3 \times 13.59 \dfrac{\text{g}}{\text{cm}^3} = 1.03 \times 10^3 \text{ g} \times \dfrac{1 \text{ kg}}{1000 \text{ g}} = 1.03 \text{ kg}$

$$F = mg = 1.03 \text{ kg} \times 9.81 \, \frac{m}{s^2} = 10.1 \, \frac{kg \, m}{s^2} = 10.1 \text{ N}$$

$$\frac{Force}{area} = \frac{10.1 \text{ N}}{cm^2} \times \left(\frac{100 \text{ cm}}{m} \right)^2 = 1.01 \times 10^5 \, \frac{N}{m^2} \text{ or } 1.01 \times 10^5 \text{ Pa}$$

(Note: 76.0 cm Hg = 1 atm = 1.01×10^5 Pa.)

To exert the same pressure, a column of water will have to contain the same mass as the 76.0 cm column of Hg. Thus, the column of water will have to be 13.59 times taller or 76.0 cm $\times$ 13.59 = 1.03×10^3 cm = 10.3 m.

4.   The pressure is proportional to the mass of the fluid. The mass is proportional to the volume of the column of fluid (or to the height of the column assuming the area of the column of fluid is constant).

$d = \dfrac{mass}{volume}$;   In this case the volume of silicon oil is the same as the volume of Hg in problem 5.2.

$$V = \frac{m}{d}; \quad V_1 = V_2; \quad \frac{m_1}{d_1} = \frac{m_2}{d_2}, \quad m_2 = \frac{m_1 d_2}{d_1}$$

Since P is proportional to the mass:

$$P_{oil} = P_{Hg} \left(\frac{d_{oil}}{d_{Hg}} \right) = P_{Hg} \left(\frac{1.30}{13.6} \right) = 0.0956 \, P_{Hg}$$

This conversion applies only to the column of silicon oil.

a. P_{flask} = 760. torr - (118 $\times$ 0.0956) torr = 760. - 11.3 = 749 torr

749 torr $\times$ $\dfrac{1 \text{ atm}}{760 \text{ torr}}$ = 0.986 atm; 0.986 atm $\times$ $\dfrac{1.013 \times 10^5 \text{ Pa}}{atm}$ = 9.99×10^4 Pa

b. P_{flask} = 760. torr + (215 $\times$ 0.0956) torr = 760. + 20.6 = 781 torr

781 torr $\times$ $\dfrac{1 \text{ atm}}{760 \text{ torr}}$ = 1.03 atm; 1.03 atm $\times$ $\dfrac{1.013 \times 10^5 \text{ Pa}}{atm}$ = 1.04×10^5 Pa

5.   If we are measuring the same pressure, the height of the silicon oil column would be 13.6 ÷ 1.30 = 10.5 times the height of a mercury column. The advantage of using a less dense fluid than mercury is in measuring small pressures. The height difference measured will be larger for the less dense fluid. Thus, the measurement will be more precise.

Gas Laws

6.   a.   PV = nRT, n and T constant

         PV = Constant

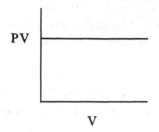

    b.   PV = nRT

$$P = \left(\frac{nR}{V}\right) T = \text{Const} \times T$$

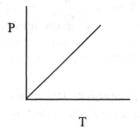

    c.   PV = nRT

$$T = \left(\frac{P}{nR}\right) V = \text{Const} \times V$$

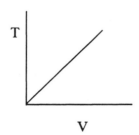

    d.   PV = nRT

$$P = \frac{nRT}{V} = \frac{\text{Const}}{V}, \quad PV = \text{Const}$$

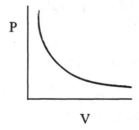

    e. $P = \dfrac{nRT}{V} = \dfrac{\text{Const}}{V} = \text{Const}\left(\dfrac{1}{V}\right)$

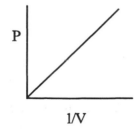

    f.   PV = nRT

$$\frac{PV}{T} = nR = \text{Const}$$

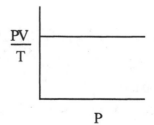

Note: The equation for a straight line is y = mx + b where y is the y-axis and x is the x-axis. Any equation that has this form will produce a straight line with slope equal to m and y-intercept equal to b. Plots b, c, and e have this straight line form.

7.   Treat each gas separately and use the relationship $P_1V_1 = P_2V_2$, since for each gas, n and T are constant.

For H_2: $P_2 = \dfrac{P_1V_1}{V_2} = 475 \text{ torr} \times \dfrac{2.00 \text{ L}}{3.00 \text{ L}} = 317 \text{ torr}$

For N_2: $P_2 = 0.200 \text{ atm} \times \dfrac{1.00 \text{ L}}{3.00 \text{ L}} = 0.0667 \text{ atm};\ \ 0.0667 \text{ atm} \left(\dfrac{760 \text{ torr}}{\text{atm}}\right) = 50.7 \text{ torr}$

$P_{total} = P_{H_2} + P_{N_2} = 317 + 50.7 = 368 \text{ torr}$

8.   Begin with the ideal gas law, $PV = nRT$. From the information given, we know that the volume is constant, i.e. the gas is in a steel cylinder, and P, n, and T are changing. So, we can say that:

$\dfrac{nT}{P} = \dfrac{V}{R} = \text{a constant; or } \dfrac{n_1T_1}{P_1} = \dfrac{n_2T_2}{P_2}$

Thus, $n_2 = \dfrac{n_1T_1P_2}{T_2P_1} = 150 \text{ mol} \times \dfrac{298 \text{ K}}{290. \text{ K}} \times \dfrac{1.2 \text{ MPa}}{7.5 \text{ MPa}} = 25 \text{ mol Ar}$

$25 \text{ mol Ar} \times \dfrac{39.95 \text{ g}}{\text{mol}} = 1.0 \times 10^3 \text{ g Ar}$

9. $PV = nRT$; Assume n is constant.

$\dfrac{PV}{T} = nR = \text{constant};\ \ \dfrac{P_1V_1}{T_1} = \dfrac{P_2V_2}{T_2}$

$\dfrac{V_2}{V_1} = \dfrac{T_2P_1}{T_1P_2} = \dfrac{(273 + 15) \text{ K} \times 720. \text{ torr}}{(273 + 25) \text{ K} \times 605 \text{ torr}} = 1.15$

$V_2 = 1.15 \text{ V};$ The volume has increased by 15%.

10. $P = \dfrac{nRT}{V} = \dfrac{\left(22.0 \text{ g} \times \dfrac{1 \text{ mol}}{44.01 \text{ g}}\right) \times \dfrac{0.08206 \text{ L atm}}{\text{mol K}} \times 300. \text{ K}}{4.00 \text{ L}} = 3.08 \text{ atm}$

With air present, the partial pressure of CO_2 will still be 3.08 atm. The total pressure will be the sum of the partial pressures.

$P_{tot} = 3.08 \text{ atm} + \left(740. \text{ torr} \times \dfrac{1 \text{ atm}}{760 \text{ torr}}\right) = 3.08 + 0.974 = 4.05 \text{ atm}$

11. $\dfrac{PV}{T} = nR = \text{a constant};\ \ \dfrac{P_1V_1}{T_1} = \dfrac{P_2V_2}{T_2}$

$$P_2 = \frac{P_1 V_1 T_2}{V_2 T_1} = 710 \text{ torr} \times \frac{5.0 \times 10^2 \text{ mL}}{25 \text{ mL}} \times \frac{(273 + 820) \text{ K}}{(273 + 30.) \text{ K}} = 5.1 \times 10^4 \text{ torr}$$

12. $n = \dfrac{PV}{RT} = \dfrac{145 \text{ atm} \times 75.0 \times 10^{-3} \text{ L}}{\dfrac{0.08206 \text{ L atm}}{\text{mol K}} \times 295 \text{ K}} = 0.449 \text{ mol O}_2$

13.   a. $PV = nRT$, n and V are constant.

$$\frac{P}{T} = \frac{nR}{V} \text{ or } \frac{P_1}{T_1} = \frac{P_2}{T_2} \,; P_2 = \frac{P_1 T_2}{T_1} = 40.0 \text{ atm} \times \frac{398 \text{ K}}{273 \text{ K}} = 58.3 \text{ atm}$$

   b. $\dfrac{P_1}{T_1} = \dfrac{P_2}{T_2} \,; T_2 = \dfrac{T_1 P_2}{P_1} = 273 \text{ K} \times \dfrac{150. \text{ atm}}{40.0 \text{ atm}} = 1.02 \times 10^3 \text{ K}$

   c. $T_2 = \dfrac{T_1 P_2}{P_1} = 273 \text{ K} \times \dfrac{25.0 \text{ atm}}{40.0 \text{ atm}} = 171 \text{ K}$

14.   a. $175 \text{ g Ar} \times \dfrac{1 \text{ mol Ar}}{39.95 \text{ g Ar}} = 4.38 \text{ mol Ar}$

$$T = \frac{PV}{nR} = \frac{2.50 \text{ L} \times 5.00 \text{ atm}}{4.38 \text{ mol} \times \dfrac{0.08206 \text{ L atm}}{\text{mol K}}} = 34.8 \text{ K}$$

   b. $P = \dfrac{nRT}{V} = \dfrac{4.38 \text{ mol} \times \dfrac{0.08206 \text{ L atm}}{\text{mol K}} \times 195 \text{ K}}{2.50 \text{ L}} = 28.0 \text{ atm}$

15. $PV = nRT$, V and n constant; $\dfrac{P}{T} = \dfrac{nR}{V} = \text{constant}$; $\dfrac{P_1}{T_1} = \dfrac{P_2}{T_2}$

$$P_2 = \frac{P_1 T_2}{T_1} = 13.7 \text{ MPa} \times \frac{(273 + 950) \text{ K}}{(273 + 23) \text{ K}} = 56.6 \text{ MPa}$$

16. $PV = nRT$, n constant; $\dfrac{PV}{T} = nR = \text{constant}$; $\dfrac{P_1 V_1}{T_1} = \dfrac{P_2 V_2}{T_2}$

$$V_2 = 1.040 \, V_1, \text{ so } \frac{V_1}{V_2} = \frac{1.000}{1.040}$$

$$P_2 = \frac{P_1 V_1 T_2}{V_2 T_1} = 100. \text{ psi} \times \frac{1.000}{1.040} \times \frac{(273 + 58) \text{ K}}{(273 + 19) \text{ K}} = 109 \text{ psi}$$

17. $PV = nRT$, P constant; $\dfrac{nT}{V} = \dfrac{P}{R} \,; \dfrac{n_1 T_1}{V_1} = \dfrac{n_2 T_2}{V_2}$

$$\frac{n_2}{n_1} = \frac{T_1 V_2}{T_2 V_1} = \frac{294 \text{ K}}{335 \text{ K}} \times \frac{4.20 \times 10^3 \text{ m}^3}{4.00 \times 10^3 \text{ m}^3} = 0.921$$

18. $P_{total} = P_{O_2} + P_{H_2O}$, $P_{O_2} = P_{tot} - P_{H_2O} = 745$ torr - 23.8 torr = 721 torr

$$PV = nRT; \; n = \frac{PV}{RT} = \frac{\dfrac{721}{760} \text{ atm} \times 0.1500 \text{ L}}{\dfrac{0.08206 \text{ L atm}}{\text{mol K}} \times 298 \text{ K}} = 5.82 \times 10^{-3} \text{ mol O}_2$$

$$5.82 \times 10^{-3} \text{ mol} \times \frac{32.00 \text{ g}}{\text{mol}} = 0.186 \text{ g O}_2$$

19.   We can use the ideal gas law to calculate the partial pressure of each gas or to calculate the total pressure. There will be less math if we calculate the total pressure from the ideal gas law.

$$n_{O_2} = 1.80 \times 10^2 \text{ mg O}_2 \times \frac{1 \text{ g}}{1000 \text{ mg}} \times \frac{1 \text{ mol O}_2}{32.00 \text{ g O}_2} = 5.63 \times 10^{-3} \text{ mol O}_2$$

$$n_{NO} = \frac{1000 \text{ cm}^3}{\text{L}} \times \frac{9.00 \times 10^{18} \text{ molecules NO}}{\text{cm}^3} \times \frac{1 \text{ mol NO}}{6.022 \times 10^{23} \text{ molecules NO}}$$

$$= 1.49 \times 10^{-2} \text{ mol NO/L}$$

$$n_{total} = n_{N_2} + n_{O_2} + n_{NO} = 2.00 \times 10^{-2} + 5.63 \times 10^{-3} + 1.49 \times 10^{-2} = 4.05 \times 10^{-2} \text{ mol}$$

$$P_{total} = \frac{nRT}{V} = \frac{4.05 \times 10^{-2} \text{ mol} \times \dfrac{0.08206 \text{ L atm}}{\text{mol K}} \times 273 \text{ K}}{1.00 \text{ L}} = 0.907 \text{ atm}$$

$$P_{N_2} = \chi_{N_2} P_{tot} = \frac{2.00 \times 10^{-2} \text{ mol N}_2}{4.05 \times 10^{-2} \text{ mol total}} \times 0.907 \text{ atm} = 0.448 \text{ atm}$$

$$P_{O_2} = \frac{5.63 \times 10^{-3}}{4.05 \times 10^{-2}} \times 0.907 \text{ atm} = 0.126 \text{ atm}; \quad P_{NO} = \frac{1.49 \times 10^{-2}}{4.05 \times 10^{-2}} \times 0.907 \text{ atm} = 0.334 \text{ atm}$$

20.   At constant temperature and pressure, volume is proportional to moles of gas. Therefore, $Br_2 + 3 F_2 \rightarrow 2 X$; two moles of X must contain two moles of Br and 6 moles of F; X must have the formula BrF_3.

21. $P_{total} = 1.00$ atm $= 760.$ torr $= P_{N_2} + P_{H_2O} = P_{N_2} + 17.5$ torr, $P_{N_2} = 743$ torr

$$PV = nRT; \; n = \frac{PV}{RT} = \frac{(743 \text{ torr}) (2.50 \times 10^2 \text{ mL})}{\left(\dfrac{0.08206 \text{ L atm}}{\text{mol K}} \right) (293 \text{ K})} \times \frac{1 \text{ atm}}{760 \text{ torr}} \times \frac{1 \text{ L}}{1000 \text{ mL}}$$

$$n = 1.02 \times 10^{-2} \text{ mol N}_2; \; 1.02 \times 10^{-2} \text{ mol N}_2 \times \frac{28.02 \text{ g N}_2}{\text{mol N}_2} = 0.286 \text{ g N}_2$$

22. $P_{He} + P_{H_2O} = 1.00$ atm $= 760.$ torr $= P_{He} + 23.8$ torr, $P_{He} = 736$ torr

$P_{tot} V_{tot} = n_{tot}RT;$ $V_{tot} = V_{He} + V_{H_2O} = V_{wet}$

$P_{He} = \chi_{He}P_{tot} = \dfrac{n_{He}}{n_{tot}}(760.) = 736;$ $n_{He} = 0.586$ g $\times \dfrac{1\ mol}{4.003\ g} = 0.146$ mol

$736 = \dfrac{0.146 \times 760.}{n_{tot}},$ $n_{tot} = 0.151$ mol

$V_{wet} = \dfrac{0.151\ mol \times \dfrac{0.08206\ L\ atm}{mol\ K} \times 298\ K}{1.00\ atm} = 3.69$ L

23. 1.00 g $H_2 \times \dfrac{1\ mol\ H_2}{2.016\ g\ H_2} = 0.496$ mol $H_2;$ 1.00 g He $\times \dfrac{1\ mol\ He}{4.003\ g\ He} = 0.250$ mol He

$P_{H_2} = \chi_{H_2}P_{tot} = \left(\dfrac{0.496\ mol\ H_2}{0.250\ mol\ He + 0.496\ mol\ H_2} \right) \times 0.480$ atm $= 0.319$ atm

$P_{H_2} + P_{He} = 0.480$ atm, $P_{He} = 0.480 - 0.319 = 0.161$ atm

24.   a. $\chi_{CH_4} = \dfrac{P_{CH_4}}{P_{tot}} = \dfrac{0.175}{0.175 + 0.250} = 0.412;$ $\chi_{O_2} = 1.000 - 0.412 = 0.588$

          b. $PV = nRT;$ $n = \dfrac{PV}{RT} = \dfrac{0.425\ atm \times 2.00\ L}{\dfrac{0.08206\ L\ atm}{mol\ K} \times 338\ K} = 0.0306$ mol

          c. $n_{CH_4} = \chi_{CH_4}(n_{tot}) = 0.412\ (0.0306\ mol) = 1.26 \times 10^{-2}$ mol CH_4

          1.26×10^{-2} mol $CH_4 \times \dfrac{16.04\ g\ CH_4}{mol\ CH_4} = 0.202$ g CH_4

          $n_{O_2} = 0.588\ (0.0306\ mol) = 1.80 \times 10^{-2}$ mol O_2

          1.80×10^{-2} mol $O_2 \times \dfrac{32.00\ g\ O_2}{mol\ O_2} = 0.576$ g O_2

Gas Density, Molar Mass, and Reaction Stoichiometry

25. $PV = nRT$; $\dfrac{n}{V} = \dfrac{P}{RT}$; Let M = molar mass and d = density, then:

$$\frac{nM}{V} = \frac{PM}{RT} \text{ where } \frac{nM}{V} = d; \ d = \frac{PM}{RT}$$

For $SiCl_4$, $M = 28.09 + 4(35.45) = \dfrac{169.89 \text{ g}}{\text{mol}}$

$$d = \frac{758 \text{ torr} \times \dfrac{169.89 \text{ g}}{\text{mol}}}{\dfrac{0.08206 \text{ L atm}}{\text{mol K}} \times 358 \text{ K}} \times \frac{1 \text{ atm}}{760 \text{ torr}} = 5.77 \text{ g/L for } SiCl_4$$

For $SiHCl_3$, $M = 28.09 + 1.008 + 3(35.45) = \dfrac{135.45 \text{ g}}{\text{mol}}$

$$d = \frac{PM}{RT} = \frac{758 \text{ torr} \times \dfrac{135.45 \text{ g}}{\text{mol}}}{\dfrac{0.08206 \text{ L atm}}{\text{mol K}} \times 358 \text{ K}} \times \frac{1 \text{ atm}}{760 \text{ torr}} = 4.60 \text{ g/L for } SiHCl_3$$

26.   Out of 100.0 g of compound, there are:

$$87.4 \text{ g N} \times \frac{1 \text{ mol N}}{14.01 \text{ g N}} = 6.24 \text{ mol N}; \ \frac{6.24}{6.24} = 1$$

$$12.6 \text{ g H} \times \frac{1 \text{ mol H}}{1.008 \text{ g H}} = 12.5 \text{ mol H}; \ \frac{12.5}{6.24} = 2$$

Empirical formula is NH_2.

$PM = dRT$ where M = molar mass and d = density in g/L.

$$M = \frac{dRT}{P} = \frac{\dfrac{0.977 \text{ g}}{\text{L}} \times \dfrac{0.08206 \text{ L atm}}{\text{mol K}} \times 373 \text{ K}}{710. \text{ torr}} \times \frac{760 \text{ torr}}{\text{atm}} = 32.0 \text{ g/mol}$$

Molar mass of $NH_2 \approx 16$ g. Therefore, molecular formula is N_2H_4.

27. $PV = \dfrac{gRT}{M}$ where M = molar mass

$$M = \frac{gRT}{PV} = \frac{0.800 \text{ g} \times \dfrac{0.08206 \text{ L atm}}{\text{mol K}} \times 373 \text{ K}}{750. \text{ torr} \times 0.256 \text{ L}} \times \frac{760 \text{ torr}}{\text{atm}} = 96.9 \text{ g/mol}$$

Molar mass of CHCl $= 12.0 + 1.0 + 35.5 = 48.5$, so molecular formula is $C_2H_2Cl_2$.

28.   If Be is trivalent and formula is $Be(C_5H_7O_2)_3$:

$$M \approx 13.5 + 15(12) + 21(1) + 6(16) = 311 \text{ g/mol}$$

If Be is divalent and formula is $Be(C_5H_7O_2)_2$:

$$M \approx 9.0 + 10(12) + 14(1) + 4(16) = 207 \text{ g/mol}$$

Data Set I:

$$M = \frac{gRT}{PV} = \frac{0.2022 \text{ g} \times \dfrac{0.08206 \text{ L atm}}{\text{mol K}} \times 286 \text{ K}}{22.6 \times 10^{-3} \text{ L} \times 765.2 \text{ torr}} \times \frac{760 \text{ torr}}{\text{atm}} = 209 \text{ g/mol}$$

Data Set II:

$$M = \frac{gRT}{PV} = \frac{0.2224 \text{ g} \times \dfrac{0.08206 \text{ L atm}}{\text{mol K}} \times 290. \text{ K}}{26.0 \times 10^{-3} \text{ L} \times 764.6 \text{ torr}} \times \frac{760 \text{ torr}}{\text{atm}} = 202 \text{ g/mol}$$

These results are close to the expected value of 207 g/mol for $Be(C_5H_7O_2)_2$. Thus, we conclude from these data that beryllium is a divalent element with an atomic weight of 9.0 g/mol.

29.   We assume that 28.01 g/mol is the true value for the molar mass of N_2. The value of 28.15 g/mol is the average of the amount of N_2 and Ar in air. Let $x = \%$ of the number of moles that are N_2 molecules. Then $100 - x = \%$ of the number of moles that are Ar atoms. Solving:

$$28.15 = \frac{x(28.01) + (100 - x)(39.95)}{100}$$

$$2815 = 28.01\, x + 3995 - 39.95\, x; \quad 11.94\, x = 1180.$$

$$x = 98.83\% \text{ N}_2; \quad \% \text{ Ar} = 100.00 - x = 1.17\% \text{ Ar}$$

Ratio of moles of Ar to moles of $N_2 = \dfrac{1.17}{98.83} = 1.18 \times 10^{-2}$.

30. $M = \dfrac{dRT}{P}$ where M = molar mass

$$M = \frac{\left(\dfrac{0.70902 \text{ g}}{L}\right)\left(\dfrac{0.08206 \text{ L atm}}{\text{mol K}}\right)(273.2 \text{ K})}{1.000 \text{ atm}} = 15.90 \text{ g/mol}$$

15.90 g/mol is the average molar mass of the mixture of methane and helium. Let x = mol % of CH_4. This is also the volume % of CH_4 since T and P are constant.

$$15.90 = \frac{x\,(16.04) + (100 - x)\,(4.003)}{100}$$

$$1590. = 16.04\,x + 400.3 - 4.003\,x; \quad 1190. = 12.04\,x$$

$$x = 98.84\ \%\ CH_4 \text{ by volume}; \quad \%\ He = 100 - x = 1.16\ \%\ He \text{ by volume}$$

31. $$1.00 \times 10^3 \text{ kg Mo} \times \frac{1000 \text{ g}}{\text{kg}} \times \frac{1 \text{ mol Mo}}{95.94 \text{ g Mo}} = 1.04 \times 10^4 \text{ mol Mo}$$

$$1.04 \times 10^4 \text{ mol Mo} \times \frac{1 \text{ mol MoO}_3}{\text{mol Mo}} \times \frac{7/2 \text{ mol O}_2}{\text{mol MoO}_3} = 3.64 \times 10^4 \text{ mol O}_2$$

$$V_{O_2} = \frac{nRT}{P} = \frac{\left(3.64 \times 10^4 \text{ mol}\right)\left(\dfrac{0.08206 \text{ L atm}}{\text{mol K}}\right)(290.\,K)}{1.00 \text{ atm}} = 8.66 \times 10^5 \text{ L of O}_2$$

$$8.66 \times 10^5 \text{ L O}_2 \times \frac{100 \text{ L air}}{21 \text{ L O}_2} = 4.1 \times 10^6 \text{ L air}$$

$$1.04 \times 10^4 \text{ mol Mo} \times \frac{3 \text{ mol H}_2}{\text{mol Mo}} = 3.12 \times 10^4 \text{ mol H}_2$$

$$V_{H_2} = \frac{\left(3.12 \times 10^4 \text{ mol}\right)\left(\dfrac{0.08206 \text{ L atm}}{\text{mol K}}\right)(290.\,K)}{1.00 \text{ atm}} = 7.42 \times 10^5 \text{ L of H}_2$$

32. $$n_{H_2} = \frac{PV}{RT} = \frac{(1.0 \text{ atm})\left[4800 \text{ m}^3 \times \left(\dfrac{100 \text{ cm}}{\text{m}}\right)^3 \times \dfrac{1 \text{ L}}{1000 \text{ cm}^3}\right]}{\left(\dfrac{0.08206 \text{ L atm}}{\text{mol K}}\right)(273 \text{ K})} = 2.1 \times 10^5 \text{ mol}$$

2.1×10^5 mol H_2 are in the balloon. This is 80.% of the total amount of H_2 that had to be generated:

$$0.80 \text{ (total mol H}_2) = 2.1 \times 10^5, \text{ total mol H}_2 = 2.6 \times 10^5 \text{ mol H}_2$$

$$2.6 \times 10^5 \text{ mol H}_2 \times \frac{1 \text{ mol Fe}}{\text{mol H}_2} \times \frac{55.85 \text{ g Fe}}{\text{mol Fe}} = 1.5 \times 10^7 \text{ g Fe}$$

$$2.6 \times 10^5 \text{ mol H}_2 \times \frac{1 \text{ mol H}_2SO_4}{\text{mol H}_2} \times \frac{98.09 \text{ g H}_2SO_4}{\text{mol H}_2SO_4} \times \frac{100 \text{ g reagent}}{98 \text{ g H}_2SO_4} = 2.6 \times 10^7 \text{ g of 98\%} \atop \text{sulfuric acid}$$

33. For ammonia (in one minute):

$$n_{NH_3} = \frac{PV}{RT} = \frac{(90.\text{ atm})(500.\text{ L})}{\left(\dfrac{0.08206 \text{ L atm}}{\text{mol K}}\right)(496 \text{ K})} = 1.1 \times 10^3 \text{ mol}$$

NH_3 flows into the reactor at a rate of 1.1×10^3 mol/min.

For CO_2 (in one minute):

$$n_{CO_2} = \frac{PV}{RT} = \frac{(45 \text{ atm})(600. \text{ L})}{\left(\dfrac{0.08206 \text{ L atm}}{\text{mol K}}\right)(496 \text{ K})} = 6.6 \times 10^2 \text{ mol}$$

CO_2 flows into the reactor at 6.6×10^2 mol/min.

To react completely with 1.1×10^3 mol NH_3/min, need:

$$\frac{1.1 \times 10^3 \text{ mol NH}_3}{\text{min}} \times \frac{1 \text{ mol CO}_2}{2 \text{ mol NH}_3} = 5.5 \times 10^2 \text{ mol CO}_2\text{/min}$$

Since 660 mol CO_2/min are present, ammonia is the limiting reagent.

$$\frac{1.1 \times 10^3 \text{ mol NH}_3}{\text{min}} \times \frac{1 \text{ mol urea}}{2 \text{ mol NH}_3} \times \frac{60.06 \text{ g urea}}{\text{mol urea}} = 3.3 \times 10^4 \text{ g urea/min}$$

34.   The unbalanced reaction is: $C_8H_{18} + O_2 \rightarrow CO_2 + H_2O$

The balanced equation is: $2 C_8H_{18} + 25 O_2 \rightarrow 16 CO_2 + 18 H_2O$

$$125 \text{ g C}_8\text{H}_{18} \times \frac{1 \text{ mol C}_8\text{H}_{18}}{114.2 \text{ g C}_8\text{H}_{18}} \times \frac{25 \text{ mol O}_2}{2 \text{ mol C}_8\text{H}_{18}} = 13.7 \text{ mol O}_2$$

$$V = \frac{nRT}{P} = \frac{13.7 \text{ mol} \times \dfrac{0.08206 \text{ L atm}}{\text{mol K}} \times 273 \text{ K}}{1.00 \text{ atm}} = 307 \text{ L of O}_2$$

Or we can make use of the fact that at STP one mole of an ideal gas occupies a volume of 22.42 L. This can be calculated using the ideal gas law.

So: $13.7 \text{ mol O}_2 \times \dfrac{22.42 \text{ L}}{\text{mol}} = 307 \text{ L}$

35. $4.10 \text{ g HgO} \times \dfrac{1 \text{ mol HgO}}{216.6 \text{ g HgO}} \times \dfrac{1 \text{ mol O}_2}{2 \text{ mol HgO}} = 9.46 \times 10^{-3} \text{ mol O}_2$

$$V = \frac{nRT}{P} = \frac{9.46 \times 10^{-3} \text{ mol} \times \dfrac{0.08206 \text{ L atm}}{\text{mol K}} \times 303 \text{ K}}{725 \text{ torr}} \times \frac{760 \text{ torr}}{\text{atm}} = 0.247 \text{ L}$$

36. $10.10 \text{ atm} - 7.62 \text{ atm} = 2.48 \text{ atm}$ is the pressure of the amount of F_2 reacted.

$PV = nRT$, V and T are constant; $\dfrac{P}{n}$ = constant, $\dfrac{P_1}{n_1} = \dfrac{P_2}{n_2}$ or $\dfrac{P_1}{P_2} = \dfrac{n_1}{n_2}$

$$\dfrac{\text{moles } F_2 \text{ reacted}}{\text{moles Xe reacted}} = \dfrac{2.48 \text{ atm}}{1.24 \text{ atm}} = 2; \quad \text{So Xe} + 2 \, F_2 \rightarrow XeF_4$$

37. $P_{total} = P_{N_2} + P_{H_2O}$, $P_{N_2} = 726 \text{ torr} - 23.8 \text{ torr} = 702 \text{ torr} \times \dfrac{1 \text{ atm}}{760 \text{ torr}} = 0.924 \text{ atm}$

$$n = \dfrac{PV}{RT} = \dfrac{0.924 \text{ atm} \times 31.8 \times 10^{-3} \text{ L}}{\dfrac{0.08206 \text{ L atm}}{\text{mol K}} \times 298 \text{ K}} = 1.20 \times 10^{-3} \text{ mol } N_2$$

Mass of N in compound $= 1.20 \times 10^{-3} \text{ mol} \times \dfrac{28.02 \text{ g } N_2}{\text{mol}} = 3.36 \times 10^{-2} \text{ g}$

$$\% \, N = \dfrac{3.36 \times 10^{-2} \text{ g}}{0.253 \text{ g}} \times 100 = 13.3\% \, N$$

38. $0.2766 \text{ g } CO_2 \times \dfrac{12.011 \text{ g C}}{44.009 \text{ g } CO_2} = 7.549 \times 10^{-2} \text{ g C}$

$$\% \, C = \dfrac{7.549 \times 10^{-2} \text{ g}}{0.1023 \text{ g}} \times 100 = 73.79\% \, C$$

$0.0991 \text{ g } H_2O \times \dfrac{2.016 \text{ g H}}{18.02 \text{ g } H_2O} = 1.11 \times 10^{-2} \text{ g H}$

$$\% \, H = \dfrac{1.11 \times 10^{-2} \text{ g}}{0.1023 \text{ g}} \times 100 = 10.9\% \, H$$

$PV = nRT; \quad n = \dfrac{PV}{RT} = \dfrac{1.00 \text{ atm} \times 27.6 \times 10^{-3} \text{ L}}{\dfrac{0.08206 \text{ L atm}}{\text{mol K}} \times 273 \text{ K}} = 1.23 \times 10^{-3} \text{ mol } N_2$

$1.23 \times 10^{-3} \text{ mol } N_2 \times \dfrac{28.02 \text{ g } N_2}{\text{mol } N_2} = 3.45 \times 10^{-2} \text{ g nitrogen}$

$$\% \, N = \dfrac{3.45 \times 10^{-2} \text{ g}}{0.4831 \text{ g}} \times 100 = 7.14\% \, N$$

$\% \, O = 100.00 - (73.79 + 10.9 + 7.14) = 8.2\% \, O$

Out of 100.00 g of compound, there are:

$73.79 \text{ g C} \times \dfrac{1 \text{ mol}}{12.011 \text{ g}} = 6.144 \text{ mol C}; \quad 7.14 \text{ g N} \times \dfrac{1 \text{ mol}}{14.01 \text{ g}} = 0.510 \text{ mol N}$

$10.9 \text{ g H} \times \dfrac{1 \text{ mol}}{1.008 \text{ g}} = 10.8 \text{ mol H}; \quad 8.2 \text{ g O} \times \dfrac{1 \text{ mol}}{16.00 \text{ g}} = 0.51 \text{ mol O}$

Divide all values by 0.51. We get the empirical formula $C_{12}H_{21}NO$.

$$M = \frac{dRT}{P} = \frac{\dfrac{4.02\ g}{L} \times \dfrac{0.08206\ L\ atm}{mol\ K} \times 400.\ K}{256\ torr} \times \frac{760\ torr}{atm} = 392\ g/mol$$

Molar mass of $C_{12}H_{21}NO \approx 195$ g/mol and $\dfrac{392}{195} \approx 2$.

Thus, the molecular formula is $C_{24}H_{42}N_2O_2$

39.   a. $PV = nRT$, P & T constant; The number of moles is directly proportional to the volume.
              From the reaction we get 4 moles of NO for every 4 moles of NH_3. Since the moles are
              equal, the volumes of gas should be equal. We will get 10.0 L of NO.

       b. 10.0×10^3 g $NH_3 \times \dfrac{1\ mol\ NH_3}{17.03\ g\ NH_3} \times \dfrac{5\ mol\ O_2}{4\ mol\ NH_3} = 734$ mol of O_2

$$V = \frac{nRT}{P}\quad \frac{(734\ mol)\left(\dfrac{0.08206\ L\ atm}{mol\ K}\right)(273\ K)}{1.00\ atm} = 1.64 \times 10^4\ L\ O_2$$

       c. $n = \dfrac{PV}{RT} = \dfrac{(3.00\ atm)(5.00 \times 10^2\ L)}{\left(\dfrac{0.08206\ L\ atm}{mol\ K}\right)(273 + 250.)K} = 35.0$ mol NH_3

              35.0 mol $NH_3 \times \dfrac{4\ mol\ NO}{4\ mol\ NH_3} \times \dfrac{30.01\ g\ NO}{mol\ NO} = 1.05 \times 10^3$ g NO

       d. $4\ NH_3(g) + 5\ O_2(g) \rightarrow 4\ NO(g) + 6\ H_2O(g)$

              65.0 L $NH_3 \times \dfrac{6\ L\ H_2O}{4\ L\ NH_3} = 97.5$ L H_2O; 75.0 L $O_2 \times \dfrac{6\ L\ H_2O}{5\ L\ O_2} = 90.0$ L H_2O

              The O_2 is limiting and 90.0 L H_2O are produced. At STP, the volume of one mole of an ideal
              gas is 22.42 L.

              90.0 L $H_2O \times \dfrac{1\ mol\ H_2O}{22.42\ L\ H_2O} \times \dfrac{18.02\ g\ H_2O}{mol\ H_2O} = 72.3$ g H_2O

       e.   90.0 L $H_2O \times \dfrac{4\ L\ NO}{6\ L\ H_2O} \times \dfrac{1\ mol\ NO}{22.42\ L\ NO} = 2.68$ mol NO

Kinetic Molecular Theory and Real Gases

40.   Boyle's Law: $P \propto 1/V$ at constant n and T

In the kinetic molecular theory (kmt), P is proportional to the collision frequency which is proportional to 1/V. As the volume increases there will be fewer collisions with the walls of the container and pressure decreases..

Charles's Law: V $\propto$ T at constant n and P

The pressure in kmt is proportional to the product of the momentum (mu) and collision frequency (nu/V). Thus,

$$P \propto \frac{nmu^2}{V}, \quad V \propto \frac{nmu^2}{P}$$

The kinetic energy is equal to $1/2\ mu^2$ and is proportional to T. Thus, V $\propto$ T at constant n and P.

Dalton's Law: $P_{tot} = P_1 + P_2 + ...$

One of the postulates of the kmt is that gas particles do not interact with each other. If this is so, then we can treat each gas in a mixture of gases independently.

41. $(KE)_{avg} = 3/2\ RT$; KE depends only on temperature. At each temperature CH_4 and N_2 will have the same average KE. For energy units of joules (J), use R = 8.3145 J mol^{-1} K^{-1}.

at 273 K: $(KE)_{avg} = \frac{3}{2} \times \frac{8.3145\ J}{mol\ K} \times 273\ K = 3.40 \times 10^3$ J/mol

at 546 K: $(KE)_{avg} = \frac{3}{2} \times \frac{8.3145\ J}{mol\ K} \times 546\ K = 6.81 \times 10^3$ J/mol

42. $u_{rms} = \left(\dfrac{3RT}{M} \right)^{1/2}$, where R = $\dfrac{8.3145\ J}{mol\ K}$ and M = molar mass in kg

For CH_4, M = 1.604×10^{-2} kg and for N_2, M = 2.802×10^{-2} kg.

For CH_4 at 273 K:

$$u_{rms} = \left(\frac{\dfrac{3 \times 8.3145\ J}{mol\ K} \times 273\ K}{1.604 \times 10^{-2}\ kg/mol} \right)^{1/2} = 652\ m/s$$

Similarly u_{rms} for CH_4 at 546 K is 921 m/s.

For N_2, u_{rms}= 493 m/s at 273 K and 697 m/s at 546 K.

43.   No. There is a distribution of energies with an average kinetic energy equal to 3/2 RT.

44.   a.   They will all have the same average kinetic energy since they are all at the same temperature.

b. Flask C: Lightest molecules have the greatest root mean square velocity (at constant T).

c. Flask A: Collision frequency is proportional to avg. velocity × n/V. Moles are proportional to pressure at constant T and V, and average velocity is proportional to $(1/M)^{1/2}$.

	n (relative)	$u_{avg.}$ (relative)	Coll. Freq. (relative)
A	1.0	1.0	1.0
B	0.33	1.0	0.33
C	0.13	3.7	0.49

45. $\dfrac{R_1}{R_2} = \left(\dfrac{M_2}{M_1}\right)^{1/2}$; $\dfrac{^{12}C\,^{17}O}{^{12}C\,^{18}O} = \left(\dfrac{30.0}{29.0}\right)^{1/2} = 1.02$; $\dfrac{^{12}C\,^{16}O}{^{12}C\,^{18}O} = \left(\dfrac{30.0}{28.0}\right)^{1/2} = 1.04$

The relative rates of effusion of $^{12}C^{16}O$: $^{12}C^{17}O$: $^{12}C^{18}O$ are 1.04: 1.02: 1.00.

Advantages: CO_2 isn't as toxic as CO.

Major disadvantages of using CO_2 instead of CO:

1. Get a mixture of oxygen isotopes in CO_2.
2. Some species, e.g., $^{12}C^{16}O^{18}O$ and $^{12}C^{17}O_2$ would effuse at about the same rate. Thus, at some points no separation occurs.

46.

	a	b	c	d
avg. KE	inc	dec	same (KE $\propto$ T)	same
u_{rms}	inc	dec	same ($u_{rms}^2 \propto$ T)	same
coll. freq. gas	inc	dec	inc	inc
coll. freq. wall	inc	dec	inc	inc
impact E	inc	dec	same (impact E $\propto$ KE $\propto$ T)	same

Both collision frequencies are proportional to the root mean square velocity (as velocity increases it takes less time to move to the next collision) and the quantity n/V (as molecules per volume increases, collision frequency increases).

47. $\dfrac{R_1}{R_2} = \left(\dfrac{M_2}{M_1}\right)^{1/2}$ where R = rate of diffusion and M = molar mass of unknown

$\dfrac{R_1}{R_2} = \left(\dfrac{M_2}{M_1}\right)^{1/2}$; $\dfrac{31.50}{30.50} = \left(\dfrac{31.998}{M}\right)^{1/2} = 1.033$

$\dfrac{31.998}{M} = 1.067$, M = 29.99; Of the choices, the gas would be NO, nitric oxide.

48. $R_1 = \dfrac{24.0 \text{ mL}}{\text{min}}$, $R_2 = \dfrac{47.8 \text{ mL}}{\text{min}}$, $M_2 = \dfrac{16.04 \text{ g}}{\text{mol}}$ and $M_1 = ?$

$$\dfrac{24.0}{47.8} = \left(\dfrac{16.04}{M_1}\right)^{1/2} = 0.502, \quad 16.04 = (0.502)^2 \, M_1, \quad M_1 = \dfrac{16.04}{0.252} = \dfrac{63.7 \text{ g}}{\text{mol}}$$

49. $\dfrac{\text{Rate (1)}}{\text{Rate (2)}} = \left(\dfrac{M_2}{M_1}\right)^{1/2}$; Let Gas (1) = He, Gas (2) = NF_3

$$\dfrac{\dfrac{1.0 \text{ L}}{4.5 \text{ min}}}{\dfrac{1.0 \text{ L}}{t}} = \left(\dfrac{71.0}{4.00}\right)^{1/2}, \quad \dfrac{t}{4.5 \text{ min}} = 4.21, \quad t = 19 \text{ min}$$

50.   a. $PV = nRT$

$$P = \dfrac{nRT}{V} = \dfrac{0.5000 \text{ mol} \left(\dfrac{0.08206 \text{ L atm}}{\text{mol K}}\right)(25.0 + 273.2) \text{ K}}{1.0000 \text{ L}} = 12.24 \text{ atm}$$

   b. $\left[P + a\left(\dfrac{n}{V}\right)^2\right](V - nb) = nRT$; For N_2: $a = 1.39$ atm L^2/mol^2 and $b = 0.0391$ L/mol

$$\left[P + 1.39\left(\dfrac{0.5000}{1.0000}\right)^2 \text{ atm}\right](1.0000 \text{ L} - 0.5000 \times 0.0391 \text{ L}) = 12.24 \text{ L atm}$$

   $(P + 0.348 \text{ atm})(0.9805 \text{ L}) = 12.24 \text{ L atm}$

   $P = \dfrac{12.24 \text{ L atm}}{0.9805 \text{ L}} - 0.348 \text{ atm} = 12.48 - 0.348 = 12.13 \text{ atm}$

   c.   The ideal gas law is high by 0.11 atm or $\dfrac{0.11}{12.13} \times 100 = 0.91\%$.

51.   a. $PV = nRT$

$$P = \dfrac{nRT}{V} = \dfrac{0.5000 \text{ mol} \times \dfrac{0.08206 \text{ L atm}}{\text{mol K}} \times 298.2 \text{ K}}{10.000 \text{ L}} = 1.224 \text{ atm}$$

   b. $\left[P + a\left(\dfrac{n}{V}\right)^2\right](V - nb) = nRT$; For N_2: $a = 1.39$ atm L^2/mol^2 and $b = 0.0391$ L/mol

$$\left[P + 1.39\left(\dfrac{0.5000}{10.000}\right)^2 \text{ atm}\right](10.000 \text{ L} - 0.5000 \times 0.0391 \text{ L}) = 12.24 \text{ L atm}$$

$(P + 0.00348 \text{ atm})(10.000 \text{ L} - 0.0196 \text{ L}) = 12.24 \text{ L atm}$

$P + 0.00348 \text{ atm} = \dfrac{12.24 \text{ L atm}}{9.980 \text{ L}} = 1.226 \text{ atm}, \quad P = 1.226 - 0.00348 = 1.223 \text{ atm}$

c.   The results agree to ± 0.001 atm (0.08%).

d.   In 5.50 the pressure is relatively high and there is a significant disagreement. In 5.51 the pressure is around 1 atm and both gas laws show better agreement. The ideal gas law is valid at relatively low pressures.

52.   Van der Waals' equation: $P = \dfrac{nRT}{V - nb} - a\left(\dfrac{n}{V}\right)^2$

P is the measured pressure and V is the volume of the container.

For NH_3: $a = 4.17 \text{ L}^2 \text{ atm/mol}^2$ and $b = 0.0371 \text{ L/mol}$

For the first experiment (assuming four significant figures in all values):

$$P = \dfrac{nRT}{V - nb} - a\left(\dfrac{n}{V}\right)^2 = \dfrac{1.000 \times 0.08206 \times 273.2}{172.1 - 0.0371} - 4.17\left(\dfrac{1.000}{172.1}\right)^2$$

$P = 0.1303 - 0.0001 = 0.1302$ atm. In Example 5.1, $P_{obs} = 0.1300$ atm. The difference is less than 0.1%. The ideal gas law also gives 0.1302 atm. At low pressures, the van der Waals equation agrees with the ideal gas law.

For experiment 6, the measured pressure is 1.000 atm. The ideal gas law gives $P = 1.015$ atm and the van der Waals equation gives $P = 1.017 - 0.086 = 1.008$ atm. The van der Waals equation accounts for approximately one-half of the deviation from ideal behavior. Note: As pressure increases, deviation from the ideal gas law increases.

53. $\left(P + \dfrac{an^2}{V^2}\right)(V - nb) = nRT$

$PV + \dfrac{an^2 V}{V^2} - nbP - \dfrac{an^3 b}{V^2} = nRT; \quad PV + \dfrac{an^2}{V} - nbP - \dfrac{an^3 b}{V^2} = nRT$

At low P and high T, the molar volume of a gas will be relatively large. The $\dfrac{an^2}{V}$ and $\dfrac{an^3 b}{V^2}$ terms become negligible because V is large.

Since nb is the actual volume of the gas molecules themselves, then nb << V and -nbP is negligible compared to PV. Thus, PV = nRT.

Atmospheric Chemistry

54. $\chi_{NO} = 5 \times 10^{-7}$ from Table 5.4

$P_{NO} = \chi_{NO}P_{total} = 5 \times 10^{-7} \times 1.0 \text{ atm} = 5 \times 10^{-7} \text{ atm}$

$$PV = nRT; \quad \frac{n}{V} = \frac{P}{RT} = \frac{5 \times 10^{-7} \text{ atm}}{\left(\dfrac{0.08206 \text{ L atm}}{\text{mol K}}\right)(273 \text{ K})} = \frac{2 \times 10^{-8} \text{ mol}}{L}$$

$$\frac{2 \times 10^{-8} \text{ mol}}{L} \times \frac{1 \text{ L}}{1000 \text{ cm}^3} \times \frac{6.02 \times 10^{23} \text{ molecules}}{\text{mol}} = \frac{1 \times 10^{13} \text{ molecules NO}}{\text{cm}^3}$$

55.   a.   If we have 10^6 L of air, there are 3.0×10^2 L of CO. Since n ∝ V:

$$P_{CO} = \chi_{CO}P_{tot} = \frac{3.0 \times 10^2}{1.0 \times 10^6} \times 628 \text{ torr} = 0.19 \text{ torr}$$

b. $n_{CO} = \dfrac{P_{CO}V}{RT}$ Assume 1.0 m^3 air, 1 m^3 = 1000 L.

$$n_{CO} = \frac{\dfrac{0.19}{760} \text{ atm} \times (1.0 \times 10^3 \text{ L})}{\dfrac{0.08206 \text{ L atm}}{\text{mol K}} \times 273 \text{ K}} = 1.1 \times 10^{-2} \text{ mol}$$

$$1.1 \times 10^{-2} \text{ mol} \times \frac{6.022 \times 10^{23} \text{ molecules}}{\text{mol}} = 6.6 \times 10^{21} \text{ CO molecules in 1.0 } m^3 \text{ of air}$$

c. $\dfrac{6.6 \times 10^{21} \text{ molecules}}{m^3} \times \left(\dfrac{1 \text{ m}}{100 \text{ cm}}\right)^3 = \dfrac{6.6 \times 10^{15} \text{ molecules CO}}{\text{cm}^3}$

56.   At 100. km, T ≈ - 90° C and P ≈ $10^{-5.5}$ ≈ 3×10^{-6} atm

$$PV = nRT; \quad \frac{PV}{T} = nR = \text{Const.}; \quad \frac{P_1V_1}{T_1} = \frac{P_2V_2}{T_2}$$

$$V_2 = \frac{V_1P_1T_2}{T_1P_2} = \frac{10.0 \text{ L} \times 3 \times 10^{-6} \text{ atm} \times 273 \text{ K}}{183 \text{ K} \times 1.0 \text{ atm}} = 4 \times 10^{-5} \text{ L} = 0.04 \text{ mL}$$

57. $2 \text{ HNO}_3(aq) + \text{CaCO}_3(s) \rightarrow \text{Ca(NO}_3)_2(aq) + \text{H}_2\text{O}(l) + \text{CO}_2(g)$

$\text{H}_2\text{SO}_4(aq) + \text{CaCO}_3(s) \rightarrow \text{CaSO}_4(aq) + \text{H}_2\text{O}(l) + \text{CO}_2(g)$

58.   Total volume = 5.0×10^3 ft $\left(5.0 \text{ mi} \times \dfrac{5280 \text{ ft}}{\text{mi}}\right)\left(10.0 \text{ mi} \times \dfrac{5280 \text{ ft}}{\text{mi}}\right) = 7.0 \times 10^{12} \text{ ft}^3$

$$7.0 \times 10^{12} \text{ ft}^3 \left(\frac{12 \text{ in}}{\text{ft}} \times \frac{2.54 \text{ cm}}{\text{in}} \right)^3 \left(\frac{1 \text{ L}}{1000 \text{ cm}^3} \right) = 2.0 \times 10^{14} \text{ L}$$

$$\text{ppmv NO}_2 = 0.20 = \frac{V_{NO_2}}{2.0 \times 10^{14} \text{ L}} \times 10^6$$

$$V_{NO_2} = \frac{(0.20)(2.0 \times 10^{14} \text{ L})}{10^6} = 4.0 \times 10^7 \text{ L NO}_2$$

$$PV = nRT; \ n = \frac{PV}{RT} = \frac{(1.0 \text{ atm}) \ (4.0 \times 10^7 \text{ L})}{\left(\dfrac{0.08206 \text{ L atm}}{\text{mol K}} \right) (273 \text{ K})} = 1.8 \times 10^6 \text{ mol NO}_2$$

$$1.8 \times 10^6 \text{ mol NO}_2 \times \frac{46.0 \text{ g NO}_2}{\text{mol NO}_2} = 8.3 \times 10^7 \text{ g NO}_2$$

59.   For benzene:

$$89.6 \times 10^{-9} \text{ g} \times \frac{1 \text{ mol}}{78.11 \text{ g}} = 1.15 \times 10^{-9} \text{ mol}$$

$$V_{ben} = \frac{nRT}{P} = \frac{1.15 \times 10^{-9} \text{ mol} \times \dfrac{0.08206 \text{ L atm}}{\text{mol K}} \times 296 \text{ K}}{748 \text{ torr}} \times \frac{760 \text{ torr}}{\text{atm}} = 2.84 \times 10^{-8} \text{ L}$$

$$\text{Mixing ratio} = \frac{2.84 \times 10^{-8} \text{ L}}{3.00 \text{ L}} \times 10^6 = 9.47 \times 10^{-3} \text{ ppmv}$$

$$\text{or ppbv} = \frac{\text{vol. of X} \times 10^9}{\text{total vol.}} = \frac{2.84 \times 10^{-8} \text{ L}}{3.00 \text{ L}} \times 10^9 = 9.47 \text{ ppbv}$$

$$\frac{1.15 \times 10^{-9} \text{ mol benzene}}{3.00 \text{ L}} \times \frac{1 \text{ L}}{1000 \text{ cm}^3} \times \frac{6.022 \times 10^{23} \text{ molecules}}{\text{mol}}$$

$$= 2.31 \times 10^{11} \text{ molecules benzene/cm}^3$$

For toluene:

$$153 \times 10^{-9} \text{ g C}_7\text{H}_8 \times \frac{1 \text{ mol}}{92.13 \text{ g}} = 1.66 \times 10^{-9} \text{ mol toluene}$$

$$V_{tol} = \frac{nRT}{P} = \frac{1.66 \times 10^{-9} \text{ mol} \times \dfrac{0.08206 \text{ L atm}}{\text{mol K}} \times 296 \text{ K}}{748 \text{ torr}} \times \frac{760 \text{ torr}}{\text{atm}} = 4.10 \times 10^{-8} \text{ L}$$

$$\text{Mixing ratio} = \frac{4.10 \times 10^{-8} \text{ L}}{3.00 \text{ L}} \times 10^6 = 1.37 \times 10^{-2} \text{ ppmv (or 13.7 ppbv)}$$

$$\frac{1.66 \times 10^{-9} \text{ mol toluene}}{3.00 \text{ L}} \times \frac{1 \text{ L}}{1000 \text{ cm}^3} \times \frac{6.022 \times 10^{23} \text{ molecules}}{\text{mol}}$$

$$= 3.33 \times 10^{11} \text{ molecules toluene/cm}^3$$

Additional Exercises

60.   a. Heating the can will increase the pressure of the gas inside the can. $P \propto T$, V & n const. As the pressure increases it may be enough to rupture the can.

        b. As you draw a vacuum in your mouth, atmospheric pressure pushes the liquid up the straw.

        c. The external atmospheric pressure pushes on the can. Since there is no opposing pressure from the air in the inside, the can collapses.

        d. How "hard" the tennis ball is depends on the difference between the pressure of the air inside the tennis ball and atmospheric pressure. A "sea level" ball will be much "harder" at high altitude. A high altitude ball will be "soft" at sea level.

61. $750 \text{ mL juice} \times \dfrac{12 \text{ mL alcohol}}{100 \text{ mL juice}} = 90. \text{ mL alcohol present}$

        $90. \text{ mL alcohol} \times \dfrac{0.79 \text{ g}}{\text{mL}} = 71 \text{ g } C_2H_5OH$

        $71 \text{ g } C_2H_5OH \times \dfrac{1 \text{ mol } C_2H_5OH}{46.1 \text{ g } C_2H_5OH} \times \dfrac{2 \text{ mol } CO_2}{2 \text{ mol } C_2H_5OH} = 1.5 \text{ mol } CO_2$

        The CO_2 will occupy (825 - 750. =) 75 mL not occupied by the liquid (headspace).

$$P = \frac{nRT}{V} = \frac{1.5 \text{ mol} \times \dfrac{0.08206 \text{ L atm}}{\text{mol K}} \times 298 \text{ K}}{75 \times 10^{-3} \text{ L}} = 490 \text{ atm}$$

        Actually, enough CO_2 will dissolve in the wine to lower the pressure of CO_2 to a much more reasonable value.

62. $PV = nRT; \quad \dfrac{nT}{P} = \dfrac{V}{R} = \text{Const.}; \quad \dfrac{n_1T_1}{P_1} = \dfrac{n_2T_2}{P_2}; \quad \text{mol} \times \dfrac{g}{\text{mol}} = g$

        $\dfrac{n_1MT_1}{P_1} = \dfrac{n_2MT_2}{P_2}; \quad \dfrac{g_1T_1}{P_1} = \dfrac{g_2T_2}{P_2}$

        $g_2 = \dfrac{g_1T_1P_2}{T_2P_1} = \dfrac{1.00 \times 10^3 \text{ g} \times 291 \text{ K} \times 650. \text{ psi}}{299 \text{ K} \times 2050. \text{ psi}} = 309 \text{ g}$

63.   a.   Volume of hot air: $V = \dfrac{4}{3}\pi r^3 = \dfrac{4}{3}\pi(2.50)^3 = 65.4 \text{ m}^3$

(Note: radius = diameter/2 = 5.00/2 = 2.50 m)

$$65.4 \text{ m}^3 \times \left(\dfrac{10 \text{ dm}}{\text{m}}\right)^3 \times \dfrac{1 \text{ L}}{\text{dm}^3} = 6.54 \times 10^4 \text{ L}$$

$$n = \dfrac{PV}{RT} = \dfrac{\left(745 \text{ torr} \times \dfrac{1 \text{ atm}}{760 \text{ torr}}\right) \times 6.54 \times 10^4 \text{ L}}{\dfrac{0.08206 \text{ L atm}}{\text{mol K}} \times (273 + 65) \text{ K}} = 2.31 \times 10^3 \text{ mol}$$

$$\text{Mass} = 2.31 \times 10^3 \text{ mol} \times \dfrac{29.0 \text{ g}}{\text{mol}} = 6.70 \times 10^4 \text{ g}$$

Mass of air displaced:

$$n = \dfrac{PV}{RT} = \dfrac{\dfrac{745}{760} \text{ atm} \times 6.54 \times 10^4 \text{ L}}{\dfrac{0.08206 \text{ L atm}}{\text{mol K}} \times (273 + 21) \text{ K}} = 2.66 \times 10^3 \text{ mol}$$

$$\text{Mass} = 2.66 \times 10^3 \text{ mol} \times \dfrac{29.0 \text{ g}}{\text{mol}} = 7.71 \times 10^4 \text{ g}$$

Lift = 7.71×10^4 g - 6.70×10^4 g = 1.01×10^4 g

b.   Mass displaced is the same, 7.71×10^4 g. Moles of He in balloon will be the same as moles of air displaced, 2.66×10^3 mol, since P, V and T are the same.

$$\text{Mass of He} = 2.66 \times 10^3 \text{ mol} \times \dfrac{4.003 \text{ g}}{\text{mol}} = 1.06 \times 10^4 \text{ g}$$

Lift = 7.71×10^4 g - 1.06×10^4 g = 6.65×10^4 g

c.   Mass of hot air:

$$n = \dfrac{PV}{RT} = \dfrac{\dfrac{630.}{760} \text{ atm} \times 6.54 \times 10^4 \text{ L}}{\dfrac{0.08206 \text{ L atm}}{\text{mol K}} \times 338 \text{ K}} = 1.95 \times 10^3 \text{ mol}$$

$$1.95 \times 10^3 \text{ mol} \times \dfrac{29.0 \text{ g}}{\text{mol}} = 5.66 \times 10^4 \text{ g of hot air}$$

Mass of air displaced:

$$n = \frac{PV}{RT} = \frac{\dfrac{630.}{760}\text{ atm} \times 6.54 \times 10^4 \text{ L}}{\dfrac{0.08206 \text{ L atm}}{\text{mol K}} \times 294 \text{ K}} = 2.25 \times 10^3 \text{ mol}$$

$$2.25 \times 10^3 \text{ mol} \times \frac{29.0 \text{ g}}{\text{mol}} = 6.53 \times 10^4 \text{ g}$$

Lift = 6.53×10^4 g - 5.66×10^4 g = 8.7×10^3 g

d.   mass of hot air = 6.70×10^4 g (from part a)

Mass of air displaced:

$$n = \frac{PV}{RT} = \frac{\dfrac{745}{760}\text{ atm} \times 6.54 \times 10^4 \text{ L}}{\dfrac{0.08206 \text{ L atm}}{\text{mol K}} \times 265 \text{ K}} = 2.95 \times 10^3 \text{ mol}$$

$$2.95 \times 10^3 \text{ mol} \times \frac{29.0 \text{ g}}{\text{mol}} = 8.56 \times 10^4 \text{ g}$$

Lift = 8.56×10^4 g - 6.70×10^4 g = 1.86×10^4 g

64.   PV = nRT, V and T are constant. $\dfrac{P_1}{n_1} = \dfrac{P_2}{n_2}, \dfrac{P_2}{P_1} = \dfrac{n_2}{n_1}$

Let's calculate the partial pressure of C_3H_3N that can be produced from each of the starting materials:

$$0.500 \text{ MPa} \times \frac{2 \text{ mol } C_3H_3N}{2 \text{ mol } C_3H_6} = 0.500 \text{ MPa if } C_3H_6 \text{ is limiting.}$$

$$0.800 \text{ MPa} \times \frac{2 \text{ mol } C_3H_3N}{2 \text{ mol } NH_3} = 0.800 \text{ MPa if } NH_3 \text{ is limiting.}$$

$$1.500 \text{ MPa} \times \frac{2 \text{ mol } C_3H_3N}{3 \text{ mol } O_2} = 1.000 \text{ MPa if } O_2 \text{ is limiting.}$$

Thus, C_3H_6 is limiting. The partial pressure of C_3H_3N after the reaction is:

$$0.500 \times 10^6 \text{ Pa} \times \frac{1 \text{ atm}}{1.013 \times 10^5 \text{ Pa}} = 4.94 \text{ atm}$$

$$n = \frac{PV}{RT} = \frac{4.94 \text{ atm} \times 150 \text{ L}}{\dfrac{0.08206 \text{ L atm}}{\text{mol K}} \times 298 \text{ K}} = 30.3 \text{ mol } C_3H_3N$$

$$30.3 \text{ mol} \times \frac{53.06 \text{ g}}{\text{mol}} = 1.61 \times 10^3 \text{ g}$$

65.   The partial pressure of CO_2 that reacted is 740. - 390. = 350. torr. Thus, the number of moles of CO_2 that reacts is given by:

$$n = \frac{PV}{RT} = \frac{\dfrac{350.}{760} \text{ atm} \times 3.00 \text{ L}}{\dfrac{0.08206 \text{ L atm}}{\text{mol K}} \times 293 \text{ K}} = 5.75 \times 10^{-2} \text{ mol } CO_2$$

$$5.75 \times 10^{-2} \text{ mol } CO_2 \times \frac{1 \text{ mol MgO}}{1 \text{ mol } CO_2} \times \frac{40.31 \text{ g MgO}}{\text{mol HgO}} = 2.32 \text{ g MgO}$$

$$\% \text{ MgO} = \frac{2.32 \text{ g}}{2.85 \text{ g}} \times 100 = 81.4\% \text{ MgO}$$

66. $$P_{total} = 97.0 \text{ kPa} = 97.0 \times 10^3 \text{ Pa} \times \frac{1 \text{ atm}}{1.013 \times 10^5 \text{ Pa}} \times \frac{760 \text{ torr}}{\text{atm}} = 728 \text{ torr}$$

$$P_{total} = P_{N_2O} + P_{H_2O}, \quad P_{N_2O} = 728 - 15 = 713 \text{ torr} \times \frac{1 \text{ atm}}{760 \text{ torr}} = 0.938 \text{ atm}$$

$$1.55 \text{ g } NH_4NO_3 \times \frac{1 \text{ mol } NH_4NO_3}{80.05 \text{ g } NH_4NO_3} \times \frac{1 \text{ mol } N_2O}{\text{mol } NH_4NO_3} = 1.94 \times 10^{-2} \text{ mol}$$

$$V = \frac{nRT}{P} = \frac{1.94 \times 10^{-2} \text{ mol} \times \dfrac{0.08206 \text{ L atm}}{\text{mol K}} \times 291 \text{ K}}{0.938 \text{ atm}} = 0.494 \text{ L or } 494 \text{ mL}$$

67.   A concentration of 1.0 ppbv means there is 1.0 L of CH_2O for every 1.0×10^9 L of air measured at the same temperature and pressure. The molar volume of an ideal gas is 22.42 at STP.

$$\frac{1.0 \text{ L } CH_2O}{1.0 \times 10^9 \text{ L}} \times \frac{1 \text{ mol } CH_2O}{22.42 \text{ L}} \times \frac{6.022 \times 10^{23} \text{ molecules}}{\text{mol}} \times \frac{1 \text{ L}}{1000 \text{ cm}^3} = \frac{2.7 \times 10^{10} \text{ molecules}}{\text{cm}^3}$$

$$V = \left(18.0 \text{ ft} \times \frac{12 \text{ in}}{\text{ft}} \times \frac{2.54 \text{ cm}}{\text{in}} \right) \left(24.0 \text{ ft} \times \frac{12 \text{ in}}{\text{ft}} \times \frac{2.54 \text{ cm}}{\text{in}} \right)$$

$$\times \left(8.0 \text{ ft} \times \frac{12 \text{ in}}{\text{ft}} \times \frac{2.54 \text{ cm}}{\text{in}} \right) = 9.8 \times 10^7 \text{ cm}^3$$

$$9.8 \times 10^7 \text{ cm}^3 \times \frac{2.7 \times 10^{10} \text{ molecules}}{\text{cm}^3} \times \frac{1 \text{ mol}}{6.022 \times 10^{23} \text{ molecules}} \times \frac{30.03 \text{ g}}{\text{mol}} = 1.3 \times 10^{-4} \text{ g } CH_2O$$

68.   The unbalanced equation is: $CaC_2 + H_2O \rightarrow C_2H_2 + Ca(OH)_2$

Balancing, we get: $CaC_2(s) + 2 \, H_2O(l) \rightarrow C_2H_2(g) + Ca(OH)_2(aq)$

$$2.50 \text{ g CaC}_2 \times \frac{1 \text{ mol CaC}_2}{64.10 \text{ g CaC}_2} \times \frac{1 \text{ mol C}_2\text{H}_2}{\text{mol CaC}_2} = 3.90 \times 10^{-2} \text{ mol C}_2\text{H}_2$$

$$P_{C_2H_2} = 715 - 23.8 = 691 \text{ torr}; \quad \chi_{C_2H_2} = \frac{P_{C_2H_2}}{P_{tot}} = \frac{691}{715} = 0.966$$

$$\chi_{C_2H_2} = 0.966 = \frac{n_{C_2H_2}}{n_{tot}} = \frac{3.90 \times 10^{-2} \text{ mol}}{n_{tot}}, \quad n_{tot} = 4.04 \times 10^{-2} \text{ mol}$$

$$V_{wet} = V_{tot} = \frac{n_{tot}RT}{P_{tot}} = \frac{4.04 \times 10^{-2} \text{ mol} \times \dfrac{0.08206 \text{ L atm}}{\text{mol K}} \times 298 \text{ K}}{715/760 \text{ atm}} = 1.05 \text{ L}$$

69. $P_1V_1 = P_2V_2$; The total volume is 1.00 L + 1.00 L + 2.00 L = 4.00 L.

For He: $P_2 = \dfrac{P_1V_1}{V_2} = 200. \text{ torr} \times \dfrac{1.00 \text{ L}}{4.00 \text{ L}} = 50.0 \text{ torr He}$

For Ne: $P_2 = 0.400 \text{ atm} \times \dfrac{1.00 \text{ L}}{4.00 \text{ L}} = 0.100 \text{ atm}$

$$0.100 \text{ atm} \times \frac{760 \text{ torr}}{\text{atm}} = 76.0 \text{ torr Ne}$$

For Ar: $P_2 = 24.0 \text{ kPa} \times \dfrac{2.00 \text{ L}}{4.00 \text{ L}} = 12.0 \text{ kPa}$

$$12.0 \text{ kPa} \times \frac{1 \text{ atm}}{101.3 \text{ kPa}} \times \frac{760 \text{ torr}}{\text{atm}} = 90.0 \text{ torr Ar}$$

$P_{total} = 50.0 + 76.0 + 90.0 = 216.0 \text{ torr}$

70. $n = \dfrac{PV}{RT} = \dfrac{\left(1.00 \times 10^{-6} \text{ torr} \times \dfrac{1 \text{ atm}}{760 \text{ torr}} \right) \times 1.00 \text{ L}}{\dfrac{0.08206 \text{ L atm}}{\text{mol K}} \times 295 \text{ K}} = 5.44 \times 10^{-11} \text{ mol}$

$$5.44 \times 10^{-11} \text{ mol} \times \frac{6.022 \times 10^{23} \text{ molecules}}{\text{mol}} = 3.28 \times 10^{13} \text{ molecules}$$

$$\frac{3.28 \times 10^{13} \text{ molecules}}{1.00 \times 10^3 \text{ cm}^3} = 3.28 \times 10^{10} \text{ molecules/cm}^3$$

71.   Out of 100.00 g compounds, there are:

$$58.51 \text{ g C} \times \frac{1 \text{ mol C}}{12.011 \text{ g C}} = 4.871 \text{ mol C}; \quad \frac{4.871}{2.436} = 2$$

$$7.37 \text{ g H} \times \frac{1 \text{ mol H}}{1.008 \text{ g H}} = 7.31 \text{ mol H}; \quad \frac{7.31}{2.436} = 3$$

$$34.12 \text{ g N} \times \frac{1 \text{ mol N}}{14.007 \text{ g N}} = 2.436 \text{ mol N}$$

Empirical formula: C_2H_3N

$$\frac{\text{Rate (1)}}{\text{Rate (2)}} = \left(\frac{M_2}{M_1}\right)^{1/2}; \quad \text{Let Gas (1)} = \text{He}; \quad 3.20 = \left(\frac{M_2}{4.003}\right)^{1/2}, \quad M_2 = 41.0$$

Mass of C_2H_3N: $2(12.0) + 3(1.0) + 1(14.0) = 41.0$

So molecular formula is also C_2H_3N.

72. $33.5 \text{ mg CO}_2 \times \dfrac{12.01 \text{ mg C}}{44.01 \text{ mg CO}_2} = 9.14 \text{ mg C}; \quad \% \text{ C} = \dfrac{9.14 \text{ mg}}{35.0 \text{ mg}} \times 100 = 26.1\%$

$41.1 \text{ mg H}_2\text{O} \times \dfrac{2.016 \text{ mg H}}{18.02 \text{ mg H}_2\text{O}} = 4.60 \text{ mg H}; \quad \% \text{ H} = \dfrac{4.60 \text{ mg}}{35.0 \text{ mg}} \times 100 = 13.1\% \text{ H}$

$$n_{N_2} = \frac{PV}{RT} = \frac{\dfrac{740.}{760} \text{ atm} \times 35.6 \times 10^{-3} \text{ L}}{\dfrac{0.08206 \text{ L atm}}{\text{mol K}} \times 298 \text{ K}} = 1.42 \times 10^{-3} \text{ mol N}_2$$

$1.42 \times 10^{-3} \text{ mol N}_2 \times \dfrac{28.02 \text{ g N}_2}{\text{mol N}_2} = 3.98 \times 10^{-2} \text{ g N} = 39.8 \text{ mg N}$

$\% \text{ N} = \dfrac{39.8 \text{ mg}}{65.2 \text{ mg}} \times 100 = 61.0\% \text{ N}$

Or we can get % N by difference: % N = 100.0 - (26.1 + 13.1) = 60.8%

Out of 100.0 g:

$$26.1 \text{ g C} \times \frac{1 \text{ mol}}{12.01 \text{ g}} = 2.17 \text{ mol}; \quad \frac{2.17}{2.17} = 1$$

$$13.1 \text{ g H} \times \frac{1 \text{ mol}}{1.008 \text{ g}} = 13.0 \text{ mol}; \quad \frac{13.0}{2.17} \approx 6$$

$$60.8 \text{ g N} \times \frac{1 \text{ mol}}{14.01 \text{ g}} = 4.34 \text{ mol}; \quad \frac{4.34}{2.17} = 2$$

Empirical formula is CH_6N_2.

$$\left(\frac{M}{39.95}\right)^{1/2} = \frac{26.4}{24.6} = 1.07, \ M = (1.07)^2 (39.95) = 45.7$$

Mass of CH_6N_2: $12.0 + 6.0 + 28.0 = 46.0$

Thus, molecular formula is CH_6N_2.

73. $PV = nRT = Const.; \ P_1V_1 = P_2V_2$

Let condition (1) correspond to He from tank that can be used to fill balloons. We must leave 1.0 atm of He in the tank, so $P_1 = 200.$ atm - $1.00 = 199$ atm and $V_1 = 15.0$ L. Condition (2) will correspond to the filled balloons with $P_2 = 1.00$ atm and $V_2 = N(2.00$ L) where N is the number of filled balloons.

(199) (15.0) = (1.00) (2.00) N; N = 1492.5; We can't fill 0.5 of a balloon. So N = 1492 balloons or to 3 significant figures, 1490 balloons.

74.   The pressure will increase because H_2 will effuse into container A faster than air will escape.

75.   The van der Waals' constant b is a measure of the size of the molecule. Thus, C_3H_8 should have the largest value of b.

76.   The values of a are: $H_2 \ \dfrac{0.244 \ L^2 \ atm}{mol^2}$, CO_2 3.59, N_2 1.39, CH_4 2.25

Since a is a measure interparticle attractions, the attractions are greatest for CO_2.

77. $M = \dfrac{dRT}{P}$, P & M constant; $dT = \dfrac{PM}{R} = const$

d = const (1/T) or $d_1T_1 = d_2T_2$, where T in kelvin (K).

$T = x + °C; \ 1.2930(x + 0.0) = 0.9460(x + 100.0)$

$1.2930 \ x = 0.9460 \ x + 94.60, \ 0.3470 \ x = 94.60, \ x = 272.6$

From these data absolute zero would be -272.6°C. Actual value is -273.15°C.

78. $PV = nRT$, n and P are constant; $\dfrac{V}{T} = \dfrac{nR}{P} = constant$

$$\frac{V_1}{T_1} = \frac{V_2}{T_2}; \ V_2 = \frac{V_1T_2}{T_1} = \frac{12.3 \ L \times 300. \ K}{450. \ K} = 8.20 \ L$$

79.   If we had 100.0 g of the gas, we would have 50.0 g He and 50.0 g Xe.

$$\chi_{He} = \frac{n_{He}}{n_{He} + n_{Xe}} = \frac{\dfrac{50.0}{4.003}}{\dfrac{50.0}{4.003} + \dfrac{50.0}{131.3}} = \frac{12.5}{12.5 + 0.381} = 0.970; \ \chi_{Xe} = 0.030$$

$P_{He} = 0.970 \times 600.$ torr $= 582$ torr; $P_{Xe} = 600. - 582 = 18$ torr

80.   Apply Boyle's Law to each gas: $P_1V_1 = P_2V_2$

N_2: $P_2 = \dfrac{V_1P_1}{V_2} = \dfrac{3.00 \text{ L} \times 0.250 \text{ atm}}{10.00 \text{ L}} = 0.0750$ atm

Ar: $P_2 = \dfrac{V_1P_1}{V_2} = \dfrac{7.00 \text{ L} \times 0.500 \text{ atm}}{10.00 \text{ L}} = 0.350$ atm

$P_{tot} = 0.0750 + 0.350 = 0.425$ atm

81.   The corrected (ideal) volume is the volume accessible to the gas molecules. For a real gas, this volume is less than the container volume because of the space occupied by the gas molecules.

82.   The force per impact, which is proportional to Δmu, is greater for He because it depends on $\sqrt{M}$.

$\dfrac{\text{Impact Force (H}_2)}{\text{Impact Force (He)}} = \sqrt{\dfrac{2.016}{4.003}} = 0.7097$

83.   Velocity $\propto (1/M)^{1/2}$; The H_2 molecules will have the largest average velocity since they have the smallest molar mass.

84. $n_i = \dfrac{P_iV}{RT} =$ initial moles of $CO_2 = \dfrac{\dfrac{750.}{760} \text{ atm} \times 1.50 \text{ L}}{\dfrac{0.08206 \text{ L atm}}{\text{mol K}} \times 303.2 \text{ K}} = 0.0595$ mol CO_2

$n_f = \dfrac{P_fV}{RT} =$ final moles of $CO_2 = \dfrac{230. \text{ torr} \times 1.50 \text{ L}}{\dfrac{0.08206 \text{ L atm}}{\text{mol K}} \times 303.2 \text{ K}} \times \dfrac{1 \text{ atm}}{760 \text{ torr}} = 0.0182$ mol CO_2

$0.0595 - 0.0182 = 0.0413$ mol CO_2 reacted.

Since each metal reacts 1:1 with CO_2, then the mixture contains 0.0413 mol of BaO and CaO.

Let x = g BaO and y = g CaO, so:

$x + y = 5.14$ g and $\dfrac{x}{153.3} + \dfrac{y}{56.08} = 0.0413$ mol reacted

Solving by simultaneous equations:

$$\begin{array}{r} x + 2.734 \text{ y} = 6.33 \\ \underline{-x -y = -5.14} \\ 1.734 \text{ y} = 1.19 \end{array}$$

y = 0.686 g CaO and 5.14 - y = x = 4.45 g BaO

% BaO = $\dfrac{4.45 \text{ g BaO}}{5.14 \text{ g}}$ × 100 = 86.6% BaO; % CaO = 100.0 - 86.6 = 13.4% CaO

85. $\dfrac{PV}{nRT} = 1 + \beta P;\ \dfrac{n}{V}(M) = d$ Where M = molar mass;

$\dfrac{M}{RT} \times \dfrac{P}{d} = 1 + \beta P,\ \dfrac{P}{d} = \dfrac{RT}{M} + \dfrac{\beta RTP}{M};$

This is in the equation for a straight line: y = b + mx. If we plot P/d vs P and extrapolate to P = 0, we get a y-intercept = b = 1.3980 = RT/M.

at 0.00°C, M = $\dfrac{0.08206 \times 273.15}{1.3980}$ = 16.03

86. $f(u) = 4\pi \left(\dfrac{m}{2\pi k_B T}\right)^{3/2} u^2 e^{(-mu^2/2k_B T)}$

As u → 0, $e^{(-mu^2/2k_B T)} \to e^0 \to 1$; At small values of u, the u^2 term causes the function to increase. At large values of u, the exponent $-mu^2/2k_B T$ is a large negative number and e raised to a negative number causes the function to decrease. As u → ∞, $e^{-\infty} \to 0$.

87. 0.050 mL × $\dfrac{1.149 \text{ g}}{\text{mL}}$ × $\dfrac{1 \text{ mol}}{32.00 \text{ g}}$ = 1.8 × 10⁻³ mol O₂

$V = \dfrac{nRT}{P} = \dfrac{1.8 \times 10^{-3} \text{ mol} \times \dfrac{0.08206 \text{ L atm}}{\text{mol K}} \times 310.\text{ K}}{1.0 \text{ atm}}$ = 4.6 10⁻² L = 46 mL

88. $u_{rms} = \left(\dfrac{3RT}{M}\right)^{1/2} = \left[\dfrac{3\left(\dfrac{8.3145 \text{ kg m}^2}{s^2 \text{ mol K}}\right)(227°C + 273)K}{28.02 \times 10^{-3} \text{ kg/mol}}\right]^{1/2}$ = 667 m/s

$u_{mp} = \left(\dfrac{2RT}{M}\right)^{1/2} = \left[\dfrac{2\left(\dfrac{8.3145 \text{ kg m}^2}{s^2 \text{ mol K}}\right)(500.\text{ K})}{28.02 \times 10^{-3} \text{ kg/mol}}\right]^{1/2}$ = 545 m/s

$u_{avg} = \left(\dfrac{8RT}{\pi M}\right)^{1/2} = \left[\dfrac{8\left(\dfrac{8.3145 \text{ kg m}^2}{s^2 \text{ mol K}}\right)(500.\text{ K})}{\pi(28.02 \times 10^{-3} \text{ kg/mol})}\right]^{1/2}$ = 615 m/s

89. $\Delta(mu) = 2mu$ = change in momentum per impact

$$\Delta(mu)_{O_2} = 2M_{O_2}\left(\frac{3RT}{M_{O_2}}\right)^{1/2} \qquad \Delta(mu)_{He} = 2M_{He}\left(\frac{3RT}{M_{He}}\right)^{1/2}$$

$$\frac{\Delta(mu)_{O_2}}{\Delta(mu)_{He}} = \frac{2M_{O_2}\left(\frac{3RT}{M_{O_2}}\right)^{1/2}}{2M_{He}\left(\frac{3RT}{M_{He}}\right)^{1/2}} = \frac{M_{O_2}}{M_{He}}\left(\frac{M_{He}}{M_{O_2}}\right)^{1/2} = \frac{31.998}{4.003}\left(\frac{4.003}{31.998}\right)^{1/2} = 2.827$$

The change in momentum per impact is 2.827 times larger for O_2 molecules than for He atoms.

$$Z = A\frac{N}{V}\left(\frac{RT}{2\pi M}\right)^{1/2} = \text{collision rate}$$

$$\frac{Z_{O_2}}{Z_{He}} = \frac{A\left(\frac{N}{V}\right)\left(\frac{RT}{2\pi M_{O_2}}\right)^{1/2}}{A\left(\frac{N}{V}\right)\left(\frac{RT}{2\pi M_{He}}\right)^{1/2}} = \frac{\left(\frac{1}{M_{O_2}}\right)^{1/2}}{\left(\frac{1}{M_{He}}\right)^{1/2}} = \left(\frac{M_{He}}{M_{O_2}}\right)^{1/2} = 0.3537; \quad \frac{Z_{He}}{Z_{O_2}} = 2.827$$

There are 2.827 times as many impacts per second for He as compared to O_2.

90. $$\frac{\Delta(mu)_{77}}{\Delta(mu)_{27}} = \frac{2m_{77}\left(\frac{3R \times 350.\,K}{M}\right)^{1/2}}{2m_{27}\left(\frac{3R \times 300.\,K}{M}\right)^{1/2}} = \left(\frac{350.}{300.}\right)^{1/2} = 1.08$$

The change in momentum is 1.08 times greater for Ar at 77°C than for Ar at 27°C.

$$\frac{Z_{77}}{Z_{27}} = \left(\frac{T_{77}}{T_{27}}\right)^{1/2} \rightarrow \text{from } Z = A\frac{N}{V}\left(\frac{RT}{2\pi M}\right)^{1/2}$$

$$\frac{Z_{77}}{Z_{27}} = \left(\frac{350.}{300.}\right)^{1/2} = 1.08$$

There are 1.08 times as many impacts per second for Ar at 77°C than for Ar at 27°C.

91.   a)   Ideal Gas Equation: $PV = nRT$

$$P = \frac{nRT}{V} = \frac{1.00 \text{ mol} \times \frac{0.08206 \text{ L atm}}{\text{mol K}} \times 310. \text{ K}}{1.00 \text{ L}} = 25.4 \text{ atm}$$

b)   van der Waals Equation: $\left[P + a\left(\frac{n}{V}\right)^2\right][V - nb] = nRT$

$$P_{obs} = \frac{nRT}{V - nb} - a\left(\frac{n}{V}\right)^2$$

For CO_2: $a = 3.59$ atm L^2 mol^{-2}; $b = 0.0427$ L/mol

$$P_{obs} = \frac{1.00 \text{ mol} \times \frac{0.08206 \text{ L atm}}{\text{mol K}} \times 310. \text{ K}}{1.00 \text{ L} - 1.00 \text{ mol} \times 0.0427 \text{ L/mol}} - \frac{3.59 \text{ atm L}^2}{\text{mol}^2} \times \frac{1.00 \text{ mol}^2}{L^2}$$

$P_{obs} = 26.6$ atm $- 3.59$ atm $= 23.0$ atm

92. $n_{Ar} = \dfrac{228 \text{ g}}{39.95 \text{ g/mol}} = 5.71$ mol Ar

$$\chi_{CH_4} = \frac{n_{CH_4}}{n_{CH_4} + n_{Ar}} = 0.650 = \frac{n_{CH_4}}{n_{CH_4} + (5.71)}$$

$0.650 \, (n_{CH_4} + 5.71) = n_{CH_4}$, $3.71 = 0.350 \, n_{CH_4}$, $n_{CH_4} = 10.6$ mol CH_4

$KE_{avg} = \dfrac{3}{2} RT$ for 1 mol

So $KE_{tot} = (10.6 + 5.71 \text{ mol}) \times 3/2 \times 8.3145$ J mol^{-1} K$^{-1} \times 298$ K $= 6.06 \times 10^4$ J $= 60.6$ kJ

93. $Z = A\left(\dfrac{N}{V}\right)\left(\dfrac{RT}{2\pi M}\right)^{1/2}$

$$\frac{Z_1}{Z_2} = \frac{\left(\dfrac{T_1}{M_1}\right)^{1/2}}{\left(\dfrac{T_2}{M_2}\right)^{1/2}} = \left(\frac{M_2 T_1}{M_1 T_2}\right)^{1/2} = 1.00; \text{ So, } M_1 T_2 = M_2 T_1$$

$$\frac{T_2}{T_1} = \frac{M_2}{M_1}; \quad \frac{T_{UF_6}}{T_{He}} = \frac{M_{UF_6}}{M_{He}} = \frac{352.0}{4.003} = 87.93$$

94.   a) $\dfrac{PV}{n} = \alpha + \beta P$

   (straight line, y = b + mx)

b) $\dfrac{\Delta(mu)}{\text{impact}} = 2\,mu = 2M\left(\dfrac{3RT}{M}\right)^{1/2} = 2\sqrt{3RTM}$

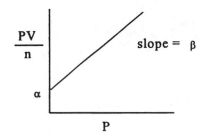

slope = β

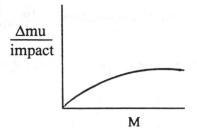

c) $P = \dfrac{nRT}{V}$, $PV = $ const

d) $P = \dfrac{nRT}{V}$; $P = $ const(T)

   (straight line, y = mx + b)

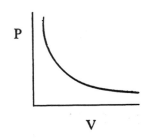

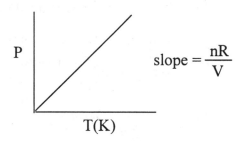

slope $= \dfrac{nR}{V}$

95. $CH_3OH + 3/2\ O_2 \rightarrow CO_2 + 2\ H_2O$ or $2\ CH_3OH(l) + 3\ O_2(g) \rightarrow 2\ CO_2(g) + 4\ H_2O(g)$

$50.0\ mL \times \dfrac{0.850\ g}{mL} \times \dfrac{1\ mol}{32.04\ g} = 1.33\ mol\ CH_3OH(l)$ available

$n_{O_2} = \dfrac{PV}{RT} = \dfrac{2.00\ atm \times 22.8\ L}{\dfrac{0.08206\ L\ atm}{mol\ K} \times 300.\ K} = 1.85\ mol\ O_2$ available

$1.33\ mol\ CH_3OH \times \dfrac{3\ mol\ O_2}{2\ mol\ CH_3OH} = 2.00\ mol\ O_2$ required for complete reaction

We only have 1.85 mol O_2, so O_2 is limiting. $1.85\ mol\ O_2 \times \dfrac{4\ mol\ H_2O}{3\ mol\ O_2} = 2.47\ mol\ H_2O$

96.   Balanced equations:

   $4\ NH_3 + 5\ O_2 \rightarrow 4\ NO + 6\ H_2O$ and $4\ NH_3 + 7\ O_2 \rightarrow 4\ NO_2 + 6\ H_2O$

Let 4x = number of mol of NO formed and let 4y = number of mol of NO_2 formed.   Then:

$4x$ NH$_3$ + $5x$ O$_2$ → $4x$ NO + $6x$ H$_2$O and $4y$ NH$_3$ + $7y$ O$_2$ → $4y$ NO$_2$ + $6y$ H$_2$O

All NH$_3$ reacted, so $4x + 4y = 2.00$. $10.00 - 6.75 = 3.25$ mol O$_2$ reacted, so $5x + 7y = 3.25$.

Solving:

$$20\,x + 28\,y = 13.0$$
$$\underline{-20\,x - 20\,y = -10.0}$$
$$8\,y = 3.0, \quad y = 0.38; \quad 4x + 4 \times 0.38 = 2.00, \quad x = 0.12$$

mol NO = $4x = 4 \times 0.12 = 0.48$ mol NO formed

97.   C$_6$H$_{14}$ + 19/2 O$_2$ → 6 CO$_2$ + 7 H$_2$O or 2 C$_6$H$_{14}$ + 19 O$_2$ → 12 CO$_2$ + 14 H$_2$O

C$_3$H$_8$ + 5 O$_2$ → 3 CO$_2$ + 4 H$_2$O

$$0.8339 \text{ g CO}_2 \times \frac{1 \text{ mol CO}_2}{44.009 \text{ g CO}_2} = 1.895 \times 10^{-2} \text{ mol CO}_2$$

Let x = g C$_6$H$_{14}$ and y = g C$_3$H$_8$, so $x + y = 0.2759$ g.

1.895×10^{-2} mol CO$_2$ = 6 × mol C$_6$H$_{14}$ + 3 × mol C$_3$H$_8$

$$1.895 \times 10^{-2} = \frac{6x}{86.177} + \frac{3y}{44.096}; \text{ Rearranging:}$$

$$x + \frac{3y}{44.096} \times \frac{86.177}{6} = 1.895 \times 10^{-2} \times \frac{86.177}{6}, \quad x + 0.9772\,y = 0.2722$$

Solving:

$$x + \quad\quad y = 0.2759$$
$$\underline{-x - 0.9772\,y = -0.2722}$$
$$0.0288\,y = 0.0037, \quad y = 0.16$$

% C$_3$H$_8$ = $\dfrac{0.16 \text{ g}}{0.2759 \text{ g}} \times 100 = 58\%$ C$_3$H$_8$; % C$_6$H$_{14}$ = 42%

98.   The reactions are:

C(s) + 1/2 O$_2$(g) → CO(g) and C(s) + O$_2$(g) → CO$_2$(g)

PV = nRT; $\quad P = n\left(\dfrac{RT}{V}\right) = n$ (const)

Since the pressure has increased by 17.0%, the number of moles of gas has also increased by 17.0%.

$n_{final} = 1.170 \; n_{initial} = 1.170 \, (5.00) = 5.85 \text{ mol gas} = n_{O_2} + n_{CO} + n_{CO_2}$

$n_{CO} + n_{CO_2} = 5.00$ (balancing moles of C)

If all C was converted to CO_2, no O_2 would be left. If all C was converted to CO, we would get 5 mol CO and 2.5 mol excess O_2 in the reaction mixture. In the final mixture: $n_{CO} = 2n_{O_2}$

$$
\begin{array}{rl}
n_{O_2} + n_{CO} + n_{CO_2} & =\; 5.85 \\
-(n_{CO} + n_{CO_2} & =\; 5.00) \\
\hline
n_{O_2} \qquad\qquad & =\; 0.85
\end{array}
$$

$n_{CO} = 2n_{O_2} = 1.70 \text{ mol CO};\; 1.70 + n_{CO_2} = 5.00, \; n_{CO_2} = 3.30 \text{ mol } CO_2$

$$\chi_{CO} = \frac{1.70}{5.85} = 0.291; \qquad \chi_{CO_2} = \frac{3.30}{5.85} = 0.564; \qquad \chi_{O_2} = \frac{0.85}{5.85} = 0.145 \approx 0.15$$

99.   The reaction is 1:1 between ethene and hydrogen. A greater volume of H_2 and thus, more moles of H_2 are flowing into the reactor. Ethene is the limiting reagent.

In one minute:

$$n_{C_2H_4} = \frac{PV}{RT} = \frac{(25.0 \text{ atm})\,(1000. \text{ L})}{\left(\dfrac{0.08206 \text{ L atm}}{\text{mol K}}\right)(573 \text{ K})} = 532 \text{ mol } C_2H_4 \text{ reacted}$$

Theoretical yield:

$$\frac{532 \text{ mol } C_2H_4}{\text{min}} \times \frac{1 \text{ mol } C_2H_6}{\text{mol } C_2H_4} \times \frac{30.07 \text{ g } C_2H_6}{\text{mol } C_2H_6} \times \frac{1 \text{ kg}}{1000 \text{ g}} = 16.0 \text{ kg } C_2H_6/\text{min}$$

$$\% \text{ yield} = \frac{15.0 \text{ kg/min}}{16.0 \text{ kg/min}} \times 100 = 93.8\%$$

100.   a.   The reaction is: $CH_4(g) + 2\,O_2(g) \rightarrow CO_2(g) + 2\,H_2O(g)$

$$PV = nRT; \quad \frac{PV}{n} = RT = \text{constant}; \quad \frac{P_{CH_4} V_{CH_4}}{n_{CH_4}} = \frac{P_{O_2} V_{O_2}}{n_{O_2}}$$

For three fold excess of O_2: $n_{O_2} = 6\, n_{CH_4}, \quad \dfrac{n_{O_2}}{n_{CH_4}} = 6$

$P_{O_2} = 0.21\, P_{air} = 0.21 \text{ atm}$

In one minute:

$$V_{O_2} = V_{CH_4} \times \frac{n_{O_2}}{n_{CH_4}} \times \frac{P_{CH_4}}{P_{O_2}} = 200. \text{ L (6)} \left(\frac{1.50 \text{ atm}}{0.21 \text{ atm}} \right) = 8.6 \times 10^3 \text{ L O}_2$$

We need: $\dfrac{8.6 \times 10^3 \text{ L O}_2}{\text{min}} \times \dfrac{100 \text{ L air}}{21 \text{ L O}_2} = 4.1 \times 10^4$ L air/min

b. If n moles of CH_4 were reacted, then 6 n mol O_2 were added, producing 0.950 n mol CO_2 and 0.050 n mol of CO. In addition, 2 n mol H_2O must be produced to balance the hydrogens.

$$CH_4 + 2 O_2 \rightarrow CO_2 + 2 H_2O; \quad CH_4 + 3/2 O_2 \rightarrow CO + 2 H_2O$$

Amount O_2 reacted:

$$0.950 \text{ n mol CO}_2 \times \frac{2 \text{ mol O}_2}{\text{mol CO}_2} = 1.90 \text{ n mol O}_2$$

$$0.050 \text{ n mol CO} \times \frac{1.5 \text{ mol O}_2}{\text{mol CO}} = 0.075 \text{ n mol O}_2$$

Amount of O_2 left in reaction mixture = 6.00 n - 1.90 n - 0.075 n = 4.03 n mol O_2

Amount of N_2 = 6.00 n mol $O_2 \times \dfrac{79 \text{ mol N}_2}{21 \text{ mol O}_2} = 22.6$ n ≈ 23 n mol N_2

The reaction mixture contains:

$$0.950 \text{ n mol CO}_2 + 0.050 \text{ n mol CO} + 4.03 \text{ n mol O}_2 + 2.00 \text{ n mol H}_2\text{O}$$

$$+ 23 \text{ n mol N}_2 = 30. \text{ n mol of gas total}$$

$$\chi_{CO} = \frac{0.050}{30.} = 0.0017; \quad \chi_{CO_2} = \frac{0.950}{30.} = 0.032 \quad ; \quad \chi_{O_2} = \frac{4.03}{30.} = 0.13;$$

$$\chi_{H_2O} = \frac{2.00}{30.} = 0.067; \quad \chi_{N_2} = \frac{23}{30.} = 0.77$$

c. The partial pressures are given by $P = \chi P_{tot}$. Since P_{tot} = 1.00 atm, then P_{CO} = 0.0017 atm, P_{CO_2} = 0.032 atm, P_{O_2} = 0.13 atm, P_{H_2O} = 0.067 atm and P_{N_2} = 0.77 atm.

101. For 1 mol of gas: KE = 3/2 RT

For 1 molecule of gas: $KE_{ave} = \dfrac{3RT}{2N_A} = \dfrac{3}{2} k_B T$, $k_B = 1.38 \times 10^{-23}$ J/K

$$KE_{ave} = \frac{3}{2}(1.38 \times 10^{-23} \text{ J/K}) (400. \text{ K}) = 8.28 \times 10^{-21} \text{ J}$$

102. n_{tot} = total # of mol of gas that have effused into the container.

$$n_{tot} = \frac{PV}{RT} = \frac{1.20 \times 10^{-6}\ atm \times 1.00\ L}{\dfrac{0.08206\ L\ atm}{mol\ K} \times 300.\ K} = 4.87 \times 10^{-8}\ mol$$

This amount has entered over a time span of 24 hours:

$$24\ hr \times \frac{60\ min}{1\ hr} \times \frac{60\ s}{1\ min} = 8.64 \times 10^{4}\ s$$

So: $\dfrac{4.87 \times 10^{-8}\ mol}{8.64 \times 10^{4}\ s} = 5.64 \times 10^{-13}$ mol/s have entered the container.

$$\frac{5.64 \times 10^{-13}\ mol}{s} \times \frac{6.022 \times 10^{23}}{1\ mol} = 3.40 \times 10^{11}\ molecules/s$$

The frequency of collisions of the gas with a given area is:

$$Z = A\left(\frac{N}{V}\right)\left(\frac{RT}{2\pi M}\right)^{1/2};\quad Z_{tot} = \frac{3.40 \times 10^{11}\ molecules}{s} = Z_{N_2} + Z_{O_2}$$

$$\frac{n}{V} = \frac{P}{RT} = \frac{1.00\ atm}{\dfrac{0.08206\ L\ atm}{mol\ K} \times 300.\ K} = 4.06 \times 10^{-2}\ mol/L$$

$$\frac{N}{V} = \frac{4.06 \times 10^{-2}\ mol}{L} \times \frac{6.022 \times 10^{23}\ molecules}{mol} \times \frac{1000\ L}{m^{3}} = 2.44 \times 10^{25}\ molecules/m^{3}$$

For N_2: $\left(\dfrac{N}{V}\right) = (0.78)(2.44 \times 10^{25}) = 1.9 \times 10^{25}\ molecules/m^{3}$

For O_2: $\left(\dfrac{N}{V}\right) = (0.22)(2.44 \times 10^{25}) = 5.4 \times 10^{24}\ molecules/m^{3}$

$Z_{tot} = 3.40 \times 10^{11}$ molecules/s $= Z_{N_2} + Z_{O_2}$

$$3.40 \times 10^{-11} = A\left[1.9 \times 10^{25}\left(\frac{8.3145 \times 300.}{2\pi(28.0 \times 10^{-3})}\right)^{1/2} + 5.4 \times 10^{24}\left(\frac{8.3145 \times 300.}{2\pi(32.0 \times 10^{-3})}\right)^{1/2}\right]$$

$$\frac{3.40 \times 10^{11}\ molecules}{s} = A\left[\frac{2.3 \times 10^{27}\ molecules}{m^{2}\ s} + \frac{6.0 \times 10^{26}\ molecules}{m^{2}\ s}\right]$$

$$A = \frac{3.40 \times 10^{11}}{2.9 \times 10^{27}}\ m^{2} = 1.2 \times 10^{-16}\ m^{2} = \pi r^{2}$$

$$r = \left(\frac{1.2 \times 10^{-16}\ m^{2}}{\pi}\right)^{1/2} = 6.2 \times 10^{-9}\ m = 6.2\ nm$$

diameter $= 2r = 2(6.2 \times 10^{-9} \text{ m}) = 1.2 \times 10^{-8} \text{ m} = 12 \text{ nm}$

103.   Each stage will give an enrichment of:

$$\frac{\text{Diff. Rate } ^{12}\text{CO}_2}{\text{Diff. Rate } ^{13}\text{CO}_2} = \left(\frac{\text{M}(^{13}\text{CO}_2)}{\text{M}(^{12}\text{CO}_2)} \right)^{1/2} = \left(\frac{45.001}{43.998} \right)^{1/2} = 1.0113$$

Since $^{12}\text{CO}_2$ moves faster, each successive stage will have less $^{13}\text{CO}_2$.

$$\frac{99.90\ ^{12}\text{CO}_2}{0.10\ ^{13}\text{CO}_2} \times 1.0113^N = \frac{99.990\ ^{12}\text{CO}_2}{0.010\ ^{13}\text{CO}_2}$$

$$1.0113^N = \frac{9,999.0}{999.00} = 10.009 \quad \text{(carry extra digits)}$$

$N \log(1.0113) = \log(10.009)$, $\quad N = \dfrac{1.000391}{4.88 \times 10^{-3}} = 2.05 \times 10^2 \approx 2.1 \times 10^2$ stages are needed.

104.   collision frequency $= Z = 4\dfrac{N}{V} d^2 \left(\dfrac{\pi RT}{M} \right)^{1/2}$ where d = diameter of He atom

$$\frac{n}{V} = \frac{P}{RT} = \frac{3.0 \text{ atm}}{\dfrac{0.08206 \text{ L atm}}{\text{mol K}} \times 300.\text{ K}} = 0.12 \text{ mol/L}$$

$$\frac{N}{V} = \frac{0.12 \text{ mol}}{L} \times \frac{6.022 \times 10^{23} \text{ molecules}}{\text{mol}} \times \frac{1000 \text{ L}}{\text{m}^3} = \frac{7.2 \times 10^{25} \text{ molecules}}{\text{m}^3}$$

$$Z = 4 \times \frac{7.2 \times 10^{25} \text{ molecules}}{\text{m}^3} \times (50. \times 10^{-12} \text{ m})^2 \times \left(\frac{\pi(8.3145)(300.)}{4.00 \times 10^{-3}} \right)^{1/2}$$

$Z = 1.0 \times 10^9$ collisions/s

mean free path $= \lambda = \dfrac{u_{avg}}{Z}$; $u_{avg} = \left(\dfrac{8RT}{\pi M} \right)^{1/2} = 1260 \text{ m/s}$; $\lambda = \dfrac{1260 \text{ m/s}}{1.0 \times 10^9 \text{ s}^{-1}} = 1.3 \times 10^{-6} \text{ m}$

105.   After the hole develops, assume each He that collides with the hole goes into the Rn side and that each Rn that collides with the hole goes into the He side. Assume no molecules return to the side in which they began. Initial moles of each gas:

$$n = \frac{PV}{RT} = \frac{(2.00 \times 10^{-6} \text{ atm}) (1.00 \text{ L})}{\dfrac{0.08206 \text{ L atm}}{\text{mol K}} \times 300.\text{ K}} = 8.12 \times 10^{-8} \text{ mol}$$

$$Z_{He} = A \times \frac{N}{V} \times \left(\frac{RT}{2\pi M} \right)^{1/2}, \quad \frac{N}{V} = \frac{P}{RT} \times N_A \times 1000 \text{ L/m}^3 \text{ and } A = \pi r^2$$

$$Z_{He} = \pi (1.00 \times 10^{-6} \text{ m})^2 \times \frac{2.00 \times 10^{-6}}{0.08206 \times 300.} \times (6.022 \times 10^{23}) \times 1000$$

$$\times \left(\frac{8.3145 \times 300.}{2\pi (4.003 \times 10^{-3})} \right)^{1/2} = 4.84 \times 10^{10} \text{ collisons/s}$$

Therefore, 4.84×10^{10} atoms/s leave He side.

$$10.0 \text{ hr} \times \frac{60 \text{ min}}{1 \text{ hr}} \times \frac{60 \text{ s}}{1 \text{ min}} \times \frac{4.84 \times 10^{10} \text{ atoms}}{s} = 1.74 \times 10^{15} \text{ atoms}$$

or $\quad \dfrac{1.74 \times 10^{15} \text{ atoms}}{6.022 \times 10^{23} \text{ atoms/mol}} = 2.89 \times 10^{-9}$ mol He leave in 10.0 hr

$$Z_{Rn} = \pi (1.00 \times 10^{-6} \text{ m})^2 \times \frac{2.00 \times 10^{-6}}{0.08206 \times 300.} \times (6.022 \times 10^{23}) \times 1000$$

$$\times \left(\frac{8.3145 \times 300.}{2\pi (222 \times 10^{-3})} \right)^{1/2} = 6.50 \times 10^9 \text{ collisions/s}$$

6.50×10^9 atoms/s leave Rn side.

$$3.60 \times 10^4 \text{ s} \times \frac{6.50 \times 10^9 \text{ atoms}}{s} \times \frac{1 \text{ mol}}{6.022 \times 10^{23} \text{ atoms}} = 3.89 \times 10^{-10} \text{ mol Rn leave in 10.0 hr.}$$

Side that began with He now contains:

$$8.12 \times 10^{-8} - 2.89 \times 10^{-9} = 7.83 \times 10^{-8} \text{ mol He} + 3.89 \times 10^{-10} \text{ mol Rn} = 7.87 \times 10^{-8} \text{ mol total}$$

The pressure in the He side is:

$$P = \frac{nRT}{V} = \frac{(7.87 \times 10^{-8} \text{ mol}) \times 0.08206 \times 300. \text{ K}}{1.00 \text{ L}} = 1.94 \times 10^{-6} \text{ atm}$$

We can get the pressure in the Rn chamber two ways. No gas has escaped. Since the initial P's were equal and the P in one of the sides decreased by 0.06×10^{-6} atm, then the P in the second side must increase by 0.06×10^{-6} atm. So the pressure on the side that originally contained Rn is 2.06×10^{-6} atm. Or we can calculate the P the same way as with He. The Rn side contains:

$$8.12 \times 10^{-8} - 3.89 \times 10^{-10} = 8.08 \times 10^{-8} \text{ mol Rn} + 2.89 \times 10^{-9} \text{ mol He} = 8.37 \times 10^{-8} \text{ mol total}$$

$$P = \frac{nRT}{V} = \frac{(8.37 \times 10^{-8} \text{ mol}) \times 0.08206 \times 300. \text{ K}}{1.00 \text{ L}} = 2.06 \times 10^{-6} \text{ atm}$$

106. $\dfrac{\text{diffusion rate } ^{235}UF_6}{\text{diffusion rate } ^{238}UF_6} = 1.0043$ (See Section 5.7 of the text.)

$\dfrac{^{235}UF_6}{^{238}UF_6} \times (1.0043)^{100} = \dfrac{1526}{1.000 \times 10^5 - 1526}$; $\dfrac{^{235}UF_6}{^{238}UF_6} \times (1.5358) = \dfrac{1526}{98500}$

$\dfrac{^{235}UF_6}{^{238}UF_6} = 1.01 \times 10^{-2} = \text{initial } ^{235}U \text{ to } ^{238}U \text{ atom ratio}$

107.   mol of He removed $= \dfrac{PV}{RT} = \dfrac{1.00 \text{ atm} (1.75 \times 10^{-3} \text{ L})}{\dfrac{0.08206 \text{ L atm}}{\text{mol K}} \times 298 \text{ K}} = 7.16 \times 10^{-5} \text{ mol}$

In the original flask, 7.16×10^{-5} mol of He exerted a partial pressure of $1.960 - 1.710 = 0.250$ atm.

$V = \dfrac{nRT}{P} = \dfrac{(7.16 \times 10^{-5} \text{ mol}) \times 0.08206 \times 298 \text{ K}}{0.250 \text{ atm}} = 7.00 \times 10^{-3} \text{ L} = 7.00 \text{ mL}$

108.   a.   Out of 100.00 g of Z, we have:

$34.38 \text{ g Ni} \times \dfrac{1 \text{ mol}}{58.69 \text{ g}} = 0.5858 \text{ mol Ni}$

$28.13 \text{ g C} \times \dfrac{1 \text{ mol}}{12.011 \text{ g}} = 2.342 \text{ mol C}$; $\dfrac{2.342}{0.5858} = 4$

$37.48 \text{ g O} \times \dfrac{1 \text{ mol}}{15.999 \text{ g}} = 2.343 \text{ mol O}$; $\dfrac{2.343}{0.5858} = 4$

The empirical formula is NiC_4O_4.

b. $\dfrac{\text{rate Z}}{\text{rate Ar}} = \left(\dfrac{M_{Ar}}{M_Z} \right)^{1/2} = \left(\dfrac{39.95}{M_Z} \right)^{1/2}$

Since initial mol Ar = mol Z, then:

$0.4837 = \left(\dfrac{39.95}{M_Z} \right)^{1/2}$, $M_Z = 170.8 \text{ g/mol}$

c. NiC_4O_4:   M $= 58.69 + 4(12.01) + 4(16.00) = 170.73$ g/mol

Molecular formula is NiC_4O_4.

d.   Each effusion step changes the concentration of Z in the gas by a factor of 0.4837. The original concentration of Z molecules to Ar atoms is a 1:1 ratio. After 5 stages:

$$n_Z/n_{Ar} = (0.4837)^5 = 2.648 \times 10^{-2}$$

109.   For O_2, n and T are constant, so $P_1V_1 = P_2V_2$.

$$P_1 = \frac{P_2V_2}{V_1} = 785 \text{ torr} \times \frac{1.94 \text{ L}}{2.00 \text{ L}} = 761 \text{ torr} = P_{O_2}$$

$$P_{tot} = P_{O_2} + P_{H_2O}, \quad P_{H_2O} = 785 - 761 = 24 \text{ torr}$$

110. $KE_{ave} = 3/2 \text{ RT per mol}; KE_{ave} = 3/2 \text{ } k_BT \text{ per molecule}$

$$KE_{(tot)} = 3/2 \times (1.3807 \times 10^{-23} \text{ J/K}) \times 300. \text{ K} \times (1.00 \times 10^{20} \text{ molecules}) = 0.621 \text{ J}$$

111.   a. $2 CH_4(g) + 2 NH_3(g) + 3 O_2(g) \rightarrow 2 HCN(g) + 6 H_2O(g)$

b.   Volumes of gases are proportional to moles. So methane and ammonia are in stoichiometric amounts and oxygen is in excess. For 1 second:

$$n_{CH_4} = \frac{PV}{RT} = \frac{(1.00 \text{ atm}) (20.0 \text{ L})}{(0.08206) (423 \text{ K})} = 0.576 \text{ mol CH}_4$$

$$\frac{0.576 \text{ mol CH}_4}{s} \times \frac{2 \text{ mol HCN}}{2 \text{ mol CH}_4} \times \frac{27.03 \text{ g HCN}}{\text{mol HCN}} = 15.6 \text{ g HCN/s}$$

CHAPTER SIX

CHEMICAL EQUILIBRIUM

Characteristics of Chemical Equilibrium

1.   a.   The rates of the forward and reverse reactions are equal.

        b.   There is no net change in the composition.

2.   False. For example consider two cases:

        i. $k_f = 10^6 \text{ s}^{-1}$ $k_r = 10^8 \text{ s}^{-1}$ fast reaction

            $K = k_f/k_r = 10^{-2}$ small K

        ii. $k_f = 10^{-5} \text{ s}^{-1}$ $k_r = 10^{-9} \text{ s}^{-1}$ slow reaction

            $K = k_f/k_r = 10^4$ large K

The equilibrium constant is the ratio of the two rate constants. This ratio does not necessarily get larger as the reactions get faster. The two examples illustrate this. If one starts with a mixture not at equilibrium, equilibrium will be reached faster if the rates of the forward and reverse reactions are fast.

3.   No, equilibrium is a dynamic process. Both reactions:

$$H_2O + CO \rightarrow H_2 + CO_2 \text{ and } H_2 + CO_2 \rightarrow H_2O + CO$$

are occurring. Thus, ^{14}C atoms will be distributed between CO and CO_2.

4.   No, it doesn't matter which direction the equilibrium position is reached. Both mixtures will give the same equilibrium position.

The Equilibrium Constant

5.   The equilibrium constant is a number that tells us the relative concentrations (pressures) of reactants and products at equilibrium.

An equilibrium position is a set of concentrations that satisfy the equilibrium constant expression. More than one equilibrium position can satisfy the same equilibrium constant expression.

6.   For the reaction; a A + b B ⇌ c C + d D,

the equilibrium constant expression is: $K_{eq} = \dfrac{[C]^c\,[D]^d}{[A]^a\,[B]^b}$

The reaction quotient has the same form: $Q = \dfrac{[C]^c\,[D]^d}{[A]^a\,[B]^b}$

The difference is that in the expressions for K_{eq} we use equilibrium concentrations, i.e., [A], [B], [C], and [D] that are in equilibrium with each other. Any set of concentrations can be plugged into the reaction quotient expression. Typically, we compare a value of Q to K_{eq} to see how far we are from equilibrium.

7.   The units for both reactions are:

$$\frac{(\text{molecules/cm}^3)}{(\text{molecules/cm}^3)\,(\text{molecules/cm}^3)} = \frac{\text{cm}^3}{\text{molecules}}$$

a. $K = \dfrac{1.26 \times 10^{-11}\ \text{cm}^3}{\text{molecules}} \times \dfrac{1\ \text{L}}{1000\ \text{cm}^3} \times \dfrac{6.022 \times 10^{23}\ \text{molecules}}{\text{mol}} = 7.59 \times 10^9\ \text{L/mol}$

$K_P = K(RT)^{\Delta n}$, where Δn = moles gaseous products - moles gaseous reactants
$\Delta n = 1 - 2 = -1$

$K_p = \dfrac{7.59 \times 10^9\ \text{L/mol}}{\left(\dfrac{0.08206\ \text{L atm}}{\text{mol K}}\right) \times 300.\ \text{K}} = 3.08 \times 10^8\ \text{atm}^{-1}$

b. $K = \dfrac{2.09 \times 10^{-12}\ \text{cm}^3}{\text{molecules}} \times \dfrac{1\ \text{L}}{1000\ \text{cm}^3} \times \dfrac{6.022 \times 10^{23}\ \text{molecules}}{\text{mol}} = 1.26 \times 10^9\ \text{L/mol}$

$K_P = K(RT)^{\Delta n}, \quad \Delta n = -1$

$K_p = \dfrac{1.26 \times 10^9\ \text{L/mol}}{\left(\dfrac{0.08206\ \text{L atm}}{\text{mol K}}\right) \times 300.\ \text{K}} = 5.12 \times 10^7\ \text{atm}^{-1}$

c. $K^* = \dfrac{[HO_2NO_2]}{[HO_2][NO_2]} = 1.26 \times 10^{-11}\ \text{cm}^3/\text{molecules}$

$[HO_2NO_2] = (1.26 \times 10^{-11})\,(1.65 \times 10^{10})\,(6.00 \times 10^{12}) = 1.25 \times 10^{12}\ \text{molecules/cm}^3$

8. $K_p = K(RT)^{\Delta n}$, $\Delta n = 4 - 2 = 2$

$$K_p = \frac{2.6 \times 10^{-5} \text{ mol}^2}{\text{L}^2}\left(\frac{0.08206 \text{ L atm}}{\text{mol K}} \times 400.\text{ K}\right)^2 = 2.8 \times 10^{-2} \text{ atm}^2$$

9. $K = 278 = \dfrac{[SO_3]^2}{[SO_2]^2[O_2]}$ for $2\ SO_2(g) + O_2(g) \rightleftharpoons 2\ SO_3(g)$

   a. $SO_2(g) + 1/2\ O_2(g) \rightleftharpoons SO_3(g)$; $K_{eq} = \dfrac{[SO_3]}{[SO_2][O_2]^{1/2}} = K^{1/2} = 16.7$

   b. $2\ SO_3(g) \rightleftharpoons 2\ SO_2(g) + O_2(g)$; $K_{eq} = \dfrac{[SO_2]^2[O_2]}{[SO_3]^2} = \dfrac{1}{K} = 3.60 \times 10^{-3}$

   c. $SO_3(g) \rightleftharpoons SO_2(g) + 1/2\ O_2(g)$; $K_{eq} = \dfrac{[SO_2][O_2]^{1/2}}{[SO_3]} = \left(\dfrac{1}{K}\right)^{1/2} = 6.00 \times 10^{-2}$

   d. $4\ SO_2(g) + 2\ O_2(g) \rightleftharpoons 4\ SO_3(g)$; $K_{eq} = \dfrac{[SO_3]^4}{[SO_2]^4[O_2]^2} = K^2 = 7.73 \times 10^4$

10. $CO_2(g) + H_2(g) \rightleftharpoons CO(g) + H_2O(g)$

$$K = \frac{[CO][H_2O]}{[CO_2][H_2]} = \frac{(5.9 \text{ mol/L}) (12 \text{ mol/L})}{(18 \text{ mol/L}) (20. \text{ mol/L})} = 0.20$$

11. $H_2(g) + I_2(g) \rightleftharpoons 2\ HI(g)$

$$K = \frac{[HI]^2}{[H_2][I_2]} = \frac{\left(\dfrac{3.50 \text{ mol}}{3.00 \text{ L}}\right)^2}{\left(\dfrac{4.10 \text{ mol}}{3.00 \text{ L}}\right)\left(\dfrac{0.30 \text{ mol}}{3.00 \text{ L}}\right)} = 10.$$

12. $C(s) + CO_2(g) \rightleftharpoons 2\ CO(g)$ $K_p = \dfrac{P_{CO}^2}{P_{CO_2}} = \dfrac{(2.6 \text{ atm})^2}{2.9 \text{ atm}} = 2.3 \text{ atm}$

13. $NH_4Cl(s) \rightleftharpoons NH_3(g) + HCl(g)$

   $K_p = P_{NH_3} \times P_{HCl}$; At equilibrium, $P_{total} = P_{NH_3} + P_{HCl}$ and $P_{NH_3} = P_{HCl}$.

   $P_{total} = 4.4 \text{ atm} = 2\,P_{NH_3}$; Thus, $P_{NH_3} = P_{HCl} = 2.2 \text{ atm}$.

   $K_p = (2.2 \text{ atm})(2.2 \text{ atm}) = 4.8 \text{ atm}^2$

14. $PCl_5(g)$ $\rightleftharpoons$ $PCl_3(g)$ + $Cl_2(g)$

Initial   0.50 atm   0   0
                  x atm PCl_5 reacts to reach equilibrium
Change $-x$ $\rightarrow$ $+x$ $+x$
Equil. $0.50 - x$ x x

$$0.84 \text{ atm} = P_{total} = P_{PCl_5} + P_{PCl_3} + P_{Cl_2} = 0.50 - x + x + x = 0.50 + x$$

$$0.84 \text{ atm} = 0.50 + x, \ x = 0.34 \text{ atm}$$

$$P_{PCl_5} = 0.16 \text{ atm}; \ P_{PCl_3} = P_{Cl_2} = 0.34 \text{ atm}$$

$$K_p = \frac{P_{PCl_3} \times P_{Cl_2}}{P_{PCl_5}} = \frac{(0.34)(0.34)}{(0.16)} = 0.72 \text{ atm}$$

$$K = \frac{K_p}{(RT)^{\Delta n}} = \frac{0.72}{(0.08206)(523)} = 0.017 \text{ mol/L}$$

15. $SO_2(g)$ + $NO_2(g)$ $\rightleftharpoons$ $SO_3(g)$ + $NO(g)$

Initial $\dfrac{2.00 \text{ mol}}{V}$ $\dfrac{2.00 \text{ mol}}{V}$ 0   0

Change $\dfrac{-1.30 \text{ mol}}{V}$ $\dfrac{-1.30 \text{ mol}}{V}$ $\rightarrow$ $\dfrac{+1.30 \text{ mol}}{V}$ $\dfrac{+1.30 \text{ mol}}{V}$

Equil. $\dfrac{0.70 \text{ mol}}{V}$ $\dfrac{0.70 \text{ mol}}{V}$ $\dfrac{1.30 \text{ mol}}{V}$ $\dfrac{1.30 \text{ mol}}{V}$

$$K = \frac{[SO_3][NO]}{[SO_2][NO_2]} = \frac{(1.30)(1.30)}{(0.70)(0.70)} = 3.4 \qquad \text{(Volume cancels.)}$$

Equilibrium Calculations

16. $H_2O(g) + Cl_2O(g) \leftrightharpoons 2 \ HOCl(g)$ $K_p = K = 0.0900 = \dfrac{[HOCl]^2}{[H_2O][Cl_2O]}$

a. $Q = \dfrac{(21.0)^2}{(200.)(49.8)} = 4.43 \times 10^{-2} < K$

Reaction will proceed to the right to reach equilibrium.

$H_2O(g) + Cl_2O(g) \rightarrow 2 \ HOCl(g)$

b. $Q = \dfrac{(20.0)^2}{(296)(15.0)} = 0.0901 \approx K$ (at equilibrium)

c. $Q = \dfrac{\left(\dfrac{0.084 \text{ mol}}{2.0 \text{ L}}\right)^2}{\left(\dfrac{0.98 \text{ mol}}{2.0 \text{ L}}\right)\left(\dfrac{0.080 \text{ mol}}{2.0 \text{ L}}\right)} = \dfrac{(0.084)^2}{(0.98)(0.080)} = 0.090 = K$ (at equilibrium)

d. $Q = \dfrac{\left(\dfrac{0.25 \text{ mol}}{3.0 \text{ L}}\right)^2}{\left(\dfrac{0.56 \text{ mol}}{3.0 \text{ L}}\right)\left(\dfrac{0.0010 \text{ mol}}{3.0 \text{ L}}\right)} = \dfrac{(0.25)^2}{(0.56)(0.0010)} = 110 > K$

Reaction will proceed to the left to reach equilibrium.

$2 \, HOCl(g) \rightarrow H_2O(g) + Cl_2O(g)$ or $H_2O(g) + Cl_2O(g) \leftarrow 2 \, HOCl(g)$

17. $CH_3CO_2H + C_2H_5OH \rightleftharpoons CH_3CO_2C_2H_5 + H_2O$

$$K = \frac{[CH_3CO_2C_2H_5]\,[H_2O]}{[CH_3CO_2H]\,[C_2H_5OH]} = 2.2$$

a. $Q = \dfrac{(0.22)\,(0.10)}{(0.010)\,(0.010)} = 220 > K$, not at equilibrium

b. $Q = \dfrac{(0.22)\,(0.0020)}{(0.0020)\,(0.10)} = 2.2 = K$, at equilibrium

c. $Q = \dfrac{(0.88)\,(0.12)}{(0.044)\,(6.0)} = 0.40 < K$, not at equilibrium

d. $Q = \dfrac{(4.4)\,(4.4)}{(0.88)\,(10.0)} = 2.2 = K$, at equilibrium

The conditions in (a) and (d) correspond to equilibrium positions.

18. $K = 2.2 = \dfrac{[CH_3CO_2C_2H_5]\,[H_2O]}{[CH_3CO_2H]\,[C_2H_5OH]}$

$2.2 = \dfrac{(2.0)\,[H_2O]}{(0.10)\,(5.0)}$, $[H_2O] = 0.55 \, M$

We can neglect the solvent in writing equilibrium constant expressions because the solvent concentration is large and essentially constant. This is why we do not include water in the expression for K in reactions that occur in AQUEOUS solution. This is not the case for this reaction; water is not the solvent. Water is a product of the reaction in some other solvent and $[H_2O]$ must appear in the expression for K.

19. $CaCO_3(s) \rightleftharpoons CaO(s) + CO_2(g)$

$K_P = P_{CO_2} = 1.04$ atm

We only need to calculate the partial pressure of CO_2. At this temperature all CO_2 will be in the gas phase.

a.   PV = nRT

$$P = \frac{nRT}{V} = \frac{\dfrac{58.4\ g}{44.01\ g/mol}\left(\dfrac{0.08206\ L\ atm}{mol\ K}\right)1173\ K}{(50.0\ L)} = 2.55\ atm > K_p$$

Reaction will proceed to the left; the mass of CaO will decrease.

b. $P = \dfrac{(23.76)\,(0.08206)\,(1173)}{(44.009)\,(50.0)} = 1.04$ atm $= K_p$

At equilibrium; mass of CaO will not change.

c.   Mass of CO_2 is the same as in part b. P = 1.04 atm $= K_P$. At equilibrium; mass of CaO will not change.

d. $P = \dfrac{(4.82)\,(0.08206)\,(1173)}{(44.01)\,(50.0)} = 0.211$ atm $< K_p$

Reaction will go to the right; the mass of CaO will increase.

20. $H_2O(g) + Cl_2O(g) \rightleftharpoons 2\ HOCl(g) \qquad K = 0.090 = \dfrac{[HOCl]^2}{[H_2O]\,[Cl_2O]}$

a.   The initial concentrations of H_2O and Cl_2O are:

$$\frac{1.0\ g\ H_2O}{1.0\ L} \times \frac{1\ mol}{18.0\ g} = 5.6 \times 10^{-2}\ mol/L$$

$$\frac{2.0\ g\ Cl_2O}{1.0\ L} \times \frac{1\ mol}{86.9\ g} = 2.3 \times 10^{-2}\ mol/L$$

	$H_2O(g)$	+	$Cl_2O(g)$	$\rightleftharpoons$	$2\ HOCl(g)$
Initial	$5.6 \times 10^{-2}\ M$		$2.3 \times 10^{-2}\ M$		0
	x mol/L H_2O reacts to reach equilibrium				
Change	$-x$		$-x$	$\rightarrow$	$+2x$
Equil.	$5.6 \times 10^{-2} - x$		$2.3 \times 10^{-2} - x$		$2x$

$$K = 0.090 = \frac{(2x)^2}{(5.6 \times 10^{-2} - x)\,(2.3 \times 10^{-2} - x)}$$

$1.16 \times 10^{-4} - 7.11 \times 10^{-3} \, x + 0.090 \, x^2 = 4 \, x^2$

$3.91 \, x^2 + 7.11 \times 10^{-3} x - 1.16 \times 10^{-4} = 0$

For a quadratic equation of the form $ax^2 + bx + c = 0$, the solutions are:

$$x = \frac{-b \pm (b^2 - 4ac)^{1/2}}{2a}$$ See Appendix A1.4 of the text.

Note: Rounding off intermediate answers in solving the quadratic formula and in the approximations method described below, leads to excessive round off error. In these problems we will discontinue the usual practice in this Solutions Guide and carry extra significant figures and round at the end.

For this equation:

$$x = \frac{-7.11 \times 10^{-3} \pm (5.06 \times 10^{-5} + 1.81 \times 10^{-3})^{1/2}}{7.82}$$

$$x = \frac{-7.11 \times 10^{-3} \pm (1.86 \times 10^{-3})^{1/2}}{7.82} = \frac{-7.11 \times 10^{-3} \pm 4.31 \times 10^{-2}}{7.82}$$

$x = 4.6 \times 10^{-3}$ or -6.4×10^{-3}

A negative answer makes no physical sense; we can't have less than nothing.
So $x = 4.6 \times 10^{-3} \, M$.

$[HOCl] = 2x = 9.2 \times 10^{-3} \, M$

$[H_2O] = 5.6 \times 10^{-2} - x = 5.6 \times 10^{-2} - 0.46 \times 10^{-2} = 5.1 \times 10^{-2} \, M$

$[Cl_2O] = 2.3 \times 10^{-2} - x = 0.023 - 0.0046 = 1.8 \times 10^{-2} \, M$

There is an easier way to solve this problem in terms of algebra. There are a lot of manipulations necessary to get a solution using the quadratic formula. If we make an early mistake, it takes a lot of work to get the final answer and find out it is wrong. As a case in point it took me (KCB) about 20 minutes to find the mistake I made when initially solving this problem. In an early step I copied an exponent wrong. We can use our chemical sense to simplify the algebra. For this problem we are trying to solve the equation:

$$\frac{4x^2}{(5.6 \times 10^{-2} - x)(2.3 \times 10^{-2} - x)} = 0.090$$

However, the value of the equilibrium constant, 0.090, is not very large; not much H_2O and Cl_2O will react to reach equilibrium. In this case, if we assume:

$0.056 - x \approx 0.056$ and $0.023 - x \approx 0.023$, then the equation becomes:

$$\frac{4x^2}{(0.056)(0.023)} = 0.090$$

We can get a value for x with much less effort. Solving: $x = 5.4 \times 10^{-3}$

If we check how good our assumptions were, we get:

$0.056 - x = 0.056 - 0.0054 = 0.0506$

$0.023 - x = 0.023 - 0.0054 = 0.0176$

These aren't very good assumptions. All is not lost! We can use 0.0506 and 0.0176 as refined approximations and solve:

$$\frac{4x^2}{(0.0506)(0.0176)} = 0.090, \quad x = 4.5 \times 10^{-3}$$

We use this value of x to get yet another approximation for $[H_2O]$ and $[Cl_2O]$.

$0.056 - 0.0045 = 0.0515$ and $0.023 - 0.0045 = 0.0185$

$$\frac{4x^2}{(0.0515)(0.0185)} = 0.090, \quad x = 4.6 \times 10^{-3}$$

Doing one more iteration:

$0.056 - 0.0046 = 0.0514$ and $0.023 - 0.0046 = 0.0184$

$$\frac{4x^2}{(0.0514)(0.0184)} = 0.090, \quad x = 4.6 \times 10^{-3}$$

We just got the same answer in consecutive iterations; we have converged on the true answer. It is the same answer to two significant figures that we determined exactly by solving using the quadratic formula. This method, called the method of successive approximations (see Appendix 1.4 in the text), at first appears to be more laborious. However, any one iteration is less time consuming than using the quadratic formula; it is easier to make an arithmetic mistake using the quadratic formula; and we can discover mistakes in a shorter period of time using successive approximations. Even more compelling, we're using our chemical sense to make approximations and, after all, this is a chemistry text. Try the method; once you get the hang of it you'll prefer successive approximation to the quadratic formula.

b.

	$H_2O(g)$	$+$	$Cl_2O(g)$	$\rightleftharpoons$	$2\ HOCl(g)$
Initial	0		0		1.0 mol/2.0 L $= 0.50\ M$
		$2x$ mol/L HOCl reacts to reach equilibrium			
Change	$+x$		$+x$	$\leftarrow$	$-2x$
Equil.	x		x		$0.50 - 2x$

$$K = 0.090 = \frac{[HOCl]^2}{[H_2O][Cl_2O]} = \frac{(0.50 - 2x)^2}{x^2}$$

We can solve this exactly very quickly. The expression is a perfect square, so we can take the square root of each side:

$$0.30 = \frac{0.50 - 2x}{x}, \quad 0.30\,x = 0.50 - 2x, \quad 2.30\,x = 0.50$$

$x = 0.217$ (carry extra significant figures)

$x = [H_2O] = [Cl_2O] = 0.217 = 0.22\ M;\ [HOCl] = 0.50 - 2x = 0.50 - 0.434 = 0.07\ M$

21.   a. $I^-(aq)\ +\ I_2(aq)\ \rightleftharpoons\ I_3^-(aq)$ $K = 710$

Before   0.10 M 0.10 M 0
            Let 0.10 mol/L I^- react completely.   From the large value of K (= 710),
Change   -0.10   -0.10 $\rightarrow$ +0.10   we know the equilibrium lies far to
                                                        the right.
After   0   0   0.10   New initial conditions to use in
                                                        solving the equilibrium problem.

            x mol/L I_3^- reacts to reach equilibrium
Change   +x +x $\leftarrow$ -x
Equil. x x 0.10 - x

$$K = \frac{[I_3^-]}{[I_2][I^-]} = 710 = \frac{0.10 - x}{x^2}$$

From the large size of K, x should be a small value. If $0.10 - x \approx 0.10$, then:

$$710 \approx \frac{0.10}{x^2}, \quad x = 0.012$$

Checking the assumption:

$$0.10 - 0.012 = 0.088, \quad \frac{0.012(100)}{0.10} = 12\%$$

The assumption introduces an error of 12%. If the error is less than 5%, then the assumption will be valid (called the 5% rule). Since the error is greater than 5%, we must solve the problem by different means, either using the quadratic formula to solve exactly or using the method of successive approximations (see Appendix 1.4 of the text). Using the method of successive approximations:

$$710 = \frac{0.10 - 0.012}{x^2}, \quad x = 0.011$$

$$710 = \frac{0.10 - 0.011}{x^2}, \quad x = 0.011$$

We have converged on the value for x that the quadratic formula would have given.

$$x = [I_2] = 0.011\ M;\ \ [I_3^-] = 0.10 - 0.011 = 0.09\ M;\ \ \dfrac{[I_3^-]}{[I_2]} = \dfrac{0.09\ M}{0.011\ M} = 8$$

b.

	I^-(aq)	+	I_2(aq)	$\rightleftharpoons$	I_3^-(aq)	
Before	0.50 M		0.10 M		0	

Let 0.10 mol/L I_2 react completely.

Change	-0.10		-0.10	$\rightarrow$	+0.10	React completely
After	0.40		0		0.10	New initial conditions

x mol/L I_3^- reacts to reach equilibrium

Change	+x		+x	$\leftarrow$	-x	
Equil.	0.40 + x		x		0.10 - x	

$$K = \dfrac{0.10 - x}{(0.40 + x)(x)} = 710;\ \text{Assume } x \text{ is small. Then:}$$

$$710 \approx \dfrac{0.10}{0.40\ x},\ \ x = 3.5 \times 10^{-4}$$

x is 0.35% of 0.10 so the assumptions hold by the 5% rule.

$$\dfrac{[I_3^-]}{[I_2]} = \dfrac{0.10\ M}{3.5 \times 10^{-4}\ M} = 290$$

c.

	I^-(aq)	+	I_2(aq)	$\rightleftharpoons$	I_3^-(aq)
Before	2.00 M		0.10 M		0

Let 0.10 mol/L I_2 react completely.

Change	-0.10		-0.10 M	$\rightarrow$	+0.10	React completely
After	1.90		0		0.10	New initial conditions

x mol/L I_3^- reacts to reach equilibrium

Change	+x		+x	$\leftarrow$	-x
Equil.	1.90 + x		x		0.10 - x

$$K = \dfrac{0.10 - x}{(1.90 + x)(x)} = 710 \approx \dfrac{0.10}{1.90\ x}\ \ \ (\text{Assuming } x \text{ is small.})$$

$x = [I_2] = 7.4 \times 10^{-5}$ Assumptions good by 5% rule.

$$\dfrac{[I_3^-]}{[I_2]} = \dfrac{0.10\ M}{7.4 \times 10^{-5}\ M} = 1.4 \times 10^3$$

22.   a.

	SO_2(g)	+	NO_2(g)	$\rightleftharpoons$	SO_3(g)	+	NO(g)	$K = 2.5$
Initial	1.00 atm		1.00 atm		0		0	

x atm of SO_2 reacts to reach equilibrium

Change	-x		-x	$\rightarrow$	+x		+x	
Equil.	1.00 - x		1.00 - x		x		x	

$$\frac{x^2}{(1.00 - x)^2} = 2.5 \quad \text{This is a perfect square.}$$

Taking the square root of both sides: $\dfrac{x}{1.00 - x} = 1.6$

$x = 1.6 - 1.6\,x; \;\; 2.6\,x = 1.6; \;\; x = 0.62$

$P_{SO_3} = P_{NO} = x = 0.62 \text{ atm}; \;\; P_{SO_2} = P_{NO_2} = 1.00 - x = 0.38 \text{ atm}$

b. $Q = 1.00$ which is less than K. Reaction shifts to the right.

$$SO_2(g) \quad + \quad NO_2(g) \; \rightleftharpoons \quad SO_3 \quad + \quad NO$$

Initial	1.00 atm	1.00 atm	1.00 atm	1.00 atm

x atm of SO_2 reacts to reach equilibrium

Change	$-x$	$-x$ $\rightarrow$	$+x$	$+x$
Equil.	$1.00 - x$	$1.00 - x$	$1.00 + x$	$1.00 + x$

$$\frac{(1.00 + x)^2}{(1.00 - x)^2} = 2.5; \text{ Taking square root of both sides:}$$

$$\frac{1.00 + x}{1.00 - x} = 1.6; \;\; 1.00 + x = 1.6 - 1.6\,x, \;\; 2.6\,x = 0.6, \;\; x = 0.2$$

$P_{SO_3} = P_{NO} = 1.00 + x = 1.2 \text{ atm}; \;\; P_{SO_2} = P_{NO_2} = 1.00 - x = 0.8 \text{ atm}$

23.

$$N_2(g) \quad + \quad O_2(g) \; \rightleftharpoons \quad 2\,NO(g) \quad K = K_p = 0.050$$

Initial	0.80 atm	0.20 atm	0

x atm of N_2 reacts to reach equilibrium

Change	$-x$	$-x$ $\rightarrow$	$+2x$
Equil.	$0.80 - x$	$0.20 - x$	$2x$

$$K = 0.050 = \frac{P_{NO}^2}{P_{N_2}P_{O_2}} = \frac{(2x)^2}{(0.80 - x)(0.20 - x)}$$

If $0.80 - x \approx 0.80$ and $0.20 - x \approx 0.20$, then:

$$0.050 = \frac{4x^2}{(0.80)(0.20)}, \;\; x = 0.045; \;\; x \text{ is } 22.5\% \text{ of } 0.20.$$

Assumptions are not valid by the 5% rule. Using successive approximations:

$$0.80 - x = 0.80 - 0.045 = 0.755, \;\; 0.20 - x = 0.20 - 0.045 = 0.155$$

$$\frac{4x^2}{(0.755)(0.155)} = 0.050, \;\; x = 0.038; \quad \frac{4x^2}{(0.762)(0.162)} = 0.050, \;\; x = 0.039$$

$$\frac{4x^2}{(0.761)\,(0.161)} = 0.050, \quad x = 0.039 \quad \text{consistent answer}$$

$$P_{NO} = 2x = 2(0.039) = 7.8 \times 10^{-2} \text{ atm}$$

24.   a.

	2 NOCl(g)	$\rightleftharpoons$	2 NO(g)	+	Cl$_2$(g)	K = 1.6 × 10^{-5}

Initial $\dfrac{2.0 \text{ mol}}{2.0 \text{ L}} = 1.0\ M$ 0   0

2x mol/L NOCl reacts to reach equilibrium

Change	-2x	$\rightarrow$	+2x	+x
Equil.	1.0 - 2x		2x	x

$$K = 1.6 \times 10^{-5} = \frac{(2x)^2\,(x)}{(1.0 - 2x)^2}$$

If 1.0 - 2x ≈ 1 (from the size of K, we know that not much reaction will occur), then

$$1.6 \times 10^{-5} = 4x^3, \quad x = 1.6 \times 10^{-2}$$

1.0 - 2x = 1.0 - 2(0.016) = 0.97 = 1.0 (to proper sig. fig.)

Our error is about 3%, i.e., 2x is 3.2% of 1.0 M. Usually if the error we introduce by making simplifying assumptions is less than 5%, we go no further, the assumption is valid (called the 5% rule).

[NO] = 2x = 0.032 M; [Cl$_2$] = x = 0.016 M; [NOCl] = 1.0 - 2x = 0.97 M ≈ 1.0 M

Note that the equation is a cubic equation. I have no idea how to solve a cubic equation exactly. If we aren't satisfied with the first answer, we need to use successive approximations.

$$\frac{4x^3}{0.97} = 1.6 \times 10^{-5}, \quad x = 1.57 \times 10^{-2} = 1.6 \times 10^{-2}$$

The 5% rule of thumb is very useful. The first approximation, to the proper number of significant figures, gave the same answer as the second try.

b.

	2 NOCl(g)	$\rightleftharpoons$	2 NO(g)	+	Cl$_2$(g)	K = 1.6 × 10^{-5}

Before	0		2.0 M	1.0 M

Let 1.0 mol/L Cl$_2$ react completely.   (K is small, reactants dominate.)

Change	+2.0	$\leftarrow$	-2.0	-1.0	React completely
After	2.0		0	0	New initial conditions

2x mol/L NOCl reacts to reach equilibrium

Change	-2x	$\rightarrow$	+2x	+x
Equil.	2.0 - 2x		2x	x

$$K = 1.6 \times 10^{-5} = \frac{(2x)^2\,(x)}{(2.0 - 2x)^2} \approx \frac{4x^3}{2.0^2} \quad \text{(Assuming } 2.0 - 2x \approx 2.0)$$

$x^3 = 1.6 \times 10^{-5}, \ x = 2.5 \times 10^{-2}$ Assumption good (2.5% error).

[NOCl] = 2.0 - 0.050 = 1.95 M = 2.0 M; [NO] = 0.050 M; [Cl$_2$] = 0.025 M

Note: If we do not break this problem into two parts, (a stoichiometric part and an
equilibrium part), we are faced with solving a cubic equation.

	2 NOCl	$\rightleftharpoons$	2 NO	+	Cl$_2$

Initial	0		2.0 M		1.0 M
Change	+2y	$\leftarrow$	-2y		-y
Equil.	2y		2.0 - 2y		1.0 - y

$$1.6 \times 10^{-5} = \frac{(2.0 - 2y)^2\,(1.0 - y)}{(2y)^2}$$

If we say that y is small to simplify the problem, then:

$$1.6 \times 10^{-5} = \frac{2.0^2}{4y^2}$$

We get $y = 250$.   THIS IS IMPOSSIBLE!

To solve this equation we cannot make any simplifying assumptions; we have to find a way
to solve a cubic equation. Or, use some chemical common sense and solve the problem the
easier way.

c.   2 NOCl(g) $\rightleftharpoons$ 2 NO(g)   +   Cl$_2$(g)

Initial	1.0 M		1.0 M		0
	2x mol/L NOCl reacts to reach equilibrium				
Change	-2x	$\rightarrow$	+2x		+x
Equil.	1.0 - 2x		1.0 + 2x		x

$$1.6 \times 10^{-5} = \frac{(1.0 + 2x)^2\,(x)}{(1.0 - 2x)^2} \approx \frac{(1.0)^2(x)}{(1.0)^2} \qquad \text{(Assuming } x \text{ is small.)}$$

$x = 1.6 \times 10^{-5}$ Assumptions are good.

[Cl$_2$] = 1.6 $\times 10^{-5}$ M and [NOCl] = [NO] = 1.0 M

d. $2 NOCl(g) \rightleftharpoons 2 NO(g) + Cl_2(g)$

Before   0   3.0 M 1.0 M
                       Let 1.0 mol/L Cl_2 react completely.
Change   +2.0 $\leftarrow$ -2.0   -1.0   React completely
After   2.0   1.0   0   New Initial
                       $2x$ mol/L NOCl reacts to reach equilibrium
Change   -2x $\rightarrow$ +2x   +x
Equil.   2.0 - 2x   1.0 + 2x   x

$$1.6 \times 10^{-5} = \frac{(1.0 + 2x)^2\,(x)}{(2.0 - 2x)^2} \approx \frac{x}{4.0}; \text{ Solving: } x = 6.4 \times 10^{-5}$$

Assumptions good. $[Cl_2] = 6.4 \times 10^{-5}\ M$; [NOCl] = 2.0 M; [NO] = 1.0 M

e. $2 NOCl(g) \rightleftharpoons 2 NO(g) + Cl_2(g)$

Before   2.0 M 2.0 M 1.0 M
                       Let 1.0 mol/L Cl_2 react completely.
Change   +2.0 $\leftarrow$ -2.0   -1.0   React completely
After   4.0   0   0   New Initial
                       $2x$ mol/L NOCl reacts to reach equilibrium
Change   -2x $\rightarrow$ +2x   +x
Equil.   4.0 - 2x   2x   x

$$1.6 \times 10^{-5} = \frac{(2x)^2\,(x)}{(4.0 - 2x)^3} \approx \frac{4x^3}{16}, \ x = 4.0 \times 10^{-2}, \text{ Assumption good (2\% error).}$$

$[Cl_2] = 0.040\ M$; [NO] = 0.080 M; [NOCl] = 4.0 - 2(0.040) = 3.92 $M \approx$ 3.9 M

f. $2 NOCl(g) \rightleftharpoons 2 NO(g) + Cl_2(g)$

Before   1.00 M 1.00 M 1.00 M
                       Let 1.00 mol/L NO react completely (limiting reagent).
Change   +1.00 $\leftarrow$ -1.00   -0.50   React completely
After   2.00   0   0.50   New Initial
                       $2x$ mol/L NOCl reacts to reach equilibrium
Change   -2x $\rightarrow$ +2x   +x
Equil.   2.00 - 2x   2x   0.50 + x

$$\frac{(2x)^2\,(0.50 + x)}{(2.00 - 2x)^2} \approx \frac{4x^2\,(0.50)}{(2.00)^2} = 1.6 \times 10^{-5}, \ x = 5.7 \times 10^{-3}, \text{ Assumptions good.}$$

[NO] = 2x = $1.1 \times 10^{-2}\ M$; $[Cl_2]$ = 0.50 + 0.0057 = 0.51 M; [NOCl] = 2.00 - 2(0.0057) = 1.99 M

25. $2 SO_2(g)$ + $O_2(g)$ $\rightleftharpoons$ $2 SO_3(g)$ $K_p = 0.25$

Initial   0.50 atm   0.50 atm   0
                $2x$ atm SO_2 reacts to reach equilibrium
Change $-2x$ $-x$ $\rightarrow$ $+2x$
Equil. $0.50 - 2x$ $0.50 - x$ $2x$

$$\frac{P_{SO_3}^2}{P_{SO_2}^2 P_{O_2}} = \frac{(2x)^2}{(0.50 - 2x)^2(0.50 - x)} \approx \frac{4x^2}{(0.50)^3} = 0.25 \quad \text{(Assuming } x \text{ is small.)}$$

$x = 0.088$; Assumptions not good (x is 18% of 0.50). Using the method of successive approximations (Appendix 1.4):

$$\frac{4x^2}{[0.50 - 2(0.088)]^2 [0.50 - (0.088)]} = \frac{4x^2}{(0.324)^2(0.412)} = 0.25, \ x = 0.052$$

$$\frac{4x^2}{[0.50 - 2(0.052)]^2 [0.50 - (0.052)]} = \frac{4x^2}{(0.396)^2(0.448)} = 0.25, \ x = 0.066$$

$$\frac{4x^2}{(0.368)^2(0.434)} = 0.25, \ x = 0.061; \quad \frac{4x^2}{(0.378)^2(0.439)} = 0.25, \ x = 0.063$$

The next trials converge at 0.062.

$P_{SO_2} = 0.50 - 2x = 0.376 = 0.38$ atm; $P_{SO_3} = 2x = 0.124 = 0.12$ atm

$P_{O_2} = 0.50 - x = 0.438 = 0.44$ atm

26.   a. $H_2(g)$ + $F_2(g)$ $\rightleftharpoons$ $2 HF(g)$ $K = 1.0 \times 10^2$

Before   2.0 M 2.0 M 0
                Let 2.0 mol/L H_2 react completely (K is large, products dominate).
Change -2.0 -2.0 $\rightarrow$ $+4.0$ React completely
After   0   0   4.0   New Initial
                $2x$ mol/L HF reacts to reach equilibrium
Change $+x$ $+x$ $\leftarrow$ $-2x$
Equil. x x $4.0 - 2x$

$$K = 1.0 \times 10^2 = \frac{[HF]^2}{[H_2][F_2]} = \frac{(4.0 - 2x)^2}{x^2}$$

The equation is a perfect square. Taking the square root:

$$10. = \frac{4.0 - 2x}{x}, \ 10.(x) = 4.0 - 2x, \ 12x = 4.0, \ x = 0.33$$

$x = [H_2] = [F_2] = 0.33 \ M$; $[HF] = 4.0 - 2x = 3.3 \ M$

Alternate Solution:

$$H_2(g) \quad + \quad F_2(g) \; \rightleftharpoons \; 2\,HF(g)$$

Initial	2.0 M	2.0 M	0

y mol/L H_2 reacts to reach equilibrium

Change	-y	-y	$\rightarrow$	+2y
Equil.	2.0 - y	2.0 - y		2y

$$\frac{4y^2}{(2.0-y)^2} = 1.0 \times 10^2; \quad \text{This is a perfect square. Solving:}$$

$$\frac{2y}{2.0-y} = 10., \quad 2y = 20. - 10.(y), \quad 12y = 20., \quad y = 1.67$$

Carrying an extra significant figure, $2y = [HF] = 3.3\ M$; $2.0 - y = [H_2] = [F_2] = 0.3\ M$

b.
$$H_2 \quad + \quad F_2 \; \rightleftharpoons \; 2\,HF$$

Initial	0.33 M	0.33 M	3.3 M (From part a.)

Add 0.50 mol H_2 to the 1.0 L flask

New Initial	0.83	0.33		3.3
Change	-x	+ -x	$\rightarrow$	+2x
Equil.	0.83 - x	0.33 - x		3.3 + 2x

$$\frac{(3.3+2x)^2}{(0.83-x)(0.33-x)} = 1.0 \times 10^2, \quad 10.9 + 13.2\,x + 4\,x^2 = 27.4 - 116\,x + 100\,x^2$$

$96\,x^2 - 129.2\,x + 16.5 = 0$; Solving using quadratic formula (Appendix 1.4):

$$x = \frac{129.2 \pm [(-129.2)^2 - 4(96)\,(16.5)]^{1/2}}{2(96)} = \frac{129.2 \pm 101.8}{192}$$

$x = 0.14$ or $x = 1.2$; Only $x = 0.14$ makes sense.

$[H_2] = 0.83 - x = 0.69\ M;$ $[F_2] = 0.33 - x = 0.19\ M;$ $[HF] = 3.3 + 2x = 3.6\ M$

27.   a. $K_p = K(RT)^{\Delta n} = \dfrac{4.5 \times 10^9\ L}{mol} \left(\dfrac{0.08206\ L\ atm}{mol\ K} \times 373\ K \right)^{-1}$ $\Delta n = 1 - 2 = -1$

$K_p = 1.5 \times 10^8\ atm^{-1}$

b. $CO(g) + Cl_2(g) \rightleftharpoons COCl_2(g)$ $K_p = 1.5 \times 10^8$

K_p is so large that at equilibrium we will have almost all $COCl_2$. Assume $P_{tot} \approx P_{COCl_2} \approx 5.0$.

Initial	0	0	5.0 atm

x atm $COCl_2$ reacts to reach equilibrium

Change	+x	+x	←	-x
Equil.	x	x		5.0 - x

$$1.5 \times 10^8 = \frac{5.0 - x}{x^2} \approx \frac{5.0}{x^2} \quad \text{(Assuming } 5.0 - x \approx 5.0)$$

Solving: $x = 1.8 \times 10^{-4}$ atm. Check assumptions: $5.0 - x = 5.0 - 1.8 \times 10^{-4} = 5.0$ atm.
Assumptions are good (well within the 5% rule).

$P_{CO} = P_{Cl_2} = 1.8 \times 10^{-4}$ atm and $P_{COCl_2} = 5.0$ atm

28. $Fe^{3+}(aq) + SCN^-(aq) \rightleftharpoons FeSCN^{2+}(aq)$ $K = 1.1 \times 10^3$

Before	0.020 M	0.10 M		0

Let 0.020 mol/L Fe^{3+} react completely (K is large, products dominate).

Change	-0.020	-0.020	→	+0.020	React completely
After	0	0.08		0.020	New Initial

x mol/L $FeSCN^{2+}$ reacts to reach equilibrium

Change	+x	+x	←	-x
Equil.	x	0.08 + x		0.020 - x

$$K = 1.1 \times 10^3 = \frac{[FeSCN^{2+}]}{[Fe^{3+}][SCN^-]} = \frac{0.020 - x}{(x)(0.08 + x)} \approx \frac{0.020}{0.08x}$$

$x = 2 \times 10^{-4}$; x is 1% of 0.020. Assumptions good by the 5% rule.

$x = [Fe^{3+}] = 2 \times 10^{-4}\ M$; $[SCN^-] = 0.08 + 2 \times 10^{-4} = 0.08\ M$

$[FeSCN^{2+}] = 0.020 - 2 \times 10^{-4} = 0.020\ M$

Le Chatelier's Principle

29.   We can do two things in the choice of solvent. First we must avoid water. Any extra water we
add from the solvent tends to push the equilibrium to the left. This eliminates water and 95%
ethanol as solvent choices. Of the remaining two solvents, acetonitrile will not take part in the
reaction, whereas ethanol is a reactant. If we use ethanol as the solvent it will drive the
equilibrium to the right, thereby reducing the concentrations of the objectionable butyric acid to
a minimum. Thus, the best solvent is 100% ethanol.

30.   When we change the pressure by adding an unreactive gas, we do not change the partial pressure of any of the substances in equilibrium with each other. In this case the equilibrium will not shift. If we change the pressure by changing the volume, we will change the partial pressures of all the substances in equilibrium by the same factor. If there are unequal number of gaseous particles on the two sides of the equation, the equilibrium will shift.

31.   A change in volume will change the partial pressure of all reactants and products by the same factor. The shift in equilibrium depends on the number of gaseous particles on each side. An increase in volume will shift the equilibrium to the side with the greater number of particles in the gas phase. A decrease in volume will favor the side with the fewer number of gas phase particles. If there are the same number of gas phase particles on each side of the reaction, then a change in volume will not shift the equilibrium.

32.   a.   Doubling the volume will decrease all concentrations by a factor of one-half.

$$Q = \frac{\frac{1}{2}[FeSCN^{2+}]_{eq}}{\left(\frac{1}{2}[Fe^{3+}]_{eq}\right)\left(\frac{1}{2}[SCN^-]_{eq}\right)} = 2\,K, \;\; Q > K$$

Equilibrium will shift to the left.

b.   Adding Ag^+ will remove SCN^- through the formation of $AgSCN(s)$. Reaction will shift to the left.

c.   Remove Fe^{3+} will shift the reaction to left.

d.   Reaction shifts to the right.

33.   a.   right   b.   right   c.   no effect ($\Delta n = 0$)

d.   Since reaction is exothermic, heat is a product:

$$CO(g) + H_2O(g) \rightarrow H_2(g) + CO_2(g) + Heat$$

Increasing T will add heat. The equilibrium shifts to the left.

34. $CoCl_2(s) + 6\,H_2O(g) \rightleftharpoons CoCl_2 \cdot 6\,H_2O(s)$

If rain is imminent, there would be a lot of water vapor in the air. Reaction would shift to the right and the indicator would turn pink (due to $CoCl_2 \cdot 6H_2O$ formation).

35. $2\,SO_3(g) \rightleftharpoons 2\,SO_2(g) + O_2(g)$

a.   Increase

b.   Increase $K_p = \dfrac{P_{SO_2}^2 \times P_{O_2}}{P_{SO_3}^2}$

Halving the volume will double each partial pressure.

$$Q = \frac{(2\,P_{SO_2})^2 \times 2P_{O_2}}{(2\,P_{SO_3})^2} = 2\,K_p, \quad Q > K_p$$

Equilibrium shifts to the left, increasing the number of moles of sulfur trioxide.

c. No effect. The partial pressures of sulfur trioxide, sulfur dioxide, and oxygen are unchanged.

d. Increase. Heat $+ 2\,SO_3 \rightleftharpoons 2\,SO_2 + O_2$

Decreasing T will remove heat, shifting reaction to the left.

e. Decrease

36.   a. right   b. left   c. right   d. no effect   e. no effect ($\Delta n = 0$)

f. An increase in temperature will shift the equilibrium to the right.

Additional Exercises

37.   a. $N_2(g) + O_2(g) \rightleftharpoons 2\,NO(g)$

$$K_p = 1 \times 10^{-31} = \frac{P_{NO}^2}{P_{N_2}P_{O_2}} = \frac{P_{NO}^2}{(0.8)\,(0.2)}, \quad P_{NO} = 1 \times 10^{-16}\ atm$$

$$n = \frac{PV}{RT} = \frac{(1 \times 10^{-16}\ atm)\,(1.0 \times 10^{-3}\ L)}{\left(\dfrac{0.08206\ L\ atm}{mol\ K}\right)(298\ K)} = 4 \times 10^{-21}\ mol\ NO$$

$$\frac{4 \times 10^{-21}\ mol}{cm^3} \times \frac{6.02 \times 10^{23}\ molecules}{mol} = \frac{2 \times 10^3\ molecules\ NO}{cm^3}$$

b. There is more NO in the atmosphere than we would expect from the value of K. The answer must lie in the rates of the reaction. At 25°C the rates of both reactions:

$$N_2 + O_2 \rightarrow 2\,NO \quad \text{and} \quad 2\,NO \rightarrow N_2 + O_2$$

are essentially zero. Very strong bonds must be broken; the activation energy is very high. Nitric oxide is produced in high energy or high temperature environments. In nature some NO is produced by lightening. The primary manmade source of NO is from automobiles. The production of NO is endothermic. At high temperature, K will increase and the rates of the reaction increases; NO is produced. Once the NO gets into a more normal environment in the atmosphere, it doesn't go back to N_2 and O_2 because of the slow rate.

c. $K_p = \dfrac{P_{NO}^2}{P_{N_2}P_{O_2}}$

To convert from partial pressures to concentrations in molecules/cm³, we will have to do the same conversion to all concentrations. All of these conversions will cancel. Therefore, $K^* = K_p = 1 \times 10^{-31}$. The equilibrium constant for this reaction is unitless.

38. a.

$$Na_2O(s) \rightleftharpoons 2\,Na(l) + 1/2\,O_2(g) \qquad K_1$$
$$2\,Na(l) + O_2(g) \rightleftharpoons Na_2O_2(s) \qquad\qquad 1/K_3$$

$$Na_2O(s) + 1/2\,O_2(g) \rightleftharpoons Na_2O_2(s) \qquad K_{eq} = (K_1)(1/K_3)$$

$$K_{eq} = \dfrac{2 \times 10^{-25}}{5 \times 10^{-29}} = 4 \times 10^3$$

b.

$$NaO(g) \rightleftharpoons Na(l) + 1/2\,O_2(g) \qquad\qquad K_2$$
$$Na_2O(s) \rightleftharpoons 2\,Na(l) + 1/2\,O_2(g) \qquad\quad K_1$$
$$2\,Na(l) + O_2(g) \rightleftharpoons Na_2O_2(s) \qquad\qquad 1/K_3$$

$$NaO(g) + Na_2O(s) \rightleftharpoons Na_2O_2(s) + Na(l)$$

$$K_{eq} = \dfrac{K_1 K_2}{K_3} = 8 \times 10^{-2}$$

c.

$$2\,NaO(g) \rightleftharpoons 2\,Na(l) + O_2(g) \qquad\qquad K_2^2$$
$$2\,Na(l) + O_2(g) \rightleftharpoons Na_2O_2(s) \qquad\qquad 1/K_3$$

$$2\,NaO(g) \rightleftharpoons Na_2O_2(s)$$

$$K_{eq} = \dfrac{K_2^2}{K_3} = 8 \times 10^{18}$$

39.

$$O(g) + NO(g) \rightleftharpoons NO_2(g) \qquad\qquad K = 1/6.8 \times 10^{-49} = 1.5 \times 10^{48}$$

$$NO_2(g) + O_2(g) \rightleftharpoons NO(g) + O_3(g) \qquad K = 1/5.8 \times 10^{-34} = 1.7 \times 10^{33}$$

$$O_2(g) + O(g) \rightleftharpoons O_3(g) \qquad\qquad K = (1.5 \times 10^{48})(1.7 \times 10^{33}) = 2.6 \times 10^{81}$$

40.

$$H_2(g) \quad + \quad S(s) \quad \rightleftharpoons \quad H_2S(g)$$

	$H_2(g)$	$S(s)$		$H_2S(g)$
Initial	0.15 M	-		0
Change	$-x$	-	$\rightarrow$	$+x$
Equil.	$0.15 - x$	-		x

$$K = 6.8 \times 10^{-2} = \frac{[H_2S]}{[H_2]} = \frac{x}{0.15 - x}, \quad 0.0102 - 0.068\,x = x \qquad \text{(Carrying extra sig. figs.)}$$

$$x = \frac{0.0102}{1.068} = 9.6 \times 10^{-3} \text{ mol/L}$$

$$P_{H_2S} = \frac{nRT}{V} = \frac{9.6 \times 10^{-3} \text{ mol} \left(\dfrac{0.08206 \text{ L atm}}{\text{mol K}} \right) \times 363 \text{ K}}{1.0 \text{ L}} = 0.29 \text{ atm}$$

41.   When SO_2 is added, the equilibrium will shift to the left, producing more SO_2Cl_2 and some energy. The energy increases the temperature of the reaction mixture.

42.   a. $2\,NO(g) \quad + \quad Br_2(g) \quad \rightleftharpoons \quad 2\,NOBr(g)$

Initial	98.4 torr	41.3 torr	0
Change	$-2x$	$-x$ →	$+2x$
Equil.	$98.4 - 2x$	$41.3 - x$	$2x$

$$P_{total} = (98.4 - 2x) + (41.3 - x) + 2x = 139.7 - x$$

$$P_{total} = 110.5 = 139.7 - x, \quad x = 29.2 \text{ torr}$$

$$P_{NO} = 98.4 - 2(29.2) = 40.0 \text{ torr} = 0.0526 \text{ atm}$$

$$P_{Br_2} = 41.3 - 29.2 = 12.1 \text{ torr} = 0.0159 \text{ atm}$$

$$P_{NOBr} = 2(29.2) = 58.4 \text{ torr} = 0.0768 \text{ atm}$$

$$K_p = \frac{P_{NOBr}^2}{P_{NO}^2 \, P_{Br_2}} = \frac{(0.0768 \text{ atm})^2}{(0.0526 \text{ atm})^2 \, (0.0159 \text{ atm})} = 134 \text{ atm}^{-1}$$

       b. $2\,NO(g) \quad + \quad Br_2(g) \quad \rightleftharpoons \quad 2\,NOBr(g)$

Initial	0.30 atm	0.30 atm	0
Change	$-2x$	$-x$ →	$+2x$
Equil.	$0.30 - 2x$	$0.30 - x$	$2x$

This would yield a cubic equation. There is no way we can simplify the task. K_P is pretty large, so let us approach equilibrium in two steps; assume the reaction goes to completion then solve the back equilibrium problem.

                               $2\,NO \quad + \quad Br_2 \quad \rightleftharpoons \quad 2\,NOBr$

Initial	0.30 atm	0.30 atm	0	
Change	-0.30	-0.15 →	$+0.30$	React completely
After	0	0.15	0.30	New initial
Change	$+2y$	$+y$ ←	$-2y$	
Equil.	$2y$	$0.15 + y$	$0.30 - 2y$	

$$\frac{(0.30 - 2y)^2}{(2y)^2(0.15 + y)} = 134, \quad \frac{(0.30 - 2y)^2}{(0.15 + y)} = 134 \times 4\,y^2 = 536\,y^2$$

If y is small: $\dfrac{(0.30)^2}{0.15} = 536\,y^2$ and $y = 0.034$, Assumptions are poor (y is 23% of 0.15).

Use 0.034 as approximation for y and solve by successive approximations:

$$\frac{(0.30 - 0.068)^2}{0.15 + 0.034} = 536\,y^2, \quad y = 0.023$$

$$\frac{(0.30 - 0.046)^2}{0.173} = 536\,y^2, \quad y = 0.026$$

$$\frac{(0.30 - 0.052)^2}{0.176} = 536\,y^2, \quad y = 0.026$$

So: $P_{NO} = 2y = 0.052$ atm; $P_{Br_2} = 0.15 + y = 0.18$ atm; $P_{NOBr} = 0.30 - 2y = 0.25$ atm

43. a. $2NaHCO_3(s)$ $\rightleftharpoons$ $Na_2CO_3(s)$ + $CO_2(g)$ + $H_2O(g)$

Initial   -   -   0   0

NaHCO_3(s) decomposes to form x atm of $CO_2(g)$ and $H_2O(g)$ at equilibrium.

Change   - $\rightarrow$ -   +x   +x
Equil.   -   -   x   x

$0.25 = K_p = P_{CO_2} \times P_{H_2O};\;\; 0.25 = x^2,\; x = P_{CO_2} = P_{H_2O} = 0.50$ atm

b. $n_{CO_2} = \dfrac{PV}{RT} = \dfrac{(0.50\ \text{atm})\,(1.00\ \text{L})}{(0.08206\ \text{L atm mol}^{-1}\,\text{K}^{-1})(398\ \text{K})} = 1.5 \times 10^{-2}\ \text{mol } CO_2$

Mass of Na_2CO_3 produced:

$$1.5 \times 10^{-2}\ \text{mol } CO_2 \times \frac{1\ \text{mol } Na_2CO_3}{\text{mol } CO_2} \times \frac{106.0\ \text{g } Na_2CO_3}{\text{mol } Na_2CO_3} = 1.6\ \text{g } Na_2CO_3$$

Mass of $NaHCO_3$ reacted:

$$1.5 \times 10^{-2}\ \text{mol } CO_2 \times \frac{2\ \text{mol } NaHCO_3}{1\ \text{mol } CO_2} \times \frac{84.01\ \text{g } NaHCO_3}{\text{mol}} = 2.5\ \text{g } NaHCO_3$$

Mass of $NaHCO_3$ left = 10.0 - 2.5 = 7.5 g

c. $10.0 \text{ g NaHCO}_3 \times \dfrac{1 \text{ mol NaHCO}_3}{84.01 \text{ g NaHCO}_3} \times \dfrac{1 \text{ mol CO}_2}{2 \text{ mol NaHCO}_3} = 5.95 \times 10^{-2} \text{ mol CO}_2$

When all of the $NaHCO_3$ has just been consumed, we will have 5.95×10^{-2} mol CO_2 gas at a pressure of 0.50 atm (from a).

$$V = \frac{nRT}{P} = \frac{(5.95 \times 10^{-2} \text{ mol}) (0.08206 \text{ L atm mol}^{-1} \text{ K}^{-1}) (398 \text{ K})}{0.50 \text{ atm}} = 3.9 \text{ L}$$

44. a. shift left

b. Since the reaction is endothermic, an increase in temperature will shift the equilibrium to the right.

c. no effect   d. shift right

45. $P_4(g) \rightleftharpoons 2 P_2(g)$ $K_p = 1.00 \times 10^{-1} \text{ atm} = \dfrac{P_{P_2}^2}{P_{P_4}}$

$P_{P_4} + P_{P_2} = P_{total} = 1.00 \text{ atm},\ \ P_{P_4} = 1.00 - P_{P_2}$

Let $y = P_{P_2}$ at equilibrium, then $\dfrac{y^2}{1.00 - y} = 0.100$

Solving: $y = 0.270 \text{ atm} = P_{P_2};\ \ P_{P_4} = 0.73 \text{ atm}$

$$P_4(g) \quad\rightleftharpoons\quad 2 P_2(g)$$

Initial	P_0	0	P_0 = initial pressure of P_4
Change	$-x$	$\rightarrow$ $+2x$	
Equil.	$P_0 - x$	$2x$	

$P_{total} = P_0 - x + 2x = 1.00 \text{ atm} = P_0 + x$

Solving: $0.270 \text{ atm} = P_{P_2} = 2x,\ x = 0.135 \text{ atm};\ P_0 = 1.00 - 0.135 = 0.87 \text{ atm}$

% Dissociation $= \dfrac{x}{P_0} = \dfrac{0.135}{0.87} \times 100 = 16\%$

46. $PV = nRT;\ \dfrac{n}{V} = \dfrac{P}{RT};\ \dfrac{nM}{V} = d = \dfrac{PM}{RT}$

$$0.168 \text{ g/L} = \frac{P_{O_2} (32.00) + P_{O_3} (48.00)}{\dfrac{0.08206 \text{ L atm}}{\text{mol K}} \times 448 \text{ K}},\ \ 32.00\ P_{O_2} + 48.00\ P_{O_3} = 6.18 \text{ (P in atm)}$$

$$P_{total} = P_{O_2} + P_{O_3} = 128 \text{ torr} \times \frac{1 \text{ atm}}{760 \text{ torr}} = 0.168 \text{ atm}$$

Solving using simultaneous equations:

$$32.00 \; P_{O_2} + 48.00 \; P_{O_3} = 6.18$$
$$-32.00 \; P_{O_2} - 32.00 \; P_{O_3} = -5.38$$
$$\overline{\hspace{3cm} 16.00 \, P_{O_3} = 0.80}$$

$$P_{O_3} = \frac{0.80}{16.00} = 0.050 \text{ atm and } P_{O_2} = 0.118 \text{ atm}$$

$$K_p = \frac{P_{O_3}^2}{P_{O_2}^3} = \frac{(0.050)^2}{(0.118)^3} = 1.5 \text{ atm}^{-1}$$

47.   2 NOBr (g) $\rightleftharpoons$ 2 NO(g) + Br$_2$(g)

Initial	P_0	0	0	P_0 = initial pressure of NOBr
Equil.	$P_0 - 2x$	$2x$	x	Note: $P_{NO} = 2P_{Br_2}$

$$P_{total} = 0.0515 \text{ atm} = (P_0 - 2x) + (2x) + (x) = P_0 + x; \; 0.0515 \text{ atm} = P_{NOBr} + 3 \, P_{Br_2}$$

$$d = \frac{PM}{RT} = 0.1861 \text{ g/L} = \frac{P_{NOBr} \, (109.9) + 2P_{Br_2} \, (30.01) + P_{Br_2} \, (159.8)}{0.08206 \times 298}$$

$$4.55 = 109.9 \; P_{NOBr} + 219.8 \; P_{Br_2}$$

Solving using simultaneous equations:

$$0.0515 = P_{NOBr} + 3 \, P_{Br_2}$$
$$-0.0414 = -P_{NOBr} - 2.000 \; P_{Br_2}$$
$$\overline{\hspace{1cm} 0.0101 = \hspace{3cm} P_{Br_2}}$$

$$P_{Br_2} = 1.01 \times 10^{-2} \text{ atm}; \; P_{NO} = 2 \, P_{Br_2} = 2.02 \times 10^{-2} \text{ atm};$$

$$P_{NOBr} = 0.0515 - 3(1.01 \times 10^{-2}) = 2.12 \times 10^{-2} \text{ atm}$$

$$K_p = \frac{(P_{Br_2}) \, (P_{NO})^2}{(P_{NOBr})^2} = \frac{(1.01 \times 10^{-2}) \, (2.02 \times 10^{-2})^2}{(2.12 \times 10^{-2})^2} = 9.17 \times 10^{-3} \text{ atm}$$

48. $N_2(g)$ + $3 H_2(g)$ $\rightleftharpoons$ $2 NH_3(g)$

Initial   0   0 P_0 P_0 = initial pressure of NH_3
Change   +x +3x $\leftarrow$ -2x
Equil. x 3x P_0 - 2x

From problem, $P_0 - 2x = \dfrac{P_0}{2.00}$, so $P_0 = 4.00\,x$

$$K_p = \frac{(4x - 2x)^2}{(x)\,(3x)^3} = \frac{(2x)^2}{(x)\,(3x)^3} = \frac{4x^2}{27x^4} = \frac{4}{27x^2} = 5.3 \times 10^5\ atm^{-2},\ x = 5.3 \times 10^{-4}\ atm$$

$P_0 = 4.00\,x = 2.1 \times 10^{-3}$ atm

49.   a. $P = \dfrac{nRT}{V} = \dfrac{0.100\ mol \times \dfrac{0.08206\ L\ atm}{mol\ K} \times 480.\ K}{12.0\ L} = 0.328$ atm

                    $PCl_5(g)$ $\rightleftharpoons$ $PCl_3(g)$ + $Cl_2(g)$ $K_p = 0.267$ atm

Initial   0.328 atm   0   0
Change   -x $\rightarrow$ +x +x
Equil.   0.328 -x x x

$$\frac{x^2}{0.328 - x} = 0.267$$

Solving using the quadratic formula: $x = 0.191$ atm

$P_{PCl_3} = P_{Cl_2} = 0.191$ atm; $P_{PCl_5} = 0.328 - 0.191 = 0.137$ atm

b. $PCl_5(g)$ $\rightleftharpoons$ $PCl_3(g)$ + $Cl_2(g)$

Initial P_0 0   0 P_0 = initial pressure of PCl_5
Change   -x $\rightarrow$ +x +x
Equil. P_0 - x x x

$P_{total} = 2.00$ atm $= (P_0 - x) + x + x = P_0 + x$, $P_0 = 2.00 - x$

$$\frac{x^2}{P_0 - x} = 0.267;\quad \frac{x^2}{2.00 - 2x} = 0.267,\ x^2 = 0.534 - 0.534\,x$$

$x^2 + 0.534\,x - 0.534 = 0$ Solving using the quadratic formula:

$$x = \frac{-0.534 \pm \sqrt{(0.534)^2 + 4(0.534)}}{2} = 0.511\ atm$$

$P_0 = 2.00 - x = 2.00 - 0.511 = 1.49$ atm; The initial pressure of PCl_5 was 1.49 atm.

$$n = \frac{PV}{RT} = \frac{(1.49 \text{ atm}) (5.00 \text{ L})}{(0.08206 \text{ L atm mol}^{-1} \text{ K}^{-1}) (480. \text{ K})} = 0.189 \text{ mol}$$

0.189 mol $PCl_5 \times 208.22$ g PCl_5/mol = 39.4 g PCl_5

50. $P_0(O_2) = 4.49$ atm   (Calculated from the ideal gas equation.)

	$CH_4(g)$	+	$2 O_2(g)$	→	$CO_2(g)$	+	$2 H_2O(g)$
Change	$-x$		$-2x$	→	$+x$		$+2x$

	$CH_4(g)$	+	$3/2 O_2(g)$	→	$CO(g)$	+	$2 H_2O(g)$
Change	$-y$		$-3/2\, y$	→	$+y$		$+2y$

Amount of O_2 reacted = 4.49 atm - 0.326 atm = 4.16 atm O_2

$2\, x + 3/2\, y = 4.16$ atm O_2 and $2\, x + 2\, y = 4.45$ atm H_2O

Solving using simultaneous equations:

$$\begin{array}{rl} 2\, x + \ 2\, y & = \ 4.45 \\ -2\, x - 3/2\, y & = -4.16 \\ \hline 0.50\, y & = \ 0.29, \ \ y = 0.58 \text{ atm} = P_{CO} \end{array}$$

$2\, x + 2(0.58) = 4.45, \ \ x = \dfrac{4.45 - 1.16}{2} = 1.65$ atm $= P_{CO_2}$

51.   a. $PCl_5(g) \ \rightleftharpoons \ PCl_3(g) \ + \ Cl_2(g)$

Initial	P_0	0	0	P_0 = initial PCl_5 pressure
Change	$-x$	$+x$	$+x$	
Equil.	$P_0 - x$	x	x	

$P_{total} = P_0 - x + x + x = P_0 + x = 358.7$ torr

$$P_0 = \frac{nRT}{V} = \frac{\dfrac{2.4156 \text{ g}}{208.22 \text{ g mol}^{-1}} \times \dfrac{0.08206 \text{ L atm}}{\text{mol K}} \times 523.2 \text{ K}}{2.000 \text{ L}} = 0.2490 \text{ atm or } 189.2 \text{ torr}$$

$x = P_{total} - P_0 = 358.7 - 189.2 = 169.5$ torr

$P_{PCl_3} = P_{Cl_2} = 169.5.$ torr = 0.2230 atm

$P_{PCl_5} = 189.2 - 169.5 = 19.7$ torr = 0.0259 atm

$K_p = \dfrac{(0.2230)^2}{0.0259} = 1.92$ atm

b. $P = \dfrac{nRT}{V} = \dfrac{0.250 \times 0.08206 \times 523.2}{2.000} = 5.37$ atm Cl_2

	$PCl_5(g)$	$\rightleftharpoons$	$PCl_3(g)$	$+$	$Cl_2(g)$	
Initial	0.0259 atm		0.2230 atm	0.2230 atm		(from a)

Add 0.250 mol Cl_2, increases P_{Cl_2} by 5.37 atm

Initial'	0.0259		0.2230	5.59	
Change	+0.2230	$\leftarrow$	-0.2230	-0.2230	React completely
After	0.2489		0	5.37	New initial
Change	-x	$\rightarrow$	+x	+x	
Equil.	0.2489 -x		x	5.37 + x	

$\dfrac{(5.37 + x)\,(x)}{(0.2489 - x)} = 1.92;\ x^2 + 7.29\,x - 0.478 = 0$

Using the quadratic formula: $x = 0.0650$ atm

$P_{PCl_3} = 0.0650$ atm; $P_{PCl_5} = 0.2489 - 0.0650 = 0.1839$ atm; $P_{Cl_2} = 5.37 + 0.0650 = 5.44$ atm

52.

	$SO_2Cl_2(g)$	$\rightleftharpoons$	$Cl_2(g)$	$+$	$SO_2(g)$	
Initial	P_0		0	0		P_0 = initial pressure of SO_2Cl_2
Change	-x	$\rightarrow$	+x	+x		
Equil.	P_0 - x		x	x		

$P_{total} = 0.900$ atm $= P_0 - x + x + x = P_0 + x$

$\dfrac{x}{P_0} \times 100 = 12.5,\ P_0 = 8.00\,x$

Solving: $0.900 = P_0 + x = 9.00\,x,\ x = 0.100$

$x = 0.100$ atm $= P_{Cl_2} = P_{SO_2};\ P_0 - x = 0.800 - 0.100 = 0.700$ atm $= P_{SO_2Cl_2}$

$K_p = \dfrac{P_{Cl_2} \times P_{SO_2}}{P_{SO_2Cl_2}} = \dfrac{(0.100)^2}{0.700} = 1.43 \times 10^{-2}$ atm

53.

	$N_2(g)$	$+$	$O_2(g)$	$\rightleftharpoons$	$2\ NO(g)$	Let:
Equil.	3.7 p		p		x	equilibrium $P_{O_2} = p$

equilibrium $P_{O_2} = p$
equilibrium $P_{N_2} = 78/21\ P_{O_2} = 3.7\ p$
equilibrium $P_{NO} = x$
equilibrium $P_{NO_2} = y$

$K_p = 1.5 \times 10^{-4} = \dfrac{P_{NO}^2}{P_{O_2} \times P_{N_2}}$

$$N_2 \quad + \quad 2\,O_2 \quad \rightleftharpoons \quad 2\,NO_2$$

Equil.   3.7 p   p y

$$K_p = 1.0 \times 10^{-5} = \frac{P_{NO_2}^2}{P_{O_2}^2 \times P_{N_2}}$$

We want $P_{NO_2} = P_{NO}$ at equilibrium, so $x = y$.

Taking the ratio of the two K_p expressions:

$$\frac{\dfrac{P_{NO}^2}{P_{O_2} \times P_{N_2}}}{\dfrac{P_{NO_2}^2}{P_{O_2}^2 \times P_{N_2}}} = \frac{1.5 \times 10^{-4}}{1.0 \times 10^{-5}} \; ; \; \text{Since } P_{NO} = P_{NO_2}, \; P_{O_2} = \frac{1.5 \times 10^{-4}}{1.0 \times 10^{-5}} = 15 \text{ atm}$$

Air is 21 mol % O_2, so:

$$P_{O_2} = 0.21\,P_{total}, \; P_{total} = \frac{15 \text{ atm}}{0.21} = 71 \text{ atm}$$

To solve for the equilibrium concentrations of all gases (not required in question), solve one of the K_p expressions where $p = P_{O_2} = 15$ atm.

$$1.5 \times 10^{-4} = \frac{x^2}{15[3.7(15)]}, \; x = P_{NO} = P_{NO_2} = 0.35 \text{ atm}$$

Equilibrium pressures:

$P_{O_2} = 15$ atm

$P_{N_2} = 3.7(15) = 55.5 = 56$ atm

$P_{NO} = P_{NO_2} = 0.35$ atm

54. $N_2O_4(g) \rightleftharpoons 2\,NO_2(g)$ $K_p = \dfrac{P_{NO_2}^2}{P_{N_2O_4}} = \dfrac{(1.2)^2}{0.33} = 4.4$

Doubling the volume decreases each partial pressure by a factor of 2 (P= nRT/V).

$P_{NO_2} = 0.60$ atm and $P_{N_2O_4} = 0.17$ atm are the new partial pressures.

$$Q = \frac{(0.60)^2}{0.17} = 2.1, \; Q < K \qquad \text{Equilibrium will shift to the right.}$$

$$N_2O_4(g) \quad \rightleftharpoons \quad 2\,NO_2(g)$$

Initial   0.17 atm   0.60 atm
Equil.   0.17 - x 0.60 + 2x

$$K_p = 4.4 = \frac{(0.60 + 2x)^2}{(0.17 - x)}, \quad 4x^2 + 6.8\,x - 0.39 = 0$$

Solving using the quadratic formula: $x = 0.056$

$$P_{NO_2} = 0.60 + 2(0.056) = 0.71 \text{ atm}$$

$$P_{N_2O_4} = 0.17 - 0.056 = 0.11 \text{ atm}$$

55.   a. $P = \dfrac{nRT}{V} = \dfrac{\dfrac{2.450\text{ g }PCl_5}{208.22\text{ g/mol}} \times \dfrac{0.08206\text{ L atm}}{\text{mol K}} \times 600.\text{ K}}{0.500\text{ L}} = 1.16 \text{ atm}$

b. $PCl_5(g) \quad \rightleftharpoons \quad PCl_3(g) \quad + \quad Cl_2(g)$

Initial   1.16 atm   0   0
Equil.   1.16 - x x x

$$\frac{x^2}{1.16 - x} = 11.5, \quad x^2 + 11.5\,x - 13.3 = 0$$

Using quadratic formula: $x = 1.06$ atm

$$P_{PCl_5} = 1.16 - 1.06 = 0.10 \text{ atm}$$

c. $P_{PCl_3} = P_{Cl_2} = 1.06$ atm; $P_{PCl_5} = 0.10$ atm

$$P_{tot} = 1.06 + 1.06 + 0.10 = 2.22 \text{ atm}$$

d.   % dissociation = $\dfrac{x}{1.16} \times 100 = \dfrac{1.06}{1.16} \times 100 = 91.4\%$

56. $CCl_4(g) \quad \rightleftharpoons \quad C(s) \quad + \quad 2\,Cl_2(g)$

Initial	P_0	-	0	P_0 = initial pressure of CCl_4
Change	-x	$\rightarrow$ -	+2x	
Equil.	P_0- x	-	2x	

$$P_{total} = P_0 - x + 2x = P_0 + x = 1.20 \text{ atm}$$

$$\frac{(2x)^2}{P_0 - x} = 0.76, \quad 4x^2 = 0.76\,P_0 - 0.76\,x, \quad P_0 = \frac{4x^2 + 0.76x}{0.76}; \text{ Substituting into } P_0 + x = 1.20:$$

$$\frac{4x^2}{0.76} + x + x = 1.20 \text{ atm}, \quad 5.3\,x^2 + 2x - 1.20 = 0; \quad \text{Solving using quadratic formula:}$$

$$x = \frac{-2 \pm (4 + 25.4)^{1/2}}{2(5.3)} = 0.32 \text{ atm}; \quad P_0 + 0.32 = 1.20, \quad P_0 = 0.88 \text{ atm}$$

57.   a. $N_2O_4(g)$ $\rightleftharpoons$ $2NO_2(g)$

Initial	4.5 atm		0
Change	-x	$\rightarrow$	+2x
Equil.	4.5 - x		2x

$$K_p = \frac{P_{NO_2}^2}{P_{N_2O_4}} = \frac{(2x)^2}{4.5 - x} = 0.25$$

Solving using successive approximations or the quadratic formula: $x = 0.50$

$$P_{NO_2} = 2x = 1.0 \text{ atm}; \quad P_{N_2O_4} = 4.5 - x = 4.0 \text{ atm}$$

   b. $N_2O_4(g)$ $\rightleftharpoons$ $2\,NO_2(g)$

Initial	0		9.0 atm
Change	+x	$\leftarrow$	-2x
Equil.	x		9.0 - 2x

$$\frac{(9.0 - 2x)^2}{x} = 0.25$$

Solving using quadratic formula: $x = 4.0$ or 5.1 Only 4.0 makes sense.

$$P_{N_2O_4} = x = 4.0 \text{ atm}; \quad P_{NO_2} = 9.0 - 2x = 1.0 \text{ atm}$$

   c.   No, we get the same equilibrium position starting with either pure N_2O_4 or pure NO_2 in stoichiometric amounts.

   d.   Decreasing volume shifts the reaction left to the side with fewer gas molecules. Each equilibrium partial pressure will increase by a factor of 2.

                              $N_2O_4(g)$ $\rightleftharpoons$ $2\,NO_2(g)$

Initial	4.0 atm		1.0 atm
New Initial	8.0		2.0
Change	+x	$\leftarrow$	-2x
Equil.	8.0 + x		2.0 - 2x

$$\frac{(2.0 - 2x)^2}{8.0 + x} = 0.25$$

Solving using quadratic formula: $x = 1.8$ or $x = 0.28$ Only 0.28 makes sense.

$P_{N_2O_4} = 8.0 + x = 8.3$ atm; $P_{NO_2} = 2.0 - 2x = 1.4$ atm

e. Tripling volume shifts reaction to the right and decreases the partial pressures by a
    factor of 3.

$$N_2O_4(g) \rightleftharpoons 2\,NO_2(g)$$

	$N_2O_4(g)$		$2\,NO_2(g)$
Initial	4.0 atm		1.0 atm
New Initial	$\dfrac{4.0}{3} = 1.3$		$\dfrac{1.0}{3} = 0.33$
Change	$-x$	$\rightarrow$	$+2x$
Equil.	$1.3 - x$		$0.33 + 2x$

$$\frac{(0.33 + 2x)^2}{1.3 - x} = 0.25$$

Solving using quadratic formula: $x = 0.11$

$P_{N_2O_4} = 1.3 - x = 1.2$ atm; $P_{NO_2} = 0.33 + 2x = 0.55$ atm

58. $SO_3(g) \rightleftharpoons SO_2(g) + \tfrac{1}{2} O_2(g)$

	$SO_3(g)$		$SO_2(g)$	$\tfrac{1}{2} O_2(g)$	
Initial	P_0		0	0	$P_0 =$ initial pressure of SO_3
Change	$-x$	$\rightarrow$	$+x$	$+x/2$	
Equil.	$P_0 - x$		x	$x/2$	

Average molar mass of the mixture is:

$$M = \frac{dRT}{P} = \frac{(1.60\text{ g/L})\,(0.08206\text{ L atm mol}^{-1}\text{ K}^{-1})\,(873\text{ K})}{1.80\text{ atm}} = 63.7\text{ g/mol}$$

The average molar mass is determined by:

$$\frac{n_{SO_3}\,(80.07\text{ g/mol}) + n_{SO_2}\,(64.07\text{ g/mol}) + n_{O_2}\,(32.00\text{ g/mol})}{n_{total}}$$

The partial pressure of each gas is proportional to the number of moles of each gas, so:

$$63.7\text{ g/mol} = \frac{P_{SO_3}\,(80.07) + P_{SO_2}\,(64.07) + P_{O_2}\,(32.00)}{P_{tot}}$$

$P_{total} = P_0 - x + x + x/2 = P_0 + x/2 = 1.80$ atm, $P_0 = 1.80 - x/2$

$$63.7 = \frac{(P_0 - x)(80.07) + x(64.07) + \frac{x}{2}(32.00)}{1.80}$$

$$63.7 = \frac{(1.80 - 3/2x)(80.07) + x(64.07) + \frac{x}{2}(32.00)}{1.80}$$

$$115 = 144 - 120.1\,x + 64.07\,x + 16.00\,x, \quad 40.0\,x = 29, \quad x = 0.73$$

$$P_{SO_3} = P_0 - x = 1.80 - 3/2\,x = 0.71 \text{ atm}; \quad P_{SO_2} = 0.73 \text{ atm}; \quad P_{O_2} = x/2 = 0.37 \text{ atm}$$

$$K_p = \frac{(0.73)(0.37)^{1/2}}{(0.71)} = 0.63 \text{ atm}^{1/2}$$

59.   The first reaction produces equal amounts of SO_3 and SO_2. Using the second reaction, calculate the SO_3, SO_2 and O_2 partial pressures at equilibrium.

$$SO_3(g) \quad \rightleftharpoons \quad SO_2(g) \quad + \quad 1/2\,O_2(g)$$

Initial	P_0		P_0	0
Change	$-x$	$\rightarrow$	$+x$	$+x/2$
Equil.	$P_0 - x$		$P_0 + x$	$x/2$

P_0 = initial pressure of SO_3 and SO_2 after first reaction occurs.

$$P_{total} = P_0 - x + P_0 + x + x/2 = 2\,P_0 + x/2 = 0.836 \text{ atm}$$

$$P_{O_2} = x/2 = 0.0275 \text{ atm}, \quad x = 0.0550 \text{ atm}$$

$$2\,P_0 + x/2 = 0.836 \text{ atm}; \quad 2\,P_0 = 0.836 - 0.0275 = 0.809 \text{ atm}, \quad P_0 = 0.405 \text{ atm}$$

$$P_{SO_3} = P_0 - x = 0.405 - 0.0550 = 0.350 \text{ atm}; \quad P_{SO_2} = P_0 + x = 0.405 + 0.0550 = 0.460 \text{ atm}$$

For $2\,FeSO_4(s) \rightleftharpoons Fe_2O_3(s) + SO_3(g) + SO_2(g)$:

$$K_p = P_{SO_2} \times P_{SO_3} = (0.460)(0.350) = 0.161 \text{ atm}^2$$

For $SO_3(g) \rightleftharpoons SO_2(g) + 1/2\,O_2(g)$:

$$K_p = \frac{P_{SO_2} \times P_{O_2}^{1/2}}{P_{SO_3}} = \frac{(0.460)(0.0275)^{1/2}}{0.350} = 0.218 \text{ atm}^{1/2}$$

60.   a. $2\,AsH_3(g) \quad \rightleftharpoons \quad 2\,As(s) \quad + \quad 3\,H_2(g)$

Initial	392.0 torr	-	0
Equil.	392.0 - 2x	-	3x

$$P_{total} = 488.0 \text{ torr} = 392.0 - 2x + 3x, \quad x = 96.0 \text{ torr}$$

$P_{H_2} = 3x = 3(96.0) = 288$ torr; $P_{AsH_3} = 392.0 - 2(96.0) = 200.0$ torr

b. $K_p = \dfrac{(P_{H_2})^3}{(P_{AsH_3})^2} = \dfrac{(288)^3}{(200.0)^2} = 597$ torr $\times \dfrac{1 \text{ atm}}{760 \text{ torr}} = 0.786$ atm

61. $N_2(g) + 3\,H_2\,(g) \rightleftharpoons 2\,NH_3(g)$ $K_p = \dfrac{(P_{NH_3})^2}{(P_{N_2})\,(P_{H_2})^3} = 6.5 \times 10^{-3}$ atm^{-2}

<u>1.0 atm</u> $N_2(g)$ + $3\,H_2(g)$ $\rightleftharpoons$ $2\,NH_3(g)$

Initial   0.25 atm   0.75 atm   0
Equil.   0.25 - x   0.75 - 3x   2x

$\dfrac{(2x)^2}{(0.75 - 3x)^3\,(0.25 - x)} = 6.5 \times 10^{-3}$; Using successive approximations:

$x = 1.2 \times 10^{-2}$; $P_{NH_3} = 2x = 0.024$ atm

<u>10 atm</u> $N_2(g)$ + $3\,H_2(g)$ $\rightleftharpoons$ $2\,NH_3(g)$

Initial   2.5 atm   7.5 atm   0
Equil.   2.5 - x   7.5 - 3x   2x

$\dfrac{(2x)^2}{(7.5 - 3x)^3\,(2.5 - x)} = 6.5 \times 10^{-3}$; Using successive approximations:

$x = 0.69$ atm; $P_{NH_3} = 1.4$ atm

<u>100 atm</u>   Using the same setup as above: $\dfrac{4x^2}{(75 - 3x)^3\,(25 - x)} = 6.5 \times 10^{-3}$

Solving by successive approximations: $x = 16$ atm; $P_{NH_3} = 32$ atm

<u>1000 atm</u>

                        $N_2(g)$ + $3\,H_2(g)$ $\rightleftharpoons$ $2\,NH_3(g)$

Initial   250 atm   750 atm   0
          Let 250 atm N_2 react completely.
New Initial   0   0 5.0×10^2
Equil.   x   3x 5.0×10^2 - 2x

$$\frac{(5.0 \times 10^2 - 2x)^2}{(3x)^3 x} = 6.5 \times 10^{-3}; \quad \text{Assume } x \text{ is small, then:}$$

$$\frac{(5.0 \times 10^2)^2}{(3x)^3 x} \approx 6.5 \times 10^{-3}, \quad x = 35$$

Assumption is poor (14% error).

Solving by successive approximations:

$$x = 32$$

$$P_{NH_3} = 5.0 \times 10^2 - 2x = 440 \text{ atm}$$

The results are plotted as
$\log P_{NH_3}$ vs. $\log P_{tot}$. The
plot will asymptotically
approach the line given by
$\log P_{NH_3} = \log P$ at the
upper limit.

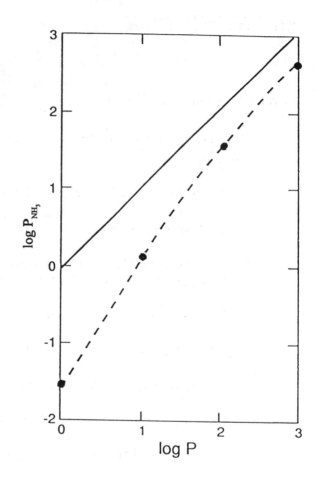

CHAPTER SEVEN

ACIDS AND BASES

Nature of Acids and Bases

1. In deciding whether a substance is an acid or a base, strong or weak, we should keep in mind three things:

    1. There are only a few important strong acids and bases.

    2. All other acids and bases are weak.

    3. Structural features to look for to identify a substance as an acid or a base: $-\overset{\displaystyle \|}{\underset{\displaystyle O}{C}}-OH$ groups (acid) or the presence of N (base) in organic compounds.

        a. weak acid   b. weak acid   c. weak base

        d. strong base   e. weak base   f. weak acid

        g. weak acid   h. weak base   i. strong acid

2. HNO_3: strong acid; $HOCl$: $K_a = 3.5 \times 10^{-8}$; H_2O: $K_a = K_w = 10^{-14}$; NH_4^+: $K_a = 5.6 \times 10^{-10}$

    $HNO_3 > HOCl > NH_4^+ > H_2O$; (As K_a increases, acid strength increases.)

3. Except for water, these are the conjugate bases of the acids in the previous exercise. The stronger the acid, the weaker the conjugate base. NO_3^- is the conjugate base of a strong acid. It is a terrible base (worse than water).

    $NH_3 > OCl^- > H_2O > NO_3^-$

4. a. HI is a strong acid and $K_a = K_w = 10^{-14}$ for H_2O

       HI is a stronger acid than H_2O.

    b. H_2O, $K_a = K_w = 10^{-14}$; $HClO_2$, $K_a = 1.2 \times 10^{-2}$

       $HClO_2$ is a stronger acid than H_2O.

c. HF, $K_a = 7.2 \times 10^{-4}$; HCN, $K_a = 6.2 \times 10^{-10}$

HF is a stronger acid than HCN.

5. a. H_2O b. ClO_2^- c. CN^-

Note: The stronger the acid, the weaker the conjugate base (K_b decreases).

6. a. $PO_4^{3-} + H_2O \rightleftharpoons HPO_4^{2-} + OH^-$ $\qquad K_b = \dfrac{[OH^-][HPO_4^{2-}]}{[PO_4^{3-}]}$

b. $HPO_4^{2-} + H_2O \rightleftharpoons H_2PO_4^- + OH^-$ $\qquad K_b = \dfrac{[OH^-][H_2PO_4^-]}{[HPO_4^{2-}]}$

c. $H_2PO_4^- + H_2O \rightleftharpoons H_3PO_4 + OH^-$ $\qquad K_b = \dfrac{[OH^-][H_3PO_4]}{[H_2PO_4^-]}$

d. $NH_3 + H_2O \rightleftharpoons NH_4^+ + OH^-$ $\qquad K_b = \dfrac{[OH^-][NH_4^+]}{[NH_3]}$

e. $CN^- + H_2O \rightleftharpoons HCN + OH^-$ $\qquad K_b = \dfrac{[OH^-][HCN]}{[CN^-]}$

f. $C_5H_5N + H_2O \rightleftharpoons C_5H_5NH^+ + OH^-$ $\qquad K_b = \dfrac{[OH^-][C_5H_5NH^+]}{[C_5H_5N]}$

g. $H_2NCH_2CO_2H + H_2O \rightleftharpoons {}^+H_3NCH_2CO_2H + OH^-$ $\qquad K_b = \dfrac{[OH^-][{}^+H_3NCH_2CO_2H]}{[H_2NCH_2CO_2H]}$

h. $CH_3CH_2NH_2 + H_2O \rightleftharpoons CH_3CH_2NH_3^+ + OH^-$ $\qquad K_b = \dfrac{[OH^-][CH_3CH_2NH_3^+]}{[CH_3CH_2NH_2]}$

i. $C_6H_5NH_2 + H_2O \rightleftharpoons C_6H_5NH_3^+ + OH^-$ $\qquad K_b = \dfrac{[OH^-][C_6H_5NH_3^+]}{[C_6H_5NH_2]}$

j. $(CH_3)_2NH + H_2O \rightleftharpoons (CH_3)_2NH_2^+ + OH^-$ $\qquad K_b = \dfrac{[OH^-][(CH_3)_2NH_2^+]}{[(CH_3)_2NH]}$

7. $H^-(aq) + H_2O(l) \rightarrow H_2(g) + OH^-(aq)$; $OCH_3^-(aq) + H_2O(l) \rightarrow CH_3OH(aq) + OH^-(aq)$

8. If we add an acid HX that is a stronger acid than H_3O^+, then the reaction

$$HX + H_2O \rightarrow H_3O^+ + X^-$$

would go to 100% completion because H_2O is a much stronger base than X^-. Thus, strong acids are "leveled" in water. They all generate H_3O^+. If a base stronger than OH^- is added to water, the reaction

$$\text{Base} + H_2O \rightarrow (HBase)^+ + OH^-$$

would go to 100% completion because H_2O is a much stronger acid than $(HBase)^+$. The reactions in Exercise 7 are examples of two bases stronger than hydroxide being leveled to OH^- in aqueous solution.

9.   a. H_2O and $CH_3CO_2^-$

     b. $CH_3CO_2^-$, since the equilibrium favors acetic acid ($K_a < 1$).

     c.   Acetate ion is a better base than water and produces basic solutions in water. When we put acetate ion into solution as the only major basic species, the reaction is:

$$CH_3CO_2^- + H_2O \rightleftharpoons CH_3CO_2H + OH^-$$

Now the competition is between $CH_3CO_2^-$ and OH^- for the proton. Hydroxide ion is the strongest base possible in water. The above equilibrium favors acetate plus water resulting in a K_b value less than one. Those species we specifically call weak bases ($10^{-14} < K_b < 1$) lie between H_2O and OH^- in base strength. Weak bases are stronger bases than water but are weaker bases than OH^-.

10.   The NH_4^+ ion is a weak acid because it lies between H_2O and H_3O^+ in terms of acid strength. Weak acids are better acids than water, thus their aqueous solutions are acidic. They are weak acids because they are not as strong as H_3O^+. Weak acids only partially dissociate in water and have K_a values between 10^{-14} and 1.

Autoionization of Water and pH Scale

11.   a.   Since the value of the equilibrium constant increases as the temperature increases, the reaction is endothermic.

     b. $H_2O \rightleftharpoons H^+ + OH^-$ $K_w = 5.47 \times 10^{-14} = [H^+][OH^-]$

          In pure water $[H^+] = [OH^-]$, so $5.47 \times 10^{-14} = [H^+]^2$, $[H^+] = 2.34 \times 10^{-7}\ M$

          $pH = -\log [H^+] = -\log (2.34 \times 10^{-7}) = 6.631$

          A neutral solution of water at 50.°C has:

          $[H^+] = [OH^-];\ [H^+] = 2.34 \times 10^{-7}\ M;\ pH = 6.631$

          Obviously, the condition that $[H^+] = [OH^-]$ is the most general definition of a neutral solution.

c.

Temp (°C)	Temp (K)	1/T	K_w	$\ln K_w$
0	273	3.66×10^{-3}	1.14×10^{-15}	-34.408
25	298	3.36×10^{-3}	1.00×10^{-14}	-32.236
35	308	3.25×10^{-3}	2.09×10^{-14}	-31.499
40.	313	3.19×10^{-3}	2.92×10^{-14}	-31.165
50.	323	3.10×10^{-3}	5.47×10^{-14}	-30.537

From the graph: 37°C = 310. K; $1/T = 3.23 \times 10^{-3}$

$$\ln K_w = -31.38; \quad K_w = 2.35 \times 10^{-14}$$

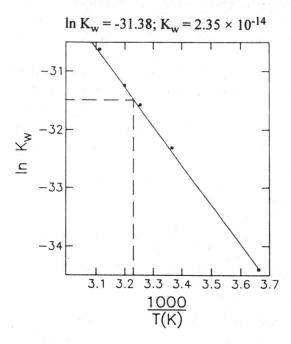

d.   At 37°C, $2.35 \times 10^{-14} = [H^+][OH^-] = [H^+]^2$, $[H^+] = 1.53 \times 10^{-7}$; pH = 6.815

12.   a.   For a straight line, y = mx + b. The equation for the straight line in Exercise 11 is:

$$\ln K_w = -6.91 \times 10^3 \left(\frac{1}{T} \right) - 9.09$$

T = 374°C + 273 = 647 K, so $\ln K_w = -6.91 \times 10^3 \left(\dfrac{1}{647} \right) - 9.09 = -19.8$

$K_w = e^{-19.8} = 2.52 \times 10^{-9}$

b.   At this temperature, $[H^+] = [OH^-] = 5.0 \times 10^{-5}$ M. Supercritical water has more H_3O^+ and OH^- available to react with substances than at lower temperatures.

c.   Steam lines are at a higher temperature than hot water pipes. Because K_w increases with temperature, the concentrations of H_3O^+ and OH^- ions are higher in steam pipes. Many corrosion reactions involve reactions of metals with acid or base. These reactions will occur more readily when in contact with a higher concentration of H_3O^+ or OH^- ions.

Solutions of Acids

13.   a. $H^+(aq)$, $Cl^-(aq)$, and H_2O (HCl is a strong acid.) $[H^+] = 0.250\ M$

pH = -log(0.250) = 0.602

b. $H^+(aq)$, $Br^-(aq)$, and H_2O (HBr is a strong acid.) pH = 0.602

c. H_2O, $H^+(aq)$, $ClO_4^-(aq)$ ($HClO_4$ is a strong acid.) pH = 0.602

d. H_2O, $H^+(aq)$, $NO_3^-(aq)$ (HNO_3 is a strong acid.) pH = 0.602

e. $HNO_2(aq)$ and H_2O are the major species. HNO_2 is a weak acid. It the major source of H^+. As is true for most weak acid solutions, we will ignore the H^+ contribution from water.

$$HNO_2 \quad \rightleftharpoons \quad H^+ \quad + \quad NO_2^-$$

Initial	0.250 M	0 (10^{-7})	0
	x mol/L HNO_2 dissociates to reach equilibrium		
Change	$-x$ $\rightarrow$	$+x$	$+x$
Equil.	0.250 $-x$	x	x

$$K_a = \frac{[H^+][NO_2^-]}{[HNO_2]} = 4.0 \times 10^{-4} = \frac{x^2}{0.250 - x} \approx \frac{x^2}{0.250} \quad \text{(Assuming } x \ll 0.250)$$

$x = 0.010\ M$; Check assumptions: $0.250 - x = 0.240$, 4% error, i.e., $\dfrac{x}{0.250} = 0.04$

Assumptions are good. H^+ contribution from water ($10^{-7}\ M$) is negligible and x is small compared to 0.250. In general, if the calculated pH for an acid in water is less than 6.0, we will assume the H^+ contribution from water is negligible. Also, if the percent error is less than 5% for an assumption, we will consider it a valid assumption (called the 5% rule). Finishing the problem: $x = 0.010\ M = [H^+]$; pH = -log(0.010) = 2.00

f. $CH_3CO_2H(aq)$ and H_2O are the major species. We will consider CH_3CO_2H as the major source of H^+.

$$CH_3CO_2H \quad \rightleftharpoons \quad H^+ \quad + \quad CH_3CO_2^-$$

Initial	0.250	0(10^{-7})	0
	x mol/L CH_3CO_2H dissociates to reach equilibrium		
Change	$-x$ $\rightarrow$	$+x$	$+x$
Equil.	0.250 $- x$	x	x

$$K_a = \frac{[H^+][CH_3CO_2^-]}{[CH_3CO_2H]} = 1.8 \times 10^{-5} = \frac{x^2}{0.250 - x} \approx \frac{x^2}{0.250}$$

$x = 2.1 \times 10^{-3}\ M;\ 0.250 - x = 0.248$, Assumptions good ($\sim 1\%$ error).

$[H^+] = x = 2.1 \times 10^{-3}\ M;\ pH = -\log(2.1 \times 10^{-3}) = 2.68$

g. $NH_4^+(aq)$, $Cl^-(aq)$ and H_2O are the major species. Cl^- has no acidic or basic properties. The major equilibrium is the dissociation of NH_4^+ ion.

$$NH_4^+ \rightleftharpoons H^+ + NH_3$$

Initial   0.250 M 0(10^{-7})   0
             x mol/L NH_4^+ dissociates to reach equilibrium
Change $-x$ $\rightarrow$ $+x$ $+x$
Equil.   0.250 - x x x

$$K_a = 5.6 \times 10^{-10} = \frac{[H^+][NH_3]}{[NH_4^+]} = \frac{x^2}{0.250 - x} \approx \frac{x^2}{0.250}$$

$x = [H^+] = 1.2 \times 10^{-5}$; Assumptions good. $pH = -\log(1.2 \times 10^{-5}) = 4.92$

h. $HCN(aq)$ and H_2O are the major species. HCN is the major source of H^+.

$$HCN \rightleftharpoons H^+ + CN^-$$

Initial   0.250 M 0(10^{-7})   0
             x mol/L HCN dissociates to reach equilibrium
Change $-x$ $\rightarrow$ $+x$ $+x$
Equil.   0.250 - x x x

$$K_a = 6.2 \times 10^{-10} = \frac{[H^+][CN^-]}{[HCN]} = \frac{x^2}{0.250 - x} \approx \frac{x^2}{0.250}$$

$x = [H^+] = 1.2 \times 10^{-5}$; Assumptions good. $pH = -\log(1.2 \times 10^{-5}) = 4.92$

14. If we just take the negative log of 10^{-12} we get $pH = 12$. This can't be. We can't add a tiny amount of acid to a neutral solution and get a strongly basic ($pH = 12$) solution. There are two sources of H^+: water and the HCl. So far we've considered the acid we add to water as the major source of H^+. In this case that is not true. The major source of H^+ is water ($10^{-7}\ M$) and not the HCl ($10^{-12}\ M$). So the $[H^+] \approx 10^{-7} + 10^{-12} \approx 10^{-7}\ M$ and $pH = 7.00$. This makes sense. If we add a minute amount of acid to a neutral solution we would still expect to have a roughly neutral solution.

15. 50.0 mL con. HCl soln $\times \dfrac{1.19\ g}{mL} \times \dfrac{38\ g\ HCl}{100\ g\ con.\ HCl\ soln} \times \dfrac{1\ mol\ HCl}{36.5\ g} = 0.62$ mol HCl

20.0 mL con. HNO_3 soln $\times \dfrac{1.42\ g}{mL} \times \dfrac{70.\ g\ HNO_3}{100\ g\ soln} \times \dfrac{1\ mol\ HNO_3}{63.0\ g\ HNO_3} = 0.32$ mol HNO_3

$HCl + H_2O \rightarrow H_3O^+ + Cl^-$ and $HNO_3 + H_2O \rightarrow H_3O^+ + NO_3^-$ (Both are strong acids.)

So we will have $0.62 + 0.32 = 0.94$ mol of H_3O^+ in the final solution.

$$[H^+] = \frac{0.94 \text{ mol}}{1.00 \text{ L}} = 0.94 \; M; \quad pH = -\log 0.94 = 0.027$$

$$[OH^-] = \frac{K_w}{[H^+]} = \frac{1.0 \times 10^{-14}}{0.94} = 1.1 \times 10^{-14} \; M$$

16.   In all four parts of this problem, the major species are the weak acid and water. We will consider the weak acid to be the major source of H^+, i.e., ignore H^+ contribution from H_2O.

a. $HC_2H_3O_2 \; + \; H_2O \;\; \rightleftharpoons \;\; H_3O^+ \; + \; C_3H_3O_2^-$

Initial	0.20 M		$0(10^{-7})$	0
	x mol/L $HC_2H_3O_2$ dissociates to reach equilibrium			
Change	$-x$	$\rightarrow$	$+x$	$+x$
Equl.	0.20 - x		x	x

$$K_a = 1.8 \times 10^{-5} = \frac{[H^+][C_2H_3O_2^-]}{[HC_2H_3O_2]} = \frac{x^2}{0.20 - x} \approx \frac{x^2}{0.20}$$

$$x = [H^+] = 1.9 \times 10^{-3} \; M$$

We have made two assumptions which we must check.

1. $0.20 - x \approx 0.20$

     $0.20 - x = 0.20 - 0.002 = 0.198 = 0.20$; Good assumption (1% error). If the percent error in the assumption is < 5%, assumption is valid.

2.   Acetic acid is the major source of H^+, i.e., we can ignore $10^{-7} \; M$ H^+ already present in neutral H_2O.

     $[H^+]$ from $HC_2H_3O_2 = 2 \times 10^{-3} \gg 10^{-7}$;   This assumption is valid.

     In future problems we will always begin the problem solving process by making these assumptions and we will always check them. However, we may not explicitly state that the assumptions are valid. We will <u>always</u> state when the assumptions are <u>not</u> valid and we have to use other techniques to solve the problem. Remember, anytime we make an assumption, we must check its validity before the solution to the problem is complete. Answering the question:

     $$[H^+] = [C_2H_3O_2^-] = 1.9 \times 10^{-3} \; M; \quad [OH^-] = 5.3 \times 10^{-12} \; M$$

$[HC_2H_3O_2] = 0.20 - x = 0.198 \approx 0.20\ M$

$pH = -\log (1.9 \times 10^{-3}) = 2.72$

b. $\qquad\qquad$ HNO$_2$ $\rightleftharpoons$ $\qquad$ H$^+$ $\quad$ + $\quad$ NO$_2^-$ $\ $ K$_a$ = 4.0 × 10^{-4}

Initial $\qquad$ 1.5 M $\qquad\qquad$ 0(10^{-7}) $\qquad\qquad$ 0
$\qquad\qquad\quad$ x mol/L HNO$_2$ dissociates to reach equilibrium
Change $\qquad$ -x $\qquad \rightarrow \qquad$ +x $\qquad\quad$ +x
Equl. $\qquad$ 1.5 - x $\qquad\qquad\quad$ x $\qquad\qquad$ x

$$K_a = 4.0 \times 10^{-4} = \frac{[H^+][NO_2^-]}{[HNO_2]} = \frac{x^2}{1.5 - x} \approx \frac{x^2}{1.5}\ \ (\text{assuming } x \ll 1.5)$$

$x = [H^+] = 2.4 \times 10^{-2}\ M$; Assumptions good: $10^{-7} \ll 2.4 \times 10^{-2} \ll 1.5$

$[H^+] = [NO_2^-] = 2.4 \times 10^{-2}\ M$; $[OH^-] = 4.2 \times 10^{-13}\ M$

$[HNO_2] = 1.5 - x = 1.48 \approx 1.5\ M$; pH = 1.62

c. $\qquad\qquad\qquad$ HF $\quad \rightleftharpoons \qquad$ H$^+$ $\ $ + $\ $ F$^-$ $\qquad$ K$_a$ = 7.2 × 10^{-4}

Initial $\qquad$ 0.020 M $\qquad\qquad$ 0(10^{-7}) $\qquad\quad$ 0
$\qquad\qquad\quad$ x mol/L HF dissociates to reach equilibrium
Change $\qquad$ -x $\qquad \rightarrow \qquad$ +x $\qquad$ +x
Equl. $\qquad$ 0.020 - x $\qquad\qquad$ x $\qquad\quad$ x

$$K_a = 7.2 \times 10^{-4} = \frac{[H^+][F^-]}{[HF]} = \frac{x^2}{0.020 - x} \approx \frac{x^2}{0.020}\ \ (\text{assuming } x \ll 0.020)$$

$x = [H^+] = 3.8 \times 10^{-3}$; Check assumptions.

$0.020 - x = 0.020 - 0.004 = 0.016$, x is 20% of 0.020.

The assumption $x \ll 0.020$ is not good (x is more than 5% of 0.020). We must continue. We can solve the equation by using the quadratic formula or by the method of successive approximations (see Appendix 1.4 of text). Using the successive approximations, we let 0.016 M be a new approximation for [HF]. That is, try $x = 0.004$, so 0.020 - 0.004 = 0.016, then solve for a new value of x.

$$\frac{x^2}{0.020 - x} \approx \frac{x^2}{0.016} = 7.2 \times 10^{-4}, \ \ x = 3.4 \times 10^{-3}$$

We use this new value of x to further refine our estimate of [HF], i.e., 0.020 - x = 0.020 - 0.0034 = 0.0166 (carry extra significant figure).

$$\frac{x^2}{0.020 - x} \approx \frac{x^2}{0.0166} = 7.2 \times 10^{-4}, \ x = 3.5 \times 10^{-3}$$

We repeat, until we get a self-consistent answer. In this case it will be: $x = 3.5 \times 10^{-3}$

So: $[H^+] = [F^-] = x = 3.5 \times 10^{-3} \ M; \ [OH^-] = 2.9 \times 10^{-12} \ M$

$[HF] = 0.020 - x = 0.020 - 0.0035 = 0.017 \ M; \ pH = 2.46$

d.

OH   OH
| |
$CH_3CHCO_2H \ \rightleftharpoons \ CH_3CHCO_2^- \ + \ H^+$

HLac   Lac$^-$

$$HLac \ \rightleftharpoons \ Lac^- \ + \ H^+ \quad K_a = 1.38 \times 10^{-4}$$

Initial	0.10 M	0	0(10^{-7})

x mol/L HLac dissociates to reach equilibrium

Change	-x $\rightarrow$	+x	+x
Equil.	0.10 - x	x	x

$$K_a = 1.38 \times 10^{-4} = \frac{[H^+][Lac^-]}{[HLac]} = \frac{x^2}{0.10 - x} \approx \frac{x^2}{0.10}$$

$x = [H^+] = 3.7 \times 10^{-3};$ Assumptions good (3.7% error): $10^{-7} << 3.7 \times 10^{-3} << 0.10$

$[H^+] = [Lac^-] = 3.7 \times 10^{-3} \ M; \ [OH^-] = 2.7 \times 10^{-12} \ M$

$[HLac] = 0.10 - 3.7 \times 10^{-3} = 0.10 \ M; \ pH = 2.43$

17.
$$HCO_2H \ + \ H_2O \ \rightleftharpoons \ H_3O^+ \ + \ HCO_2^- \qquad K_a = 1.8 \times 10^{-4}$$

Initial	0.025 M	0(10^{-7})	0

x mol/L HCO$_2$H dissociates to reach equilibrium

Change	-x $\rightarrow$	+x	+x
Equil.	0.025 - x	x	x

$$K_a = 1.8 \times 10^{-4} = \frac{[H^+][HCO_2^-]}{[HCO_2H]} = \frac{x^2}{0.025 - x} \approx \frac{x^2}{0.025}$$

$x = [H^+] = 2.1 \times 10^{-3};$ Check assumptions: $0.025 - x = 0.025 - 0.002 = 0.023$

Assumption $x << 0.025$ is not good (~8% error). Solving using method of successive approximations:

$$\frac{x^2}{0.025 - x} \approx \frac{x^2}{0.023} = 1.8 \times 10^{-4}, \ x = 2.0 \times 10^{-3} \text{ which we get consistently.}$$

$$x = [H^+] = 2.0 \times 10^{-3} \ M; \ \text{pH} = 2.70$$

18.   Major species: HIO_3, H_2O; Major source of H^+: HIO_3

$$HIO_3 \ \rightleftharpoons \ H^+ \ + \ IO_3^-$$

Initial	0.20 M	$0(10^{-7})$	0

x mol/L HIO_3 dissociates to reach equilibrium

Change	$-x$ $\rightarrow$	$+x$	$+x$
Equil.	0.20 - x	x	x

$$K_a = 0.17 = \frac{[H^+][IO_3^-]}{[HIO_3]} = \frac{x^2}{0.20 - x} \approx \frac{x^2}{0.20}, \ x = 0.18; \quad \text{Check assumption.}$$

Assumption is horrible (90% error). When the assumption is this poor, it is generally quickest to solve exactly using the quadratic formula (see Appendix 1.4 in text). The method of successive approximations will require many trials to finally converge on the answer. For this problem, 5 trials were required. Using the quadratic formula:

$$0.17 = \frac{x^2}{0.20 - x}, \ x^2 = 0.17 \ (0.20 - x), \ x^2 + 0.17 \ x - 0.034 = 0$$

$$x = \frac{-0.17 \pm [(0.17)^2 - 4(1)(-0.034)]^{1/2}}{2(1)} = \frac{-0.17 \pm 0.406}{2}, \ x = 0.12 \text{ or } -0.29$$

Only $x = 0.12$ makes sense. $x = 0.12 \ M = [H^+]$; pH $= -\log (0.12) = 0.92$

19. $B(OH)_3 \ + \ H_2O \ \rightleftharpoons \ B(OH)_4^- \ + \ H^+$

Initial	0.50 M	0	$0(10^{-7})$

x mol/L $B(OH)_3$ reacts to reach equilibrium

Change	$-x$ $\rightarrow$	$+x$	$+x$
Equil.	0.50 - x	x	x

$$K_a = 5.8 \times 10^{-10} = \frac{[H^+][B(OH)_4^-]}{[B(OH)_3]} = \frac{x^2}{0.50 - x} \approx \frac{x^2}{0.50}$$

$$x = [H^+] = 1.7 \times 10^{-5} \ M; \ \text{pH} = 4.77 \quad \text{Assumptions good.}$$

20. $0.56 \text{ g } C_6H_5CO_2H(HBz) \times \dfrac{1 \text{ mol HBz}}{122.1 \text{ g}} = 4.6 \times 10^{-3} \text{ mol}; \quad [HBz]_o = 4.6 \times 10^{-3} \text{ } M$

$$HBz \quad \rightleftharpoons \quad H^+ \quad + \quad Bz^-$$

Initial $4.6 \times 10^{-3} \text{ } M$ $0(10^{-7})$ 0
    x mol/L HBz dissociates to reach equilibrium
Change $-x \quad \rightarrow \quad +x \qquad +x$
Equil. $4.6 \times 10^{-3} - x \qquad\quad x \qquad\quad x$

$K_a = 6.4 \times 10^{-5} = \dfrac{[H^+][Bz^-]}{[HBz]} = \dfrac{x^2}{4.6 \times 10^{-3} - x} \approx \dfrac{x^2}{4.6 \times 10^{-3}}$

$x = [H^+] = 5.4 \times 10^{-4};$ Check assumptions.

$0.0046 - x = 0.0046 - 0.0005 = 0.0041$

Assumption is not good (0.0005 is greater than 5% of 0.0046). When assumption(s) fail, we must solve exactly using the quadratic formula or the method of successive approximations (see Appendix 1.4 of text). Using successive approximations:

$$\dfrac{x^2}{0.0041} = 6.4 \times 10^{-5}, \quad x = 5.1 \times 10^{-4}$$

$$\dfrac{x^2}{(0.0046 - 0.00051)} = \dfrac{x^2}{0.0041} = 6.4 \times 10^{-5}, \quad x = 5.1 \times 10^{-4}$$

So: $x = [H^+] = [Bz^-] = 5.1 \times 10^{-4} \text{ } M; \quad [HBz] = 4.6 \times 10^{-3} - x = 4.1 \times 10^{-3} \text{ } M$

$pH = 3.29; \quad pOH = 10.71; \quad [OH^-] = 10^{-10.71} = 1.9 \times 10^{-11} \text{ } M$

21. $HBz \quad \rightleftharpoons \quad H^+ \quad + \quad Bz^-$

Initial C $0(10^{-7})$ 0 $C = [HBz]_o$
    x mol/L HBz dissociates to reach equilibrium
Change $-x \quad \rightarrow \quad +x \qquad +x$
Equil. $C - x \qquad\qquad x \qquad\quad x$

$K_a = \dfrac{[H^+][Bz^-]}{[HBz]} = 6.4 \times 10^{-5} = \dfrac{x^2}{C - x}, \text{ where } x = [H^+]$

$6.4 \times 10^{-5} = \dfrac{[H^+]^2}{C - [H^+]}; \quad pH = 2.80; \quad [H^+] = 10^{-2.80} = 1.6 \times 10^{-3}$

$C - 1.6 \times 10^{-3} = \dfrac{(1.6 \times 10^{-3})^2}{6.4 \times 10^{-5}} = 4.0 \times 10^{-2}, \quad C = 4.0 \times 10^{-2} + 1.6 \times 10^{-3} = 4.2 \times 10^{-2} \text{ } M$

The molar solubility is 4.2×10^{-2} mol/L.

$$\frac{4.2 \times 10^{-2} \text{ mol}}{L} \times \frac{122.1 \text{ g}}{\text{mol}} \times 0.100 \text{ L} = 0.51 \text{ g per } 100. \text{ mL}$$

22.   Let HX symbolize the weak acid.

$$\text{HX} \rightleftharpoons \text{H}^+ + \text{X}^-$$

Initial   0.15 $0(10^{-7})$ 0
               x mol/L HX dissociates to reach equilibrium
Change $-x$ $\rightarrow$ $+x$ $+x$
Equil. $0.15 - x$ x x

If the acid is 3.0% dissociated, then x is 3.0% of 0.15. $x = 0.030 (0.15) = 4.5 \times 10^{-3}$

$$K_a = \frac{[\text{H}^+][\text{X}^-]}{[\text{HX}]} = \frac{(4.5 \times 10^{-3})^2}{0.15 - 4.5 \times 10^{-3}} = 1.4 \times 10^{-4}$$

23.   20.0 mL glacial acetic acid $\times \dfrac{1.05 \text{ g}}{\text{mL}} \times \dfrac{1 \text{ mol}}{60.05 \text{ g}} = 0.350 \text{ mol}$

Initial concentration of $HC_2H_3O_2 = \dfrac{0.350 \text{ mol}}{0.2500 \text{ L}} = 1.40 \ M$

A common abbreviation used by chemists for acetic acid, $HC_2H_3O_2$, is HOAc. The abbreviation for $C_2H_3O_2^-$ is OAc$^-$.

$$\text{HOAc} \rightleftharpoons \text{H}^+ + \text{OAc}^- \quad K_a = 1.8 \times 10^{-5}$$

Initial   1.40 M $0(10^{-7})$ 0
               x mol/L $HC_2H_3O_2$ dissociates to reach equilibrium
Change $-x$ $\rightarrow$ $+x$ $+x$
Equil. $1.40 - x$ x x

$$K_a = 1.8 \times 10^{-5} = \frac{[\text{H}^+][\text{OAc}^-]}{[\text{HOAc}]} = \frac{x^2}{1.40 - x} \approx \frac{x^2}{1.40}$$

$x = [\text{H}^+] = 5.0 \times 10^{-3} \ M;\ \ \text{pH} = 2.30$ Assumptions good.

24.   a. HCl is a strong acid. It will produce 0.10 M H$^+$. HOCl is a weak acid. Let's consider the
           equilibrium:

$$\text{HOCl} \quad \rightleftharpoons \quad \text{H}^+ \quad + \quad \text{OCl}^-$$

Initial   0.10 M 0.10 M 0

x mol/L HOCl dissociates to reach equilibrium

Change $-x$ $\rightarrow$ $+x$ $+x$

Equil.   0.10 - x 0.10 + x x

$$K_a = 3.5 \times 10^{-8} = \frac{[\text{H}^+][\text{OCl}^-]}{[\text{HOCl}]} = \frac{(0.10 + x)(x)}{0.10 - x} \approx x, \; x = 3.5 \times 10^{-8}$$

Assumptions are good. We are really assuming that HCl is the only important source of H$^+$, which it is. The [H$^+$] contribution from the HOCl is negligible. Therefore,

[H$^+$] = 0.10 M; pH = 1.00

b. HNO$_3$ is a strong acid, giving an initial concentration of H$^+$ equal to 0.050 M. Consider the equilibrium:

$$\text{HC}_2\text{H}_3\text{O}_2 \quad \rightleftharpoons \quad \text{H}^+ \quad + \quad \text{C}_2\text{H}_3\text{O}_2^- \qquad K_a = 1.8 \times 10^{-5}$$

Initial   0.50 M 0.050 M 0

x mol/L HC$_2$H$_3$O$_2$ dissociates to reach equilibrium

Change $-x$ $\rightarrow$ $+x$ $+x$

Equil.   0.50 - x 0.050 + x x

$$K_a = 1.8 \times 10^{-5} = \frac{[\text{H}^+][\text{OAc}^-]}{[\text{HOAc}]} = \frac{(0.050 + x)x}{(0.50 - x)} \approx \frac{0.050\,x}{0.50}$$

$x = 1.8 \times 10^{-4}$; 0.050 + x = 0.050; Assumptions are good (well within the 5% rule).

[H$^+$] = 0.050 M and pH = 1.30.

25.   Both are strong acids.

0.050 mmol/mL $\times$ 50.0 mL = 2.5 mmol HCl

0.10 mmol/mL $\times$ 150.0 mL = 15 mmol HNO$_3$

$$[\text{H}^+] = \frac{2.5 \text{ mmol} + 15 \text{ mmol}}{200.0 \text{ mL}} = 0.088 \; M; \quad [\text{OH}^-] = \frac{K_w}{[\text{H}^+]} = 1.1 \times 10^{-13} \; M$$

$$[\text{Cl}^-] = \frac{2.5 \text{ mmol}}{200.0 \text{ mL}} = 0.0125 \text{ mol/L} \approx 0.013 \; M; \quad [\text{NO}_3^-] = \frac{15 \text{ mmol}}{200.0 \text{ mL}} = 0.075 \; M$$

26.   For H_2SO_4 the first dissociation occurs to completion. Hydrogen sulfate ion, HSO_4^-, is a weak
        acid with $K_a = 1.2 \times 10^{-2}$. We will consider the equilibrium:

$$HSO_4^- \quad \rightleftharpoons \quad H^+ \quad + \quad SO_4^{2-}$$

Initial	0.0010 M	0.0010 M	0

x mol/L HSO_4^- dissociates to reach equilibrium

Change	$-x$ $\rightarrow$	$+x$	$+x$
Equil.	0.0010 - x	0.0010 + x	x

$$K_a = 0.012 = \frac{(0.0010 + x)\,(x)}{(0.0010 - x)} \approx x, \ x = 0.012; \quad \text{Assumption is horrible (200\% error).}$$

Using the quadratic formula:

$$1.2 \times 10^{-5} - 0.012\,x = x^2 + 0.0010\,x, \ x^2 + 0.013\,x - 1.2 \times 10^{-5} = 0$$

$$x = \frac{-0.013 \pm (1.69 \times 10^{-4} + 4.80 \times 10^{-5})^{1/2}}{2} = \frac{-0.013 \pm 0.0147}{2}, \ x = 8.5 \times 10^{-4}$$

$$[H^+] = 0.0010 + x = 0.0010 + 0.00085 = 0.0019 \ M; \ pH = 2.72$$

27.   The conductivity of the solution is a measure of the number of ions. In addition the colligative
        properties, which will be discussed in Chapter 17, depend on the number of particles. So
        measurements of osmotic pressure, vapor pressure lowering, freezing point depression or boiling
        point elevation will also allow us to determine the extent of ionization.

28. $HF \rightleftharpoons H^+ + F^-$; From normal weak acid problem: $[H^+] = [F^-]$ and $[HF] = 0.100 - [H^+]$

$$[H^+] = [F^-] = 0.081[HF]_o = (0.081)(0.100) = 8.1 \times 10^{-3} \ M$$

$$[HF] = 0.100 - 8.1 \times 10^{-3} = 0.092 \ M; \ K_a = \frac{[H^+]\,[F^-]}{[HF]} = \frac{(8.1 \times 10^{-3})^2}{(9.2 \times 10^{-2})} = 7.1 \times 10^{-4}$$

29.   a.
$$HOAc \quad \rightleftharpoons \quad H^+ \quad + \quad OAc^-$$

Initial	0.50 M	0(10^{-7})	0

x mol/L HOAc ($HC_2H_3O_2$) dissociates to reach equilibrium

Change	$-x$ $\rightarrow$	$+x$	$+x$
Equil.	0.50 - x	x	x

$$K_a = 1.8 \times 10^{-5} = \frac{[H^+]\,[OAc^-]}{[HOAc]} = \frac{x^2}{(0.50 - x)} \approx \frac{x^2}{0.50}$$

$$x = [H^+] = [OAc^-] = 3.0 \times 10^{-3} \ M \qquad \text{Assumptions good.}$$

$$\% \ \text{ionized} = \frac{[OAc^-]}{0.50 \ M} \times 100 = \frac{3.0 \times 10^{-3}}{0.50} \times 100 = 0.60\%$$

b. The set-up for (b) and (c) is similar to (a) except the final equation is slightly different.

$$K_a = 1.8 \times 10^{-5} = \frac{x^2}{0.050 - x} \approx \frac{x^2}{0.050}$$

$x = [H^+] = [OAc^-] = 9.5 \times 10^{-4}\ M$ Assumptions good.

$$\% \text{ ionized} = \frac{9.5 \times 10^{-4}}{0.050} \times 100 = 1.9\%$$

c. $K_a = 1.8 \times 10^{-5} = \dfrac{x^2}{0.0050 - x} \approx \dfrac{x^2}{0.0050}$

$x = [H^+] = [OAc^-] = 3.0 \times 10^{-4}\ M;$ Check assumptions.

Assumption that x is negligible is borderline (6% error). We should solve exactly. Using the method of successive approximations (Appendix 1.4 of text):

$$1.8 \times 10^{-5} = \frac{x^2}{0.0047}, \quad x = 2.9 \times 10^{-4};\quad \text{Next trial gives } x = 2.9 \times 10^{-4}.$$

$$\% \text{ ionized} = \frac{2.9 \times 10^{-4}}{5.0 \times 10^{-3}} \times 100 = 5.8\%$$

d. As we dilute a solution, all concentrations are decreased by the same function. Dilution will shift the equilibrium to the side with the greater number of particles. For example, suppose we double the volume of an equilibrium mixture of a weak acid by adding water, then:

$$Q = \frac{\left(\dfrac{[H^+]_{eq}}{2}\right)\left(\dfrac{[X^-]_{eq}}{2}\right)}{\left(\dfrac{[HX]_{eq}}{2}\right)} = \frac{1}{2} K_a$$

$Q < K_a$, equilibrium shifts to the right or towards a greater percent dissociation.

e. $[H^+]$ depends on the initial concentration of weak acid and on how much weak acid dissociates. For solutions a-c the initial concentration of acid decreases more rapidly than the percent dissociation increases. Thus, $[H^+]$ decreases.

30. $HClO_2 \quad \rightleftharpoons \quad H^+ \quad + \quad ClO_2^- \quad K_a = 1.2 \times 10^{-2}$

Initial	0.22 M	0(10^{-7})	0
	x mol/L $HClO_2$ dissociates to reach equilibrium		
Change	$-x$ $\rightarrow$	$+x$	$+x$
Equil.	0.22 - x	x	x

$$K_a = 1.2 \times 10^{-2} = \frac{[H^+][ClO^-]}{[HClO_2]} = \frac{x^2}{0.22 - x} \approx \frac{x^2}{0.22}, \quad x = 5.1 \times 10^{-2}$$

The assumption that x is small is not good (23% error). Using the method of successive approximations:

$$\frac{x^2}{0.169} = 1.2 \times 10^{-2}, \quad x = 4.5 \times 10^{-2}$$

$$\frac{x^2}{0.175} = 1.2 \times 10^{-2}, \quad x = 4.6 \times 10^{-2} \quad \text{We can stop here } (4.6 \times 10^{-2} \text{ will repeat}).$$

$$[H^+] = [ClO_2^-] = x = 4.6 \times 10^{-2} \; M; \quad \% \text{ ionized} = \frac{4.6 \times 10^{-2}}{0.22} \times 100 = 21\%$$

31. $HOBr \rightleftharpoons H^+ + OBr^-$

From normal weak acid problem: $[H^+] = [OBr^-]$ and $[HOBr] = 0.063 - [H^+]$

$pH = 4.95, \; [H^+] = 10^{-4.95} = 1.1 \times 10^{-5} \; M$

$$K_a = \frac{[H^+][OBr^-]}{[HOBr]} = \frac{(1.1 \times 10^{-5})^2}{(0.063 - 1.1 \times 10^{-5})} = 1.9 \times 10^{-9}$$

32. $CCl_3CO_2H \rightleftharpoons H^+ + CCl_3CO_2^-; \quad [H^+] = [CCl_3CO_2^-]$ and $[CCl_3CO_2H] = 0.050 - [H^+]$

$pH = 1.40, \; [H^+] = 10^{-1.40} = 4.0 \times 10^{-2} \; M$

$$K_a = \frac{[H^+][CCl_3CO_2^-]}{[CCl_3CO_2H]} = \frac{(4.0 \times 10^{-2})^2}{0.050 - 0.040} = 0.16$$

33.   The reactions are:

$$H_3PO_4 \rightleftharpoons H^+ + H_2PO_4^- \quad K_{a_1} = 7.5 \times 10^{-3}$$

$$H_2PO_4^- \rightleftharpoons H^+ + HPO_4^{2-} \quad K_{a_2} = 6.2 \times 10^{-8}$$

$$HPO_4^{2-} \rightleftharpoons H^+ + PO_4^{3-} \quad K_{a_3} = 4.8 \times 10^{-13}$$

We will deal with the reactions in order of importance, beginning with the largest K_a.

$$H_3PO_4 \quad \rightleftharpoons \quad H^+ \; + \; H_2PO_4^- \quad K_{a_1} = 7.5 \times 10^{-3} = \frac{[H^+][H_2PO_4^-]}{[H_3PO_4]}$$

Initial	0.100 M	0(10^{-7})	0
Equil.	0.100 - x	x	x

$$7.5 \times 10^{-3} = \frac{x^2}{0.100 - x}, \; x = 2.4 \times 10^{-2} \; M \qquad \text{(By using successive approximation or the quadratic formula.)}$$

$[H^+] = [H_2PO_4^-] = 2.4 \times 10^{-2} \; M; \; [H_3PO_4] = 0.100 - 0.024 = 0.076 \; M$

Since $K_{a_2} = \dfrac{[H^+][HPO_4^{2-}]}{[H_2PO_4^-]} = 6.2 \times 10^{-8}$ is much smaller than the K_{a_1} value, very little of the second (and third) reactions occur as compared to the first reaction. Therefore, $[H^+]$ and $[H_2PO_4^-]$ will not change significantly by the second reaction. Using the above concentrations of H^+ and $H_2PO_4^-$ to calculate the concentration of HPO_4^{2-}:

$$6.2 \times 10^{-8} = \frac{(2.4 \times 10^{-2})[HPO_4^{2-}]}{2.4 \times 10^{-2}}, \; [HPO_4^{2-}] = 6.2 \times 10^{-8} \; M$$

Assumption that second reaction does not change $[H^+]$ and $[H_2PO_4^-]$ is good. We repeat the process using K_{a_3} to get $[PO_4^{3-}]$.

$$K_{a_3} = 4.8 \times 10^{-13} = \frac{[H^+][PO_4^{3-}]}{[HPO_4^{2-}]} = \frac{(2.4 \times 10^{-2})[PO_4^{3-}]}{(6.2 \times 10^{-8})}$$

$[PO_4^{3-}] = 1.2 \times 10^{-18} \; M$ Assumptions good.

So in $0.100 \; M$ analytical concentration of H_3PO_4:

$\quad [H_3PO_4] = 7.6 \times 10^{-2} \; M; \; [H^+] = [H_2PO_4^-] = 2.4 \times 10^{-2} \; M$

$\quad [HPO_4^{2-}] = 6.2 \times 10^{-8} \; M; \; [PO_4^{3-}] = 1.2 \times 10^{-18} \; M$

$\quad [OH^-] = 1.0 \times 10^{-14}/[H^+] = 4.2 \times 10^{-13} \; M$

34.   The relevant reactions are:

$\quad H_2CO_3 \rightleftharpoons H^+ + HCO_3^- \qquad K_{a_1} = 4.3 \times 10^{-7}$

$\quad HCO_3^- \rightleftharpoons H^+ + CO_3^{2-} \qquad K_{a_2} = 5.6 \times 10^{-11}$

Initially, we deal only with the first reaction (since $K_{a_1} \gg K_{a_2}$) and then let those results control values of concentrations in the second reaction.

$$H_2CO_3 \rightleftharpoons H^+ + HCO_3^-$$

Initial $0.010\ M$ $0(10^{-7})$ 0
Equil. $0.010 - x$ x x

$$K_{a_1} = 4.3 \times 10^{-7} = \frac{[H^+][HCO_3^-]}{[H_2CO_3]} = \frac{x^2}{0.010 - x} \approx \frac{x^2}{0.010}$$

$x = 6.6 \times 10^{-5}\ M = [H^+] = [HCO_3^-]$ Assumptions good.

$$HCO_3^- \rightleftharpoons H^+ + CO_3^{2-}$$

Initial $6.6 \times 10^{-5}\ M$ $6.6 \times 10^{-5}\ M$ 0
Equil. $6.6 \times 10^{-5} - y$ $6.6 \times 10^{-5} + y$ y

If y is small, then $[H^+] = [HCO_3^-]$ and $K_{a_2} = 5.6 \times 10^{-11} = \dfrac{[H^+][CO_3^{2-}]}{[HCO_3^-]} \approx y$

$y = [CO_3^{2-}] = 5.6 \times 10^{-11}\ M$ Assumptions good.

The amount of H^+ from the second dissociation is $5.6 \times 10^{-11}\ M$ or

$$\frac{5.6 \times 10^{-11}}{6.6 \times 10^{-5}} \times 100 = 8.5 \times 10^{-5}\ \%$$

This result justifies our treating the equilibria separately. If the second dissociation contributed a significant amount of H^+ we would have to treat both equilibria simultaneously. The reaction that occurs when acid is added to a solution of HCO_3^- is

$$HCO_3^- + H^+ \rightarrow (H_2CO_3) \rightarrow H_2O + CO_2$$

The bubbles are $CO_2(g)$ and are formed by the breakdown of unstable H_2CO_3 molecules. We should write $H_2O(l) + CO_2(aq)$ or $CO_2(aq)$ for what we call carbonic acid. It is for convenience, however, that we write $H_2CO_3(aq)$.

Solutions of Bases

35. NO_3^-: $K_b \approx 0$ since HNO_3 is a strong acid. H_2O: $K_b = K_w = 10^{-14}$

NH_3: $K_b = 1.8 \times 10^{-5}$; CH_3NH_2: $K_b = 4.38 \times 10^{-4}$

$CH_3NH_2 > NH_3 > H_2O > NO_3^-$ (As K_b increases, base strength increases.)

36.   Excluding water these are the conjugate acids of the bases in the previous exercise: the stronger the base, the weaker the conjugate acid.

$HNO_3 > NH_4^+ > CH_3NH_3^+ > H_2O$

37.   a. NH_3 b. NH_3 c. OH^- d. CH_3NH_2

The base with the largest K_b value is the strongest base.

38.   a. HNO_3 b. NH_4^+ c. NH_4^+

39. $2.48 \text{ g TlOH} \times \dfrac{1 \text{ mol TlOH}}{221.4 \text{ g}} = 1.12 \times 10^{-2} \text{ mol}$

TlOH is a strong base, so $[OH^-] = \dfrac{1.12 \times 10^{-2} \text{ mol}}{L}$

$pOH = -\log[OH^-] = 1.951$; $pH = 14.000 - pOH = 12.049$

40.   a. $NaOH(aq) \rightarrow Na^+(aq) + OH^-(aq)$; NaOH is a strong base and dissociates completely.

$[OH^-] = 0.25 \ M$; $pOH = -\log[OH^-] = 0.60$; $pH = 14.00 - pOH = 13.40$

b. $Ba(OH)_2 \rightarrow Ba^{2+} + 2 \ OH^-$; $Ba(OH)_2$ is a strong base.

$[OH^-] = 2(0.00040) = 8.0 \times 10^{-4} \ M$; $pOH = 3.10$; $pH = 10.90$

c. $\dfrac{25 \text{ g KOH}}{L} \times \dfrac{1 \text{ mol}}{56.1 \text{ g}} = 0.45 \text{ mol/L}$

KOH is a strong base, so $[OH^-] = 0.45 \ M$; $pOH = 0.35$; $pH = 13.65$

d. $\dfrac{150.0 \text{ g NaOH}}{L} \times \dfrac{1 \text{ mol}}{40.00 \text{ g}} = 3.750 \ M$; NaOH is a strong base, so $[OH^-] = 3.750 \ M$.

$pOH = -0.5740$ and $pH = 14.5740$

Although we are justified in calculating the answer to four significant figures, in reality the pH can only be measured to ± 0.01 pH units.

41.   Nitrogen

42.   a.

$$C_2H_5NH_2 \ + \ H_2O \ \rightleftharpoons \ C_2H_5NH_3^+ \ + \ OH^- \qquad K_b = 5.6 \times 10^{-4}$$

Initial	0.200 M		0	0(10⁻⁷)

x mol/L $C_2H_5NH_2$ reacts with H_2O to reach equilibrium

Change	$-x$	$\rightarrow$	$+x$	$+x$
Equil.	$0.200 - x$		x	x

$$K_b = \frac{[C_2H_5NH_3^+][OH^-]}{[C_2H_5NH_2]} = \frac{x^2}{0.200-x} \approx \frac{x^2}{0.200}$$

$x = 1.1 \times 10^{-2}$; $0.200 - 0.011 = 0.189$ Assumption borderline (~5% error).

Using successive approximations:

$$\frac{x^2}{0.189} = 5.6 \times 10^{-4}, \ x = 1.0 \times 10^{-2} \ M \qquad \text{(consistent answer)}$$

$$x = [OH^-] = 1.0 \times 10^{-2} \ M; \ [H^+] = \frac{K_w}{[OH^-]} = \frac{1.0 \times 10^{-14}}{1.0 \times 10^{-2}} = 1.0 \times 10^{-12} \ M; \ pH = 12.00$$

Note: Assumption that OH⁻ contribution from water is negligible is good. This assumption will hold for basic solutions when pH > 8.0.

b. $Et_2NH + H_2O \rightleftharpoons Et_2NH_2^+ + OH^-$ Et = -C$_2$H$_5$ $K_b = 1.3 \times 10^{-3}$

Initial   0.200 M 0   0(10⁻⁷)
                  x mol/L (C$_2$H$_5$)$_2$NH reacts with H$_2$O to reach equilibrium
Change $-x$ $\rightarrow$ $+x$ $+x$
Equil.   0.200 - x x x

$$K_b = 1.3 \times 10^{-3} = \frac{[Et_2NH_2^+][OH^-]}{[Et_2NH]} = \frac{x^2}{0.200-x} \approx \frac{x^2}{0.200}$$

$x = 1.6 \times 10^{-2}$ Assumption bad (x is 8% of 0.20).

Using successive approximations:

$$\frac{x^2}{0.184} = 1.3 \times 10^{-3}, \ x = 1.55 \times 10^{-2} \ \text{(carry extra significant figure)}$$

$$\frac{x^2}{0.185} = 1.3 \times 10^{-3}, \ x = 1.55 \times 10^{-2}$$

$[OH^-] = x = 1.55 \times 10^{-2} \ M$; $[H^+] = 6.45 \times 10^{-13} \ M$; To correct significant figures:

$[OH^-] = 1.6 \times 10^{-2} \ M$; $[H^+] = 6.5 \times 10^{-13} \ M$; pH = 12.19

c. $Et_3NH + H_2O \rightleftharpoons Et_3NH^+ + OH^-$ $K_b = 4.0 \times 10^{-4}$

Equil.   0.200 - x x x

$$4.0 \times 10^{-4} = \frac{x^2}{0.200-x} \approx \frac{x^2}{0.200}, \ x = 8.9 \times 10^{-3}; \ \text{Assumptions good.}$$

$[OH^-] = 8.9 \times 10^{-3} \ M$; $[H^+] = 1.1 \times 10^{-12} \ M$; pH = 11.96

d. $C_6H_5NH_2$ + H_2O $\rightleftharpoons$ $C_6H_5NH_3^+$ + OH^- $K_b = 3.8 \times 10^{-10}$

Equil. 0.200 - x x x

$3.8 \times 10^{-10} = \dfrac{x^2}{0.200 - x} \approx \dfrac{x^2}{0.200}$, $x = [OH^-] = 8.7 \times 10^{-6}\ M$; Assumptions good.

$[H^+] = 1.1 \times 10^{-9}\ M$; pH = 8.96

e. C_5H_5N + H_2O $\rightleftharpoons$ $C_5H_5NH^+$ + OH^- $K_b = 1.7 \times 10^{-9}$

Equil. 0.200 - x x x

$K_b = 1.7 \times 10^{-9} = \dfrac{x^2}{0.200 - x} \approx \dfrac{x^2}{0.200}$, $x = 1.8 \times 10^{-5}\ M$; Assumptions good.

$[OH^-] = 1.8 \times 10^{-5}\ M$; $[H^+] = 5.6 \times 10^{-10}\ M$; pH = 9.25

f. $HONH_2$ + H_2O $\rightleftharpoons$ $HONH_3^+$ + OH^- $K_b = 1.1 \times 10^{-8}$

Equil. 0.200 - x x x

$K_b = 1.1 \times 10^{-8} = \dfrac{x^2}{0.200 - x} \approx \dfrac{x^2}{0.200}$, $x = [OH^-] = 4.7 \times 10^{-5}\ M$ Assumptions good.

$[H^+] = 2.1 \times 10^{-10}\ M$; pH = 9.68

43.   The N-atom is protonated in each case.

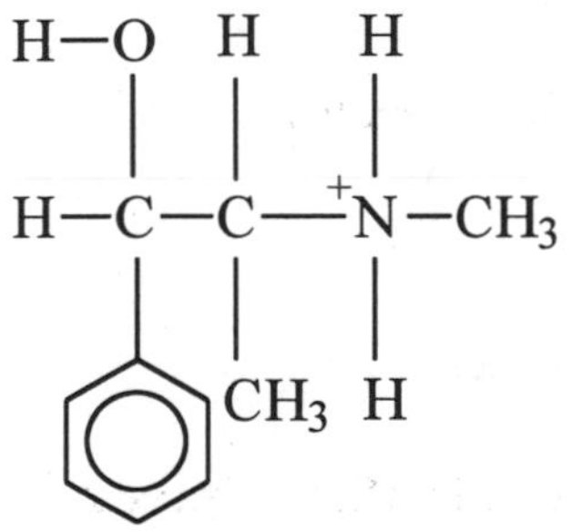

conjugate acid
of ephedrine

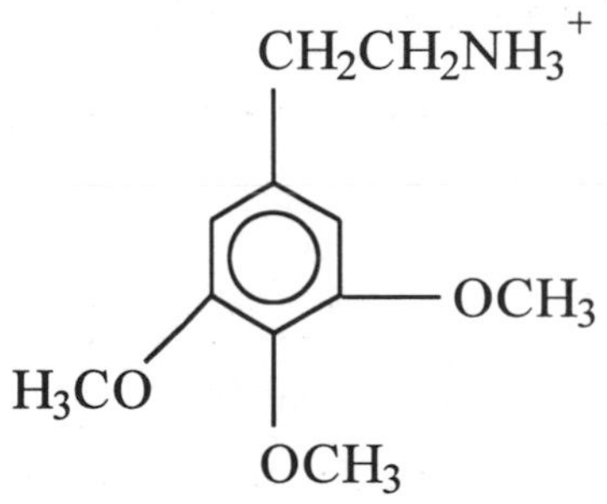

conjugate acid
of mescaline

44. $\dfrac{5.0\ mg}{10.0\ mL} \times \dfrac{1\ mmol}{299.4\ mg} = 1.7 \times 10^{-3}\ \dfrac{mmol}{mL} = 1.7 \times 10^{-3}\ M = [\text{codeine}]$

cod = codeine, $C_{18}H_{21}NO_3$

$$\text{cod} + H_2O \rightleftharpoons \text{codH}^+ + OH^- \qquad K_b = 10^{-6.05} = 8.9 \times 10^{-7}$$

Initial	$1.7 \times 10^{-3}\ M$	0	$0(10^{-7})$

x mol/L codeine reacts with H_2O to reach equilibrium

Change	$-x$ $\rightarrow$	$+x$	$+x$
Equil.	$1.7 \times 10^{-3} - x$	x	x

$$K_b = 8.9 \times 10^{-7} = \frac{x^2}{1.7 \times 10^{-3} - x} \approx \frac{x^2}{1.7 \times 10^{-3}}, \quad x = 3.9 \times 10^{-5} \qquad \text{Assumptions good.}$$

$[OH^-] = 3.9 \times 10^{-5}\ M;\ [H^+] = 2.6 \times 10^{-10}\ M;\ pH = 9.59$

45.   Codeine: $C_{18}H_{21}NO_3$;   Codeine sulfate = $C_{36}H_{44}N_2O_{10}S$

The formula for codeine sulfate works out to (codeine H^+)$_2$ SO_4^{2-} where codeine H^+ = $HC_{18}H_{21}NO_3^+$. Two codeine molecules are protonated by H_2SO_4, forming the conjugate acid of codeine. The SO_4^{2-} then acts as the counter ion to give a neutral compound. Codeine sulfate is an ionic compound which is more soluble in water than codeine, allowing more of the drug into the bloodstream.

46.   a. $NH_3 + H_2O \rightleftharpoons NH_4^+ + OH^- \qquad K_b = 1.8 \times 10^{-5}$

Initial	0.10 M	0	$0(10^{-7})$
Equil.	0.10 - x	x	x

$$1.8 \times 10^{-5} = \frac{x^2}{0.10 - x} \approx \frac{x^2}{0.10}, \quad x = [NH_4^+] = 1.3 \times 10^{-3}\ M$$

$$\% \text{ ionized} = \frac{1.3 \times 10^{-3}\ M}{0.10\ M} \times 100 = 1.3\% \qquad \text{Assumptions good.}$$

b. $NH_3 + H_2O \rightleftharpoons NH_4^+ + OH^-$

Equil.	0.010 - x	x	x

$$1.8 \times 10^{-5} = \frac{x^2}{0.010 - x} \approx \frac{x^2}{0.010}, \quad x = [NH_4^+] = 4.2 \times 10^{-4};\ \text{Assumptions good.}$$

$$\% \text{ ionized} = \frac{4.2 \times 10^{-4}}{0.010} \times 100 = 4.2\%$$

c. $CH_3NH_2 + H_2O \rightleftharpoons CH_3NH_3^+ + OH^- \qquad K_b = 4.38 \times 10^{-4}$

Equil.	0.10 - x	x	x

$$4.38 \times 10^{-4} = \frac{x^2}{0.10 - x} \approx \frac{x^2}{0.10}, \quad x = 6.6 \times 10^{-3};\ \text{Assumption fails the 5\% rule (6.6\% error).}$$

Using successive approximations:

$$\frac{x^2}{0.093} = 4.38 \times 10^{-4}, \quad x = 6.4 \times 10^{-3}$$

$$\frac{x^2}{0.094} = 4.38 \times 10^{-4}, \quad x = 6.4 \times 10^{-3} \, M = [CH_3NH_3^+]$$

$$\% \text{ ionized} = \frac{6.4 \times 10^{-3}}{0.10} \times 100 = 6.4\%$$

47. $H_2NNH_2 \; + \; H_2O \;\; \rightleftharpoons \;\; H_2NNH_3^+ \; + \; OH^- \quad K_b = 3.0 \times 10^6$

Initial	2.0 M		0	0(10^{-7})

x mol/L H_2NNH_2 reacts with H_2O to reach equilibrium

Change	$-x$	$\rightarrow$	$+x$	$+x$
Equil.	2.0 - x		x	x

$$K_b = 3.0 \times 10^{-6} = \frac{[H_2NNH_3^+][OH^-]}{[H_2NNH_2]} = \frac{x^2}{2.0 - x} \approx \frac{x^2}{2.0}$$

$x = [OH^-] = 2.4 \times 10^{-3} \, M; \quad pOH = 2.62; \quad pH = 11.38 \quad$ Assumptions good.

48. $\dfrac{1.0 \text{ g}}{1.9 \text{ L}} \times \dfrac{1 \text{ mol quinine}}{324.4 \text{ g}} = 1.6 \times 10^{-3} \, M$ in quinine

                    $Q \; + \; H_2O \;\; \rightleftharpoons \;\; QH^+ \; + \; OH^- \quad K_b = 10^{-5.1} = 8 \times 10^{-6}$

Initial	$1.6 \times 10^{-3} \, M$		0	0(10^{-7})

x mol/L quinine reacts with H_2O to reach equilibrium

Change	$-x$	$\rightarrow$	$+x$	$+x$
Equil.	$1.6 \times 10^{-3} - x$		x	x

$$K_b = 8 \times 10^{-6} = \frac{[QH^+][OH^-]}{[Q]} = \frac{x^2}{1.6 \times 10^{-3} - x} \approx \frac{x^2}{1.6 \times 10^{-3}}$$

$x = 1 \times 10^{-4};$ Assumption fails 5% rule (~7% error). Using successive approximations:

$$\frac{x^2}{1.5 \times 10^{-3}} = 8 \times 10^{-6}, \quad x = 1 \times 10^{-4}$$

$x = [OH^-] = 1 \times 10^{-4} \, M; \quad pOH = 4.0; \quad pH = 10.0$

49.   PT = p-toluidine

$PT + H_2O \rightleftharpoons PTH^+ + OH^-$ $K_b = \dfrac{[PTH^+][OH^-]}{[PT]}$

From normal setup: $[PTH^+] = [OH^-]$; $[PT] = 0.016 - [OH^-]$

$pH = 8.60$; $pOH = 5.40$; $[OH^-] = 4.0 \times 10^{-6}$

$K_b = \dfrac{(4.0 \times 10^{-6})^2}{0.016 - 4.0 \times 10^{-6}} = 1.0 \times 10^{-9}$

50.   pyr = pyrrolidine, C_4H_8NH

$pyr + H_2O \rightleftharpoons pyrH^+ + OH^-$ $K_b = \dfrac{[OH^-][pyrH^+]}{[pyr]}$

From normal setup: $[OH^-] = [pyrH^+]$; $[pyr] = 1.00 \times 10^{-3} - [OH^-]$

$pH = 10.82$; $pOH = 3.18$; $[OH^-] = 10^{-3.18} = 6.6 \times 10^{-4}$

$K_b = \dfrac{(6.6 \times 10^{-4})^2}{1.00 \times 10^{-3} - 6.6 \times 10^{-4}} = 1.3 \times 10^{-3}$

Acid-Base Properties of Salts

51. $K_a K_b = K_w$; $-\log K_a K_b = -\log K_w$

$-\log K_a - \log K_b = -\log K_w$; $pK_a + pK_b = pK_w = 14.00$ (at 25°C)

52.   a. $KCl \rightarrow K^+ + Cl^-$ Neither K^+ nor Cl^- can change the pH of water. neutral

        In general, +1 and +2 metal ions have no acidic/basic properties and all conjugate bases of strong acids have no basic properties.

        b. $NaNO_3 \rightarrow Na^+ + NO_3^-$ neutral; Neither species have any acidic/basic properties.

        c. $NaNO_2 \rightarrow Na^+ + NO_2^-$ basic

        NO_2^- is a weak base. It is the conjugate base of the weak acid, HNO_2. Ignore Na^+.

        $NO_2^- + H_2O \rightleftharpoons HNO_2 + OH^-$

        d. $NH_4NO_3 \rightarrow NH_4^+ + NO_3^-$ acidic

        NH_4^+, weak acid: $NH_4^+ \rightleftharpoons H^+ + NH_3$. NO_3^- has no basic (or acidic) properties.

e. $NH_4NO_2 \rightarrow NH_4^+ + NO_2^-$

NH_4^+ is a weak acid and NO_2^- is a weak base.

$NH_4^+ \rightleftharpoons NH_3 + H^+$ $K_a = \dfrac{K_w}{K_b} = \dfrac{1.0 \times 10^{-14}}{1.8 \times 10^{-5}} = 5.6 \times 10^{-10}$

$NO_2^- + H_2O \rightleftharpoons HNO_2 + OH^-$ $K_b = \dfrac{K_w}{K_a} = \dfrac{1.0 \times 10^{-14}}{4.0 \times 10^{-4}} = 2.5 \times 10^{-11}$

NH_4^+ is stronger as an acid in water than NO_2^- is as a base. Therefore, the solution is acidic.

f. $NaHCO_3 \rightarrow Na^+ + HCO_3^-$

HCO_3^- can be either an acid or a base. Ignore Na^+.

$HCO_3^- \rightleftharpoons H^+ + CO_3^{2-}$ $K_{a_2} = 5.6 \times 10^{-11}$

$HCO_3^- + H_2O \rightleftharpoons H_2CO_3 + OH^-$ $K_b = \dfrac{K_w}{K_{a_1}} = 2.3 \times 10^{-8}$

HCO_3^- is a stronger base than an acid. The solution is basic.

g. $NH_4C_2H_3O_2 \rightleftharpoons NH_4^+ + C_2H_3O_2^-$
     weak   weak
     acid   base

$NH_4^+ \rightleftharpoons NH_3 + H^+$ $K_a = \dfrac{K_w}{K_b} = \dfrac{1.0 \times 10^{-14}}{1.8 \times 10^{-5}} = 5.6 \times 10^{-10}$

$C_2H_3O_2^- + H_2O \rightleftharpoons HC_2H_3O_2 + OH^-$ $K_b = \dfrac{K_w}{K_a} = \dfrac{1.0 \times 10^{-14}}{1.8 \times 10^{-5}} = 5.6 \times 10^{-10}$

The acid and base are equal in strengths. The solution is neutral.

h. $NaF \rightarrow Na^+ + F^-$; F^- = weak base; $F^- + H_2O \rightleftharpoons HF + OH^-$; Solution is basic.

53. a. $CH_3NH_3Cl \rightarrow CH_3NH_3^+ + Cl^-$; Methylammonium ion is a weak acid. Cl^- is the conjugate base of a strong acid. Cl^- has no basic (or acidic) properties.

$CH_3NH_3^+ \rightleftharpoons CH_3NH_2 + H^+$

$K_a = \dfrac{[CH_3NH_2][H^+]}{[CH_3NH_3^+]} = \dfrac{[CH_3NH_2][H^+][OH^-]}{[CH_3NH_3^+][OH^-]} = \dfrac{K_w}{K_b} = \dfrac{1.00 \times 10^{-14}}{4.38 \times 10^{-4}} = 2.28 \times 10^{-11}$

$$CH_3NH_3^+ \rightleftharpoons CH_3NH_2 + H^+$$

Initial	0.10 *M*		0	0(10^{-7})

x mol/L $CH_3NH_3^+$ dissociates to reach equilibrium

Change	-x	$\rightarrow$	+x	+x
Equil.	0.10 - x		x	x

$$2.28 \times 10^{-11} = \frac{x^2}{0.10 - x} \approx \frac{x^2}{0.10}$$

$x = [H^+] = 1.5 \times 10^{-6} \, M;$ pH = 5.82   Assumptions good.

b. $NaCN \rightarrow Na^+ + CN^-$ Cyanide ion is a weak base. Na^+ has no acidic (or basic) properties.

$$CN^- + H_2O \rightleftharpoons HCN + OH^- \quad K_b = \frac{K_w}{K_a} = \frac{1.0 \times 10^{-14}}{6.2 \times 10^{-10}} = 1.6 \times 10^{-5}$$

Initial	0.050 *M*		0	0(10^{-7})

x mol/L CN^- reacts with H_2O to reach equilibrium

Change	-x	$\rightarrow$	+x	+x
Equil.	0.050 - x		x	x

$$K_b = 1.6 \times 10^{-5} = \frac{[HCN][OH^-]}{[CN^-]} = \frac{x^2}{0.050 - x} \approx \frac{x^2}{0.050}$$

$x = [OH^-] = 8.9 \times 10^{-4} \, M;$ pOH = 3.05; pH = 10.95   Assumptions good.

c. $Na_2CO_3 \rightarrow 2 \, Na^+ + CO_3^{2-}$ CO_3^{2-} is a weak base. Ignore Na^+.

$$CO_3^{2-} + H_2O \rightleftharpoons HCO_3^- + OH^- \quad K_b = \frac{K_w}{K_{a_2}} = \frac{1.0 \times 10^{-14}}{5.6 \times 10^{-11}} = 1.8 \times 10^{-4}$$

Initial	0.20 *M*		0	0(10^{-7})
Equil.	0.20 - x		x	x

$$K_b = 1.8 \times 10^{-4} = \frac{[HCO_3^-][OH^-]}{[CO_3^{2-}]} = \frac{x^2}{0.20 - x} \approx \frac{x^2}{0.20}$$

$x = 6.0 \times 10^{-3} \, M = [OH^-];$ pOH = 2.22; pH = 11.78   Assumptions good.

d. $NaNO_2 \rightarrow Na^+ + NO_2^-$ NO_2^- is a weak base. Ignore Na^+.

$$NO_2^- + H_2O \rightleftharpoons HNO_2 + OH^- \quad K_b = \frac{K_w}{K_a} = \frac{1.0 \times 10^{-14}}{4.0 \times 10^{-4}} = 2.5 \times 10^{-11}$$

Initial	0.12 *M*		0	0(10^{-7})
Equil.	0.12 - x		x	x

$$K_b = 2.5 \times 10^{-11} = \frac{[OH^-][HNO_2]}{[NO_2^-]} = \frac{x^2}{0.12 - x} \approx \frac{x^2}{0.12}$$

$x = [OH^-] = 1.7 \times 10^{-6}~M$; pOH = 5.77; pH = 8.23   Assumptions good.

e. $NaOCl \rightarrow Na^+ + OCl^-$ OCl^- is a weak base. Ignore Na^+.

$$OCl^- + H_2O \rightleftharpoons HOCl + OH^- \qquad K_b = \frac{K_w}{K_a} = \frac{1.0 \times 10^{-14}}{3.5 \times 10^{-8}} = 2.9 \times 10^{-7}$$

Initial 0.45 M 0 $0(10^{-7})$
Equil.   0.45 - x x x

$$K_b = 2.9 \times 10^{-7} = \frac{[HOCl][OH^-]}{[OCl^-]} = \frac{x^2}{0.45 - x} \approx \frac{x^2}{0.45}$$

$x = [OH^-] = 3.6 \times 10^{-4}~M$; pOH = 3.44; pH = 10.56   Assumptions good.

54.   KOH: strong base; KBr: neutral, K^+ and Br^- have no acidic/basic properties.

KCN: CN^- is weak base, $K_b = 1.0 \times 10^{-14}/6.2 \times 10^{-10} = 1.6 \times 10^{-5}$.

NH_4Br: NH_4^+ is weak acid, $K_a = 5.6 \times 10^{-10}$.

NH_4CN: slightly basic, CN^- is a stronger base compared to NH_4^+ as an acid.

HCN: weak acid, $K_a = 6.2 \times 10^{-10}$

Most acidic → most basic: $HCN > NH_4Br > KBr > NH_4CN > KCN > KOH$

55. HNO_2, $K_a = 4.0 \times 10^{-4}$; NO_2^-, $K_b = 2.5 \times 10^{-11}$

NH_4^+, $K_a = 10^{-14}/K_b = 5.6 \times 10^{-10}$; HNO_3, Strong acid

Most acidic → most basic: $HNO_3 > HNO_2 > NH_4NO_3 > NH_4NO_2 > H_2O > KNO_2$

56.   Solution is acidic from $HSO_4^- \rightleftharpoons H^+ + SO_4^{2-}$.

$$HSO_4^- \rightleftharpoons H^+ + SO_4^{2-} \qquad K_a = 1.2 \times 10^{-2}$$

Initial   0.10 M $0(10^{-7})$ 0
Equil.   0.10 - x x x

$$1.2 \times 10^{-2} = \frac{[H^+][SO_4^{2-}]}{[HSO_4^-]} = \frac{x^2}{0.10 - x} \approx \frac{x^2}{0.10}, ~ x = 0.035$$

Assumption is not good (35% error). Using successive approximations:

$$\frac{x^2}{0.10 - x} \approx \frac{x^2}{0.10 - 0.035} = 1.2 \times 10^{-2}, ~ x = 0.028$$

$$\frac{x^2}{0.10 - 0.028} = 1.2 \times 10^{-2}, \ x = 0.029$$

$$\frac{x^2}{0.10 - 0.029} = 1.2 \times 10^{-2}, \ x = 0.029 \qquad \text{(consistent answers)}$$

$x = [H^+] = 0.029 \ M; \ pH = 1.54$

If we add Na_2CO_3 to a solution of $NaHSO_4$, the base CO_3^{2-} can react with the acid HSO_4^-. Depending on relative amounts, two reactions are possible.

$$CO_3^{2-}(aq) \ + \ HSO_4^-(aq) \rightleftharpoons HCO_3^-(aq) + SO_4^{2-}(aq)$$
or
$$CO_3^{2-}(aq) \ + \ 2 \ HSO_4^-(aq) \rightleftharpoons 2 \ SO_4^{2-}(aq) + H_2O(l) + CO_2(g)$$

57. $NaN_3 \rightarrow Na^+ + N_3^-$; Azide, N_3^-, is a weak base. Ignore Na^+.

$$N_3^- \ + \ H_2O \ \rightleftharpoons \ HN_3 \ + \ OH^- \quad K_b = \frac{K_w}{K_a} = \frac{1.0 \times 10^{-14}}{1.9 \times 10^{-5}} = 5.3 \times 10^{-10}$$

	N_3^-		HN_3	OH^-
Initial	0.010 M		0	0(10^{-7})
Equil.	0.010 - x		x	x

$$K_b = \frac{[HN_3] [OH^-]}{[N_3^-]} = 5.3 \times 10^{-10} = \frac{x^2}{0.010 - x} \approx \frac{x^2}{0.010}$$

$$x = [OH^-] = 2.3 \times 10^{-6} \ M; \ [H^+] = \frac{1.0 \times 10^{-14}}{2.3 \times 10^{-6}} = 4.3 \times 10^{-9} \ M \quad \text{Assumptions good.}$$

$[HN_3] = [OH^-] = 2.3 \times 10^{-6} \ M; \ [Na^+] = 0.010 \ M; \ [N_3^-] = 0.010 - 2.3 \times 10^{-6} = 0.010 \ M$

58.   From the K_a values, acetic acid is a stronger acid than hypochlorous acid. Conversely, the conjugate base of acetic acid, $C_2H_3O_2^-$, is a weaker base than the conjugate base of hypochlorous acid, OCl^-. Thus, the hypochlorite ion, OCl^-, is a stronger base than the acetate ion, $C_2H_3O_2^-$.

59.   From exercise 45, codeine sulfate = $(HC_{18}H_{21}NO_3)_2SO_4$. $HC_{18}H_{21}NO_3^+$ is the conjugate acid of codeine and is abbreviated codH$^+$. In a solution of codeine sulfate, the major species are: codH$^+$, SO_4^{2-}, and H_2O. CodH$^+$ is a weak acid and is the major source of H^+ in this solution. SO_4^{2-} is such a weak base ($K_b = \frac{K_w}{K_{a_2}} = 8.3 \times 10^{-13}$) that we will ignore its basic properties.

$$1.0 \text{ g } (codH)_2SO_4 \times \frac{1 \text{ mol } (codH)_2SO_4}{696.8 \text{ g}} \times \frac{2 \text{ mol codH}^+}{\text{mol } (codH)_2SO_4} = 2.9 \times 10^{-3} \text{ mol codH}^+$$

$$[codH^+] = \frac{2.9 \times 10^{-3} \text{ mol}}{30. \times 10^{-3} \text{ L}} = 0.097 \ M$$

From exercise 44, pK_b for cod = 6.05, so pK_a for codH$^+$ = 14.00 - 6.05 = 7.95.

$$codH^+ + H_2O \rightleftharpoons cod + H_3O^+ \qquad K_a = 10^{-7.95} = 1.1 \times 10^{-8}$$

Initial	0.097 M		0	0(10^{-7})
Equil.	0.097 - x		x	x

$$\frac{x^2}{0.097 - x} = 1.1 \times 10^{-8} \approx \frac{x^2}{0.097}$$

$x = [H^+] = 3.3 \times 10^{-5}\ M; \ pH = 4.48$ \qquad Assumptions good.

Additional Exercises

60. a. An acid will generate NH_4^+ in liquid ammonia and a base will generate NH_2^-. NH_4^+ corresponds to H^+ and NH_2^- corresponds to OH^-.

b. $[NH_4^+] = [NH_2^-]$

c. $2\ Na(s) + 2\ NH_3(l) \rightarrow 2\ Na^+ + 2\ NH_2^- + H_2(g)$

d. Ammonia is more basic than water. The acidity of substances is enhanced in liquid ammonia. For example, acetic acid is a strong acid in ammonia

61. $H_3Cit + HCO_3^- \rightarrow H_2Cit^- + H_2CO_3\ (H_2O + CO_2)$

$$K_{eq} = \frac{[H_2Cit^-][H_2CO_3]}{[H_3Cit][HCO_3^-]} \times \frac{[H^+]}{[H^+]} = \frac{K_{a_1}\ (citric\ acid)}{K_{a_1}\ (carbonic\ acid)}$$

$$K_{eq} = \frac{8.4 \times 10^{-4}}{4.3 \times 10^{-7}} = 1950 = 2.0 \times 10^3$$

$H_3Cit + 3\ HCO_3^- \rightleftharpoons Cit^{3-} + 3\ H_2CO_3\ (3\ H_2O + 3\ CO_2)$

$$K_{eq} = \frac{[Cit^{3-}][H_2CO_3]^3}{[H_3Cit][HCO_3^-]^3} \times \frac{[H^+]^3}{[H^+]^3} = \frac{K_{a_1}K_{a_2}K_{a_3}\ (citric\ acid)}{K_{a_1}^3\ (carbonic\ acid)}$$

$$K_{eq} = \frac{(8.4 \times 10^{-4})(1.8 \times 10^{-5})(4.0 \times 10^{-6})}{(4.3 \times 10^{-7})^3} = 7.6 \times 10^5$$

62. $pH = 2.77, \ [H^+] = 10^{-2.77} = 1.7 \times 10^{-3}$

$HOCN \rightleftharpoons H^+ + OCN^-$; From normal weak acid setup:

$[H^+] = [OCN^-] = 1.7 \times 10^{-3} \ M; \ \ [HOCN] = 0.0100 - 0.0017 = 0.0083 \ M$

$$K_a = \frac{[H^+][OCN^-]}{[HOCN]} = \frac{(1.7 \times 10^{-3})^2}{8.3 \times 10^{-3}} = 3.5 \times 10^{-4}$$

63.   a. $HCO_3^- + HCO_3^- \rightleftharpoons H_2CO_3 + CO_3^{2-}$

$$K_{eq} = \frac{[H_2CO_3][CO_3^{2-}]}{[HCO_3^-][HCO_3^-]} \times \frac{[H^+]}{[H^+]} = \frac{K_{a_2}}{K_{a_1}} = \frac{5.6 \times 10^{-11}}{4.3 \times 10^{-7}} = 1.3 \times 10^{-4}$$

    b. $[H_2CO_3] = [CO_3^{2-}]$ if the reaction in (a) is considered to be the only reaction.

    c. $H_2CO_3 \rightleftharpoons 2\ H^+ + CO_3^{2-}$ $K_{eq} = \dfrac{[H^+]^2[CO_3^{2-}]}{[H_2CO_3]} = K_{a_1}K_{a_2}$

       Since, $[H_2CO_3] = [CO_3^{2-}]$ from part b, $[H^+]^2 = K_{a_1}K_{a_2}$

       $[H^+] = (K_{a_1}K_{a_2})^{1/2}; \ \ pH = \dfrac{pK_{a_1} + pK_{a_2}}{2}$

    d. $[H^+] = [(4.3 \times 10^{-7})(5.6 \times 10^{-11})]^{1/2}, \ [H^+] = 4.9 \times 10^{-9}; \ pH = 8.31$

64.   a. $NH_3 + H_3O^+ \rightleftharpoons NH_4^+ + H_2O$

$$K_{eq} = \frac{[NH_4^+]}{[NH_3][H^+]} = \frac{1}{K_a \text{ for } NH_4^+} = \frac{K_b}{K_w} = \frac{1.8 \times 10^{-5}}{1.0 \times 10^{-14}} = 1.8 \times 10^9$$

    b. $NO_2^- + H_3O^+ \rightleftharpoons H_2O + HNO_2$

$$K_{eq} = \frac{[HNO_2]}{[NO_2^-][H^+]} = \frac{1}{K_a} = \frac{1}{4.0 \times 10^{-4}} = 2.5 \times 10^3$$

    c. $NH_4^+ + CH_3CO_2^- \rightleftharpoons NH_3 + CH_3CO_2H$ $K_{eq} = \dfrac{[NH_3][CH_3CO_2H]}{[NH_4^+][CH_3CO_2^-]} \times \dfrac{[H^+]}{[H^+]}$

$$K_{eq} = \frac{K_a \text{ for } NH_4^+}{K_a \text{ for } HOAc} = \frac{K_w}{(K_b \text{ for } NH_3)(K_a \text{ for } HOAc)} = \frac{1.0 \times 10^{-14}}{(1.8 \times 10^{-5})(1.8 \times 10^{-5})}$$

       $K_{eq} = 3.1 \times 10^{-5}$

    d. $H_3O^+ + OH^- \rightleftharpoons 2\ H_2O$ $K_{eq} = \dfrac{1}{K_w} = 1.0 \times 10^{14}$

e. $NH_4^+ + OH^- \rightleftharpoons NH_3 + H_2O$ $K_{eq} = \dfrac{1}{K_b \text{ for } NH_3} = 5.6 \times 10^4$

f. $HNO_2 + OH^- \rightleftharpoons H_2O + NO_2^-$

$$K_{eq} = \frac{[NO_2^-]}{[HNO_2][OH^-]} \times \frac{[H^+]}{[H^+]} = \frac{K_a \text{ for } HNO_2}{K_w} = \frac{4.0 \times 10^{-4}}{1.0 \times 10^{-14}} = 4.0 \times 10^{10}$$

65.   a.   In the lungs, there is a lot of O_2 and the equilibrium favors $Hb_2(O_2)_4$. In the cells there is a deficiency of O_2 and the equilibrium favors HbH_4^{4+}.

   b. CO_2 is a weak acid, $CO_2 + H_2O \rightleftharpoons HCO_3^- + H^+$. Removing CO_2 essentially decreases H^+. $Hb(O_2)_4$ is then favored and O_2 is not released by hemoglobin in the cells. Breathing into a paper bag increases CO_2 in the blood.

   c. CO_2 builds up in the blood and it becomes too acidic, driving the equilibrium to the left. Hemoglobin can't bind O_2 as strongly in the lungs. Bicarbonate ion acts as a base in water and neutralizes the excess acidity.

66.   Can't neglect OH^- contribution from H_2O.

Charge balance: $[Na^+] + [H^+] = [OH^-]$

$$1.0 \times 10^{-7} + \frac{1.0 \times 10^{-14}}{[OH^-]} = [OH^-], \ [OH^-]^2 - 1.0 \times 10^{-7}[OH^-] - 1.0 \times 10^{-14} = 0$$

Using the quadratic formula: $[OH^-] = 1.6 \times 10^{-7}$; $pOH = 6.80$; $pH = 7.20$

67. $HBrO$ $\rightleftharpoons$ H^+ + BrO^- $K_a = 2 \times 10^{-9}$

Initial $1.0 \times 10^{-6} \ M$ $0(10^{-7})$ 0
            x mol/L $HBrO$ dissociates to reach equilibrium
Change $-x$ $\rightarrow$ $+x$ $+x$
Equil. $1.0 \times 10^{-6} - x$ x x

$$\frac{x^2}{1.0 \times 10^{-6} - x} \approx \frac{x^2}{1.0 \times 10^{-6}} = 2 \times 10^{-9}$$

$x = [H^+] = 4 \times 10^{-8}$; $pH = 7.4$ Check assumptions.

This answer is impossible. We can't add a small amount of a very weak acid to a neutral solution and get a basic solution. In a correct solution, we would have to take into account the autoionization of water. For a weak acid dissolved in water, the H^+ contribution from water should be considered when the pH is greater than 6.0.

68. CH_2ClCO_2H $\rightleftharpoons$ H^+ + $CH_2ClCO_2^-$

Initial   0.0200 M 0(10^{-7})   0
Equil.   0.0200 - x x x

$$K_a = 1.39 \times 10^{-3} = \frac{[H^+][CH_2ClCO_2^-]}{[CH_2ClCO_2H]} = \frac{x^2}{(0.0200 - x)} \approx \frac{x^2}{0.0200}$$

$x = 5.27 \times 10^{-3}$; Check 5% Rule: $\dfrac{5.27 \times 10^{-3}}{0.020} \times 100 = 26\%$; Assumption is poor.

Using the quadratic formula: $x^2 + (1.39 \times 10^{-3})x - 2.78 \times 10^{-5} = 0$

$x = 4.62 \times 10^{-3} = [H^+]$; pH = -log $[H^+]$ = -log (4.62×10^{-3}) = 2.335

69. HIO_3 $\rightleftharpoons$ H^+ + IO_3^-

Initial   0.010 M 0(10^{-7})   0
Equil.   0.010 - x x x

$$\frac{x^2}{(0.010 - x)} = 1.7 \times 10^{-1} \approx \frac{x^2}{0.0100}, \, x = 0.041; \text{ Assumption is horrible.}$$

Using the quadratic formula: $x^2 + (1.7 \times 10^{-1})x - 1.7 \times 10^{-3} = 0$

$x = 9.5 \times 10^{-3}\ M = [H^+]$; pH = -log $[H^+]$ = 2.02

70.   Can't neglect $[H^+]$ contribution from H_2O.

[positive charge] = [negative charge]   (called charge balance)

$[H^+] = [Cl^-] + [OH^-]$; $[H^+] = 7.00 \times 10^{-7} + \dfrac{K_w}{[H^+]}$

$\dfrac{[H^+]^2 - K_w}{[H^+]} = 7.00 \times 10^{-7}$, $[H^+]^2 - 7.00 \times 10^{-7}\,[H^+] - 1.00 \times 10^{-14} = 0$

Using the quadratic formula: $[H^+] = 7.14 \times 10^{-7}\ M$; pH = 6.146

Alternative method: Consider the effect of added H^+ on the H_2O equilibrium:

                          H_2O $\rightleftharpoons$ H^+ + OH^-

Initial $(1.00 + 7.00) \times 10^{-7}\ M$ $1.00 \times 10^{-7}\ M$
Equil. $8.00 \times 10^{-7} - x$ $1.00 \times 10^{-7} - x$

$$K_w = [H^+][OH^-] = 1.00 \times 10^{-14} = (8.00 \times 10^{-7} - x)(1.00 \times 10^{-7} - x)$$

$$x^2 - (9.00 \times 10^{-7})x + 7.00 \times 10^{-14} = 0, \quad x = 8.60 \times 10^{-8} \text{ (from quadratic formula)}$$

$$[H^+] = 8.00 \times 10^{-7} - 8.60 \times 10^{-8} = 7.14 \times 10^{-7} \, M; \quad pH = 6.146$$

71. B^- is a weak base.

$$B^- \;\; + \;\; H_2O \;\; \rightleftharpoons \;\; HB \;\; + \;\; OH^-$$

Initial   0.050 M 0   0(10^{-7})
Equil.   0.050 - x x x

pH = 9.00,   pOH = 5.00, and $[OH^-] = 1.0 \times 10^{-5} = x$

$$K_b = \frac{[HB][OH^-]}{[B^-]} = \frac{(1.0 \times 10^{-5})^2}{(0.050 - 1.0 \times 10^{-5})} = 2.0 \times 10^{-9}$$

Alternate Method: Mass balance for B^-

$0.050 = [B^-] + [HB]$ and $[HB] = [OH^-]$ from the reaction

$$K_b = \frac{[HB][OH^-]}{[B^-]} = \frac{[OH^-]^2}{0.050 - [OH^-]} = \frac{(1.00 \times 10^{-5})^2}{0.050 - 1.0 \times 10^{-5}} = 2.0 \times 10^{-9}$$

Since B^- is a weak base, then HB is a weak acid.

$$HB \;\; \rightleftharpoons \;\; H^+ \;\; + \;\; B^-$$

Initial   0.010 M 0(10^{-7})   0
Equil.   0.010 - x x x

$$K_a = \frac{K_w}{K_b} = \frac{(1.0 \times 10^{-14})}{2.0 \times 10^{-9}} = 5.0 \times 10^{-6} = \frac{x^2}{(0.010 - x)} \approx \frac{x^2}{0.010}$$

$x = [H^+] = 2.2 \times 10^{-4} \, M; \quad pH = 3.66$ Assumptions good.

72.   Major Species: H^+, Cl^-, CH_3CO_2H, H_2O

$$HOAc \;\; \rightleftharpoons \;\; H^+ \;\; + \;\; OAc^-$$

Initial   1.0 M 1.0 $\times 10^{-3}$ M 0
Equil.   1.0 - x 1.0 $\times 10^{-3}$ + x x

$$\frac{x(1.0 \times 10^{-3} + x)}{(1.0 - x)} = 1.8 \times 10^{-5}$$

If $x \ll 1.0 \times 10^{-3}$, then $x(1.0 \times 10^{-3}) = 1.8 \times 10^{-5}$, $x = 1.8 \times 10^{-2}$ Assumption is horrible.

If $x \ll 1.0$, then: $1.8 \times 10^{-5} = x(1.0 \times 10^{-3} + x)$, $x^2 + 1.0 \times 10^{-3}\, x - 1.8 \times 10^{-5} = 0$

Using the quadratic formula: $x = 3.8 \times 10^{-3}$ This assumption is good (0.4% error).

So $[H^+] = 1.0 \times 10^{-3} + x = 4.8 \times 10^{-3}\ M$; pH = 2.32; $[OAc^-] = 3.8 \times 10^{-3}\ M$

73. $d = \dfrac{PM}{RT}$, $M = \dfrac{dRT}{P} = \dfrac{5.11\ g/L \times \dfrac{0.08206\ L\ atm}{mol\ K} \times 298\ K}{1.00\ atm} = 125\ g/mol$

$\dfrac{1.50\ g \times \dfrac{1\ mol}{125\ g}}{0.100\ L} = 0.120\ M$; pH = 1.80; $[H^+] = 10^{-1.80} = 1.6 \times 10^{-2}\ M$

$HA \rightleftharpoons H^+ + A^-$; $[H^+] = [A^-]$, $0.120 = [HA] + [A^-]$ (mass balance)

$K_a = \dfrac{[H^+][A^-]}{[HA]} = \dfrac{(1.6 \times 10^{-2})^2}{0.120 - 0.016} = 2.5 \times 10^{-3}$

74.   For 0.0010% dissociation: $[NH_4^+] = 1.0 \times 10^{-5}(0.050) = 5.0 \times 10^{-7}\ M$

$NH_3 + H_2O \rightleftharpoons NH_4^+ + OH^-$ $K_b = \dfrac{(5.0 \times 10^{-7})[OH^-]}{0.050 - 5.0 \times 10^{-7}} = 1.8 \times 10^{-5}$

Solving: $[OH^-] = 1.8\ M$; Assuming no volume change:

$1.0\ L \times \dfrac{1.8\ mol\ NaOH}{L} \times \dfrac{40.00\ g\ NaOH}{mol\ NaOH} = 72\ g\ of\ NaOH$

75.   For this problem we will abbreviate $CH_2{=}CHCO_2H$ as Hacr and $CH_2{=}CHCO_2^-$ as acr$^-$.

   a.   Hacr $\rightleftharpoons$ H^+ +   acr$^-$

	Hacr	H^+	acr$^-$
Initial	0.10 M	0(10^{-7})	0
Equil.	0.10 - x	x	x

$\dfrac{x^2}{0.10 - x} = 5.6 \times 10^{-5} \approx \dfrac{x^2}{0.10}$, $x = [H^+] = 2.4 \times 10^{-3}\ M$; pH = 2.62

Assumptions good.

b.   % dissociation = $\dfrac{2.4 \times 10^{-3}}{0.10} \times 100 = 2.4\%$

c.   For 0.010% dissociation: $[acr^-] = 1.0 \times 10^{-4}\,(0.10) = 1.0 \times 10^{-5}\ M$

$$K_a = \frac{[H^+]\,[acr^-]}{[Hacr]} = 5.6 \times 10^{-5} = \frac{[H^+]\,(1.0 \times 10^{-5})}{0.10 - 1.0 \times 10^{-5}},\quad [H^+] = 0.56\ M$$

d. acr^- is a weak base and the major source of OH^- in this solution.

$$acr^- \ +\ H_2O \ \rightleftharpoons\ Hacr \ +\ OH^- \qquad K_b = \frac{K_w}{K_a} = \frac{1.0 \times 10^{-14}}{5.6 \times 10^{-5}}$$

Initial	0.050 M		0	0(10^{-7})
Equil.	0.050 - x		x	x

$K_b = 1.8 \times 10^{-10}$

$$K_b = \frac{[OH^-]\,[Hacr]}{[acr^-]} = 1.8 \times 10^{-10} = \frac{x^2}{0.050 - x} \approx \frac{x^2}{0.050}$$

$x = [OH^-] = 3.0 \times 10^{-6}$;   pOH = 5.52;   pH = 8.48   Assumptions good.

76.   a.
$$Fe(H_2O)_6^{3+} \ +\ H_2O \ \rightleftharpoons\ Fe(H_2O)_5(OH)^{2+} \ +\ H_3O^+$$

Initial	0.10 M	0	0(10^{-7})
Equil.	0.10 - x	x	x

$$K_a = \frac{[H_3O^+]\,[Fe(H_2O)_5(OH)^{2+}]}{[Fe(H_2O)_6^{3+}]} = 6.0 \times 10^{-3} = \frac{x^2}{0.10 - x} \approx \frac{x^2}{0.10}$$

$x = 2.4 \times 10^{-2}$;   Assumption is poor (24% error).

Using successive approximations:

$$\frac{x^2}{0.10 - 0.024} = 6.0 \times 10^{-3},\ x = 0.021$$

$$\frac{x^2}{0.10 - 0.021} = 6.0 \times 10^{-3},\ x = 0.022;\quad \frac{x^2}{0.10 - 0.022} = 6.0 \times 10^{-3},\ x = 0.022$$

$x = [H^+] = 0.022\ M$;   pH = 1.66

b. $\dfrac{[Fe(H_2O)_5(OH)^{2+}]}{[Fe(H_2O)_6^{3+}]} = \dfrac{0.0010}{0.9990}$; $K_a = 6.0 \times 10^{-3} = \dfrac{[H^+]\,(0.0010)}{(0.9990)}$

Solving: $[H^+] = 6.0\ M$;   pH = -log 6.0 = -0.78

c. Because of the lower charge, $Fe^{2+}(aq)$ will not be as strong an acid as $Fe^{3+}(aq)$. A solution of iron(II) nitrate will be less acidic (have a higher pH) than a solution with the same concentration of iron(III) nitrate.

77. Major species: BH^+, X^-, H_2O; If BH^+ is the best acid and X^- is the best base in solution, then the principal equilibrium is:

$$BH^+ + X^- \rightleftharpoons B + HX$$

$$K_{eq} = \frac{K_a(BH^+)}{K_a(HX)} = \frac{[B][HX]}{[BH^+][X^-]} \text{ where } [B] = [HX] \text{ and } [BH^+] = [X^-]$$

$$\frac{K_a(BH^+)}{K_a(HX)} = \frac{[HX]^2}{[X^-]^2}; \quad K_a(HX) = \frac{[H^+][X^-]}{[HX]}, \quad \frac{[HX]}{[X^-]} = \frac{[H^+]}{K_a(HX)}$$

$$\frac{K_a(BH^+)}{K_a(HX)} = \frac{[HX]^2}{[X^-]^2} = \left(\frac{[H^+]}{K_a(HX)}\right)^2, \quad [H^+]^2 = K_a(BH^+) \times K_a(HX)$$

Taking the -log of both sides: $pH = \dfrac{pK_a(BH^+) + pK_a(HX)}{2}$

$$K_b(B) = 1.0 \times 10^{-3}; \quad K_a(BH^+) = \frac{K_w}{K_b} = 1.0 \times 10^{-11}$$

$$8.00 = \frac{11.00 + pK_a}{2}, \quad pK_a = 5.00 \text{ and } K_a(HX) = 1.0 \times 10^{-5}$$

78. The major contribution of H^+ is from HCl. $[H^+] = 1.0$ M and pH = 0.00

Now, let's check this assumption with the HCN equilibrium.

	HCN	$\rightleftharpoons$	H^+	+	CN^-
Initial	1.0 M		1.0 M		0
Equil.	1.0 - x		1.0 + x		x

$$K_a = \frac{x(1.0 + x)}{(1.0 - x)} = 6.2 \times 10^{-10} \approx x$$

H^+ contribution from HCN <<< H^+ contribution from HCl. Assumptions good.

79. At a pH = 0.00, the $[H^+] = 10^{-0.00} = 1.0$ M. Begin with 1.0 L $\times$ 2.0 mol/L NaOH = 2.0 mol OH^-. We will need 2.0 mol HCl to neutralize the OH^- plus an additional 1.0 mol excess to reduce to a pH of 0.00. Need 3.0 mol HCl total.

80.   0.0500 M HCO$_2$H (HA), K$_a$ = 1.77 × 10^{-4}; 0.150 M CH$_3$CH$_2$CO$_2$H (HB), K$_a$ = 1.34 × 10^{-5}

Since two comparable weak acids are present, each contributes to the total pH.

Charge balance: [H$^+$] = [A$^-$] + [B$^-$] + [OH$^-$] = [A$^-$] + [B$^-$] + K$_w$/[H$^+$]

Mass balance for HA and HB: 0.0500 = [HA] + [A$^-$] and 0.150 = [HB] + [B$^-$]

$$\frac{[H^+][A^-]}{[HA]} = 1.77 \times 10^{-4}; \qquad \frac{[H^+][B^-]}{[HB]} = 1.34 \times 10^{-5}$$

We have 5 equations and 5 unknowns. Manipulate the equations to solve.

[H$^+$] = [A$^-$] + [B$^-$] + K$_w$/[H$^+$]; [H$^+$]^2 = [H$^+$][A$^-$] + [H$^+$][B$^-$] + K$_w$

[H$^+$][A$^-$] = [HA](1.77 × 10^{-4}) = (1.77 × 10^{-4}) (0.0500 - [A$^-$])

If [A$^-$] << 0.0500, then [H$^+$][A$^-$] ≈ (1.77 × 10^{-4}) (0.0500) = 8.85 × 10^{-6}

Similarly, [H$^+$][B$^-$] ≈ (1.34 × 10^{-5})(0.150) = 2.01 × 10^{-6}

[H$^+$]^2 = 8.85 × 10^{-6} + 2.01 × 10^{-6} + 1.00 × 10^{-14}, [H$^+$] = 3.30 × 10^{-3} mol/L

Check assumptions: [H$^+$][A$^-$] ≈ 8.85 × 10^{-6}, [A$^-$] ≈ $\dfrac{8.85 \times 10^{-6}}{3.30 \times 10^{-3}}$ ≈ 2.68 × 10^{-3}

Assumed 0.050 - [A$^-$] ≈ 0.050, 0.050 - 0.0027 = 0.0473. This assumption is borderline (5.4% error).   The HB assumption is good (0.4% error).

Using successive approximations: [H$^+$] = 3.22 × 10^{-3} M, pH = -log (3.22 × 10^{-3}) = 2.492.

Note: If we treat each acid separately:

H$^+$ from HA = 2.9 × 10^{-3}
H$^+$ from HB = 1.4 × 10^{-3}
                 4.3 × 10^{-3} M = [H$^+$] total, assuming the acids did not suppress each others
ionization. They do and we expect the [H$^+$] to be less than 4.3 × 10^{-3}. We get such an answer.

81.   HA   ⇌   H$^+$ +   A$^-$

Initial	C	0(10^{-7})	0	C = [HA]$_o$
Equil.	(C-10^{-4})	10^{-4}	10^{-4}	(Write 10^{-4} for brevity, really 1.00 × 10^{-4}.)

$$\frac{(10^{-4})^2}{C - 10^{-4}} = 1.00 \times 10^{-6}, \; C = 0.0101 \, M$$

Solution initially contains 0.0101 mmol/mL × 50.0 mL = 0.505 mmol HA. Dilute to total volume, V. The resulting pH = 5.000, $[H^+] = 1.00 \times 10^{-5}$. From normal setup:

$$HA \quad \rightleftharpoons \quad H^+ \quad + \quad A^-$$

Equil. $(0.505/V - 1.00 \times 10^{-5})$ 1.00×10^{-5} 1.00×10^{-5}

$$\frac{(1.00 \times 10^{-5})^2}{0.505/V - 1.00 \times 10^{-5}} = 1.00 \times 10^{-6}, \; 1.00 \times 10^{-4} = 0.505/V - 1.00 \times 10^{-5}$$

V = 4590 mL; 50.0 mL are present, so we need to add 4540 mL of water.

82.   From Exercise 7.77: $pH = \dfrac{pK_a(NH_4^+) + pK_a(HCN)}{2}$

HCN: $K_a = 6.2 \times 10^{-10}$, $pK_a = 9.21$

NH_4^+: $K_a = \dfrac{K_w}{K_b \text{ for } NH_3}$, $pK_a = - \log \left(\dfrac{1.0 \times 10^{-14}}{1.8 \times 10^{-5}} \right) = 9.26$

$pH = \dfrac{9.21 + 9.26}{2} = 9.24$

83. $\dfrac{0.135 \, \text{mol } CO_2}{2.50 \, L} = 5.40 \times 10^{-2} \, \text{mol } CO_2/L = 5.40 \times 10^{-2} \, M \, H_2CO_3$; $0.105 \, M \, CO_3^{2-}$

$$H_2CO_3 + CO_3^{2-} \rightarrow 2 \, HCO_3^- \qquad K = \frac{1}{1.3 \times 10^{-4}} = 7.7 \times 10^3 \;\; \text{(See Exercise 63)}$$

From large size of K, assume all CO_2 (H_2CO_3) is converted into HCO_3^-, i.e., 5.40×10^{-2} mol/L CO_3^{2-} converted into HCO_3^-.

$[HCO_3^-] = 2(5.40 \times 10^{-2}) = 0.108 \, M$; $[CO_3^{2-}] = 0.105 - 0.0540 = 0.051 \, M$

To solve for the $[H^+]$ in equilibrium with HCO_3^- and CO_3^{2-}, use the K_a expression for HCO_3^-.

$HCO_3^- \leftarrow H^+ + CO_3^{2-}$ $K_a = 5.6 \times 10^{-11}$

$$5.6 \times 10^{-11} = \frac{[H^+][CO_3^{2-}]}{[HCO_3^-]} \approx [H^+] \left(\frac{0.051}{0.108} \right)$$

$[H^+] = 1.2 \times 10^{-10}$; pH = 9.92   Assumptions good.

84.   Since such a dilute solution of strong acid, H_2O contribution to $[H^+]$ must be considered.

Charge balance: $[H^+] = [NO_3^-] + [OH^-]$, $[H^+] = [NO_3^-] + K_w/[H^+]$

$[H^+]^2 - 1.0 \times 10^{-14} = [H^+](5.0 \times 10^{-8})$, $[H^+]^2 - (5.0 \times 10^{-8})[H^+] - (1.0 \times 10^{-14}) = 0$

Using the quadratic formula: $[H^+] = 1.3 \times 10^{-7}$; pH = 6.89

85.   HCN $\rightleftharpoons$ H^+ + CN^-

Initial $5.0 \times 10^{-4}\ M$ $0(10^{-7})$ 0
Equil. $5.0 \times 10^{-4} - x$ x x

$$\frac{x^2}{5.0 \times 10^{-4} - x} \approx \frac{x^2}{5.0 \times 10^{-4}} = 6.2 \times 10^{-10}, \ x = 5.6 \times 10^{-7}\ M \quad \text{Check assumptions.}$$

The assumption that the H^+ contribution from water is negligible is poor. Whenever the calculated pH is greater than 6.0 for a weak acid, water contribution to $[H^+]$ must be considered. From Section 7.9 in text:

if $\dfrac{[H^+]^2 - K_w}{[H^+]} \ll [HCN]_0 = 5.0 \times 10^{-4}$ then we can use: $[H^+] = (K_a[HCN]_o + K_w)^{1/2}$.

Using this formula: $[H^+] = [(6.2 \times 10^{-10})(5.0 \times 10^{-4}) + (1.0 \times 10^{-14})]^{1/2}$, $[H^+] = 5.7 \times 10^{-7}\ M$

Checking assumptions: $\dfrac{[H^+]^2 - K_w}{[H^+]} = 5.5 \times 10^{-7} \ll 5.0 \times 10^{-4}$

Assumptions good. pH $= -\log 5.7 \times 10^{-7} = 6.24$

86.   phOH $\rightleftharpoons$ phO^- + H^+ phOH = phenol

Equil. $4.0 \times 10^{-5} - x$ x x

$$\frac{x^2}{4.0 \times 10^{-5}} \approx 1.6 \times 10^{-10}, \ x = [H^+] \approx 8.0 \times 10^{-8} \qquad \text{Check assumptions.}$$

One of the assumptions is poor. We have to consider the H^+ contribution from the H_2O. From Section 7.9 in text, try $[H^+] = (K_a[HA]_o + K_w)^{1/2}$

$[H^+] = [(1.6 \times 10^{-10})(4.0 \times 10^{-5}) + (1.0 \times 10^{-14})]^{1/2} = 1.3 \times 10^{-7}\ M$

This will work if $4.0 \times 10^{-5} \gg \dfrac{[H^+]^2 - K_w}{[H^+]} = 5.3 \times 10^{-8}$. Assumption good.

$[H^+] = 1.3 \times 10^{-7}\ M$; pH = 6.89

87.   Major species: H_2O, Na^+, NO_2^-; NO_2^- is a weak base.

$NO_2^- + H_2O \rightleftharpoons HNO_2 + OH^-$

Since this is a very dilute solution of a weak base, OH^- contribution from H_2O must be considered. The analogous weak base equations from the weak acid equations are:

For $A^- + H_2O \rightleftharpoons HA + OH^-$

I. $K_b = \dfrac{[OH^-]^2 - K_w}{[A^-]_0 - \dfrac{[OH^-]^2 - K_w}{[OH^-]}}$

II. When $[A^-]_0 \gg \dfrac{[OH^-]^2 - K_w}{[OH^-]}$, then $K_b = \dfrac{[OH^-]^2 - K_w}{[A^-]_0}$

and $[OH^-] = (K_b[A^-]_0 + K_w)^{1/2}$

Try $[OH^-] = \left(\dfrac{1.0 \times 10^{-14}}{4.0 \times 10^{-4}} \times (6.0 \times 10^{-4}) + 1.0 \times 10^{-14} \right)^{1/2} = 1.6 \times 10^{-7} M$

Checking assumption: $6.0 \times 10^{-4} \gg \dfrac{(1.6 \times 10^{-7})^2 - 1.0 \times 10^{-14}}{1.6 \times 10^{-7}} = 9.8 \times 10^{-8}$

Assumption good. $[OH^-] = 1.6 \times 10^{-7} M$; $pOH = 6.80$; $pH = 7.20$

88.   Ammonia and hydrochloric acid will react to completion since HCl is a strong acid. Taking into account the change in volume, we initially have a $5.0 \times 10^{-5} M$ solution of NH_4Cl.

	NH_4^+	$\rightleftharpoons$	H^+	$+$	NH_3	$K_a = \dfrac{K_w}{K_b} = 5.6 \times 10^{-10}$
Initial	$5.0 \times 10^{-5} M$		$0(10^{-7})$		0	
Equil.	$5.0 \times 10^{-5} - x$		x		x	

$\dfrac{x^2}{5.0 \times 10^{-5} - x} \approx \dfrac{x^2}{5.0 \times 10^{-5}} = 5.6 \times 10^{-10}$, $x = 1.7 \times 10^{-7} M$ Check assumptions.

We cannot neglect $[H^+]$ that comes from H_2O. Assume extent of reaction is small compared to 5.0×10^{-5}.

$[H^+] = (K_a[HA]_0 + K_w)^{1/2} = 1.9 \times 10^{-7} M$

$$\frac{[H^+]^2 - K_w}{[H^+]} = 1.4 \times 10^{-7} \ll 5.0 \times 10^{-5} \quad \text{Assumption good.}$$

So $[H^+] = 1.9 \times 10^{-7}\ M$; pH $= 6.72$

89. $0.50\ M$ HA, $K_a = 1.0 \times 10^{-3}$; $0.20\ M$ HB, $K_a = 1.0 \times 10^{-10}$; $0.10\ M$ HC, $K_a = 1.0 \times 10^{-12}$

Major source of H^+ is HA since its K_a value is significantly larger than other K_a values.

$$\text{HA} \quad \rightleftharpoons \quad H^+ \quad + \quad A^-$$

Initial	$0.50\ M$	$0(10^{-7})$	0
Equil.	$0.50 - x$	x	x

$$\frac{x^2}{0.50 - x} = 1.0 \times 10^{-3} \approx \frac{x^2}{0.50}, \quad x \approx 0.022\ M = [H^+], \quad \frac{0.022}{0.50} \times 100 = 4.4\ \%\ \text{error}$$

Assumptions good. Let's check out the assumption that only HA is an important source of H^+.

For HB: $1.0 \times 10^{-10} = \dfrac{(0.022)\ [B^-]}{(0.20)}$, $[B^-] = 9.1 \times 10^{-10}\ M$

At most, HB will produce an additional $9.1 \times 10^{-10}\ M\ H^+$. Even less will be produced by HC. Thus, our original assumption was good. $[H^+] = 0.022\ M$.

90. $[HA]_0 = \dfrac{1.0\ \text{mol}}{2.0\ \text{L}} = 0.50\ \text{mol/L}$

$[HA]_0 = [HA] + [A^-]$ (mass balance), $[A^-] = 0.50 - 0.45 = 0.05\ M$

$[H^+] = [A^-] + [OH^-] \approx [A^-] \approx 0.05\ M$ (charge balance)

$K_a = \dfrac{[H^+]\ [A^-]}{[HA]} = \dfrac{(0.05)\ (0.05)}{(0.45)} = 5.6 \times 10^{-3} = 6 \times 10^{-3}$ Assumption good.

91. $Ca(OH)_2\ (s) \rightarrow Ca^{2+}\ (aq) + 2\ OH^-\ (aq)$; $[OH^-]_0 = 6.0 \times 10^{-7}\ M$

We can't ignore the OH^- contribution from H_2O. From dissociation of $Ca(OH)_2$ alone, $2[Ca^{2+}] = [OH^-]$. Including H_2O dissociation, the overall charge balance is:

$$2[Ca^{2+}] + [H^+] = [OH^-]$$

$6.0 \times 10^{-7}\ M + K_w/[OH^-] = [OH^-]$, $[OH^-]^2 = 6.0 \times 10^{-7}\ [OH^-] + K_w$

$[OH^-]^2 - (6.0 \times 10^{-7})\ [OH^-] - 1.0 \times 10^{-14} = 0$; Using quadratic formula: $[OH^-] = 6.2 \times 10^{-7}\ M$

92. NH_3 + H_2O $\rightleftharpoons$ NH_4^+ + OH^- $K_b = 1.8 \times 10^{-5}$

Initial   15.0 M 0   0.0100 M (Assume no volume change.)
Equil.   15.0 - x x 0.0100 + x

$$1.8 \times 10^{-5} = \frac{x(0.0100 + x)}{15.0 - x} \approx \frac{x(0.0100)}{15.0}, \ x = 0.027; \text{ Assumption is horrible (270\% error).}$$

Using the quadratic formula:

$$1.8 \times 10^{-5}(15.0 - x) = 0.0100\, x + x^2, \ x^2 + 0.0100\, x - 2.7 \times 10^{-4} = 0$$

$$x = 1.2 \times 10^{-2}, \ [OH^-] = 1.2 \times 10^{-2} + 0.0100 = 0.022 \ M$$

93. $[H^+]_o = 1.0 \times 10^{-2} + 1.0 \times 10^{-2} = 2.0 \times 10^{-2} \ M$ from strong acids HCl and H_2SO_4.

HSO_4^- is a good weak acid ($K_a = 0.012$). However, HCN is a poor weak acid ($K_a = 6.2 \times 10^{-10}$) and can be ignored. Calculating the H^+ contribution from HSO_4^-:

                 HSO_4^- $\rightleftharpoons$ H^+ + SO_4^{2-} $K_a = 0.012$

Initial   0.010 M 0.020 M 0
Equil.   0.010 - x 0.020 + x x

$$\frac{x(0.020 + x)}{(0.010 - x)} = 0.012 \approx \frac{x(0.020)}{(0.010)}, \ x = 0.0060; \text{ Assumption poor (60\% error).}$$

Using the quadratic formula: $x^2 + 0.032\, x - 1.2 \times 10^{-4} = 0, \ x = 3.4 \times 10^{-3}$

$[H^+] = 0.020 + x = 0.020 + 3.4 \times 10^{-3} = 0.023 \ M; \ \text{pH} = 1.64$

94. NH_3 + H_2O $\rightleftharpoons$ OH^- + NH_4^+ $K_b = 1.8 \times 10^{-5}$

Initial   15.0 M $0(10^{-7})$ 0
Equil.   15.0 - x x x

$$\frac{x^2}{15.0 - x} = 1.8 \times 10^{-5} \approx \frac{x^2}{15.0}, \ x = 1.6 \times 10^{-2}; \text{ Assumptions good.}$$

$[OH^-] = 1.6 \times 10^{-2} \ M; \ \text{pOH} = 1.80; \ \text{pH} = 12.20$

95.   a.   HOAc $\rightleftharpoons$ H^+ + OAc^-

Initial   0.100 M $5.00 \times 10^{-4} \ M$ 0
Equil.   0.100 - x $5.00 \times 10^{-4} + x$ x

$$1.8 \times 10^{-5} = \frac{x(5.00 \times 10^{-4} + x)}{(0.100 - x)} \approx \frac{x(5.00 \times 10^{-4})}{0.100}$$

$x = 3.6 \times 10^{-3}$; Assumption is horrible. Using the quadratic formula:

$x = 1.1 \times 10^{-3}\ M$; $[H^+] = 5.00 \times 10^{-4} + x = 1.6 \times 10^{-3}\ M$; pH = 2.80

b. $x = [OAc^-] = 1.1 \times 10^{-3}\ M$

96.   a. $NH_4(HCO_3) \rightarrow NH_4^+ + HCO_3^-$

$$K_a(NH_4^+) = \frac{1.0 \times 10^{-14}}{1.8 \times 10^{-5}} = 5.6 \times 10^{-10};\ \ K_b(HCO_3^-) = \frac{K_w}{K_{a_1}} = \frac{1.0 \times 10^{-14}}{4.3 \times 10^{-7}} = 2.3 \times 10^{-8}$$

Solution is basic since HCO_3^- is a stronger base than NH_4^+ is as an acid. The acidic properties of HCO_3^- were ignored since K_{a_2} is small (5.6×10^{-11}).

b. $NaH_2PO_4 \rightarrow Na^+ + H_2PO_4^-$; Ignore Na^+.

$$K_{a_2}(H_2PO_4^-) = 6.2 \times 10^{-8};\ \ K_b(H_2PO_4^-) = \frac{K_w}{K_{a_1}} = \frac{1.0 \times 10^{-14}}{7.5 \times 10^{-3}} = 1.3 \times 10^{-12}$$

Solution is acidic since $K_a > K_b$.

c. $Na_2HPO_4 \rightarrow 2\ Na^+ + HPO_4^{2-}$; Ignore Na^+.

$$K_{a_3}(HPO_4^{2-}) = 4.8 \times 10^{-13};\ \ K_b(HPO_4^{2-}) = \frac{K_w}{K_{a_2}} = \frac{1.0 \times 10^{-14}}{6.2 \times 10^{-8}} = 1.6 \times 10^{-7}$$

Solution is basic since $K_b > K_a$.

d. $NH_4(H_2PO_4) \rightarrow NH_4^+ + H_2PO_4^-$

NH_4^+ is weak acid and $H_2PO_4^-$ is also acidic (see b). Solution with both ions present will be acidic.

e. $NH_4(HCO_2) \rightarrow NH_4^+ + HCO_2^-$

$$K_a(NH_4^+) = 5.6 \times 10^{-10};\ \ K_b(HCO_2^-) = \frac{K_w}{K_a} = \frac{1.0 \times 10^{-14}}{1.8 \times 10^{-4}} = 5.6 \times 10^{-11}$$

Solution is acidic since NH_4^+ is a stronger acid than HCO_2^- is as a base.

CHAPTER EIGHT

APPLICATIONS OF AQUEOUS EQUILIBRIA

Common Ions: Buffers

1. A buffered solution must contain both a weak acid and a weak base. Buffer solutions are useful for controlling the pH of a solution since they resist pH change.

2. $CO_3^{2-} + H^+ \rightarrow HCO_3^-$; $HCO_3^- + OH^- \rightarrow H_2O + CO_3^{2-}$

3. The capacity of a buffer is a measure of how much strong acid or base the buffer can neutralize. All the buffers listed have the same pH. The 1.0 M buffer has the greatest capacity; the 0.01 M buffer the least capacity.

4. $NH_3 + H_2O \rightleftharpoons NH_4^+ + OH^-$ $K_b = \dfrac{[NH_4^+][OH^-]}{[NH_3]}$

   $-\log K_b = -\log [OH^-] - \log \dfrac{[NH_4^+]}{[NH_3]}$, $-\log [OH^-] = -\log K_b + \log \dfrac{[NH_4^+]}{[NH_3]}$

   $pOH = pK_b + \log \dfrac{[NH_4^+]}{[NH_3]}$ or $pOH = pK_b + \log \dfrac{[Acid]}{[Base]}$

5. a. This is a weak acid problem. Let $CH_3CH_2CO_2H$ = HOPr and $CH_3CH_2CO_2^-$ = OPr⁻.

   $$HOPr \rightleftharpoons H^+ + OPr^- \qquad K_a = 1.3 \times 10^{-5}$$

	HOPr	H⁺	OPr⁻
Initial	0.100 M	~0	0

   x mol/L HOPr dissociates to reach equilibrium

   | Change | -x $\rightarrow$ | +x | +x |
   | Equil. | 0.100 - x | x | x |

   $K_a = 1.3 \times 10^{-5} = \dfrac{[H^+][OPr^-]}{[HOPr]} = \dfrac{x^2}{0.100 - x} \approx \dfrac{x^2}{0.100}$

   $x = [H^+] = 1.1 \times 10^{-3}\ M$; pH = 2.96 Assumptions good by the 5% rule.

b. This is a weak base problem.

$$OPr^- \;+\; H_2O \;\rightleftharpoons\; HOPr \;+\; OH^- \qquad K_b = \dfrac{K_w}{K_a} = 7.7 \times 10^{-10}$$

Initial	0.100 M	0	0
	x mol/L OPr^- reacts with H_2O to reach equilibrium		
Change	$-x$ $\rightarrow$	$+x$	$+x$
Equil.	0.100 $- x$	x	x

$$K_b = 7.7 \times 10^{-10} = \dfrac{[HOPr]\,[OH^-]}{[OPr^-]} = \dfrac{x^2}{0.100 - x} \approx \dfrac{x^2}{0.100}$$

$x = [OH^-] = 8.8 \times 10^{-6}\ M;\ \ pOH = 5.06;\ \ pH = 8.94$ \qquad Assumptions good.

c. pure H_2O, $[H^+] = [OH^-] = 1.0 \times 10^{-7}\ M;\ \ pH = 7.00$

d. This solution contains a weak acid and its conjugate base. This is a buffer solution. Solve for the pH through the normal set-up.

$$HOPr \;\rightleftharpoons\; H^+ \;+\; OPr^- \qquad K_a = 1.3 \times 10^{-5}$$

Initial	0.100 M	~0	0.100 M
	x mol/L HOPr dissociates to reach equilibrium		
Change	$-x$ $\rightarrow$	$+x$	$+x$
Equil.	0.100 $- x$	x	0.100 $+ x$

$$1.3 \times 10^{-5} = \dfrac{(0.100 + x)\,(x)}{0.100 - x} \approx \dfrac{(0.100)\,(x)}{0.100} = x = [H^+]$$

$[H^+] = 1.3 \times 10^{-5}\ M;\ \ pH = 4.89$ \quad Assumptions good.

Alternately, we can use the Henderson-Hasselbalch equation to calculate the pH of buffer solutions.

$$pH = pK_a + \log \dfrac{[Base]}{[Acid]} = pK_a + \log \dfrac{(0.100)}{(0.100)} = pK_a = -\log\,(1.3 \times 10^{-5}) = 4.89$$

The Henderson-Hasselbalch equation will be valid when an assumption of the type, $0.1 + x \approx 0.1$, that we just made in this problem is valid. From a practical standpoint, this will almost always be true for useful buffer solutions. If the assumption is not valid, the solution will have such a low buffering capacity it will not be of any use to control the pH. Note: The Henderson-Hasselbalch equation can only be used to solve for the pH of buffer solutions.

6. a.   HOPr $\rightleftharpoons$ H^+ + OPr^- $K_a = 1.3 \times 10^{-5}$

Initial   0.100 M 0.020 M   0

x mol/L HOPr dissociates to reach equilibrium

Change $-x$ $\rightarrow$ $+x$ $+x$

Equil.   0.100 - x 0.020 + x x

$[H^+] = 0.020 + x \approx 0.020\ M$; pH = 1.70   Assumption good ($x = 6.5 \times 10^{-5}$).

Note: H^+ contribution from the weak acid HOPr was negligible. The pH of the solution
can be determined by only considering the amount of strong acid present.

b. H^+ reacts completely with the best base present, OPr^-.

OPr^- + H^+ $\rightarrow$ HOPr

Before   0.100 M 0.020 M 0

Change   -0.020   -0.020 $\rightarrow$ +0.020   Reacts completely

After   0.080   0   0.020 M

After reaction, a weak acid, HOPr , and its conjugate base, OPr^-, are present. This is a buffer
solution. Using the Henderson-Hasselbalch equation:

$$pH = pK_a + \log \frac{[\text{Base}]}{[\text{Acid}]} = 4.89 + \log \frac{(0.080)}{(0.020)} = 5.49 \qquad \text{Assumptions good.}$$

c. $[H^+] = 0.020\ M$; pH = 1.70

d. Added H^+ reacts completely with best base present, OPr^-.

OPr^- + H^+ $\rightarrow$ HOPr

Before   0.100 M 0.020 M 0.100 M

Change   -0.020   -0.020 $\rightarrow$ +0.020   Reacts completely

After   0.080   0   0.120

A buffer solution results (weak acid and conjugate base). Using Henderson-Hasselbalch
equation:

$$pH = pK_a + \log \frac{[\text{Base}]}{[\text{Acid}]} = 4.89 + \log \frac{(0.080)}{(0.120)} = 4.71$$

7.   a.   OH⁻ will react completely with the best acid present, HOPr.

HOPr   +   OH⁻   →   OPr⁻   + H_2O

Before   0.100 M 0.020 M 0
Change   -0.020   -0.020   →   +0.020   Reacts completely
After   0.080   0   0.020

A buffer solution results after the reaction. Using the Henderson-Hasselbalch equation:

$$pH = pK_a + \log \frac{[Base]}{[Acid]} = 4.89 + \log \frac{(0.020)}{(0.080)} = 4.29$$

b.   OPr⁻   + H_2O ⇌   HOPr   +   OH⁻ $K_b = 7.7 \times 10^{-10}$

Before   0.100 M 0   0.020 M
    x mol/L OPr⁻ reacts with H_2O to reach equilibrium
Change   -x →   +x +x
Equil.   0.100 - x x 0.020 + x

[OH⁻] = 0.020 + x ≈ 0.020 M; pOH = 1.70; pH = 12.30   Assumption good.

Note: OH⁻ contribution from the weak base OPr⁻ was negligible ($x = 3.9 \times 10^{-9}$). pH can be determined by only considering the amount of strong base present.

c.   [OH⁻] = 0.020 M; pOH = 1.70; pH = 12.30

d.   OH⁻ will react completely with HOPr, the best acid present.

HOPr   +   OH⁻   →   OPr⁻   + H_2O

Initial   0.100 M 0.020 M 0.100 M
Change   -0.020   -0.020   →   +0.020   Reacts completely
After   0.080   0   0.120

Using Henderson-Hasselbalch equation to solve for the pH of the resulting buffer solution:

$$pH = pK_a + \log \frac{[Base]}{[Acid]} = 4.89 + \log \frac{(0.120)}{(0.080)} = 5.07$$

8.   Consider all of the results to Exercises 8.5, 6, and 7.

Solution	Initial pH	after added acid	after added base
a	2.96	1.70	4.29
b	8.94	5.49	12.30
c	7.00	1.70	12.30
d	4.89	4.71	5.07

The solution in (d) is a buffer; it contains both a weak acid (HOPr) and a weak base (OPr⁻). It resists a change in pH when strong acid or strong base is added.

9.   a. $HNO_2 \rightleftharpoons H^+ + NO_2^-$ $K_a = 4.0 \times 10^{-4}$

This is a buffer solution. Using the Henderson-Hasselbalch equation:

$$pH = pK_a + \log \frac{[Base]}{[Acid]} = 3.40 + \log \frac{(0.15)}{(0.10)} = 3.40 + 0.18 = 3.58$$

b. $25.0 \text{ g } CH_3CO_2H \times \dfrac{1 \text{ mol}}{60.05 \text{ g}} = 0.416 \text{ mol HOAc}; \quad [HOAc] = \dfrac{0.416 \text{ mol}}{0.500 \text{ L}} = 0.832 \, M$

Remember we write acetic acid several different ways. Acetic acid: HOAc, CH_3CO_2H, $HC_2H_3O_2$; acetate ion: OAc^-, $CH_3CO_2^-$, $C_2H_3O_2^-$

$40.0 \text{ g } CH_3CO_2Na \times \dfrac{1 \text{ mol}}{82.03 \text{ g}} = 0.488 \text{ mol } OAc^-; \quad [OAc^-] = \dfrac{0.488 \text{ mol}}{0.500 \text{ L}} = 0.976 \, M$

$pH = pK_a + \log \dfrac{[Base]}{[Acid]}; \quad K_a = 1.8 \times 10^{-5}; \quad pK_a = -\log 1.8 \times 10^{-5} = 4.74$

$pH = 4.74 + \log \dfrac{(0.976)}{(0.832)} = 4.74 + 0.0693 = 4.81$

c. $50.0 \text{ mL} \times \dfrac{1.0 \text{ mmol}}{\text{mL}} = 50. \text{ mmol HOCl}; \quad 30.0 \text{ mL} \times \dfrac{0.80 \text{ mmol NaOH}}{\text{mL}} = 24 \text{ mmol } OH^-$

The strong base reacts with the best acid available, HOCl. The reaction goes to completion. Whenever strong base or strong acid reacts, the reaction is always assumed to go to completion.

	HOCl	+	OH⁻	→	OCl⁻	+	H₂O
Before	50. mmol		24 mmol		0		

24 mmol OH⁻ reacts completely.

	HOCl	+	OH⁻	→	OCl⁻	+	H₂O
Change	-24		-24	→	+24		Reacts completely
After	26 mmol		0		24 mmol		

After reaction, the solution contains 26 mmol HOCl and 24 mmol OCl⁻ in 250 mL of solution. This is a buffer solution. Using the Henderson-Hasselbalch equation:

$pH = pK_a + \log \dfrac{[Base]}{[Acid]}; \quad K_a = 3.5 \times 10^{-8}; \quad pK_a = 7.46$

$pH = 7.46 + \log \dfrac{\dfrac{24 \text{ mmol OCl}^-}{250 \text{ mL}}}{\dfrac{26 \text{ mmol HOCl}}{250 \text{ mL}}} = 7.46 - 0.035 = 7.43$

d. $100.0 \text{ g NH}_4\text{Cl} \times \dfrac{1 \text{ mol}}{53.49 \text{ g}} = 1.870 \text{ mol NH}_4^+$; $[\text{NH}_4^+] = 1.87 \, M$ (Carry extra S.F.)

$65.0 \text{ g NaOH} \times \dfrac{1 \text{ mol}}{40.00 \text{ g}} = 1.63 \text{ mol NaOH}$; $[\text{OH}^-] = 1.63 \, M$

OH⁻ reacts completely with best acid present, NH_4^+.

$$\text{NH}_4^+ \quad + \quad \text{OH}^- \qquad \rightarrow \qquad \text{NH}_3 \quad + \quad \text{H}_2\text{O}$$

Before	1.87 M	1.63 M	0	
Change	-1.63	-1.63 $\rightarrow$	+1.63	Reacts completely
After	0.24 M	0	1.63 M	

After reaction, a buffer solution results.

$$\text{pH} = \text{pK}_a + \log \frac{[\text{NH}_3]}{[\text{NH}_4^+]}; \quad \text{K}_a = \frac{\text{K}_w}{\text{K}_b} = \frac{1.0 \times 10^{-14}}{1.8 \times 10^{-5}} = 5.6 \times 10^{-10}; \quad \text{pK}_a = 9.25$$

$$\text{pH} = 9.25 + \log \frac{1.63}{0.24} = 9.25 + 0.83 = 10.08 \approx 10.1$$

e. $26.4 \text{ g CH}_3\text{CO}_2\text{Na} \times \dfrac{1 \text{ mol}}{82.03 \text{ g}} = 0.322 \text{ mol NaOAc}$

$50.0 \text{ mL HCl} \times \dfrac{1 \text{ L}}{1000 \text{ mL}} \times \dfrac{6.00 \text{ mol HCl}}{\text{L}} = 0.300 \text{ mol HCl}$

H⁺ reacts completely with best base present, OAc⁻. (Strong acids always react to completion.)

$$\text{H}^+ \quad + \quad \text{OAc}^- \qquad \rightarrow \qquad \text{HOAc}$$

Before	0.300 mol	0.322 mol	0	
Change	-0.300	-0.300 $\rightarrow$	+0.300	Reacts completely
After	0	0.022 mol	0.300 mol	

After reaction, a buffer solution results. Using the Henderson-Hasselbalch equation:

$$\text{pH} = \text{pK}_a + \log \frac{[\text{OAc}^-]}{[\text{HOAc}]} = 4.74 + \log \frac{(0.022 \text{ mol}/0.500 \text{ L})}{(0.300 \text{ mol}/0.500 \text{ L})}, \quad \text{pH} = 4.74 - 1.13 = 3.61$$

10. $\text{NH}_4^+ \rightleftharpoons \text{H}^+ + \text{NH}_3$ $\text{K}_a = \dfrac{\text{K}_w}{\text{K}_b} = 5.6 \times 10^{-10}$; $\text{pK}_a = -\log (5.6 \times 10^{-10}) = 9.25$

a. $9.00 = 9.25 + \log \dfrac{[NH_3]}{[NH_4^+]}$ b. $8.80 = 9.25 + \log \dfrac{[NH_3]}{[NH_4^+]}$

$\log \dfrac{[NH_3]}{[NH_4^+]} = -0.25$ $\dfrac{[NH_3]}{[NH_4^+]} = 10^{-0.45} = 0.35$

$\dfrac{[NH_3]}{[NH_4^+]} = 10^{-0.25} = 0.56$

c. $10.00 = 9.25 + \log \dfrac{[NH_3]}{[NH_4^+]}$ d. $9.60 = 9.25 + \log \dfrac{[NH_3]}{[NH_4^+]}$

$\dfrac{[NH_3]}{[NH_4^+]} = 10^{0.75} = 5.6$ $\dfrac{[NH_3]}{[NH_4^+]} = 10^{0.35} = 2.2$

11.   When OH⁻ is added, it converts HOAc into OAc⁻: $HOAc + OH^- \rightarrow OAc^- + H_2O$

The moles of OAc⁻ produced <u>equals</u> the moles of OH⁻ added. Since the volume is 1.0 L (assuming no volume change), then:

   $[OAc^-] + [HOAc] = 2.0$ *M* and $[OAc^-]$ produced $= [OH^-]$ added.

a. $pH = pK_a + \log \dfrac{[OAc^-]}{[HOAc]}$; For $pH = pK_a$, $\log \dfrac{[OAc^-]}{[HOAc]} = 0$

   Therefore, $\dfrac{[OAc^-]}{[HOAc]} = 1.0$ and $[OAc^-] = [HOAc]$

   Since $[OAc^-] + [HOAc] = 2.0$, then $[OAc^-] = [HOAc] = 1.0$ *M*

   To produce a 1.0 *M* OAc⁻ solution we need to add 1.0 mol of NaOH to 1.0 L of the HOAc solution.

   1.0 mol of NaOH $\times \dfrac{40.00 \text{ g}}{\text{mol NaOH}} = 40.$ g NaOH

b. $4.00 = 4.74 + \log \dfrac{[OAc^-]}{[HOAc]}$; $\dfrac{[OAc^-]}{[HOAc]} = 10^{-0.74} = 0.18$

   $[OAc^-] = 0.18$ $[HOAc]$ or $[HOAc] = 5.6$ $[OAc^-]$; $[OAc^-] + [HOAc] = 2.0$ *M*

   $[OAc^-] + 5.6 [OAc^-] = 2.0$ *M*, $[OAc^-] = \dfrac{2.0}{6.6} = 0.30$ *M*

   We need to add 0.30 mol of NaOH to 1.0 L of HOAc solution to produce 0.30 *M* OAc⁻.

   0.30 mol $\times \dfrac{40.0 \text{ g}}{\text{mol}} = 12$ g NaOH

c. $5.00 = 4.74 + \log \dfrac{[OAc^-]}{[HOAc]}, \quad \dfrac{[OAc^-]}{[HOAc]} = 10^{0.26} = 1.8$

$1.8\,[HOAc] = [OAc^-]$ or $[HOAc] = 0.56\,[OAc^-]$; $[HOAc] + [OAc^-] = 2.0\ M$

$1.56\,[OAc^-] = 2.0\ M$; $[OAc^-] = 1.3\ M$

We need to add 1.3 mol of NaOH or $1.3\ \text{mol} \times \dfrac{40.0\ \text{g}}{\text{mol}} = 52$ g NaOH

12. $75.0\ \text{g CH}_3\text{CO}_2\text{Na} \times \dfrac{1\ \text{mol}}{82.03\ \text{g}} = 0.914$ mol NaOAc; $[OAc^-] = \dfrac{0.914\ \text{mol}}{0.5000\ \text{L}} = 1.83\ M$

$pH = pK_a + \log \dfrac{[OAc^-]}{[HOAc]} = 4.74 + \log\left(\dfrac{1.83}{0.64}\right) = 5.20$

13. $[H^+]$ added $= \dfrac{0.010\ \text{mol}}{0.25\ \text{L}} = 0.040\ M$; The added H^+ converts NH_3 to NH_4^+.

a.

	NH_4^+	$\rightleftharpoons$	NH_3	+	H^+	
Before	0.15 M		0.050 M		0.040 M	
Change	+0.040	$\leftarrow$	-0.040		-0.040	Reacts completely
After	0.19		0.010		0	

A buffer solution still exists. Using the Henderson-Hasselbalch equation:

$$pH = pK_a + \log \dfrac{[NH_3]}{[NH_4^+]} = 9.25 + \log\left(\dfrac{0.010}{0.19}\right) = 7.97$$

b.

	NH_4^+	$\rightleftharpoons$	NH_3	+	H^+	
Before	1.50 M		0.50 M		0.040 M	
Change	+0.040	$\leftarrow$	-0.040		-0.040	Reacts completely
After	1.54		0.46		0	

$$pH = pK_a + \log \dfrac{[NH_3]}{[NH_4^+]} = 9.25 + \log\left(\dfrac{0.46}{1.54}\right) = 8.73$$

14. $pH = pK_a + \log \dfrac{[\text{Base}]}{[\text{Acid}]}, \quad 7.41 = 6.37 + \log \dfrac{[HCO_3^-]}{[H_2CO_3]}$

$\dfrac{[HCO_3^-]}{[H_2CO_3]} = 10^{1.04} = 11, \quad \dfrac{[CO_2]}{[HCO_3^-]} = 0.091$

15.   The reaction $H^+ + NH_3 \rightarrow NH_4^+$ goes to completion. After this reaction occurs there must be both NH_3 and NH_4^+ in solution for it to be a buffer. After the reaction the solutions contain:

   a.   0.05 M NH_4^+ and 0.05 M H^+ b.   0.05 M NH_4^+

   c.   0.05 M NH_4^+ and 0.05 M H^+ d.   0.05 M NH_4^+ and 0.05 M NH_3

   Only the combination in (d) results in a buffer.

16.   a.   No, A strong acid and neutral salt do not constitute a buffer solution.

   b.   Yes, HNO_2 = weak acid, NO_2^- = weak base

   c.   Yes; 250 mL × 0.10 mmol/mL = 25 mmol CH_3CO_2H

        500 mg KOH × 1 mmol/56.1 mg = 9 mmol OH^-

        $OH^- + CH_3CO_2H \rightarrow CH_3CO_2^- + H_2O$ Reacts completely. After this reaction goes to completion, both a weak acid and weak base are present: 16 mmol CH_3CO_2H and 9 mmol $CH_3CO_2^-$ ion. This is a buffer solution.

   d.   No, both the CO_3^{2-} and PO_4^{3-} ions are weak bases. No weak acid is present.

Acid-Base Titrations

17.

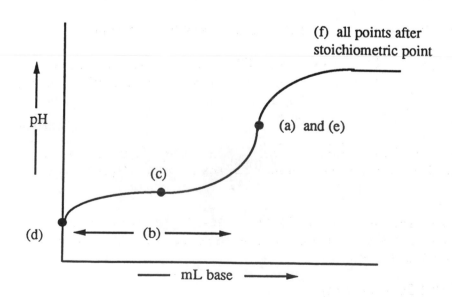

18.   At the beginning of the titration, only the weak acid HLac is present:

$$HLac \quad \rightleftharpoons \quad H^+ \quad + \quad Lac^- \qquad K_a = 1.4 \times 10^{-4} \quad HLac: HC_3H_5O_3$$
$$Lac^-: C_3H_5O_3^-$$

Initial   0.100 M ~0   0
            x mol/L HLac dissociates to reach equilibrium
Change $-x$ $\rightarrow$ $+x$ $+x$
Equil.   0.100 - x x x

$1.4 \times 10^{-4} = \dfrac{x^2}{0.100 - x} = \dfrac{x^2}{0.100}$, $x = [H^+] = 3.7 \times 10^{-3}$ M; pH = 2.43   Assumptions good.

Up to the stoichiometric point, we calculate the pH using the Henderson-Hasselblach equation. This is the buffer region. For example, 4.0 mL of NaOH added:

$$\text{initial mmol HLac} = 25.0 \text{ mL} \times \frac{0.100 \text{ mmol}}{\text{mL}} = 2.50 \text{ mmol HLac}$$

$$\text{mmol OH}^- \text{ added} = 4.0 \text{ mL} \times \frac{0.100 \text{ mmol}}{\text{mL}} = 0.40 \text{ mmol OH}^-$$

Note: The units mmol are usually easier numbers to work with. The units for molarity are moles/L but are also equal to mmoles/mL.

The OH⁻ converts 0.40 mmoles HLac to 0.40 mmoles Lac⁻ according to the equation:

$$HLac + OH^- \rightarrow Lac^- + H_2O \qquad \text{Reacts completely}$$

mmol HLac remaining = 2.50 - 0.40 = 2.10 mmol; mmol Lac⁻ produced = 0.40 mmol

We have a buffer solution. Using the Henderson-Hasselbalch equation:

$$pH = pK_a + \log \frac{[Lac^-]}{[HLac]} = 3.86 + \log \frac{(0.40)}{(2.10)} \qquad \text{(Total volume cancels, so we can use the ratio of moles.)}$$

pH = 3.86 - 0.72 = 3.14

Other points in the buffer region are calculated in a similar fashion. Do a stoichiometry problem first, followed by a buffer problem. The buffer region includes all points up to 24.9 mL OH⁻ added.

At the stoichiometric point, we have added enough OH⁻ to convert all of the HLac (2.50 mmol) into its conjugate base, Lac⁻. All that is present is a weak base. To determine the pH, we perform a weak base calculation.

$$[Lac^-] = \frac{2.50 \text{ mmol}}{50.0 \text{ mL}} = 0.0500 \text{ } M$$

$$\text{Lac}^- + H_2O \rightleftharpoons \text{HLac} + \text{OH}^- \qquad K_b = \frac{1.0 \times 10^{-14}}{1.4 \times 10^{-4}} = 7.1 \times 10^{-11}$$

Initial	$0.0500\ M$		0	0

x mol/L Lac$^-$ reacts with H_2O to reach equilibrium

Change	$-x$	$\rightarrow$	$+x$	$+x$
Equil.	$0.0500 - x$		x	x

$$\frac{x^2}{0.0500 - x} = \frac{x^2}{0.0500} = 7.1 \times 10^{-11}$$

$x = [\text{OH}^-] = 1.9 \times 10^{-6}\ M$; $\text{pOH} = 5.72$; $\text{pH} = 8.28$ Assumptions good.

Past the stoichiometric point, we have added more than 2.50 mmol of NaOH. The pH will be determined by the excess OH$^-$ ion present.

At 25.1 mL: OH$^-$ added $= 25.1\ \text{mL} \times \dfrac{0.100\ \text{mmol}}{\text{mL}} = 2.51$ mmol OH$^-$

excess OH$^-$ = 2.51 - 2.50 = 0.01 mmol

$[\text{OH}^-] = \dfrac{0.01\ \text{mmol}}{50.1\ \text{mL}} = 2 \times 10^{-4}\ M$; $\text{pOH} = 3.7$; $\text{pH} = 10.3$

All results are listed in Table 8.1 at the end of the solution to Exercise 8.21.

19.   At beginning of the titration, only the weak base NH$_3$ is present:

$$NH_3 + H_2O \rightleftharpoons NH_4^+ + \text{OH}^- \qquad K_b = 1.8 \times 10^{-5}$$

Equil.	$0.100 - x$		x	x

$$\frac{x^2}{0.100 - x} = \frac{x^2}{0.100} = 1.8 \times 10^{-5}$$

$x = [\text{OH}^-] = 1.3 \times 10^{-3}\ M$; $\text{pOH} = 2.89$; $\text{pH} = 11.11$ Assumptions good.

In the buffer region (4 - 24.9 mL), use $\text{pH} = \text{p}K_a + \log \dfrac{[\text{Base}]}{[\text{Acid}]}$.

$$K_a = \frac{1.0 \times 10^{-14}}{1.8 \times 10^{-5}} = 5.6 \times 10^{-10}; \quad \text{p}K_a = 9.25; \quad \text{pH} = 9.25 + \log \frac{[NH_3]}{[NH_4^+]}$$

For example, after 8.0 mL HCl added:

initial mmol NH$_3$ = $25.0\ \text{mL} \times \dfrac{0.100\ \text{mmol}}{\text{mL}} = 2.50$ mmol

mmol H$^+$ added = $8.0\ \text{mL} \times \dfrac{0.100\ \text{mmol}}{\text{mL}} = 0.80$ mmol

Added H^+ reacts with NH_3 to completion: $NH_3 + H^+ \rightarrow NH_4^+$

mmol NH_3 remaining = 2.50 - 0.80 = 1.70 mmol; mmol NH_4^+ produced = 0.80 mmol

$pH = 9.25 + \log \dfrac{1.70}{0.80} = 9.58$ (Mole ratios can be used since total volume cancels.)

At the stoichiometric point, enough HCl has been added to convert all of the weak base (NH_3) into its conjugate acid, NH_4^+. Perform a weak acid calculation. $[NH_4^+]_0 = \dfrac{2.50 \text{ mmol}}{50.0 \text{ mL}} = 0.0500 \ M$

$$NH_4^+ \quad \rightleftharpoons \quad H^+ \ + \ NH_3 \qquad K_a = 5.6 \times 10^{-10}$$

Equil.   0.0500 - x x x

$5.6 \times 10^{-10} = \dfrac{x^2}{0.0500 - x} = \dfrac{x^2}{0.0500}$

$x = [H^+] = 5.3 \times 10^{-6} \ M$; pH = 5.28   Assumptions good.

Beyond the stoichiometric point, the pH is determined by the excess H^+.

At 28.0 mL: H^+ added = 28.0 mL $\times \dfrac{0.100 \text{ mmol}}{\text{mL}} = 2.80$ mmol H^+

Excess H^+ = 2.80 mmol - 2.50 mmol = 0.30 mmol excess H^+

$[H^+] = \dfrac{0.30 \text{ mmol}}{53.0 \text{ mL}} = 5.7 \times 10^{-3} \ M$; pH = 2.24

All results are in Table 8.1 after Exercise 8.21.

20.   Initially, a weak base problem:

$$py \ + \ H_2O \quad \rightleftharpoons \quad Hpy^+ \ + \ OH^- \qquad \text{py is pyridine}$$

Equil.   0.100 - x x x

$K_b = \dfrac{[Hpy^+][OH^-]}{[py]} = \dfrac{x^2}{0.100 - x} = \dfrac{x^2}{0.100} = 1.7 \times 10^{-9}$

$x = [OH^-] = 1.3 \times 10^{-5}$; pOH = 4.89; pH = 9.11   Assumptions good.

Buffer region (4-24.9) mL: Added acid converts: $py + H^+ \rightarrow Hpy^+$. Determine moles of py and Hpy^+ after reaction and use the Henderson-Hasselbalch equation to solve for the pH.

$K_a = \dfrac{K_w}{K_b} = \dfrac{1.0 \times 10^{-14}}{1.7 \times 10^{-9}} = 5.9 \times 10^{-6}$; $pK_a = 5.23$; $pH = 5.23 + \log \dfrac{[py]}{[Hpy^+]}$

At stoichiometric point, a weak acid problem:

$$Hpy^+ \quad \rightleftharpoons \quad py \quad + \quad H^+ \qquad K_a = 5.9 \times 10^{-6}$$

Equil.   0.0500 - x x x

$$5.9 \times 10^{-6} = \frac{x^2}{0.0500 - x} = \frac{x^2}{0.0500}, \quad x = [H^+] = 5.4 \times 10^{-4}; \quad pH = 3.27 \quad \text{Assumptions good.}$$

Beyond the equivalence point, the pH is the same as in Exercise 8.19. All results are summarized in Table 8.1 after Exercise 8.21.

21. For the titration of a strong base with a strong acid, we calculate the excess $[OH^-]$ or $[H^+]$ and then calculate the pH.

Initial (0 mL H^+ added): $[OH^-] = 0.100$ M; pOH = 1.000; pH = 13.000

4.0 mL: initial mmol $OH^- = 25.0$ mL $\times \dfrac{0.100 \text{ mmol}}{\text{mL}} = 2.50$ mmol OH^-

mmol H^+ added = 4.0 mL $\times \dfrac{0.100 \text{ mmol}}{\text{mL}} = 0.40$ mmol H^+

Added H^+ reacts with OH^- to completion: $OH^- + H^+ \rightarrow H_2O$

mmol OH^- remaining = 2.50 - 0.40 = 2.10 mmol excess OH^-

$[OH^-] = \dfrac{2.10 \text{ mmol}}{29.0 \text{ mL}} = 7.24 \times 10^{-2}$ M; pOH = 1.140; pH = 12.860

Equivalence point: neutral solution, pH = 7.000

Past the equivalence point, excess H^+ is added. pH determined by amount (molarity) of excess H^+. All results are summarized in Table 8.1.

Table 8.1: Summary of pH Results for Exercises 8.18 - 8.21   (Graph follows)

titrant mL	Exercise 8.18	Exercise 8.19	Exercise 8.20	Exercise 8.21
0.0	2.43	11.11	9.11	13.000
4.0	3.14	9.97	5.95	12.860
8.0	3.53	9.58	5.56	12.712
12.5	3.86	9.25	5.23	12.523
20.0	4.46	8.65	4.63	12.05
24.0	5.24	7.87	3.85	11.31
24.5	5.6	7.6	3.5	11.0
24.9	6.3	6.9	-	10.3
25.0	8.28	5.28	3.27	7.000
25.1	10.3	3.7	-	3.7
26.0	11.29	2.71	2.71	2.71
28.0	11.75	2.25	2.25	2.25
30.0	11.96	2.04	2.04	2.04

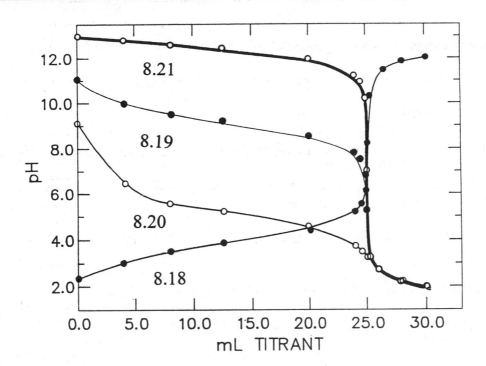

22.   Between the starting point of the titration and the equivalence point, we are dealing with a buffer solution. The Henderson-Hasselbalch equation can be used to determine pH.

$$pH = pK_a + \log \frac{[Base]}{[Acid]}$$

At the halfway point to equivalence:

[Base] = [Acid] so $pH = pK_a + \log 1 = pK_a$

23.   Yes, at any point this can be done. In Chapter 7 we calculated the K_a from a solution of only the weak acid. In the buffer region, we can calculate the ratio of the base to acid form of the weak acid and use the Henderson-Hasselbalch equation to determine the K_a value.

$$pH = pK_a + \log \frac{[Base]}{[Acid]}$$

24.   a. $0.104 \text{ g NaC}_2\text{H}_3\text{O}_2 \times \dfrac{1 \text{ mol NaC}_2\text{H}_3\text{O}_2}{82.03 \text{ g}} = 1.27 \times 10^{-3} \text{ mol} = 1.27 \text{ mmol NaC}_2\text{H}_3\text{O}_2$

$1.27 \text{ mmol NaC}_2\text{H}_3\text{O}_2 \times \dfrac{1 \text{ mmol HCl}}{\text{mmol NaC}_2\text{H}_3\text{O}_2} \times \dfrac{1 \text{ mL}}{0.0996 \text{ mmol HCl}} = 12.8 \text{ mL}$

Added H^+ reacts with the acetate ion to completion: $H^+ + C_2H_3O_2^- \rightarrow HC_2H_3O_2$. The titration converts acetate ion into acetic acid. At the equivalence point, there is a solution of acetic acid with a concentration of:

$$\frac{1.27 \text{ mmol}}{25.0 + 12.8 \text{ mL}} = 3.36 \times 10^{-2} \ M$$

The acetic acid partially dissociates, making the equivalence point acidic.

$$HOAc \rightleftharpoons H^+ + OAc^-$$

Initial	0.0336 M	0	0
	x mol/L HOAc dissociates to reach equilibrium		
Change	$-x$ $\rightarrow$	$+x$	$+x$
Equil.	0.0336 $- x$	x	x

$$K_a = 1.8 \times 10^{-5} = \frac{[H^+][OAc^-]}{[HOAc]} = \frac{x^2}{0.0336 - x} \approx \frac{x^2}{0.0336}$$

$x = [H^+] = 7.8 \times 10^{-4}$ M; pH $= 3.11$ Assumptions good.

b. $50.00 \text{ mL} \times \dfrac{0.0426 \text{ mmol HOCl}}{\text{mL}} = 2.13 \text{ mmol HOCl}$

$2.13 \text{ mmol HOCl} \times \dfrac{1 \text{ mmol NaOH}}{\text{mmol HOCl}} \times \dfrac{1 \text{ mL}}{0.1028 \text{ mmol NaOH}} = 20.7 \text{ mL of NaOH soln.}$

Added OH^- reacts with HOCl to completion: $OH^- + HOCl \rightarrow OCl^- + H_2O$. The titration converts HOCl into OCl^- ion. At the equivalence point, there are 2.13 mmol of OCl^- in 70.7 mL of solution (50.0 mL + 20.7 mL). The concentration of OCl^- is:

$$\frac{2.13 \text{ mmol}}{70.7 \text{ mL}} = 3.01 \times 10^{-2} M$$

The OCl^- reacts with H_2O to produce a basic solution.

$$OCl^- + H_2O \rightleftharpoons HOCl + OH^-$$

Initial	0.0301 M	0	0
	x mol/L OCl^- reacts with H_2O to reach equilibrium		
Change	$-x$	$\rightarrow$ $+x$	$+x$
Equil.	0.0301 $- x$	x	x

$$K_b = \frac{K_w}{K_a} = \frac{1.0 \times 10^{-14}}{3.5 \times 10^{-8}} = 2.9 \times 10^{-7} = \frac{[HOCl][OH^-]}{[OCl^-]} = \frac{x^2}{0.0301 - x}$$

$2.9 \times 10^{-7} \approx \dfrac{x^2}{0.0301}$, $x = [OH^-] = 9.3 \times 10^{-5}$ M; pOH $= 4.03$; pH $= 9.97$ Assumptions good.

c. HBr is a strong acid; NaOH is a strong base. At the equivalence point we will have an aqueous solution of NaBr. Neither Na^+ nor Br^- reacts with water; pH $= 7.00$.

25.   The pH will be less than about 0.5.

26.   pH > 5 for bromcresol green to be blue. pH < 8 for thymol blue to be yellow. pH is between 5 and 8.

27.   a. yellow   b. orange   c. blue   d. bluish-green

28.   a. yellow   b. yellow

       c. green (Both yellow and blue forms are present.)   d. colorless

29.   No, since there are three colored forms there must be two proton transfer reactions. Thus, there must be at least two acidic protons in the acid (orange) form of thymol blue.

30.   Equivalence point: moles acid = moles base. End point: indicator changes color. We want the indicator to tell us when we have reached the equivalence point. We can detect the end point and assume it is the equivalence point for doing stoichiometric calculations. They don't have to be as close as 0.01 pH units, since at the equivalence point the pH is changing very rapidly with added titrant. The range over which an indicator changes color only needs to be close to the pH of the equivalence point.

31.   The two forms of an indicator are different colors. To see only one color, that form must be in ten fold excess or greater over the other. When the ratio of the two forms is less than 10, both colors are present. To go from [HIn]/[In$^-$] = 10 to [HIn]/[In$^-$] = 0.1 requires a change of 2 pH units as the indicator changes from the acid color to the base color.

32.
Exercise	pH at eq. pt.	Indicator
8.18	8.28	phenolphthalein
8.20	3.27	2-4-dinitrophenol

(The Titration in 8.20 is not feasible. The pH break is too small.)

33.
Exercise	pH at eq. pt.	Indicator
8.19	5.28	bromcresol green
8.21	7.00	bromthymol blue

34.   Color will change over the range of pH = pK$_a$ ± 1 = 5.3 ± 1. Useful pH range of methyl red would be about 4.3 (red) - 6.3 (yellow). In titrating a weak acid with base the color would change from red to yellow at pH ~ 4.3. In titrating a weak base with acid the color change would be from yellow to red at pH ~ 6.3. Methyl red would be best for the titration in Exercise 19, since the equivalence point will occur at a pH within the range that methyl red changes color.

35.   From results of the bromcresol green indicator, pH ≈ 5.

       HX ⇌ H$^+$ + X$^-$; From normal weak acid setup: [H$^+$] = [X$^-$] ≈ 1 × 10^{-5} M and [HX] ≈ 0.01 M

$$K_a = \frac{[H^+][X^-]}{[HX]} \approx \frac{(1 \times 10^{-5})^2}{0.01} = 1 \times 10^{-8}$$

Solubility Equilibria

36.   a. $CaC_2O_4(s)$ $\rightleftharpoons$ $Ca^{2+}(aq) + C_2O_4^{2-}(aq)$

s = solubility $\rightarrow$ s   s
in mol/L

$s = [Ca^{2+}] = [C_2O_4^{2-}] = 4.8 \times 10^{-5}\ M$; $K_{sp} = [Ca^{2+}][C_2O_4^{2-}] = (4.8 \times 10^{-5})^2 = 2.3 \times 10^{-9}$

   b. $PbBr_2(s)$ $\rightleftharpoons$ $Pb^{2+}(aq) + 2\ Br^-(aq)$

s = solubility $\rightarrow$ s   2s
in mol/L

$s = 2.14 \times 10^{-2}\ M$; $K_{sp} = [Pb^{2+}][Br^-]^2 = (2.14 \times 10^{-2})(4.28 \times 10^{-2})^2 = 3.92 \times 10^{-5}$

   c. $BiI_3(s)$ $\rightleftharpoons$ $Bi^{3+} + 3\ I^-$

s = solubility $\rightarrow$ s   3s
in mol/L

$s = 1.32 \times 10^{-5}\ M$; $K_{sp} = [Bi^{3+}][I^-]^3 = (1.32 \times 10^{-5})(3.96 \times 10^{-5})^3 = 8.20 \times 10^{-19}$

   d. $FeC_2O_4(s)$ $\rightleftharpoons$ $Fe^{2+} + C_2O_4^{2-}$

s = solubility $\rightarrow$ s   s
in mol/L

$$s = \frac{65.9 \times 10^{-3}\ g}{L} \times \frac{1\ mol}{143.9\ g} = 4.58 \times 10^{-4}\ mol/L$$

$[Fe^{2+}] = [C_2O_4^{2-}] = 4.58 \times 10^{-4}\ M$; $K_{sp} = [Fe^{2+}][C_2O_4^{2-}] = (4.58 \times 10^{-4})^2 = 2.10 \times 10^{-7}$

   e. $Cu(IO_4)_2(s)$ $\rightleftharpoons$ $Cu^{2+} + 2\ IO_4^-$

s = solubility $\rightarrow$ s   2s
in mol/L

$$s = \frac{0.146\ g\ Cu(IO_4)_2}{0.100\ L} \times \frac{1\ mol}{445.4\ g} = 3.28 \times 10^{-3}\ mol/L$$

$K_{sp} = [Cu^{2+}][IO_4^-]^2 = (3.28 \times 10^{-3})(6.56 \times 10^{-3})^2 = 1.41 \times 10^{-7}$

   f. $Li_2CO_3(s)$ $\rightleftharpoons$ $2\ Li^+ + CO_3^{2-}$

s = solubility $\rightarrow$ 2s   s
in mol/L

d. $CaCO_3(s)$ $\rightleftharpoons$ Ca^{2+} + CO_3^{2-}

Initial   0   0
Equil.   s   s

$K_{sp} = 8.7 \times 10^{-9} = [Ca^{2+}][CO_3^{2-}] = s^2$

$s = 9.3 \times 10^{-5}$ mol/L; $\dfrac{9.3 \times 10^{-5} \text{ mol}}{L} \times \dfrac{100.1 \text{ g}}{mol} = 9.3 \times 10^{-3}$ g/L

Note: This ignores any reaction between CO_3^{2-} and H_2O to form HCO_3^{-}.

e. $Mg(NH_4)PO_4(s)$ $\rightleftharpoons$ Mg^{2+} + NH_4^{+} + PO_4^{3-}

Initial   0   0   0
Equil.   s   s   s

$K_{sp} = 3 \times 10^{-13} = [Mg^{2+}][NH_4^{+}][PO_4^{3-}] = s^3$

$s = 7 \times 10^{-5}$ mol/L; $\dfrac{7 \times 10^{-5} \text{ mol}}{L} \times \dfrac{137.3 \text{ g}}{mol} = 1 \times 10^{-2}$ g/L

Note: We are ignoring any reaction between NH_4^{+} and PO_4^{3-}.

f. $Hg_2Cl_2(s)$ $\rightleftharpoons$ Hg_2^{2+} +   2 Cl^{-}

Initial   0   0
Equil.   s   2s

$K_{sp} = 1.1 \times 10^{-18} = [Hg_2^{2+}][Cl^{-}]^2 = (s)(2s)^2 = 4s^3,\ \ s = 6.5 \times 10^{-7}$ mol/L

$\dfrac{6.5 \times 10^{-7} \text{ mol}}{L} \times \dfrac{472.1 \text{ g}}{mol} = \dfrac{3.1 \times 10^{-4} \text{ g}}{L}$

g. $SrSO_4(s)$ $\rightleftharpoons$ Sr^{2+} + SO_4^{2-}

s mol/L $\rightarrow$ s   s
dissolves

$K_{sp} = 3.2 \times 10^{-7} = [Sr^{2+}][SO_4^{2-}] = s^2,\ \ s = 5.7 \times 10^{-4}$ mol/L

$\dfrac{5.7 \times 10^{-4} \text{ mol}}{L} \times \dfrac{183.7 \text{ g}}{mol} = \dfrac{0.10 \text{ g}}{L}$

Note: This ignores any reaction between SO_4^{2-} and H_2O.

h. $Ag_2CO_3(s) \rightleftharpoons 2\,Ag^+ + CO_3^{2-}$ (ignoring $CO_3^{2-} + H_2O \rightleftharpoons HCO_3^- + OH^-$)

  s mol/L $\rightarrow$ 2s   s
  dissolves

$K_{sp} = 8.1 \times 10^{-12} = [Ag^+]^2\,[CO_3^{2-}] = (2s)^2\,(s) = 4s^3$

$s = 1.3 \times 10^{-4}$ mol/L; $\dfrac{1.3 \times 10^{-4}\,\text{mol}}{L} \times \dfrac{275.8\,g}{\text{mol}} = \dfrac{3.6 \times 10^{-2}\,g}{L}$

i. $Ag_2Cl_2O_7(s) \rightleftharpoons 2\,Ag^+ + Cl_2O_7^{2-}$

  s mol/L $\rightarrow$ 2s   s
  dissolves

$K_{sp} = 2 \times 10^{-7} = [Ag^+]^2\,[Cl_2O_7^{2-}] = (2s)^2\,(s) = 4s^3$

$s = 3.7 \times 10^{-3}$ mol/L $\approx 4 \times 10^{-3}$ mol/L; $\dfrac{4 \times 10^{-3}\,\text{mol}}{L} \times \dfrac{398.7\,g}{\text{mol}} = 1.6$ g/L ≈ 2 g/L

j. $Cu_2S(s) \rightleftharpoons 2\,Cu^+ + S^{2-}$ (ignoring any HS^- formation)

  s mol/L $\rightarrow$ 2s   s
  dissolves

$K_{sp} = 2 \times 10^{-47} = [Cu^+]^2\,[S^{2-}] = (2s)^2(s) = 4s^3$

$s = 1.7 \times 10^{-16}$ mol/L $\approx 2 \times 10^{-16}$ mol/L; $\dfrac{2 \times 10^{-16}\,\text{mol}}{L} \times \dfrac{159.2\,g}{\text{mol}} = 3 \times 10^{-14}$ g/L

38. If the number of ions in the two salts is the same, the K_{sp}'s can be compared, i.e., 1:1 electrolytes (1 cation:1 anion) can be compared to each other; 2:1 electrolytes can be compared to each other, etc.

39. a. $Fe(OH)_3(s) \rightleftharpoons Fe^{3+} + 3\,OH^-$

  Initial   0 $1 \times 10^{-7}\,M$
       s mol/L dissolves to reach equilibrium
  Change -s $\rightarrow$ +s   +3s
  Equil.   s $3s + 1 \times 10^{-7}$

$K_{sp} = 4 \times 10^{-38} = [Fe^{3+}]\,[OH^-]^3 = (s)\,(3s + 1 \times 10^{-7})^3 \approx s(1 \times 10^{-7})^3$

$s = 4 \times 10^{-17}$ mol/L Assumption good ($s \ll 1 \times 10^{-7}$).

b. $Fe(OH)_3(s) \rightleftharpoons$ Fe^{3+} + $3\ OH^-$ pH = 5.0 so $[OH^-] = 1 \times 10^{-9}\ M$

Initial		0	$1 \times 10^{-9}\ M$ (buffered)

s mol/L dissolves to reach equilibrium

Change	-s	$\rightarrow$ +s	----- (assume no pH change in buffer)
Equil.		s	1×10^{-9}

$K_{sp} = 4 \times 10^{-38} = [Fe^{3+}][OH^-]^3 = (s)(1 \times 10^{-9})^3$, s = 4×10^{-11} mol/L

c. $Fe(OH)_3(s) \rightleftharpoons$ Fe^{3+} + $3\ OH^-$ pH = 11.0 so $[OH^-] = 1 \times 10^{-3}\ M$

Initial		0	0.001 M (buffered)

s mol/L dissolves to reach equilibrium

Change	-s	$\rightarrow$ +s	----- (assume no pH change)
Equil.		s	0.001

$K_{sp} = 4 \times 10^{-38} = [Fe^{3+}][OH^-]^3 = (s)(0.001)^3$, s = 4×10^{-29} mol/L

40.   If the anion in the salt can act as a base in water, the solubility will increase as the solution becomes more acidic. This is true for $Al(OH)_3$, $CaCO_3$, $Mg(NH_4)PO_4$, Ag_2CO_3, and Cu_2S.

$Al(OH)_3(s) + 3\ H^+ \rightarrow Al^{3+} + 3\ H_2O$

$CaCO_3(s) + H^+ \rightarrow Ca^{2+} + HCO_3^- \xrightarrow{H^+} Ca^{2+} + H_2O + CO_2(g)$

$Mg(NH_4)PO_4(s) + H^+ \rightarrow Mg^{2+} + NH_4^+ + HPO_4^{2-}$ and

$\quad HPO_4^{2-} \xrightarrow{H^+} H_2PO_4^- \xrightarrow{H^+} H_3PO_4$

$Ag_2CO_3(s) + H^+ \rightarrow 2\ Ag^+ + HCO_3^- \xrightarrow{H^+} 2\ Ag^+ + H_2O + CO_2(g)$

$Cu_2S(s) + H^+ \rightarrow 2\ Cu^+ + HS^- \xrightarrow{H^+} 2\ Cu^+ + H_2S$

$CaSO_4$ and $SrSO_4$ will be slightly more soluble in acid since $MSO_4(s) + H^+ \rightarrow M^{2+} + HSO_4^-$ occurs. However, K_b for SO_4^{2-} is only about 10^{-12}, and the effect of acid on these salts is not as great as on the others.

41. $Ca_5(PO_4)_3(OH)\ (s)$ $\rightleftharpoons$ $5\ Ca^{2+}$ + $3\ PO_4^{3-}$ + OH^-

s = solubility (mol/L) $\rightarrow$ 5s   3s $s + 1.0 \times 10^{-7} \approx s$

$K_{sp} = 6.8 \times 10^{-37} = [Ca^{2+}]^5 [PO_4^{3-}]^3 [OH^-] = (5s)^5(3s)^3(s)$

$6.8 \times 10^{-37} = (3125)(27)s^9$, s = 2.7×10^{-5} mol/L   Assumption good.

The solubility of hydroxyapatite will increase as a solution gets more acidic, since both phosphate and hydroxide can react with H^+.

$$Ca_5(PO_4)_3(F)(s) \quad \rightleftharpoons \quad 5\,Ca^{2+} \;+\; 3\,PO_4^{3-} \;+\; F^-$$

solubility = s(mol/L) $\rightarrow$ 5s   3s   s

$$1 \times 10^{-60} = (5s)^5 (3s)^3 (s) = (3125)\,(27)s^9, \quad s = 6 \times 10^{-8}\ \text{mol/L}$$

The hydroxyapatite in the tooth enamel is converted to the less soluble fluorapatite by fluoride treated water. The less soluble fluorapatite will then be more difficult to remove, making teeth less susceptible to decay.

42.   a. $PbI_2(s) \quad \rightleftharpoons \quad Pb^{2+} \;+\; 2\,I^-$

s mol/L $\rightarrow$ s   2s
dissolves

$$K_{sp} = 1.4 \times 10^{-8} = [Pb^{2+}]\,[I^-]^2 = 4s^3, \quad s = 1.5 \times 10^{-3}\ \text{mol/L}$$

b. $PbI_2(s) \rightleftharpoons \qquad Pb^{2+} \;+\; 2\,I^-$

Initial $\rightarrow$ 0.10 M 0
Equil.   0.10 + s   2s

$$1.4 \times 10^{-8} = (0.10 + s)\,(2s)^2 \approx (0.10)\,(2s)^2 = 0.40s^2$$

$s = 1.9 \times 10^{-4}$ mol/L   Assumption good.

c. $PbI_2(s) \rightleftharpoons \qquad Pb^{2+} \;+\; 2\,I^-$

Initial $\rightarrow$ 0   0.010 M
Equil.   s   0.010 + 2s

$$1.4 \times 10^{-8} = (s)\,(0.010 + 2s)^2 \approx (s)\,(0.010)^2, \quad s = 1.4 \times 10^{-4}\ \text{mol/L} \qquad \text{Assumption good.}$$

43.

a.

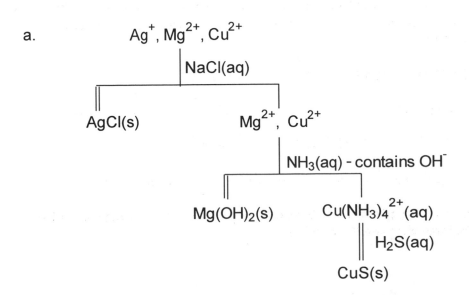

b.

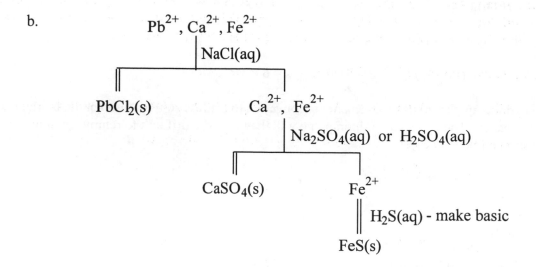

c.

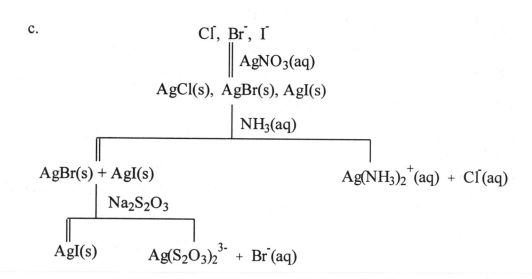

d.

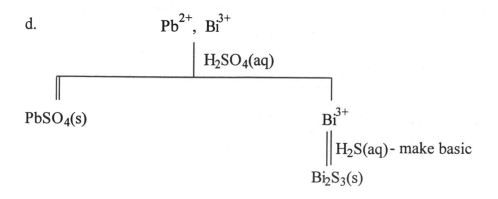

44.   It will be more soluble in base. Addition of OH^- will react with H^+ to produce H_2O. This drives the equilibrium $H_4SiO_4 \rightleftharpoons H_3SiO_4^- + H^+$ to the right. This, in turn, drives the solubility reaction $SiO_2(s) + 2 H_2O \rightleftharpoons H_4SiO_4$ to the right, increasing the solubility of the silica.

45.   After mixing: $[Pb^{2+}]_0 = [Cl^-]_0 = 0.010\ M$

$[Pb^{2+}]_0[Cl^-]_0^2 = (0.010)(0.010)^2 = 1.0 \times 10^{-6} < K_{sp}\ (1.6 \times 10^{-5})$; No precipitate will form.

Thus, $[Pb^{2+}] = [Na^+] = [Cl^-] = 0.010\ M$; $[NO_3^-] = 2(0.010) = 0.020\ M$

46. $[Ba^{2+}]_0 = \dfrac{75.0\ \text{mL} \times \dfrac{0.020\ \text{mmol}}{\text{mL}}}{200.\ \text{mL}} = 7.5 \times 10^{-3}\ M$

$[SO_4^{2-}]_0 = \dfrac{125\ \text{mL} \times \dfrac{0.040\ \text{mmol}}{\text{mL}}}{200.\ \text{mL}} = 2.5 \times 10^{-2}\ M$

$[Ba^{2+}]_0[SO_4^{2-}]_0 = (7.5 \times 10^{-3})(2.5 \times 10^{-2}) = 1.9 \times 10^{-4} > K_{sp}\ (1.5 \times 10^{-9})$

A precipitate of $BaSO_4(s)$ will form.

	$BaSO_4(s)$	$\rightleftharpoons$	Ba^{2+}	$+$	SO_4^{2-}	
Before			0.0075 M		0.025 M	
		0.0075 mol/L Ba^{2+} reacts with SO_4^{2-} to completion.				
Change		$\leftarrow$	-0.0075		-0.0075	Reacts completely
After			0		0.0175	New initial (carry extra S.F.)
		s mol/L $BaSO_4$ dissolves to reach equilibrium				
Change	-s	$\rightarrow$	+s		+s	
Equil.			s		0.0175 + s	

$K_{sp} = 1.5 \times 10^{-9} = [Ba^{2+}][SO_4^{2-}] = (s)(0.0175 + s) \approx s(0.0175)$

$s = 8.6 \times 10^{-8}$; $[Ba^{2+}] = 8.6 \times 10^{-8}\ M$; $[SO_4^{2-}] = 0.018\ M$ Assumption good.

Complex Ion Equilibria

47. $\begin{array}{ll} Mn^{2+} + C_2O_4^{2-} \rightleftharpoons MnC_2O_4 & K_1 = 7.9 \times 10^3 \\ MnC_2O_4 + C_2O_4^{2-} \rightleftharpoons Mn(C_2O_4)_2^{2-} & K_2 = 7.9 \times 10^1 \end{array}$

$\overline{\quad Mn^{2+} + 2\ C_2O_4^{2-} \rightleftharpoons Mn(C_2O_4)_2^{2-} \qquad K_f = K_1K_2 = 6.2 \times 10^5 \quad}$

48.   An ammonia solution is basic. Initially the reaction that occurs is $Cu^{2+}(aq) + 2\ OH^-(aq) \rightarrow$ $Cu(OH)_2(s)$. As the concentration of NH_3 increases the complex ion $Cu(NH_3)_4^{2+}$ forms: $Cu(OH)_2(s) + 4\ NH_3(aq) \rightleftharpoons Cu(NH_3)_4^{2+}(aq) + 2\ OH^-(aq)$

49. $$\frac{65\ g\ KI}{0.500\ L} \times \frac{1\ mol}{166.0\ g} = \frac{0.78\ mol}{L}$$

	Hg^{2+}	+	$4\ I^-$	$\rightleftharpoons$	HgI_4^{2-}	$K_f = 1.0 \times 10^{30}$
Before	0.010 M		0.78 M		0	
Change	-0.010		-0.040	$\rightarrow$	+0.010	Reacts completely (K large)
After	0		0.74		0.010	New initial

x mol/L HgI_4^{2-} dissociates to reach equilibrium

| Change | +x | | +4x | $\leftarrow$ | -x | |
| Equil. | x | | 0.74 + 4x | | 0.010 - x | |

$$K_f = 1.0 \times 10^{30} = \frac{[HgI_4^{2-}]}{[Hg^{2+}][I^-]^4} = \frac{(0.010 - x)}{(x)(0.74 + 4x)^4}$$

$$1.0 \times 10^{30} \approx \frac{(0.010)}{(x)(0.74)^4}, \quad x = [Hg^{2+}] = 3.3 \times 10^{-32}\ M \quad \text{Assumptions good.}$$

Note: 3.3×10^{-32} mol/L corresponds to one Hg^{2+} ion per 5×10^7 L. It is very reasonable to approach the equilibrium in two steps. The reaction essentially goes to completion.

50.

	Ni^{2+}	+	$6\ NH_3$	$\rightleftharpoons$	$Ni(NH_3)_6^{2+}$	$K_f = 5.5 \times 10^8$
Initial	0		3.0 M		0.10 mol/0.50 L = 0.20 M	

x mol/L $Ni(NH_3)_6^{2+}$ dissociates to reach equilibrium

| Change | +x | | +6x | $\leftarrow$ | -x | |
| Equil. | x | | 3.0 + 6x | | 0.20 - x | |

$$K_f = 5.5 \times 10^8 = \frac{[Ni(NH_3)_6^{2+}]}{[Ni^{2+}][NH_3]^6} = \frac{(0.20 - x)}{(x)(3.0 + 6x)^6}, \quad 5.5 \times 10^8 \approx \frac{(0.20)}{(x)(3.0)^6}$$

$x = [Ni^{2+}] = 5.0 \times 10^{-13}\ M;\quad [Ni(NH_3)_6^{2+}] = 0.20\ M \qquad \text{Assumptions good.}$

51. $Ag^+(aq) + Cl^-(aq) \rightleftharpoons AgCl(s)$, white ppt.; $AgCl(s) + 2\ NH_3(aq) \rightleftharpoons Ag(NH_3)_2^+(aq) + Cl^-(aq)$

$Ag(NH_3)_2^+(aq) + Br^-(aq) \rightleftharpoons AgBr(s) + 2\ NH_3(aq)$, pale yellow ppt.

$AgBr(s) + 2\ S_2O_3^{2-}(aq) \rightleftharpoons Ag(S_2O_3)_2^{3-}(aq) + Br^-(aq)$

$Ag(S_2O_3)_2^{3-}(aq) + I^-(aq) \rightleftharpoons AgI(s) + 2\ S_2O_3^{2-}(aq)$, yellow ppt.

K_{sp} (AgCl) > K_{sp} (AgBr) > K_{sp} (AgI); K_f [Ag(S$_2$O$_3$)$_2$$^{3-}$] > K_f [Ag(NH$_3$)$_2$$^+$]

52.

$$AgBr(s) \rightleftharpoons Ag^+ + Br^- \qquad K_{sp} = 5.0 \times 10^{-13}$$
$$Ag^+ + 2\ S_2O_3^{2-} \rightleftharpoons Ag(S_2O_3)_2^{3-} \qquad K_f = 2.9 \times 10^{13}$$

$AgBr(s) + 2\ S_2O_3^{2-} \rightleftharpoons Ag(S_2O_3)_2^{3-} + Br^-$ $K = K_{sp}K_f = 14.5$ (Carry extra S.F.)

$$AgBr(s) \ + \ 2\ S_2O_3^{3-} \quad \rightleftharpoons \quad Ag(S_2O_3)_2^{3-} \ + \ Br^-$$

Initial		0.500 M	0	0

s mol/L AgBr dissolves to reach equilibrium

Change	-s	-2s	→	+s	+s
Equil.		0.500 - 2s		s	s

$\dfrac{s^2}{0.500 - 2s} = 14.5$; Using the quadratic equation since s is not small:

$s^2 = 7.25 - 29.0\ s$, $s^2 + 29.0\ s - 7.25 = 0$

$$s = \frac{-29.0 + \sqrt{(29.0)^2 + 4(7.25)}}{2} = 0.248\ mol/L$$

$$1.00\ L \times \frac{0.248\ mol\ AgBr}{L} \times \frac{187.8\ g\ AgBr}{mol\ AgBr} = 46.6\ g\ AgBr = 47\ g\ AgBr$$

53.

$$AgBr(s) \rightleftharpoons Ag^+ + Br^- \qquad K_{sp} = 5.0 \times 10^{-13}$$
$$Ag^+ + 2\ NH_3 \rightleftharpoons Ag(NH_3)_2^+ \qquad K_f = 1.7 \times 10^7$$

$AgBr(s) + 2\ NH_3 \rightleftharpoons Ag(NH_3)_2^+ + Br^-$ $K = 8.5 \times 10^{-6}$

$$AgBr(s) \ + \ 2\ NH_3 \quad \rightleftharpoons \quad Ag(NH_3)_2^+ \ + \ Br^-$$

Initial		3.0 M	0	0

If 0.10 mol AgBr dissolves in 1.0 L:

Change	-0.10	-0.20	→	+0.10	+0.10
After		2.8		0.10	0.10

If all the AgBr(s) dissolves these would be the concentrations after dissolution. To check if all AgBr(s) dissolves, use the reaction quotient.

$$Q = \frac{(0.10)^2}{(2.8)^2} = 1.3 \times 10^{-3}$$

Since Q > K (8.5×10^{-6}), the reaction will shift left, producing AgBr(s). No, 0.10 mol AgBr will not dissolve in 1.0 L of 3.0 M NH$_3$.

Additional Exercises

54. $1.0 \text{ mL} \times \dfrac{1.0 \text{ mmol}}{\text{mL}} = 1.0 \text{ mmol Cd}^{2+}$ added to the ammonia solution

Thus, $[\text{Cd}^{2+}]_0 = 1.0 \times 10^{-3}$ mol/L. We will treat each equilibrium separately, considering the one with the largest K first.

$$\text{Cd}^{2+} \quad + \quad 4\,\text{NH}_3 \quad \rightleftharpoons \quad \text{Cd(NH}_3)_4^{2+}$$

Before	$1.0 \times 10^{-3}\,M$	$5.0\,M$	0
Change	-1.0×10^{-3}	-4.0×10^{-3} $\rightarrow$	$+1.0 \times 10^{-3}$	Reacts completely
After	0	$4.996 \approx 5.0$	1.0×10^{-3}	New initial

x mol/L Cd(NH$_3$)$_4^{2+}$ dissociates to reach equilibrium

Change	$+x$	$+4x$ $\leftarrow$	$-x$
Equil.	x	$5.0 + 4x$	$0.0010 - x$

$$K_f = 1.0 \times 10^7 = \frac{(0.0010 - x)}{(x)(5.0 + 4x)^4} \approx \frac{(0.0010)}{(x)(5.0)^4}$$

$x = [\text{Cd}^{2+}] = 1.6 \times 10^{-13}\,M$; Assumptions good. This is the maximum $[\text{Cd}^{2+}]$ possible. If a precipitate forms, $[\text{Cd}^{2+}]$ will be less than this value. In 5.0 M NH$_3$ we can calculate the pH:

$$\text{NH}_3 + \text{H}_2\text{O} \quad \rightleftharpoons \quad \text{NH}_4^+ + \text{OH}^- \qquad K_b = 1.8 \times 10^{-5}$$

Initial	$5.0\,M$	0	~ 0
Equil.	$5.0 - y$	y	y

$$K_b = 1.8 \times 10^{-5} = \frac{[\text{NH}_4^+][\text{OH}^-]}{[\text{NH}_3]} = \frac{y^2}{5.0 - y} \approx \frac{y^2}{5.0}, \quad y = [\text{OH}^-] = 9.5 \times 10^{-3}\,M; \quad \text{Assumptions good.}$$

We now calculate the value of the solubility quotient:

$$Q = [\text{Cd}^{2+}][\text{OH}^-]^2 = (1.6 \times 10^{-13})(9.5 \times 10^{-3})^2$$

$$Q = 1.4 \times 10^{-17} < K_{sp}(5.9 \times 10^{-15}); \quad \text{Therefore, no precipitate forms.}$$

55.   a. $\dfrac{28.6 \text{ g borax}}{\text{L}} \times \dfrac{1 \text{ mol}}{381.4 \text{ g}} = \dfrac{0.0750 \text{ mol borax}}{\text{L}}$

$$\frac{0.0750 \text{ mol borax}}{\text{L}} \times \frac{2 \text{ mol B(OH)}_3}{\text{mol borax}} = 0.150\,M$$

The concentration of B(OH)$_4^-$ is also 0.150 M.

$$\text{B(OH)}_3 + \text{H}_2\text{O} \rightleftharpoons \text{B(OH)}_4^- + \text{H}^+ \qquad K_a = 5.8 \times 10^{-10}$$

$$\text{pH} = \text{p}K_a + \log \frac{[\text{Base}]}{[\text{Acid}]} = \text{p}K_a + \log 1.00 = \text{p}K_a = 9.24$$

b. $100. \text{ mL} \times \dfrac{0.10 \text{ mmol}}{\text{mL}} = 10. \text{ mmol NaOH added or } 0.010 \text{ mol NaOH}$

For $B(OH)_3$: $\dfrac{0.150 \text{ mol } B(OH)_3}{L} \times 1.0 \text{ L} = 0.15 \text{ mol } B(OH)_3$

Added OH^- converts $B(OH)_3$ to $B(OH)_4^-$: $B(OH)_3 + OH^- \rightarrow B(OH)_4^-$

mol $B(OH)_3 = 0.15 - 0.010 = 0.14$ mol; mol $B(OH)_4^- = 0.15 + 0.010 = 0.16$ mol

$[B(OH)_3] = \dfrac{0.14 \text{ mol}}{1.1 \text{ L}}$; Similarly $[B(OH)_4^-] = \dfrac{0.16 \text{ mol}}{1.1 \text{ L}}$

$pH = pK_a + \log \dfrac{[Base]}{[Acid]} = 9.24 + \log \dfrac{[B(OH)_4^-]}{[B(OH)_3]} = 9.24 + \log \dfrac{(0.16)}{(0.14)} = 9.30$

56.   a. The optimum pH for a buffer is when $pH = pK_a$. At this pH a buffer will have equal neutralization capacity for both added acid and base.

$K_b = 1.19 \times 10^{-6}$; $K_a = K_w/K_b = 8.40 \times 10^{-9}$; $pH = pK_a = 8.076$

b. $pH = pK_a + \log \dfrac{[TRIS]}{[TRISH^+]}$, $7.00 = 8.076 + \log \dfrac{[TRIS]}{[TRISH^+]}$

$\dfrac{[TRIS]}{[TRISH^+]} = 10^{-1.08} = 0.083$ (at pH = 7.00)

$9.00 = 8.076 + \log \dfrac{[TRIS]}{[TRISH^+]}$, $\dfrac{[TRIS]}{[TRISH^+]} = 10^{0.92} = 8.3$ (at pH = 9.00)

c. $\dfrac{50.0 \text{ g TRIS}}{2.0 \text{ L}} \times \dfrac{1 \text{ mol}}{121.1 \text{ g}} = 0.206 \, M = 0.21 \, M = [TRIS]$

$\dfrac{65.0 \text{ g TRISHCl}}{2.0 \text{ L}} \times \dfrac{1 \text{ mol}}{157.6 \text{ g}} = 0.206 \, M = 0.21 \, M = [TRISHCl] = [TRISH^+]$

$pH = pK_a + \log \dfrac{[TRIS]}{[TRISH^+]} = 8.076 + \log \dfrac{(0.21)}{(0.21)} = 8.076 = 8.08$

Amount of H^+ added from HCl is: $0.50 \times 10^{-3} \text{ L} \times \dfrac{12 \text{ mol}}{L} = 6.0 \times 10^{-3} \text{ mol } H^+$

The H^+ from HCl will convert TRIS into $TRISH^+$. The reaction is:

	TRIS	+	H$^+$	→	TRISH$^+$	
Before	0.21 M		$\dfrac{6.0 \times 10^{-3}}{0.2005} = 0.030\ M$		0.21 M	
Change	-0.030		-0.030	→	+0.030	Reacts completely
After	0.18		0		0.24	

Now use the Henderson-Hasselbalch equation to solve the buffer problem.

$$pH = 8.076 + \log \frac{0.18}{0.24} = 7.95$$

57. $pH = pK_a + \log \dfrac{[(CH_3)_2AsO_2^-]}{[(CH_3)_2AsO_2H]}$, $6.60 = 6.19 + \log \dfrac{[Cac^-]}{[HCac]}$

$\dfrac{[Cac^-]}{[HCac]} = 10^{0.41} = 2.6$, $[Cac^-] = 2.6\ [HCac]$

$[Cac^-] + [HCac] = 0.25$; $3.6[HCac] = 0.25$, $[HCac] = 0.069\ M$ and $[Cac^-] = 0.18\ M$

$$0.500\ L \times \frac{0.069\ mol\ (CH_3)_2AsO_2H}{L} \times \frac{138.0\ g}{mol} = 4.8\ g\ cacodylic\ acid$$

$$0.500\ L \times \frac{0.18\ mol\ (CH_3)_2AsO_2^-}{L} \times \frac{160.0\ g\ (CH_3)_2AsO_2Na}{mol} = 14\ g\ sodium\ cacodylate$$

58.   cacodylic acid $pK_a = 6.19$
       TRISHCl   8.08
       benzoic acid   4.19
       acetic acid   4.74
       HF   3.14
       NH$_4$Cl   9.26

Best buffer is when pH = pK$_a$. Choose combinations that yield a buffer where pH ≈ pK$_a$.

a.   potassium fluoride + HCl   b.   Benzoic acid + NaOH

c.   sodium acetate + acetic acid

d. $(CH_3)_2AsO_2Na$ + HCl: This is the best choice to produce a conjugate acid/base pair.
     Actually the best choice is an equimolar mixture of ammonium chloride and sodium acetate.
     NH_4^+ is a weak acid ($K_a = 5.6 \times 10^{-10}$). $C_2H_3O_2^-$ is a weak base ($K_b = 5.6 \times 10^{-10}$). A mixture
     of the two will give a buffer at pH = 7 since the weak acid and weak base are the same
     strengths. $NH_4C_2H_3O_2$ is commercially available and its solutions are used as pH = 7 buffers.

e.   ammonium chloride + NaOH

59.   a. $pH = pK_a = 4.19$

b.   [Bz⁻] will increase to 0.120 M. [HBz] will decrease to 0.080 M.

$$pH = pK_a + \log \frac{[Bz^-]}{[HBz]}, \quad pH = 4.19 + \log \frac{(0.120)}{(0.080)} = 4.37$$

c.   Bz⁻   +   H₂O   ⇌   HBz   +   OH⁻

Initial   0.120 M 0.080 M 0
Equil.   0.120 - x 0.080 + x x

$$K_b = \frac{K_w}{K_a} = \frac{1.0 \times 10^{-14}}{6.4 \times 10^{-5}} = \frac{(0.080 + x)\,(x)}{(0.120 - x)} \approx \frac{(0.080)\,(x)}{0.120}$$

$x = [OH^-] = 2.34 \times 10^{-10}\ M$ (carry extra S.F.); Assumptions good.

$pOH = 9.63$; $pH = 4.37$

d.   We get the same answer. Both equilibria involve the two major species, benzoic acid and benzoate anion. Both equilibria must hold true. K_b is related to K_a by K_w. [OH⁻] is also related to [H⁺] by K_w.

60.   NaOH added = 50.0 mL × $\dfrac{0.500\ \text{mmol}}{\text{mL}}$ = 25.0 mmol NaOH

NaOH left unreacted = 31.92 mL HCl × $\dfrac{0.289\ \text{mmol}}{\text{mL}}$ × $\dfrac{1\ \text{mmol NaOH}}{\text{mmol HCl}}$ = 9.22 mmol NaOH

NaOH reacted with aspirin = 25.0 - 9.22 = 15.8 mmol NaOH

15.8 mmol NaOH × $\dfrac{1\ \text{mmol aspirin}}{2\ \text{mmol NaOH}}$ × $\dfrac{180.2\ \text{mg}}{\text{mmol}}$ = 1420 mg = 1.42 g aspirin

Purity = $\dfrac{1.42\ \text{g}}{1.427\ \text{g}}$ × 100 = 99.5%

A strong base is being titrated with a strong acid. The equivalence point is at pH = 7. Bromothymol blue would be the best indicator since it changes color at pH ≈ 7.

61.   a.   The predominate equilibrium is:

$$NH_4^+ + X^- \rightleftharpoons NH_3 + HX \qquad K_{eq} = \frac{[NH_3]\,[HX]}{[NH_4^+]\,[X^-]} = \frac{K_a(NH_4^+)}{K_a(HX)}$$

From the reaction, [NH₄⁺] = [X⁻] and [NH₃] = [HX].

Therefore, $K_{eq} = \dfrac{K_a(NH_4^+)}{K_a(HX)} = \dfrac{[HX]^2}{[X^-]^2}$

The K_a expression for HX is: $K_a(HX) = \dfrac{[H^+][X^-]}{[HX]}, \quad \dfrac{[HX]}{[X^-]} = \dfrac{[H^+]}{K_a(HX)}$

Substituting into the K_{eq} expression: $K_{eq} = \dfrac{K_a(NH_4^+)}{K_a(HX)} = \dfrac{[HX]^2}{[X^-]^2} = \left(\dfrac{[H^+]}{K_a(HX)}\right)^2$

Rearranging: $[H^2]^2 = K_a(NH_4^+) \times K_a(HX)$ or taking the -log of both sides:

$$pH = \dfrac{pK_a(NH_4^+) + pK_a(HX)}{2}$$

b. Ammonium formate = $NH_4(HCO_2)$

$K_a(NH_4^+) = \dfrac{1.0 \times 10^{-14}}{1.8 \times 10^{-5}} = 5.6 \times 10^{-10}; \; K_a(HCO_2H) = 1.8 \times 10^{-4}$

$pH = \dfrac{pK_a(NH_4^+) + pK_a(HCO_2H)}{2} = \dfrac{9.25 + 3.74}{2} = 6.50$

Ammonium acetate = $NH_4(CH_3CO_2); \; pK_a(CH_3CO_2H) = 4.74$

$pH = \dfrac{9.25 + 4.74}{2} = 7.00$

Ammomium bicarbonate = $NH_4(HCO_3); \; K_a(H_2CO_3) = 4.3 \times 10^{-7}; \; pK_a = 6.37$

$pH = \dfrac{9.25 + 6.37}{2} = 7.81$

c. $NH_4^+ + OH^- \rightarrow NH_3 + H_2O; \; CH_3CO_2^- + H^+ \rightarrow CH_3CO_2H$

62.   At 4.0 mL NaOH added: $\left|\dfrac{\Delta pH}{\Delta mL}\right| = \left|\dfrac{2.43 - 3.14}{0 - 4.0}\right| = 0.18$

See following table for the other points. As can be seen from the plot, the advantage of this approach is that it is much easier to accurately determine the location of the equivalence point.

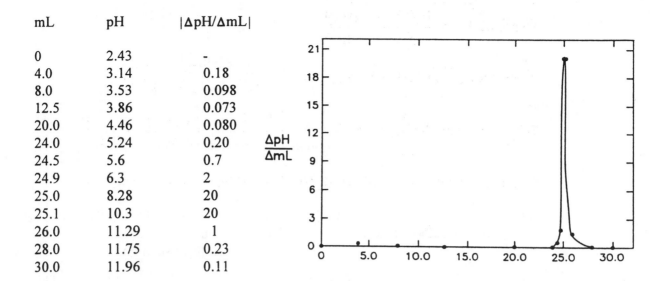

mL	pH	$\|\Delta pH/\Delta mL\|$
0	2.43	-
4.0	3.14	0.18
8.0	3.53	0.098
12.5	3.86	0.073
20.0	4.46	0.080
24.0	5.24	0.20
24.5	5.6	0.7
24.9	6.3	2
25.0	8.28	20
25.1	10.3	20
26.0	11.29	1
28.0	11.75	0.23
30.0	11.96	0.11

63.   At the equivalence point, $C_6H_4(CO_2)_2^{2-}$ = P^{2-} is the major species. It is a weak base in water.

$$P^{2-} \quad + \quad H_2O \quad \rightleftharpoons \quad HP^- \quad + \quad OH^-$$

Initial $\dfrac{0.5 \text{ g}}{0.1 \text{ L}} \times \dfrac{1 \text{ mol}}{204.2 \text{ g}} = 0.024 \, M$ $\qquad$ 0 $\qquad$ 0 $\qquad$ (carry extra sig. fig.)

Equil. $0.024 - x$ $\qquad\qquad\qquad\qquad$ x $\qquad$ x

$$K_b = \frac{[HP^-][OH^-]}{[P^{2-}]} = \frac{K_w}{K_{a_2}} = \frac{1.0 \times 10^{-14}}{10^{-5.51}} = 3.2 \times 10^{-9} = \frac{x^2}{0.024 - x} \approx \frac{x^2}{0.024}$$

$x = [OH^-] = 8.8 \times 10^{-6}$; pH = 8.9   Assumptions good.

Phenolphthalein would be the best indicator for this titration since it changes color at pH ~ 9.

64. $$Cr^{3+} \quad + \quad H_2EDTA^{2-} \quad \rightleftharpoons \quad CrEDTA^- \quad + \quad 2 \, H^+$$

Before	0.0010 M	0.050 M	0	$1.0 \times 10^{-6} \, M$	(buffer-[H⁺] constant)
Change	-0.0010	-0.0010	$\rightarrow$ +0.0010	No change	Reacts completely
After	0	0.049	0.0010	1.0×10^{-6}	New initial

x mol/L CrEDTA⁻ dissociates to reach equilibrium

Change	$+x$	$+x$	$\leftarrow$ $-x$	---	
Equil.	x	$0.049 + x$	$0.0010 - x$	1.0×10^{-6}	(buffer)

$$K_f = 1.0 \times 10^{23} = \frac{[CrEDTA^-][H^+]^2}{[Cr^{3+}][H_2EDTA^{2-}]} = \frac{(0.0010 - x)(1.0 \times 10^{-6})^2}{(x)(0.049 + x)}$$

$1.0 \times 10^{23} \approx \dfrac{(0.0010)(1.0 \times 10^{-12})}{x(0.049)}$, $\quad x = [Cr^{3+}] = 2.0 \times 10^{-37} \, M$ $\quad$ Assumptions good.

65.   a. $Pb(OH)_2(s)$ $\rightleftharpoons$ Pb^{2+} + $2\ OH^-$

s mol/L $\rightarrow$ s   2s   (ignore OH^- from H_2O)
dissolves

$K_{sp} = 1.2 \times 10^{-15} = [Pb^{2+}]\,[OH^-]^2 = 4s^3$

$s = [Pb^{2+}] = 6.7 \times 10^{-6}\ M$;   Assumption to ignore water is good.

b. $Pb(OH)_2(s)$ $\rightleftharpoons$ Pb^{2+} + $2\ OH^-$

Initial   0   0.10 M pH = 13.00, $[OH^-]$ = 0.10 M
       s mol/L $Pb(OH)_2$ dissolves to reach equilibrium
Equil.   s   0.10 (buffered solution)

$1.2 \times 10^{-15} = (s)\,(0.10)^2$, $s = [Pb^{2+}] = 1.2 \times 10^{-13}\ M$

c.   Calculate the Pb^{2+} concentration in equilibrium with $EDTA^{4-}$.

Pb^{2+} + $EDTA^{4-}$ $\rightleftharpoons$ $PbEDTA^{2-}$ $K_f = 1.1 \times 10^{18}$

Before   0.010 M 0.050 M 0
        0.010 mol/L Pb^{2+} reacts completely (large K)
Change   -0.010   -0.010 $\rightarrow$ +0.010   Reacts completely
After   0   0.040   0.010   New initial
        x mol/L $PbEDTA^{2-}$ dissociates to reach equilibrium
Equil. x $0.040 + x$ $0.010 - x$

$1.1 \times 10^{18} = \dfrac{(0.010 - x)}{(x)\,(0.040 + x)} \approx \dfrac{(0.010)}{x(0.040)}$, $x = [Pb^{2+}] = 2.3 \times 10^{-19}\ M$ Assumptions good.

Calculate the solubility quotient for $Pb(OH)_2$ to see if precipitation occurs:

$Q = [Pb^{2+}]\,[OH^-]^2 = (2.3 \times 10^{-19})\,(0.10)^2 = 2.3 \times 10^{-21} < K_{sp}\,(1.2 \times 10^{-15})$

$Pb(OH)_2(s)$ will not form.

66. $Fe(OH)_3 \rightleftharpoons Fe^{3+} + 3\ OH^-$ $K_{sp} = [Fe^{3+}]\,[OH^-]^3 = 4 \times 10^{-38}$

pH = 7.41; pOH = 6.59; $[OH^-] = 2.6 \times 10^{-7}\ M$; $[Fe^{3+}] = \dfrac{4 \times 10^{-38}}{(2.6 \times 10^{-7})^3} = 2 \times 10^{-18}\ M$

The lowest level of iron in serum: $\dfrac{60 \times 10^{-6}\ g}{0.1\ L} \times \dfrac{1\ mol}{55.85\ g} = 1 \times 10^{-5}\ M$

The actual concentration of iron is much greater than expected when considering only the solubility of $Fe(OH)_3$. There must be complexing agents present to increase the solubility.

67. $Cu(OH)_2 \rightleftharpoons Cu^{2+} + 2\ OH^-$ $K_{sp} = 1.6 \times 10^{-19}$
              $Cu^{2+} + 4\ NH_3 \rightleftharpoons Cu(NH_3)_4^{2+}$ $K_f = 1.0 \times 10^{13}$

$$\overline{Cu(OH)_2 + 4\ NH_3 \rightleftharpoons Cu(NH_3)_4^{2+} + 2\ OH^-} \qquad K = K_{sp}K_f = 1.6 \times 10^{-6}$$

68. $Cu(OH)_2(s)\ +\ 4\ NH_3\ \rightleftharpoons\ Cu(NH_3)_4^{2+}\ +\ 2\ OH^-$

Initial   5.0 M 0   0.010 M
              s mol/L $Cu(OH)_2$ dissolves to reach equilibrium
Equil.   5.0 - 4s s 0.010 + 2s

$$1.6 \times 10^{-6} = \frac{[Cu(NH_3)_4^{2+}]\,[OH^-]^2}{[NH_3]^4} = \frac{s(0.010 + 2s)^2}{(5.0 - 4s)^4}$$

If s is small: $1.6 \times 10^{-6} = \dfrac{s(0.010)^2}{(5.0)^4}$, $s = 10.$ mol/L

Assumptions are not good. We will solve the problem by successive approximations.

$$s_{calc} = \frac{1.6 \times 10^{-6}\,(5.0 - 4s_{guess})^4}{(0.010 + 2s_{guess})^2},\ \ \text{The results from six trials are:}$$

  s_{guess}:   0.10, 0.050, 0.060, 0.055, 0.058, 0.056

  s_{calc}: 1.6×10^{-2}, 0.070, 0.049, 0.058, 0.052, 0.056

Thus, the solubility is 0.056 mol/L.

69.   a. AgF   b. PbO   c. $Sr(NO_2)_2$ d. $Ni(CN)_2$

All of the above have anions that are weak bases. The other choices have anions (and cations) with no basic (or acidic) properties.

70.   The solubility of SiO_2 will equal the sum of the concentrations of H_4SiO_4 and $H_3SiO_4^-$.

$SiO_2(s) + 2\ H_2O \rightleftharpoons H_4SiO_4(aq)$ $K = 2 \times 10^{-3}\ M = [H_4SiO_4]$

$H_4SiO_4 \rightleftharpoons H_3SiO_4^- + H^+$ $K_a = 10^{-9.46} = 3.5 \times 10^{-10} = \dfrac{[H_3SiO_4^-]\,[H^+]}{[H_4SiO_4]}$

At pH = 7.0: $3.5 \times 10^{-10} = \dfrac{[H_3SiO_4^-]\,(1 \times 10^{-7})}{2 \times 10^{-3}}$, $[H_3SiO_4^-] = 7 \times 10^{-6}\ M$

Solubility = $2 \times 10^{-3} + 7 \times 10^{-6} \approx 2 \times 10^{-3}$ mol/L at pH = 7.0

At pH = 10.0: $3.5 \times 10^{-10} = \dfrac{[H_3SiO_4^-] \, (1 \times 10^{-10})}{2 \times 10^{-3}}$, $[H_3SiO_4^-] = 7 \times 10^{-3} \, M$

Solubility $= 2 \times 10^{-3} + 7 \times 10^{-3} = 9 \times 10^{-3}$ mol/L at pH = 10.0

71. $100. \text{ mL} \times \dfrac{0.200 \text{ mmol } K_2C_2O_4}{\text{mL}} = 20.0 \text{ mmol } K_2C_2O_4$

$150. \text{ mL} \times \dfrac{0.250 \text{ mmol } BaBr_2}{\text{mL}} = 37.5 \text{ mmol } BaBr_2$

$$Ba^{2+}(aq) \; + \; C_2O_4^{2-}(aq) \; \rightleftharpoons \; BaC_2O_4(s) \qquad K_{sp} = 2.3 \times 10^{-8}$$

	Ba^{2+}	$C_2O_4^{2-}$		BaC_2O_4	
Before	37.5 mmol	20.0 mmol		0	
Change	-20.0	-20.0	$\rightarrow$	+20.0	Reacts completely (K_{sp} small)
After	17.5	0		20.0	

New initial concentrations are: $[Ba^{2+}] = \dfrac{17.5 \text{ mmol}}{250. \text{ mL}} = 7.00 \times 10^{-2} \, M$

$[K^+] = \dfrac{40.0 \text{ mmol}}{250. \text{ mL}} = 0.160 \, M$; $[Br^-] = \dfrac{75.0 \text{ mmol}}{250. \text{ mL}} = 0.300 \, M$

For K^+ and Br^-, these are also the final concentrations. For Ba^{2+} and $C_2O_4^{2-}$:

$$BaC_2O_4(s) \; \rightleftharpoons \; Ba^{2+}(aq) \; + \; C_2O_4^{2-}(aq)$$

	Ba^{2+}	$C_2O_4^{2-}$
Initial	0.0700 M	0
	s mol/L BaC_2O_4 dissolves to reach equilibrium	
Equil.	0.0700 + s	s

$K_{sp} = 2.3 \times 10^{-8} = [Ba^{2+}] \, [C_2O_4^{2-}] = (0.0700 + s) \, (s) \approx 0.0700 \, s$

$s = [C_2O_4^{2-}] = 3.3 \times 10^{-7}$ mol/L; $[Ba^{2+}] = 0.0700 \, M$ Assumptions good.

72. $AgC_2H_3O_2(s) \rightleftharpoons Ag^+ + C_2H_3O_2^-$ $K_{sp} = 2.5 \times 10^{-3}$; $HC_2H_3O_2 \rightleftharpoons H^+ + C_2H_3O_2^-$ $K_a = 1.8 \times 10^{-5}$

Solubility $= [Ag^+] =$ total $C_2H_3O_2^-$ concentration $= [C_2H_3O_2^-] + [HC_2H_3O_2]$

$K_a = \dfrac{(1.0 \times 10^{-3}) \, [C_2H_3O_2^-]}{[HC_2H_3O_2]} = 1.8 \times 10^{-5}$

$\dfrac{[C_2H_3O_2^-]}{[HC_2H_3O_2]} = 1.8 \times 10^{-2}$, $[HC_2H_3O_2] = 56 \, [C_2H_3O_2^-]$

$[Ag^+] = [C_2H_3O_2^-] + [HC_2H_3O_2] = [C_2H_3O_2^-] + 56 \, [C_2H_3O_2^-]$, $[C_2H_3O_2^-] = \dfrac{[Ag^+]}{57}$

$$K_{sp}' = [Ag^+][C_2H_3O_2^-] = 2.5 \times 10^{-3} = [Ag^+] \left(\frac{[Ag^+]}{57} \right), \ [Ag^+] = solubility = 0.38 \ mol/L$$

0.38 mol of silver acetate will dissolve per liter of pH = 3.00 buffer.

73.   a. $SrF_2(s) \ \rightleftharpoons \ Sr^{2+}(aq) \ + \ 2 \ F^-(aq)$

Initial   0   0
            s mol/L SrF_2 dissolves to reach equilibrium
Equil. s $2s$

$[Sr^{2+}][F^-]^2 = K_{sp} = 7.9 \times 10^{-10} = 4s^3, \ s = 5.8 \times 10^{-4} \ mol/L$

b.   Greater, because some of the F⁻ would react with water:

$$F^- + H_2O \rightleftharpoons HF + OH^- \quad K_b = \frac{K_w}{K_a(HF)} = 1.4 \times 10^{-11}$$

This lowers the concentration of F⁻, forcing more SrF_2 to dissolve.

c. $SrF_2(s) \rightleftharpoons Sr^{2+} + 2 \ F^- \quad K_{sp} = 7.9 \times 10^{-10} = [Sr^{2+}][F^-]^2$

Let s = solubility = $[Sr^{2+}]$, then 2s = $2[Sr^{2+}]$ = total F⁻ concentration.

Since F⁻ is a weak base, then 2s = $[F^-] + [HF]$.

$$HF \rightleftharpoons H^+ + F^- \quad K_a = 7.2 \times 10^{-4} = \frac{[H^+][F^-]}{[HF]} = \frac{1.0 \times 10^{-2} [F^-]}{[HF]}$$

$7.2 \times 10^{-2} = \dfrac{[F^-]}{[HF]}, \ [HF] = 14 \ [F^-]; \ $ Solving:

$[Sr^{2+}] = s; \ 2s = [F^-] + [HF] = [F^-] + 14 \ [F^-], \ 2s = 15 \ [F^-], \ [F^-] = 2s/15$

$$7.9 \times 10^{-10} = [Sr^{2+}][F^-]^2 = (s) \left(\frac{2s}{15} \right)^2, \ s = 3.5 \times 10^{-3} \ mol/L$$

74.   a. $Al(OH)_3(s) \rightleftharpoons Al^{3+} + 3 \ OH^-; \ Al(OH)_3(s) + OH^- \rightleftharpoons Al(OH)_4^-$

S = solubility = total Al^{3+} concentration = $[Al^{3+}] + [Al(OH)_4^-]$

$$[Al^{3+}] = \frac{K_{sp}}{[OH^-]^3} = K_{sp} \times \frac{[H^+]^3}{K_w^3}$$

$$\frac{[Al(OH)_4^-]}{[OH^-]} = K; \quad [OH^-] = \frac{K_w}{[H^+]}; \quad [Al(OH)_4^-] = K[OH^-] = \frac{KK_w}{[H^+]}$$

$$S = [Al^{3+}] + [Al(OH)_4^-] = [H^+]^3 K_{sp}/K_w^3 + KK_w/[H^+]$$

b. $K_{sp} = 2 \times 10^{-32}; \quad K_w = 1.0 \times 10^{-14}; \quad K = 40.0$

$$S = \frac{[H^+]^3 (2 \times 10^{-32})}{(1.0 \times 10^{-14})^3} + \frac{40.0 (1.0 \times 10^{-14})}{[H^+]} = [H^+]^3 (2 \times 10^{10}) + \frac{4.0 \times 10^{-13}}{[H^+]}$$

pH	solubility	log S
4.0	2×10^{-2}	-1.7
5.0	2×10^{-5}	-4.7
6.0	4.2×10^{-7}	-6.38
7.0	4.0×10^{-6}	-5.40
8.0	4.0×10^{-5}	-4.40
9.0	4.0×10^{-4}	-3.40
10.0	4.0×10^{-3}	-2.40
11.0	4.0×10^{-2}	-1.40
12.0	4.0×10^{-1}	-0.40

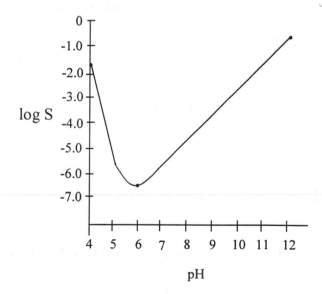

As expected, the solubility of $Al(OH)_3(s)$ is increased by very acidic solutions and by very basic solutions.

75.   For HOCl, $K_a = 3.5 \times 10^{-8}; \quad pH = pK_a + \log \frac{[OCl^-]}{[HOCl]}$

$$8.00 = 7.46 + \log \frac{[OCl^-]}{[HOCl]}, \quad \frac{[OCl^-]}{[HOCl]} = 10^{0.54} = 3.5$$

$1.00 \text{ L} \times 0.0500 \ M = 0.0500$ mol HOCl initially. Added OH^- converts HOCl into OCl^-.

$n_{OCl^-} + n_{HOCl} = 0.0500$ and $n_{OCl^-} = 3.5 \ n_{HOCl}$

$4.5 \ n_{HOCl} = 0.0500$, $n_{HOCl} = 0.011$ mol; $n_{OCl^-} = 0.039$ mol

Need to add 0.039 mol NaOH to produce 0.039 mol OCl^-.

$0.039 \text{ mol} = 0.0100 \ M \times \text{V}$, V = 3.9 L NaOH

76. $350.0 \text{ mL} \times \dfrac{0.200 \text{ mmol}}{\text{mL}} = 70.0$ mmol HCl; $100.0 \text{ mL} \times \dfrac{0.400 \text{ mmol}}{\text{mL}} = 40.0$ mmol NaF

The reaction that occurs is $HCl + F^- \rightarrow HF + Cl^-$. After this reaction goes to completion (F^- is limiting):

$$[HF]_0 = \frac{40.0 \text{ mmol}}{450.0 \text{ mL}} = 0.0889 \ M; \ [H^+]_0 = \frac{70.0 - 40.0}{450.0} = 0.0667 \ M$$

If we assume $[H^+]$ is the major source of H^+: $[H^+] = 0.0667 \ M$, pH = 1.176

Let's check assumption: $7.2 \times 10^{-4} = \dfrac{(0.0667) \ [F^-]}{0.0889}$, $[F^-] = 9.6 \times 10^{-4} \ M$

At most, an additional $9.6 \times 10^{-4} \ M \ H^+$ will be generated by HF ($\approx 1.4\%$ error). Assumption good. pH = 1.176.

77. a. $NH_3 + H_2O \rightleftharpoons OH^- + NH_4^+$; pH = 8.95; pOH = 5.05

$$K_b = 1.8 \times 10^{-5} = \frac{(10^{-5.05}) \ (x)}{(0.500)}, \ x = [NH_4^+] = 1.0 \ M$$

b. $4.00 \text{ g NaOH} \times \dfrac{1 \text{ mol}}{40.00 \text{ g}} = 0.100$ mol; OH^- converts NH_4^+ into NH_3.

$NH_4^+ + OH^- \rightarrow NH_3 + H_2O$; After this reaction goes to completion, a buffer solution still exists.

$$\text{pH} = 9.26 + \log \frac{0.600}{0.9} = 9.26 - 0.18 = 9.08 = 9.1$$

78. a. $HA^- \rightleftharpoons H^+ + A^{2-}$ $K_a = 1 \times 10^{-8}$; When $[HA^-] = [A^{2-}]$, pH = pK_{a2} = 8.00.

The titration reaction is $A^{2-} + H^+ \rightarrow HA^-$ (goes to completion). Begin with $100.0 \text{ mL} \times 0.200$ mmol/mL = 20.0 mmol A^{2-}. We need to convert 10.0 mmol A^{2-} into HA^- by adding 10.0 mmol H^+. This will produce a solution where pH = pK_{a2} = 8.00.

10.0 mmol = 1.00 mmol/mL × V, V = 10.0 mL HCl

b. At the 2nd stoichiometric point, all A^{2-} is converted into H_2A. This requires 40.0 mmol HCl which is 40.0 mL of 1.00 M HCl.

$$[H_2A]_0 = \frac{20.0 \text{ mmol}}{140.0 \text{ mL}} = 0.143 \ M; \text{ Since } K_{a1} >> K_{a2}, H_2A \text{ is the major source of } H^+.$$

$$H_2A \quad \rightleftharpoons \quad H^+ \quad + \quad HA^-$$

Initial	0.143 M	0	0
Equil.	0.143 - x	x	x

$$\frac{x^2}{0.143 - x} = 1.0 \times 10^{-3} \approx \frac{x^2}{0.143}, \ x = 0.012 \ M \quad \text{Check assumptions:}$$

$$\frac{0.012}{0.143} \times 100 = 8.4\%; \text{ Can't neglect } x. \text{ Using successive approximations, } x = 0.0115.$$

$[H^+] = 0.0115 \ M; \ pH = 1.94$

79. a. 1.00 L × 0.100 mol/L = 0.100 mol HCl added to reach stoichiometric point.

The 10.0 g sample must have contained 0.100 mol of NaA. $\dfrac{10.0 \text{ g}}{0.100 \text{ mol}} = 100. \text{ g/mol}$

b. 500.0 mL of HCl added represents the halfway point to equivalence. So, pH = pK_a = 5.00 and $K_a = 1.0 \times 10^{-5}$.

$$[HA]_0 = \frac{0.100 \text{ mol}}{1.10 \text{ L}} = 0.0909 \ M$$

$$HA \quad \rightleftharpoons \quad H^+ \quad + \quad A^- \qquad\qquad K_a = 1.0 \times 10^{-5}$$

Initial	0.0909 M	0	0
Equil.	0.0909 - x	x	x

$$1.0 \times 10^{-5} = \frac{x^2}{0.0909 - x} \approx \frac{x^2}{0.0909}$$

$x = 9.5 \times 10^{-4} \ M = [H^+]; \ pH = 3.02 \quad \text{Assumptions good.}$

80. a. $C_2H_5NH_3^+ \rightleftharpoons H^+ + C_2H_5NH_2 \quad K_a = \dfrac{K_w}{K_b} = \dfrac{1.0 \times 10^{-14}}{5.6 \times 10^{-4}} = 1.8 \times 10^{-11}; \ pK_a = 10.74$

$$pH = pK_a + \log \frac{[C_2H_5NH_2]}{[C_2H_5NH_3^+]} = 10.74 + \log \frac{0.10}{0.20} = 10.74 - 0.30 = 10.44$$

b. $C_2H_5NH_3^+ + OH^- \rightleftharpoons C_2H_5NH_2$; After this reaction goes to completion, a buffer solution still exists where $[C_2H_5NH_3^+] = [C_2H_5NH_2] = 0.15 \ M$.

$$pH = 10.74 + \log \frac{0.15}{0.15} = 10.74$$

81.   Aniline = $C_6H_5NH_2$; $K_b = 3.8 \times 10^{-10}$; Titration reaction: $C_6H_5NH_2 + H^+ \rightleftharpoons C_6H_5NH_3^+$

100.0 mL $\times$ 0.250 mmol/mL = 25.0 mmol aniline; Need 25.0 mmol HCl to reach the stoichiometric point.

$$V_{HCl} = \frac{25.0 \text{ mmol}}{0.500 \text{ mmol/mL}} = 50.0 \text{ mL}$$

At stoichiometric point:

$$C_6H_5NH_3^+ \quad \rightleftharpoons \quad H^+ \; + \; C_6H_5NH_2$$

Initial	$\dfrac{25.0 \text{ mmol}}{150.0 \text{ mL}} = 0.167\ M$	0	0
Equil.	$0.167 - x$	x	x

$$K_a = \frac{K_w}{3.8 \times 10^{-10}} = 2.6 \times 10^{-5} = \frac{x^2}{0.167 - x} \approx \frac{x^2}{0.167}$$

$x = [H^+] = 2.1 \times 10^{-3}\ M$;   pH = 2.68 Assumptions good.

82.   HOAc $\rightleftharpoons$ H$^+$ + OAc$^-$; Let C_o = initial concentration of HOAc.

From normal setup: $K_a = 1.8 \times 10^{-5} = \dfrac{[H^+][OAc^-]}{[HOAc]} = \dfrac{[H^+]^2}{C_o - [H^+]}$

$[H^+] = 10^{-2.68} = 2.1 \times 10^{-3}\ M$; $1.8 \times 10^{-5} = \dfrac{4.4 \times 10^{-6}}{C_o - 2.1 \times 10^{-3}}$, $C_o = 0.25\ M$

25.0 mL $\times$ 0.25 mmol/mL = 6.3 mmol acetic acid

Need 6.3 mmol KOH = 0.0975 mmol/mL $\times$ V, $V_{KOH} = 65$ mL

83.   HX $\rightleftharpoons$ H$^+$ + X$^-$; At equil., $[H^+] = [X^-]$ and $[HX] = C_o - [H^+]$ where $C_o = [HX]_o$.

$K_a = \dfrac{[H^+][X^-]}{[HX]} = 1.5 \times 10^{-4} = \dfrac{[H^+]^2}{C_o - [H^+]}$; $[H^+] = 10^{-3.58} = 2.6 \times 10^{-4}\ M$

$1.5 \times 10^{-4} = \dfrac{(2.6 \times 10^{-4})^2}{C_o - 2.6 \times 10^{-4}}$, $C_o = 7.1 \times 10^{-4}\ M$

75.0 mL $\times$ 7.1 $\times$ 10^{-4} mmol/mL = 5.3 $\times$ 10^{-2} mmol HX

Need 5.3×10^{-2} mmol NaOH $= 9.5 \times 10^{-2}$ mmol/mL $\times$ V, $V_{NaOH} = 0.56$ mL

84.   a)   50.0 mL $\times$ 1.0 mmol Na_3PO_4/mL = 50. mmol Na_3PO_4

           270.0 mL $\times$ 0.50 mmol HCl/mL = 135 mmol HCl   (carry extra S.F.)

           100.0 mL $\times$ 0.10 mmol NaCN/mL = 10. mmol NaCN

           110. mmol of HCl reacts to convert all CN^- to HCN and all PO_4^{3-} to $H_2PO_4^-$. At this point 10. mmol HCN, 50. mmol $H_2PO_4^-$ and 25 mmol HCl are in solution. The remaining HCl reacts completely with $H_2PO_4^-$, converting 25 mmol to H_3PO_4. Final solution contains: 25 mmol H_3PO_4, 25 mmol $H_2PO_4^-$ and 10. mmol HCN. HCN is a much weaker acid than either H_3PO_4 or $H_2PO_4^-$, so ignore it. Principle equilibrium reaction that is a source of H^+ is:

$$H_3PO_4 \quad \rightleftharpoons \quad H^+ \quad + \quad H_2PO_4^-$$

Equil. $\dfrac{25}{420.0} - x$ x $\dfrac{25}{420.0} + x$

$$K_{a1} = 7.5 \times 10^{-3} = \frac{x\left(\dfrac{25}{420.0} + x\right)}{\left(\dfrac{25}{420.0} - x\right)} = \frac{x(0.060 + x)}{(0.060 - x)}$$

Normal assumptions don't hold here. Using the quadratic formula:

$$x^2 + 0.0675\, x - 4.5 \times 10^{-4} = 0, \quad x = 0.0061\ M = [H^+];\ \ pH = 2.21 = 2.2$$

b) $[HCN] = \dfrac{10.\ \text{mmol}}{420.0\ \text{mL}} = 2.4 \times 10^{-2}\ M$; HCN dissociation will be minimal.

85.   In the final solution: $[H^+] = 10^{-2.15} = 7.1 \times 10^{-3}\ M$

Beginning mmol HCl = 500.0 mL $\times$ 0.200 mmol/mL = 100. mmol HCl

Amount of HCl that reacts with NaOH = 1.50×10^{-2} mmol/mL $\times$ V

$$\frac{7.1 \times 10^{-3}\ \text{mmol}}{\text{mL}} = \frac{\text{final mmol H}^+}{\text{total vol soln}} = \frac{100. - 0.0150\ V}{500.0 + V}$$

$3.6 + 7.1 \times 10^{-3}\ V = 100. - 1.50 \times 10^{-2}\ V,\ \ 2.21 \times 10^{-2}\ V = 100. - 3.6$

$V = 4.36 \times 10^3$ mL = 4.36 L = 4.4 L NaOH

86. $V_{HCl} = 1/2\ V_{NH_3}$; 0.400 mol/L $\times V_{NH_3}$ = mol NH_3 = mol NH_4^+ after reaction with HCl

$$C_{NH_4^+} = \frac{0.400 \times V_{NH_3}}{1.50\, V_{NH_3}} = 0.267\ M;\ \text{At equivalence point:}$$

$$NH_4^+ \rightleftharpoons H^+ + NH_3$$

	NH_4^+	H^+	NH_3
Initial	$0.267\ M$	0	0
Equil.	$0.267 - x$	x	x

$$K_a = \frac{K_w}{1.8 \times 10^{-5}} = 5.6 \times 10^{-10} = \frac{x^2}{0.267 - x} \approx \frac{x^2}{0.267}$$

$$x = [H^+] = 1.2 \times 10^{-5}\ M;\ pH = 4.92 \qquad \text{Assumptions good.}$$

87.   50.0 mL $\times$ 0.100 M = 5.00 mmol H_2SO_4;   30.0 mL $\times$ 0.10 M = 3.0 mmol HOCl

25.0 mL $\times$ 0.20 M = 5.0 mmol NaOH; 25.0 mL $\times$ 0.10 M = 2.5 mmol $Ca(OH)_2$ = 5.00 mmol OH^-

10.0 mL $\times$ 0.15 M = 1.5 mmol KOH; We've added 11.5 mmol OH^- total.

10.0 mmol OH^- + 5.00 mmol $H_2SO_4 \rightarrow$ 5.0 mmol H_2O + 5.0 mmol SO_4^{2-}

The remaining 1.5 mmol OH^- will convert 1.5 mmol HOCl to OCl^-, leaving 1.5 mmol OCl^- and 1.5 mmol HOCl. Major species at this point: HOCl, OCl^-, SO_4^{2-}, H_2O plus cations that don't affect pH. SO_4^{2-} is an extremely weak base ($K_b = 8.3 \times 10^{-13}$). Major equilibrium affecting pH: HOCl $\rightleftharpoons H^+ + OCl^-$. Since [HOCl] = [$OCl^-$]:

$$[H^+] = K_a = 3.5 \times 10^{-8}\ M;\ pH = 7.46$$

88.   a.   HA $\rightleftharpoons H^+ + A^-$ $K_a = 5.0 \times 10^{-10}$; $[HA]_0 = 1.0 \times 10^{-4}\ M$

Since this is a dilute solution of a weak acid, H_2O cannot be ignored as a source of H^+.

Try: $[H^+] = (K_a[HA]_o + K_w)^{1/2} = 2.4 \times 10^{-7}$; $pH = 6.62$

Check assumption:

$$\frac{[H^+]^2 - K_w}{[H^+]} = 2.0 \times 10^{-7} << 1.0 \times 10^{-4} \qquad \text{Assumption good.}\ pH = 6.62$$

b.   100.0 mL $\times$ $(1.00 \times 10^{-4}\ M) = 1.00 \times 10^{-2}$ mmol HA

5.00 mL $\times$ $(1.0 \times 10^{-3}\ M) = 5.00 \times 10^{-3}$ mmol NaOH added; Reacts completely with HA.

After reaction 5.00×10^{-3} mmol HA and 5.00×10^{-3} mmol A^- are in 105.0 mL. [A^-] = [HA] = 5.0×10^{-3} mmol/105.0 mL = $4.76 \times 10^{-5}\ M$

$$A^- + H_2O \rightleftharpoons HA + OH^- \qquad K_b = 2.0 \times 10^{-5}$$

	$A^- + H_2O$		HA +	OH^-
Initial	$4.76 \times 10^{-5}\,M$		$4.76 \times 10^{-5}\,M$	0
Equil.	$4.76 \times 10^{-5} - x$		$4.76 \times 10^{-5} + x$	x

$$2.0 \times 10^{-5} = \frac{(4.76 \times 10^{-5} + x)x}{(4.76 \times 10^{-5} - x)}; \quad x \text{ will not be small.}$$

Using the quadratic formula: $x = [OH^-] = 1.2 \times 10^{-5}\,M$; pOH = 4.92; pH = 9.08

c) At stoichiometric point, all the HA is converted into A^-.

$$A^- + H_2O \rightleftharpoons HA + OH^- \qquad K_b = 2.0 \times 10^{-5}$$

	A^- + H_2O		HA +	OH^-
Initial	$\dfrac{1.00 \times 10^{-2}}{110.0} = 9.09 \times 10^{-5}\,M$		0	0
Equil.	$9.09 \times 10^{-5} - x$		x	x

$$2.0 \times 10^{-5} = \frac{x^2}{9.09 \times 10^{-5} - x} \approx \frac{x^2}{9.09 \times 10^{-5}}, \quad x = [OH^-] = 4.3 \times 10^{-5}$$

Assumption is not good. Using the quadratic formula or successive approximations:

$$x = 3.4 \times 10^{-5}\,M = [OH^-]; \quad pOH = 4.47; \quad pH = 9.53$$

Assumption to ignore H_2O contribution to OH^- is good.

89. $50.0 \text{ mL} \times 0.100\,M = 5.00 \text{ mmol NaOH initially}$

at pH = 10.50, pOH = 3.50, $[OH^-] = 10^{(-3.50)} = 3.2 \times 10^{-4}\,M$

mmol OH^- remaining = 3.2×10^{-4} mmol/mL $\times$ 73.75 mL = 2.4×10^{-2} mmol

mmol OH^- that reacted = 5.00 - 0.024 = 4.98 mmol

Assuming the weak acid is monoprotic, then 23.75 mL of the weak acid solution contains 4.98 mmol HA.

$$C_o(HA) = \frac{4.98 \text{ mmol}}{23.75 \text{ mL}} = 0.210\,M$$

90. H_3A: $pK_{a1} = 3.00$, $pK_{a2} = 7.30$, $pK_{a3} = 11.70$

$$pH = 9.50 = \frac{pK_{a2} + pK_{a3}}{2} = \frac{7.30 + 11.70}{2} = \text{2nd stoichiometric point where } HA^{2-} \text{ dominates.}$$
See Section 8.6 on calculating the pH of amphoteric species like HA^{2-} or H_2A^-.

$$s = \frac{5.48 \text{ g}}{\text{L}} \times \frac{1 \text{ mol}}{73.89 \text{ g}} = 7.42 \times 10^{-2} \text{ mol/L}; \quad [Li^+] = 1.48 \times 10^{-1} \, M; \quad [CO_3^{2-}] = 7.42 \times 10^{-2} \, M$$

$$K_{sp} = [Li^+]^2 [CO_3^{2-}] = (1.48 \times 10^{-1})^2 (7.42 \times 10^{-2}) = 1.63 \times 10^{-3}$$

37.   In all setups, s = solubility of solid in mol/L. Note: Since solids do not appear in the K_{sp} expression, we do not need to worry about their initial or equilibrium amounts.

a. $Al(OH)_3(s) \rightleftharpoons Al^{3+} + 3 \, OH^-$

Initial		0	$10^{-7} \, M$ from water

s mol/L dissolves to reach equilibrium

Change	-s	$\rightarrow$	+s	+3s
Equil.			s	$10^{-7} + 3s$

$$K_{sp} = 2 \times 10^{-32} = [Al^{3+}][OH^-]^3 = (s)(10^{-7} + 3s)^3 \approx s(10^{-7})^3$$

$$s = \frac{2 \times 10^{-32}}{1 \times 10^{-21}} = 2 \times 10^{-11} \text{ mol/L} \qquad \text{Assumption good } (10^{-7} + 3s \approx 10^{-7}).$$

$$\frac{2 \times 10^{-11} \text{ mol}}{\text{L}} \times \frac{78.00 \text{ g}}{\text{mol}} = 2 \times 10^{-9} \text{ g/L}$$

b. $Be(IO_4)_2(s) \rightleftharpoons Be^{2+} + 2 \, IO_4^-$

Initial			0	0

s mol/L dissolves to reach equilibrium

Change	-s	$\rightarrow$	+s	+2s
Equil.			s	2s

$$K_{sp} = 1.57 \times 10^{-9} = [Be^{2+}][IO_4^-]^2 = (s)(2s)^2 = 4s^3, \quad s = \left(\frac{1.57 \times 10^{-9}}{4}\right)^{\frac{1}{3}}$$

$$s = 7.32 \times 10^{-4} \text{ mol/L}; \quad \frac{7.32 \times 10^{-4} \text{ mol}}{\text{L}} \times \frac{390.8 \text{ g}}{\text{mol}} = 0.286 \text{ g/L}$$

c. $CaSO_4(s) \rightleftharpoons Ca^{2+} + SO_4^{2-}$

Initial			0	0
Equil.			s	s

$$K_{sp} = 6.1 \times 10^{-5} = [Ca^{2+}][SO_4^{2-}] = s^2, \quad s = (6.1 \times 10^{-5})^{1/2}$$

$$s = 7.8 \times 10^{-3} \text{ mol/L}; \quad \frac{7.8 \times 10^{-3} \text{ mol}}{\text{L}} \times \frac{136.2 \text{ g}}{\text{mol}} = 1.1 \text{ g/L}$$

Note: This ignores any reaction between SO_4^{2-} (a very weak base) and water.

100.0 mL × 0.0500 M = 5.00 mmol H_3A initially. To reach the 2nd stiochiometric point, need 10.0 mmol OH^- = 1.0 mmol/mL × V. Solving for V:

V_{NaOH} = 10.0 mL (to reach pH = 9.50)

pH = 4.00 is between the first half way point to equivalence (pH = pK_{a1} = 3.00) and the first stoichiometric point (pH = $\dfrac{pK_{a1} + pK_{a2}}{2}$ = 5.15).

This is the buffer region controlled by: $H_3A \rightleftharpoons H_2A^- + H^+$

$pH = pK_{a1} + \log \dfrac{[H_2A^-]}{[H_3A]}$, $4.00 = 3.00 + \log \dfrac{[H_2A^-]}{[H_3A]}$, $\dfrac{[H_2A^-]}{[H_3A]} = 10.$

Since both species are in the same volume, the mole ratio also equals 10.

$\dfrac{n_{H_2A^-}}{n_{H_3A}} = 10.$ and $n_{H_2A^-} + n_{H_3A} = 5.00$ mmol (mass balance)

11 n_{H_3A} = 5.00, n_{H_3A} = 0.45 mmol; $n_{H_2A^-}$ = 4.55 mmol

We need to add 4.55 mmol OH^- to get 4.55 mmol H_2A^- from the original H_3A present.

4.55 mmol = 1.00 mmol/mL × V, V = 4.55 mL of NaOH (to reach pH = 4.00)

91. a. Since $K_{a1} > K_{a2}$ and K_{a3}, then initial pH is determined by H_3A. Consider only the first dissociation:

$$H_3A \rightleftharpoons H^+ + H_2A^-$$

Initial	0.100 M	~0	0
Equil.	0.100 - x	x	x

$K_{a1} = \dfrac{[H^+][H_2A^-]}{[H_3A]} = \dfrac{x^2}{0.100 - x} = 1.5 \times 10^{-4} \approx \dfrac{x^2}{0.100}$, $x = 3.9 \times 10^{-3}$

$[H^+] = 3.9 \times 10^{-3}$ M; pH = 2.41   Assumptions good.

b. $H_3A \rightleftharpoons 3\,H^+ + A^{3-}$ $K = K_{a1} \cdot K_{a2} \cdot K_{a3} = 2.3 \times 10^{-23} = \dfrac{[H^+]^3 [A^{3-}]}{[H_3A]}$

From (a), $[H^+] = 3.9 \times 10^{-3}$ M and $[H_3A] = 0.100 - 3.9 \times 10^{-3}$ M, so

$[A^{3-}] = \dfrac{2.3 \times 10^{-23}\,(0.100 - 3.9 \times 10^{-3})}{(3.9 \times 10^{-3})^3} = 3.7 \times 10^{-17}$ M

c. 10.0 mL × 1.00 M = 10.0 mmol NaOH. Began with 100.0 mL × 0.100 M = 10.0 mmol H_3A. This is at the 1st stoichiometric point where H_2A^- is the major species present.

$$pH = \frac{pK_{a1} + pK_{a2}}{2} = \frac{3.82 + 7.52}{2} = 5.67$$

d. 25.0 mL × 1.00 M = 25.0 mmol NaOH added. Mixture contains 5.00 mmol A^{3-} and 5.00 mmol HA^{2-}.

$$K_{a3} = \frac{[H^+][A^{3-}]}{[HA^{2-}]};\ \text{Since } [A^{3-}] = [HA^{2-}], \text{ then } [H^+] = K_{a3};\ pH = pK_{a3} = 11.30$$

Assumptions good.

92.   100.0 mL × 0.0500 M = 5.00 mmol H_3X initially

a. Since $K_{a1} > K_{a2} >> K_{a3}$, pH initially determined by H_3X.

$$H_3X \quad \rightleftharpoons \quad H^+ \ + \ H_2X^-$$

Initial	0.0500 M	~0	0
Equil.	0.0500 - x	x	x

$$1.0 \times 10^{-3} = \frac{x^2}{0.0500 - x} \approx \frac{x^2}{0.0500},\ x = 7.1 \times 10^{-3} \quad \text{Assumption poor.}$$

Using the quadratic formula: $x = 6.6 \times 10^{-3}\ M = [H^+];\ pH = 2.18$

b. 1.00 mmol OH^- added which converts H_3X into H_2X^-. After reaction, 4.00 mmol H_3X and 1.00 mmol H_2X^- are in a total volume of 110.0 mL. Solving the buffer problem:

$$H_3X \quad \rightleftharpoons \quad H^+ \ + \ H_2X^-$$

Initial	0.0364 M	~0	0.00909 M
Equil.	0.0364 - x	x	0.00909 + x

$$1.0 \times 10^{-3} = \frac{x(0.00909 + x)}{0.0364 - x}$$

Using the quadratic formula: $x = 2.8 \times 10^{-3}\ M = [H^+];\ pH = 2.55$

c. 2.50 mmol OH^- added, resulting in 2.50 mmol H_3X and 2.50 mmol H_2X^-. This is the first halfway point to equivalence. pH = pK_{a1} = 3.00; Assumptions good (5% error).

d. 5.00 mmol OH^- added, resulting in 5.00 mmol H_2X^-. This is the 1st stoichiometric point.

$$pH = \frac{pK_{a1} + pK_{a2}}{2} = \frac{3.00 + 7.00}{2} = 5.00$$

e.   6.0 mmol OH⁻ added, resulting in 1.00 mmol HX^{2-} and 4.00 mmol H_2X^-.

Using $H_2X^- \rightleftharpoons H^+ + HX^{2-}$ equation:

$$pH = pK_{a2} + \log \frac{[HX^{2-}]}{[H_2X^-]} = 7.00 - \log (1.00/4.00) = 6.40 \qquad \text{Assumptions good.}$$

f.   7.50 mmol KOH added, resulting in 2.50 mmol HX^{2-} and 2.50 mmol H_2X^-. This is the second halfway point to equivalence.

$pH = pK_{a2} = 7.00$ Assumptions good.

g.   10.0 mmol OH⁻ added, resulting in 5.0 mmol HX^{2-}. This is the 2nd stoichiometric point.

$$pH = \frac{pK_{a2} + pK_{a3}}{2} = \frac{7.00 + 12.00}{2} = 9.50$$

h.   12.5 mmol OH⁻ added, resulting in 2.5 mmol X^{3-} and 2.5 mmol HX^{2-}. This is the third halfway point to equivalence. Usually $pH = pK_{a3}$ but normal assumptions don't hold. Must solve for the pH exactly.

$[X^{3-}] = [HX^{2-}] = 2.5 \text{ mmol}/225.0 \text{ mL} = 1.1 \times 10^{-2} M$

$$X^{3-} + H_2O \rightleftharpoons HX^{2-} + OH^- \qquad K_b = \frac{K_w}{K_{a3}} = 1.0 \times 10^{-2}$$

Initial	0.011 M	0.011 M	0
Equil.	0.011 - x	0.011 + x	x

$1.0 \times 10^{-2} = \dfrac{x(0.011 + x)}{(0.011 - x)}$; Using quadratic formula:

$x^2 + 2.1 \times 10^{-2} x - 1.1 \times 10^{-4} = 0,\ x = 4.3 \times 10^{-3} M = [OH^-]$; $pH = 11.63$

i.   15.0 mmol OH⁻ added, resulting in 5.0 mmol X^{3-}. This is the 3rd stoichiometric point.

$$X^{3-} + H_2O \rightleftharpoons HX^{2-} + OH^- \qquad K_b = \frac{K_w}{K_{a3}} = 1.0 \times 10^{-2}$$

Initial	$\dfrac{5.0 \text{ mmol}}{250.0 \text{ mL}} = 0.020\ M$	0	0
Equil.	0.020 - x	x	x

$\dfrac{x^2}{0.020 - x} = 1.0 \times 10^{-2} \approx \dfrac{x^2}{0.020}$, $x = 1.4 \times 10^{-2}$ Assumption poor.

Using the quadratic formula: $x^2 + 1.0 \times 10^{-2} x - 2.0 \times 10^{-4} = 0$

$x = [OH^-] = 1.0 \times 10^{-2} M$; $pH = 12.00$

j. 20.0 mmol OH⁻ added, resulting in 5.0 mmol X³⁻ and 5.0 mmol OH⁻ excess. Since K_b for X³⁻ is fairly large for a weak base, we have to worry about the OH⁻ contribution from X³⁻.

$$[X^{3-}] = [OH^-] = \frac{5.0 \text{ mmol}}{300.0 \text{ mL}} = 1.7 \times 10^{-2} M$$

$$X^{3-} + H_2O \rightleftharpoons OH^- + HX^{2-}$$

Initial $1.7 \times 10^{-2} M$ $1.7 \times 10^{-2} M$ 0

Equil. $1.7 \times 10^{-2} - x$ $1.7 \times 10^{-2} + x$ x

$$K_b = \frac{[OH^-][HX^{2-}]}{[X^{3-}]} = 1.0 \times 10^{-2} = \frac{(1.7 \times 10^{-2} + x)x}{(1.7 \times 10^{-2} - x)}$$

Using the quadratic formula: $x^2 + 2.7 \times 10^{-2} x - 1.7 \times 10^{-4} = 0$, $x = 5.3 \times 10^{-3} M$

$[OH^-] = 1.7 \times 10^{-2} + x = 1.7 \times 10^{-2} + 5.3 \times 10^{-3} = 2.2 \times 10^{-2} M$; pH = 12.34

k. pH = 10.00: From calculation g and h, this solution will be a mixture of HX²⁻ and X³⁻, i.e., between 100.0 and 125.0 mL NaOH added.

$$X^{3-} + H_2O \rightarrow HX^{2-} + OH^- \qquad K_b = \frac{K_w}{K_{a3}} = 1.0 \times 10^{-2}; \; [OH^-] = 1.0 \times 10^{-4} M$$

$$1.0 \times 10^{-2} = \frac{[OH^-][HX^{2-}]}{[X^{3-}]} = (1.0 \times 10^{-4}) \frac{[HX^{2-}]}{[X^{3-}]}, \; \frac{[HX^{2-}]}{[X^{3-}]} = 1.0 \times 10^2 = \frac{n_{HX^{2-}}}{n_{X^{3-}}}$$

$n_{HX^{2-}} = 1.0 \times 10^2 n_{X^{3-}}$ and $n_{HX^{2-}} + n_{X^{3-}} = 5.00$; $101 n_{X^{3-}} = 5.00$, $n_{X^{3-}} = 4.95 \times 10^{-2}$ mmol

We need to add 5.0×10^{-2} mmol OH⁻ beyond the 2nd stoichiometric point (100.0 mL OH⁻).

5.0×10^{-2} mmol = 0.100 mmol/mL × V, V = 0.50 mL beyond 2nd stoichiometric point

$V_{tot} = 100.0 + 0.50 = 100.5$ mL NaOH

93.   50.0 mL × 0.00200 M = 0.100 mmol Ag⁺; 50.0 mL × 0.0100 M = 0.500 mmol IO₃⁻

Assume AgIO₃ precipitates completely. After reaction, 0.400 mmol IO₃⁻ is remaining. Now, let some AgIO₃ dissolve in solution with excess IO₃⁻ to reach equilibrium.

$$AgIO_3 \rightleftharpoons Ag^+ + IO_3^-$$

Initial   0 $\frac{0.400 \text{ mmol}}{100.0 \text{ mL}} = 4.00 \times 10^{-3} M$

Equil.   s $4.00 \times 10^{-3} + s$

$K_{sp} = [Ag^+][IO_3^-] = 3.0 \times 10^{-8} = s(4.00 \times 10^{-3} + s) \approx s(4.00 \times 10^{-3})$

$s = 7.5 \times 10^{-6} \ M = [Ag^+]$ Assumptions good.

94.   a. $CuCl(s) \rightleftharpoons Cu^+ + Cl^-$

Equil.   s   s

$K_{sp} = 1.2 \times 10^{-6} = [Cu^+][Cl^-] = s^2$, $s = 1.1 \times 10^{-3}$ mol/L

b. $CuCl(s) \rightarrow Cu^+ (aq) + Cl^- (aq)$ $K_{sp} = 1.2 \times 10^{-6}$
         $Cu^+ (aq) + 2 Cl^- (aq) \rightarrow CuCl_2^-(aq)$ $K_f = 8.7 \times 10^4$

$\overline{\qquad CuCl (s) + Cl^-(aq) \rightarrow CuCl_2^-(aq) \qquad}$ $K = 1.0 \times 10^{-1}$

Initial   0.10 M 0
Equil.   0.10 - s   s

$0.10 = \dfrac{[CuCl_2^-]}{[Cl^-]} = \dfrac{s}{0.10 - s}$

$1.0 \times 10^{-2} - 0.10 \ s = s$, $1.10 \ s = 1.0 \times 10^{-2}$, $s = 9.1 \times 10^{-3} \ M$

95. $MX (s) \rightarrow M^+ + X^-$

$X^- + H_2O \rightarrow HX + OH^-$ $K_b = \dfrac{1.00 \times 10^{-14}}{1.00 \times 10^{-15}} = 10.0$

$K_b = 10.0 = \dfrac{1.00 \times 10^{-14} \ [HX]}{[X^-]}$, $[X^-] = 1.00 \times 10^{-15} \ [HX]$

$[M^+]$ = total X^- concentration = $[HX] + [X^-] \approx [HX]$ (since K_b is large)

Thus $[X^-] = 1.00 \times 10^{-15} \ [M^+]$ and $[M^+]$ = solubility = $3.17 \times 10^{-8} \ M$.

$[X^-] = (1.00 \times 10^{-15})(3.17 \times 10^{-8}) = 3.17 \times 10^{-23} \ M$

$K_{sp} = [M^+][X^-] = (3.17 \times 10^{-8})(3.17 \times 10^{-23}) = 1.00 \times 10^{-30}$

When MX dissolves in H_2O, the reaction is:

    $MX(s) + H_2O \rightarrow M^+ + HX + OH^-$ $K = K_{sp} (K_b) = 1.00 \times 10^{-30} (10.) = 1.00 \times 10^{-29}$

Equil.   s   s $1.00 \times 10^{-7} + s$

$1.00 \times 10^{-29} = [M^+][HX][OH^-] = (s)(s)(1.00 \times 10^{-7} + s) \approx 1.00 \times 10^{-7} \ s^2$

$s^2 = \dfrac{1.00 \times 10^{-29}}{1.00 \times 10^{-7}} = 1.00 \times 10^{-22}$, $s = 1.00 \times 10^{-11}$ mol/L   Assumption good.

96.

$$AgBr(s) \rightleftharpoons Ag^+ + Br^- \qquad\qquad K_{sp} = 5.0 \times 10^{-13}$$
$$Ag^+ + 2\,NH_3 \rightleftharpoons Ag(NH_3)_2^+ \qquad K_f = K_1K_2 = 1.7 \times 10^7$$

$$AgBr(s) + 2\,NH_3 \rightleftharpoons Ag(NH_3)_2^+ + Br^- \qquad K = 8.5 \times 10^{-6}$$

	$AgBr(s)$	$+$	$2\,NH_3$	$\rightleftharpoons$	$Ag(NH_3)_2^+$	$+$	Br^-
Initial			0.200 M		0		0
Equil.			0.200 - 2s		s		s

$$\frac{s^2}{(0.200 - 2s)^2} = 8.5 \times 10^{-6} \approx \frac{s^2}{(0.200)^2}, \quad s = 5.8 \times 10^{-4}\ \text{mol/L} \qquad \text{Assumption good.}$$

97.   In 2.0 M NH_3, the soluble complex ion $Ag(NH_3)_2^+$ forms which increases the solubility of AgCl. The reaction is: $AgCl\,(s) + 2\,NH_3 \rightleftharpoons Ag(NH_3)_2^+ + Cl^-$. In 2.0 M NH_4NO_3, NH_3 is only formed by the dissocaition of the weak acid NH_4^+. There is not enough NH_3 produced by this reaction to dissolve AgCl.

98.   Since $pK_{a2} = 7.00$, an equimolar mixture of H_2A^- and HA^{2-} will produce a solution with pH = $pK_{a2} = 7.00$.

$$20.8\ \text{g Na}_3\text{A} \times \frac{1\ \text{mol}}{208\ \text{g}} = 0.100\ \text{mol Na}_3\text{A} = 100.\ \text{mmol Na}_3\text{A}$$

To convert 100. mmol Na_3A to an equimolar solution of H_2A^- and HA^{2-}, we must add 100. mmol H^+ + 50. mmol H^+ = 150. mmol H^+. 100. mmol H^+ converts all A^{3-} into HA^{2-}. The next 50. mmol H^+ converts half of the HA^{2-} into H_2A^-: Only answer c adds 150. mmol H^+ (= 1500 mL $\times$ 0.100 mmol/mL).

99.   50.0 mL $\times$ 1.0 M = 50. mmol CH_3NH_2 initially

a.   50.0 mL $\times$ 0.50 M = 25. mmol HCl added. This is the halfway point to equivalence where $[CH_3NH_2] = [CH_3NH_3^+]$.

$$pH = pK_a + \log \frac{[CH_3NH_2]}{[CH_3NH_3^+]} = pK_a; \quad K_a = \frac{1.0 \times 10^{-14}}{4.0 \times 10^{-4}} = 2.5 \times 10^{-11}$$

$$pH = pK_a = -\log (2.5 \times 10^{-11}) = 10.60$$

b.   It will take 100. mL of HCl solution to reach the stoichiometric point.

$$[CH_3NH_3^+] = \frac{50.\ \text{mmol}}{150.\ \text{mL}} = 0.33\ M$$

$$CH_3NH_3^+ \rightleftharpoons H^+ + CH_3NH_2 \qquad K_a = \frac{K_w}{K_b} = 2.5 \times 10^{-11}$$

	$CH_3NH_3^+$	H^+	CH_3NH_2
Initial	$0.33\ M$	0	0
Equil.	$0.33 - x$	x	x

$$2.5 \times 10^{-11} = \frac{x^2}{0.33 - x} \approx \frac{x^2}{0.33}, \quad x = 2.9 \times 10^{-6}\ M;\ \ pH = 5.54 \qquad \text{Asumptions good.}$$

100. $100.0\ mL \times 0.100\ M = 10.0$ mmol H_3A initially

a. $100.0\ mL \times 0.0500$ mmol/mL $= 5.00$ mmol OH^- added

This is the first halfway point to equivalence where $[H_3A] = [H_2A^-]$ and $pII = pK_{a1}$.

pH $= -\log (5.0 \times 10^{-4}) = 3.30$ Assumptions good.

b.   Since $pK_{a2} = 8.00$, a buffer mixture of H_2A^- and HA^{2-} can produce a pH $= 8.67$ solution.

$$8.67 = 8.00 + \log \frac{[HA^{2-}]}{[H_2A^-]}, \quad \frac{[HA^{2-}]}{[H_2A^-]} = 10^{+0.67} = 4.7$$

$$\frac{n_{HA^{2-}}}{n_{H_2A^-}} = 4.7, \quad n_{HA^{2-}} = 4.7\ n_{H_2A^-}; \quad n_{HA^{2-}} + n_{H_2A^-} = 10.0 \text{ mmol (mass balance)}$$

$$5.7\ n_{H_2A^-} = 10.0 \text{ mmol}, \quad n_{H_2A^-} = 1.8 \text{ mmol}; \quad n_{HA^{2-}} = 8.2 \text{ mmol}$$

To reach this point, must add a total of 18.2 mmol NaOH. 10.0 mmol OH^- converts all of the 10.0 mmol H_3A into H_2A^-. The next 8.2 mmol OH^- converts 8.2 mmol H_2A^- into 8.2 mmol HA^{2-}, leaving 1.8 mmol H_2A^-.

18.2 mmol $= 0.0500\ M \times V$, V $= 364$ mL NaOH

101. $100.0\ mL \times 0.100\ M = 10.0$ mmol NaF; $100.0\ mL \times 0.025\ M = 2.5$ mmol HCl

2.5 mmol H^+ converts 2.5 mmol F^- into 2.5 mmol HF. A buffer solution results containing 2.5 mmol HF and $(10.0 - 2.5 =)$ 7.5 mmol F^- in 200.0 mL.

$$pH = pK_a + \log \frac{[F^-]}{[HF]} = 3.14 + \log 3.0 = 3.62 \qquad \text{Assumptions good.}$$

102. $[X^-]_0 = 5.00\ M$ and $[Cu^+]_0 = 1.0 \times 10^{-3}\ M$ since equal volumes of each reagent.

Assume the reaction goes completely to CuX_3^{2-}, then solve equilibrium problem.

$$Cu^+ + 3\ X^- \rightleftharpoons CuX_3^{2-} \qquad K = K_1 \cdot K_2 \cdot K_3 = 1.0 \times 10^9$$

	Cu^+	$3\ X^-$	CuX_3^{2-}
Before	$1.0 \times 10^{-3}\ M$	$5.00\ M$	0
After	0	$5.00 - 3(10^{-3}) \approx 5.00$	1.0×10^{-3}
Equil.	x	$5.00 + 3x$	$1.0 \times 10^{-3} - x$

$$\frac{(1.0 \times 10^{-3} - x)}{x(5.00 + 3x)^3} = 1.0 \times 10^9 \approx \frac{1.0 \times 10^{-3}}{x(5.00)^3}, \quad x = [Cu^+] = 8.0 \times 10^{-15} \ M \quad \text{Assumptions good.}$$

$$[CuX_3^{2-}] = 1.0 \times 10^{-3} - 8.0 \times 10^{-15} = 1.0 \times 10^{-3} \ M$$

$$K_3 = \frac{[CuX_3^{2-}]}{[CuX_2^-][X^-]} = 1.0 \times 10^3 = \frac{(1.0 \times 10^{-3})}{[CuX_2^-](5.00)}, \quad [CuX_2^-] = 2.0 \times 10^{-7} \ M$$

Summarizing:

$[CuX_3^{2-}] = 1.0 \times 10^{-3} \ M$ (answer a)

$[CuX_2^-] = 2.0 \times 10^{-7} \ M$ (answer b)

$[Cu^{2+}] = 8.0 \times 10^{-15} \ M$ (answer c)

103.   a. Na^+ is present in all solutions.

     A. CO_3^{2-}, H_2O B. CO_3^{2-}, HCO_3^-, H_2O, Cl^-

     C. HCO_3^-, H_2O, Cl^- D. HCO_3^-, $CO_2(H_2CO_3)$, H_2O, Cl^-

     E. $CO_2 (H_2CO_3)$, H_2O, Cl^- F. H_3O^+ (excess), $CO_2(H_2CO_3)$, H_2O, Cl^-

   b.   <u>Point A</u>:

$$CO_3^{2-} + H_2O \rightleftharpoons HCO_3^- + OH^- \quad K_b(CO_3^{2-}) = \frac{K_w}{K_{a2}} = \frac{1.0 \times 10^{-14}}{5.6 \times 10^{-11}} = 1.8 \times 10^{-4}$$

Initial   0.100 M 0   ~0

Equil.   0.100 - x x x

$$1.8 \times 10^{-4} = \frac{[HCO_3^-][OH^-]}{[CO_3^{2-}]} = \frac{x^2}{0.100 - x} \approx \frac{x^2}{0.100}$$

$x = 4.2 \times 10^{-3} = [OH^-]; \ pH = 11.62$ Assumptions good.

<u>Point B</u>:   The first halfway point where $[CO_3^{2-}] = [HCO_3^-]$.

    $pH = pK_{a2} = -\log (5.6 \times 10^{-11}) = 10.25$ Assumptions good.

<u>Point C</u>:   First equivalence point. The amphoteric HCO_3^- dominates.

$$pH = \frac{pK_{a1} + pK_{a2}}{2}; \ pK_{a1} = -\log (4.3 \times 10^{-7}) = 6.37$$

$$pH = \frac{6.37 + 10.25}{2} = 8.31$$

Point D: The second halfway point where $[HCO_3^-] = [H_2CO_3]$.

pH = pK_{a1} = 6.37   Assumptions good.

Point E: 50.0 mL HCl added. $[H_2CO_3]$ = 2.50 mmol/75.0 mL = 0.0333 M

$$H_2CO_3 \;\rightleftharpoons\; H^+ \;+\; HCO_3^- \qquad K_{a1} = 4.3 \times 10^{-7}$$

Initial	0.0333 M	0	0
Equil.	0.0333 - x	x	x

$$4.3 \times 10^{-7} = \frac{x^2}{0.0333 - x} \approx \frac{x^2}{0.0333}$$

$x = [H^+]$ = 1.2×10^{-4} M;   pH = 3.92   Assumptions good.

c. V_1 corresponds to $CO_3^{2-} + H^+ \rightarrow HCO_3^-$; V_2 corresponds to $HCO_3^- + H^+ \rightarrow H_2CO_3$.

There are two sources of HCO_3^-: $NaHCO_3$ and Na_2CO_3.   So, $V_2 > V_1$

d. V_1 corresponds to $OH^- + H^+ \rightarrow H_2O$ and $CO_3^{2-} + H^+ \rightarrow HCO_3^-$.

V_2 corresponds to $HCO_3^- + H^+ \rightarrow H_2CO_3$;   So, $V_1 > V_2$

e.   0.100 mmol HCl/mL × 18.9 mL = 1.89 mmol H^+

Since the first stoichiometric point only involves the titration of Na_2CO_3 by H^+, then 1.89 mmol of CO_3^{2-} has been converted into HCO_3^-. The sample contains 1.89 mmol Na_2CO_3 × 105.99 mg/mmol = 2.00×10^2 mg = 0.200 g Na_2CO_3.

The second stoichiometric point involves the titration of HCO_3^- by H^+.

$$\frac{0.100 \text{ mmol } H^+}{mL} \times 36.7 \text{ mL} = 3.67 \text{ mmol } H^+ = 3.67 \text{ mmol } HCO_3^-$$

1.89 mmol $NaHCO_3$ came from the first stoichiometric point of the Na_2CO_3 titration. 3.67 - 1.89 = 1.78 mmol HCO_3^- came from $NaHCO_3$.

1.78 mmol $NaHCO_3$ × 84.01 mg $NaHCO_3$/mmol = 1.50×10^2 mg $NaHCO_3$ = 0.150 g $NaHCO_3$

$$\% \text{ Na}_2CO_3 = \frac{0.200 \text{ g}}{0.350 \text{ g}} \times 100 = 57.1 \% \text{ Na}_2CO_3 \text{ by mass}$$

$$\% \text{ NaHCO}_3 = \frac{0.150 \text{ g}}{0.350 \text{ g}} \times 100 = 42.9 \% \text{ NaHCO}_3 \text{ by mass}$$

104. K_{a3} is so small that a break is not seen at the third stoichiometric point.

105.   We will see only the first stoichiometric point in the titration of salicylic acid because K_{a2} is so small. For adipic acid the K_a's are fairly close to each other. Both protons will be titrated almost simultaneously, giving us only one break. The stoichiometric points will occur when 1 mol of H^+ is added per mol of salicylic acid present and when 2 mol of H^+ is added per mol of adipic acid present. Thus, the 25.00 mL volume corresponded to the titration of salicylic acid and the 50.00 mL volume corresponded to the titration of adipic acid.

106.   We will abbreviate $C_6H_5CO_2H$ as HBz and $C_6H_5CO_2^-$ as Bz^-. In order to do part b, we need to calculate the K_{sp} value for calcium benzoate.

$$Ca(Bz)_2 \cdot 3H_2O(s) \quad \rightleftharpoons \quad Ca^{2+} \quad + \quad 2\,Bz^- \quad + \quad 3\,H_2O$$

Initial   0   0
                s mol/L $Ca(Bz)_2 \cdot 3H_2O$ dissolves to reach equilibrium
Equil.   s   2s

s $\doteq$ 0.080 mol/L (from problem); $K_{sp} = 4s^3 = 4(0.080)^3 = 2.0 \times 10^{-3}$

Note: The $[Bz^-]$ will be slightly less than 2s since Bz^- is a weak base. However, the effect on the K_{sp} calculation is negligible. At equilibrium, $[Bz^-] \approx 2s$ as will be seen below.

a.   From set-up above, $[Ca^{2+}] = s = 8.0 \times 10^{-2}$ M and $[Bz^-] = 2s = 0.16$ M.

Bz^- is a weak base.

$$Bz^- \quad + \quad H_2O \quad \rightleftharpoons \quad HBz \quad + \quad OH^- \qquad K_b = \frac{1.0 \times 10^{-14}}{6.4 \times 10^{-5}} = 1.6 \times 10^{-10}$$

Initial   0.16 M 0   ~0
Equil.   0.16 - x x x

$1.6 \times 10^{-10} = \dfrac{x^2}{0.16 - x} \approx \dfrac{x^2}{0.16}$, $x = 5.1 \times 10^{-6}$ All assumptions good.

$[Ca^{2+}] = 0.080$ M; $[C_6H_5CO_2^-] = 0.16 - x = 0.16$ M

$[OH^-] = [C_6H_5CO_2H] = x = 5.1 \times 10^{-6}$ M; $[H^+] = 2.0 \times 10^{-9}$ M

b. $Ca(Bz)_2 \cdot 3H_2O(s) \rightleftharpoons Ca^{2+} + 2\,Bz^- + 3\,H_2O$ $K_{sp} = 2.0 \times 10^{-3}$
        $2\,Bz^- + 2\,H_2O \rightleftharpoons 2\,HBz + 2\,OH^-$ $K = K_b^2 = 2.6 \times 10^{-20}$

$\overline{}$

$Ca(Bz)_2 \cdot 3H_2O(s) \rightleftharpoons Ca^{2+} + 2\,HBz + 2\,OH^- + H_2O$ $K = K_{sp} \cdot K_b^2 = 5.2 \times 10^{-23}$

Equil.   s   2s 1.0×10^{-10} M (buffer)

$K = 5.2 \times 10^{-23} = [Ca^{2+}]\,[HBz]^2\,[OH^-]^2 = (s)\,(2s)^2\,(1.0 \times 10^{-10})^2$

$5.2 \times 10^{-3} = 4s^3$, s = 0.11 mol/L

c. H$^+$ reacts with the benzoate anion (a weak base) to form benzoic acid. As benzoate is removed, more calcium benzoate can dissolve (solubility equilibrium shifts right).

107. At equivalence point: 16.00 mL × 0.125 mmol/mL = 2.00 mmol OH$^-$ added; There must be 2.00 mmol HX present initially.

2.00 mL NaOH added = 2.00 mL × 0.125 mmol/mL = 0.250 mmol OH$^-$

0.250 mmol of OH$^-$ added will convert 0.250 mmol HX into 0.250 mmol X$^-$.

Remaining HX = 2.00 - 0.250 = 1.75 mmol HX

This is a buffer solution where [H$^+$] = $10^{-6.912}$ = 1.22 × 10^{-7} M.   Since total volume cancels:

$$K_a = \frac{[H^+][X^-]}{[HX]} = \frac{1.22 \times 10^{-7}(0.250)}{1.75} = 1.74 \times 10^{-8}$$

CHAPTER NINE

ENERGY, ENTHALPY AND THERMOCHEMISTRY

Potential and Kinetic Energy

1. Ball A: $PE = mgz = 5.0 \text{ kg} \times \dfrac{9.8 \text{ m}}{s^2} \times 5.0 \text{ m} = \dfrac{245 \text{ kg m}^2}{s^2} = 245 \text{ J} = 250 \text{ J}$

    At Point I: All of this energy is transferred to Ball B. All of B's energy is kinetic energy at this point. $E_{tot} = KE = 250 \text{ J}$

    At Point II: $PE = mgz = 1.0 \text{ kg} \times \dfrac{9.8 \text{ m}}{s^2} \times 2.0 \text{ m} = 19.6 \text{ J} \approx 20. \text{ J}$

    $KE = E_{tot} - PE = 250 \text{ J} - 20. \text{ J} = 230 \text{ J}$

2. $KE = \dfrac{1}{2} mv^2 = \dfrac{1}{2} (2.0 \text{ kg}) \left(\dfrac{1.0 \text{ m}}{s} \right)^2 = 1.0 \text{ J}; \ \ KE = \dfrac{1}{2} (1.0 \text{ kg}) \left(\dfrac{2.0 \text{ m}}{s} \right)^2 = 2.0 \text{ J}$

    The 1.0 kg object with a velocity of 2.0 m/s has the greater kinetic energy.

3. $q = \dfrac{20.8 \text{ J}}{°\text{C mol}} \times 39.1 \text{ mol} \times (38.0 - 0.0) \text{ }°\text{C} = 30,900 \text{ J} = 30.9 \text{ kJ}$

    $w = -P\Delta V = -1.00 \text{ atm } (998 \text{ L} - 876 \text{ L}) = -122 \text{ L atm}$

    $w = -122 \text{ L atm} \times \dfrac{101.3 \text{ J}}{\text{L atm}} = -12,400 \text{ J} = -12.4 \text{ kJ}$

    $\Delta E = q + w = 30.9 \text{ kJ} + (-12.4 \text{ kJ}) = 18.5 \text{ kJ}$

4. In this problem q = w = -950. J

    $-950. \text{ J} \times \dfrac{1 \text{ L atm}}{101.3 \text{ J}} = - 9.38 \text{ L atm of work done by the gases.}$

    $w = -P\Delta V, \ \ -9.38 \text{ L atm} = \dfrac{-650.}{760} \text{ atm } (V_f - 0.040 \text{ L})$

    $V_f - 0.040 = 11.0 \text{ L}, \ \ V_f = 11.0 \text{ L}$

5. $313 \text{ g He} \times \dfrac{1 \text{ mol He}}{4.003 \text{ g He}} = 78.2 \text{ mol He}$

$q = 78.2 \text{ mol He} \times \dfrac{20.8 \text{ J}}{\text{mol °C}} \times (-15 \text{ °C}) = -2.4 \times 10^4 \text{ J or } -24 \text{ kJ}$

$w = -P\Delta V = -1.00 \text{ atm } (1814 - 1910.) \text{ L} = +96 \text{ L atm}$

$w = +96 \text{ L atm} \times \dfrac{101.3 \text{ J}}{\text{L atm}} = +9.7 \times 10^3 \text{ J or } +9.7 \text{ kJ}$

$\Delta E = q + w = -24 \text{ kJ} + 9.7 \text{ kJ} = -14.3 \text{ kJ} = -14 \text{ kJ}$

Properties of Enthalpy

6.   One should try to cool the reaction mixture or provide some means of removing heat. The reaction is very exothermic. The H_2SO_4(aq) will get very hot and possibly boil.

7.   a. $1.00 \text{ g CH}_4 \times \dfrac{1 \text{ mol CH}_4}{16.04 \text{ g CH}_4} \times \dfrac{-891 \text{ kJ}}{\text{mol CH}_4} = -55.5 \text{ kJ}$

       b. $n = \dfrac{PV}{RT} = \dfrac{\dfrac{740.}{760} \text{ atm} \times (1.00 \times 10^3 \text{ L})}{\dfrac{0.08206 \text{ L atm}}{\text{mol K}} \times 298 \text{ K}} = 39.8 \text{ mol}; \quad 39.8 \text{ mol} \times \dfrac{-891 \text{ kJ}}{\text{mol}} = -3.55 \times 10^4 \text{ kJ}$

8. $S(s) + O_2(g) \rightarrow SO_2(g) \quad \Delta H° = -296 \text{ kJ/mol}$

       a. $275 \text{ g S} \times \dfrac{1 \text{ mol S}}{32.07 \text{ g S}} \times \dfrac{-296 \text{ kJ}}{\text{mol S}} = -2.54 \times 10^3 \text{ kJ}$

       b. $25 \text{ mol S} \times \dfrac{-296 \text{ kJ}}{\text{mol S}} = -7.4 \times 10^3 \text{ kJ}$

       c. $150. \text{ g SO}_2 \times \dfrac{1 \text{ mol SO}_2}{64.07 \text{ g SO}_2} \times \dfrac{-296 \text{ kJ}}{\text{mol SO}_2} = -693 \text{ kJ}$

9.   The combustion of phosphorus is exothermic. Thus the product, P_4O_{10}, is lower in energy than either red or white phosphorus. Since the conversion of white phosphorus to red phosphorus is exothermic, red phosphorus is lower in energy than white phosphorus. Thus, white phosphorus will release more heat when burned in air since there is a larger energy difference between white phosphorus and products.

The Thermodynamics of Ideal Gases

10.   Calculate the constant volume process first.

$$n = 1.00 \times 10^3 \text{ g} \times \frac{1 \text{ mol}}{30.07 \text{ g}} = 33.3 \text{ mol } C_2H_6; \quad C_v = \frac{44.60 \text{ J}}{K \text{ mol}} = \frac{44.60 \text{ J}}{°C \text{ mol}}$$

$$\Delta E = nC_v\Delta T = (33.3 \text{ mol}) (44.60 \text{ J } °C^{-1} \text{ mol}^{-1}) (75.0 - 25.0°C) = 74,300 \text{ J} = 74.3 \text{ kJ}$$

$$\Delta E = q + w; \quad \text{Since } \Delta V = 0, \ w = 0; \quad \Delta E = q_v = 74.3 \text{ kJ}$$

$$\Delta H = \Delta E + \Delta PV = \Delta E + nR\Delta T$$

$$\Delta H = 74.3 \text{ kJ} + (33.3 \text{ mol})(8.3145 \text{ J mol}^{-1} \text{ K}^{-1}) (50.0 \text{ K})(1 \text{ kJ}/1000 \text{ J})$$

$$\Delta H = 74.3 \text{ kJ} + 13.8 \text{ kJ} = 88.1 \text{ kJ}$$

Now consider the constant pressure process.

$$q_p = \Delta H = nC_p\Delta T = (33.3 \text{ mol}) (52.92 \text{ J mol}^{-1} \text{ K}^{-1}) (50.0 \text{ K})$$

$$q_p = 88,100 \text{ J} = 88.1 \text{ kJ} = \Delta H$$

$$w = -P\Delta V = -nR\Delta T = -(33.3 \text{ mol}) (8.3145 \text{ J mol}^{-1} \text{ K}^{-1}) (50.0 \text{ K}) = -13,800 \text{ J} = -13.8 \text{ kJ}$$

$$\Delta E = q + w = 88.1 \text{ kJ} - 13.8 \text{ kJ} = 74.3 \text{ kJ}$$

Summary:	Constant V	Constant P
q	74.3 kJ	88.1 kJ
ΔE	74.3 kJ	74.3 kJ
ΔH	88.1 kJ	88.1 kJ
w	0	-13.8 kJ

11. $88.0 \text{ g } N_2O \times \dfrac{1 \text{ mol } N_2O}{44.02 \text{ g } N_2O} = 2.00 \text{ mol } N_2O$

At constant pressure, $q_p = \Delta H$

$$\Delta H = nC_p\Delta T = (2.00 \text{ mol}) (38.70 \text{ J mol}^{-1} \text{ °C}^{-1}) (55°C - 165°C)$$

$$\Delta H = -8510 \text{ J} = -8.51 \text{ kJ} = q_p$$

$$w = -P\Delta V = -nR\Delta T = -(2.00 \text{ mol}) (8.3145 \text{ J mol}^{-1} \text{ K}^{-1}) (-110. \text{ K}) = 1830 \text{ J} = 1.83 \text{ kJ}$$

$$\Delta E = q + w = -8.51 \text{ kJ} + 1.83 \text{ kJ} = -6.68 \text{ kJ}$$

12.   Pathway I:

Step 1: (5.00 mol, 3.00 atm, 15.0 L) → (5.00 mol, 3.00 atm, 55.0 L)

$w = -P\Delta V = -(3.00 \text{ atm}) (55.0 - 15.0 \text{ L}) = -120. \text{ L atm}$

$w = -120. \text{ L atm} \times \dfrac{101.3 \text{ J}}{\text{L atm}} \times \dfrac{1 \text{ kJ}}{1000 \text{ J}} = -12.2 \text{ kJ}$

$\Delta H = q_p = nC_p\Delta T = nC_p \ \dfrac{\Delta(PV)}{nR} = \dfrac{C_p\Delta(PV)}{R}; \ \ \Delta(PV) = (P_2V_2 - P_1V_1)$

For an ideal monatomic gas: $C_p = \dfrac{5}{2} R$

$\Delta H = q_p = \dfrac{5}{2}\Delta(PV) = \dfrac{5}{2}(165 - 45.0) \text{ L atm} = 300. \text{ L atm}$

$\Delta H = q_p = 300. \text{ L atm} \times \dfrac{101.3 \text{ J}}{\text{L atm}} \times \dfrac{1 \text{ kJ}}{1000 \text{ J}} = 30.4 \text{ kJ}$

$\Delta E = q + w = 30.4 \text{ kJ} - 12.2 \text{ kJ} = 18.2 \text{ kJ}$

Step 2: (5.00 mol, 3.00 atm, 55.0 L) → (5.00 mol, 6.00 atm, 20.0 L)

$\Delta E = nC_v\Delta T = n\left(\dfrac{3}{2} R\right)\left(\dfrac{\Delta(PV)}{nR}\right) = \dfrac{3}{2}\Delta PV$

$\Delta E = \dfrac{3}{2}(120. - 165) \text{ L atm} = -67.5 \text{ L atm}$ (Carry extra significant figure.)

$\Delta E = -67.5 \text{ L atm} \times \dfrac{101.3 \text{ J}}{\text{L atm}} \times \dfrac{1 \text{ kJ}}{1000 \text{ J}} = -6.8 \text{ kJ}$

$\Delta H = nC_p\Delta T = n\left(\dfrac{5}{2} R\right)\left(\dfrac{\Delta(PV)}{nR}\right) = \dfrac{5}{2}\Delta(PV)$

$\Delta H = \dfrac{5}{2}(-45 \text{ L atm}) = -113 \text{ L atm}$ (Carry extra significant figure.)

$\Delta H = -113 \text{ L atm} \times \dfrac{101.3 \text{ J}}{\text{L atm}} \times \dfrac{1 \text{ kJ}}{1000 \text{ J}} = -11 \text{ kJ}$

$w = -P_{ext}\Delta V = -(6.00 \text{ atm}) (20.0 - 55.0)\text{L} = 210. \text{ L atm}$

$w = 210. \text{ L atm} \times \dfrac{101.3 \text{ J}}{\text{L atm}} \times \dfrac{1 \text{ kJ}}{1000 \text{ J}} = 21.3 \text{ kJ}$

$\Delta E = q + w$, -6.8 kJ $= q + 21.3$ kJ, $q = -28.1$ kJ

Summary:

	Path I	Step 1	Step 2	Total
q		30.4 kJ	-28.1 kJ	2.3 kJ
w		-12.2 kJ	21.3 kJ	9.1 kJ
ΔE		18.2 kJ	-6.8 kJ	11.4 kJ
ΔH		30.4 kJ	-11 kJ	19 kJ

Pathway II:

Step 3: (5.00 mol, 3.00 atm, 15.0 L) $\rightarrow$ (5.00 mol, 6.00 atm, 15.0 L)

$$\Delta E = q_v = \frac{3}{2}\Delta(PV) = \frac{3}{2}(90.0 - 45.0)\text{L atm} = 67.5 \text{ L atm}$$

$$\Delta E = q_v = 67.5 \text{ L atm} \times \frac{101.3 \text{ J}}{\text{L atm}} \times \frac{1 \text{ kJ}}{1000 \text{ J}} = 6.84 \text{ kJ}$$

$w = -P\Delta V = 0$ since $\Delta V = 0$

$\Delta H = \Delta E + \Delta(PV) = 67.5$ L atm $+ 45.0$ L atm $= 112.5$ L atm $= 11.40$ kJ

Step 4: (5.00 mol, 6.00 atm, 15.0 L) $\rightarrow$ (5.00 mol, 6.00 atm, 20.0 L)

$$\Delta H = q_p = nC_p\Delta T = n\left(\frac{5}{2}R\right)\left(\frac{\Delta(PV)}{nR}\right) = \frac{5}{2}\Delta PV$$

$$\Delta H = \frac{5}{2}(120. - 90.0) \text{ L atm} = 75 \text{ L atm}$$

$$\Delta H = q_p = 75 \text{ L atm} \times \frac{101.3 \text{ J}}{\text{L atm}} \times \frac{1 \text{ kJ}}{1000 \text{ J}} = 7.6 \text{ kJ}$$

$w = -P\Delta V = -(6.00 \text{ atm})(20.0 - 15.0)\text{L} = -30.$ L atm

$$w = -30. \text{ L atm} \times \frac{101.3 \text{ J}}{\text{L atm}} \times \frac{1 \text{ kJ}}{1000 \text{ J}} = -3.0 \text{ kJ}$$

$\Delta E = q + w = 7.6$ kJ $- 3.0$ kJ $= 4.6$ kJ

Summary: Path II   Step 3   Step 4   Total

q   6.84 kJ   7.6 kJ   14.4 kJ

w   0   -3.0 kJ   -3.0 kJ

ΔE 6.84 kJ   4.6 kJ   11.4 kJ

ΔH 11.40 kJ   7.6 kJ   19.0 kJ

Note that ΔE and ΔH are the same for both pathways. They are state functions.

13.   Step 1: (2.00 mol, 10.0 atm, 10.0 L) → (2.00 mol, 10.0 atm, 5.0 L)

$$\Delta H = q_p = nC_p\Delta T = n\left(\frac{5}{2}R\right)\left(\frac{\Delta(PV)}{nR}\right) = \frac{5}{2}\Delta(PV)$$

$\Delta H = q_p = \dfrac{5}{2}(50. - 100.) = -125$ L atm $= -12.7$ kJ   (We will carry all calculations to 0.1 kJ.)

$w = -P\Delta V = -(10.0\ \text{atm})(5.0 - 10.0)\text{L} = 50.$ L atm $= 5.1$ kJ

$\Delta E = q + w = -12.7 + 5.1 = -7.6$ kJ

Step 2: (2.00 mol, 10.0 atm, 5.0 L) → (2.00 mol, 20.0 atm, 5.0 L)

$\Delta E = q_v = \dfrac{3}{2}\Delta(PV) = \dfrac{3}{2}(100 - 50.) = 75$ L atm $= 7.6$ kJ; $w = 0$ since $\Delta V = 0$

$\Delta H = \Delta E + \Delta(PV) = 75$ L atm $+ 50$ L atm $= 125$ L atm $= 12.7$ kJ

Step 3: (2.00 mol, 20.0 atm, 5.0 L) → (2.00 mol, 20.0 atm, 25.0 L)

$\Delta H = q_p = \dfrac{5}{2}\Delta(PV) = \dfrac{5}{2}(500. - 100) = 1.0 \times 10^3$ L atm $= 101.3$ kJ

$w = -P\Delta V = -(20.0\ \text{atm})(25.0 - 5.0)\text{L} = -400.$ L atm $= -40.5$ kJ

$\Delta E = q + w = 101.3 - 40.5 = 60.8$ kJ

Summary: Path I   Step 1   Step 2   Step 3   Total

q   -12.7 kJ   7.6 kJ   101.3 kJ   96.2 kJ

w   5.1 kJ   0   -40.5 kJ   -35.4 kJ

ΔE -7.6 kJ   7.6 kJ   60.8 kJ   60.8 kJ

ΔH -12.7 kJ   12.7 kJ   101.3 kJ   101.3 kJ

Calorimetry and Heat Capacity

14.   A solution calorimeter is at constant (atmospheric) pressure. The heat at constant P is ΔH. A bomb calorimeter is at constant volume. The heat at constant volume is ΔE.

15.   a. $\Delta H = q = \dfrac{0.900 \text{ J}}{\text{g }^\circ\text{C}} \times 850. \text{ g} \times (94.6 - 22.8)^\circ\text{C} = 5.49 \times 10^4 \text{ J or } 54.9 \text{ kJ}$

       b. $\dfrac{0.900 \text{ J}}{\text{g }^\circ\text{C}} \times \dfrac{26.98 \text{ g}}{\text{mol}} = \dfrac{24.3 \text{ J}}{\text{mol }^\circ\text{C}}$

16.   Specific heat capacity $= \dfrac{78.2 \text{ J}}{45.6 \text{ g} \times 13.3^\circ\text{C}} = \dfrac{0.129 \text{ J}}{\text{g }^\circ\text{C}}$

       Molar heat capacity $= \dfrac{0.129 \text{ J}}{\text{g }^\circ\text{C}} \times \dfrac{207.2 \text{ g}}{\text{mol}} = \dfrac{26.7 \text{ J}}{\text{mol }^\circ\text{C}}$

17.   Heat gained by water = Heat lost by nickel

       Heat gain $= \dfrac{4.18 \text{ J}}{\text{g }^\circ\text{C}} \times 150.0 \text{ g} \times (25.0^\circ\text{C} - 23.5^\circ\text{C}) = 940 \text{ J}$

       Note: A temperature <u>change</u> of one Kelvin is the same as a temperature change of one degree Celsius.

       Heat loss = S × m × ΔT where S = specific heat capacity. Keeping all quantities positive to avoid sign error:

       $940 \text{ J} = \text{S} \times 28.2 \text{ g} \times (99.8 - 25.0)^\circ\text{C}, \ \text{S} = \dfrac{940 \text{ J}}{28.2 \text{ g} \times 74.8 \ ^\circ\text{C}} = \dfrac{0.45 \text{ J}}{\text{g }^\circ\text{C}}$

18.   Heat lost by solution = Heat gained by KBr

       Note: Sign errors are common with calorimetry problems. However, the correct sign for ΔH can easily be obtained from the ΔT data. When working calorimetry problems, keep all quantities positive (ignore signs). When finished, deduce the correct sign for ΔH. For this problem, T decreases so ΔH is positive.

       Heat lost by solution $= \dfrac{4.18 \text{ J}}{\text{g }^\circ\text{C}} \times 136 \text{ g} \times 3.1^\circ\text{C} = 1800 \text{ J}$

       Heat gained by KBr $= \dfrac{1800 \text{ J}}{10.5 \text{ g}} = \dfrac{170 \text{ J}}{\text{g}} = \Delta H$

       $\Delta H = \dfrac{170 \text{ J}}{\text{g KBr}} \times \dfrac{119.0 \text{ g KBr}}{\text{mol KBr}} \times \dfrac{1 \text{ kJ}}{1000 \text{ J}} = \dfrac{20. \text{ kJ}}{\text{mol}}$

19. $NaOH(aq) + HCl(aq) \rightarrow NaCl(aq) + H_2O(l)$

We have a stoichiometric mixture. All of the NaOH and HCl will react.

$0.1000 \text{ L} \times \dfrac{1.0 \text{ mol}}{\text{L}} = 0.10$ mol of HCl is neutralized.

Heat lost by chemicals = heat gained by solution.

Heat gain $= \dfrac{4.18 \text{ J}}{\text{g }^\circ\text{C}}\left(200.0 \text{ mL} \times \dfrac{1.0 \text{ g}}{\text{mL}}\right)(31.3 - 24.6)^\circ\text{C}\ (1 \text{ kJ}/1000 \text{ J}) = 5.6 \text{ kJ}$

Heat loss = 5.6 kJ; This is the heat released by the neutralization of 0.10 mol HCl (0.1000 L HCl × 1.0 mol/L). Since the temperature increased, the sign for ΔH must be negative, i.e., the reaction is exothermic. For calorimetry problems, keep all quantities positive until the end of the calculation, then decide the sign for ΔH.

$\Delta H = \dfrac{-5.6 \text{ kJ}}{0.10 \text{ mol}} = -56 \text{ kJ/mol}$

20. $50.0 \times 10^{-3} \text{ L} \times 0.100 \text{ mol/L} = 5.00 \times 10^{-3}$ mol of both $AgNO_3$ and HCl are reacted. Thus, 5.00×10^{-3} mol of AgCl will be produced.

Heat lost by chemicals = Heat gained by water

Heat gain $= \dfrac{4.18 \text{ J}}{\text{g }^\circ\text{C}} \times 100.0 \text{ g} \times 0.80^\circ\text{C} = 330 \text{ J}$

Heat loss = 330 J; This is the heat evolved (exothermic reaction) when 5.00×10^{-3} mol of AgCl is produced. So ΔH is negative with a value of:

$\Delta H = \dfrac{-330 \text{ J}}{5.00 \times 10^{-3} \text{ mol}} \times \dfrac{1 \text{ kJ}}{1000 \text{ J}} = -66 \text{ kJ/mol}$

21.   Heat lost by camphor = Heat gained by calorimeter

Heat lost by combustion of camphor $= \dfrac{5903.6 \text{ kJ}}{\text{mol}} \times \dfrac{1 \text{ mol}}{152.23 \text{ g}} \times 0.1204 \text{ g} = 4.669 \text{ kJ}$

Heat gain by calorimeter $= C_{cal} \times \Delta T$; $C_{cal}(2.28^\circ\text{C}) = 4.669 \text{ kJ}$, $C_{cal} = 2.05 \text{ kJ/}^\circ\text{C}$

22.   Heat gain by calorimeter $= \dfrac{1.56 \text{ kJ}}{^\circ\text{C}} \times 3.2^\circ\text{C} = 5.0 \text{ kJ}$ = heat loss by quinone

Heat loss = 5.0 kJ which is the heat evolved (exothermic reaction) by the combustion of 0.1964 g of quinone.

$\Delta E_{comb} = \dfrac{-5.0 \text{ kJ}}{0.1964 \text{ g}} = -25 \text{ kJ/g}$; $\dfrac{-25 \text{ kJ}}{\text{g}} \times \dfrac{108.1 \text{ g}}{\text{mol}} = -2700 \text{ kJ/mol}$

23.   First, we need to get the heat capacity of the calorimeter from the combustion of benzoic acid.

Heat lost by combustion = Heat gained by calorimeter

Heat loss = $\dfrac{26.42 \text{ kJ}}{\text{g}} \times 0.1584 \text{ g} = 4.185 \text{ kJ}$

Heat gain = 4.185 kJ = $C_{cal} \times \Delta T$, $C_{cal} = \dfrac{4.185 \text{ kJ}}{2.54°C} = 1.65 \text{ kJ/°C}$

Now we can calculate the heat of combustion of vanillin.

Heat gain by calorimeter = $\dfrac{1.65 \text{ kJ}}{°C} \times 3.25°C = 5.36 \text{ kJ}$

Heat loss = 5.36 kJ which is the heat evolved by the combustion of 0.2130 g vanillin.

$\Delta E_{comb} = \dfrac{-5.36 \text{ kJ}}{0.2130 \text{ g}} = -25.2 \text{ kJ/g};$ $\dfrac{-25.2 \text{ kJ}}{\text{g}} \times \dfrac{152.1 \text{ g}}{\text{mol}} = -3830 \text{ kJ/mol}$

24. $V = 10.0 \text{ m} \times 4.0 \text{ m} \times 3.0 \text{ m} = 1.2 \times 10^2 \text{ m}^3 \times (100 \text{ cm/m})^3 = 1.2 \times 10^8 \text{ cm}^3$

Mass of water = 1.2×10^8 g since the density of water is 1.0 g/cm^3.

Heat required = $\dfrac{4.18 \text{ J}}{\text{g °C}} \times (1.2 \times 10^8 \text{ g}) \times 9.8°C = 4.9 \times 10^9 \text{ J} = 4.9 \times 10^6 \text{ kJ}$

25.   The specific heat of water is 4.18 J g^{-1} $°C^{-1}$, which is equal to 4.18 kJ kg^{-1} $°C^{-1}$. We have 1.00 kg of H_2O:

$$1.00 \text{ kg} \times \dfrac{4.18 \text{ kJ}}{\text{kg °C}} = 4.18 \text{ kJ/°C}$$

This is the portion of the heat capacity that can be attributed to H_2O.

Total heat capacity = $C_{cal} + C_{H_2O}$, $C_{cal} = 10.84 - 4.18 = 6.66 \text{ kJ/°C}$

26.   Heat released = 26.42 kJ/g $\times$ 1.056 g = 27.90 kJ

Heat gained by calorimeter = 27.90 kJ = $\dfrac{4.18 \text{ kJ}}{\text{kg °C}} \times 0.987 \text{ kg} \times \Delta T + \dfrac{6.66 \text{ kJ}}{°C} \times \Delta T$

27.90 = (4.13 + 6.66) ΔT = 10.79 ΔT, $\Delta T = 2.586 °C$

2.586 °C = T_f - 23.32 °C, $T_f = 25.91 °C$

Hess's Law

27. $S + 3/2\ O_2 \rightarrow SO_3$ $\Delta H° = -395.2\ kJ$
              $SO_3 \rightarrow SO_2 + 1/2\ O_2$ $\Delta H° = +198.2\ kJ/2 = +99.1\ kJ$

          _____

          $S(s) + O_2(g) \rightarrow SO_2(g)$ $\Delta H° = -296.1\ kJ$

28. $2\ NO_2 \rightarrow N_2 + 2\ O_2$ $\Delta H° = -67.7\ kJ$
          $N_2 + 2\ O_2 \rightarrow N_2O_4$ $\Delta H° = 9.7\ kJ$

          _____

          $2\ NO_2(g) \rightarrow N_2O_4(g)$ $\Delta H° = -58.0\ kJ$

29. $4\ HNO_3 \rightarrow 2\ N_2O_5 + 2\ H_2O$ $\Delta H° = 2(+76.6\ kJ)$
          $2\ N_2 + 6\ O_2 + 2\ H_2 \rightarrow 4\ HNO_3$ $\Delta H° = 4(-174.1\ kJ)$
                      $2\ H_2O \rightarrow 2\ H_2 + O_2$ $\Delta H° = 2(+285.8\ kJ)$

          _____

          $2\ N_2(g) + 5\ O_2(g) \rightarrow 2\ N_2O_5(g)$ $\Delta H° = 28.4\ kJ$

30.   To avoid fractions, let's first calculate ΔH for the reaction:

          $6\ FeO(s) + 6\ CO(g) \rightarrow 6\ Fe(s) + 6\ CO_2(g)$

              $6\ FeO + 2\ CO_2 \rightarrow 2\ Fe_3O_4 + 2\ CO$ $\Delta H° = 2(-18\ kJ)$
              $2\ Fe_3O_4 + CO_2 \rightarrow 3\ Fe_2O_3 + CO$ $\Delta H° = +39\ kJ$
              $3\ Fe_2O_3 + 9\ CO \rightarrow 6\ Fe + 9\ CO_2$ $\Delta H° = 3(-23\ kJ)$

          _____

          $6\ FeO(s) + 6\ CO(g) \rightarrow 6\ Fe(s) + 6\ CO_2(g)$ $\Delta H° = -66\ kJ$

          So for: $FeO(s) + CO(g) \rightarrow Fe(s) + CO_2(g)$ $\Delta H° = \dfrac{-66\ kJ}{6} = -11\ kJ$

31.   Information given:

              $C(s) + O_2(g) \rightarrow CO_2(g)$ $\Delta H° = -393.7\ kJ$
          $CO(g) + 1/2\ O_2(g) \rightarrow CO_2(g)$ $\Delta H° = -283.3\ kJ$

          So:

              $2\ C + 2\ O_2 \rightarrow 2\ CO_2$ $\Delta H° = 2(-393.7\ kJ)$
                  $2\ CO_2 \rightarrow 2\ CO + O_2$ $\Delta H° = 2(+283.3\ kJ)$

          _____

          $2\ C(s) + O_2(g) \rightarrow 2\ CO(g)$ $\Delta H° = -220.8\ kJ$

32. $2\ C + 2\ O_2 \rightarrow 2\ CO_2$ $\Delta H° = 2(-394\ kJ)$
          $H_2 + 1/2\ O_2 \rightarrow H_2O$ $\Delta H° = -286\ kJ$
          $2\ CO_2 + H_2O \rightarrow C_2H_2 + 5/2\ O_2$ $\Delta H° = +1300.\ kJ$

          _____

          $2\ C(s) + H_2(g) \rightarrow C_2H_2(g)$ $\Delta H° = 226\ kJ$

33. $NO + O_3 \rightarrow NO_2 + O_2$ $\quad\quad\quad$ $\Delta H° = -199$ kJ
$\quad\quad$ $3/2\ O_2 \rightarrow O_3$ $\quad\quad\quad\quad\quad$ $\Delta H° = +427$ kJ/2
$\quad\quad\quad$ $O \rightarrow 1/2\ O_2$ $\quad\quad\quad\quad$ $\Delta H° = -495$ kJ/2

$\overline{}$
$NO(g) + O(g) \rightarrow NO_2(g)$ $\quad\quad\quad\quad$ $\Delta H° = -233$ kJ

34. $C_6H_4(OH)_2(aq) \rightarrow C_6H_4O_2(aq) + H_2(g)$ $\quad\quad$ $\Delta H° = +177.4$ kJ
$\quad\quad\quad$ $H_2O_2(aq) \rightarrow H_2(g) + O_2(g)$ $\quad\quad\quad$ $\Delta H° = +191.2$ kJ
$\quad\quad$ $2\ H_2(g) + O_2(g) \rightarrow 2\ H_2O(g)$ $\quad\quad\quad$ $\Delta H° = 2(-241.8$ kJ$)$
$\quad\quad\quad\quad$ $2\ H_2O(g) \rightarrow 2\ H_2O(l)$ $\quad\quad\quad\quad$ $\Delta H° = 2(-43.8$ kJ$)$

$\overline{}$
$C_6H_4(OH)_2(aq) + H_2O_2(aq) \rightarrow C_6H_4O_2(aq) + 2\ H_2O(l)$ $\quad$ $\Delta H° = -202.6$ kJ

35. $1/2\ O_2 + 1/2\ H_2 \rightarrow OH$ $\quad\quad\quad$ $\Delta H° = 77.9$ kJ/2
$\quad\quad\quad$ $O \rightarrow 1/2\ O_2$ $\quad\quad\quad\quad\quad$ $\Delta H° = -495$ kJ/2
$\quad\quad\quad$ $H \rightarrow 1/2\ H_2$ $\quad\quad\quad\quad\quad$ $\Delta H° = -435.9$ kJ/2

$\overline{}$
$\quad\quad$ $O(g) + H(g) \rightarrow OH(g)$ $\quad\quad\quad\quad$ $\Delta H° = -427$ kJ

36.   It will be easier to calculate $\Delta H°$ for combustion of four moles of N_2H_4.

$\quad\quad$ $9\ H_2(g) + 9/2\ O_2(g) \rightarrow 9\ H_2O(l)$ $\quad\quad\quad$ $\Delta H° = 9(-286$ kJ$)$
$\quad\quad$ $3\ N_2H_4(l) + 3\ H_2O(l) \rightarrow 3\ N_2O(g) + 9\ H_2(g)$ $\quad$ $\Delta H° = 3(317$ kJ$)$
$\quad\quad$ $2\ NH_3(g) + 3\ N_2O(g) \rightarrow 4\ N_2(g) + 3\ H_2O(l)$ $\quad$ $\Delta H° = -1010.$ kJ
$\quad\quad$ $N_2H_4(l) + H_2O(l) \rightarrow 2\ NH_3(g) + 1/2\ O_2(g)$ $\quad$ $\Delta H° = +143$ kJ

$\overline{}$
$\quad\quad$ $4\ N_2H_4(l) + 4\ O_2(g) \rightarrow 4\ N_2(g) + 8\ H_2O(l)$ $\quad\quad$ $\Delta H° = -2490.$ kJ

$\quad\quad$ $N_2H_4(l) + O_2(g) \rightarrow N_2(g) + 2\ H_2O(l)$ $\quad\quad$ $\Delta H° = \dfrac{-2490\ \text{kJ}}{4} = -623$ kJ

37. $2\ N_2(g) + 6\ H_2(g) \rightarrow 4\ NH_3(g)$ $\quad\quad\quad$ $\Delta H° = 4(-46$ kJ$)$
$\quad\quad$ $6\ H_2O(g) \rightarrow 6\ H_2(g) + 3\ O_2(g)$ $\quad\quad\quad$ $\Delta H° = 3(484$ kJ$)$

$\overline{}$
$\quad\quad$ $2\ N_2(g) + 6\ H_2O(g) \rightarrow 3\ O_2(g) + 4\ NH_3(g)$ $\quad\quad$ $\Delta H° = +1268$ kJ

No, since the reaction is very endothermic (requires a lot of heat), it would not be a practical way of making ammonia.

Standard Enthalpies of Formation

38.   The enthaply change for the formation of one mole of a compound at 25 °C from its elements, with all substances in their standard states.

$\quad\quad$ $Na(s) + 1/2\ Cl_2(g) \rightarrow NaCl(s);$ $\quad$ $H_2(g) + 1/2\ O_2(g) \rightarrow H_2O(l)$

$\quad\quad$ $6\ C(\text{graphite}) + 6\ H_2(g) + 3\ O_2(g) \rightarrow C_6H_{12}O_6(s);$ $\quad$ $Pb(s) + S(s) + 2\ O_2(g) \rightarrow PbSO_4(s)$

39.   In general: $\Delta H° = \Sigma \Delta H_f°(\text{products}) - \Sigma \Delta H_f°(\text{reactants})$

a. $2 NH_3(g) + 3 O_2(g) + 2 CH_4(g) \rightarrow 2 HCN(g) + 6 H_2O(g)$

$\Delta H° = 2 \text{ mol HCN} \times \Delta H_f°(\text{HCN}) + 6 \text{ mol } H_2O(g) \times \Delta H_f°(H_2O,g)$

$- [2 \text{ mol } NH_3 \times \Delta H_f°(NH_3) + 2 \text{ mol } CH_4 \times (CH_4)]$

$\Delta H° = 2(135.1) + 6(-242) - [2(-46) + 2(-75)] = -940. \text{ kJ}$

b. $\Delta H° = 3 \text{ mol } CaSO_4 \left(\dfrac{-1433 \text{ kJ}}{\text{mol}} \right) + 2 \text{ mol } H_3PO_4(l) \left(\dfrac{-1267 \text{ kJ}}{\text{mol}} \right)$

$- \left[1 \text{ mol } Ca_3(PO_4)_2 \left(\dfrac{-4126 \text{ kJ}}{\text{mol}} \right) + 3 \text{ mol } H_2SO_4(l) \left(\dfrac{-814 \text{ kJ}}{\text{mol}} \right) \right]$

$\Delta H° = -6833 \text{ kJ} - (-6568 \text{ kJ}) = -265 \text{ kJ}$

c. $\Delta H° = 1 \text{ mol } NH_4Cl \times \Delta H_f°(NH_4Cl) - [1 \text{ mol } NH_3 \times \Delta H_f°(NH_3) + 1 \text{ mol } HCl \times \Delta H_f°(HCl)]$

$\Delta H° = 1 \text{ mol} \left(\dfrac{-314 \text{ kJ}}{\text{mol}} \right) - \left[1 \text{ mol} \left(\dfrac{-46 \text{ kJ}}{\text{mol}} \right) + 1 \text{ mol} \left(\dfrac{-92 \text{ kJ}}{\text{mol}} \right) \right]$

$\Delta H° = -314 \text{ kJ} + 138 \text{ kJ} = -176 \text{ kJ}$

d. $SiCl_4(l) + 2 H_2O(l) \rightarrow SiO_2(s) + 4 HCl(aq)$

Since $HCl(aq)$ is $H^+(aq) + Cl^-(aq)$, $\Delta H_f° = 0 - 167 = -167 \text{ kJ/mol}$

$\Delta H° = [4 \text{ mol}(-167 \text{ kJ/mol}) + 1 \text{ mol}(-911 \text{ kJ/mol})]$

$- [1 \text{ mol}(-687 \text{ kJ/mol}) + 2 \text{ mol}(-286 \text{ kJ/mol})]$

$\Delta H° = -1579 \text{ kJ} - (-1259 \text{ kJ}) = -320. \text{ kJ}$

e. $MgO(s) + H_2O(l) \rightarrow Mg(OH)_2(s)$

$\Delta H° = 1 \text{ mol}(-925 \text{ kJ/mol}) - [1 \text{ mol}(-602 \text{ kJ/mol}) + 1 \text{ mol}(-286 \text{ kJ/mol})]$

$\Delta H° = -925 \text{ kJ} - (-888 \text{ kJ}) = -37 \text{ kJ}$

40.   a. $4 NH_3(g) + 5 O_2(g) \rightarrow 4 NO(g) + 6 H_2O(g)$

$\Delta H° = \left[4 \text{ mol} \left(\dfrac{90 \text{ kJ}}{\text{mol}} \right) + 6 \text{ mol} \left(\dfrac{-242 \text{ kJ}}{\text{mol}} \right) \right] - 4 \text{ mol} \left(\dfrac{-46 \text{ kJ}}{\text{mol}} \right) = -908 \text{ kJ}$

$$2 \text{ NO(g)} + \text{O}_2(g) \rightarrow 2 \text{ NO}_2(g)$$

$$\Delta H° = 2 \text{ mol} \left(\frac{34 \text{ kJ}}{\text{mol}} \right) - 2 \text{ mol} \left(\frac{90 \text{ kJ}}{\text{mol}} \right) = -112 \text{ kJ}$$

$$3 \text{ NO}_2(g) + \text{H}_2\text{O(l)} \rightarrow 2 \text{ HNO}_3(aq) + \text{NO(g)}$$

$$\Delta H° = 2 \text{ mol} \left(\frac{-207 \text{ kJ}}{\text{mol}} \right) + 1 \text{ mol} \left(\frac{90 \text{ kJ}}{\text{mol}} \right)$$

$$- \left[3 \text{ mol} \left(\frac{34 \text{ kJ}}{\text{mol}} \right) + 1 \text{ mol} \left(\frac{-286 \text{ kJ}}{\text{mol}} \right) \right] = -140. \text{ kJ}$$

b. $12 \text{ NH}_3(g) + 15 \text{ O}_2(g) \rightarrow 12 \text{ NO(g)} + 18 \text{ H}_2\text{O(g)}$
      $12 \text{ NO(g)} + 6 \text{ O}_2(g) \rightarrow 12 \text{ NO}_2(g)$
      $12 \text{ NO}_2(g) + 4 \text{ H}_2\text{O(l)} \rightarrow 8 \text{ HNO}_3(aq) + 4 \text{ NO(g)}$
      $\phantom{12 \text{ NO}_2(g)} 4 \text{ H}_2\text{O(g)} \rightarrow 4 \text{ H}_2\text{O(l)}$
      _____

      $12 \text{ NH}_3(g) + 21 \text{ O}_2(g) \rightarrow 8 \text{ HNO}_3(aq) + 4 \text{ NO(g)} + 14 \text{ H}_2\text{O(g)}$

The overall reaction is exothermic since each step is exothermic.

41. $4 \text{ Na(s)} + \text{O}_2(g) \rightarrow 2 \text{ Na}_2\text{O(s)}$ $\Delta H° = 2 \text{ mol} \left(\frac{-416 \text{ kJ}}{\text{mol}} \right) = -832 \text{ kJ}$

$$\text{Ia(s)} + 2 \text{ H}_2\text{O(l)} \rightarrow 2 \text{ NaOH(aq)} + \text{H}_2(g)$$

$$\Delta H° = 2 \text{ mol} \left(\frac{-470 \text{ kJ}}{\text{mol}} \right) - 2 \text{ mol} \left(\frac{-286 \text{ kJ}}{\text{mol}} \right) = -368 \text{ kJ}$$

$$2\text{Na(s)} + \text{CO}_2(g) \rightarrow \text{Na}_2\text{O(s)} + \text{CO(g)}$$

$$\Delta H° = \left[1 \text{ mol} \left(\frac{-416 \text{ kJ}}{\text{mol}} \right) + 1 \text{ mol} \left(\frac{-110.5 \text{ kJ}}{\text{mol}} \right) \right] - 1 \text{ mol} \left(\frac{-393.5 \text{ kJ}}{\text{mol}} \right) = -133 \text{ kJ}$$

In both cases sodium metal reacts with the "extinguishing agent." Both reactions are exothermic and each reaction produces a flammable gas, H_2 and CO respectively.

42. $3 \text{ Al(s)} + 3 \text{ NH}_4\text{ClO}_4(s) \rightarrow \text{Al}_2\text{O}_3(s) + \text{AlCl}_3(s) + 3 \text{ NO(g)} + 6 \text{ H}_2\text{O(g)}$

$$\Delta H° = \left[6 \text{ mol} \left(\frac{-242 \text{ kJ}}{\text{mol}} \right) + 3 \text{ mol} \left(\frac{90 \text{ kJ}}{\text{mol}} \right) + 1 \text{ mol} \left(\frac{-704 \text{ kJ}}{\text{mol}} \right) + 1 \text{ mol} \left(\frac{-1676 \text{ kJ}}{\text{mol}} \right) \right]$$

$$- \left[3 \text{ mol} \left(\frac{-295 \text{ kJ}}{\text{mol}} \right) \right] = -2677 \text{ kJ}$$

43. $5 N_2O_4(l) + 4 N_2H_3CH_3(l) \rightarrow 12 H_2O(g) + 9 N_2(g) + 4 CO_2(g)$

$\Delta H° = \left[12 \text{ mol} \left(\dfrac{-242 \text{ kJ}}{\text{mol}}\right) + 4 \text{ mol} \left(\dfrac{-393.5 \text{ kJ}}{\text{mol}}\right)\right]$

$- \left[5 \text{ mol} \left(\dfrac{-20 \text{ kJ}}{\text{mol}}\right) + 4 \text{ mol} \left(\dfrac{54 \text{ kJ}}{\text{mol}}\right)\right] = -4594 \text{ kJ}$

44.   For exercise 42 a mixture of 3 mol Al and 3 mol NH_4ClO_4 yields 2677 kJ of energy. The mass of the stoichiometric reactant mixture is:

$3 \text{ mol} \times \dfrac{26.98 \text{ g}}{\text{mol}} + 3 \text{ mol} \times \dfrac{117.49 \text{ g}}{\text{mol}} = 433.41 \text{ g}$

For 1.000 kg of fuel: $1.000 \times 10^3 \text{ g} \times \dfrac{-2677 \text{ kJ}}{433.41 \text{ g}} = -6177 \text{ kJ}$

In Exercise 43 we get 4594 kJ of energy from 5 mol of N_2O_4 and 4 mol of $N_2H_3CH_3$. The mass is: $5 \text{ mol} \times \dfrac{92.02 \text{ g}}{\text{mol}} + 4 \text{ mol} \times \dfrac{46.08 \text{ g}}{\text{mol}} = 644.42 \text{ kJ}$

For 1.000 kg of fuel: $1.000 \times 10^3 \text{ g} \times \dfrac{-4594 \text{ kJ}}{644.42 \text{ g}} = -7129 \text{ kJ}$

Thus, we get more energy per kg from the $N_2O_4/N_2H_3CH_3$ mixture.

45.   a. $\Delta H° = 3 \text{ mol} (227 \text{ kJ/mol}) - 1 \text{ mol} (49 \text{ kJ/mol}) = 632 \text{ kJ}$

b.   Since 3 $C_2H_2(g)$ is higher in energy than $C_6H_6(l)$, then acetylene will release more energy per gram when burned in air.

46.   a. $C_2H_4(g) + O_3(g) \rightarrow CH_3CHO(g) + O_2(g)$ $\Delta H° = -166 \text{ kJ} - [143 \text{ kJ} + 52 \text{ kJ}] = -361 \text{ kJ}$

b. $O_3(g) + NO(g) \rightarrow NO_2(g) + O_2(g)$ $\Delta H° = 34 \text{ kJ} - [90 \text{ kJ} + 143 \text{ kJ}] = -199 \text{ kJ}$

c. $SO_3(g) + H_2O(l) \rightarrow H_2SO_4(aq)$ $\Delta H° = -909 \text{ kJ} - [-396 \text{ kJ} + (-286 \text{ kJ})] = -227 \text{ kJ}$

d. $2 NO(g) + O_2(g) \rightarrow 2 NO_2(g)$ $\Delta H° = 2(34) \text{ kJ} - 2(90) \text{ kJ} = -112 \text{ kJ}$

47. $2 ClF_3(g) + 2 NH_3(g) \rightarrow N_2(g) + 6 HF(g) + Cl_2(g)$

$\Delta H° = 6 \ \Delta H_f^° (HF) - [2 \ \Delta H_f^° (ClF_3) + 2 \ \Delta H_f^° (NH_3)]$

$-1196 \text{ kJ} = 6 \text{ mol} \left(\dfrac{-271 \text{ kJ}}{\text{mol}}\right) - 2 \ \Delta H_f^° - 2 \text{ mol} \left(\dfrac{-46 \text{ kJ}}{\text{mol}}\right)$

$$-1196 \text{ kJ} = -1626 \text{ kJ} - 2 \; \Delta H_f^{\circ} + 92 \text{ kJ}, \quad \Delta H_f^{\circ} = \frac{(-1626 + 92 + 1196) \text{ kJ}}{2 \text{ mol}} = \frac{-169 \text{ kJ}}{\text{mol}}$$

48. $C_2H_4(g) + 3 \; O_2(g) \rightarrow 2 \; CO_2(g) + 2 \; H_2O(l)$

$\Delta H^{\circ} = -1411.1 \text{ kJ} = 2(-393.5) \text{ kJ} + 2(-285.9) \text{ kJ} - \Delta H_f^{\circ}$

$-1411.1 \text{ kJ} = -1358.8 \text{ kJ} - \Delta H_f^{\circ}, \quad \Delta H_f^{\circ} = 52.3 \text{ kJ/mol}$

Energy Consumption and Sources

49. $CH_4(g) + O_2(g) \rightarrow CO_2(g) + 2 \; H_2O(l)$ $\Delta H^{\circ} = -891 \text{ kJ}$

Need 4.9×10^6 kJ of energy (from Exercise 24).

$4.9 \times 10^6 \text{ kJ} \times \dfrac{1 \text{ mol CH}_4}{891 \text{ kJ}} = 5.5 \times 10^3 \text{ mol CH}_4; \quad 5.5 \times 10^3 \text{ mol} \times \dfrac{22.4 \text{ L}}{\text{mol}} \text{ at STP} = 1.2 \times 10^5 \text{ L}$

Or using the ideal gas law: $V = \dfrac{5.5 \times 10^3 \text{ mol} \times \dfrac{0.08206 \text{ L atm}}{\text{mol K}} \times 273 \text{ K}}{1.0 \text{ atm}} = 1.2 \times 10^5 \text{ L}$

50. $C_2H_5OH(l) + 3 \; O_2(g) \rightarrow 2 \; CO_2(g) + 3 \; H_2O(l)$

$\Delta H^{\circ} = [2 \; (-393.5 \text{ kJ}) + 3(-286 \text{ kJ})] - (-278 \text{ kJ}) = -1367 \text{ kJ/mol ethanol}$

$\dfrac{-1367 \text{ kJ}}{\text{mol}} \times \dfrac{1 \text{ mol}}{46.068 \text{ g}} = -29.67 \text{ kJ/g}$

51. $CO(g) + 2 \; H_2(g) \rightarrow CH_3OH(l) \quad \Delta H^{\circ} = (-239 \text{ kJ}) - (-110.5 \text{ kJ}) = -129 \text{ kJ}$

52.   If there are fewer forests, then less CO_2 will be removed from the atmosphere by photosynthesis. The greenhouse effect is a result of increased levels of CO_2 in the atmosphere, and thus will be more severe with forest destruction.

53. $C_3H_8(g) + 5 \; O_2(g) \rightarrow 3 \; CO_2(g) + 4 \; H_2O(l)$

$\Delta H^{\circ} = [3(-393.5 \text{ kJ}) + 4(-286 \text{ kJ})] - (-104 \text{ kJ}) = -2221 \text{ kJ/mol C}_3H_8$

$\dfrac{-2221 \text{ kJ}}{\text{mol}} \times \dfrac{1 \text{ mol}}{44.096 \text{ g}} = \dfrac{-50.37 \text{ kJ}}{\text{g}}$ vs. -47.7 kJ/g for octane (Example 9.8)

The fuel values of the fuels are very close. An advantage of propane is that it burns more cleanly. The boiling point of propane is -42°C. Thus, it is more difficult to store propane and there are extra safety hazards associated in using high pressure compressed gas tanks.

54. $NaH(s) + H_2O(l) \rightarrow NaOH(aq) + H_2(g)$; If water gets into the storage container, flammable hydrogen gas can form.

55.   Energy needed $= \dfrac{20. \times 10^3 \text{ g } C_{12}H_{22}O_{11}}{\text{hr}} \times \dfrac{1 \text{ mol } C_{12}H_{22}O_{11}}{342.3 \text{ g } C_{12}H_{22}O_{11}} \times \dfrac{5640 \text{ kJ}}{\text{mol}} = 3.3 \times 10^5 \text{ kJ/hr}$

Energy from sun $= 1.0 \text{ kW/m}^2 = 1000 \text{ W/m}^2 = \dfrac{1000 \text{ J}}{\text{s m}^2} = \dfrac{1.0 \text{ kJ}}{\text{s m}^2}$

$10,000 \text{ m}^2 \times \dfrac{1.0 \text{ kJ}}{\text{s m}^2} \times \dfrac{60 \text{ s}}{\text{min}} \times \dfrac{60 \text{ min}}{\text{hr}} = 3.6 \times 10^7 \text{ kJ/hr}$

% efficiency $= \dfrac{\text{Energy used per hour}}{\text{Total energy per hour}} \times 100 = \dfrac{3.3 \times 10^5 \text{ kJ}}{3.6 \times 10^7 \text{ kJ}} \times 100 = 0.92\%$

56.   Energy needed $= 40. \text{ kWh} = \dfrac{40. \text{ kJ h}}{\text{s}} \times \dfrac{3600 \text{ s}}{\text{h}} = 1.4 \times 10^5 \text{ kJ}$

Energy from the sun in 8.0 hours $= \dfrac{1.0 \text{ kJ}}{\text{s m}^2} \times \dfrac{60 \text{ s}}{\text{min}} \times \dfrac{60 \text{ min}}{\text{h}} \times 8.0 \text{ h} = 2.9 \times 10^4 \text{ kJ/m}^2$

Only 13% of the sunlight is converted into electricity:

$(0.13) (2.9 \times 10^4 \text{ kJ/m}^2) (\text{Area}) = 1.4 \times 10^5 \text{ kJ}$,   Area $= 37 \text{ m}^2$

Additional Exercises

57.

Pathway 1   w = complete shaded area
$w = -P\Delta V = -5.0(4.0 - 1.0) = -15 \text{ L atm}$

Pathway 2   w = area of cross hatched rectangle
$w = -P\Delta V = -2.0(4.0 - 1.0) = -6.0 \text{ L atm}$

Sign is (-) because system is doing work on the surroundings

No, since work depends on pathway it cannot be a state function.

58. $KE = 1/2 \, mv^2 = \dfrac{1}{2}\left(7.8 \text{ g} \times \dfrac{1 \text{ kg}}{1000 \text{ g}} \right)\left(\dfrac{5.0 \times 10^4 \text{ cm}}{\text{sec}} \times \dfrac{1 \text{ m}}{100 \text{ cm}} \right)^2 = 980 \, \dfrac{\text{kg m}^2}{\text{s}^2} = 980 \text{ J}$

Heat gained by bullet and wood $= 980 \text{ J}$.

$$980 \text{ J} = 7.8 \text{ g} \times \frac{0.13 \text{ J}}{\text{g °C}} \times \Delta T + 1.00 \times 10^3 \text{ g} \times \frac{2.1 \text{ J}}{\text{g °C}} \times \Delta T = \frac{1.0 \text{ J}}{°C} \times \Delta T + \frac{2100 \text{ J}}{°C} \times \Delta T$$

$$\Delta T = 0.47°C; \quad \Delta T = T_f - T_i, \quad 0.47 = T_f - 25.0, \quad T_f = 25.5°C$$

59. $$0.100 \text{ L} \times \frac{0.500 \text{ mol HCl}}{\text{L}} \times \frac{1 \text{ mol BaCl}_2}{2 \text{ mol HCl}} = 2.50 \times 10^{-2} \text{ mol BaCl}_2$$

$$0.300 \text{ L} \times \frac{0.500 \text{ mol Ba(OH)}_2}{\text{L}} \times \frac{1 \text{ mol BaCl}_2}{1 \text{ mol Ba(OH)}_2} = 1.50 \times 10^{-1} \text{ mol BaCl}_2$$

Thus, HCl is limiting.

$$2.50 \times 10^{-2} \text{ mol BaCl}_2 \times \frac{-118 \text{ kJ}}{\text{mol BaCl}_2} = -2.95 \text{ kJ of heat is evolved.}$$

60. $$400 \text{ kcal} \times \frac{4.18 \text{ kJ}}{\text{kcal}} = 1.67 \times 10^3 \text{ kJ} \approx 2 \times 10^3 \text{ kJ}$$

$$PE = mgz = \left(180 \text{ lb} \times \frac{1 \text{ kg}}{2.205 \text{ lb}} \right) \left(\frac{9.8 \text{ m}}{\text{s}^2} \right) \left(8 \text{ in} \times \frac{2.54 \text{ cm}}{\text{in}} \times \frac{1 \text{ m}}{100 \text{ cm}} \right) = 160 \text{ J} \approx 200 \text{ J}$$

200 J of energy are needed to climb one step.

$$2 \times 10^6 \text{ J} \times \frac{1 \text{ step}}{200 \text{ J}} = 1 \times 10^4 \text{ steps}$$

61.   If Hess's law were not true it would be possible to create energy by reversing a reaction using a different series of steps. This violates the law of conservation of energy (first law). Thus, Hess's Law is another statement of the law of conservation of energy.

62. $$N_2H_4(l) + O_2(g) \rightarrow N_2(g) + 2 H_2O(g) \quad \Delta H° = 2(-242 \text{ kJ}) - (51 \text{ kJ}) = -535 \text{ kJ}$$

$$2 N_2H_4(l) + N_2O_4(l) \rightarrow 3 N_2(g) + 4 H_2O(g) \quad \Delta H° = 4(-242 \text{ kJ}) - [2(51 \text{ kJ}) + (-20 \text{ kJ})] = -1050. \text{ kJ}$$

For hydrazine plus oxygen the stoichiometric reactant mixture contains 1 mol N_2H_4 for every mol O_2. This is a total mass of 64.05 g or 0.06405 kg.

$$\Delta H = \frac{-535 \text{ kJ}}{0.06405 \text{ kg}} = -8.35 \times 10^3 \text{ kJ/kg for } N_2H_4(l) + O_2(g)$$

For hydrazine plus N_2O_4 the optimum mixture contains 2 mol of N_2H_4 for every mol N_2O_4. This is a total mass of 156.12 g or 0.15612 kg.

$$\Delta H = \frac{-1050. \text{ kJ}}{0.15612 \text{ kg}} = -6.726 \times 10^3 \text{ kJ/kg for } 2 N_2H_4(l) + N_2O_4(l)$$

From Exercise 44, we calculated -7129 kJ/kg for methyl hydrazine ($N_2H_3CH_3$) plus N_2O_4. The hydrazine plus oxygen is the most efficient fuel.

63.   a.   Using Hess's Law and the equation $\Delta H = nC_p\Delta T$:

$CH_3Cl(248°C) + H_2(248°C) \rightarrow CH_4(248°C) + HCl(248°C)$ $\Delta H_1 = -83.3$ kJ
$CH_3Cl(25°C) \rightarrow CH_3Cl(248°C)$ $\Delta H_2 = 1$ mol (48.5 J °C^{-1} mol^{-1}) (223°C)
                                          $\Delta H_2 = 10,800$ J $= 10.8$ kJ
$H_2(25°C) \rightarrow H_2(248°C)$ $\Delta H_3 = 1(28.9) (223) (1$ kJ/1000 J) $= 6.44$ kJ
$CH_4(248°C) \rightarrow CH_4(25°C)$ $\Delta H_4 = 1(41.3) (-223) (1/1000) = -9.21$ kJ
$HCl(248°C) \rightarrow HCl(25°C)$ $\Delta H_5 = 1(29.1) (-223) (1/1000) = -6.49$ kJ

$CH_3Cl(25°C) + H_2(25°C) \rightarrow CH_4(25°C) + HCl(25°C)$ $\Delta H° = \Delta H_1 + \Delta H_2 + \Delta H_3 + \Delta H_4 + \Delta H_5$

$\Delta H° = -83.3$ kJ $+ 10.8$ kJ $+ 6.44$ kJ $- 9.21$ kJ $- 6.49$ kJ $= -81.8$ kJ

b. $\Delta H° = [\Delta H_f°(CH_4) + \Delta H_f°(HCl)] - [\Delta H_f°(CH_3Cl) + \Delta H_f°(H_2)]$

-81.8 kJ $= -75$ kJ $- 92$ kJ $- [\Delta H_f°(CH_3Cl) + 0]$, $\Delta H_f°(CH_3Cl) = -85$ kJ/mol

64.   a. $C_{12}H_{22}O_{11}(s) + 12$ $O_2(g) \rightarrow 12$ $CO_2(g) + 11$ $H_2O(l)$

b.   A bomb calorimeter is at constant volume, so heat released $= q_v = \Delta E°$:

$$\Delta E° = \frac{-24.00 \text{ kJ}}{1.46 \text{ g}} \times \frac{342.3 \text{ g}}{\text{mol}} = -5630 \text{ kJ/mol } C_{12}H_{22}O_{11}$$

c. $\Delta H° = \Delta E° + \Delta(PV) = \Delta E° + \Delta(nRT) = \Delta E° + RT\Delta n_{gas}$

For this reaction $\Delta n_{gas} = 12-12 = 0$, so $\Delta H° = \Delta E° = -5630$ kJ/mol.

65. $w = -P\Delta V = -RT\Delta n_{gas}$, $\Delta n_{gas} =$ gaseous product moles - gaseous reactant moles

$\Delta n < 0$, $w > 0$ and surroundings does work on system (a, c, and d).

$\Delta n > 0$, $w < 0$ and system does work on surroundings (e and f).

$\Delta n = 0$, $w = 0$ and no work is done (b).

66. $C_3H_8(g) + 5$ $O_2(g) \rightarrow 3$ $CO_2(g) + 4$ $H_2O(l)$

$\Delta H° = 4(-286$ kJ$) + 3(-393.5$ kJ$) - (-104$ kJ$) = -2221$ kJ

We need 1.3×10^8 J $= 1.3 \times 10^5$ kJ of energy. The mass of C_3H_8 required is:

$$1.3 \times 10^5 \text{ kJ} \times \frac{1 \text{ mol } C_3H_8}{2221 \text{ kJ}} \times \frac{44.1 \text{ g } C_3H_8}{\text{mol}} = 2600 \text{ g of } C_3H_8$$

67.   A state function is a function whose change depends only on the initial and final states and **not** on how one got from the initial to the final state. If H and E were not state functions, the law of conservation of energy (first law) would not be true.

68.   a. $2 Al(s) + 3/2 O_2(g) \rightarrow Al_2O_3(s)$

       b. $C_2H_5OH(l) + 3 O_2(g) \rightarrow 2 CO_2(g) + 3 H_2O(l)$

       c. $Ba(OH)_2(aq) + 2 HCl(aq) \rightarrow 2 H_2O(l) + BaCl_2(aq)$

       d. $2 C(graphite) + 3/2 H_2(g) + 1/2 Cl_2(g) \rightarrow C_2H_3Cl(g)$

       e. $C_6H_6(l) + 15/2 O_2(g) \rightarrow 6 CO_2(g) + 3 H_2O(l)$

       Note: ΔH_{comb} values generally assume 1 mol of compound combusted.

       f. $NH_4Br(s) \rightarrow NH_4^+(aq) + Br^-(aq)$

69. $n = \dfrac{PV}{RT} = \dfrac{125 \text{ atm} \times 50.0 \times 10^{-3} \text{ L}}{\dfrac{0.08206 \text{ L atm}}{\text{mol K}} \times 295 \text{ K}} = 0.258 \text{ mol O}_2$

       The balanced reaction for the combustion of benzoic acid is:

       $$2 C_7H_6O_2(s) + 15 O_2(g) \rightarrow 14 CO_2(g) + 6 H_2O(l)$$

       We want 25 times as much oxygen as needed, so only 0.258/25 moles of oxygen will react.

       $$\dfrac{0.258 \text{ mol O}_2}{25} \times \dfrac{2 \text{ mol C}_7H_6O_2}{15 \text{ mol O}_2} \times \dfrac{122.1 \text{ g C}_7H_6O_2}{\text{mol C}_7H_6O_2} = 0.168 \text{ g} = 0.17 \text{ g C}_7H_6O_2$$

70.   The specific heat capacities are: Al, $\dfrac{0.89 \text{ J}}{\text{g °C}}$; Fe, $\dfrac{0.45 \text{ J}}{\text{g °C}}$

       Al would be the better choice. It has a higher heat capacity and a lower density than Fe. Using Al, the same amount of heat could be dissipated by a smaller mass, keeping the mass of the amplifier down.

CHAPTER TEN

SPONTANEITY, ENTROPY AND FREE ENERGY

Spontaneity and Entropy

1.   A spontaneous process is one that occurs without any outside intervention.

2.   (a), (b), (d), and (g) are spontaneous. Therefore, (c), (e), and (f) require an external source of energy in order to occur.

3.   We draw all of the possible arrangements of the two particles in the three levels.

2 kJ	—	—	x	—	x	xx
1 kJ	—	x	—	xx	x	—
0 kJ	xx	x	x	—	—	—
Total E =	0 kJ	1 kJ	2 kJ	2 kJ	3 kJ	4 kJ

The most likely total energy is 2 kJ.

4.
2 kJ	—	—	AB	—	—	B	A	B	A
1 kJ	—	AB	—	B	A	—	—	A	B
0 kJ	AB	—	—	A	B	A	B	—	—
E_T =	0 kJ	2 kJ	4 kJ	1 kJ	1 kJ	2 kJ	2 kJ	3 kJ	3 kJ

The most likely total energy is 2 kJ.

5.   a.   Entropy is a measure of disorder.

      b.   The quantity $T\Delta S$ has units of energy.

6.   Of the three phases, solids are most ordered and gases are most disordered. Thus, a, b, c, e, and g involve an increase in entropy. All increase disorder.

7.   There are more ways to roll a seven. We can consider all of the possible throws by constructing a table.

one die	1	2	3	4	5	6	
1	2	3	4	5	6	7	
2	3	4	5	6	7	8	
3	4	5	6	7	8	9	sum of the two dice
4	5	6	7	8	9	10	
5	6	7	8	9	10	11	
6	7	8	9	10	11	12	

There are six ways to get a seven, more than any other number. The seven is not favored by energy; rather it is favored by probability. To change the probability we would have to expend energy (do work).

8.   Arrangement I: $S = k \ln \Omega$; $\Omega = 1$; $S = k \ln 1 = 0$

Arrangement II: $\Omega = 4$; $S = k \ln 4 = 1.38 \times 10^{-23}$ J/K $\ln 4$, $S = 1.91 \times 10^{-23}$ J/K

Arrangement III: $\Omega = 6$; $S = k \ln 6 = 2.47 \times 10^{-23}$ J/K

9. $S = k \ln \Omega$; S has units of J K^{-1} mol^{-1} and k has units of J/K

To make units match: S (J K^{-1} mol^{-1}) $= N_A k \ln \Omega$ when N_A = Avogadro's number

$$189 \text{ J } K^{-1} \text{ mol}^{-1} = 8.31 \text{ J } K^{-1} \text{ mol}^{-1} \ln \Omega_g$$
$$70. \text{ J } K^{-1} \text{ mol}^{-1} = 8.31 \text{ J } K^{-1} \text{ mol}^{-1} \ln \Omega_l$$

Subtracting: 119 J K^{-1} mol^{-1} $= 8.31$ J K^{-1} mol^{-1} (ln Ω_g - ln Ω_l)

$14.3 = \ln(\Omega_g/\Omega_l)$, $\dfrac{\Omega_g}{\Omega_l} = e^{14.3} = 1.6 \times 10^6$

10. a.   Entropy increases; there is a greater volume accessible to the randomly moving gas molecules which increases disorder.

b.   The positional entropy doesn't change. There is no change in volume and thus, no change in the numbers of positions of the molecules. The total entropy increases because the increase in temperature increases the energy disorder.

c.   Entropy decreases; volume decreases.

11.   a. N_2O; It is the more complex molecule, i.e., more parts, more disorder.

       b. H_2 at 100°C and 0.5 atm; Higher temperature and lower pressure means greater volume and hence, greater positional entropy.

       c. N_2 at STP has the greater volume.   d. $H_2O(l)$ is more disordered.

Entropy and the Second Law of Thermodynamics: Free Energy

12.   a.   The system is the portion of the universe in which we are interested.

       b.   The surroundings are everything else in the universe.

       c.   A closed system can only exchange energy with its surroundings. Matter is not exchanged.

       d.   An open system can exchange both matter and energy with its surroundings.

13.   Living organisms need an external source of energy to carry out these processes. Green plants use the energy from sunlight to produce glucose from carbon dioxide and water by photosynthesis. The energy released from the metabolism of glucose helps drive the synthesis of proteins. A living organism is an open system. For all processes combined, ΔS_{univ} must be greater than zero (2nd law).

14.   No, living organisms need an outside source of matter (food) to survive.

15.   a. $C_{12}H_{22}O_{11}$; Larger molecule, more parts, more disorder

       b. H_2O (0°C); Higher temp.; $S° = 0$ at 0 K

       c. $H_2S(g)$; A gas has greater disorder than a liquid.   d.   He (10 K); $S = 0$ at 0 K

       e. N_2O; More complicated molecule   f.   HCl; More electrons in HCl

16.   a.   Decrease in disorder; ΔS (-)   b.   Increase in disorder; ΔS (+)

       c.   Decrease in disorder ($\Delta n < 0$); ΔS (-)   d.   Decrease in disorder ($\Delta n < 0$); ΔS (-)

       e.   HCl(g) is more disordered. One mole of a gas will still be more disordered than two moles of ions in solution; ΔS (-)

       f.   Increase in disorder; ΔS (+)

17.   a. $H_2(g) + 1/2\ O_2(g) \rightarrow H_2O(g)$; $\Delta S° = S°[H_2O(g)] - [S°(H_2) + 1/2\ S°(O_2)]$

             $\Delta S° = 1\ mol(189\ J\ K^{-1}\ mol^{-1}) - [1\ mol(131\ J\ K^{-1}\ mol^{-1}) + 1/2\ mol(205\ J\ K^{-1}\ mol^{-1})]$

$\Delta S° = 189 \text{ J/K} - 234 \text{ J/K} = -45 \text{ J/K}$

b. $3 \text{ O}_2(g) \rightarrow 2 \text{ O}_3(g); \quad \Delta S° = 2 \text{ mol}(239 \text{ J K}^{-1} \text{ mol}^{-1}) - [3 \text{ mol}(205 \text{ J K}^{-1} \text{ mol}^{-1})] = -137 \text{ J/K}$

c. $\text{N}_2(g) + \text{O}_2(g) \rightarrow 2 \text{ NO}(g); \quad \Delta S° = 2(211) - (192 + 205) = +25 \text{ J/K}$

18. $\text{CS}_2(g) + 3 \text{ O}_2(g) \rightarrow \text{CO}_2(g) + 2 \text{ SO}_2(g); \quad \Delta S° = S°(\text{CO}_2) + 2 \, S°(\text{SO}_2) - [3 \, S°(\text{O}_2) + S°(\text{CS}_2)]$

$-143 \text{ J/K} = 214 \text{ J/K} + 2(248 \text{ J/K}) - 3(205 \text{ J/K}) - (1 \text{ mol})S°, \quad S° = 238 \text{ J K}^{-1} \text{ mol}^{-1}$

19. $2 \text{ Al}(s) + 3 \text{ Br}_2(l) \rightarrow 2 \text{ AlBr}_3(s)$

$-144 \text{ J/K} = 2 \text{ mol}[S°(\text{AlBr}_3)] - [2(28 \text{ J/K}) + 3(152 \text{ J/K})], \quad S° = 184 \text{ J K}^{-1} \text{ mol}^{-1}$

20. At the boiling point $\Delta G = 0$, so $\Delta H = T\Delta S$.

$$\Delta S = \frac{\Delta H}{T} = \frac{31.4 \text{ kJ/mol}}{(273.2 + 61.7)\text{K}} = 9.38 \times 10^{-2} \text{ kJ mol}^{-1} \text{ K}^{-1} = 93.8 \text{ J K}^{-1} \text{ mol}^{-1}$$

21. $\Delta G = 0$ so $\Delta H = T\Delta S, \quad T = \dfrac{\Delta H}{\Delta S} = \dfrac{58.51 \times 10^3 \text{ J/mol}}{92.92 \text{ J K}^{-1} \text{ mol}^{-1}} = 629.7 \text{ K}$

22. $\Delta G = 0$ so $\Delta H = T\Delta S, \quad \Delta S = \dfrac{\Delta H}{T} = \dfrac{35.2 \times 10^3 \text{ J/mol}}{3680 \text{ K}} = 9.57 \text{ J K}^{-1} \text{ mol}^{-1}$

23. a. $\text{NH}_3(s) \rightarrow \text{NH}_3(l); \quad \Delta G = \Delta H - T\Delta S = 5650 \text{ J/mol} - 200. \text{ K}(28.9 \text{ J K}^{-1} \text{ mol}^{-1})$

$\Delta G = 5650 \text{ J/mol} - 5780 \text{ J/mol} = -130 \text{ J/mol}$

Yes, NH_3 will melt since $\Delta G < 0$ at this temperature.

b. At the melting point, $\Delta G = 0$ so $T = \dfrac{\Delta H}{\Delta S} = \dfrac{5650 \text{ J/mol}}{28.9 \text{ J K}^{-1} \text{ mol}^{-1}} = 196 \text{ K}$

24. $\text{P}_4(s, \alpha) \rightarrow \text{P}_4(s, \beta)$

a. At $T < -76.9°\text{C}$, this reaction is spontaneous and the sign of ΔG is (-). At $76.9°\text{C}$, $\Delta G = 0$ and above $-76.9 °\text{C}$, the sign of ΔG is (+). This is consistent with ΔH (-) and ΔS (-).

b. Since the sign of ΔS is negative, then the β form has the more ordered structure.

Free Energy and Chemical Reactions

25.   a. $CH_4(g)$ + $2 O_2(g)$ → $CO_2(g)$ + $2 H_2O(g)$

ΔH_f°	-75 kJ/mol	0	-393.5	-242
ΔG_f°	-51 kJ/mol	0	-394	-229 Data from Appendix 4
S°	186 J K^{-1} mol^{-1}	205	214	189

ΔH° = 2 mol(-242 kJ/mol) + 1 mol (-393.5 kJ/mol) - [1 mol(-75 kJ/mol)] = -803 kJ

ΔS° = 2 mol(189 J K^{-1} mol^{-1}) + 1 mol(214 J K^{-1} mol^{-1})

$\qquad$ - [1 mol(186 J K^{-1} mol^{-1}) + 2 mol(205 J K^{-1} mol^{-1})] = -4 J/K

There are two ways to get ΔG°:

$\qquad \Delta G^\circ = \Delta H^\circ - T\Delta S^\circ$ = -803 × 10^3 J - 298 K(-4 J/K) = -8.018 × 10^5 J = -802 kJ

or from ΔG_f° , we get:

$\qquad \Delta G^\circ$ = 2 mol(- 229 kJ/mol) + 1 mol(-394 kJ/mol) - [1 mol(- 51 kJ/mol)]

$\qquad \Delta G^\circ$ = -801 kJ   (Answers are the same within round off error)

b. $6 CO_2(g)$ + $6 H_2O(l)$ → $C_6H_{12}O_6(s)$ + $6 O_2(g)$

ΔH_f°	-393.5 kJ/mol	-286	-1275	0
S°	214 J K^{-1} mol^{-1}	70	212	205

ΔH° = -1275 - [6(-286) + 6(-393.5)] = +2802 kJ

ΔS° = 6(205) + 212 - [6(214) + 6(70)] = -262 J/K

ΔG° = 2802 kJ - 298 K(-0.262 kJ/K) = 2880. kJ

c. $P_4O_{10}(s) + 6 H_2O(l) \rightarrow 4 H_3PO_4(s)$

ΔH_f° (kJ/mol)	-2984	-286	-1279
S° (J K^{-1} mol^{-1})	229	70	110

ΔH° = 4 mol (-1279 kJ/mol) - [1 mol(-2984 kJ/mol) + 6 mol(-286 kJ/mol)] = -416 kJ

ΔS° = 4(110) - [229 + 6(70)] = -209 J/K

$\Delta G^{\circ} = \Delta H^{\circ} - T\Delta S^{\circ}$ = -416 kJ - 298 K (-0.209 kJ/K) = -354 kJ

d. $HCl(g) + NH_3(g) \rightarrow NH_4Cl(s)$

ΔH_f° (kJ/mol)	-92	-46	-314
S° (J K^{-1} mol^{-1})	187	193	96

ΔH° = -314 - [-92 - 46] = -176 kJ; ΔS° = 96 - [187 + 193] = -284 J/K

$\Delta G^{\circ} = \Delta H^{\circ} - T\Delta S^{\circ}$ = -176 kJ - (298 K) (-0.284 kJ/K) = -91 kJ

26. $SF_4(g) + F_2(g) \rightarrow SF_6(g)$; -374 kJ = -1105 kJ - $\Delta G_f^{\circ}(SF_4)$, ΔG_f° = -731 kJ/mol

27. $2 Al(OH)_3(s) \rightarrow Al_2O_3(s) + 3 H_2O(g)$

7 kJ = 3(-229 kJ) - 1582 kJ - (2 mol)ΔG_f°, ΔG_f° = -1138 kJ/mol

28.   a.

$$CH_2\!\!-\!\!CH_2(g) + HCN(g) \longrightarrow CH_2=CHCN(g) + H_2O(l)$$
$$\underset{O}{\diagdown\diagup}$$

ΔH_f° (kJ/mol)	-53	135.1	185.0	-286
S° (J K^{-1} mol^{-1})	242	202	274	70

$\Delta H° = 185.0 - 286 - (-53 + 135.1) = -183 \text{ kJ}; \ \Delta S° = 274 + 70 - (242 + 202) = -100. \text{ J/K}$

$\Delta G° = \Delta H° - T\Delta S° = -183 \text{ kJ} - 298 \text{ K}(-0.100 \text{ kJ/K}) = -153 \text{ kJ}$

b. $HC\equiv CH(g) + HCN(g) \rightarrow CH_2=CHCN(g)$

For $C_2H_2(g)$: $\Delta H_f° = 227 \text{ kJ/mol}$ and $S° = 201 \text{ J K}^{-1} \text{ mol}^{-1}$

$\Delta H° = 185.0 - [135.1 + 227] = -177 \text{ kJ}; \ \Delta S° = 274 - [202 + 201] = -129 \text{ J/K}$

$T = 70.°C = 343 \text{ K}; \ \Delta G° = \Delta H° - T\Delta S° = -177 \text{ kJ} - 343 \text{ K}(-0.129 \text{ kJ/K})$

$\Delta G° = -177 \text{ kJ} + 44 \text{ kJ} = -133 \text{ kJ}$

c. $4 \ CH_2=CHCH_3(g) \ + \ 6 \ NO(g) \rightarrow 4 \ CH_2=CHCN(g) + 6 \ H_2O(g) \ + \ N_2(g)$

$\Delta H_f°$ (kJ/mol)	20.9	90	185.0	-242	0
$S°$ (J K^{-1} mol^{-1})	266.9	211	274	189	192

$\Delta H° = 6(-242) + 4(185.0) - [4(20.9) + 6(90)] = -1336 \text{ kJ}$

$\Delta S° = 192 + 6(189) + 4(274) - [6(211) + 4(266.9)] = 88 \text{ J/K}$

$T = 700.°C = 973 \text{ K}; \ \Delta G° = \Delta H° - T\Delta S° = -1336 \text{ kJ} - 973 \text{ K}(0.088 \text{ kJ/K})$

$\Delta G° = -1336 \text{ kJ} - 86 \text{ kJ} = -1422. \text{ kJ}$

29. $C_2H_4(g) + H_2O(g) \rightarrow CH_3CH_2OH(l)$

$\Delta H° = -278 - (52 - 242) = -88 \text{ kJ}; \ \Delta S° = 161 - (219 + 189) = -247 \text{ J/K}$

When $\Delta G° = 0$, $\Delta H° = T\Delta S°$, $T = \dfrac{\Delta H°}{\Delta S°} = \dfrac{-88 \times 10^3 \text{ J}}{-247 \text{ J/K}} = 360 \text{ K}$

Since the signs of $\Delta H°$ and $\Delta S°$ are both negative, this reaction will be spontaneous at temperatures below 360 K (where the $\Delta H°$ term will dominate).

$C_2H_6(g) + H_2O(g) \rightarrow CH_3CH_2OH(l) + H_2(g)$

$\Delta H° = -278 - (-84.7 - 242) = +49 \text{ kJ}; \ \Delta S° = 131 + 161 - (229.5 + 189) = -127 \text{ J/K}$

This reaction will never be spontaneous because of the signs of $\Delta H°$ and $\Delta S°$.

Thus the reaction $C_2H_4(g) + H_2O(g) \rightarrow C_2H_5OH(l)$ would be preferred.

30.   Enthalpy is not favorable, so ΔS must provide the driving force for the change. Thus, ΔS is positive. There is an increase in disorder, so the original enzyme has the more ordered structure.

31.   a.   A bond is broken: ΔH (+); Increase in disorder ($\Delta n > 0$): ΔS (+)

       b. $\Delta G = \Delta H - T\Delta S$; For the reaction to be spontaneous, the entropy term must dominate. The reaction will be spontaneous at higher temperatures.

32. $2\ NH_3(g)\ +\ 3\ O_2(g)\ +\ 2\ CH_4(g) \rightarrow 2\ HCN(g)\ +\ 6\ H_2O(g)$

ΔH_f° (kJ/mol)	-46	0	-75	135.1	-242
S° ($J\ K^{-1}\ mol^{-1}$)	193	205	186	202	189

$\Delta H^{\circ} = 6(-242) + 2(135.1) - [2(-75) + 2(-46)] = -940.\ kJ$

$\Delta S^{\circ} = 2(202) + 6(189) - [2(193) + 3(205) + 2(186)] = +165\ J/K$

From the signs of ΔH° and ΔS°, this reaction is spontaneous at all temperatures. It will cost money to heat the reaction mixture. Since there is no thermodynamic reason to do this, then the purpose of the elevated temperature must be to increase the rate of the reaction.

33. $\Delta G^{\circ} = -RT \ln K = -\dfrac{8.3145\ J}{K\ mol}\ (298\ K)\ \ln(1.00 \times 10^{-14}) = 7.99 \times 10^4\ J = 79.9\ kJ$

34.   a. ΔH_f° (kJ/mol) S° ($J\ K^{-1}\ mol^{-1}$)

	ΔH_f° (kJ/mol)	S° ($J\ K^{-1}\ mol^{-1}$)
$NH_3(g)$	-46	193
$O_2(g)$	0	205
$NO(g)$	90	211
$H_2O(g)$	-242	189
$NO_2(g)$	34	240
$HNO_3(l)$	-174	156
$H_2O(l)$	-286	70

$4\ NH_3(g) + 5\ O_2(g) \rightarrow 4\ NO(g) + 6\ H_2O(g)$

$\Delta H^{\circ} = 6(-242) + 4(90) - [4(-46)] = -908\ kJ$

$\Delta S^{\circ} = 4(211) + 6(189) - [4(193) + 5(205)] = 181\ J/K$

$\Delta G° = -908$ kJ $- 298$ K $(0.181$ kJ/K$) = -962$ kJ

$\Delta G° = -$ RT ln K, ln K $= \dfrac{-\Delta G°}{RT} = \left(\dfrac{+962 \times 10^3 \text{ J}}{8.3145 \text{ J K}^{-1} \text{ mol}^{-1} \times 298 \text{ K}} \right) = 388$

ln K $= 2.303$ log K, log K $= 168$, K $= 10^{168}$

2 NO(g) $+ O_2$(g) $\rightarrow 2$ NO$_2$(g)

$\Delta H° = 2(34) - [2(90)] = -112$ kJ; $\Delta S° = 2(240) - [2(211) + (205)] = -147$ J/K

$\Delta G° = -112$ kJ $- (298$ K$)(-0.147$ kJ/K$) = -68$ kJ

K $= \exp \dfrac{-\Delta G°}{RT} = \exp \left(\dfrac{+68,000 \text{ J}}{8.3145 \text{ J K}^{-1} \text{ mol}^{-1} (298 \text{ K})} \right) = e^{27.44} = 8.3 \times 10^{11}$

Note: When determining exponents, we will round off after the calculation is complete.

3 NO$_2$(g) $+ H_2$O(l) $\rightarrow 2$ HNO$_3$(l) $+$ NO(g)

$\Delta H° = 2(-174) + (90) - [3(34) + (-286)] = -74$ kJ

$\Delta S° = 2(156) + (211) - [3(240) + (70)] = -267$ J/K

$\Delta G° = -74$ kJ $- (298$ K$)(-0.267$ kJ/K$) = +6$ kJ

K $= \exp \dfrac{-\Delta G°}{RT} = \exp \left(\dfrac{-6000 \text{ J}}{8.3145 \text{ J K}^{-1} \text{ mol}^{-1} (298 \text{ K})} \right) = e^{-2.4} = 9 \times 10^{-2}$

b. $\Delta G° = -$RT ln K; T $= 825°$C $= (825 + 273)$K $= 1098$ K

$\Delta G° = -908$ kJ $- (1098$ K$)(0.181$ kJ/K$) = -1107$ kJ

K $= \exp \dfrac{-\Delta G°}{RT} = \exp \left(\dfrac{1.107 \times 10^6 \text{ J}}{8.3145 \text{ J K}^{-1} \text{ mol}^{-1} (1098 \text{ K})} \right) = e^{121.258} = 4.589 \times 10^{52}$

c. There is no thermodynamic reason for the elevated temperature since $\Delta H°$ is negative and $\Delta S°$ is positive. Thus, the purpose for the high temperature must be to increase the rate of the reaction.

35.   NO(g) $+ O_3$(g) $\leftrightarrow$ NO$_2$(g) $+ O_2$(g); $\Delta G° = \Sigma \Delta G_f°$ (Products) $- \Sigma \Delta G_f°$ (Reactants)

$\Delta G° = 1$ mol$(52$ kJ/mol$) - [1$ mol$(87$ kJ/mol$) + 1$ mol$(163$ kJ/mol$)] = -198$ kJ

$\Delta G° = -$RT ln K, K $= \exp \dfrac{-\Delta G°}{RT} = \exp \left(\dfrac{+1.98 \times 10^5 \text{ J}}{8.3145 \text{ J K}^{-1} \text{ mol}^{-1} (298 \text{ K})} \right) = e^{79.912} = 5.07 \times 10^{34}$

36. $2\ H_2S(g)\ +\ SO_2(g)\ \leftharpoons\ 3\ S(s)\ +\ 2\ H_2O(g)$

ΔG_f° (kJ/mol)	-34	-300	0	-229

ΔG° = 2 mol (-229 kJ/mol) - [2 mol (-34 kJ/mol) + 1 mol (-300 kJ/mol)] = -90. kJ

$$K = \exp\ \frac{-\Delta G^\circ}{RT} = \exp\left(\frac{90.\times 10^3\ J}{(8.3145)\,(298\ K)}\right)\ =\ e^{36.32} = 5.9 \times 10^{15}$$

Since there is a decrease in the number of moles of gaseous particles, ΔS° is negative. Since ΔG° is negative, then ΔH° must be negative. The reaction will be spontaneous at low temperatures (the ΔH° term dominates at low temperatures).

37.   a. ΔG° = - RT ln K, K = exp (-ΔG°/RT) = $\exp\left(\dfrac{30,500\ J}{8.3145\ J\ K^{-1}\ mol^{-1} \times 298\ K}\right)$ = 2.22×10^5

       b. $C_6H_{12}O_6(s) + 6\ O_2(g) \rightarrow 6\ CO_2(g) + 6\ H_2O(l)$

             ΔG° = 6 mol (-394 kJ/mol) + 6 mol (-237 kJ/mol) - 1 mol (-911 kJ/mol) = -2875 kJ

             $\dfrac{2875\ kJ}{mol\ glucose} \times \dfrac{1\ mol\ ATP}{30.5\ kJ}$ = 94.3 mol ATP; 94.3 molecules ATP/molecule glucose

             This is an overstatement. The assumption that all of the free energy goes into this reaction is false. Actually only 38 moles of ATP are produced by metabolism of one mole of glucose.

38. ΔG°(CO) = -RT ln K_{CO} and ΔG°(O_2) = -RT ln K_{O_2}

       ΔG°(CO) - ΔG°(O_2) = -RT ln K_{CO} + RT ln K_{O_2} = -RT ln (K_{CO}/K_{O_2})

       ΔG°(CO) - ΔG°(O_2) = -(8.3145 J K^{-1} mol^{-1}) (298 K) ln 210 = -13,000 J/mol = -13 kJ/mol

39.   a. ΔG° = -RT ln K

             $\ln K = \dfrac{-\Delta G^\circ}{RT} = \dfrac{-14,000\ J}{(8.3145\ J\ K^{-1}\ mol^{-1})\,(298\ K)}$ = -5.65, K = $e^{-5.65}$ = 3.5×10^{-3}

       b.

Glutamic acid + NH_3 → Glutamine + H_2O	ΔG° = 14 kJ
ATP + H_2O → ADP + $H_2PO_4^-$	ΔG° = -30.5 kJ
Glutamic acid + ATP + NH_3 → Glutamine + ADP + $H_2PO_4^-$	ΔG° = 14 - 30.5 = -17 kJ

$$\ln K = \frac{-\Delta G^\circ}{RT} = \frac{-(-17,000 \text{ J})}{8.3145 \text{ J K}^{-1}\text{mol}^{-1}(298 \text{ K})} = 6.86, \quad K = e^{6.86} = 9.5 \times 10^2$$

40.   a. $\Delta G^\circ_{298} = 3(191.2) - 78.2 = 495.4 \text{ kJ}; \quad \Delta H^\circ = 3(241.3) - 132.8 = 591.1 \text{ kJ}$

$$\Delta S^\circ = \frac{\Delta H^\circ - \Delta G^\circ}{T} = \frac{591.1 \text{ kJ} - 495.4 \text{ kJ}}{298 \text{ K}} = 0.321 \text{ kJ/K} = 321 \text{ J/K}$$

b. $\Delta G^\circ = -RT \ln K, \quad \ln K = \frac{-\Delta G^\circ}{RT} = \frac{-495,400 \text{ J}}{(8.3145 \text{ J K}^{-1}\text{mol}^{-1})(298 \text{ K})} = -199.942$

$$K = e^{-199.942} = 1.47 \times 10^{-87}$$

c.   Assuming ΔH° and ΔS° are temperature independent:

$$\Delta G^\circ_{3000} = 591.1 \text{ kJ} - 3000. \text{ K} (0.321 \text{ kJ/K}) = -372 \text{ kJ}$$

$$\ln K = \frac{-(-372,000 \text{ J})}{(8.3145 \text{ J K}^{-1}\text{mol}^{-1})(3000. \text{K})} = 14.914, \quad K = e^{14.914} = 3.00 \times 10^6$$

Free Energy and Pressure

41.   From Exercise 10.35 we get $\Delta G^\circ = -198 \text{ kJ}. \quad \Delta G = \Delta G^\circ + RT \ln \dfrac{P_{NO_2} P_{O_2}}{P_{NO} P_{O_3}}$

$$\Delta G = -198 \text{ kJ} + \frac{8.3145 \text{ J K}^{-1}\text{mol}^{-1}}{1000 \text{ J/kJ}} (298 \text{ K}) \ln \frac{(1.00 \times 10^{-7})(1.00 \times 10^{-3})}{(1.00 \times 10^{-6})(2.00 \times 10^{-6})}$$

$$\Delta G = -198 \text{ kJ} + 9.69 \text{ kJ} = -188 \text{ kJ}$$

42. $\Delta G = \Delta G^\circ + RT \ln \dfrac{P^2_{H_2O}}{P^2_{H_2S} P_{SO_2}} = -90. \text{ kJ} + \dfrac{(8.3145)(298)}{1000} \text{ kJ} \left[\ln \dfrac{(0.030)^2}{(1.0 \times 10^{-4})^2 (0.010)} \right]$

$$\Delta G = -90. \text{ kJ} + 39.7 \text{ kJ} = -50. \text{ kJ}$$

43. $N_2(g) + 3 \text{ H}_2(g) \rightleftharpoons 2 \text{ NH}_3(g)$

$\Delta H^\circ = 2 \Delta H^\circ_f (NH_3) = 2(-46) = -92 \text{ kJ}; \quad \Delta G^\circ = 2 \Delta G^\circ_f (NH_3) = 2(-17) = -34 \text{ kJ}$

$\Delta S^\circ = 2(193 \text{ J/K}) - [192 \text{ J/K} + 3(131 \text{ J/K})] = -199 \text{ J/K}$

$$K = \exp \frac{-\Delta G^\circ}{RT} = \exp \left(\frac{34,000 \text{ J}}{(8.3145 \text{ J K}^{-1}\text{mol}^{-1})(298 \text{ K})} \right) = e^{13.72} = 9.1 \times 10^5$$

a. $\Delta G = \Delta G^\circ + RT \ln \dfrac{P_{NH_3}^2}{P_{N_2} P_{H_2}^3} = -34 \text{ kJ} + \dfrac{(8.3145 \text{ J K}^{-1} \text{mol}^{-1})(298 \text{ K})}{1000 \text{ J/kJ}} \ln \dfrac{(50.)^2}{(200.)(200.)^3}$

$\Delta G = -34 \text{ kJ} - 33 \text{ kJ} = -67 \text{ kJ}$

b. $\Delta G = -34 \text{ kJ} + \dfrac{(8.3145 \text{ J K}^{-1} \text{mol}^{-1})(298 \text{ K})}{1000 \text{ J/kJ}} \ln \dfrac{(200.)^2}{(200.)(600.)^3}$

$\Delta G = -34 \text{ kJ} - 34.4 \text{ kJ} = -68 \text{ kJ}$

c. Assume ΔH° and ΔS° are temperature independent.

$\Delta G_{100}^\circ = \Delta H^\circ - T\Delta S^\circ, \quad \Delta G_{100}^\circ = -92 \text{ kJ} - (100. \text{ K})(-0.199 \text{ kJ/K}) = -72 \text{ kJ}$

$\Delta G_{100} = \Delta G_{100}^\circ + RT \ln Q = -72 \text{ kJ} + \dfrac{(8.3145 \text{ J K}^{-1} \text{mol}^{-1})(100. \text{ K})}{1000 \text{ J/kJ}} \ln \dfrac{(10.)^2}{(50.)(200.)^3}$

$\Delta G_{100} = -72 \text{ kJ} - 13 \text{ kJ} = -85 \text{ kJ}$

d. $\Delta G_{700}^\circ = -92 \text{ kJ} - (700. \text{ K})(-0.199 \text{ kJ/K}) = 47 \text{ kJ}$

$\Delta G_{700} = 47 \text{ kJ} + \dfrac{(8.3145 \text{ J K}^{-1} \text{mol}^{-1})(700. \text{ K})}{1000 \text{ J/kJ}} \ln \dfrac{(10.)^2}{(50.)(200.)^3} = 47 \text{ kJ} - 88 \text{ kJ} = -41 \text{ kJ}$

Additional Exercises

44.   No, when using Appendix 4 ΔG_f° values, we have specified a temperature of 25°C. Further, if gases or solutions are involved, we have specified partial pressures of 1 atm and solute concentration of 1 molar. At other temperatures and compositions the reaction may not be spontaneous. A negative ΔG° means the reaction is spontaneous under standard conditions.

45.   The more widely something is dispersed, the greater the disorder. We must do work to overcome this disorder. In terms of the 2nd law it would be more advantageous to prevent contamination of the environment rather than clean it up later.

46.   It appears that the sum of the two processes has no net change. This is not so, ΔS_{univ} has increased even though it looks as if we have gone through a cyclic process.

47.   If we graph ln K vs 1/T we get a straight line (y = mx + b). For an endothermic process the slope is negative (m = slope = $-\Delta H^\circ/R$).

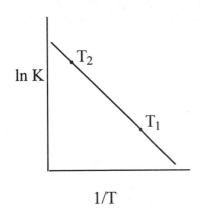

As we increase the temperature from T_1 to T_2 $(1/T_2 < 1/T_1)$, we can see that ln K and, hence, K increases. A larger value of K means some more reactants can be converted to products (reaction shifts right).

48.   a.

Temp (°C)	T(K)	1000/T (K⁻¹)	K	ln K
0	273	3.66	1.14×10^{-15}	-34.408
25	298	3.36	1.00×10^{-14}	-32.236
35	308	3.25	2.09×10^{-14}	-31.499
40.	313	3.19	2.92×10^{-14}	-31.165
50.	323	3.10	5.47×10^{-14}	-30.537

A graph of lnK vs. 1/T will yield a straight line with slope equal to $-\Delta H°/R$ and y-intercept equal to $\Delta S°/R$ (y = mx + b). See Exercise 7.11 for the plot.

The straight line equation is: $\ln K = -6.91 \times 10^3 \left(\dfrac{1}{T} \right) - 9.09$

Slope $= -6.91 \times 10^3 = \dfrac{-\Delta H°}{R}$, $\Delta H° = 5.75 \times 10^4$ J $= 57.5$ kJ

y-intercept $= -9.09 = \dfrac{\Delta S°}{R}$, $\Delta S° = -75.6$ J/K

b.   From part a, $\Delta H° = 57.4$ kJ and $\Delta S° = -75.6$ J/K. Assuming $\Delta H°$ and $\Delta S°$ are temperature independent:

$\Delta G° = 57,500$ J $- 647$ K $(-75.6$ J/K$) = 57,500 + 48,900 = 106,400$ J $= 106.4$ kJ

c.   Both $\Delta H°$ and $\Delta S°$ are unfavorable. Thus, $\Delta G°$ will always be positive and K will always be less than 1. There is no temperature at which K can equal 1. This assumes that pressure has no effect on the equilibrium constant and that $\Delta H°$ and $\Delta S°$ are independent of temperature. Neither of these assumptions are valid under the conditions in which H_2O might be an ionic solid.

49. $3 O_2(g) \rightleftharpoons 2 O_3(g)$

ΔG_f° 0   163 kJ/mol

ΔH_f° 0   143 kJ/mol

$\Delta H^{\circ} = 2(143) = 286$ kJ; $\Delta G^{\circ} = 2(163) = 326$ kJ

$$\ln K = \frac{-\Delta G^{\circ}}{RT} = \frac{-326 \times 10^3 \text{ J}}{(8.3145 \text{ J K}^{-1} \text{mol}^{-1})(298 \text{ K})} = -131.573, \ K = e^{-131.573} = 7.22 \times 10^{-58}$$

We need the value of K at 230. K. From Exercise 47: $\ln K = \dfrac{-\Delta H^{\circ}}{R}\left(\dfrac{1}{T}\right) + \dfrac{\Delta S^{\circ}}{R}$

For two sets of K and T:

$$\ln K_1 = \frac{-\Delta H^{\circ}}{R}\left(\frac{1}{T_1}\right) + \frac{\Delta S^{\circ}}{R}; \ \ \ln K_2 = \frac{-\Delta H^{\circ}}{R}\left(\frac{1}{T_2}\right) + \frac{\Delta S^{\circ}}{R}$$

Subtracting the first expression from the second:

$$\ln K_2 - \ln K_1 = \frac{\Delta H^{\circ}}{R}\left(\frac{1}{T_1} - \frac{1}{T_2}\right) = \frac{\Delta H^{\circ}}{R}\left(\frac{T_2 - T_1}{T_2 T_1}\right)$$

Let $K_2 = 7.22 \times 10^{-58}$, $T_2 = 298$; $K_1 = K_{230}$, $T_1 = 230.$ K; $\Delta H^{\circ} = 286 \times 10^3$ J

$$\ln 7.22 \times 10^{-58} - \ln K_{230} = \frac{286 \times 10^3}{(8.3145)}\left(\frac{68}{(298)(230.)}\right)$$

$\ln K_{230} = -132 - 34 = -166, \ K_{230} = e^{-166} = 8.07 \times 10^{-73}$

$$8.07 \times 10^{-73} = \frac{P_{O_3}^2}{P_{O_2}^3} = \frac{P_{O_3}^2}{(1.0 \times 10^{-3} \text{ atm})^3}, \ \ P_{O_3} = 2.8 \times 10^{-41} \text{ atm}$$

The volume occupied by one molecule of ozone is:

$$V = \frac{nRT}{P} = \frac{(1/6.022 \times 10^{23} \text{ mol})(0.08206 \text{ L atm mol}^{-1} \text{K}^{-1})(230. \text{ K})}{(2.8 \times 10^{-41} \text{ atm})}, \ V = 1.1 \times 10^{18} \text{ L}$$

Equilibrium is probably not maintained under these conditions. When only two ozone molecules are in a volume of 1.1×10^{18} L, the reaction is not at equilibrium. Under these conditions, Q > K and the reaction shifts left. But with only 2 ozone molecules in a very large volume, it is extremely unlikely that they will collide with each other. At these conditions, the concentration of ozone is not large enough to maintain equilibrium.

50.   S (rhombic) $\rightarrow$ S (monoclinic); $\Delta H° = 0.30$ kJ; $\Delta S° = 32.55 - 31.88 = 0.67$ J/K

For $\Delta G° = 0 = \Delta H° - T\Delta S°$, $\Delta H° = T\Delta S°$, $T = \dfrac{\Delta H°}{\Delta S°} = \dfrac{3.0 \times 10^2 \text{ J}}{0.67 \text{ J/K}} = 450$ K

51. $CH_4(g) + CO_2(g) \rightarrow CH_3CO_2H(l)$

$\Delta H° = -484 - [-75 + (-393.5)] = -16$ kJ; $\Delta S° = 160 - [186 + 214] = -240.$ J/K

$\Delta G° = \Delta H° - T\Delta S° = -16$ kJ $- (298$ K$)(-0.240$ kJ/K$) = +56$ kJ

This reaction is spontaneous only at a temperature below $T = \Delta H°/\Delta S° = 67$ K. This is not practical. Substances will be in condensed phases and rates will be very slow.

$CH_3OH(g) + CO(g) \rightarrow CH_3CO_2H(l)$

$\Delta H° = -484 - [-110.5 + (-201)] = -173$ kJ; $\Delta S° = 160 - [198 + 240] = -278$ J/K

$\Delta G° = -173$ kJ $- (298$ K$)(-0.278$ kJ/K$) = -90.$ kJ

This reaction is spontaneous at a temperature below $T = \Delta H°/\Delta S° = 622$ K. The reaction of CH_3OH and CO will be preferred. It is spontaneous at high enough temperatures that the rates of reaction should be reasonable.

52.   From our answer to Exercise 10.33, we get $\Delta G° = 79.9$ kJ.

$\Delta G = \Delta G° + RT \ln [H^+] [OH^-] = 79.9$ kJ $+ \dfrac{(8.3145 \text{ J K}^{-1} \text{ mol}^{-1}) (298 \text{ K})}{1000 \text{ J/kJ}} \ln [H^+] [OH^-]$

a. $\Delta G = 79.9 - 79.9 = 0$, At equilibrium

b. $\Delta G = 0$, At equilibrium   c. $\Delta G = -34$ kJ, Shifts right

d. $\Delta G = 45.7$ kJ, Shifts left   e. $\Delta G = 79.9$ kJ, Shifts left

Le Chatelier's principle gives the same results.

53. K^+ (blood) $\rightleftharpoons K^+$ (muscle) $\Delta G° = 0$; $\Delta G = RT \ln \left(\dfrac{[K^+]_m}{[K^+]_b} \right)$; $\Delta G = w_{max}$

$\Delta G = \dfrac{8.3145 \text{ J}}{\text{K mol}} (310. \text{ K}) \ln \left(\dfrac{0.15}{0.0050} \right)$, $\Delta G = 8.8 \times 10^3$ J/mol $= 8.8$ kJ/mol

At least 8.8 kJ of work must be applied. $\dfrac{8.8 \text{ kJ}}{\text{mol K}^+} \times \dfrac{1 \text{ mol ATP}}{30.5 \text{ kJ}} = 0.29$ mol ATP

Other ions will have to be transported in order to maintain electroneutrality. Either anions **must** be transported into the cells, or cations (Na^+) in the cell must be transported to the blood. The latter is what happens: $[Na^+]$ in blood is greater than $[Na^+]$ in cells as a result of this pumping.

54.   As any process occurs, ΔS_{univ} will increase; ΔS_{univ} cannot decrease. Time also goes in one direction, just as ΔS_{univ} goes in one direction.

55.   As ΔS_{univ} continually increases, there will be less energy available for work. When no work **can** be done, the world ends.

56.   The introduction of mistakes is an effect of entropy. The purpose of redundant information **is to** provide a control to check the "correctness" of the transmitted information.

57. $q_v = \Delta E = C_v \Delta T = (12.47 \text{ J K}^{-1} \text{ mol}^{-1})\,(56.5 \text{ K}) = 705 \text{ J/mol}$

$$15.0 \text{ g He} \times \frac{1 \text{ mol}}{4.003 \text{ g}} \times \frac{705 \text{ J}}{\text{mol}} = 2640 \text{ J} = 2.64 \text{ kJ}$$

58. $\Delta E = q + w;\ \ w = -P\Delta V;\ \ \Delta E = q - P\Delta V = 4.53 \text{ kJ} - 2.74 \text{ kJ} = 1.79 \text{ kJ}$

59.   Heat gained by water $= 4.18 \text{ J g}^{-1} \text{ }^{\circ}\text{C}^{-1} \times 250.0 \text{ g} \times 2.5 ^{\circ}\text{C} = 2600 \text{ J}$

Heat lost by Al = Heat gained by H_2O

$2600 \text{ J} = S_{Al}\,(50.0 \text{ g})\,(85.0 - 27.5)^{\circ}\text{C},\ \ S_{Al} = 0.90 \text{ J g}^{-1}\text{ }^{\circ}\text{C}^{-1}$

60. $q_v = \Delta E = C_v \Delta T = (44.60 \text{ J mol}^{-1} \text{ K}^{-1})\,(48.4 \text{ K}) = 2160 \text{ J/mol}$

$$\frac{2160 \text{ J}}{\text{mol}} \times 1.00 \times 10^3 \text{ g} \times \frac{1 \text{ mol}}{30.07 \text{ g}} = 7.18 \times 10^4 \text{ J} = 71.8 \text{ kJ}$$

At constant volume, 71.8 J of energy are required and $\Delta E = 71.8 \text{ kJ}$.

At constant pressure (assuming ethane acts as an ideal gas):

$C_p = C_v + R = 44.60 + 8.31 = 52.91 \text{ J mol}^{-1} \text{ K}^{-1}$

Energy required $= q_p = \Delta H = C_p \Delta T = (52.91 \text{ J mol}^{-1} \text{ K}^{-1})\,(48.4 \text{ K}) = 2560 \text{ J/mol}$

$$\frac{2560 \text{ J}}{\text{mol}} \times 1.00 \times 10^3 \text{ g} \times \frac{1 \text{ mol}}{30.07 \text{ g}} = 8.51 \times 10^4 \text{ J} = 85.1 \text{ kJ} = q_p$$

For the constant pressure process, $\Delta E = 71.8 \text{ kJ}$ as calculated previously (ΔE is unchanged).

61.   a. $q_v = \Delta E = C_v \Delta T = (28.95 \text{ J mol}^{-1} \text{ K}^{-1})\,(350.0 - 298.0 \text{ K}) = 1.51 \times 10^3 \text{ J/mol} = 1.51 \text{ kJ/mol}$

$q_p = \Delta H = C_p \Delta T = 37.27\,(350.0 - 298.0) = 1.94 \times 10^3 \text{ J/mol} = 1.94 \text{ kJ/mol}$

b. $\Delta S = S_{350} - S_{298} = nC_p \ln (T_2/T_1)$

$S_{350} - 213.64$ J/K $= (1.000$ mol$) (37.27$ J mol^{-1} K$^{-1}) \ln (350.0/298.0)$

$S_{350} = 213.64$ J/K $+ 5.994$ J/K $= 219.63$ J/K $=$ molar entropy at 350.0 K and 1.000 atm

c. $\Delta S = nR \ln (V_2/V_1),\; V = nRT/P,\; \Delta S = nR \ln (P_1/P_2) = S_{(350,\; 1.174)} - S_{(350,\; 1.000)}$

$\Delta S = S_{(350,\; 1.174)} - 219.63$ J/K $= (1.000$ mol$) (8.3145$ J mol^{-1} K$^{-1}) \ln (1.000/1.174)$

$\Delta S = -1.334$ J/K $= S - 219.63,\quad S = 218.30$ J K^{-1} mol$^{-1} = S_{(350,\; 1.174)}$

62.   It takes $nC_p\Delta T$ amount of energy to carry out this process. The internal energy of the system increases by $nC_v\Delta T$. So the fraction that goes into raising the internal energy is:

$$\frac{nC_v\Delta T}{nC_p\Delta T} = \frac{C_v}{C_p} = \frac{20.8}{29.1} = 0.715$$

The remainder of the energy ($nR\Delta T$) goes into expanding the gas against the constant pressure.

$$100.0 \text{ g N}_2 \times \frac{1 \text{ mol}}{28.014 \text{ g}} = 3.570 \text{ mol}$$

$q_v = \Delta E = nC_v\Delta T = 3.570$ mol $(20.8$ J mol^{-1} K$^{-1}) (60.0$ K$) = 4.46 \times 10^3$ J $= 4.46$ kJ

63.   Using LeChatelier's principle: A decrease in pressure (volume increases) will favor the side with the greater number of particles. Thus, 2 I(g) will be favored at low pressure.

Looking at ΔG: $\Delta G = \Delta G° + RT \ln (P_I^2/P_{I_2})$; $\ln (P_I^2/P_{I_2}) > 0$ for $P_I = P_{I_2} = 10$ atm and ΔG is positive (not spontaneous). But at $P_I = P_{I_2} = 0.10$ atm, the logarithm term is negative. If $|RT \ln Q| > \Delta G°$, then ΔG becomes negative and the reaction is spontaneous.

64.   Assuming $\Delta S°$ and $\Delta H°$ are temperature independent:

$$\Delta S° = \frac{\Delta H°}{T} = \frac{13.8 \times 10^3 \text{ J/mol}}{121 \text{ K}} = 114 \text{ J mol}^{-1} \text{ K}^{-1}$$

65.   a.   He(g, 0.100 mol, 25°C, 1.00 atm) $\rightarrow$ He(g, 0.100 mol, 25°C, 5.00 L)

$\Delta S = nR \ln (V_2/V_1) = S_{final} - S_{initial};\; S_i = 0.100$ mol $(126.1$ J mol^{-1} K$^{-1}) = 12.6$ J/K

$$V_1 = \frac{nRT}{P_1} = \frac{0.100 \text{ mol} \times \dfrac{0.08206 \text{ L atm}}{\text{mol K}} \times 298 \text{ K}}{1.00 \text{ atm}} = 2.45 \text{ L}$$

$S_f - 12.6$ J/K $= (0.100$ mol$) (8.3145$ J mol^{-1} K$^{-1}) \ln (5.00$ L/2.45 L$)$

S_f = 12.6 J/K + 0.593 J/K = 13.2 J/K

b. He (3.00 mol, 25°C, 1.00 atm) → He (3.00 mol, 25°C, 3000.0 L)

$\Delta S = nR \ln (V_2/V_1) = S_{final} - S_{initial}$; S_i = 3.00 mol (126.1 J mol^{-1} K^{-1}) = 378 J/K

$$V_1 = \frac{nRT}{P_1} = \frac{3.00 \text{ mol} \times \dfrac{0.08206 \text{ L atm}}{\text{mol K}} \times 298 \text{ K}}{1.00 \text{ atm}} = 73.4 \text{ L}$$

S_f - 378 J/K = (3.00 mol) (8.3145 J mol^{-1} K^{-1}) ln $\left(\dfrac{3000.0}{73.4} \right)$ = 92.6 J/K

S_f = 378 + 92.6 = 471 J/K

66.   Rhombic → Monoclinic; ΔH is (+) and ΔG is (-) above 95.3°C, thus ΔS must be positive.

At 95.3°C, $\Delta G = 0$; $\Delta S = \dfrac{\Delta H}{T} = \dfrac{(0.400 \times 10^3 \text{ J/mol})}{(95.3 + 273.2)\text{K}}$ = 1.09 J mol^{-1} K^{-1}

67. $\Delta S = \dfrac{q_{rev}}{T} = \dfrac{\Delta H_{vap}}{T}$; For methane: $\Delta S = \dfrac{8.20 \times 10^3 \text{ J/mol}}{112 \text{ K}}$ = 73.2 J mol^{-1} K^{-1}

For hexane: $\Delta S = \dfrac{28.9 \times 10^3 \text{ J/mol}}{342 \text{ K}}$ = 84.5 J mol^{-1} K^{-1}

$V_{met} = \dfrac{nRT}{P} = \dfrac{1.00 \text{ mol} (0.08206) (112 \text{ K})}{1.00 \text{ atm}} = 9.19 \text{ L}$; $V_{hex} = \dfrac{nRT}{P} = R(342 \text{ K}) = 28.1 \text{ L}$

$\Delta S_{hex} - \Delta S_{met}$ = 84.5 - 73.2 = 11.3 J mol^{-1} K^{-1}; R ln (V_{hex}/V_{met}) = 9.29 J mol^{-1} K^{-1}

As the molar volume of a gas increases, ΔS_{vap} also increases. In the case of hexane and methane, the difference in molar volume accounts for 82% of the difference in the entropies.

68.   The volumes for each step are:

a. P_1 = 5.00 atm, n = 1.00 mol, T = 350. K; $V_1 = \dfrac{nRT}{P_1}$ = 5.74 L

b. P_2 = 2.24 atm, $V_2 = \dfrac{nRT}{P_2}$ = 12.8 L   c. P_3 = 1.00 atm, V_3 = 28.7 L

The process can be carried out in the following steps:

$(P_1, V_1) \rightarrow (P_2, V_1)$ w = -PΔV = 0

$(P_2, V_1) \rightarrow (P_2, V_2)$ w = -(2.24 atm)(12.8 - 5.74)L = -16 L atm

$(P_2, V_2) \rightarrow (P_3, V_2)$ w = 0

$(P_3, V_2) \rightarrow (P_3, V_3)$ $w = -(1.00 \text{ atm})(28.7 - 12.8)\text{L} = -15.9 \text{ L atm}$

$w_{tot} = -16 - 15.9 = -32 \text{ L atm};$ $-32 \text{ L atm} \times \dfrac{101.3 \text{ J}}{1 \text{ L atm}} = -3200 \text{ J} = \text{total work}$

$w_{rev} = -nRT \ln (P_1/P_2) = -(1.00 \text{ mol})(8.3145 \text{ J mol}^{-1} \text{ K}^{-1})(350. \text{ K}) \ln (5.00/1.00) = -4680 \text{ J}$

69. $(3.00 \text{ mol})(75.3 \text{ J mol}^{-1} \text{ °C}^{-1})(T_f - 0°C) = (1.00 \text{ mol})(75.3 \text{ J mol}^{-1} \text{ °C}^{-1})(100.°C - T_f)$

Solving: $T_f = 25°C = 298 \text{ K};$ $\Delta S = nC_p \ln (T_2/T_1)$

Heat 3 mol H_2O: $\Delta S_1 = (3.00 \text{ mol})(75.3 \text{ J mol}^{-1} \text{ K}^{-1}) \ln (298/273) = 19.8 \text{ J/K}$

Cool 1 mol H_2O: $\Delta S_2 = (1.00 \text{ mol})(75.3 \text{ J mol}^{-1} \text{ K}^{-1}) \ln (298/373) = -16.9 \text{ J/K}$

$\Delta S_{tot} = 19.8 - 16.9 = 2.9 \text{ J/K}$

70. $18.02 \text{ g ice} = 1.000 \text{ mol ice};$ $54.05 \text{ g } H_2O = 3.000 \text{ mol } H_2O$

Heat gained by the ice:

$(1.000 \text{ mol})(37.5 \text{ J mol}^{-1} \text{ °C}^{-1})(10.0 \text{ °C}) + 6.01 \times 10^3 \text{ J} + (1.000 \text{ mol})(75.3 \text{ J mol}^{-1} \text{ °C}^{-1})(T_f - 0.0)$

Heat lost by $H_2O(l) = 3.000 \text{ mol } (75.3 \text{ J mol}^{-1} \text{ °C}^{-1})(100.0 \text{ °C} - T_f)$

Heat gain by ice = Heat lost by H_2O; $375 \text{ J} + 6010 \text{ J} + 75.3 \text{ T}_f = 22,600 \text{ J} - 226 \text{ T}_f$

Solving: $T_f = 53.8°C;$ $\Delta S = nC_p \ln (T_2/T_1)$ or for a phase change, $\Delta S = \Delta H/T$

$\Delta S_{ice} = (1.000 \text{ mol}) (37.5 \text{ J mol}^{-1} \text{ K}^{-1}) \ln \left(\dfrac{273.2 \text{ K}}{263.2 \text{ K}} \right) + \dfrac{6010 \text{ J}}{273.2 \text{ K}}$

$+ (1.000 \text{ mol}) (75.3 \text{ J mol}^{-1} \text{ K}^{-1}) \ln \left(\dfrac{327.0}{273.2} \right)$

$\Delta S_{ice} = 1.40 \text{ J/K} + 22.0 \text{ J/K} + 13.5 \text{ J/K} = 36.9 \text{ J/K}$

$\Delta S_{water} = (3.000 \text{ mol})(75.3 \text{ J mol}^{-1} \text{ K}^{-1}) \ln (327.0/373.2) = -29.9 \text{ J/K}$

$\Delta S_{tot} = 36.9 - 29.9 = 7.0 \text{ J/K}$

71. $q = (1.000 \text{ mol})(37.5 \text{ J mol}^{-1} \text{ K}^{-1})(30.0 \text{ K}) + 6010 \text{ J} + (1.000 \text{ mol})(75.3 \text{ J mol}^{-1} \text{ K}^{-1})(100.0 \text{ K})$

$+ 40,700 \text{ J} + (1.000 \text{ mol})(36.4 \text{ J mol}^{-1} \text{ K}^{-1})(40.0 \text{ K})$

$q = 1130 \text{ J} + 6010 \text{ J} + 7530 \text{ J} + 40,700 \text{ J} + 1460 \text{ J} = 56,800 \text{ J} = 56.8 \text{ kJ}$

At constant pressure: $q_p = \Delta H = 56.8 \text{ kJ}$

at 100.0°C: $V = \dfrac{nRT}{P} = \dfrac{1.000 \text{ mol} \times \dfrac{0.08206 \text{ L atm}}{\text{mol K}} \times 373.2 \text{ K}}{1.00 \text{ atm}} = 30.6 \text{ L}$

at 140.0°C: $V = \dfrac{nR(413.2 \text{ K})}{P} = 33.9 \text{ L}$

Work is only done when vaporization occurs and when the vapor expands as T is increased from 100.0°C to 140.0°C.

$w = -P\Delta V = -1.00 \text{ atm} (30.6 \text{ L} - 0.018 \text{ L}) - 1.00 \text{ atm} (33.9 - 30.6 \text{ L})$

$w = -30.6 \text{ L atm} - 3.3 \text{ L atm} = -33.9 \text{ L atm}; \; -33.9 \text{ L atm} \times \dfrac{101.3 \text{ J}}{\text{L atm}} = -3430 \text{ J} = -3.43 \text{ kJ}$

$\Delta E = q + w = 56.8 \text{ kJ} - 3.43 \text{ kJ} = 53.4 \text{ kJ}; \; \Delta S = nC_p \ln (T_2/T_1) \text{ or } \Delta S = \Delta H/T \text{ (phase change)}$

$\Delta S = (1.000 \text{ mol})(37.5 \text{ J mol}^{-1} \text{ K}^{-1}) \ln \left(\dfrac{273.2}{243.2} \right) + \dfrac{6010 \text{ J}}{273.2 \text{ K}}$

$\qquad + (1.000 \text{ mol}) (75.3 \text{ J mol}^{-1} \text{ K}^{-1}) \ln \left(\dfrac{373.2}{273.2} \right) + \dfrac{40{,}700 \text{ J}}{372.2 \text{ K}}$

$\qquad + (1.000 \text{ mol}) (36.4 \text{ J mol}^{-1} \text{ K}^{-1}) \ln \left(\dfrac{413.2}{373.2} \right)$

$\Delta S = 4.36 \text{ J/K} + 22.0 \text{ J/K} + 23.5 \text{ J/K} + 109 \text{ J/K} + 3.71 \text{ J/K} = 163 \text{ J/K}$

Summary: $q = 56.8 \text{ kJ}; \; \Delta H = 56.8 \text{ kJ}; \; w = -3.43 \text{ kJ}; \; \Delta E = 53.4 \text{ kJ}; \; \Delta S = 163 \text{ J/K}$

72. $1000 \text{ gal } H_2O \times \dfrac{4 \text{ qt}}{1 \text{ gal}} \times \dfrac{1 \text{ L}}{1.06 \text{ qt}} \times \dfrac{1000 \text{ g}}{1 \text{ L}} \times \dfrac{1 \text{ mol}}{18.0 \text{ g}} = 2 \times 10^5 \text{ mol } H_2O$

There is such an excess of water, let's assume the final temperature is very close to 0°C. Therefore, $\Delta S_{H_2O} \approx 0$.

$\Delta S_{Fe} = \left(\dfrac{111.7 \text{ g Fe}}{55.85 \text{ g/mol}} \right) (25.1 \text{ J mol}^{-1} \text{ K}^{-1}) \ln \left(\dfrac{273 \text{ K}}{373 \text{ K}} \right) = -15.7 \text{ J/K}$

73. $P_i = \dfrac{1.00 \text{ mol} \times \dfrac{0.08206 \text{ L atm}}{\text{mol K}} \times 300. \text{ K}}{30.0 \text{ L}} = 0.821 \text{ atm}; \; P_f = \dfrac{nRT}{40.0 \text{ L}} = 0.615 \text{ atm}$

a.   free expansion

$\quad w = 0 \text{ since } P_{ext} = 0; \; \Delta E = nC_v\Delta T, \text{ since } \Delta T = 0, \Delta E = 0$

$\Delta E = q + w$, $q = 0$; $\Delta H = nC_p\Delta T$, since $\Delta T = 0$, $\Delta H = 0$

$$\Delta S = nR \ln\left(\frac{V_2}{V_1}\right) = (1.00 \text{ mol}) (8.3145 \text{ J mol}^{-1} \text{ K}^{-1}) \ln\left(\frac{40.0}{30.0}\right) = 2.39 \text{ J/K}$$

$\Delta G = \Delta H - T\Delta S = 0 - 300. \text{ K } (2.39 \text{ J/K}) = -717 \text{ J}$

b. reversible expansion

$\Delta E = 0$; $\Delta H = 0$; $\Delta S = 2.39 \text{ J/K}$; $\Delta G = -717 \text{ J}$; These are state functions.

$$w_{rev} = -nRT \ln\left(\frac{V_2}{V_1}\right) = -(1.00 \text{ mol})(8.3145 \text{ J mol}^{-1} \text{ K}^{-1}) (300. \text{ K}) \ln\left(\frac{40.0}{30.0}\right) = -718 \text{ J}$$

$\Delta E = 0 = q + w$, $q_{rev} = -w_{rev} = 718 \text{ J}$

Summary: a) free expansion   b) reversible

	a) free expansion	b) reversible
q	0	718 J
w	0	-718 J
ΔE	0	0
ΔH	0	0
ΔS	2.39 J/K	2.39 J/K
ΔG	-717 J	-717 J

74.   Step 1: $\Delta E = 0$ and $\Delta H = 0$ since $\Delta T = 0$

$$w = -P\Delta V = -(9.87 \times 10^{-3} \text{ atm}) \Delta V; \quad V = \frac{nRT}{P}, R = 0.08206 \text{ L atm mol}^{-1} \text{ K}^{-1}$$

$$\Delta V = V_f - V_i = nRT\left(\frac{1}{P_f} - \frac{1}{P_i}\right) = \left(\frac{nRT}{9.87 \times 10^{-3}} - \frac{nRT}{2.45 \times 10^{-2}}\right) = 1480 \text{ L}$$

$w = -(9.87 \times 10^{-3} \text{ atm})(1480 \text{ L}) = -14.6 \text{ L atm } (101.3 \text{ J L}^{-1} \text{ atm}^{-1}) = -1480 \text{ J}$

$\Delta E = q + w = 0$, $q = -w = +1480 \text{ J}$

$$\Delta S = nR \ln\left(\frac{V_2}{V_1}\right) = 1.00 \text{ mol } (8.3145 \text{ J mol}^{-1} \text{ K}^{-1}) \ln\left(\frac{P_1}{P_2}\right), \quad \Delta S = 7.56 \text{ J/K}$$

$\Delta G = \Delta H - T\Delta S = 0 - 298 \text{ K}(7.56 \text{ J/K}) = -2250 \text{ J}$

Step 2: $\Delta E = 0$, $\Delta H = 0$

$$w = -(4.93 \times 10^{-3} \text{ atm})\left(\frac{nRT}{4.93 \times 10^{-3}} - \frac{nRT}{9.87 \times 10^{-3}} \right) L \times \frac{101.3 \text{ J}}{L \text{ atm}} = -1240 \text{ J}$$

$$q = -w = 1240 \text{ J}; \quad \Delta S = nR \ln\left(\frac{9.87 \times 10^{-3} \text{ atm}}{4.93 \times 10^{-3} \text{ atm}} \right) = 5.77 \text{ J/K}$$

$$\Delta G = 0 - 298 \text{ K}(5.77 \text{ J/K}) = -1720 \text{ J}$$

Step 3: $\Delta E = 0$, $\Delta H = 0$

$$w = -(2.45 \times 10^{-3} \text{ atm})\left(\frac{nRT}{2.45 \times 10^{-3}} - \frac{nRT}{4.93 \times 10^{-3}} \right) L \times \frac{101.3 \text{ J}}{L \text{ atm}} = -1250 \text{ J}$$

$$q = -w = 1250 \text{ J}; \quad \Delta S = nR \ln\left(\frac{4.93 \times 10^{-3} \text{ atm}}{2.45 \times 10^{-3} \text{ atm}} \right) = 5.81 \text{ J/K}$$

$$\Delta G = 0 - 298 \text{ K} (5.81 \text{ J/K}) = -1730 \text{ J}$$

	q	w	ΔE	ΔS	ΔH	ΔG
Step 1	1480 J	-1480 J	0	7.56 J/K	0	-2250 J
Step 2	1240 J	-1240 J	0	5.77 J/K	0	-1720 J
Step 3	1250 J	-1250 J	0	5.81 J/K	0	-1730 J
Total	3970 J	-3970 J	0	19.14 J/K	0	-5.70×10^3 J

75.   a.   Isothermal: $\Delta E = 0$, $\Delta H = 0$; $PV = nRT$, $R = 0.08206$ L atm K^{-1} mol^{-1}

$$w = -P\Delta V = -2.45 \times 10^{-3} \text{ atm}\left(\frac{nRT}{2.45 \times 10^{-3}} - \frac{nRT}{2.45 \times 10^{-2}} \right) L \times \frac{101.3 \text{ J}}{L \text{ atm}} = -2230 \text{ J}$$

$$q = -w = 2230 \text{ J}$$

$$\Delta S = nR \ln\left(\frac{P_1}{P_2} \right) = (1.00 \text{ mol})(8.3145 \text{ J mol}^{-1} \text{ K}^{-1}) \ln\left(\frac{2.45 \times 10^{-2} \text{ atm}}{2.45 \times 10^{-3} \text{ atm}} \right) = 19.1 \text{ J/K}$$

$$\Delta G = \Delta H - T\Delta S = 0 - (298 \text{ K})(19.1 \text{ J/K}) = -5.69 \times 10^3 \text{ J} = -5.69 \text{ kJ}$$

   b. $\Delta E = 0$; $\Delta H = 0$; $\Delta S = 19.1$ J/K; $\Delta G = -5.69 \times 10^3$ J; Same as (a) since these are all state functions.

$$\Delta S = \frac{q_{rev}}{T}, \quad q_{rev} = T\Delta S = 5.69 \times 10^3 \text{ J} = 5.69 \text{ kJ}$$

$\Delta E = 0 = q + w$, $w_{rev} = -5.69 \times 10^3 \text{ J} = -5.69 \text{ kJ}$

c. $\Delta E = 0$; $\Delta H = 0$; $\Delta S = -19.1 \text{ J/K}$; $\Delta G = 5690 \text{ J}$ (Opposite signs of a.)

$$w = -2.45 \times 10^{-2} \text{ atm} \left(\frac{nRT}{2.45 \times 10^{-2}} - \frac{nRT}{2.45 \times 10^{-3}} \right) \text{L} \times \frac{101.3 \text{ J}}{\text{L atm}} = 22,300 \text{ J} = 22.3 \text{ kJ}$$

$\Delta E = q + w = 0$, $q = -22.3 \text{ kJ}$

d.

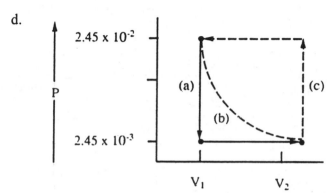

e. $\Delta S_{surr} = \dfrac{-q_{actual}}{T}$; $\Delta S_{surr, a} = \dfrac{-2230 \text{ J}}{298 \text{ K}} = -7.48 \text{ J/K}$

$\Delta S_{surr, b} = -\Delta S = -19.1 \text{ J/K}$; $\Delta S_{surr, c} = \dfrac{22,300 \text{ J}}{298 \text{ K}} = 74.8 \text{ J/K}$

76. $\Delta G_{298}^{\bullet} = -58.03 \text{ kJ} - 298 \text{ K} (-0.1766 \text{ kJ/K}) = -5.40 \text{ kJ}$

$\Delta G^{\circ} = 0 = \Delta H^{\circ} - T\Delta S^{\circ}$, $T = \dfrac{\Delta H^{\circ}}{\Delta S^{\circ}} = \dfrac{-58.03 \text{ kJ}}{-0.1766 \text{ kJ/K}} = 328.6 \text{ K}$

ΔG° is negative below 328.6 K where ΔH° term dominates.

77. $2 \text{ SO}_2(g)$ + $\text{O}_2(g)$ → $2 \text{ SO}_3(g)$

$\Delta H_f^{\bullet}$ -297 kJ/mol   0   -396

S° 248 J mol^{-1} K^{-1} 205   257

$\Delta H_{298}^{\bullet} = 2(-396) - 2(-297) = -198 \text{ kJ}$; $\Delta S_{298}^{\bullet} = 2(257) - [205 + 2(248)] = -187 \text{ J/K}$

Set up a thermochemical cycle to convert to $T = 227°C = 500.$ K.

$2\ SO_2\ (g, 227°C) \rightarrow 2\ SO_2\ (g, 25°C)$	$\Delta H = nC_p\Delta T$
$O_2\ (g, 227°C) \rightarrow O_2\ (g, 25°C)$	$\Delta H = nC_p\Delta T$
$2\ SO_2\ (g, 25°C) + O_2\ (g, 25°C) \rightarrow 2\ SO_3\ (g, 25°C)$	$\Delta H° = -198$ kJ
$2\ SO_3\ (g, 25°C) \rightarrow 2\ SO_3\ (g, 227°C)$	$\Delta H = nC_p\Delta T$

$2\ SO_2\ (g, 227°C) + O_2\ (g, 227°C) \rightarrow 2\ SO_3\ (g, 227°C) \quad \Delta H°_{500} = ?$

$$\Delta H°_{500} = 2\ \text{mol} \times \frac{39.9\ \text{J mol}^{-1}\text{K}^{-1}}{1000\ \text{J/kJ}} \times (-202\ \text{K}) + 1 \times \frac{29.4}{1000} \times (-202) - 198\ \text{kJ} + 2 \times \frac{50.7}{1000} \times 202$$

$$\Delta H°_{500} = -16.1\ \text{kJ} - 5.94\ \text{kJ} - 198\ \text{kJ} + 20.5\ \text{kJ} = -199.5\ \text{kJ} = -200.\ \text{kJ}$$

For the same cycle using $\Delta S = nC_p \ln (T_2/T_1)$:

$$\Delta S°_{500} = 2(39.9) \ln (298\ \text{K}/500.\ \text{K}) + 1(29.4) \ln (298/500.) - 187 + 2(50.7) \ln (500./298)$$

$$\Delta S°_{500} = -41.3\ \text{J/K} - 15.2\ \text{J/K} - 187\ \text{J/K} + 52.5\ \text{J/K} = -191\ \text{J/K}$$

78.

$H_2O\ (l, 340.2\ \text{K}) \rightarrow H_2O\ (l, 373.2\ \text{K})$	$\Delta H = 2.49\ \text{kJ/mol} = C_p\Delta T/1000$
$H_2O\ (l, 373.2\ \text{K}) \rightarrow H_2O\ (g, 373.2\ \text{K})$	$\Delta H = 40.66\ \text{kJ/mol}$
$H_2O\ (g, 373.2\ \text{K}) \rightarrow H_2O\ (g, 340.2\ \text{K})$	$\Delta H = -1.20\ \text{kJ/mol} = C_p\Delta T/1000$

$H_2O\ (l, 340.2\ \text{K}) \rightarrow H_2O\ (g, 340.2\ \text{K}) \qquad \Delta H_{vap} = 41.95\ \text{kJ/mol}$

79. $\Delta G° = 2(-371) - [2(-300)] = -142\ \text{kJ}; \quad \Delta G = \Delta G° + RT \ln Q = -142\ \text{kJ} + RT \ln \left(\dfrac{P_{SO_3}^2}{P_{SO_2}^2 P_{O_2}} \right)$

$$\Delta G = -142\ \text{kJ} + \frac{(8.3145\ \text{J mol}^{-1}\text{K}^{-1})}{1000\ \text{J/kJ}}(298\ \text{K}) \ln \left(\frac{(10.0)^2}{(10.0)^2(10.0)} \right) = -148\ \text{kJ}$$

80. $H_2O\ (l, 298\ \text{K}) \rightarrow H_2O\ (g, V = 1000.\ \text{L/mol})$; Break process into 2 steps:

Step 1: $H_2O\ (l, 298\ \text{K}) \rightarrow H_2O\ (g, 298\ \text{K}, V = \dfrac{nR(298)}{1.00\ \text{atm}} = 24.5\ \text{L})$

$$\Delta S = S°_{H_2O(g)} - S°_{H_2O(l)} = 189\ \text{J/K} - 70\ \text{J/K} = 119\ \text{J/K}$$

Step 2: H_2O (g, 298 K, 24.5 L) → H_2O (g, 298 K, 1000. L)

$$\Delta S = nR \ln \frac{V_2}{V_1} = (1.00 \text{ mol})(8.3145 \text{ J mol}^{-1} \text{ K}^{-1}) \ln \left(\frac{1000. \text{ L}}{24.5 \text{ L}} \right) = 30.8 \text{ J/K}$$

$$\Delta S_{tot} = 119 + 30.8 = 150. \text{ J/K}$$

$$\Delta G = 44.02 \times 10^3 \text{ J} - 298 \text{ K} (150. \text{ J/K}) = -700 \text{ J}; \text{ Spontaneous}$$

For H_2O (l, 298 K) → H_2O (g, 298 K, V = 100. L/mol):

$$\Delta S = 119 \text{ J/K} + 8.3145 \text{ J/K} \ln \left(\frac{100. \text{ L}}{24.5 \text{ L}} \right) = 131 \text{ J/K}$$

$$\Delta G = 44.02 \times 10^3 \text{ J} - 298 \text{ K}(131 \text{ J/K}) = 5.0 \times 10^3 \text{ J}; \text{ Not spontaneous}$$

81.   Isothermal: $\Delta H = 0$ (assume ideal gas)

$$\Delta S = nR \ln \left(\frac{V_2}{V_1} \right) = (1.00 \text{ mol})(8.3145 \text{ J mol}^{-1} \text{ K}^{-1}) \ln \left(\frac{1.00 \text{ L}}{100.0 \text{ L}} \right) = -38.3 \text{ J/K}$$

$$\Delta G = \Delta H - T\Delta S = 0 - (300. \text{ K})(-38.3 \text{ J/K}) = +11,500 \text{ J} = 11.5 \text{ kJ}$$

82. $1.00 \, M \, \text{HCl} \rightarrow 0.100 \, M \, \text{HCl}$ $\Delta G° = 0; \Delta G = \Delta G° + RT \ln Q = RT \ln \dfrac{[H^+][Cl^-]}{[H^+][Cl^-]}$

$$\Delta G = (8.3145 \text{ J mol}^{-1} \text{ K}^{-1}) (298 \text{ K}) \ln \left(\frac{(0.100)^2}{(1.00)^2} \right) = -11,400 \text{ J} = -11.4 \text{ kJ}$$

The 0.1 M HCl is lower in free energy by 11.4 kJ.

83. $\ln K = \dfrac{-\Delta H°}{RT} + \dfrac{\Delta S°}{R}$,   R = 8.3145 J mol^{-1} K^{-1}

For K at two temperatures T_2 and T_1: $\ln \dfrac{K_2}{K_1} = \dfrac{\Delta H°}{R} \left(\dfrac{1}{T_1} - \dfrac{1}{T_2} \right)$

$\ln (10.0) = \dfrac{\Delta H°}{8.3145} \left(\dfrac{1}{300.0 \text{ K}} - \dfrac{1}{350.0 \text{ K}} \right)$, $2.30 = \dfrac{\Delta H°}{8.3145} (4.76 \times 10^{-4})$

$\Delta H° = 4.02 \times 10^4 \text{ J/mol} = 40.2 \text{ kJ/mol}$

84.

T(°C)	T(K)	C_p(J mol^{-1} K^{-1})	C_p/T (J mol^{-1} K^{-2})
-200.	73	12	0.16
-180.	93	15	0.16
-160.	113	17	0.15
-140.	133	19	0.14
-100.	173	24	0.14
-60.	213	29	0.14
-30.	243	33	0.14
-10.	263	36	0.14
0	273	37	0.14

Total area of C_p/T vs T plot = ΔS = I + II + III   (See plot)

ΔS = (0.16 J mol^{-1} K^{-2})(20. K) + (0.14 J mol^{-1} K^{-2})(180. K) + 1/2(0.02 J mol^{-1} K^{-2})(40. K)

ΔS = 3.2 + 25 + 0.4 = 29 J mol^{-1} K^{-1}

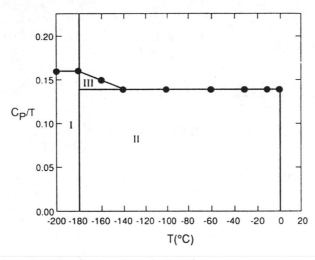

85.   We can set up 3 equations in 3 unknowns:

28.7262 = a + 300.0 b + c(300.0)^2

29.2937 = a + 400.0 b + c(400.0)^2

29.8545 = a + 500.0 b + c(500.0)^2

These can be solved by several methods. One way involves setting up a matrix and solving with a calculator such as:

$$\begin{pmatrix} 1 & 300.0 & 90,000 \\ 1 & 400.0 & 160,000 \\ 1 & 500.0 & 250,000 \end{pmatrix} \begin{pmatrix} a \\ b \\ c \end{pmatrix} = \begin{pmatrix} 28.7262 \\ 29.2937 \\ 29.8545 \end{pmatrix}$$

The solution is: $a = 26.98$; $b = 5.91 \times 10^{-3}$; $c = -3.4 \times 10^{-7}$

At 900. K: $C_p = 26.98 + 5.91 \times 10^{-3}(900.) - 3.4 \times 10^{-7}(900.)^2 = 32.02$ J K^{-1} mol^{-1}

$$\Delta S = n \int_{T_1}^{T_2} \frac{C_p dT}{T} = n \int_{T_1}^{T_2} \frac{(a + bT + cT^2)}{T} dT, \quad n = 1.00 \text{ mol}$$

$$\Delta S = a \int_{T_1}^{T_2} \frac{dT}{T} + b \int_{T_1}^{T_2} dT + c \int_{T_1}^{T_2} T dT = a \ln\left(\frac{T_2}{T_1}\right) + b(T_2 - T_1) + \frac{c(T_2^2 - T_1^2)}{2}$$

Solving using $T_2 = 900.$ K and $T_1 = 100.$ K: $\Delta S = 59.3 + 4.73 - 0.14 = 63.9$ J/K

86. Because of hydrogen bonding there is greater "structure" (or order) in liquid water than in most other liquids. Thus, there is a greater increase in disorder when water evaporates or a greater increase in entropy.

87. a. Isothermal: $\Delta E = 0$ and $\Delta H = 0$ if gas is ideal.

   $\Delta S = nR \ln(P_1/P_2) = (1.00 \text{ mol})(8.3145 \text{ J mol}^{-1} \text{ K}^{-1}) \ln(5.00/2.00) = 7.62$ J/K

   $T = \dfrac{PV}{nR} = \dfrac{5.00 \times 5.00}{1.00 \times 0.08206} = 305$ K; $\Delta G = \Delta H - T\Delta S = 0 - (305 \text{ K})(7.62 \text{ J/K}) = -2320$ J

   $w = -P\Delta V = -(2.00 \text{ atm})\Delta V$ where $V_f = \dfrac{nRT}{2.00 \text{ atm}}$ and $V_i = \dfrac{nRT}{5.00 \text{ atm}}$

   $w = -2.00 \text{ atm} \left(\dfrac{nRT}{2.00} - \dfrac{nRT}{5.00} \right) \text{L} \times (101.3 \text{ J L}^{-1} \text{ atm}^{-1}) = -1520 \text{ J} = -1500$ J

   $\Delta E = 0 = q + w$, $q = +1520 \text{ J} = 1500$ J

   b. Second law, $\Delta S_{univ} > 0$; $\Delta S_{univ} = \Delta S_{sys} + \Delta S_{surr} = \Delta S_{sys} - \dfrac{q_{actual}}{T}$

   $\Delta S_{univ} = 7.62 \text{ J/K} - \dfrac{1500 \text{ J}}{305 \text{ K}} = 7.62 - 4.9 = 2.7$ J/K; Thus, the process is spontaneous.

88. a. $\Delta G° = 2 \text{ mol}(-394 \text{ kJ/mol}) - 2 \text{ mol}(-137 \text{ kJ/mol}) = -514$ kJ

   $$K = \exp\left(\frac{-\Delta G°}{RT}\right) = \exp\left(\frac{+514{,}000 \text{ J}}{(8.3145 \text{ J mol}^{-1} \text{ K}^{-1})(298 \text{ K})}\right) = 1.24 \times 10^{90}$$

b. $\Delta H° = 2 (-393.5) - 2 (-110.5) = -566.0$ kJ; $\Delta S° = \Sigma$ S°(products) - Σ S°(reactants)

$$\Delta S° = \frac{\Delta H° - \Delta G°}{T} = \frac{-566.0 \text{ kJ} - (-514 \text{ kJ})}{298 \text{ K}} = - 0.174 \text{ kJ/K} = -174 \text{ J/K} \quad \text{(Carry extra S.F.)}$$

-174 J/K $= 2(214$ J/K$) - [S_{O_2}° + 2(198$ J/K$)]$, $S_{O_2}° = 206$ J mol^{-1} K$^{-1} = 210$ J mol^{-1} K^{-1}

c.   From Appendix 4, $\Delta S° = -173$ J/K

2 CO (1.00 atm) + O$_2$ (1.00 atm) → 2 CO$_2$ (1.00 atm)	$\Delta S° = -173$ J/K
2 CO$_2$ (1.00 atm) → 2 CO$_2$ (10.0 atm)	$\Delta S = nR \ln (P_1/P_2) = -38.3$ J/K
2 CO (10.0 atm) → 2 CO (1.00 atm)	$\Delta S = nR \ln (P_1/P_2) = 38.3$ J/K
O$_2$ (10.0 atm) → O$_2$ (1.00 atm)	$\Delta S = nR \ln (P_1/P_2) = 19.1$ J/K

2 CO (10.0 atm) + O$_2$ (10.0 atm) → CO$_2$ (10.0 atm)	$\Delta S = -173 + 19.1 = -154$ J/K

89.   a. $w_{rev} = -nRT \ln (V_2/V_1) = -(1 \text{ mol}) (8.3145 \text{ J mol}^{-1} \text{ K}^{-1}) (298 \text{ K}) \ln \left(\dfrac{20.0}{10.0} \right) = -1720$ J

For isothermal expansion: $\Delta E = 0$, so $q = +1720$ J

b. $w = -P\Delta V = -1.23$ atm $(20.0$ L $- 10.0$ L$) = -12.3$ L atm

-12.3 L atm $\times 101.3$ J L^{-1} atm$^{-1} = -1250$ J

$\Delta E = 0$ for isothermal expansion, so $q = 1250$ J.

90. ΔS is more favorable for reaction two than reaction one. In reaction one, seven particles in solution are forming one particle in solution. In reaction two, four particles are forming one which results in a smaller decrease in disorder than for reaction one.

91.   The hydration of the dissociated ions results in a more ordered "structure" in solution. The smallest ion (F$^-$) has the largest charge density and should show the most ordering.

92.   a. $\Delta G = G_B° - G_A° = 11,718 - 8996 = 2722$ J

$$K = \exp\left(\frac{-\Delta G°}{RT} \right) = \exp\left(\frac{-2722 \text{ J}}{(8.3145 \text{ J mol}^{-1} \text{ K}^{-1}) (298 \text{ K})} \right) = 0.333$$

b.   Since $Q = 1.00 > K$, reaction shifts left.

$$A(g) \quad \rightleftharpoons \quad B(g)$$

	A(g)	B(g)
Initial	1.00 atm	1.00 atm
Equil.	1.00 + x	1.00 - x

$$\frac{1.00 - x}{1.00 + x} = 0.333, \ 1.00 - x = 0.333 + 0.333\,x, \ x = 0.50 \text{ atm}$$

$$P_B = 1.00 - 0.50 = 0.50 \text{ atm}; \ P_A = 1.00 + 0.50 = 1.50 \text{ atm}$$

c. $\Delta G = \Delta G° + RT \ln Q = \Delta G° + RT \ln (P_B/P_A)$

$$\Delta G = 2722 \text{ J} + (8.3145)(298) \ln (0.50/1.50) = 2722 \text{ J} - 2722 \text{ J} = 0$$

93.   At equilibrium:

$$P_{H_2} = \frac{nRT}{V} = \frac{\left(\dfrac{1.10 \times 10^{13} \text{ molecules}}{6.022 \times 10^{23} \text{ molecules/mol}}\right)\left(\dfrac{0.08206 \text{ L atm}}{\text{mol K}}\right)(298 \text{ K})}{1.00 \text{ L}} = 4.47 \times 10^{-10} \text{ atm}$$

Essentially all of the H_2 and Br_2 has reacted. $P_{HBr} = 2.00$ atm

Since we began with the same amount of H_2 and Br_2, at equilibrium $P_{H_2} = P_{Br_2} = 4.47 \times 10^{-10}$ atm.

$$K = \frac{P_{HBr}^2}{P_{H_2} P_{Br_2}} = \frac{(2.00)^2}{(4.47 \times 10^{-10})^2} = 2.00 \times 10^{19} \quad \text{Assumptions good.}$$

$\Delta G° = -RT \ln K = -(8.3145 \text{ J mol}^{-1} \text{ K}^{-1})(298 \text{ K}) \ln (2.00 \times 10^{19}) = -1.10 \times 10^5 \text{ J/mol}$

$$\Delta S° = \frac{\Delta H° - \Delta G°}{T} = \frac{-103{,}800 \text{ J} - (-1.10 \times 10^5 \text{ J})}{298 \text{ K}} = 20 \text{ J/K}$$

94.   Isothermal: $\Delta E = 0$, $\Delta H = 0$; $V_f = nRT/5.00 \text{ atm} = 2.00$ L

$\Delta S = nR \ln (P_1/P_2) = (1.00 \text{ mol})(8.3145 \text{ J mol}^{-1} \text{ K}^{-1}) \ln (1.50/5.00) = -10.0 \text{ J/K}$

$w = -5.00 \text{ atm}(2.00 \text{ L} - 6.67 \text{ L})(101.3 \text{ J L}^{-1} \text{ atm}^{-1}) = 2370 \text{ J}$

$\Delta E = 0 = q + w$, $q = -2370 \text{ J}$; $\Delta S_{surr} = \dfrac{-q}{T} = \dfrac{2370 \text{ J}}{122 \text{ K}} = 19.4 \text{ J/K}$

$\Delta S_{univ} = \Delta S_{sys} + \Delta S_{surr} = -10.0 \text{ J/K} + 19.4 \text{ J/K} = 9.4 \text{ J/K}$

$\Delta G = \Delta H - T\Delta S = 0 - (122 \text{ K})(-10.0 \text{ J/K}) = 1220 \text{ J}$

95. $P_1 V_1 = P_2 V_2$, $P_2 = \dfrac{P_1 V_1}{V_2} = \dfrac{(5.0)(1.0)}{(2.0)} = 2.5$ atm

Gas expands isothermally against no pressure, so $\Delta E = 0$, $w = 0$ and $q = 0$.

$\Delta E = 0$, so $q_{rev} = -w_{rev} = nRT \ln (V_2/V_1)$; $T = \dfrac{PV}{nR} = 61$ K

$q_{rev} = (1.0 \text{ mol})(8.3145 \text{ J mol}^{-1} \text{ K}^{-1})(61 \text{ K}) \ln (2.0/1.0) = 350$ J

96.   For A(l, 125°C) → A(l, 75°C):

$\Delta S = nC_p \ln (T_2/T_1) = 1.00 \text{ mol } (75.0 \text{ J K}^{-1} \text{ mol}^{-1}) \ln (348 \text{ K}/398 \text{ K}) = -10.1$ J/K

For A(l, 75°C) → A(g, 155°C): $\Delta S = 75.0 \text{ J K}^{-1} \text{ mol}^{-1}$

For A(g, 155°C) → A(g,125°C):

$\Delta S = nC_p \ln (T_2/T_1) = 1.00 \text{ mol } (29.0 \text{ J K}^{-1} \text{ mol}^{-1}) \ln (398 \text{ K}/428 \text{ K}) = -2.11$ J/K

The sum of the three step gives: A(l, 125°C) → A(g, 125°C)

$\Delta S = -10.1 + 75.0 - 2.11 = 62.8$ J/K

At 125°C: $\Delta G = 0 = \Delta H - T\Delta S$, $\Delta H_{vap} = T\Delta S = 398 \text{ K } (62.8 \text{ J/K}) = 2.50 \times 10^4$ J

97.   Heat gain by He = Heat loss by N_2

$(0.400 \text{ mol})(12.5 \text{ J °C}^{-1} \text{ mol}^{-1})(T_f - 20.0 \text{ °C}) = (0.600 \text{ mol})(20.7 \text{ J °C}^{-1} \text{ mol}^{-1})(100.0 \text{ °C} - T_f)$

$5.00 \text{ T}_f - 100. = 1240 - 12.4 \text{ T}_f$, $T_f = \dfrac{1340}{17.4} = 77.0$ °C

98.   We would expect the solubility to increase for salts with an endothermic heat of solution. $\Delta H°$ refers to a solution with one molar concentration. Solubility deals with a saturated solution. In a saturated solution the concentrations are not one molar and the sign for ΔH_{sol} may be different.

99. $K_p = P_{CO_2}$; To keep Ag_2CO_3 from decomposing, P_{CO_2} must be greater than or equal to K_p.

From Exercise 10.47, $\ln K = \dfrac{-\Delta H°}{RT} + \dfrac{\Delta S°}{R}$. For two conditions of K and T, the equation reduces to:

$$\ln(K_2/K_1) = \dfrac{\Delta H°}{R} \left(\dfrac{1}{T_1} - \dfrac{1}{T_2} \right)$$

Let $T_1 = 25°C = 298$ K, $K_1 = 6.23 \times 10^{-3}$ torr; $T_2 = 110.°C = 383$ K, $K_2 = ?$

$$\ln K_2 - \ln (6.23 \times 10^{-3} \text{ torr}) = \dfrac{79.14 \times 10^3 \text{ J/mol}}{8.3145 \text{ J K}^{-1} \text{ mol}^{-1}} \left(\dfrac{1}{298 \text{ K}} - \dfrac{1}{383 \text{ K}} \right)$$

$\ln K_2 = -5.08 + 7.1 = 2.0$, $K_2 = e^{2.0} = 7.4$ torr

To prevent decomposition of Ag_2CO_3, the partial pressure of CO_2 must be equal to or **greater** than 7.4 torr.

100.   For a phase change, $\Delta S_{sys} = \dfrac{q_{rev}}{T} = \dfrac{\Delta H}{T}$ where ΔH is determined at the melting point (5.5°C).

$C_6H_6(l, 25.0°C) \rightarrow C_6H_6 (s, 25.0°C)$	$\Delta H = -10.04$ kJ
$C_6H_6(l, 5.5°C) \rightarrow C_6H_6 (l, 25.0°C)$	$\Delta H = nC_p\Delta T/1000 = 2.59$ kJ
$C_6H_6(s, 25.0°C) \rightarrow C_6H_6 (s, 5.5°C)$	$\Delta H = nC_p\Delta T/1000 = -1.96$ kJ

$C_6H_6(l, 5.5°C) \rightarrow C_6H_6 (s, 5.5°C)$	$\Delta H = -9.41$ kJ

At the melting point, $\Delta S_{sys} = \dfrac{-9.41 \times 10^3 \text{ J}}{278.7 \text{ K}} = -33.8$ J/K

For the phase change at 10.0°C (283.2 K):

$C_6H_6(l, 278.7 \text{ K}) \rightarrow C_6H_6 (s, 278.7 \text{ K})$	$\Delta S = -33.8$ J/K
$C_6H_6(l, 283.2 \text{ K}) \rightarrow C_6H_6 (l, 278.7 \text{ K})$	$\Delta S = nC_p \ln (T_2/T_1) = -2.130$ J/K
$C_6H_6(s, 278.7 \text{ K}) \rightarrow C_6H_6 (s, 283.2 \text{ K})$	$\Delta S = nC_p \ln (T_2/T_1) = 1.608$ J/K

$C_6H_6(l, 283.2 \text{ K}) \rightarrow C_6H_6 (s, 283.2 \text{ K})$	$\Delta S_{sys} = -34.3$ J/K

To calculate ΔS_{surr}, we need ΔH_{sys} at 10.0°C ($\Delta S_{surr} = \dfrac{-\Delta H_{sys}}{T}$).

$C_6H_6(l, 25.0°C) \rightarrow C_6H_6 (s, 25.0°C)$	$\Delta H = -10.04$ kJ
$C_6H_6(l, 10.0°C) \rightarrow C_6H_6 (l, 25.0°C)$	$\Delta H = nC_p\Delta T/1000 = 2.00$ kJ
$C_6H_6(s, 25.0°C) \rightarrow C_6H_6 (s, 10.0°C)$	$\Delta H = nC_p\Delta T/1000 = -1.51$ kJ

$C_6H_6(l, 10.0°C) \rightarrow C_6H_6 (s, 10.0°C)$	$\Delta H_{sys} = -9.55$ kJ

$\Delta S_{surr} = \dfrac{-\Delta H_{sys}}{T} = \dfrac{-(-9.55 \times 10^3 \text{ J})}{283.2 \text{ K}} = 33.7$ J/K

101.   a. $\Delta S°$ will be negative since there is a decrease in the number of moles of gas.

b.   Since $\Delta S°$ is negative, $\Delta H°$ must be negative for the reaction to be spontaneous at some temperatures. Therefore, ΔS_{surr} is positive.

c.

	Ni(s)	+	4 CO(g)	$\rightleftharpoons$	$Ni(CO)_4(g)$
$\Delta H°$	0		-110.5		-607 kJ/mol
$S°$	30		198		417 J K^{-1} mol^{-1}

$\Delta H° = -607 - [4(-110.5)] = -165$ kJ; $\Delta S° = 417 - [4(198) + (30)] = -405$ J/K

d. $\Delta G° = 0 = \Delta H° - T\Delta S°$; $T = \dfrac{\Delta H°}{\Delta S°} = \dfrac{-165 \times 10^3 \text{ J}}{-405 \text{ J/K}} = 407$ K or 134°C

e. $T = 50.°C + 273 = 323$ K

$\Delta G°_{323} = -165$ kJ $- (323$ K$)(-0.405$ kJ/K$) = -34$ kJ

$\ln K = \dfrac{-\Delta G°}{RT} = \dfrac{-(-34,000 \text{ J})}{(8.3145 \text{ J K}^{-1} \text{ mol}^{-1})(323 \text{ K})} = 12.66$, $K = e^{12.66} = 3.1 \times 10^5$

f. $T = 227°C + 273 = 500.$ K

$\Delta G°_{500} = -165$ kJ $- (500.$ K$)(-0.405$ kJ/K$) = 38$ kJ

$\ln K = \dfrac{-38,000}{(8.3145)(500.)} = -9.14$, $K = e^{-9.14} = 1.1 \times 10^{-4}$

g.   The temperature change causes the value of the equilibrium constant to change from a large value favoring formation of $Ni(CO)_4$ to a small value favoring the decomposition of $Ni(CO)_4$ to Ni and CO.

h. $Ni(CO)_4(l) \rightleftharpoons Ni(CO)_4(g)$ $K_p = P_{Ni(CO)_4}$

At 42°C (the boiling point): $\Delta G° = 0 = \Delta H° - T\Delta S°$

$\Delta S° = \dfrac{\Delta H°}{T} = \dfrac{29.0 \times 10^3 \text{ J}}{315 \text{ K}} = 92.1$ J/K

At 152°C: $\Delta G°_{152} = \Delta H° - T\Delta S° = 29.0 \times 10^3$ J $- 425$ K $(92.1$ J/K$) = -10,100$ J

$\Delta G° = -RT \ln K_p$, $\ln K_p = \dfrac{10,100 \text{ J}}{8.3145(425 \text{ K})} = 2.86$, $K_p = e^{2.86} = 17$ atm

A maximum pressure of 17 atm can be attained before $Ni(CO)_4(g)$ will liquify.

CHAPTER ELEVEN

ELECTROCHEMISTRY

Galvanic Cells

1. In a galvanic cell a spontaneous reaction occurs, producing an electric current. In an electrolytic cell electricity is used to force a reaction to occur that is not spontaneous.

2. The salt bridge completes the electrical circuit and allows counter ions to flow into the two cell compartments to maintain electrical neutrality. It also separates the cathode and anode compartments such that electrical energy can be extracted. If the cathode and anode compartments are not separated the cell would rapidly go to equilibrium without any useful work being done.

3.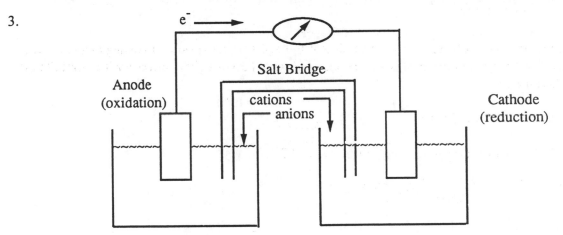

    The diagram for all cells will look like this. The contents of each half-cell will be identified for each reaction, with all concentrations at 1.0 M and partial pressures at 1.0 atm.

    a. $(Cl_2 + 2 e^- \rightarrow 2 Cl^-) \times 3$ $E° = 1.36$ V
        $7 H_2O + 2 Cr^{3+} \rightarrow Cr_2O_7^{2-} + 14 H^+ + 6 e^-$ $-E° = -1.33$ V

        $7 H_2O(l) + 2 Cr^{3+}(aq) + 3 Cl_2(g) \rightarrow Cr_2O_7^{2-}(aq) + 6 Cl^-(aq) + 14 H^+(aq)$ $E°_{cell} = 0.03$ V

        Cathode: Pt electrode; Cl_2 bubbled into solution, Cl^- in solution

        Anode: Pt electrode; Cr^{3+}, H^+, and $Cr_2O_7^{2-}$ in solution

We need a nonreactive metal to use as the electrode in each case, since all of the reactants and products are in solution. Pt is the most common choice. Other possibilities are gold and graphite.

b. $Cu^{2+} + 2\,e^- \rightarrow Cu$ $E° = 0.34$ V
                        $Mg \rightarrow Mg^{2+} + 2e^-$ $-E° = 2.37$ V
_____

       $Cu^{2+}(aq) + Mg(s) \rightarrow Cu(s) + Mg^{2+}(aq)$ $E°_{cell} = 2.71$ V

Cathode: Cu electrode; Cu^{2+} in solution

Anode: Mg electrode; Mg^{2+} in solution

c. $5\,e^- + 6\,H^+ + IO_3^- \rightarrow 1/2\,I_2 + 3\,H_2O$ $E° = 1.20$ V
                    $(Fe^{2+} \rightarrow Fe^{3+} + e^-) \times 5$ $-E° = -0.77$ V
_____

       $6\,H^+ + IO_3^- + 5\,Fe^{2+} \rightarrow 5\,Fe^{3+} + 1/2\,I_2 + 3\,H_2O$ $E°_{cell} = 0.43$ V

or $12\,H^+(aq) + 2\,IO_3^-(aq) + 10\,Fe^{2+}(aq) \rightarrow 10\,Fe^{3+}(aq) + I_2(s) + 6\,H_2O(l)$ $E°_{cell} = 0.43$ V

Cathode: Pt electrode; IO_3^-, I_2 and H_2SO_4 (H^+ source) in solution.

Anode: Pt electrode; Fe^{2+} and Fe^{3+} in solution

d. $(Ag^+ + e^- \rightarrow Ag) \times 2$ $E° = 0.80$ V
                        $Zn \rightarrow Zn^{2+} + 2\,e^-$ $-E° = 0.76$ V
_____

       $Zn(s) + 2\,Ag^+(aq) \rightarrow 2\,Ag(s) + Zn^{2+}(aq)$ $E° = 1.56$ V

Cathode: Ag electrode; Ag^+ in solution

Anode: Zn electrode; Zn^{2+} in solution

4.   The diagram will look like that for Exercise 11.3. We get the reaction by remembering that the reduction occurs at the cathode and oxidation occurs at the anode.

a. $Cl_2 + 2\,e^- \rightarrow 2\,Cl^-$
                        $2\,Br^- \rightarrow Br_2 + 2\,e^-$
_____

       $Cl_2(g) + 2\,Br^-(aq) \rightarrow Br_2(aq) + 2\,Cl^-(aq)$

Cathode: Pt electrode; $Cl_2(g)$ bubbled in, Cl^- in solution

Anode: Pt electrode; Br_2 and Br^- in solution

b. $(2\,e^- + 2\,H^+ + IO_4^- \rightarrow IO_3^- + H_2O) \times 5$
                    $(4\,H_2O + Mn^{2+} \rightarrow MnO_4^- + 8\,H^+ + 5\,e^-) \times 2$
_____

   $10\,H^+ + 5\,IO_4^- + 8\,H_2O + 2\,Mn^{2+} \rightarrow 5\,IO_3^- + 5\,H_2O + 2\,MnO_4^- + 16\,H^+$

This simplifies to:

$$3 H_2O(l) + 5 IO_4^-(aq) + 2 Mn^{2+}(aq) \rightarrow 5 IO_3^-(aq) + 2 MnO_4^-(aq) + 6 H^+(aq)$$

Cathode: Pt electrode; IO_4^-, IO_3^-, and H_2SO_4 (as a source of H^+) in solution

Anode: Pt electrode; Mn^{2+}, MnO_4^- and H_2SO_4 in solution

c. $(Ni^{2+} + 2 e^- \rightarrow Ni) \times 3$
            $(Al \rightarrow Al^{3+} + 3 e^-) \times 2$
            _____

$$3 Ni^{2+}(aq) + 2 Al(s) \rightarrow 2 Al^{3+}(aq) + 3 Ni(s)$$

Cathode: Ni electrode; Ni^{2+} in solution

Anode: Al electrode; Al^{3+} in solution

d. $Co^{3+} + e^- \rightarrow Co^{2+}$
            $Fe^{2+} \rightarrow Fe^{3+} + e^-$
            _____

$$Co^{3+}(aq) + Fe^{2+}(aq) \rightarrow Fe^{3+}(aq) + Co^{2+}(aq)$$

Cathode: Pt electrode; Co^{3+} and Co^{2+} in solution

Anode: Pt electrode; Fe^{2+} and Fe^{3+} in solution

Cell Potential, Standard Reduction Potentials, and Free Energy

5.   The half-reaction for the SCE is:

$$Hg_2Cl_2 + 2 e^- \rightarrow 2 Hg + 2 Cl^- \qquad E_{SCE} = +0.242 \text{ V}$$

For a spontaneous reaction to occur, E_{cell} must be positive. Using the standard reduction potentials in Table 11.1 and the given SCE potential, deduce which combination will produce a positive overall cell potential.

a. $Cu^{2+} + 2 e^- \rightarrow Cu \qquad E° = 0.34 \text{ V}$

$E_{cell} = 0.34 - 0.242 = 0.10 \text{ V}$; SCE is the anode.

b. $Fe^{3+} + e^- \rightarrow Fe^{2+} \qquad E° = 0.77 \text{ V}$

$E_{cell} = 0.77 - 0.242 = 0.53 \text{ V}$; SCE is the anode.

c. $AgCl + e^- \rightarrow Ag + Cl^-$ $E° = 0.22$ V

$E_{cell} = 0.242 - 0.22 = 0.02$ V; SCE is the cathode.

d. $Al^{3+} + 3 e^- \rightarrow Al$ $E° = -1.66$ V

$E_{cell} = 0.242 + 1.66 = 1.90$ V; SCE is the cathode.

e. $Ni^{2+} + 2 e^- \rightarrow Ni$ $E° = -0.23$ V

$E_{cell} = 0.242 + 0.23 = 0.47$ V; SCE is the cathode.

6.   a. $2 Ag^+ + 2 e^- \rightarrow 2 Ag$ $E° = 0.80$ V
              $Cu \rightarrow Cu^{2+} + 2 e^-$ $-E° = -0.34$ V

$$\overline{2 Ag^+ + Cu \rightarrow Cu^{2+} + 2 Ag \qquad E^{\circ}_{cell} = 0.46 \text{ V}}$$ Spontaneous ($E^{\circ}_{cell} > 0$)

b. $Zn^{2+} + 2 e^- \rightarrow Zn$ $E° = -0.76$ V
      $Ni \rightarrow Ni^{2+} + 2 e^-$ $-E° = +0.23$ V

$$\overline{Zn^{2+} + Ni \rightarrow Zn + Ni^{2+} \qquad E^{\circ}_{cell} = -0.53 \text{ V}}$$ Not spontaneous ($E^{\circ}_{cell} < 0$)

c. $(5 e^- + 8 H^+ + MnO_4^- \rightarrow Mn^{2+} + 4 H_2O) \times 2$ $E° = 1.51$ V
                    $(2 I^- \rightarrow I_2 + 2 e^-) \times 5$ $-E° = -0.54$ V

$$\overline{16 H^+ + 2 MnO_4^- + 10 I^- \rightarrow 5 I_2 + 2 Mn^{2+} + 8 H_2O \qquad E^{\circ}_{cell} = 0.97 \text{ V}}$$ Spontaneous

d. $(5 e^- + 8 H^+ + MnO_4^- \rightarrow Mn^{2+} + 4 H_2O) \times 2$ $E° = 1.51$ V
                    $(2 F^- \rightarrow F_2 + 2 e^-) \times 5$ $-E° = -2.87$ V

$$\overline{16 H^+ + 2 MnO_4^- + 10 F^- \rightarrow 5 F_2 + 2 Mn^{2+} + 8 H_2O \qquad E^{\circ}_{cell} = -1.36 \text{ V}}$$ Not spontaneous

7.   a. $2 H^+ + 2 e^- \rightarrow H_2$ $E° = 0.00$ V; $Cu \rightarrow Cu^{2+} + 2 e^-$ $-E° = -0.34$ V

$E^{\circ}_{cell} = -0.34$ V; No, H^+ cannot oxidize Cu to Cu^{2+} at standard conditions ($E^{\circ}_{cell} < 0$).

b. $2 H^+ + 2 e^- \rightarrow H_2$ $E° = 0.00$ V; $Mg \rightarrow Mg^{2+} + 2 e^-$ $-E° = -(-2.37$ V$)$

$E^{\circ}_{cell} = 2.37$ V; Yes, H^+ can oxidize Mg to Mg^{2+}.

c. $Fe^{3+} + e^- \rightarrow Fe^{2+}$ $E° = 0.77$ V; $2 I^- \rightarrow I_2 + 2 e^-$ $-E° = -0.54$ V

$E^{\circ}_{cell} = 0.77 - 0.54 = 0.23$ V; Yes, Fe^{3+} can oxidize I^- to I_2.

d. $Fe^{3+} + e^- \rightarrow Fe^{2+}$ $E° = 0.77$ V; $2 Br^- \rightarrow Br_2 + 2 e^-$ $-E° = -1.09$ V

No, for: $2 Fe^{3+} + 2 Br^- \rightarrow Br_2 + 2 Fe^{2+}$ $E°_{cell} = -0.32$ V; The reaction is not spontaneous.

8.   a. $H_2 \rightarrow 2 H^+ + 2 e^-$ $-E° = 0.00$ V; $Ag^+ + e^- \rightarrow Ag$ $E° = 0.80$ V

$E°_{cell} = 0.80$ V; Yes, H_2 can reduce Ag^+ to Ag at standard conditions ($E°_{cell} > 0$).

b. $E°_{cell} = -0.23$ V; No, H_2 cannot reduce Ni^{2+} to Ni.

c. $Fe^{2+} \rightarrow Fe^{3+} + e^-$ $-E° = -0.77$ V; $VO_2^+ + 2 H^+ + e^- \rightarrow VO^{2+} + H_2O$ $E° = 1.00$ V

$E°_{cell} = 0.23$ V; Yes

d. $Fe^{2+} \rightarrow Fe^{3+} + e^-$ $-E° = -0.77$ V; $Cr^{3+} + e^- \rightarrow Cr^{2+}$ $E° = -0.50$ V

$E°_{cell} = -1.27$ V; No

e. $Fe^{2+} \rightarrow Fe^{3+} + e^-$ $-E° = -0.77$; $Cr^{3+} + 3 e^- \rightarrow Cr$ $E° = -0.73$

$E°_{cell} = -1.50$ V; No

f. $Fe^{2+} \rightarrow Fe^{3+} + e^-$ $-E° = -0.77$; $Sn^{2+} + 2 e^- \rightarrow Sn$ $E° = -0.14$

$E°_{cell} = -0.91$; No

9.   Good oxidizing agents are easily reduced. Oxidizing agents are on the left side of the reduction half-reactions listed in Table 11.1. We look for the largest, most positive E° values to correspond to the best oxidizing agents.

$$MnO_4^- > Cl_2 > Cr_2O_7^{2-} > Fe^{3+} > Fe^{2+} > Mg^{2+}$$

E°   1.68   1.36   1.33   0.77   -0.44   -2.37

10.   Good reducing agents are easily oxidized. The reducing agents are on the right side of the reduction half-reactions listed in Table 11.1. The best reducing agents have the most negative standard reduction potentials.

$$Li > Zn > H_2 > Fe^{2+} > Cr^{3+} > F^-$$

E°   -3.05   -0.76   0.00   0.77   1.33   2.87

11. $Br_2 + 2 e^- \rightarrow 2 Br^-$ $E° = 1.09$ V $2 H^+ + 2 e^- \rightarrow H_2$ $E° = 0.00$ V
       $Cd^{2+} + 2 e^- \rightarrow Cd$ $E° = -0.40$ V $La^{3+} + 3 e^- \rightarrow La$ $E° = -2.37$ V
       $Ca^{2+} + 2 e^- \rightarrow Ca$ $E° = -2.76$ V

a. Oxidizing agents are on the left side of the reduction half-reactions. Br_2 is the best oxidizing agent (largest E°).

b. Reducing agents are on the right side of the reduction half-reactions. Ca is the best reducing agent (most negative E°).

c. $MnO_4^- + 8 H^+ + 5 e^- \rightarrow Mn^{2+} + 4 H_2O$ E° = 1.51 V

Permanganate can oxidize Br^-, H_2, Ca and Cd. When MnO_4^- is coupled with these reagents, E°_{cell} is positive.

d. $Zn \rightarrow Zn^{2+} + 2 e^-$ -E° = 0.76 V;   Zinc can reduce Br_2 and H^+ since $E^\circ_{cell} > 0$.

12.
$Ce^{4+} + e^- \rightarrow Ce^{3+}$	E° = 1.70 V	$Fe^{3+} + e^- \rightarrow Fe^{2+}$	E° = 0.77 V
$Fe^{3+} + 3 e^- \rightarrow Fe$	E° = -0.036 V	$Sn^{2+} + 2 e^- \rightarrow Sn$	E° = -0.14 V
$Ni^{2+} + 2 e^- \rightarrow Ni$	E° = -0.23 V	$Fe^{2+} + 2 e^- \rightarrow Fe$	E° = -0.44 V
$Mg^{2+} + 2 e^- \rightarrow Mg$	E° = -2.37 V		

a. Ce^{4+} is the strongest oxidizing agent (largest E°).

b. Mg is the strongest reducing agent (largest -E°).

c. Yes, Ce^{4+} will oxidize Fe to Fe^{3+} ($E^\circ_{cell} > 0$).

d. Fe, Sn, and Mg can be oxidized by H^+ ($E^\circ_{cell} > 0$).

e. Ce^{4+} and Fe^{3+} can be reduced by H_2 ($E^\circ_{cell} > 0$).

f. Sn and Mg will reduce Fe^{3+} to Fe^{2+} ($E^\circ_{cell} > 0$). Sn and Mg should also reduce Fe^{3+} to Fe. However, concentrations might be adjusted to prevent this from happening. It is easier to adjust concentrations using Sn since E°_{cell} with Sn is closer to zero. Therefore, Sn is the preferred reducing agent.

13.   a. E° > 0.80 V, oxidize Hg; E° < 0.91 V, not oxidize Hg_2^{2+}; No half-reaction in this table fits this requirement. However, by changing concentrations, Ag^+ may be able to work.

b. E° > 1.09 V, oxidize Br^-; E° < 1.36 V, not oxidize Cl^-; $Cr_2O_7^{2-}$, O_2, MnO_2, and IO_3^- are all possible.

c. $Ni^{2+} + 2 e^- \rightarrow Ni$ E° = -0.23 V; $Mn^{2+} + 2 e^- \rightarrow Mn$ E° = -1.18 V; Any oxidizing agent with -0.23 V > E° > -1.18 V will work. $PbSO_4$, Cd^{2+}, Fe^{2+}, Cr^{3+}, Zn^{2+} and H_2O will be able to do this.

14.   a. E° < 0.77 V, reduce Fe^{3+} to Fe^{2+}; E° > -0.44 V, not reduce Fe^{2+} to Fe; H_2O_2, MnO_4^{2-}, I^-, Cu, OH^-, Hg (in 1 M Cl^-), Ag (in 1 M Cl^-), H_2SO_3, H_2, Pb, Sn, Ni, and Cd are all capable of this.

b. $Ag^+ + e^- \rightarrow Ag$ $E° = 0.80$ V; $O_2 + 2 H^+ + 2 e^- \rightarrow H_2O_2$ $E° = 0.68$ V; Fe^{2+} will reduce Ag^+
to Ag but will not reduce O_2 to H_2O_2. A potential of -0.77 V is required to oxidize Fe^{2+} to
Fe^{3+}, which is large enough to oxidize Ag^+ but small enough not to oxidize O_2.

15.   Reduce I_2 to I^-, $E° < 0.54$ V; Not reduce Cu^{2+} to Cu, $E° > 0.34$ V; Any reducing agent with a
reduction potential between 0.34 V and 0.54 V will work.

16.   a. $Sn^{4+} + 2 e^- \rightarrow Sn^{2+}$ $E° = 0.15$ V; $Sn^{2+} + 2 e^- \rightarrow Sn$ $E° = -0.14$ V; Need 0.15 V $> E° > -$
0.14 V; H_2, Fe, and Pb are all possible.

b.   Look for oxidizing agents in the same range. H^+, Fe^{3+}, and Pb^{2+} are all possible.

17. $H_2O_2 + 2 H^+ + 2 e^- \rightarrow 2 H_2O$ $E° = 1.78$ V; H_2O_2 is the oxidizing agent.
      $O_2 + 2 H^+ + 2 e^- \rightarrow H_2O_2$ $E° = 0.68$ V; H_2O_2 is the reducing agent.

$$H_2O_2 + 2 H^+ + 2 e^- \rightarrow 2 H_2O \qquad\qquad E° = 1.78 \text{ V}$$
$$H_2O_2 \rightarrow O_2 + 2 H^+ + 2 e^- \qquad\qquad -E° = -0.68 \text{ V}$$

$$2 H_2O_2 \rightarrow 2 H_2O + O_2 \qquad\qquad E°_{cell} = 1.10 \text{ V}$$

18. $2 Cu^+ \rightarrow Cu + Cu^{2+}$; $E°_{cell} = 0.52$ V - 0.16 V = 0.36 V

$$\Delta G° = -nFE° = -(1 \text{ mol } e^-)(96{,}485 \text{ C/mol } e^-)(0.36 \text{ V}) = -34{,}700 \text{ C V} = -34.7 \text{ kJ} = -35 \text{ kJ}$$

$$\log K = \frac{nE°}{0.0592} = \frac{0.36}{0.0592} = 6.08; \quad K = 10^{6.08} = 1.2 \times 10^6$$

Note: When determining exponents, we will round-off when the calculation is complete.

19.   a. $(ClO_2^- \rightarrow ClO_2 + e^-) \times 2$ $-E° = -0.954$ V
             $Cl_2 + 2 e^- \rightarrow 2 Cl^-$ $E° = 1.36$ V

$$2 ClO_2^- + Cl_2 \rightarrow 2 ClO_2 + 2 Cl^- \qquad E°_{cell} = 0.41 \text{ V} = 0.41 \text{ J/C}$$

$$\Delta G° = -nFE° = -(2 \text{ mol } e^-)(96{,}485 \text{ C/mol } e^-)(0.41 \text{ J/C}) = -7.91 \times 10^4 \text{ J} = -79 \text{ kJ}$$

$$\Delta G° = -RT \ln K \text{ so } K = \exp(-\Delta G°/RT)$$

$$K = \exp[(7.9 \times 10^4 \text{ J})/(8.3145 \text{ J mol}^{-1} \text{ K}^{-1})(298 \text{ K})] = 7.0 \times 10^{13}$$

or $\log K = \dfrac{nE°}{0.0592} = \dfrac{2(0.41)}{0.0592} = 13.85$, $K = 10^{13.85} = 7.1 \times 10^{13}$

b. $(H_2O + ClO_2 \rightarrow ClO_3^- + 2 H^+ + e^-) \times 5$
        $5 e^- + 4 H^+ + ClO_2 \rightarrow Cl^- + 2 H_2O$

$$3 H_2O(l) + 6 ClO_2(g) \rightarrow 5 ClO_3^-(aq) + Cl^-(aq) + 6 H^+(aq)$$

20.   Consider the strongest oxidizing agent combined with the strongest reducing agent:

$$F_2 + 2\ e^- \rightarrow 2\ F^- \qquad\qquad E° = 2.87\ V$$
$$(Li \rightarrow Li^+ + e^-) \times 2 \qquad -E° = +3.05\ V$$

$$\overline{F_2 + 2\ Li \rightarrow 2\ Li^+ + 2\ F^- \qquad E°_{cell} = 5.92\ V}$$

The claim is impossible. The strongest oxidizing agent and reducing agent when combined only give E° of about 6 V.

21.   a.   Possible reaction: $I_2 + 2\ Cl^- \rightarrow 2\ I^- + Cl_2$ $E°_{cell} = 0.54\ V - 1.36\ V = -0.82\ V$; This reaction is not spontaneous. No reaction occurs.

        b.   Possible reaction: $Cl_2 + 2\ I^- \rightarrow I_2 + 2\ Cl^-$ $E°_{cell} = 0.82\ V$; This reaction is spontaneous. The reaction will occur.

        c.   Possible reaction: $2\ Ag + Cu^{2+} \rightarrow Cu + 2\ Ag^+$ $E°_{cell} = -0.46\ V$; No reaction occurs.

        d.   Possible reaction: $Pb + Cu^{2+} \rightarrow Cu + Pb^{2+}$ $E°_{cell} = +0.47$; spontaneous

        e.   Possible reaction: $4\ Fe^{2+} + 4\ H^+ + O_2 \rightarrow 4\ Fe^{3+} + 2\ H_2O$ $E°_{cell} = +0.46\ V$; spontaneous

22.   b. $Cl_2 + 2\ I^- \rightarrow I_2 + 2\ Cl^-$ $E°_{cell} = +0.82\ V = 0.82\ J/C$

        $$\Delta G° = -nFE° = -(2\ mol\ e^-)(96{,}485\ C/mol\ e^-)(0.82\ J/C) = -1.6 \times 10^5\ J = -160\ kJ$$

        $$E° = \frac{0.0592}{n}\ \log K, \log K = \frac{nE°}{0.0592} = \frac{2(0.82)}{0.0592} = 27.70,\ K = 10^{27.70} = 5.0 \times 10^{27}$$

        d. $Pb + Cu^{2+} \rightarrow Cu + Pb^{2+}$ $E°_{cell} = 0.47\ V$

        $$\Delta G° = -nFE° = -(2\ mol\ e^-)(96{,}485\ C/mol\ e^-)(0.47\ J/C)\ (1\ kJ/1000\ J) = -91\ kJ$$

        $$\log K = \frac{2(0.47)}{0.0592} = 15.88,\ K = 7.6 \times 10^{15}$$

        e. $(Fe^{2+} \rightarrow Fe^{3+} + e^-) \times 4$ $-E° = -0.77\ V$
                        $4\ H^+ + O_2 + 4\ e^- \rightarrow 2\ H_2O$ $E° = 1.23\ V$

        $$\overline{4\ H^+ + O_2 + 4\ Fe^{2+} \rightarrow 4\ Fe^{3+} + 2\ H_2O \qquad E°_{cell} = +0.46\ V}$$

        $$\Delta G° = -nFE° = -(4\ mol\ e^-)(96{,}485\ C/mol\ e^-)(0.46\ J/C)\ (1\ kJ/1000\ J) = -180\ kJ$$

        $$\log K = \frac{4(0.46)}{0.0592} = 31.08,\ K = 1.2 \times 10^{31}$$

23.   3a. $7 H_2O + 2 Cr^{3+} + 3 Cl_2 \rightarrow Cr_2O_7^{2-} + 6 Cl^- + 14 H^+$ $E^\circ_{cell} = 0.03 V = 0.03 J/C$

$\Delta G^\circ = -nFE^\circ = -(6 \text{ mol } e^-)(96{,}485 \text{ C/mol } e^-)(0.03 \text{ J/C}) = -1.7 \times 10^4 J = -20 kJ$

$\log K = \dfrac{nE^\circ}{0.0592} = \dfrac{6(0.03)}{0.0592} = 3.04, \ K = 10^{3.04} = 1 \times 10^3$

3b. $\Delta G^\circ = -(2 \text{ mol } e^-)(96{,}485 \text{ C/mol } e^-)(2.71 \text{ J/C}) = -5.23 \times 10^5 J = -523 kJ$

$\log K = \dfrac{2(2.71)}{0.0592} = 91.55, \ K = 3.55 \times 10^{91}$

3c. $\Delta G^\circ = -(5 \text{ mol } e^-)(96{,}485 \text{ C/mol } e^-)(0.43 \text{ J/C}) = -2.1 \times 10^5 J = -210 kJ$

$\log K = \dfrac{5(0.43)}{0.0592} = 36.32, \ K = 2.1 \times 10^{36}$

This is for the reaction: $6 H^+ + IO_3^- + 5 Fe^{2+} \rightarrow 5 Fe^{3+} + 1/2 I_2 + 3 H_2O$

3d. $\Delta G^\circ = -(2 \text{ mol } e^-)(96{,}485 \text{ C/mol } e^-)(1.56 \text{ J/C}) = -3.01 \times 10^5 J = -301 kJ$

$\log K = \dfrac{2(1.56)}{0.0592} = 52.70, \ K = 5.01 \times 10^{52}$

4a. $\Delta G^\circ = -(2 \text{ mol } e^-)(96{,}485 \text{ C/mol } e^-)(0.27 \text{ J/C}) = -5.21 \times 10^4 J = -52 kJ$

$\log K = \dfrac{2(0.27)}{0.0592} = 9.12, \ K = 1.3 \times 10^9$

4b. $\Delta G^\circ = -(10 \text{ mol } e^-)(96{,}485 \text{ C/mol } e^-)(0.09 \text{ J/C}) = -8.7 \times 10^4 J = -90 kJ$

$\log K = \dfrac{10(0.09)}{0.0592} = 15.20, \ K = 2 \times 10^{15}$

4c. $\Delta G^\circ = -(6 \text{ mol } e^-)(96{,}485 \text{ C/mol } e^-)(1.43 \text{ J/C}) = -8.28 \times 10^5 J = -828 kJ$

$\log K = \dfrac{6(1.43)}{0.0592} = 144.93 = 145, \ K \approx 10^{145}$

4d. $\Delta G^\circ = -(1 \text{ mol } e^-)(96{,}485 \text{ C/mol } e^-)(1.18 \text{ J/C}) = -1.14 \times 10^5 J = -114 kJ$

$\log K = \dfrac{(1)(1.18)}{0.0592} = 19.93, \ K = 8.51 \times 10^{19}$

24.   a. $(Mn \rightarrow Mn^{2+} + 2 e^-) \times 3$ $-E^\circ = 1.18 V$
              $(4 H^+ + NO_3^- + 3 e^- \rightarrow NO + 2 H_2O) \times 2$ $E^\circ = 0.96 V$

$3 Mn + 8 H^+ + 2 NO_3^- \rightarrow 2 NO + 4 H_2O + 3 Mn^{2+}$ $E^\circ_{cell} = 2.14 V$

$$5 \times (2\ e^- + 2\ H^+ + IO_4^- \rightarrow IO_3^- + H_2O) \qquad E° = 1.60\ V$$
$$2 \times (Mn^{2+} + 4\ H_2O \rightarrow MnO_4^- + 8\ H^+ + 5\ e^-) \qquad -E° = -1.51\ V$$

$$\overline{5\ IO_4^- + 2\ Mn^{2+} + 3\ H_2O \rightarrow 5\ IO_3^- + 2\ MnO_4^- + 6\ H^+ \quad E°_{cell} = 0.09\ V}$$

b. Nitric acid oxidation:

$$\Delta G° = -(6\ \text{mol}\ e^-)(96{,}485\ \text{C/mol}\ e^-)(2.14\ \text{J/C}) = -1.24 \times 10^6\ \text{J} = -1240\ \text{kJ}$$

$$\log K = \frac{nE°}{0.0592} = \frac{6(2.14)}{0.0592} = 217,\ K \approx 10^{217}$$

Periodate oxidation:

$$\Delta G° = -(10\ \text{mol}\ e^-)(96{,}485\ \text{C/mol}\ e^-)(0.09\ \text{J/C})\ (1\ \text{kJ}/1000\ \text{J}) = -90\ \text{kJ}$$

$$\log K = \frac{10(0.09)}{0.0592} = 15.2,\ K = 10^{15.2} = 2 \times 10^{15}$$

25. $$\Delta G° = -nFE° = \Delta H° - T\Delta S°,\quad E° = \frac{T\Delta S°}{nF} - \frac{\Delta H°}{nF}$$

If we graph E° vs. T we should get a straight line (y = mx + b). The slope of the line is equal to $\Delta S°/nF$ and the y-intercept is equal to $-\Delta H°/nF$.

26. a. $$(2\ H^+ + 2\ e^- \rightarrow H_2) \times 2 \qquad E° = 0.00\ V$$
$$2\ H_2O \rightarrow O_2 + 4\ H^+ + 4\ e^- \qquad -E° = -1.23\ V$$

$$\overline{2\ H_2O \rightarrow 2\ H_2 + O_2 \qquad\qquad E°_{cell} = -1.23\ V}$$

$$\Delta G° = -nFE° = -(4\ \text{mol}\ e^-)(96{,}485\ \text{C/mol}\ e^-)(-1.23\ \text{J/C})\ (1\ \text{kJ}/1000\ \text{J}) = 475\ \text{kJ}$$

b. $$\Delta H° = -2\ \Delta H_f°\ (H_2O) = -2\ \text{mol}\ (-286\ \text{kJ/mol}) = 572\ \text{kJ}$$

$$\Delta S° = 2\ \text{mol}\ (131\ \text{J}\ K^{-1}\ \text{mol}^{-1}) + 1(205) - 2(70) = 327\ \text{J/K}$$

c. At 90.°C (363 K): $\Delta G° = 572\ \text{kJ} - (363\ K)(0.327\ \text{kJ/K}) = 453\ \text{kJ}$

$$\Delta G° = -nFE°,\quad E° = \frac{-\Delta G°}{nF} = \frac{-4.53 \times 10^5\ \text{J}}{(4\ \text{mol}\ e^-)\ (96{,}485\ \text{C/mol}\ e^-)} = -1.17\ \text{J/C} = -1.17\ V$$

At 0°C: $\Delta G° = 572\ \text{kJ} - (273\ K)(0.327\ \text{kJ/K}) = 483\ \text{kJ}$

$$E° = \frac{-\Delta G°}{nF} = \frac{-4.83 \times 10^5\ \text{J}}{(4\ \text{mol}\ e^-)\ (96{,}485\ \text{C/mol}\ e^-)} = -1.25\ \text{J/C} = -1.25\ V$$

27. E° will have little temperature dependence for cell reactions with $\Delta S°$ close to zero. (See Exercise 25.)

28. $CH_3OH(l) + 3/2\ O_2(g) \rightarrow CO_2(g) + 2\ H_2O(l)$ $\Delta G° = 2(-237) + (-394) - [-166] = -702$ kJ

The balanced half-reactions are:

$H_2O + CH_3OH \rightarrow CO_2 + 6\ H^+ + 6\ e^-$ and $O_2 + 4\ H^+ + 4\ e^- \rightarrow 2\ H_2O$

For 3/2 mol O_2, 6 moles of electrons will be transferred (n = 6).

$\Delta G° = -nFE°,\ E° = \dfrac{-\Delta G°}{nF} = \dfrac{702,000\text{ J}}{(6\text{ mol e}^-)\,(96,485\text{ C/mol e}^-)} = 1.21$ J/C $= 1.21$ V

For this reaction: $\Delta S° = 2(70) + 214 - [127 + 3/2(205)] = -81$ J/K

$E° = \dfrac{T\Delta S°}{nF} - \dfrac{\Delta H°}{nF}$; Since $\Delta S°$ is negative, $E°$ will decrease with an increase in temperature.

29. $\Delta G° = -nFE° = -(1\text{ mol e}^-)(96,485\text{ C/mol e}^-)(0.80\text{ J/C})(1\text{ kJ}/1000\text{ J}) = -77$ kJ; $\Delta G_f°\,(e^-) = 0$

-77 kJ $= \Delta G_f°\,(Ag) - \Delta G_f°\,(Ag^+)$, -77 kJ $= 0 - \Delta G_f°\,(Ag^+)$, $\Delta G_f°\,(Ag^+) = 77$ kJ/mol

30. $Fe^{2+} + 2\ e^- \rightarrow Fe$ $E° = -0.44$ V $= -0.44$ J/C

$\Delta G° = -(2\text{ mol e}^-)(96,485\text{ C/mol e}^-)(-0.44\text{ J/C})(1\text{ kJ}/1000\text{ J}) = 85$ kJ; $\Delta G_f°\,(e^-) = 0$

85 kJ $= 0 - \Delta G_f°\,(Fe^{2+})$, $\Delta G_f°\,[Fe^{2+}(aq)] = -85$ kJ

We can get $\Delta G_f°\,(Fe^{3+})$ two ways. Consider: $Fe^{3+} + e^- \rightarrow Fe^{2+}$ $E° = 0.77$ V

$\Delta G° = -(1\text{ mol e})(96,485\text{ C/mol e}^-)(0.77\text{ J/C}) = -74,300$ J $= -74$ kJ

$Fe^{2+} \rightarrow Fe^{3+} + e^-$ $\Delta G° = +74$ kJ
$Fe \rightarrow Fe^{2+} + 2\ e^-$ $\Delta G° = -85$ kJ
_____
$Fe \rightarrow Fe^{3+} + 3\ e^-$ $\Delta G° = -11$ kJ, $\Delta G_f°\,[Fe^{3+}(aq)] = -11$ kJ/mol

or consider: $Fe^{3+} + 3\ e^- \rightarrow Fe$ $E° = -0.036$ V

$\Delta G° = -(3\text{ mol e}^-)(96,485\text{ C/mol e}^-)(-0.036\text{ J/C}) = 10,400$ J $\approx 10.$ kJ

10. kJ $= 0 - \Delta G_f°$, $\Delta G_f° = -10.$ kJ/mol

31. $2\ H_2O + 2\ e^- \rightarrow H_2 + 2\ OH^-$ $\Delta G° = 2(-157) - 2(-237) = +160.$ kJ   (For e^-, $\Delta G_f° = 0$)

$\Delta G° = -nFE°,\ E = \dfrac{-\Delta G°}{nF} = \dfrac{-1.60 \times 10^5\text{ J}}{(2\text{ mol e}^-)\,(96,485\text{ C/mol e}^-)} = -0.829$ J/C $= -0.829$ V

The two values agree (-0.83 V in Table 11.1).

32.   a. $E_{cell}^{\circ} = 0.52 - 0.16 = 0.36$ V; spontaneous

      b. $E_{cell}^{\circ} = -0.14$ V $- (0.15$ V$) = -0.29$ V; not spontaneous

      c. $E_{cell}^{\circ} = -0.44$ V $- (0.77$ V$) = -1.21$ V; not spontaneous

      d. $E_{cell}^{\circ} = 1.65 - 1.21 = 0.44$ V; spontaneous

33. $2\ Cu^{+}(aq) \rightarrow Cu^{2+}(aq) + Cu(s)$ $E_{cell}^{\circ} = 0.36$ V

$\Delta G^{\circ} = -nFE^{\circ} = -(1\ mol\ e^{-})(96,485\ C/mol\ e^{-})(0.36\ J/C) = -34,700\ J = -35\ kJ$

$$\log K = \frac{nE^{\circ}}{0.0592} = \frac{1(0.36)}{0.0592} = 6.08, \quad K = 1.2 \times 10^{6}$$

$2\ HClO_2(aq) \rightarrow ClO_3^{-}(aq) + H^{+}(aq) + HClO(aq)$ $E_{cell}^{\circ} = 0.44$ V

$\Delta G^{\circ} = -nFE^{\circ} = -(2\ mol\ e^{-})(96,485\ C/mol\ e^{-})(0.44\ J/C) = -84,900\ J = -85\ kJ$

$$\log K = \frac{nE^{\circ}}{0.0592} = \frac{2(0.44)}{0.0592} = 14.86, \quad K = 7.2 \times 10^{14}$$

34.

$$\begin{array}{ll} Ag^{+} + e^{-} \rightarrow Ag & E^{\circ} = 0.80\ V \\ Ag + 2\ S_2O_3^{2-} \rightarrow Ag(S_2O_3)_2^{3-} + e^{-} & -E^{\circ} = -0.017\ V \end{array}$$

$\overline{}$

$Ag^{+} + 2\ S_2O_3^{2-} \rightarrow Ag(S_2O_3)_2^{3-}$ $E_{cell}^{\circ} = 0.78$ V $K = ?$

$$\log K = \frac{nE^{\circ}}{0.0592} = \frac{(1)(0.78)}{0.0592} = 13.18, \quad K = 10^{13.18} = 1.5 \times 10^{13}$$

35.

$$\begin{array}{ll} CdS + 2\ e^{-} \rightarrow Cd + S^{2-} & E^{\circ} = -1.21\ V \\ Cd \rightarrow Cd^{2+} + 2\ e^{-} & -E^{\circ} = 0.402\ V \end{array}$$

$\overline{}$

$CdS(s) \rightarrow Cd^{2+} + S^{2-}$ $E_{cell}^{\circ} = -0.81$ V $K_{sp} = ?$

$$\log K_{sp} = \frac{nE^{\circ}}{0.0592} = \frac{2(-0.81)}{0.0592} = -27.36, \quad K_{sp} = 10^{-27.36} = 4.4 \times 10^{-28}$$

36.

$$\begin{array}{ll} Ag \rightarrow Ag^{+} + e^{-} & -E^{\circ} = -0.80\ V \\ e^{-} + AgI \rightarrow Ag + I^{-} & E_{AgI}^{\circ} = ? \end{array}$$

$\overline{}$

$AgI \rightarrow Ag^{+} + I^{-}$ $E_{cell}^{\circ} = -0.80 + E_{AgI}^{\circ}$ $K_{sp} = 1.5 \times 10^{-16}$

For this overall reaction:

$$E_{cell}^{\circ} = \frac{0.0592}{n}\ \log K_{sp} = \frac{0.0592}{1}\ \log(1.5 \times 10^{-16}) = -0.94\ V$$

$E_{cell}^{\circ} = -0.94 \text{ V} = -0.80 \text{ V} + E_{AgI}^{\circ}, \quad E_{AgI}^{\circ} = -0.94 + 0.80 = -0.14 \text{ V}$

The Nernst Equation

37. a. $E = E^{\circ} - \dfrac{RT}{nF} \ln Q \text{ or } E = E^{\circ} - \dfrac{0.0592}{n} \log_{10} Q \quad (\text{at } 25^{\circ} \text{ C})$

For $Cu^{2+} + 2 e^{-} \rightarrow Cu \qquad E^{\circ} = 0.34 \text{ V}$

$E = E^{\circ} - \dfrac{0.0592}{n} \log (1/[Cu^{2+}]) = E^{\circ} + \dfrac{0.0592}{n} \log [Cu^{2+}]$

$E = 0.34 \text{ V} + \dfrac{0.0592}{2} \log (0.10) = 0.34 \text{ V} - 0.030 \text{ V} = 0.31 \text{ V}$

b. $E = 0.34 + \dfrac{0.0592}{2} \log 2.0 = 0.34 \text{ V} + 8.9 \times 10^{-3} \text{ V} = 0.35 \text{ V}$

c. $E = 0.34 + \dfrac{0.0592}{2} \log (1.0 \times 10^{-4}) = 0.34 - 0.12 = 0.22 \text{ V}$

d. $5 e^{-} + 8 H^{+} + MnO_4^{-} \rightarrow Mn^{2+} + 4 H_2O \qquad E^{\circ} = 1.51 \text{ V}$

$E = E^{\circ} - \dfrac{0.0592}{5} \log \dfrac{[Mn^{2+}]}{[MnO_4^{-}] [H^{+}]^8} = 1.51 \text{ V} - \dfrac{0.0592}{5} \log \dfrac{(0.010)}{(0.10)(1.0 \times 10^{-3})^8}$

$E = 1.51 - \dfrac{0.0592}{5} (23) = 1.51 \text{ V} - 0.27 \text{ V} = 1.24 \text{ V}$

e. $E = 1.51 - \dfrac{0.0592}{5} \log \left[\dfrac{(0.010)}{(0.10)(0.10)^8} \right] = 1.51 - 0.083 = 1.43 \text{ V}$

38. a. $n = 2$ for this reaction.

$E = E^{\circ} - \dfrac{0.0592}{2} \log \left(\dfrac{1}{[H^{+}]^2 [HSO_4^{-}]^2} \right) = 2.04 \text{ V} + \dfrac{0.0592}{2} \log ([H^{+}]^2 [HSO_4^{-}]^2)$

$E = 2.04 \text{ V} + \dfrac{0.0592}{2} \log [(4.5)^2 (4.5)^2] = 2.04 \text{ V} + 0.077 \text{ V} = 2.12 \text{ V}$

b. We can calculate ΔG° from $\Delta G^{\circ} = \Delta H^{\circ} - T\Delta S^{\circ}$ and then E° from $\Delta G^{\circ} = -nFE^{\circ}$; or we can use the equation derived in Exercise 11.25.

$E^{\circ} = \dfrac{T\Delta S^{\circ} - \Delta H^{\circ}}{nF} = \dfrac{(253 \text{ K}) (263.5 \text{ J/K}) + 315.9 \times 10^3 \text{ J}}{(2 \text{ mol } e^{-}) (96,485 \text{ C/mol } e^{-})} = 1.98 \text{ J/C} = 1.98 \text{ V}$

c. $E = E° - \dfrac{RT}{nF} \ln Q = E° + \dfrac{RT}{nF} \ln ([H^+]^2 [HSO_4^-]^2)$

$E = 1.98 \text{ V} + \dfrac{(8.3145 \text{ J/K mol}) (253 \text{ K})}{(2 \text{ mol e}^-) (96,485 \text{ C/mol e}^-)} \ln [(4.5)^2 (4.5)^2]$

$E = 1.98 \text{ V} + 0.066 \text{ V} = 2.05 \text{ V}$

d.   As the temperature decreases, the cell potential decreases. Also, since oil becomes more viscous at lower temperatures, it is more difficult to start an engine on a cold day. The combination of these two factors results in batteries failing more often on cold days than on warm days.

39.   For the cell on the left, $E_l = E° = 0.80$ V since $[Ag^+] = 1.0$ M. Let's calculate the E for the half-cell on the right, and then figure out what reaction occurs. The half-cell with the smallest reduction potential will be the anode.

a. $E_r = E° = 0.80$ V; Since $E_l = E_r$, then $E_{cell} = 0$. No reaction occurs. Concentration cells only produce a voltage when the ion concentrations are <u>not</u> equal.

b.   For $Ag^+ + e^- \rightarrow Ag$: $E_r = E° - \dfrac{0.0592}{1} \log \dfrac{1}{[Ag^+]} = 0.80 \text{ V} + 0.0592 \log [Ag^+]$

$E_r = 0.80 \text{ V} + 0.0592 \log 2.0 = 0.80 \text{ V} + 0.018 \text{ V} = 0.82 \text{ V}$

The half-cell with 2.0 M Ag^+ is the cathode. Electrons flow from the anode to the cathode, to the right in the diagram.

$E_{cell} = E_r - E_l = 0.82 - 0.80 = 0.02 \text{ V}$

Note: The driving force for the reaction is to equalize the concentration of Ag^+ in the two half-cells. The half-cell with the largest Ag^+ concentration will always be the cathode.

c. $E_r = 0.80 + 0.0592 \log 0.10 = 0.80 \text{ V} - 0.059 \text{ V} = 0.74 \text{ V}$

The half-cell with 0.10 M Ag^+ is the anode. Electrons move to the cathode (to the left in the diagram).

$E_{cell} = E_l - E_r = 0.80 - 0.74 = 0.06 \text{ V}$

d. $E = 0.80 + 0.0592 \log (4.0 \times 10^{-5}) = 0.80 - 0.26 = 0.54 \text{ V}$

$E_{cell} = E_l - E_r = 0.80 - 0.54 = 0.26 \text{ V}$

The half-cell with 4.0×10^{-5} M Ag^+ is the anode. Electrons move from the anode to the cathode (to the left in the diagram).

e.   If the concentrations are the same in each half-cell, $E_l = E_r$ and no reaction occurs ($E_{cell} = 0$).

40. $5 e^- + 8 H^+ + MnO_4^- \rightarrow Mn^{2+} + 4 H_2O$ $E° = 1.51$ V
                   $(Fe^{2+} \rightarrow Fe^{3+} + e^-) \times 5$ $-E° = -0.77$ V

_____

$8 H^+ + MnO_4^- + 5 Fe^{2+} \rightarrow 5 Fe^{3+} + Mn^{2+} + 4 H_2O$ $E_{cell}° = 0.74$ V

$E = E° - \dfrac{0.0592}{n} \log Q = 0.74 \text{ V} - \dfrac{0.0592}{5} \log \dfrac{[Fe^{3+}]^5 [Mn^{2+}]}{[Fe^{2+}]^5 [MnO_4^-] [H^+]^8}$

$E = 0.74 - \dfrac{0.0592}{5} \log \dfrac{(1 \times 10^{-6})^5 (1 \times 10^{-6})}{(1 \times 10^{-3})^5 (1 \times 10^{-2}) (1 \times 10^{-4})^8}$

$E = 0.74 - \dfrac{0.0592}{5} \log (1 \times 10^{13}) = 0.74 \text{ V} - 0.15 \text{ V} = 0.59 \text{ V} = 0.6 \text{ V}$

Yes, E > 0 so reaction will occur as written.

41. $Cu^{2+} + H_2 \rightarrow 2 H^+ + Cu$ $E_{cell}° = 0.34 \text{ V} - 0.00 \text{V} = 0.34$ V; Since $P_{H_2} = 1.0$ atm and $[H^+] = 1.0$ M:

$E = E° - \dfrac{0.0592}{2} \log \dfrac{1}{[Cu^{2+}]} = 0.34 + \dfrac{0.0592}{2} \log [Cu^{2+}]$

a. $E = 0.34 \text{ V} + \dfrac{0.0592}{2} \log (2.5 \times 10^{-4}) = 0.23$ V

b. $Cu(OH)_2(s) \rightleftharpoons Cu^{2+} + 2 OH^-$ $K_{sp} = 1.6 \times 10^{-19}$

Equil.   s $0.10 + 2s$

$1.6 \times 10^{-19} = (s)(0.10 + 2s)^2 \approx s(0.10)^2$, $s = [Cu^{2+}] = 1.6 \times 10^{-17}$ M Assumption good.

$E = E° + \dfrac{0.0592}{2} \log [Cu^{2+}] = 0.34 \text{ V} + \dfrac{0.0592}{2} \log (1.6 \times 10^{-17}) = 0.34 - 0.50 = -0.16$ V

Not spontaneous, reverse reaction occurs. The Cu electrode becomes the anode and $E_{cell} = 0.16$ V for the reverse reaction.

c. $0.195 \text{ V} = 0.34 \text{ V} + \dfrac{0.0592}{2} \log [Cu^{2+}]$

$\log [Cu^{2+}] = -4.899$, $[Cu^{2+}] = 10^{-4.899} = 1.3 \times 10^{-5}$ M

d. $E = E° + (0.0592/2) \log [Cu^{2+}]$ is in the form of a straight line equation, $y = mx + b$. A graph of E vs. $\log [Cu^{2+}]$ will yield a straight line with slope equal to 0.0296 V or 29.6 mV.

42.   The potential oxidizing agents are NO_3^- and H^+. Hydrogen ion cannot oxidize Pt under any condition. Nitrate cannot oxidize Pt unless there is Cl^- in the solution. Aqua regia has both Cl^- and NO_3^-. Nitrate oxidizes Pt to Pt^{2+} which is complexed by the Cl^-.

$$12\ Cl^- + 3\ Pt \rightarrow 3\ PtCl_4^{2-} + 6\ e^- \qquad\qquad -E^\circ = -0.76\ V$$
$$2\ NO_3^- + 8\ H^+ + 6\ e^- \rightarrow 2\ NO + 4\ H_2O \qquad\qquad E^\circ = 0.96\ V$$

$$\overline{12\ Cl^- + 3\ Pt + 2\ NO_3^- + 8\ H^+ \rightarrow 3\ PtCl_4^{2-} + 2\ NO + 4\ H_2O \qquad E^\circ_{cell} = +0.20\ V}$$

43.   a. $E = E^\circ = \dfrac{0.0592}{n} \log \dfrac{[OH^-]^5}{[CrO_4^{2-}]}$; $[OH^-] = 10^{-6.60} = 2.5 \times 10^{-7}\ M$

$$E = -0.13\ V - \dfrac{0.0592}{3} \log \dfrac{(2.5 \times 10^{-7})^5}{1.0 \times 10^{-6}}$$

$$E = -0.13 - \dfrac{0.0592}{3} \log (9.8 \times 10^{-28}) = -0.13\ V + 0.53\ V = +0.40\ V$$

   b. $E = E^\circ - \dfrac{0.0592}{n} \log \dfrac{[Cr^{3+}]^2}{[Cr_2O_7^{2-}][H^+]^{14}} = 1.33\ V - \dfrac{0.0592}{6} \log \dfrac{(1.0 \times 10^{-6})^2}{(1.0 \times 10^{-6})(1.0 \times 10^{-2})^{14}}$

$$E = 1.33 - \dfrac{0.0592}{6} \log (1.0 \times 10^{22}) = 1.33\ V - 0.22\ V = 1.11\ V$$

44.   a. $Ag_2CrO_4 + 2\ e^- \rightarrow 2\ Ag + CrO_4^{2-}$ $E^\circ = 0.446\ V$

   $Hg_2Cl_2 + 2\ e^- \rightarrow 2\ Hg + 2\ Cl^-$ $E_{SCE} = 0.242\ V$

   SCE will be the oxidation half-reaction, $E_{cell} = 0.446 - 0.242 = 0.204\ V$.

   $\Delta G = -nFE = -2(96,485)(0.204)J = -3.94 \times 10^4\ J = -39.4\ kJ$

   b.   In SCE, we assume all concentrations are constant.

$$E = E^\circ - \dfrac{0.0592}{2} \log [CrO_4^{2-}] = 0.204\ V - \dfrac{0.0592}{2} \log [CrO_4^{2-}]$$

   c. $E = 0.204 - \dfrac{0.0592}{2} \log (1.00 \times 10^{-5}) = 0.204\ V + 0.148\ V = 0.352\ V$

   d. $0.504\ V = 0.204\ V - (0.0592/2) \log [CrO_4^{2-}]$

   $\log [CrO_4^{2-}] = -10.135,\ \ [CrO_4^{2-}] = 10^{-10.135} = 7.33 \times 10^{-11}\ M$

e.
$$(Ag \rightarrow Ag^+ + e^-) \times 2 \qquad\qquad -E° = -0.80 \text{ V}$$
$$Ag_2CrO_4 + 2\ e^- \rightarrow 2\ Ag + CrO_4^{2-} \qquad E° = 0.446 \text{ V}$$

$$Ag_2CrO_4(s) \rightarrow 2\ Ag^+ + CrO_4^{2-} \qquad E°_{cell} = -0.35 \text{ V} \qquad K_{sp} = ?$$

$$E°_{cell} = \frac{0.0592}{n} \log K_{sp}, \ \log K_{sp} = \frac{(-0.35 \text{ V})(2)}{0.0592} = -11.82, \ K_{sp} = 10^{-11.82} = 1.5 \times 10^{-12}$$

45.   Cathode: $O_2 + 2\ H_2O + 4\ e^- \rightarrow 4\ OH^-$ $\qquad\qquad$ $E° = 0.40 \text{ V}$
       Anode: $\qquad\qquad\qquad$ $(Fe \rightarrow Fe^{2+} + 2\ e^-) \times 2$ $\qquad$ $-E° = 0.44 \text{ V}$

$$O_2 + 2\ H_2O + 2\ Fe \rightarrow 2\ Fe^{2+} + 4\ OH^- \qquad E°_{cell} = 0.84 \text{ V}$$

$$E = 0.84 \text{ V} - \frac{0.0592}{4} \log \left(\frac{[OH^-]^4 [Fe^{2+}]^2}{P_{O_2}} \right)$$

If $[OH^-]$ is very small, say $10^{-14} = [OH^-]$ ($[H^+] = 1.0\ M$), the log term will become more positive. Thus, as the solution becomes more acidic, E_{cell} becomes more positive, and the corrosion process becomes more spontaneous. Note: This is oversimplified as $Fe(OH)_2(s)$ or $Fe_2O_3(s)$ formation has been ignored.

46.   From Exercise 11.3a: $3\ Cl_2 + 2\ Cr^{3+} + 7\ H_2O \rightleftharpoons 14\ H^+ + Cr_2O_7^{2-} + 6\ Cl^-$ $\quad$ $E°_{cell} = 0.03 \text{ V}$

$$E = E° - \frac{0.0592}{6} \log \frac{[Cr_2O_7^{2-}]\ [H^+]^{14} [Cl^-]^6}{[Cr^{3+}]^2\ P_{Cl_2}^3}$$

When $K_2Cr_2O_7$ and Cl^- are added to concentrated H_2SO_4, Q becomes a large number due to $[H^+]^{14}$ term. The log of a large number is positive. E becomes negative, which means the reverse reaction becomes spontaneous. The pungent fumes were $Cl_2(g)$.

Electrolysis

47.   a. $Al^{3+} + 3\ e^- \rightarrow Al$; 3 mol e^- are needed to produce 1 mol Al.

$$1.0 \times 10^3 \text{ g} \times \frac{1 \text{ mol Al}}{27.0 \text{ g}} \times \frac{3 \text{ mol } e^-}{\text{mol Al}} \times \frac{96{,}485 \text{ C}}{\text{mol } e^-} \times \frac{1 \text{ s}}{100.0 \text{ C}} = 1.07 \times 10^5 \text{ s} = 3.0 \times 10^1 \text{ hours}$$

b. $$1.0 \text{ g Ni} \times \frac{1 \text{ mol}}{58.7 \text{ g}} \times \frac{2 \text{ mol } e^-}{\text{mol Ni}} \times \frac{96{,}485 \text{ C}}{\text{mol } e^-} \times \frac{1 \text{ s}}{100.0 \text{ C}} = 33 \text{ s}$$

c. $$5.0 \text{ mol Ag} \times \frac{1 \text{ mol } e^-}{\text{mol Ag}} \times \frac{96{,}485 \text{ C}}{\text{mol } e^-} \times \frac{1 \text{ s}}{100.0 \text{ C}} = 4.8 \times 10^3 \text{ s} = 1.3 \text{ hours}$$

48. $$15A = \frac{15 \text{ C}}{\text{s}} \times \frac{60 \text{ s}}{\text{min}} \times \frac{60 \text{ min}}{\text{h}} = 5.4 \times 10^4 \text{ C of charge passed in 1 hour}$$

a. $5.4 \times 10^4 \text{ C} \times \dfrac{1 \text{ mol e}^-}{96,485 \text{ C}} \times \dfrac{1 \text{ mol Co}}{2 \text{ mol e}^-} \times \dfrac{58.9 \text{ g}}{\text{mol}} = 16 \text{ g Co}$

b. $5.4 \times 10^4 \text{ C} \times \dfrac{1 \text{ mol e}^-}{96,485 \text{ C}} \times \dfrac{1 \text{ mol Hf}}{4 \text{ mol e}^-} \times \dfrac{178.5 \text{ g}}{\text{mol}} = 25 \text{ g Hf}$

c. $I_2 + 2e^- \rightarrow 2I^-$; $5.4 \times 10^4 \text{ C} \times \dfrac{1 \text{ mol e}^-}{96,485 \text{ C}} \times \dfrac{1 \text{ mol I}_2}{2 \text{ mol e}^-} \times \dfrac{253.8 \text{ g I}_2}{\text{mol I}_2} = 71 \text{ g I}_2$

d.   Cr is in the +6 oxidation state in CrO_3 (n = 6).

$5.4 \times 10^4 \text{ C} \times \dfrac{1 \text{ mol e}^-}{96,485 \text{ C}} \times \dfrac{1 \text{ mol Cr}}{6 \text{ mol e}^-} \times \dfrac{52.0 \text{ g Cr}}{\text{mol Cr}} = 4.9 \text{ g Cr}$

49. $2.30 \text{ min} \times \dfrac{60 \text{ s}}{\text{min}} = 138 \text{ s}$; $138 \text{ s} \times \dfrac{2.00 \text{ C}}{\text{s}} \times \dfrac{1 \text{ mol e}^-}{96,485 \text{ C}} \times \dfrac{1 \text{ mol Ag}}{\text{mol e}^-} = 2.86 \times 10^{-3} \text{ mol Ag}$

$[\text{Ag}^+] = 2.86 \times 10^{-3} \text{ mol Ag}^+/0.250 \text{ L} = 1.14 \times 10^{-2} \text{ } M$

50.   a.   Cathode (reduction occurs): $K^+ + e^- \rightarrow K$ $E° = -2.92 \text{ V}$

Anode (oxidation occurs): $2 F^- \rightarrow F_2 + 2 e^-$ $-E° = -2.87 \text{ V}$

b.   Cathode: easier to reduce H_2O than K^+; $2 H_2O + 2 e^- \rightarrow H_2 + 2 OH^-$ $E° = -0.83 \text{ V}$

Anode: easier to oxidize H_2O than F^-; $2 H_2O \rightarrow 4 H^+ + O_2 + 4 e^-$ $-E° = -1.23 \text{ V}$

c.   Possible cathode reactions:

$2 H_2O + 2 e^- \rightarrow H_2 + 2 OH^-$	$E° = -0.83 \text{ V}$
$2 H^+ + 2 e^- \rightarrow H_2$	$E° = 0.00 \text{ V}$
$SO_4^{2-} + 4 H^+ + 2 e^- \rightarrow H_2SO_3 + H_2O$	$E° = 0.20 \text{ V}$
$H_2O_2 + 2 H^+ + 2 e^- \rightarrow 2 H_2O$	$E° = 1.78 \text{ V}$

Reduction of H_2O_2 will occur since E° is most positive.

Possible anode reactions:

$2 H_2O \rightarrow O_2 + 4 H^+ + 4 e^-$	$-E° = -1.23 \text{ V}$
$H_2O_2 \rightarrow O_2 + 2 H^+ + 2 e^-$	$-E° = -0.68 \text{ V}$

Oxidation of H_2O_2 will occur since -E° is more positive.

d.   Cathode: $Mg^{2+} + 2 e^- \rightarrow Mg$; Anode: $2 Cl^- \rightarrow Cl_2 + 2 e^-$

e.   Possible cathode reactions:

$Ag^+ + e^- \rightarrow Ag$	$E° = 0.80 \text{ V}$
$2 H_2O + 2 e^- \rightarrow H_2 + 2 OH^-$	$E° = -0.83 \text{ V}$

Ag^+ would be reduced at the cathode since E° is most positive. Note: Reduction of NO_3^- was ignored (E° = 0.96 V) since H^+ is a reactant in the NO_3^- half-reaction. Some NO_3^- may initially reduce, but will quickly stop since $[H^+]$ from water is so small (10^{-7} M).

Anode: Only possible oxidation reaction is the oxidation of water. $2 H_2O \rightarrow 4 H^+ + O_2 + 4 e^-$

51. $Fe^{2+} + 2 e^- \rightarrow Fe$ $E = E° = -0.44$ V

   $Ag^+ + e^- \rightarrow Ag$ $E = E° - \dfrac{0.0592}{1} \log (1/[Ag^+]) = 0.80 + 0.0592 \log [Ag^+]$

   $E_{Ag} = 0.80$ V $+ 0.0592 \log(0.010) = 0.68$ V; It is still easier to plate out Ag.

52. $(Au(CN)_4^- + 3e^- \rightarrow Au + 4 CN^-) \times 4$
             $(4 OH^- \rightarrow O_2 + 2 H_2O + 4 e^-) \times 3$
   _____
   $4 Au(CN)_4^- + 12 OH^- \rightarrow 4 Au + 16 CN^- + 3 O_2 + 6 H_2O$

   1.00×10^3 g Au $\times \dfrac{1 \text{ mol Au}}{197.0 \text{ g}} \times \dfrac{3 \text{ mol O}_2}{4 \text{ mol Au}} = 3.81$ mol O_2

   $V = \dfrac{nRT}{P} = \dfrac{3.81 \text{ mol} \left(\dfrac{0.08206 \text{ L atm}}{\text{mol K}} \right) (298 \text{ K})}{740. \text{ torr}} \times \dfrac{760 \text{ torr}}{\text{atm}} = 95.7 \text{ L O}_2$

53.   To begin plating out Pd: $E = 0.62$ V $- \dfrac{0.0592}{2} \log \dfrac{[Cl^-]^4}{[PdCl_4^{2-}]} = 0.62 - \dfrac{0.0592}{2} \log \dfrac{(1.0)^4}{0.020}$

   $E = 0.62$ V $- 0.050$ V $= 0.57$ V

   When 99% of Pd has plated out, $[PdCl_4^-] = \dfrac{1}{100} (0.020) = 0.00020$ M.

   $E = 0.62 - \dfrac{0.0592}{2} \log \dfrac{(1.0)^4}{2.0 \times 10^{-4}} = 0.62$ V $- 0.11$V $= 0.51$ V

   To begin Pt plating: $E = 0.73$ V $- \dfrac{0.0592}{2} \log \dfrac{(1.0)^4}{0.020} = 0.73 - 0.050 = 0.68$ V

   When 99% of Pt plated: $E = 0.73 - \dfrac{0.0592}{2} \log \dfrac{(1.0)^4}{2.0 \times 10^{-4}} = 0.73 - 0.11 = 0.62$ V

   To begin Ir plating: $E° = 0.77$ V $- \dfrac{0.0592}{3} \log \dfrac{(1.0)^4}{0.020} = 0.77 - 0.034 = 0.74$ V

   When 99% of Ir plated: $E° = 0.77 - \dfrac{0.0592}{3} \log \dfrac{(1.0)^4}{2.0 \times 10^{-4}} = 0.77 - 0.073 = 0.70$ V

Yes, since the range of potentials for plating out each metal do not overlap, we should be able to separate the three metals. The exact potential to apply depends on the oxidation reaction. The order of plating will be Ir(s) first, followed by Pt(s) and finally Pd(s) as the potential is gradually increased.

54. $600. \text{ s} \times \dfrac{5.00 \text{ C}}{\text{s}} \times \dfrac{1 \text{ mol e}^-}{96{,}485 \text{ C}} \times \dfrac{1 \text{ mol M}}{3 \text{ mol e}^-} = 1.04 \times 10^{-2} \text{ mol M}$

Atomic mass $= \dfrac{1.19 \text{ g M}}{1.04 \times 10^{-2} \text{ mol}} = \dfrac{114 \text{ g}}{\text{mol}}$; The element is indium, In. Indium forms 3+ ions.

55. $74.6 \text{ s} \times \dfrac{2.50 \text{ C}}{\text{s}} \times \dfrac{1 \text{ mol e}^-}{96{,}485 \text{ C}} \times \dfrac{1 \text{ mol M}}{2 \text{ mol e}^-} = 9.66 \times 10^{-4} \text{ mol M}$

Atomic mass $= \dfrac{0.1086 \text{ g M}}{9.66 \times 10^{-4} \text{ mol}} = \dfrac{112 \text{ g}}{\text{mol}}$

The metal is Cd. Cd has the correct atomic mass and does form Cd^{2+} ions.

56. $\dfrac{150. \times 10^3 \text{ g}}{\text{h}} \times \dfrac{1 \text{ h}}{60 \text{ min}} \times \dfrac{1 \text{ min}}{60 \text{ s}} \times \dfrac{1 \text{ mol C}_6\text{H}_8\text{N}_2}{108.1 \text{ g C}_6\text{H}_8\text{N}_2} \times \dfrac{2 \text{ mol e}^-}{\text{mol C}_6\text{H}_8\text{N}_2} \times \dfrac{96{,}485 \text{ C}}{\text{mol e}^-}$

$= 7.44 \times 10^4 \text{ C/s or current of } 7.44 \times 10^4 \text{ A.}$

57. F_2 is produced at the anode: $2 \text{ F}^- \rightarrow F_2 + 2 \text{ e}^-$

$2.00 \text{ h} \times \dfrac{60 \text{ min}}{\text{h}} \times \dfrac{60 \text{ s}}{\text{min}} \times \dfrac{10.0 \text{ C}}{\text{s}} \times \dfrac{1 \text{ mol e}^-}{96{,}485 \text{ C}} = 0.746 \text{ mol e}^-$

$0.746 \text{ mol e}^- \times \dfrac{1 \text{ mol F}_2}{2 \text{ mol e}^-} = 0.373 \text{ mol F}_2$; $V = \dfrac{nRT}{P} = \dfrac{(0.373)(0.08206)(298)}{1.00} = 9.12 \text{ L F}_2$

K is produced at the cathode: $K^+ + \text{e}^- \rightarrow K$

$0.746 \text{ mol e}^- \times \dfrac{1 \text{ mol K}}{\text{mol e}^-} \times \dfrac{39.10 \text{ g K}}{\text{mol K}} = 29.2 \text{ g K}$

58. $15 \text{ kWh} = \dfrac{15000 \text{ J h}}{\text{s}} \times \dfrac{60 \text{ s}}{\text{min}} \times \dfrac{60 \text{ min}}{\text{h}} = 5.4 \times 10^7 \text{ J or } 5.4 \times 10^4 \text{ kJ}$ (Hall process)

To melt Al requires: $1.0 \times 10^3 \text{ g Al} \times \dfrac{1 \text{ mol Al}}{27.0 \text{ g}} \times \dfrac{10.7 \text{ kJ}}{\text{mol Al}} = 4.0 \times 10^2 \text{ kJ}$

It is feasible to recycle Al by melting the metal because it takes less than 1% of the energy required to produce the same amount of Al by the Hall process.

59.   Anode reaction: $2 \text{ Cl}^- \rightarrow Cl_2 + 2 \text{ e}^-$; Cathode reaction: $2 \text{ H}_2\text{O} + 2 \text{ e}^- \rightarrow H_2 + 2 \text{ OH}^-$

If 6.00 L of H_2 are produced, then 6.00 L of Cl_2 will also be produced.

60. $(2 H_2O \rightarrow 4 H^+ + O_2 + 4 e^-) \times 1/2$
           $2 e^- + 2 H_2O \rightarrow H_2 + 2 OH^-$

           _____

                    $H_2O \rightarrow H_2 + 1/2 O_2$ n = 2 for reaction as it is written.

$15.0 \text{ min} \times \dfrac{60 \text{ s}}{\text{min}} \times \dfrac{2.50 \text{ C}}{\text{s}} \times \dfrac{1 \text{ mol e}^-}{96,485 \text{ C}} \times \dfrac{1 \text{ mol } H_2}{2 \text{ mol e}^-} = 1.17 \times 10^{-2} \text{ mol } H_2$

At STP, 1 mole of an ideal gas occupies a volume of 22.42 L (see chapter 5).

$1.17 \times 10^{-2} \text{ mol } H_2 \times \dfrac{22.42 \text{ L}}{\text{mol } H_2} = 0.262 \text{ L} = 262 \text{ mL } H_2$

$1.17 \times 10^{-2} \text{ mol } H_2 \times \dfrac{0.500 \text{ mol } O_2}{\text{mol } H_2} \times \dfrac{22.42 \text{ L}}{\text{mol } O_2} = 0.131 \text{ L} = 131 \text{ mL } O_2$

61.   a.   Species present: Na^+, SO_4^{2-} and H_2O. From the potentials, H_2O is the most easily oxidized
             and the most easily reduced species present.

             Cathode: $2 H_2O + 2 e^- \rightarrow H_2(g) + 2 OH^-$; Anode: $2 H_2O \rightarrow O_2(g) + 4 H^+ + 4 e^-$

       b.   When water is electrolyzed a significantly higher voltage than predicted is necessary to
             produce the chemical change (called overvoltage). This higher voltage is probably great
             enough to cause some SO_4^{2-} to be oxidized instead of H_2O. Thus, the volume of O_2 generated
             would be less than expected and the measured volume ratio would be greater than 2:1.

Additional Exercises

62.   a.   See text for the example of the electrorefining of copper. Essentially electrorefining is
             possible because of the selectivity of the electrode reactions. The anode is made up of the
             impure metal. The metal of interest and all more easily oxidized metals are oxidized at the
             anode. The metal of interest is the only metal plated at the cathode. The metal ions of
             metals that could plate out at the cathode in preference to the metal we are purifying will not
             be in solution, since these metals were not oxidized at the anode.

       b.   A more easily oxidized metal is placed in electrical contact with the metal we are trying to
             protect. It is oxidized in preference to the protected metal. The protected metal becomes the
             cathode; thus, cathodic protection.

63.   1.   protective coating   2. alloying   3. cathodic protection

64. $Zn^{2+} + 2 e^- \rightarrow Zn$ $E° = -0.76$ V; $Fe^{2+} + 2 e^- \rightarrow Fe$ $E° = -0.44$ V

       It is easier to oxidize Zn than Fe, so the Zn will be oxidized protecting the iron of the *Monitor's*
       hull.

65.   Fuel cells are more efficient in converting chemical energy to electrical energy; they are also less massive. The major disadvantage is that they are expensive.

66. $Al^{3+} + 3 e^- \rightarrow Al$ $E° = -1.66$ V
         $Al + 6 F^- \rightarrow AlF_6^{3-} + 3 e^-$ $-E° = -(-2.07$ V$)$

$$Al^{3+} + 6 F^- \rightarrow AlF_6^{3-} \qquad E°_{cell} = 0.41 \text{ V} \qquad K_f = ?$$

$\log K_f = \dfrac{nE°}{0.0592} = \dfrac{3(0.41)}{0.0592} = 20.78, \; K_f = 10^{20.78} = 6.0 \times 10^{20}$

67. $1/2\, O_2 + 2 H^+ + 2 e^- \rightarrow H_2O$ $E° = 1.23$ V
                      $H_2 \rightarrow 2 H^+ + 2 e^-$ $-E° = 0.00$ V

$$H_2 + 1/2\, O_2 \rightarrow H_2O \qquad E°_{cell} = 1.23 \text{ V}$$

$\Delta G° = -nFE° = -(2 \text{ mol } e^-)(96{,}485 \text{ C/mol } e^-)(1.23 \text{ J/C}) = -237{,}000 \text{ J} = -237 \text{ kJ}$

$1.00 \times 10^3 \text{ g } H_2O \times \dfrac{1 \text{ mol } H_2O}{18.02 \text{ g}} \times \dfrac{-237 \text{ kJ}}{\text{mol } H_2O} = -13{,}200 \text{ kJ} = w_{max}$

The work done can be no larger than the free energy change. The best that could happen is that all of the free energy released goes into doing work, but this does not occur in any real process.

68.   As a battery discharges, E_{cell} decreases, eventually reaching zero. A charged battery is not at equilibrium. At equilibrium $E_{cell} = 0$ and $\Delta G = 0$. We can get no work out of an equilibrium system. A battery is useful to us, because it can do work as it approaches equilibrium.

69.   a.   In base, $Co(OH)_2$ will precipitate ($K_{sp} = 2.5 \times 10^{-16}$). We need E° for $Co(OH)_2 + 2 e^- \rightarrow Co + 2 OH^-$.

         We look at the effect of the solubility on:

$$Co^{2+} + 2 e^- \rightarrow Co \quad E = -0.28 - \dfrac{0.0592}{2} \log \dfrac{1}{[Co^{2+}]}$$

Since $K_{sp} = [Co^{2+}][OH^-]^2$, $E = -0.28 - \dfrac{0.0592}{2} \log \dfrac{[OH^-]^2}{K_{sp}}$

If $[OH^-] = 1.0$ M, then the E we calculate is E° for $Co(OH)_2 + 2 e^- \rightarrow Co + 2 OH^-$.

$$E = -0.28 - \dfrac{0.0592}{2} \log \dfrac{(1.0)^2}{2.5 \times 10^{-16}} = -0.74$$

$$(2\ OH^- + Co \rightarrow Co(OH)_2 + 2\ e^-) \times 2 \qquad -E° = 0.74\ V$$
$$O_2 + 2\ H_2O + 4\ e^- \rightarrow 4\ OH^- \qquad E° = 0.40\ V$$

$$2\ Co + O_2 + 2\ H_2O \rightarrow 2\ Co(OH)_2 \qquad E° = 1.14\ V$$

Yes, the corrosion reaction is spontaneous.

b. $O_2 + 4\ H^+ + 4e^- \rightarrow 2\ H_2O \qquad E° = 1.23\ V$

This half-reaction is even more positive than the reduction of O_2 in base. Yes, corrosion will also occur in 1.0 M H^+.

c.   Using the reaction: $2\ Co + O_2 + 4\ H^+ \rightarrow 2\ H_2O + 2\ Co^{2+} \quad E° = 1.23\ V - (-0.28\ V) = 1.51\ V$

$$E = 1.51\ V - \frac{0.0592}{4} \log \frac{[Co^{2+}]^2}{P_{O_2}\ [H^+]^4},\ [H^+] = 1.0 \times 10^{-7}\ M$$

$$E = 1.51\ V - \frac{0.0592}{4} \log (1.0 \times 10^{28}) - \frac{0.0592}{4} \log \frac{[Co^{2+}]^2}{P_{O_2}}$$

$$E = 1.51\ V - 0.41\ V - \frac{0.0592}{4} \log \frac{[Co^{2+}]^2}{P_{O_2}}$$

Typically, $[Co^{2+}]$ won't be very large and $P_{O_2} \approx 0.2$ atm. E will still be greater than zero and the corrosion reaction is spontaneous at pH = 7.

70.   a. $Hg_2Cl_2 + 2e^- \rightarrow 2\ Hg + 2\ Cl^- \qquad E_{SCE} = 0.242\ V$
      $H_2 \rightarrow 2\ H^+ + 2\ e^- \qquad -E° = 0.000\ V$

$$HgCl_2 + H_2 \rightarrow 2\ H^+ + 2\ Cl^- + 2\ Hg \qquad E_{cell} = 0.242\ V$$

b.   The hydrogen half-cell is the oxidation half-reaction, thus the hydrogen electrode is the anode.

c.   Ignoring concentrations from the SCE (they are constant):

$$E = 0.242\ V - \frac{0.0592}{2} \log \frac{[H^+]^2}{P_{H_2}}$$

If we keep $P_{H_2} = 1.0$ atm: $E = 0.242\ V - \frac{0.0592}{2} \log [H^+]^2$

Since, $\log [H^+]^2 = 2 \log [H^+] = -2$ pH, $E = 0.242 - 0.0592 \log [H^+]$ or

$E = 0.242\ V + 0.0592$ pH

d.   i) $E = 0.242 - 0.0592 \log [H^+] = 0.242 - 0.0592 \log (1.0 \times 10^{-3}) = 0.42$ V

    ii) $E = 0.242 - 0.0592 \log 2.5 = 0.242 - 0.024 = 0.218$ V

    iii) $E = 0.242 - 0.0592 \log (1.0 \times 10^{-9}) = 0.242 + 0.53 = 0.77$ V

e. $E = 0.242$ V $+ 0.0592$ pH   (from c)

    $0.285 = 0.242 + 0.0592$ pH, pH $= 0.73$; $[H^+] = 0.19\ M$

f.   Primarily the reason is convenience. It is inconvenient to deal with H_2 gas and particularly to keep P_{H_2} constant. Gas cylinders are bulky and H_2 presents a fire and explosion hazard.

71.   a. $E_{cell} = E_{ref} + 0.05916$ pH, 0.480 V $= 0.250$ V $+ 0.05916$ pH

    $pH = \dfrac{0.480 - 0.250}{0.05916} = 3.888$; Uncertainty $= \pm 1$ mV $= \pm 0.001$ V

    $pH_{max} = \dfrac{0.481 - 0.250}{0.05916} = 3.905$; $pH_{min} = \dfrac{0.479 - 0.250}{0.05916} = 3.871$

    So if the uncertainty in potential is ± 0.001 V, the uncertainty in pH is ± 0.017 or about ± 0.02 pH units.

    For this measurement, $[H^+] = 1.29 \times 10^{-4}\ M$. For an error of $+ 1$ mV, $[H^+] = 1.24 \times 10^{-4}\ M$. For an error of -1 mV, $[H^+] = 1.35 \times 10^{-4}\ M$. So the uncertainty in $[H^+]$ is $\pm 0.06 \times 10^{-4}\ M$.

b.   From the previous example, we will be within ± 0.02 pH units if we measure the potential to the nearest ± 0.001 V (1 mV).

72. $NO_3^- + 4\ H^+ + 3\ e^- \rightarrow NO + 2\ H_2O$ $E° = 0.96$ V

The $NaNO_3$ solution will have $[H^+] = 1.0 \times 10^{-7}\ M$ and $[NO_3^-] = 5.0\ M$.

$E = 0.96 - \dfrac{0.0592}{3} \log \dfrac{P_{NO}}{[H^+]^4\,[NO_3^-]}$; Substituting H^+ and NO_3^- concentrations:

$E = 0.96 - 0.54 - \dfrac{0.0592}{3} \log P_{NO} = 0.42 - \dfrac{0.0592}{3} \log P_{NO}$

At pH $= 7$ it is still easier to reduce NO_3^- than water ($E = -0.42$ V). So it will be possible to produce NO from the electrolysis of the $NaNO_3$ solution.

73.   a. $E_{meas} = E_{ref} - 0.05916 \log [F^-]$, $0.4462 = 0.2420 - 0.05916 \log [F^-]$

    $\log [F^-] = -3.4517$, $[F^-] = 3.534 \times 10^{-4}\ M$

b.   pH = 9.00; pOH = 5.00; $[OH^-] = 1.0 \times 10^{-5}\ M$

   $0.4462 = 0.2420 - 0.05916\ \log\ [[F^-] + 10.0(1.0 \times 10^{-5})]$

   $\log\ ([F^-] + 1.0 \times 10^{-4}) = -3.452,\ \ [F^-] + 1.0 \times 10^{-4} = 3.532 \times 10^{-4},\ \ [F^-] = 2.5 \times 10^{-4}\ M$

   True value is 2.5×10^{-4} and by ignoring the $[OH^-]$ we would say $[F^-]$ was 3.5×10^{-4}, so:

   $$\%\ \text{Error} = \frac{1.0 \times 10^{-4}}{2.5 \times 10^{-4}} \times 100 = 40.\ \%$$

c. $[F^-] = 2.5 \times 10^{-4};\ \ \dfrac{[F^-]}{k[OH^-]} = 50. = \dfrac{2.5 \times 10^{-4}}{10.0[OH^-]}$

   $$[OH^-] = \frac{2.5 \times 10^{-4}}{10. \times 50.} = 5.0 \times 10^{-7}\ M;\ \ pOH = 6.30;\ \ pH = 7.70$$

d. $HF \rightleftharpoons H^+ + F^-$ $K_a = \dfrac{[H^+][F^-]}{[HF]} = 7.2 \times 10^{-4};$ If 99% is F^-, then $[F^-]/[HF] = 99$.

   $99[H^+] = 7.2 \times 10^{-4},\ \ [H^+] = 7.3 \times 10^{-6}\ M;\ \ pH = 5.14$

e.   The buffer controls the pH so that there is little HF present and that there is little response to OH^-. Typically a buffer of pH = 6 is used.

74.   Cathode: $M^{2+} + 2e^- \rightarrow M(s)$ $E° = 0.80\ V$
      Anode: $M(s) \rightarrow M^{2+} + 2e^-$ $-E° = -0.80\ V$

      _____

                $M^{2+}\ (\text{cathode}) \rightarrow M^{2+}\ (\text{anode})$ $E° = 0.00\ V$

   $E_{cell} = 0.44\ V = 0.00\ V - \dfrac{0.0592}{2}\ \log\ \dfrac{[M^{2+}]_{anode}}{[M^{2+}]_{cathode}},\ \ 0.44 = -\dfrac{0.0592}{2}\ \log\ \dfrac{[M^{2+}]_a}{1.0}$

   $\log\ [M^{2+}]_a = -\dfrac{2(0.44)}{0.0592} = -14.86,\ \ [M^{2+}]_{anode} = 1.4 \times 10^{-15}\ M$

   Since we started with equal numbers of moles of SO_4^{2-} and M^{2+}, then $[M^{2+}] = [SO_4^{2-}]$ at equilibrium.

   $K_{sp} = [M^{2+}][SO_4^{2-}] = (1.4 \times 10^{-15})^2 = 2.0 \times 10^{-30}$

75.   a.

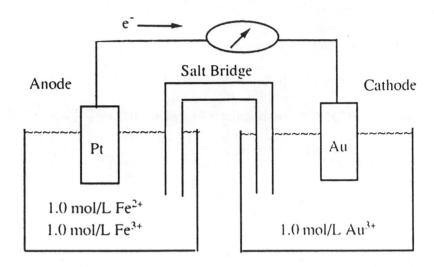

1.0 mol/L Fe^{2+}
1.0 mol/L Fe^{3+}

1.0 mol/L Au^{3+}

b. $Au^{3+} + 3\ Fe^{2+} \rightarrow 3\ Fe^{3+} + Au$ $E° = 1.50 - 0.77 = 0.73$ V

$$E_{cell} = 0.73\ V - \frac{0.0592}{3} \log \frac{[Fe^{3+}]^3}{[Au^{3+}]\ [Fe^{2+}]^3}$$

Since $[Fe^{3+}] = [Fe^{2+}] = 1.0\ M$: $0.31\ V = 0.73\ V - \dfrac{0.0592}{3} \log \dfrac{1}{[Au^{3+}]}$

$$\frac{3(-0.42)}{0.0592} = -\log \frac{1}{[Au^{3+}]}, \quad \log [Au^{3+}] = -21.28, \quad [Au^{3+}] = 10^{-21.28} = 5.2 \times 10^{-22}\ M$$

$Au^{3+} + 4\ Cl^- \rightleftharpoons AuCl_4^-$

Since the equilibrium Au^{3+} concentration is so small, assume $[AuCl_4^-] \approx [Au^{3+}]_o \approx 1.0\ M$.

$$K = \frac{[AuCl_4^-]}{[Au^{3+}]\ [Cl^-]^4} = \frac{1.0}{(5.2 \times 10^{-22})\ (0.10)^4} = 1.9 \times 10^{25}; \quad \text{Assumption good (K large)}.$$

76.   a. $3 \times (e^- + 2\ H^+ + NO_3^- \rightarrow NO_2 + H_2O)$ $E° = 0.775$ V
                         $2\ H_2O + NO \rightarrow NO_3^- + 4\ H^+ + 3\ e^-$ $-E° = -0.957$ V
       _____
            $2\ H^+(aq) + 2\ NO_3^-(aq) + NO(g) \rightarrow 3\ NO_2(g) + H_2O(l)$ $E_{cell}° = -0.182$ V $K = ?$

$$\log K = \frac{nE°}{0.0592} = \frac{3(-0.182)}{0.0592} = -9.223, \quad K = 10^{-9.223} = 5.98 \times 10^{-10}$$

b.   Let C = concentration of $HNO_3 = [H^+] = [NO_3^-]$

$$5.98 \times 10^{-10} = \frac{P_{NO_2}^3}{P_{NO}\ [H^+]^2\ [NO_3^-]^2} = \frac{P_{NO_2}^3}{P_{NO}C^4}$$

If 0.20 mol % NO_2 and P_{tot} = 1.00 atm:

$$P_{NO_2} = \frac{0.20 \text{ mol } NO_2}{100. \text{ mol total}} \times 1.00 \text{ atm} = 2.0 \times 10^{-3} \text{ atm}; \quad P_{NO} = 1.00 - 0.0020 = 1.00 \text{ atm}$$

$$5.98 \times 10^{-10} = \frac{(2.0 \times 10^{-3})^3}{(1.00) \text{ C}^4}, \quad C = 1.9 \text{ } M \text{ } HNO_3$$

77.   a. $\Delta G° = 2(-480.) + 3(86) - [3(-40)] = -582 \text{ kJ}$

$$\Delta G° = -nFE°, \quad E° = \frac{-\Delta G°}{nF} = \frac{-(-582,000 \text{ J})}{6(96,485) \text{ C}} = 1.01 \text{ V}$$

$$\log K = \frac{nE°}{0.0592} = \frac{6(1.01)}{0.0592} = 102.365, \quad K = 10^{102.365} = 2.32 \times 10^{102}$$

   b. $3 \times (2 \text{ e}^- + Ag_2S \rightarrow 2 \text{ Ag} + S^{2-})$ $E°_{Ag_2S} = ?$
          $2 \times (Al \rightarrow Al^{3+} + 3 \text{ e}^-)$ $-E° = -(-1.66)$

   _____

          $3 \text{ Ag}_2S + 2 \text{ Al} \rightarrow 6 \text{ Ag} + 3 \text{ S}^{2-} + 3 \text{ Al}^{3+}$ $E°_{cell} = 1.01 \text{ V} = E°_{Ag_2S} + 1.66$

   $E°_{Ag_2S} = 1.01 - 1.66 = -0.65 \text{ V}$

78.   a. $Zn + Cu^{2+} \rightarrow Zn^{2+} + Cu$ $E° = 1.10 \text{ V}; \quad E = 1.10 \text{ V} - \frac{0.0592}{2} \log \frac{[Zn^{2+}]}{[Cu^{2+}]}$

$$E = 1.10 \text{ V} - \frac{0.0592}{2} \log \frac{0.10}{2.50} = 1.10 \text{ V} + 0.041 \text{ V} = 1.14 \text{ V}$$

   b. $10.0 \text{ h} \times \frac{60 \text{ min}}{\text{h}} \times \frac{60 \text{ s}}{\text{min}} \times \frac{10.0 \text{ C}}{\text{s}} \times \frac{1 \text{ mol e}^-}{96,485 \text{ C}} \times \frac{1 \text{ mol Cu}}{2 \text{ mol e}^-} = 1.87 \text{ mol Cu}$

   The copper(II) concentration decreases by 1.87 mol/L and the zinc(II) concentration will increase by 1.87 mol/L.

   $[Cu^{2+}] = 2.50 - 1.87 = 0.63 \text{ } M; \quad [Zn^{2+}] = 0.10 + 1.87 = 1.97 \text{ } M$

$$E = 1.10 \text{ V} - \frac{0.0592}{2} \log \frac{1.97}{0.63} = 1.10 \text{ V} - 0.015 \text{ V} = 1.09 \text{ V}$$

   c. $1.87 \text{ mol Zn consumed} \times \frac{65.38 \text{ g Zn}}{\text{mol Zn}} = 122 \text{ g Zn}; \quad$ Mass of electrode = 200. - 122 = 78 g Zn

   $1.87 \text{ mol Cu formed} \times \frac{63.55 \text{ g Cu}}{\text{mol Cu}} = 119 \text{ g Cu}; \quad$ Mass of electrode = 200. + 119 = 319 g Cu

d. Three things could possibly cause this battery to go dead:

    1. All of the Zn is consumed.
    2. All of the Cu^{2+} is consumed.
    3. Equilibrium is reached ($E_{cell} = 0$).

We began with 2.5 mol Cu^{2+} and 200. g $\times$ 1 mol Zn/65.38 g Zn = 3.06 mol Zn. Cu^{2+} is the limiting reagent and will run out first.

$$2.50 \text{ mol } Cu^{2+} \times \frac{2 \text{ mol } e^-}{\text{mol } Cu^{2+}} \times \frac{96,485 \text{ C}}{\text{mol } e^-} \times \frac{1 \text{ s}}{10.0 \text{ C}} \times \frac{1 \text{ h}}{3600 \text{ s}} = 13.4 \text{ h}$$

For equilibrium to be reached: $E = 0 = 1.10 \text{ V} - \dfrac{0.0592}{2} \log \dfrac{[Zn^{2+}]}{[Cu^{2+}]}$

$$\frac{[Zn^{2+}]}{[Cu^{2+}]} = K = 10^{2(1.10)/0.0592} = 1.45 \times 10^{37}$$

This is such a large equilibrium constant that virtually all of the Cu^{2+} must react to reach equilibrium. So, the battery will go dead in 13.4 hours.

79.   a. $3 \text{ e}^- + 4 \text{ H}^+ + NO_3^- \rightarrow NO + 2 \text{ H}_2O$ $E° = 0.96 \text{ V}$

Nitric acid can oxidize Co to Co^{2+}, but is not strong enough to oxidize Co to Co^{3+}. Co^{2+} is the primary product.

b.   Concentrated nitric acid is about 16 mol/L. $[H^+] = [NO_3^-] = 16 \ M$; Assume $P_{NO} = 1$ atm

$$E = 0.96 \text{ V} - \frac{0.0592}{3} \log \frac{P_{NO}}{[H^+]^4 [NO_3^-]} = 0.96 - \frac{0.0592}{3} \log \frac{1}{(16)^5} = 0.96 + 0.12 = 1.08 \text{ V}$$

No, concentrated nitric acid will still only be able to oxidize Co to Co^{2+}.

c. $H_2O_2 + 2 \text{ H}^+ + 2 \text{ e}^- \rightarrow 2 \text{ H}_2O$ $E° = 1.78 \text{ V}$

Hydrogen peroxide should oxidize Co to Co^{3+}. Co^{3+} is the primary product.

CHAPTER TWELVE

QUANTUM MECHANICS AND ATOMIC THEORY

Light and Matter

1. Planck found that heated bodies only give off certain frequencies of light. Einstein's studies of the photoelectric effect.

2. $\nu = \dfrac{c}{\lambda} = \dfrac{2.998 \times 10^8 \text{ m/s}}{780. \times 10^{-9} \text{ m}} = 3.84 \times 10^{14} \text{ s}^{-1}$; $E = h\nu = 2.54 \times 10^{-19} \text{ J}$ where $h = 6.626 \times 10^{-34} \text{ J s}$

3. $\nu = \dfrac{c}{\lambda} = \dfrac{3.00 \times 10^8 \text{ m/s}}{1.0 \times 10^{-2} \text{ m}} = 3.0 \times 10^{10} \text{ s}^{-1}$

    $E = h\nu = 6.63 \times 10^{-34} \text{ J s} \times (3.0 \times 10^{10} \text{ s}^{-1}) = 2.0 \times 10^{-23} \text{ J/photon}$

    $E = 2.0 \times 10^{-23} \text{ J/photon} \times (6.02 \times 10^{23} \text{ photons/mol}) = 12 \text{ J/mol}$

4. $E_{photon} = h\nu$ and $\lambda\nu = c$. So, $\nu = \dfrac{c}{\lambda}$ and $E = \dfrac{hc}{\lambda}$

    $\lambda = \dfrac{hc}{E_{photon}} = \dfrac{(6.626 \times 10^{-34} \text{ J s})(2.998 \times 10^8 \text{ m/s})}{7.21 \times 10^{-19} \text{ J}} = 2.76 \times 10^{-7} \text{ m} = 276 \text{ nm}$

5. The energy needed to remove a single electron is:

    $\dfrac{279.7 \text{ kJ}}{\text{mol}} \times \dfrac{1 \text{ mol}}{6.022 \times 10^{23}} = 4.645 \times 10^{-22} \text{ kJ} = 4.645 \times 10^{-19} \text{ J}$

    $E = \dfrac{hc}{\lambda}$ for a single photon.

    $\lambda = \dfrac{hc}{E} = \dfrac{(6.626 \times 10^{-34} \text{ J s})(2.9979 \times 10^8 \text{ m/s})}{4.645 \times 10^{-19} \text{ J}} = 4.276 \times 10^{-7} \text{ m} = 427.6 \text{ nm}$

6. For 404.7 nm light:

    $\nu = \dfrac{c}{\lambda} = \dfrac{2.9979 \times 10^8 \text{ m/s}}{404.7 \times 10^{-9} \text{ m}} = 7.408 \times 10^{14} \text{ s}^{-1}$

$$E = h\nu = (6.626 \times 10^{-34} \text{ J s}) (7.408 \times 10^{14} \text{ s}^{-1}) = 4.909 \times 10^{-19} \text{ J}$$

$$\frac{4.909 \times 10^{-19} \text{ J}}{\text{photon}} \times \frac{6.022 \times 10^{23} \text{ photons}}{\text{mol}} = 2.956 \times 10^5 \text{ J/mol} = 295.6 \text{ kJ/mol}$$

For 435.8 nm light:

$$\nu = \frac{c}{\lambda} = \frac{2.9979 \times 10^8 \text{ m/s}}{435.8 \times 10^{-9} \text{ m}} = 6.879 \times 10^{14} \text{ s}^{-1}$$

$$E = h\nu = (6.626 \times 10^{-34} \text{ J s}) (6.879 \times 10^{14} \text{ s}^{-1}) = 4.558 \times 10^{-19} \text{ J}$$

$$\frac{4.558 \times 10^{-19} \text{ J}}{\text{photon}} \times \frac{6.022 \times 10^{23} \text{ photons}}{\text{mol}} = 2.745 \times 10^5 \text{ J/mol} = 274.5 \text{ kJ/mol}$$

7. $$\frac{890.1 \text{ kJ}}{\text{mol}} \times \frac{1 \text{ mol}}{6.022 \times 10^{23} \text{ atoms}} = \frac{1.478 \times 10^{-21} \text{ kJ}}{\text{atom}} = \frac{1.478 \times 10^{-18} \text{ J}}{\text{atom}}$$

$$E = \frac{hc}{\lambda}, \quad \lambda = \frac{hc}{E} = \frac{(6.626 \times 10^{-34} \text{ J s}) (2.9979 \times 10^8 \text{ m/s})}{1.478 \times 10^{-18} \text{ J}} = 1.344 \times 10^{-7} \text{ m} = 134.4 \text{ nm}$$

No, it will take light with a wavelength of 134.4 nm or less to ionize gold. A photon of light with a wavelength of 225 nm is longer wavelength and, thus, less energy than 134.4 nm light.

8. $$\frac{492 \text{ kJ}}{\text{mol}} \times \frac{1 \text{ mol}}{6.022 \times 10^{23}} = 8.17 \times 10^{-22} \text{ kJ} = 8.17 \times 10^{-19} \text{ J to remove 1 electron}$$

$$\lambda = \frac{hc}{E} = \frac{(6.626 \times 10^{-34} \text{ J s}) (2.998 \times 10^8 \text{ m/s})}{(8.17 \times 10^{-19} \text{ J})} = 2.43 \times 10^{-7} \text{ m} = 243 \text{ nm}$$

9.   The energy to remove a single electron is:

$$\frac{208.4 \text{ kJ}}{\text{mol}} \times \frac{1 \text{ mol}}{6.022 \times 10^{23}} = 3.461 \times 10^{-22} \text{ kJ} = 3.461 \times 10^{-19} \text{ J} = E_w$$

Energy of 254 nm light is:

$$E = \frac{hc}{\lambda} = \frac{(6.626 \times 10^{-34} \text{ J s}) (2.998 \times 10^8 \text{ m/s})}{254 \times 10^{-9} \text{ m}} = 7.82 \times 10^{-19} \text{ J}$$

$$E_{photon} = E_K + E_w, \quad E_K = 7.82 \times 10^{-19} \text{ J} - 3.461 \times 10^{-19} \text{ J} = 4.36 \times 10^{-19} \text{ J} = \text{maximum KE}$$

Hydrogen Atom: The Bohr Model

10.   In the Bohr model of the H-atom, the energy of the electron is quantized.

11.   For the H-atom: $E_n = \dfrac{-R_H}{n^2}$, $R_H = 2.178 \times 10^{-18}$ J; For a spectral transition, $\Delta E = E_f - E_i$:

$$\Delta E = \left(\frac{-R_H}{n_f^2} \right) - \left(\frac{-R_H}{n_i^2} \right) = -R_H \left(\frac{1}{n_f^2} - \frac{1}{n_i^2} \right)$$

The quantum numbers n_i and n_f are the initial and final states of the electron, respectively. If we follow this convention a positive value of ΔE will correspond to absorption of light and a negative value of ΔE will correspond to emission of light.

a. $\Delta E = -2.178 \times 10^{-18} \text{ J} \left(\frac{1}{2^2} - \frac{1}{3^2} \right) = -2.178 \times 10^{-18} \text{ J} \left(\frac{1}{4} - \frac{1}{9} \right)$

$\Delta E = -2.178 \times 10^{-18} \text{ J} (0.2500 - 0.1111) = -3.025 \times 10^{-19} \text{ J}$

The photon of light will have energy equal to 3.025×10^{-19} J.

$$|\Delta E| = E_{photon} = h\nu = \frac{hc}{\lambda} \text{ or } \lambda = \frac{hc}{|\Delta E|}$$

$$\lambda = \frac{(6.626 \times 10^{-34} \text{ J s}) (2.9979 \times 10^8 \text{ m/s})}{3.025 \times 10^{-19} \text{ J}} = 6.567 \times 10^{-7} \text{ m} = 656.7 \text{ nm}$$

b. $\Delta E = -2.178 \times 10^{-18} \text{ J} \left(\frac{1}{2^2} - \frac{1}{4^2} \right) = -4.084 \times 10^{-19} \text{ J}$

$$\lambda = \frac{hc}{|\Delta E|} = \frac{(6.626 \times 10^{-34} \text{ J s}) (2.9979 \times 10^8 \text{ m/s})}{4.084 \times 10^{-19} \text{ J}} = 4.864 \times 10^{-7} \text{ m} = 486.4 \text{ nm}$$

c. $\Delta E = -2.178 \times 10^{-18} \text{ J} \left(\frac{1}{1^2} - \frac{1}{2^2} \right) = -1.634 \times 10^{-18} \text{ J}$

$$\lambda = \frac{(6.626 \times 10^{-34} \text{ J s}) (2.9979 \times 10^8 \text{ m/s})}{1.634 \times 10^{-18} \text{ J}} = 1.216 \times 10^{-7} \text{ m} = 121.6 \text{ nm}$$

d. $\Delta E = -2.178 \times 10^{-18} \text{ J} \left(\frac{1}{3^2} - \frac{1}{4^2} \right) = -1.059 \times 10^{-19} \text{ J}$

$$\lambda = \frac{(6.626 \times 10^{-34} \text{ J s}) (2.9979 \times 10^8 \text{ m/s})}{1.059 \times 10^{-19} \text{ J}} = 1.876 \times 10^{-6} \text{ m} = 1876 \text{ nm}$$

12.   Ionization from $n = 1$ corresponds to the transition $n = 1 \rightarrow n = \infty$ where $E_\infty = 0$.

$$\Delta E = E_\infty - E_1 = -E_1 = R_H \left(\frac{1}{1^2} \right) = R_H, \ \ \Delta E = 2.178 \times 10^{-18} \text{ J} = E_{photon}$$

$$\lambda = \frac{hc}{E} = \frac{(6.626 \times 10^{-34} \text{ J s}) (2.9979 \times 10^8 \text{ m/s})}{2.178 \times 10^{-18} \text{ J}} = 9.120 \times 10^{-8} \text{ m} = 91.20 \text{ nm}$$

To ionize from n = 3, $\Delta E = 0 - E_3 = R_H\left(\dfrac{1}{3^2}\right) = 2.178 \times 10^{-18}\ J\left(\dfrac{1}{9}\right)$

$\Delta E = 2.420 \times 10^{-19}\ J;\ \lambda = 8.208 \times 10^{-7}\ m = 820.8\ nm$

13. a. False, It takes less energy to ionize an electron from n = 3 than from the ground state.

b. True

c. False, The energy difference from n = 3 → n = 2 is less than the energy difference from n = 3 → n = 1. Thus, the wavelength is larger for n = 3 → n = 2 than for n = 3 → n = 1.

d. True

e. False, n = 2 is the first excited state, n = 3 is the second excited state.

14. The longest wavelength light emitted will correspond to the transition with the lowest energy change. This is the transition from n = 6 to n = 5.

$\Delta E = -2.178 \times 10^{-18}\ J\left(\dfrac{1}{5^2} - \dfrac{1}{6^2}\right) = -2.662 \times 10^{-20}\ J$

$\lambda = \dfrac{hc}{|\Delta E|} = \dfrac{(6.626 \times 10^{-34}\ J\ s)(2.9979 \times 10^{8}\ m/s)}{2.662 \times 10^{-20}\ J} = 7.462 \times 10^{-6}\ m = 7462\ nm$

The shortest wavelength emitted will correspond to the largest ΔE; this is n = 6 → n = 1.

$\Delta E = -2.178 \times 10^{-18}\ J\left(\dfrac{1}{1^2} - \dfrac{1}{6^2}\right) = -2.118 \times 10^{-18}\ J$

$\lambda = \dfrac{hc}{|\Delta E|} = \dfrac{(6.626 \times 10^{-34}\ J\ s)(2.9979 \times 10^{8}\ m/s)}{2.118 \times 10^{-18}\ J} = 9.379 \times 10^{-8}\ m = 93.79\ nm$

15. $\Delta E = h\nu = (6.626 \times 10^{-34}\ J\ s)(1.141 \times 10^{14}\ s^{-1}) = 7.560 \times 10^{-20}\ J$

$\Delta E = -7.560 \times 10^{-20}\ J$ since light is emitted.

$\Delta E = E_4 - E_n,\ -7.560 \times 10^{-20}\ J = -2.178 \times 10^{-18}\ J\left(\dfrac{1}{4^2} - \dfrac{1}{n^2}\right)$

$3.471 \times 10^{-2} = 6.250 \times 10^{-2} - \dfrac{1}{n^2},\ \dfrac{1}{n^2} = 2.779 \times 10^{-2},\ n^2 = 35.98,\ n = 6$

Wave Mechanics and Particle in a Box

16. $\lambda = \dfrac{h}{mv}$ = wavelength of particle

    a. 5.0% of speed of light = $0.050 \times 3.00 \times 10^8$ m/s = 1.5×10^7 m/s

$$\lambda = \frac{6.63 \times 10^{-34} \text{ J s}}{1.67 \times 10^{-27} \text{ kg} \times (1.5 \times 10^7 \text{ m/s})} = 2.6 \times 10^{-14} \text{ m} = 2.6 \times 10^{-5} \text{ nm}$$

Note: For units to come out, the mass must be in kg since $1 \text{ J} = \dfrac{1 \text{ kg m}^2}{\text{s}^2}$.

    b. $\lambda = \dfrac{6.63 \times 10^{-34} \text{ J s}}{9.11 \times 10^{-31} \text{ kg} \times (0.15 \times 3.00 \times 10^8 \text{ m/s})} = 1.6 \times 10^{-11} \text{ m} = 1.6 \times 10^{-2} \text{ nm}$

    c. $m = 5.2 \text{ oz} \times \dfrac{1 \text{ lb}}{16 \text{ oz}} \times \dfrac{1 \text{ kg}}{2.205 \text{ lb}} = 0.15 \text{ kg}$

$$v = \frac{100.8 \text{ mi}}{\text{hr}} \times \frac{1 \text{ hr}}{3600 \text{ s}} \times \frac{1760 \text{ yd}}{\text{mi}} \times \frac{0.9144 \text{ m}}{\text{yd}} = 45.06 \text{ m/s}$$

$$\lambda = \frac{h}{mv} = \frac{6.63 \times 10^{-34} \text{ J s}}{0.15 \text{ kg} \times 45.06 \text{ m/s}} = 9.8 \times 10^{-35} \text{ m} = 9.8 \times 10^{-26} \text{ nm}$$

This number is so small that it is essentially zero. We cannot detect a wavelength this small. The meaning of this number is that we do not have to consider the wave properties of large objects.

17.   a. $\lambda = \dfrac{h}{mv} = \dfrac{6.626 \times 10^{-34} \text{ J s}}{(1.675 \times 10^{-27} \text{ kg}) (0.0100 \times 2.998 \times 10^8 \text{ m/s})} = 1.32 \times 10^{-13} \text{ m}$

    b. $\lambda = \dfrac{h}{mv}$, $\lambda mv = h$, $v = \dfrac{h}{\lambda m} = \dfrac{6.626 \times 10^{-34} \text{ J s}}{(75 \times 10^{-12} \text{ m}) (1.675 \times 10^{-27} \text{ kg})} = 5.3 \times 10^3$ m/s

18. $\lambda = \dfrac{h}{mv}$, $v = \dfrac{h}{\lambda m}$; For $\lambda = 1.0 \times 10^2$ nm = 1.0×10^{-7} m:

$$v = \frac{6.63 \times 10^{-34} \text{ J s}}{(9.11 \times 10^{-31} \text{ kg}) (1.0 \times 10^{-7} \text{ m})} = 7.3 \times 10^3 \text{ m/s}$$

For $\lambda = 1.0$ nm = 1.0×10^{-9} m: $v = \dfrac{6.63 \times 10^{-34} \text{ J s}}{(9.11 \times 10^{-31} \text{ kg}) (1.0 \times 10^{-9} \text{ m})} = 7.3 \times 10^5$ m/s

19. $\lambda = \dfrac{h}{mv}$ or $h = \lambda mv$; If $\lambda = 5.0$ cm, $m = 0.15$ kg, and $v = 45.06$ m/s then:

$$h = 5.0 \times 10^{-2} \text{ m} \times 0.15 \text{ kg} \times 45.06 \text{ m/s} = 0.34 \text{ J s}$$

For $\lambda = 5.0 \times 10^2$ nm: $h = 5.0 \times 10^{-7}$ m $\times 0.15$ kg $\times 45.06$ m/s $= 3.4 \times 10^{-6}$ J s

20.   1) Electrons can be diffracted like light.

        2) The electron microscope uses electrons in a fashion similar to the way in which light is used in a light microscope.

21.   Units of $\Delta E \times \Delta t = $ J $\times$ s, the same as the units of Planck's constant. Linear momentum, p, is equal to mass times velocity, $p = mv$.

Units of $\Delta p \Delta x = $ kg $\times \dfrac{m}{s} \times m = \dfrac{kg\,m^2}{s} = \dfrac{kg\,m^2}{s^2} \times s = $ J $\times$ s

22.   a. $\Delta p = m\Delta v = 9.11 \times 10^{-31}$ kg $\times 0.100$ m/s $= \dfrac{9.11 \times 10^{-32}\,kg\,m}{s}$

         $\Delta p \Delta x \geq \dfrac{h}{4\pi}$, $\Delta x = \dfrac{h}{4\pi\Delta p} = \dfrac{6.626 \times 10^{-34}\,J\,s}{4 \times 3.142 \times (9.11 \times 10^{-32}\,kg\,m/s)} = 5.79 \times 10^{-4}$ m

    b. $\Delta x = \dfrac{h}{4\pi\Delta p} = \dfrac{6.626 \times 10^{-34}\,J\,s}{4 \times 3.142 \times 0.145\,kg \times 0.100\,m/s} = 3.64 \times 10^{-33}$ m

    c. Diameter of H atom is roughly 1.0×10^{-8} cm. The uncertainty in position is much larger than the size of the atom.

    d. The uncertainty is insignificant compared to the size of a baseball.

23. $E_n = \dfrac{n^2 h^2}{8\,mL^2}$; $\Delta E = E_5 - E_1 = \dfrac{h^2}{8\,mL^2}(5^2 - 1^2)$

    $\Delta E = \dfrac{(6.626 \times 10^{-34}\,J\,s)^2}{8(9.11 \times 10^{-31}\,kg)(40.0 \times 10^{-12}\,m)^2}(24) = 9.04 \times 10^{-16}$ J

    $\Delta E = \dfrac{hc}{\lambda}$, $\lambda = \dfrac{hc}{\Delta E} = \dfrac{(6.626 \times 10^{-34}\,J{\bullet}s)(2.998 \times 10^8\,m/s)}{(9.04 \times 10^{-16}\,J)} = 2.20 \times 10^{-10}$ m $= 0.220$ nm

24. $E_n = \dfrac{n^2 h^2}{8\,mL^2}$; As L increases, E_n will decrease and the spacing between energy levels will also decrease.

25. $E_n = \dfrac{n^2 h^2}{8\,mL^2}$, $n = 1$ for ground state

    10^{-6} m box: $E_1 = \dfrac{h^2}{8\,m}(10^{12}\,m^{-2})$; 10^{-10} m box: $E_1 = \dfrac{h^2}{8\,m}(10^{20}\,m^{-2})$

    The e^- in the 10^{-6} m box has the lowest ground state energy.

26.   a. $L_x = L_y = L_z$; $E = \dfrac{h^2(n_x^2 + n_y^2 + n_z^2)}{8\,mL^2}$

$E_{111} = \dfrac{3\,h^2}{8\,mL^2}$; $E_{112} = \dfrac{h^2}{8\,mL^2}\,(1^2 + 1^2 + 2^2) = \dfrac{6\,h^2}{8\,mL^2}$

$E_{122} = \dfrac{h^2}{8\,mL^2}\,(1^2 + 2^2 + 2^2) = \dfrac{9\,h^2}{8\,mL^2}$

b. E_{111}: only a single state; E_{112}: triple degenerate, either n_x, n_y or n_z can equal 2; E_{122}: triple degenerate, either n_x, n_y or n_z can equal 1; E_{222}: single state

Orbitals and Quantum Numbers

27.   n: gives the energy (it completely specifies the energy only for the H-atom or ions with one electron) and the relative size of the orbitals.

ℓ: gives the type (shape) of orbital.

$m_ℓ$: gives information about the direction in which the orbital is pointing in space.

28.   1p: n = 1, ℓ = 1, not possible; 3f: n = 3, ℓ = 3, not possible; 2d: n = 2, ℓ = 2, not possible; In all three incorrect cases, n = ℓ. The maximum value ℓ can have is n - 1, not n.

29.   b. ℓ must be smaller than n.   d. For ℓ = 0, $m_ℓ$ = 0 only allowed value.

f. For ℓ = 3, $m_ℓ$ can range from -3 to +3; thus +4 is not allowed.

g. n cannot equal zero.   h. ℓ cannot be a negative number.

30.   No, for n = 2, the allowed values of ℓ are 0 and 1; for n = 3 the allowed values of ℓ are 0, 1, and 2.

31.   5p: three orbitals $3d_{z^2}$: one orbital   4d: five orbitals

        $n = 5$: $\ell = 0$ (1 orbital), $\ell = 1$ (3 orbitals), $\ell = 2$ (5 orbitals), $\ell = 3$ (7 orbitals), $\ell = 4$ (9 orbitals)

        Total for $n = 5$ is 25 orbitals.

        $n = 4$: $\ell = 0$ (1), $\ell = 1$ (3), $\ell = 2$ (5), $\ell = 3$ (7);   Total for $n = 4$ is 16 orbitals.

32.   The diagrams of orbitals give probabilities. We can never be 100% certain of the location of the electrons.

33.

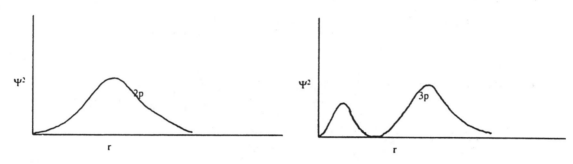

34.   The 2p orbitals differ from each other in the direction in which they point in space.

35.   The 2p and 3p orbitals differ from each other in their size and number of nodes.

36.   A nodal surface in an atomic orbital is a surface in which the probability of finding an electron is zero.

37. ψ^2 gives the probability of finding the electron at that point.

38.   A node occurs when $\psi = 0$. $\psi_{300} = 0$ when $27 - 18\sigma + 2\sigma^2 = 0$.

        Solving using the quadratic formula: $\sigma = \dfrac{18 \pm \sqrt{(18)^2 - 4(2)\,(27)}}{4} = \dfrac{18 \pm \sqrt{108}}{4}$

        $\sigma = 7.10$ or $\sigma = 1.90$; Since $\sigma = r/a_o$, the nodes occur at $r = 7.10\ a_o = 3.76 \times 10^{-10}$ m and at $r = 1.90\ a_o = 1.01 \times 10^{-10}$ m where r is the distance from the nucleus.

Polyelectronic Atoms

39.   The electrostatic energy of repulsion from Coulomb's Law will be of the form Q^2/r where Q is the charge of the electron and r is the distance between the two electrons. From the Heisenberg uncertainty principle, we cannot know precisely the position of each electron. Thus, we cannot precisely know the distance between the electrons nor the value of the electrostatic repulsions.

40. $E_n = \dfrac{-R_H Z^2}{n^2}$, $R_H = 2.178 \times 10^{-18}$ J, Z = atomic number (nuclear charge)

Normal ionization energy is for the transition $n = 1 \to n = \infty$ (E = 0). Thus, the I.E. is given by the energy of state n = 1.

a.   IE = 2.178×10^{-18} J $\left(\dfrac{1^2}{1^2}\right)$ = 2.178×10^{-18} J/atom

IE = $\dfrac{2.178 \times 10^{-18}\ \text{J}}{\text{atom}} \times \dfrac{1\ \text{kJ}}{1000\ \text{J}} \times \dfrac{6.022 \times 10^{23}\ \text{atoms}}{\text{mol}}$ = 1311.6 kJ/mol ≈ 1312 kJ/mol

For brevity, the I.E. of heavier one electron species can be given as:

IE = $\dfrac{1311.6\ \text{kJ}}{\text{mol}} \left(\dfrac{Z^2}{n^2}\right)$ (We will carry an extra significant figure.)

We get this by combining IE = 2.178×10^{-18} J $\left(\dfrac{Z^2}{n^2}\right)$ and

$\dfrac{2.178 \times 10^{-18}\ \text{J}}{\text{atom}} \times \dfrac{6.022 \times 10^{23}\ \text{atoms}}{\text{mol}} \times \dfrac{1\ \text{kJ}}{1000\ \text{J}} = \dfrac{1311.6\ \text{kJ}}{\text{mol}}$.

b. He^+:   Z = 2; IE = 1311.6 kJ/mol × 2^2 = 5246 kJ/mol   (Assume n = 1 for all.)

c. Li^{2+}:   Z = 3; IE = 1311.6 kJ/mol × 3^2 = 1.180×10^4 kJ/mol

d. C^{5+}:   Z = 6; IE = 1311.6 kJ/mol × 6^2 = 4.722×10^4 kJ/mol

e. Fe^{25+}:   Z = 26; IE = 1311.6 kJ/mol × $(26)^2$ = 8.866×10^5 kJ/mol

41.   The size of the 1s orbitals would be proportional to 1/Z, that is as Z increases, the electrons are more strongly attracted to the nucleus and will be drawn in closer. Thus the relative sizes would be:

H : He^+ : Li^{2+} : C^{5+} : Fe^{25+} → $1 : \dfrac{1}{2} : \dfrac{1}{3} : \dfrac{1}{6} : \dfrac{1}{26}$

42. $E_{photon} = \dfrac{hc}{\lambda} = \dfrac{6.626 \times 10^{-34}\ \text{J s}\ (2.9979 \times 10^8\ \text{m/s})}{253.4 \times 10^{-9}\ \text{m}} = 7.839 \times 10^{-19}$ J, $\Delta E = -7.839 \times 10^{-19}$ J

$\Delta E = -2.178 \times 10^{-18}$ J $(Z)^2 \left(\dfrac{1}{n_f^2} - \dfrac{1}{n_i^2}\right)$, Z = 4 for Be^{3+}

-7.839×10^{-19} J $= -2.178 \times 10^{-18}\ (4)^2 \left(\dfrac{1}{n_f^2} - \dfrac{1}{5^2}\right)$

$\dfrac{7.839 \times 10^{-19}}{2.178 \times 10^{-18}\ (16)} + \dfrac{1}{25} = \dfrac{1}{n_f^2}$, $\dfrac{1}{n_f^2} = 0.06249$, $n_f = 4$

This emission line corresponds to the n = 5 → n = 4 electronic transition.

43.   No, the spin is a convenient model. Since we cannot locate or "see" the electron, we cannot determine if it is spinning.

44.   a.   n = 4: ℓ can be 0, 1, 2, or 3. Thus we have s(2 e⁻), p(6 e⁻), d(10 e⁻) and f(14 e⁻) orbitals present. Total number of electrons to fill these orbitals is 32.

       b.   n = 5, m_ℓ = +1: For n = 5, ℓ = 0, 1, 2, 3, 4. For ℓ = 1, 2, 3, 4, all can have m_ℓ = +1. Four distinct orbitals, thus 8 electrons.

       c.   n = 5, m_s = +1/2: For n = 5, ℓ = 0, 1, 2, 3, 4. Number of orbitals = 1, 3, 5, 7, 9 for each value of ℓ, respectively. There are 25 orbitals with n = 5. They can hold 50 electrons and 25 of these electrons can have m_s = +1/2.

       d.   n = 3, ℓ = 2: These quantum numbers define a set of 3d orbitals. There are 5 degenerate 3d orbitals which can hold a total of 10 electrons.

       e.   n = 2, ℓ = 1; These define a set of 2p orbitals. There are 3 degenerate 2p orbitals which can hold a total of 6 electrons.

       f.   It is impossible for n = 0. Thus, no electrons can have this set of quantum numbers.

       g.   The 4 quantum numbers completely specify a single electron.

       h.   n = 3:   3s, 3p and 3d orbitals all have n = 3. These orbitals can hold up to 18 electrons.

       i.   n = 2, ℓ = 2: This combination is not possible ($\ell \neq$ 2 for n = 2). Zero electrons in an atom can have these quantum numbers.

       j.   n = 1, ℓ = 0, m_ℓ = 0: These define a 1s orbital which can hold 2 electrons.

45.   Si: $1s^2 2s^2 2p^6 3s^2 3p^2$ or $[Ne]3s^2 3p^2$; Ga: $1s^2 2s^2 2p^6 3s^2 3p^6 4s^2 3d^{10} 4p^1$ or $[Ar]4s^2 3d^{10} 4p^1$

       As: $[Ar]4s^2 3d^{10} 4p^3$; Ge: $[Ar]4s^2 3d^{10} 4p^2$; Al: $[Ne]3s^2 3p^1$; Cd: $[Kr]5s^2 4d^{10}$

       S: $[Ne]3s^2 3p^4$; Se: $[Ar]4s^2 3d^{10} 4p^4$

46.   Cu: $[Ar]4s^2 3d^9$ (using periodic table), $[Ar]4s^1 3d^{10}$ (actual)

       O: $1s^2 2s^2 2p^4$; La: $[Xe]6s^2 5d^1$; Y: $[Kr]5s^2 4d^1$; Ba: $[Xe]6s^2$

       Tl: $[Xe]6s^2 4f^{14} 5d^{10} 6p^1$; Bi: $[Xe]6s^2 4f^{14} 5d^{10} 6p^3$

47.   The following are complete electron configurations. Noble gas shorthand notation could also be used.

Sc: $1s^22s^22p^63s^23p^64s^23d^1$; Fe: $1s^22s^22p^63s^23p^64s^23d^6$

P: $1s^22s^22p^63s^23p^3$; Cs: $1s^22s^22p^63s^23p^64s^23d^{10}4p^65s^24d^{10}5p^66s^1$

Eu: $1s^22s^22p^63s^23p^64s^23d^{10}4p^65s^24d^{10}5p^66s^24f^65d^1$*

Pt: $1s^22s^22p^63s^23p^64s^23d^{10}4p^65s^24d^{10}5p^66s^24f^{14}5d^8$*

Xe: $1s^22s^22p^63s^23p^64s^23d^{10}4p^65s^24d^{10}5p^6$; Br: $1s^22s^22p^63s^23p^64s^23d^{10}4p^5$

*Note: These electron configurations were written down using only the periodic table.

Actual electron configurations are: Eu: $[Xe]6s^24f^7$ and Pt: $[Xe]6s^14f^{14}5d^9$

48.   K: $1s^22s^22p^63s^23p^64s^1$; Rb: $1s^22s^22p^63s^23p^64s^23d^{10}4p^65s^1$

Fr: $1s^22s^22p^63s^23p^64s^23d^{10}4p^65s^24d^{10}5p^66s^24f^{14}5d^{10}6p^67s^1$ or $[Rn]7s^1$

Pu: $[Rn]7s^26d^15f^5$ (expected from periodic table)

Sb: $1s^22s^22p^63s^23p^64s^23d^{10}4p^65s^24d^{10}5p^3$; Os: $[Xe]6s^24f^{14}5d^6$

Pd: $1s^22s^22p^63s^23p^64s^23d^{10}4p^65s^24d^8$ (expected from periodic table)

Pb: $[Xe]6s^24f^{14}5d^{10}6p^2$; I: $1s^22s^22p^63s^23p^64s^23d^{10}4p^65s^24d^{10}5p^5$

49.   Exceptions: Cr, Cu, Nb, Mo, Tc, Ru, Rh, Pd, Ag, Pt, Au; Tc, Ru, Rh, Pd and Pt do not correspond to the supposed extra stability of half-filled and filled subshells.

50.   a.   The smallest halogen is fluorine: $1s^22s^22p^5$ b.   K: $1s^22s^22p^63s^23p^64s^1$

c.   Be: $1s^22s^2$;   Mg: $1s^22s^22p^63s^2$;   Ca: $1s^22s^22p^63s^23p^64s^2$

d.   In: $[Kr]5s^24d^{10}5p^1$ e.   C : $1s^22s^22p^2$; Si : $1s^22s^22p^63s^23p^2$

f.   This will be element #118: $[Rn]7s^25f^{14}6d^{10}7p^6$

51.   There is a higher probability of finding the 4s electron very close to the nucleus than for the 3d electron.

52.   When atoms interact with each other, it will be the outermost electrons that are involved in those interactions. The outermost electrons are the valence electrons.

53.   O: $1s^2 2s^2 2p_x^2 2p_y^2$ ($\underline{\uparrow\downarrow}\ \underline{\uparrow\downarrow}\ \underline{\quad}$); There are no unpaired electrons in this oxygen atom. This configuration would be an excited state and in going to the ground state ($\underline{\uparrow\downarrow}\ \underline{\uparrow}\ \underline{\uparrow}$), energy would be released.

54.   Sc: $[Ar]4s^2 3d^1$, one unpaired d electron; Ti: $[Ar]4s^2 3d^2$, two unpaired d electrons

       Al: $[Ne]3s^2 3p^1$, one unpaired p electron; Sn: $[Kr]5s^2 4d^{10} 5p^2$, two unpaired p electrons

       Te: $[Kr]5s^2 4d^{10} 5p^4$, two unpaired p electrons; Br: $[Ar]4s^2 3d^{10} 4p^5$, one unpaired p electron

55.   a.   This atom has 10 electrons. Ne   b.   S

       c.   The ground state configuration is $[Kr]5s^2 4d^9$. The element is Ag.

       d.   Bi: $[Xe]6s^2 4f^{14} 5d^{10} 6p^3$

The Periodic Table and Periodic Properties

56.   a.   Be < Mg < Ca   b.   Xe < I < Te   c.   Ge < Ga < In

       d.   F < N < As   e.   F < Cl < S

57.   a.   Ca < Mg < Be   b.   Te < I < Xe   c.   In < Ga < Ge

       d.   As < N < F   e.   S < Cl < F

58.   Ge: $[Ar]4s^2 3d^{10} 4p^2$; As: $[Ar]4s^2 3d^{10} 4p^3$; Se: $[Ar]4s^2 3d^{10} 4p^4$

       There are extra electron-electron repulsions in Se because two electrons are in the same 4p orbital, resulting in a lower ionization energy.

59.   If one more electron is added to a half-filled subshell, electron-electron repulsions will increase.

60.   Size also decreases going across a period. Sc & Ti and Y & Zr are adjacent elements. There are 14 elements (the lanthanides) between La and Hf, making Hf considerably smaller.

61.   a.   Sg: $[Rn]7s^2 5f^{14} 6d^4$ b.   W   c. SgO_3 and SgO_4^{2-} (similar to Cr)

62.   a.   Uus will have 117 electrons. $[Rn]7s^2 5f^{14} 6d^{10} 7p^5$

       b.   It will be in the halogen family and most similar to astatine, At.

       c.   NaUus, $Mg(Uus)_2$, $C(Uus)_4$, $O(Uus)_2$

       d.   Assuming Uus is like the other halogens: $UusO^-$, $UusO_2^-$, $UusO_3^-$, $UusO_4^-$

63.   a. O < S, S most exothermic   b. I < Br < F < Cl, Cl most exothermic

      c. N < O < F, F most exothermic

64.   Electron-electron repulsions become more important when we try to add electrons to an atom. From the standpoint of electron-electron repulsions, larger atoms would have more favorable (more exothermic) electron affinities. Considering only electron-nucleus attractions, smaller atoms would be expected to have the more favorable (more exothermic) EA's. These trends are exactly the opposite of each other. Thus, the overall variation in EA is not as great as ionization energy in which attractions to the nucleus dominate.

65.   Yes, since it is an exothermic EA there are conditions in which we might expect Na$^-$ to exist. Such compounds were first synthesized by James L. Dye at Michigan State University.

66.   O, because electron-electron repulsions will be much more severe for O$^-$ + e$^-$ → O^{2-} than for O + e$^-$ → O$^-$.

67.   a. Li   b. P   c. O$^+$

      d. From the radii trend, Ar < Cl < S and Kr > Ar. Since variation in size down a family is greater than the variation across a period, we would predict Cl to be the smallest of the three.

      e. Cu

68.   a. Cs   b. Ga   c. Tl   d. Tl   e. O^{2-}

69.   Al (-44 kJ/mol), Si (-120), P (-74), S (-200.4), Cl (-348.7);   Based on effective nuclear charge, we would expect the EA to become more exothermic as we go from left to right in the period. Phosphorus is out of line. The reaction for the EA of P is:

$$P(g) \quad + \quad e^- \quad \rightarrow \quad P^-(g)$$
$$[Ne]3s^23p^3 \qquad\qquad\qquad [Ne]3s^23p^4$$

      The additional electron in P$^-$ will have to go into an orbital that already has one electron. There will be greater repulsion between electrons in P$^-$, causing the EA to be less favorable than predicted based solely on attractions to the nucleus.

70.   Ar   : $1s^22s^22p^63s^23p^6$:   1.527 MJ/mol : 0.98 Å

      Mg   : $1s^22s^22p^63s^2$:   0.735 MJ/mol : 1.60 Å

      K   : $1s^22s^22p^63s^23p^64s^1$:   0.419 MJ/mol : 2.35 Å

      Size: Ar < Mg < K; IE: K < Mg < Ar

71.   a.   The electron affinity of Mg^{2+} is ΔH for: $Mg^{2+}(g) + e^- \rightarrow Mg^+(g)$; This is just the reverse of the second ionization energy, or: $EA(Mg^{2+}) = -IE_2(Mg) = -1445$ kJ/mol

       b.   EA of Al^+ is ΔH for: $Al^+(g) + e^- \rightarrow Al(g)$; $EA(Al^+) = -IE_1(Al) = -580$ kJ/mol

       c.   IE of Cl^- is ΔH for: $Cl^-(g) \rightarrow Cl(g) + e^-$; $IE(Cl^-) = -EA(Cl) = +348.7$ kJ/mol

       d. $Cl(g) \rightarrow Cl^+(g) + e^-$ IE = 1255 kJ/mol (Table 12.6)

       e. $Cl^+(g) + e^- \rightarrow Cl(g)$ $\Delta H = -IE_1 = -1255$ kJ/mol $= EA(Cl^+)$

72.   No, our generalization was that the first ionization energy decreases as we go down a group. For the coinage metals the first ionization energy decreases and then increases as we go down the group.

Sc 549.5 kJ/mol	Ti 658	Fe 759.4	Ga 578.8	As 944
Y 616	Zr 660	Ru 711	In 558.3	Sb 831.6
La 538.1	Hf 654	Os 840	Tl 589.3	Bi 703.3

(Fe, Ru, Os) and (Ga, In, Tl) follow the same trend as the coinage metals. Only (As, Sb, Bi) follow the trend we would predict.

73. $EA = -IE_1$ for the neutral atoms.

       Cu^+, $EA = -745.5$ kJ/mol; Ag^+, $EA = -731.0$ kJ/mol; Au^+, $EA = -890.1$ kJ/mol

74.   As electrons are removed, the effective nuclear charge exerted on the remaining electrons increases. Since the remaining electrons are 'held' more strongly by the nucleus, the energy required to remove these electrons increases.

75.   Yes, the electron configuration is $1s^2 2s^2 2p^6 3s^2 3p^2$. There should be another big jump when the thirteenth electron is removed, i.e., when the 1s electrons begin to be removed. The 1s electrons, on the average, are closer to the nucleus than the electrons in the n = 2 orbitals.

76.   a.   More favorable EA: Li, S, B, Cl   b.   Higher IE: Li, S, N, F

       c.   Larger radius: K, Sc, B, Cl

The Alkaline Metals

77. Li^+ ions will be the smallest of the alkali metal cations, and will be most strongly attracted to the water molecules.

78.   It should be potassium peroxide, K_2O_2. K^{2+} ions are not stable; the second ionization energy of K is very large compared to the first.

79.   a. $4 \, Li(s) + O_2(g) \rightarrow 2 \, Li_2O(s)$ b. $2 \, K(s) + S(s) \rightarrow K_2S(s)$

      c. $2 \, Cs(s) + 2 \, H_2O(l) \rightarrow 2 \, CsOH(aq) + H_2(g)$ d. $2 \, Na(s) + Cl_2(g) \rightarrow 2 \, NaCl(s)$

80. $\nu = \dfrac{c}{\lambda} = \dfrac{2.9979 \times 10^8 \text{ m/s}}{455.5 \times 10^{-9} \text{ m}} = 6.582 \times 10^{14} \text{ s}^{-1}$

      $E = h\nu = (6.626 \times 10^{-34} \text{ J s})(6.582 \times 10^{14} \text{ s}^{-1}) = 4.361 \times 10^{-19} \text{ J}$

81.   a.   Carbonate ion is CO_3^{2-}. Lithium form Li^+ ions. Thus, lithium carbonate is Li_2CO_3.

      b. $\dfrac{1 \times 10^{-3} \text{ mol Li}}{L} \times \dfrac{6.9 \text{ g Li}}{\text{mol Li}} = \dfrac{7 \times 10^{-3} \text{ g Li}}{L}$

82.   a. Li_3N;   lithium nitride   b.   NaBr;   sodium bromide

      c. K_2S;   potassium sulfide   d. Li_3P;   lithium phosphide

      e.   RbH;   rubidium hydride   f.   NaH;   sodium hydride

83.   It should be element #119 with ground state electron configuration: $[Rn] \, 7s^2 5f^{14} 6d^{10} 7p^6 8s^1$

Additional Exercises

84. $\nu = \dfrac{c}{\lambda}$ and $E = h\nu$; For 589.0 nm: $\nu = \dfrac{2.9979 \times 10^8 \text{ m/s}}{589.0 \times 10^{-9} \text{ m}} = 5.090 \times 10^{14} \text{ s}^{-1}$

      $E = 6.626 \times 10^{-34} \text{ J s} \times (5.090 \times 10^{14} \text{ s}^{-1}) = 3.373 \times 10^{-19} \text{ J}$

      For 589.6 nm: $\nu = 5.085 \times 10^{14} \text{ s}^{-1}$; $E = 3.369 \times 10^{-19} \text{ J}$

      The energies in kJ/mol are:

      $3.373 \times 10^{-19} \text{ J} \times \dfrac{1 \text{ kJ}}{1000 \text{ J}} \times \dfrac{6.022 \times 10^{23}}{\text{mol}} = 203.1 \text{ kJ/mol}$

      $3.369 \times 10^{-19} \text{ J} \times \dfrac{1 \text{ kJ}}{1000 \text{ J}} \times \dfrac{6.022 \times 10^{23}}{\text{mol}} = 202.9 \text{ kJ/mol}$

85.   Each element has a characteristic spectrum. Thus, the presence of the characteristic spectral lines of an element confirms its presence in any particular sample.

86.   We get a number of unpaired electrons by looking at the incompletely filled subshells.

O: $[He]2s^22p^4$: 2 unpaired e⁻: $2p^4$:   ⇅ ↿ ↿

O⁺: $[He]2s^22p^3$: 3 unpaired e⁻: $2p^3$:   ↿ ↿ ↿

O⁻: $[He]2s^22p^5$: 1 unpaired e⁻: $2p^5$:   ⇅ ⇅ ↿

Fe: $[Ar]4s^23d^6$: 4 unpaired e⁻: $3d^6$:   ⇅ ↿ ↿ ↿ ↿

Mn: $[Ar]4s^23d^5$: 5 unpaired e⁻: $3d^5$:   ↿ ↿ ↿ ↿ ↿

S: $[Ne]3s^23p^4$: 2 unpaired e⁻: $3p^4$:   ⇅ ↿ ↿

F: $[He]2s^22p^5$: 1 unpaired e⁻: $2p^5$:   ⇅ ⇅ ↿

Ar: $[Ne]3s^23p^6$: no unpaired e⁻: $3p^6$:   ⇅ ⇅ ⇅

87.   a. n   b. n and ℓ

88.   For $r = a_o$ and $\theta = 0°$ (Z = 1 for H):

$$\psi_{2p_z} = \frac{1}{4(2\pi)^{1/2}}\left(\frac{1}{5.29 \times 10^{-11}}\right)^{3/2}(1)\,e^{-1/2}\cos 0 = 1.57 \times 10^{14};\quad \psi^2 = 2.46 \times 10^{28}$$

For $r = a_o$ and $\theta = 90°$: $\psi_{2p_z} = 0$ since $\cos 90° = 0$; $\psi^2 = 0$

89.   A large IE is usually the result of a large Z_{eff}. A large Z_{eff} can also result in a large, exothermic EA. The noble gases are an exception. The noble gases have a large IE but have an endothermic EA. Noble gases have a stable arrangement of electrons. Adding an electron disrupts this stable arrangement, resulting in unfavorable electron affinities.

90.   a. $P(g) \rightarrow P^+(g) + e^-$ b. $P(g) + e^- \rightarrow P^-(g)$

        It is important to remember that these terms are defined for the gaseous state.

91.   The IE is for removal of the electron from the atom in the gas phase. The work function is for the removal of an electron from the solid.

        $M(g) \rightarrow M^+(g) + e^-$ IE; $M(s) \rightarrow M^+(s) + e^-$ work function

92.   The electron is no longer part of that atom. The proton and electron are completely separated.

93.   a. excited state of boron   b. ground state of neon

        B ground state: $1s^22s^22p^1$

c. excited state of fluorine   d. excited state of iron

F ground state: $1s^2 2s^2 2p^5$ Fe ground state: $[Ar]4s^2 3d^6$

94.   Yes, the maximum number of unpaired electrons in any configuration corresponds to a minimum in electron-electron repulsions.

95.   Electron-electron repulsions are much greater in O⁻ than S⁻ because the electron goes into a smaller 2p orbital vs. the larger 3p orbital in sulfur. This results in a more favorable (more exothermic) EA for sulfur.

96. $E_{photon} = \dfrac{hc}{\lambda} = \dfrac{6.626 \times 10^{-34} \text{ J s} \times (2.998 \times 10^8 \text{ m/s})}{589 \times 10^{-9} \text{ m}} = 3.37 \times 10^{-19} \text{ J}$

$E = 3.37 \times 10^{-19} \text{ J} \times \dfrac{1 \text{ kJ}}{1000 \text{ J}} \times \dfrac{6.022 \times 10^{23}}{\text{mol}} = 203 \text{ kJ/mol}$

No, since $E_{photon} < IE$, this light cannot ionize sodium.

97.   Expected order: Li < Be < B < C < N < O < F < Ne

B and O are out of order. The IE of O (and also F and Ne) is lower because of the extra electron-electron repulsions present when two electrons are paired in the same orbital. B is out of order because of the different penetrating abilities of the 2p electron in B compared to the 2s electrons in Be.

98. | | n | ℓ | m_ℓ | m_s | Ti : $[Ar]4s^2 3d^2$
       |---|---|---|---|---|

| 4s | 4 | 0 | 0 | +1/2 |

| 4s | 4 | 0 | 0 | -1/2 |

| 3d | 3 | 2 | -2 | +1/2 | Only one of 10 possible combinations of m_ℓ and m_s for the first d electron. For the ground state, the second d electron should be in a

| 3d | 3 | 2 | -1 | +1/2 | different orbital with spin parallel; 4 possibilities.

99. $E = \dfrac{310. \text{ kJ}}{\text{mol}} \times \dfrac{1 \text{ mol}}{6.022 \times 10^{23}} = 5.15 \times 10^{-22} \text{ kJ} = 5.15 \times 10^{-19} \text{ J}$

$\lambda = \dfrac{hc}{E} = \dfrac{6.626 \times 10^{-34} \text{ J s} \times (2.998 \times 10^8 \text{ m/s})}{5.15 \times 10^{-19} \text{ J}} = 3.86 \times 10^{-7} \text{ m} = 386 \text{ nm}$

100.   a. n = 3: We can have 3s, 3p, and 3d orbitals. Nine orbitals can hold 18 electrons.

       b. n = 2, ℓ = 0: Specifies a 2s orbital. 2 electrons

c. $n = 2$, $\ell = 2$: Not possible; No electrons can have this combination of quantum numbers.

d. These four quantum numbers completely specify a single electron.

101. $n = 5$; $m_\ell = -4, -3, -2, -1, 0, 1, 2, 3, 4$; 18 electrons

102.   Elements in the same group have similar chemical properties.

103. a. $Cu^+(g) + e^- \rightarrow Cu(g)$ $-I_1 = -746$ kJ
          $Cu^+(g) \rightarrow Cu^{2+}(g) + e^-$ $I_2 = 1958$ kJ

          $2\ Cu^+(g) \rightarrow Cu(g) + Cu^{2+}(g)$ $\Delta H = 1212$ kJ

b. $Na^-(g) \rightarrow Na(g) + e^-$ $-EA = 52$ kJ
          $Na^+(g) + e^- \rightarrow Na(g)$ $-I_1 = -495$ kJ

          $Na^-(g) + Na^+(g) \rightarrow 2\ Na(g)$ $\Delta H = -443$ kJ

c. $Mg^{2+}(g) + e^- \rightarrow Mg^+(g)$ $-I_2 = -1445$ kJ
            $K(g) \rightarrow K^+(g) + e^-$ $I_1 = 419$ kJ

     $Mg^{2+}(g) + K(g) \rightarrow Mg^+(g) + K^+(g)$ $\Delta H = -1026$ kJ

d. $Na(g) \rightarrow Na^+(g) + e^-$ $I_1 = 495$ kJ
          $Cl(g) + e^- \rightarrow Cl^-(g)$ $EA = -348.7$ kJ

     $Na(g) + Cl(g) \rightarrow Na^+(g) + Cl^-(g)$ $\Delta H = 146$ kJ

e. $Mg(g) \rightarrow Mg^+(g) + e^-$ $I_1 = 735$ kJ
          $F(g) + e^- \rightarrow F^-(g)$ $EA = -327.8$ kJ

     $Mg(g) + F(g) \rightarrow Mg^+(g) + F^-(g)$ $\Delta H = 407$ kJ

f. $Mg^+(g) \rightarrow Mg^{2+}(g) + e^-$ $I_2 = 1445$ kJ
          $F(g) + e^- \rightarrow F^-(g)$ $EA = -327.8$ kJ

     $Mg^+(g) + F(g) \rightarrow Mg^{2+}(g) + F^-(g)$ $\Delta H = 1117$ kJ

g. From parts e and f we get:

     $Mg(g) + F(g) \rightarrow Mg^+(g) + F^-(g)$ $\Delta H = 407$ kJ
     $Mg^+(g) + F(g) \rightarrow Mg^{2+}(g) + F^-(g)$ $\Delta H = 1117$ kJ

     $Mg(g) + 2\ F(g) \rightarrow Mg^{2+}(g) + 2\ F^-(g)$ $\Delta H = 1524$ kJ

104.   a. $Se^{3+}(g) \rightarrow Se^{4+}(g) + e^-$ b. $S^-(g) + e^- \rightarrow S^{2-}(g)$ c. $Fe^{3+}(g) + e^- \rightarrow Fe^{2+}(g)$

       d. $Mg(g) \rightarrow Mg^+(g) + e^-$ e. $Mg(s) \rightarrow Mg^+(s) + e^-$

105.   a.   As: $1s^2 2s^2 2p^6 3s^2 3p^6 4s^2 3d^{10} 4p^3$

       b.   Element 116 will be below Po in the periodic table: [Rn] $7s^2 5f^{14} 6d^{10} 7p^4$

       c.   Ta: [Xe]$6s^2 4f^{14} 5d^3$ or Ir: [Xe]$6s^2 4f^{14} 5d^7$

       d.   Ti: [Ar]$4s^2 3d^2$; Ni: [Ar]$4s^2 3d^8$; Os: [Xe]$6s^2 4f^{14} 5d^6$

106.   Sb: $1s^2 2s^2 2p^6 3s^2 3p^6 4s^2 3d^{10} 4p^6 5s^2 4d^{10} 5p^3$

       a. $\ell = 1$: Designates p orbitals. There are 21 electrons in p orbitals.

       b. $m_\ell = 0$: All s electrons, 2 out of each set of 2p, 3p, 4p electrons, 2 out of each set of 3d and
            4d electrons, and one of the 5p electrons have $m_\ell = 0$. $10 + 6 + 4 + 1 = 21$ e$^-$ with $m_\ell = 0$.

       c. $m_\ell = 1$: 2 out of each set of 2p, 3p, and 4p electrons, 2 out of each set of 3d and 4d
            electrons, and one of the 5p electrons have $m_\ell = 1$. $6 + 4 + 1 = 11$ e$^-$ with $m_\ell = 1$.

107.   a.   The 4+ ion contains 20 electrons. Thus, the electrically neutral atom will contain 24
            electrons. The atomic number is 24.

       b.   The ground state electron configuration of the ion must be: $1s^2 2s^2 2p^6 3s^2 3p^6 4s^0 3d^2$; There are
            6 electrons in s orbitals.

       c.   12   d.   2

       e.   This is an isotope of $^{50}_{24}Cr$. There are 26 neutrons in the nucleus.

       f. 3.01×10^{23} atoms $\times \dfrac{49.9 \text{ amu}}{\text{atom}} \times \dfrac{1 \text{ g}}{6.022 \times 10^{23} \text{ amu}} = 24.94 \text{ g} \approx 24.9 \text{ g}$

       g. $1s^2 2s^2 2p^6 3s^2 3p^6 4s^1 3d^5$ is the ground state electron configuration for Cr.

108.   a.   As we remove succeeding electrons, the electron being removed is closer to the nucleus and
            there are fewer electrons left repelling it. In addition, we are removing a negatively charged
            particle from a positively charged particle. All of these factors go in the direction of
            requiring increasingly more energy to remove successive electrons.

       b.   Al: $1s^2 2s^2 2p^6 3s^2 3p^1$; For I_4, we begin removing an electron with n = 2. For I_3, we remove an
            electron with n = 3. In going from n = 3 to n = 2 there is a big jump in ionization energy
            because the n = 2 electrons are much closer to the nucleus, on the average, than n = 3
            electrons. Also, the n = 3 electrons are effectively shielded from the nuclear charge by all of

the n = 2 electrons while the n = 2 electrons are not as effective in shielding each other from the nuclear charge.

c. Al^{4+}; The electron affinity for Al^{4+} is ΔH for the reaction:

$$Al^{4+}(g) + e^- \rightarrow Al^{3+}(g) \quad \Delta H = -I_4 = EA = -11,600 \text{ kJ/mol}$$

d. The greater the number of electrons, the greater the size.

Size trend: $Al^{4+} < Al^{3+} < Al^{2+} < Al^+ < Al$

109. a. True for H only.   b. True for all atoms.   c. True for all atoms.

d. This is false for all atoms. In the presence of a magnetic field, the d orbitals are no longer degenerate.

110. a. Each orbital could hold 4 electrons.

b. First period: 4; Second period: 16   c. 20   d. 28

111. a. 1st period: $p = 1$, $q = 1$, $r = 0$, $s = \pm 1/2$ (2 elements)

2nd period: $p = 2$, $q = 1$, $r = 0$, $s = \pm 1/2$ (2 elements)

3rd period: $p = 3$, $q = 1$, $r = 0$, $s = \pm 1/2$ (2 elements)

$p = 3$, $q = 3$, $r = -2$, $s = \pm 1/2$ (2 elements)

$p = 3$, $q = 3$, $r = 0$, $s = \pm 1/2$ (2 elements)

$p = 3$, $q = 3$, $r = +2$, $s = \pm 1/2$ (2 elements)

4th period: $p = 4$; q and r values are the same as with $p = 3$ (8 total elements)

1							2
3							4
5	6	7	8	9	10	11	12
13	14	15	16	17	18	19	20

b. Elements 2, 4, 12 and 20 all have filled shells and will be least reactive.

c. Draw similarities to the modern periodic table.

XY could be X^+Y^-, $X^{2+}Y^{2-}$ or $X^{3+}Y^{3-}$. Possible ions for each are:

X^+ could be elements 1, 3, 5 or 13; Y^- could be 11 or 19.

X^{2+} could be 6 or 14; Y^{2-} could be 10 or 18.

X^{3+} could be 7 or 15; Y^{3-} could be 9 or 17.

$X^{4+}Y^{4-}$ probably won't form.

XY_2 will be $X^{2+}(Y^-)_2$; See above for possible ions.

X_2Y will be $(X^+)_2Y^{2-}$; See above for possible ions.

XY_3 will be $X^{3+}(Y^-)_3$; See above for possible ions.

X_2Y_3 will be $(X^{3+})_2(Y^{2-})_3$; See above for possible ions.

d. From (a), we can see that eight electrons can have p = 3.

e. p = 4, q = 3, r = 2, s = ± 1/2 (2 electrons)

f. p = 4, q = 3, r = -2 , s = ± 1/2 (2)

p = 4, q = 3, r = 0, s = ± 1/2 (2)

p = 4, q = 3, r = +2, s = ± 1/2 (2)

A total of 6 electrons can have p = 4 and q = 3.

g. p = 3, q = 0, r = 0: This is not allowed; q must be odd. Zero electrons can have these
quantum numbers.

h. p = 5, q = 1, r = 0

p = 5, q = 3, r = -2, 0, +2

p = 5, q = 5, r = -4, -2, 0, +2, +4

i. p = 6, q = 1, r = 0, s = ± 1/2 (2)

p = 6, q = 3, r = -2, 0, +2; s = ± 1/2 (6)

p = 6, q = 5, r = -4, -2, 0, +2, 4; s = ± 1/2 (10)

Eighteen electrons can have p = 6.

112.   None of the noble gases and no subatomic particles had been discovered when Mendeleev published his periodic table. Thus, there was not an element out of place in terms of reactivity. There was no reason to predict an entire family of elements. Mendeleev ordered his table by mass, he had no way of knowing there were gaps in atomic numbers (they hadn't been invented yet).

113. $\psi_{1s} = \dfrac{1}{\sqrt{\pi}} \left(\dfrac{Z}{a_0} \right)^{3/2} e^{-\sigma}$; $Z = 1$ for H, $\sigma = \dfrac{Zr}{a_0} = \dfrac{r}{a_0}$, $a_0 = 5.29 \times 10^{-11}$ m

$\psi_{1s} = \dfrac{1}{\sqrt{\pi}} \left(\dfrac{1}{a_0} \right)^{3/2} \exp\left(\dfrac{-r}{a_0} \right)$

Probability is proportional to ψ^2: $\psi_{1s}^2 = \dfrac{1}{\pi} \left(\dfrac{1}{a_0} \right)^3 \exp\left(\dfrac{-2r}{a_0} \right)$ (units of $\psi^2 = m^{-3}$)

a. ψ_{1s}^2 (at nucleus) $= \dfrac{1}{\pi} \left(\dfrac{1}{a_0} \right)^3 \exp\left[\dfrac{-2\,(0)}{a_0} \right] = 2.15 \times 10^{30}$ m^{-3}

If we assume this probability is constant throughout the 1.0×10^{-3} pm^3 volume then the total probability, p, is $\psi_{1s}^2 \times V$.

1.0×10^{-3} pm$^3 = (1.0 \times 10^{-3}$ pm$) \times (10^{-12}$ m/pm$)^3 = 1.0 \times 10^{-39}$ m^3

total probability $= p = (2.15 \times 10^{30}$ m$^{-3}) \times (1.0 \times 10^{-39}$ m$^3) = 2.2 \times 10^{-9}$

b.   For an electron that is 1.0×10^{-11} m from the nucleus:

$\psi_{1s}^2 = \dfrac{1}{\pi} \left(\dfrac{1}{5.29 \times 10^{-11}} \right)^3 \exp\left[\dfrac{-2(1.0 \times 10^{-11})}{(5.29 \times 10^{-11})} \right] = 1.5 \times 10^{30}$ m^{-3}

$V = 1.0 \times 10^{-39}$ m^3; $p = \psi_{1s}^2 \times V = 1.5 \times 10^{-9}$

c. $\psi_{1s}^2 = 2.15 \times 10^{30}$ m$^{-3} \exp\left[\dfrac{-2(53 \times 10^{-12})}{(5.29 \times 10^{-11})} \right] = 2.9 \times 10^{29}$; $V = 1.0 \times 10^{-39}$ m^3

$p = \psi_{1s}^2 \times V = 2.9 \times 10^{-10}$

d. $V = \dfrac{4}{3}\pi\,[(10.05 \times 10^{-12}$ m$)^3 - (9.95 \times 10^{-12}$ m$)^3] = 1.3 \times 10^{-34}$ m^3

We shall evaluate ψ_{1s}^2 at the middle of the shell, $r = 10.00$ pm, and assume ψ_{1s}^2 is constant from $r = 9.95$ to 10.05 pm. The concentric spheres are assumed centered about the nucleus.

$$\psi_{1s}^2 = 2.15 \times 10^{30} \text{ m}^{-3} \exp\left[\frac{-2(10.0 \times 10^{-12} \text{ m})}{(5.29 \times 10^{-11} \text{ m})}\right] = 1.47 \times 10^{30} \text{ m}^{-3}$$

$$p = (1.47 \times 10^{30} \text{ m}^{-3})(1.3 \times 10^{-34} \text{ m}^3) = 1.9 \times 10^{-4}$$

e. $V = \dfrac{4}{3}\pi\,[(52.95 \times 10^{-12} \text{ m})^3 - (52.85 \times 10^{-12} \text{ m})^3] = 4 \times 10^{-33} \text{ m}^3$

Evaluate ψ_{1s}^2 at r = 52.90 pm: $\psi_{1s}^2 = 2.15 \times 10^{30} \text{ m}^{-3}\,(e^{-2}) = 2.91 \times 10^{29} \text{ m}^{-3}$; $p = 1 \times 10^{-3}$

114.

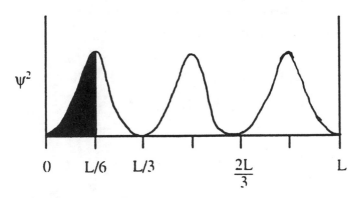

$$0 \qquad L/6 \quad L/3 \qquad\qquad \frac{2L}{3} \qquad\qquad L$$

Total Area = 1; Area of one hump = 1/3

Shaded area = 1/6 = probability of finding the electron between x = 0 and x = L/6
in a one dimensional box with n = 3.

115. $\psi_{2p_z} = \dfrac{1}{4\sqrt{2\pi}}\left(\dfrac{Z}{a_0}\right)^{3/2}\sigma\,(e^{-\sigma/2})\cos\theta,\quad \sigma = \dfrac{Zr}{a_0}$

z-axis: $\theta = 0°$, $\cos\theta = 1$ which is the maximum that the $\cos\theta$ part of ψ_{2p_z} function can be.

xy plane: $\theta = 90°$, $\cos\theta = 0$ so $\psi_{2p_z} = 0$ (a node).

116.   The 3p and 3d orbitals have the same energy as the 3s orbital in hydrogen. No, in He the 3s, 3p and 3d are at different energies. The orbital energies in polyelectronic atoms depend on n and ℓ. The orbital energies in hydrogen only depend on n.

117. $E = \dfrac{h^2(n_x^2 + n_y^2 + n_z^2)}{8\,mL^2}$; $E_{111} = \dfrac{3\,h^2}{8\,mL^2}$; $E_{112} = \dfrac{6\,h^2}{8\,mL^2}$; $\Delta E = \dfrac{3\,h^2}{8\,mL^2} = E_{112} - E_{111}$

$\Delta E = \dfrac{hc}{\lambda} = \dfrac{(6.626 \times 10^{-34} \text{ J·sec})(2.998 \times 10^8 \text{ m/s})}{9.50 \times 10^{-9} \text{ m}} = 2.09 \times 10^{-17} \text{ J}$

$$L^2 = \frac{3\,h^2}{8\,m\Delta E}, \quad L = \left(\frac{3\,h^2}{8\,m\Delta E}\right)^{1/2} = \left[\frac{3(6.626 \times 10^{-34}\ \text{J sec})^2}{8(9.11 \times 10^{-31}\ \text{kg})\,(2.09 \times 10^{-17}\ \text{J})}\right]^{1/2}$$

$L = 9.30 \times 10^{-11}$ m = 93.0 pm

The sphere that fits in this cube will touch the cube at the center of each face. The diameter of the sphere will equal the length of the cube. So:

   $2\,r = L$ and $r = 46.5$ pm

118.   At $x = 0$, the value of the square of the wave function must be zero. The particle must be inside the box. For $\psi = A\cos(Lx)$, at $x = 0$, $\cos(0) = 1$ and $\psi^2 = A^2$. This violates the boundary condition.

CHAPTER THIRTEEN

BONDING - GENERAL CONCEPTS

Chemical Bonds and Electronegativity

1. a. Electronegativity: The ability of an atom <u>in a molecule</u> to attract electrons to itself.

    Electron affinity: The energy change for $M(g) + e^- \rightarrow M^-(g)$. EA deals with isolated atoms in the gas phase.

    b. Covalent bond: Sharing of electron pair(s); Polar covalent bond: Unequal sharing of electron pair(s).

    c. Ionic bond: Electrons are no longer shared, i.e., complete transfer of electron(s) from one atom to another.

2. a. There are two attractions of the form $\dfrac{(+1)(-1)}{r}$, where $r = 1 \times 10^{-10}$ m = 0.1 nm.

    $$V = 2 \times (2.31 \times 10^{-19} \text{ J nm}) \left[\frac{(+1)(-1)}{0.1 \text{ nm}} \right] = -4.62 \times 10^{-18} \text{ J}$$

    b. There are 4 attractions of +1 and -1 charges at a distance of 0.1 nm from each other. The two negative charges and the two positive charges repel each other across the diagonal of the square. This is at a distance of $\sqrt{2} \times 0.1$ nm.

    $$V = 4 \times (2.31 \times 10^{-19}) \left[\frac{(+1)(-1)}{0.1} \right] + 2.31 \times 10^{-19} \left[\frac{(+1)(+1)}{\sqrt{2}(0.1)} \right] + 2.31 \times 10^{-19} \left[\frac{(-1)(-1)}{\sqrt{2}(0.1)} \right]$$

    $$V = -9.24 \times 10^{-18} \text{ J} + 1.63 \times 10^{-18} \text{ J} + 1.63 \times 10^{-18} \text{ J} = -5.98 \times 10^{-18} \text{ J}$$

    Note: There is a greater net attraction in arrangement b than in a.

3. Using the periodic table we expect the general trend for electronegativity to be:
    1) increase as we go from left to right across a period
    2) decrease as we go down a group

    a. $C < N < O$ b. $Se < S < Cl$ c. $Sn < Ge < Si$

    d. $Tl < Ge < S$ e. $Rb < K < Na$ f. $Ga < B < O$

4.   The most polar bond will have the greatest difference in electronegativity between the two atoms. From positions in the periodic table, we would predict:

   a. Ge-F   b. P-Cl   c. S-F   d. Ti-Cl   e. Sn-H   f. Tl-Br

5.   The general trends in electronegativity used on Exercises 13.3 and 13.4 are only rules of thumb. In this exercise we use experimental values of electronegativities and can begin to see several exceptions. The order of EN from Figure 13.3 is:

   a.   C (2.5) < N (3.0) < O (3.5)   same as predicted

   b.   Se (2.4) < S (2.5) < Cl (3.0)   same

   c.   Si = Ge = Sn (1.8)   different   d.   Tl (1.8) = Ge (1.8) < S (2.5)   different

   e.   Rb = K < Na   different   f.   Ga < B < O   same

Most polar bonds using actual EN values:

   a.   Si-F and Ge-F equal polarity (Ge-F predicted)

   b.   P-Cl (same as predicted)

   c.   S-F (same as predicted)   d.   Ti-Cl (same as predicted)

   e.   C-H (Sn-H predicted)   f.   Al-Br (Tl-Br predicted)

6.   (IE - EA)   (IE - EA)/502   EN (text)   2006/502 = 4.0

	(IE - EA)	(IE - EA)/502	EN (text)
F	2006 kJ/mol	4.0	4.0
Cl	1604	3.2	3.0
Br	1463	2.9	2.8
I	1302	2.6	2.5

   The values calculated from IE and EA show the same trend (and agree fairly closely) to the values given in the text.

Ionic Compounds

7.   a. $Cu > Cu^+ > Cu^{2+}$ b. $Pt^{2+} > Pd^{2+} > Ni^{2+}$ c. $Se^{2-} > S^{2-} > O^{2-}$

   d. $La^{3+} > Eu^{3+} > Gd^{3+} > Yb^{3+}$ e. $Te^{2-} > I^- > Xe > Cs^+ > Ba^{2+} > La^{3+}$

   For answer a, as electrons are removed from an atom, the size decreases. Answers b - d follow the radii trend. Answer e follows the trend for an isoelectronic series, i.e., the smallest ion has the most protons.

8.   a. Mg^{2+}: $1s^22s^22p^6$ Sn^{2+}: $[Kr]5s^24d^{10}$

     K$^+$: $1s^22s^22p^63s^23p^6$ Al^{3+}: $1s^22s^22p^6$

     Tl$^+$: $[Xe]6s^24f^{14}5d^{10}$ As^{3+}: $[Ar]4s^23d^{10}$

     b. N^{3-}, O^{2-} and F^-: $1s^22s^22p^6$ Te^{2-}: $[Kr]5s^24d^{10}5p^6$

     c. Be^{2+}: $1s^2$ Rb^+: $[Ar]4s^23d^{10}4p^6$

     Ba^{2+}: $[Kr]5s^24d^{10}5p^6$ Se^{2-}: $[Ar]4s^23d^{10}4p^6$

     I$^-$: $[Kr]5s^24d^{10}5p^6$

9.   a. Sc^{3+} b. Te^{2-} c. Ce^{4+} and Ti^{4+} d. Ba^{2+}

10. Isoelectronic: Same number of electrons.

     There are two variables, number of protons and number of electrons, that will determine the size of an ion. Keeping the number of electrons constant we only have to consider the number of protons to predict trends in size. The smallest ion has the most protons.

11. Se^{2-}, Br$^-$, Rb^+, Sr^{2+}, Y^{3+}, Zr^{4+} are all isoelectronic with Kr (36 electrons).

12. Lattice energy is proportional to $\dfrac{Q_1Q_2}{r}$.

     In general, charge effects are much greater than size effects.

     a.   NaCl, Na$^+$ smaller than K$^+$ b   LiF, F$^-$ smaller than Cl$^-$

     c.   MgO, O^{2-} greater charge than OH$^-$ d. $Fe(OH)_3$, Fe^{3+} greater charge than Fe^{2+}

     e. Na_2O, O^{2-} greater charge than Cl$^-$ f.   MgO, smaller ions

13. Ionic solids can be characterized as being held together by strong omnidirectional forces.

     I.   For electrical conductivity, charged species must be free to move. In ionic solids the charged ions are held rigidly in place. Once the forces are disrupted (melting or dissolution) the ions can move about (conduct).

     II. Melting and boiling disrupts the attractions of the ions for each other. If the forces are strong it will take a lot of energy (high temp.) to accomplish this.

III. If we try to bend a piece of material, the atoms/ions must slide across each other. For an ionic solid the following might happen:

strong attraction   strong repulsion

Just as the layers begin to slide, there will be very strong repulsions causing the solid to snap across a fairly clean plane.

These properties and their correlation to chemical forces will be discussed in detail in Chapter 16.

14.
$$Na(s) \rightarrow Na(g) \qquad \Delta H = 109 \text{ kJ} \quad \text{(sublimation)}$$
$$Na(g) \rightarrow Na^+(g) + e^- \qquad \Delta H = 495 \text{ kJ} \quad \text{(ionization energy)}$$
$$1/2\ Cl_2(g) \rightarrow Cl(g) \qquad \Delta H = 239/2 \text{ kJ} \quad \text{(bond energy)}$$
$$Cl(g) + e^- \rightarrow Cl^-(g) \qquad \Delta H = -349 \text{ kJ} \quad \text{(electron affinity)}$$
$$Na^+(g) + Cl^-(g) \rightarrow NaCl(s) \qquad \Delta H = -786 \text{ kJ} \quad \text{(lattice energy)}$$

$$Na(s) + 1/2\ Cl_2(g) \rightarrow NaCl(s) \qquad \Delta H_f^\circ = -412 \text{ kJ/mol}$$

15.
$$Ba(s) \rightarrow Ba(g) \qquad \Delta H = 178 \text{ kJ} \qquad \text{(sublimation)}$$
$$Ba(g) \rightarrow Ba^+(g) + e^- \qquad \Delta H = 503 \text{ kJ} \qquad (IE_1)$$
$$Ba^+(g) \rightarrow Ba^{2+}(g) + e^- \qquad \Delta H = 965 \text{ kJ} \qquad (IE_2)$$
$$Cl_2(g) \rightarrow 2\ Cl(g) \qquad \Delta H = 239 \text{ kJ} \qquad (BE)$$
$$2\ Cl(g) + 2\ e^- \rightarrow 2\ Cl^-(g) \qquad \Delta H = 2(-349) \text{ kJ} \qquad (EA)$$
$$Ba^{2+}(g) + 2\ Cl^-(g) \rightarrow BaCl_2(s) \qquad \Delta H = -2056 \text{ kJ} \qquad (LE)$$

$$Ba(s) + Cl_2(g) \rightarrow BaCl_2(s) \qquad \Delta H_f^\circ = -869 \text{ kJ/mol}$$

16. a. From the data given, it costs less energy to produce $Mg^+(g) + O^-(g)$ than to produce $Mg^{2+}(g) + O^{2-}(g)$. However, the lattice energy for $Mg^{2+}O^{2-}$ will be much larger than for Mg^+O^-. The favorable lattice energy term will dominate and $Mg^{2+}O^{2-}$ forms.

b. Mg^+ and O^- both have unpaired electrons. In Mg^{2+} and O^{2-}, there are no unpaired electrons. Hence, Mg^+O^- would be paramagnetic; $Mg^{2+}O^{2-}$ would be diamagnetic. Paramagnetism can be detected by measuring the mass of a sample in the presence and absence of a magnetic field. The apparent mass of a paramagnetic substance will be larger in a magnetic field because of the force between the unpaired electrons and the field.

17.   Let us look at the complete cycle for Li_2S.

$$
\begin{array}{ll}
2\ Li(s) \rightarrow 2\ Li(g) & 2\ \Delta H_{sub}(Li) = 2(161)\ kJ \\
2\ Li(g) \rightarrow 2\ Li^+(g) + 2\ e^- & 2\ IE = 2(520.)\ kJ \\
S(s) \rightarrow S(g) & \Delta H_{sub}(S) = 277\ kJ \\
S(g) + e^- \rightarrow S^-(g) & EA_1 = -200.\ kJ \\
S^-(g) + e^- \rightarrow S^{2-}(g) & EA_2 = ? \\
2\ Li^+(g) + S^{2-}(g) \rightarrow Li_2S(s) & LE = -2472\ kJ
\end{array}
$$

$$
\begin{array}{ll}
\hline
2\ Li(s) + S(s) \rightarrow Li_2S(s) & \Delta H_f^\circ = -500.\ kJ
\end{array}
$$

$-500. = 2\ \Delta H_{sub}(Li) + 2\ IE + \Delta H_{sub}(S) + EA_1 + EA_2 + LE$

$-500. = -1033 + EA_2,\ \ EA_2 = +533\ kJ$

For each salt: $\Delta H_f^\circ = 2\ \Delta H_{sub}(M) + 2\ IE + 277 - 200. + LE + EA_2$

Na_2S: $-365 = 2(109) + 2(495) + 277 - 200. - 2203 + EA_2,\ \ EA_2 = +553\ kJ$

K_2S: $-381 = 2(90.) + 2(419) + 277 - 200. - 2052 + EA_2,\ \ EA_2 = +576\ kJ$

Rb_2S: $-361 = 2(82) + 2(409) + 277 - 200. - 1949 + EA_2,\ \ EA_2 = +529\ kJ$

Cs_2S: $-360. = 2(78) + 2(382) + 277 - 200. - 1850. + EA_2,\ \ EA_2 = +493\ kJ$

We get values from 493 to 576 kJ.

The mean value is: $\dfrac{533 + 553 + 576 + 529 + 493}{5} = 537\ kJ$

We can represent the results as $EA_2 = 540\ kJ \pm 50\ kJ$.

18.
$$
\begin{array}{ll}
O(g) + e^- \rightarrow O^-(g) & \Delta H = -141\ kJ/mol \\
O^-(g) + e^- \rightarrow O^{2-}(g) & \Delta H = 878\ kJ/mol
\end{array}
$$

$$
\begin{array}{ll}
\hline
O(g) + 2\ e^- \rightarrow O^{2-}(g) & \Delta H = 737\ kJ/mol
\end{array}
$$

19.   The extra electron-electron repulsions are much greater than the attraction of the electron for the nucleus.

20. Ca^{2+} has greater charge than Na^+, and Se^{2-} is smaller than Te^{2-}. The effect of charge on the lattice energy is greater than the effect of size. We expect the trend from most exothermic to least exothermic to be:

$$CaSe\ >\ CaTe\ >\ Na_2Se\ >\ Na_2Te$$

$\ \ \ \ (-2862)\ \ \ \ (-2721)\ \ \ \ (-2130)\ \ \ \ (-2095)$ This is what we observe.

21.   The compounds are FeO, Fe_2O_3, $FeCl_2$ and $FeCl_3$. Lattice energy is proportional to the charge of the cation times the charge of the anion, Q_1Q_2.

Compound	Q_1Q_2	Lattice Energy
$FeCl_2$	(+2)(-1)	-2631 kJ/mol
$FeCl_3$	(+3)(-1)	-3865 kJ/mol
FeO	(+2)(-2)	-5359 kJ/mol
Fe_2O_3	(+3)(-2)	-14,744 kJ/mol

22.   a. Li^+ and N^{3-} are the expected ions. The formula of the compound would be Li_3N (lithium nitride).

      b. Ga^{3+} and O^{2-}; Ga_2O_3 [gallium(III) oxide]

      c. Rb^+ and Cl^-; RbCl (rubidium chloride)

      d. Ba^{2+} and S^{2-}; BaS (barium sulfide)

Bond Energies

23.

Bonds broken: 1 C – N (305 kJ/mol)   Bonds formed: 1 C – C (347 kJ/mol)

$\Delta H = \Sigma D_{broken} - \Sigma D_{formed}$, $\Delta H = 305 - 347 = -42$ kJ

Note: Many bonds usually remain the same between reactants and products. Only break and form bonds that are involved in the reaction.

24.   a.   H – H + Cl – Cl → 2 H – Cl

      Bonds broken: Bonds formed:

        1 H – H (432 kJ/mol)   2 H – Cl (427 kJ/mol)
        1 Cl – Cl (239 kJ/mol)

      $\Delta H = \Sigma D_{broken} - \Sigma D_{formed}$, $\Delta H = 432$ kJ + 239 kJ - 2(427) kJ = -183 kJ

b.

$$N \equiv N + 3 \text{ H}-\text{H} \longrightarrow 2 \text{ H}-\overset{\displaystyle |}{\underset{\displaystyle H}{N}}-\text{H}$$

Bonds broken: Bonds formed:

1 N ≡ N (941 kJ/mol)   6 N – H (391 kJ/mol)
3 H – H (432 kJ/mol)

ΔH = 941 kJ + 3(432) kJ - 6(391) kJ = -109 kJ

c.

$$\overset{\text{H}}{\underset{\text{H}}{\diagdown}}C=C\overset{\text{H}}{\underset{\text{H}}{\diagup}} \quad + \quad \text{Br}-\text{Br} \longrightarrow \quad \text{H}-\overset{\text{H}}{\underset{\text{Br}}{C}}-\overset{\text{H}}{\underset{\text{Br}}{C}}-\text{H}$$

Bonds broken: Bonds formed:

1 C = C (614 kJ/mol)   1 C – C (347 kJ/mol)
1 Br – Br (193 kJ/mol)   2 C – Br (276 kJ/mol)

ΔH = 614 kJ + 193 kJ - [347 kJ + 2(276 kJ)] = -92 kJ

d.

$$\overset{\text{H}}{\underset{\text{H}}{\diagdown}}C=C\overset{\text{H}}{\underset{\text{H}}{\diagup}} \quad + \quad \text{H}-\text{O}-\text{O}-\text{H} \longrightarrow \quad \text{H}-\overset{\text{OH}}{\underset{\text{H}}{C}}-\overset{\text{OH}}{\underset{\text{H}}{C}}-\text{H}$$

Bonds broken: Bonds formed:

1 C = C (614 kJ/mol)   1 C – C (347 kJ/mol)
1 O – O (146 kJ/mol)   2 C – O (358 kJ/mol)

ΔH = 614 kJ + 146 kJ - [347 kJ + 2(358 kJ)] = -303 kJ

25.   a. $\Delta H = 2 \ \Delta H_f^{\circ} (HCl) = 2 \text{ mol}(-92 \text{ kJ/mol}) = -184 \text{ kJ}$ (-183 kJ from bond energies)

b. $\Delta H = 2 \ \Delta H_f^{\circ} (NH_3) = 2 \text{ mol}(-46 \text{ kJ/mol}) = -92 \text{ kJ}$ (-109 kJ from bond energies)

Comparing the values for each reaction, bond energies seem to give a reasonably good estimate for the enthalpy change of a reaction. The estimate is especially good for gas phase reactions.

26.   a.   i. $C_6H_6N_{12}O_{12} \rightarrow 6\ CO + 6\ N_2 + 3\ H_2O + 3/2\ O_2$

The NO_2 groups have one N – O single bond and one N = O double bond and each carbon atom has one C – H single bond. We must break and form all bonds.

Bonds broken: Bonds formed:

3 C – C (347 kJ/mol)	6 C ≡ O (1072 kJ/mol)
6 C – H (413 kJ/mol)	6 N ≡ N (941 kJ/mol)
12 C – N (305 kJ/mol)	6 H – O (467 kJ/mol)
6 N – N (160 kJ/mol)	3/2 O = O (495 kJ/mol)
6 N – O (201 kJ/mol)	
6 N = O (607 kJ/mol)	$\Sigma D_{formed} = 15{,}623$ kJ

$\Sigma D_{broken} = 12{,}987$ kJ

$\Delta H = \Sigma D_{broken} - \Sigma D_{formed} = 12{,}987$ kJ - $15{,}623$ kJ = -2636 kJ

ii. $C_6H_6N_{12}O_{12} \rightarrow 3\ CO + 3\ CO_2 + 6\ N_2 + 3\ H_2O$

Note: The bonds broken will be the same for all three reactions.

Bonds formed:

3 C ≡ O (1072 kJ/mol)
6 C = O (799 kJ/mol)
6 N ≡ N (941 kJ/mol)
6 H – O (467 kJ/mol)

$\Sigma D_{formed} = 16{,}458$ kJ

$\Delta H = 12{,}987$ kJ - $16{,}458$ kJ = -3471 kJ

iii. $C_6H_6N_{12}O_{12} \rightarrow 6\ CO_2 + 6\ N_2 + 3\ H_2$

Bonds formed:

12 C = O (799 kJ/mol)
6 N ≡ N (941 kJ/mol)
3 H – H (432 kJ/mol)

$\Sigma D_{formed} = 16{,}530.$ kJ

$\Delta H = 12{,}987$ kJ - $16{,}530.$ kJ = -3543 kJ

b.   Reaction iii yields the most energy per mole of CL-20 so it will yield the most energy per kg.

$$\frac{-3543 \text{ kJ}}{\text{mol}} \times \frac{1 \text{ mol}}{438.23 \text{ g}} \times \frac{1000 \text{ g}}{\text{kg}} = -8085 \text{ kJ/kg}$$

27.

Bonds broken: Bonds formed:

    1 C = C (614 kJ/mol)   1 C – C (347 kJ/mol)
    1 O – O (146 kJ/mol)   1 C = O (799 kJ/mol)

$\Delta H = 614 + 146 - (347 + 799) = -386$ kJ

From Exercise 9.46, $\Delta H° = -361$ kJ

28.   a.   I.

Bonds broken (*):   Bonds formed (*):

    1 C – O (358 kJ)   1 O – H (467 kJ)
    1 H – C (413 kJ)   1 C – C (347 kJ)

$\Delta H_I = 358$ kJ $+ 413$ kJ $- [467$ kJ $+ 347$ kJ$] = -43$ kJ

II.

Bonds broken (*):   Bonds formed (*):

1 C – O (358 kJ/mol)   1 H – O (467 kJ/mol)
1 C – H (413 kJ/mol)   1 C = C (614 kJ/mol)
1 C – C (347 kJ/mol)

ΔH_{II} = 358 kJ + 413 kJ + 347 kJ - [467 kJ + 614 kJ] = +37 kJ

ΔH = ΔH_I + ΔH_{II} = -43 kJ + 37 kJ = -6 kJ

b.

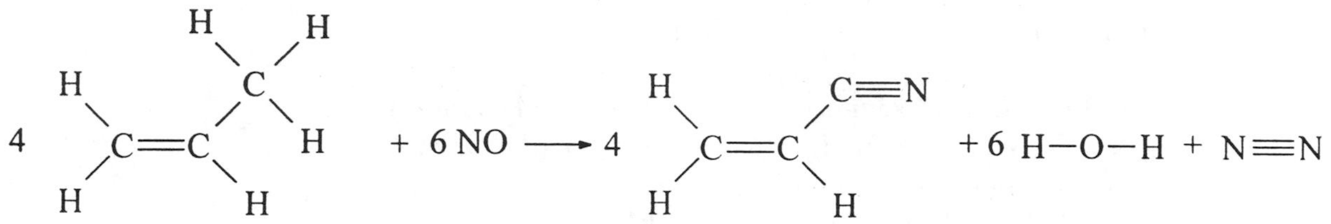

Bonds broken: Bonds formed:

4 × 3 C – H (413 kJ/mol)   4 C ≡ N (891 kJ/mol)
6 N = O (630. kJ/mol   6 × 2 H – O (467 kJ/mol)
                                           1 N ≡ N (941 kJ/mol)

ΔH = 12(413) + 6(630.) - [4(891) + 12(467) + 941] = -1373 kJ

c.

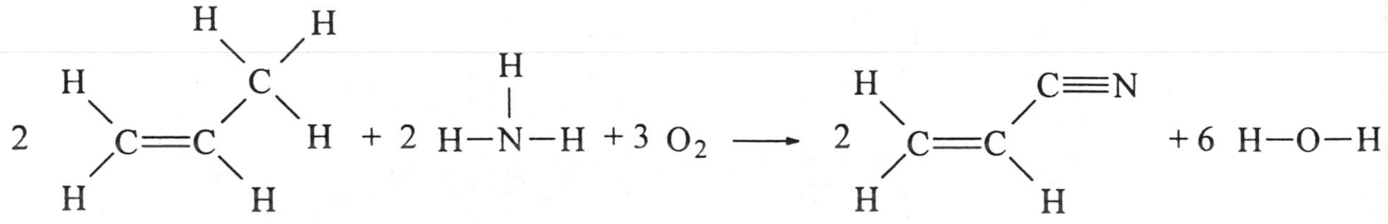

Bonds broken: Bonds formed:

2 × 3 C – H (413 kJ/mol)   2 C ≡ N (891 kJ/mol)
2 × 3 N – H (391 kJ/mol)   6 × 2 O – H (467 kJ/mol)
3 O = O (495 kJ/mol)

ΔH = 6(413) + 6(391) + 3(495) - [2(891) + 12(467)] = -1077 kJ

29. Since both reactions are highly exothermic, the high temperature is not needed to provide energy. It must be necessary for some other reason. This will be discussed in Chapter 15 on kinetics.

30.

Bonds broken:

1 C ≡ O (1072 kJ/mol)

Bonds formed:

1 C – C (347 kJ/mol)
1 C = O (799 kJ/mol)

$\Delta H = 1072 - [347 + 799] = -74$ kJ

31. a.
$HF(g) \rightarrow H(g) + F(g)$	$\Delta H = 565$ kJ
$H(g) \rightarrow H^+(g) + e^-$	$\Delta H = 1312$ kJ
$F(g) + e^- \rightarrow F^-(g)$	$\Delta H = -327.8$ kJ

$HF(g) \rightarrow H^+(g) + F^-(g)$	$\Delta H = 1549$ kJ

b.
$HCl(g) \rightarrow H(g) + Cl(g)$	$\Delta H = 427$ kJ
$H(g) \rightarrow H^+(g) + e^-$	$\Delta H = 1312$ kJ
$Cl(g) + e^- \rightarrow Cl^-(g)$	$\Delta H = -348.7$ kJ

$HCl(g) \rightarrow H^+(g) + Cl^-(g)$	$\Delta H = 1390.$ kJ

c.
$HI(g) \rightarrow H(g) + I(g)$	$\Delta H = 295$ kJ
$H(g) \rightarrow H^+(g) + e^-$	$\Delta H = 1312$ kJ
$I(g) + e^- \rightarrow I^-(g)$	$\Delta H = -295.2$ kJ

$HI(g) \rightarrow H^+(g) + I^-(g)$	$\Delta H = 1312$ kJ

d.
$H_2O(g) \rightarrow OH(g) + H(g)$	$\Delta H = 467$ kJ
$H(g) \rightarrow H^+(g) + e^-$	$\Delta H = 1312$ kJ
$OH(g) + e^- \rightarrow OH^-(g)$	$\Delta H = -180.$ kJ

$H_2O(g) \rightarrow H^+(g) + OH^-(g)$	$\Delta H = 1599$ kJ

32. a. Using SF_4 data: $SF_4(g) \rightarrow S(g) + 4 F(g)$

$\Delta H° = 4 D_{SF} = 278.8$ kJ $+ 4(79.0$ kJ$) - (-775$ kJ$) = 1370.$ kJ

$$D_{SF} = \frac{1370.\,kJ}{4\,mol\,SF\,bonds} = 342.5\,kJ/mol$$

Using SF_6 data: $SF_6(g) \rightarrow S(g) + 6\,F(g)$

$\Delta H^\circ = 6\,D_{SF} = 278.8\,kJ + 6(79.0\,kJ) - (-1209\,kJ) = 1962\,kJ$

$D_{SF} = \dfrac{1962\,kJ}{6} = 327.0\,kJ/mol$

b. The S – F bond energy in the table is 327 kJ/mol. The value in the table was based on the S – F bond in SF_6.

c. S(g) and F(g) are not the most stable form of the element at 25°C and 1 atm. The most stable forms are $S_8(s)$ and $F_2(g)$; $\Delta H_f^{\bullet} = 0$ for these two species.

33. $NH_3(g) \rightarrow N(g) + 3\,H(g)$

$\Delta H^\circ = 3\,D_{NH} = 472.7\,kJ + 3(216.0\,kJ) - (-46.1\,kJ) = 1166.8\,kJ$

$D_{NH} = \dfrac{1166.8\,kJ}{3\,mol\ NH\ bonds} = 388.93\,kJ/mol$

$D_{calc} = 389\,kJ/mol$ as compared to 391 kJ/mol in the table. There is good agreement.

34. $N_2 + 3\,H_2 \rightarrow 2\,NH_3$; $\Delta H = D_{N_2} + 3\,D_{H_2} - 6\,D_{NH}$; $\Delta H^\circ = 2(-46\,kJ) = -92\,kJ$

$-92\,kJ = 941\,kJ + 3(432\,kJ) - (6\,D_{N-H})$, $6\,D_{N-H} = 2329\,kJ$, $D_{N-H} = 388.2\,kJ/mol$

Exercise 33: 389 kJ/mol; Table in text: 391 kJ/mol

There is good agreement between all three values.

35.

$2N(g) + 4H(g) \quad \Delta H = D_{N-N} + 4\,D_{N-H} = D_{N-N} + 4(388.9)$

$\Delta H^\circ = 2\,\Delta H_f^{\bullet}(N) + 4\,\Delta H_f^{\bullet}(H) - \Delta H_f^{\bullet}(N_2H_4) = 2(472.7\,kJ) + 4(216.0\,kJ) - 95.4\,kJ$

$\Delta H^\circ = 1714.0\,kJ = D_{N-N} + 4(388.9)$

$D_{N-N} = 158.4\,kJ/mol$ (160 kJ/mol in Table 13.5)

36. $\Delta H_f^{\bullet}$ for H(g) is ΔH° for the reaction: $1/2\,H_2(g) \rightarrow H(g)$

$\Delta H_f^{\bullet}$ for H(g) equals one-half the H–H bond energy.

Lewis Structures and Resonance

37.   a.   HCN has $1 + 4 + 5 = 10$ valence electrons.

$$H—C—N \qquad\qquad H—C\equiv N\!:$$

     skeletal   complete

uses 4 e⁻: (6 e⁻ left, 3 pairs)

b. PH_3 has $5 + 3(1) = 8$ valence electrons.

$$H—P—H \qquad\qquad H—\overset{\cdot\cdot}{P}—H$$
$$\quad\;\; | \qquad\qquad\qquad\qquad\; |$$
$$\quad\;\; H \qquad\qquad\qquad\qquad\; H$$

     skeletal   complete

uses 6 e⁻: (2 e⁻ left, 1 pair)

c. $CHCl_3$ has $4 + 1 + 3(7) = 26$ valence electrons.

$$\begin{array}{c} H \\ | \\ Cl—C—Cl \\ | \\ Cl \end{array} \qquad\qquad \begin{array}{c} H \\ | \\ :\overset{\cdot\cdot}{\underset{}{Cl}}—C—\overset{\cdot\cdot}{\underset{\cdot\cdot}{Cl}}: \\ | \\ :\overset{}{\underset{\cdot\cdot}{Cl}}: \end{array}$$

     skeletal   complete

uses 8 e⁻: (18 e⁻ left, 9 pairs)

d. NH_4^+ has $5 + 4(1) - 1 = 8$ valence electrons.

$$\left[\begin{array}{c} H \\ | \\ H—N—H \\ | \\ H \end{array} \right]^+$$

e. BF_4^- has $3 + 4(7) + 1 = 32$ valence electrons.

$$\left[\begin{array}{c} :\overset{\cdot\cdot}{F}: \\ | \\ :\overset{\cdot\cdot}{F}—B—\overset{\cdot\cdot}{F}: \\ | \\ :\overset{\cdot\cdot}{F}: \end{array} \right]^-$$

f. SeF_2 has $6 + 2(7) = 20$ valence electrons.

$$:\overset{\cdot\cdot}{\underset{\cdot\cdot}{F}}—\overset{\cdot\cdot}{\underset{\cdot\cdot}{Se}}—\overset{\cdot\cdot}{\underset{\cdot\cdot}{F}}:$$

38.   a. $POCl_3$ has $5 + 6 + 3(7) = 32$ valence electrons.

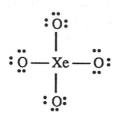

skeletal structure   complete octets

This structure uses all 32 e⁻ while satisfying the octet rule for all atoms. This is a valid Lewis structure.

SO_4^{2-} has $6 + 4(6) + 2 = 32$ valence electrons.

Note: A negatively charged ion will have additional electrons to those that come from the valence shells of the atoms.

XeO_4, $8 + 4(6) = 32$ e⁻ PO_4^{3-}, $5 + 4(6) + 3 = 32$ e⁻

ClO_4^- has $7 + 4(6) + 1 = 32$ valence electrons.

Note: All of these species have the same number of atoms and the same number of valence electrons. They also have the same Lewis structure.

    b. NF_3 has $5 + 3(7) = 26$ valence electrons.

skeletal
structure   complete
octets

SO_3^{2-}, $6 + 3(6) + 2 = 26$ e⁻ PO_3^{3-}, $5 + 3(6) + 3 = 26$ e⁻

$$\left[\ddot{:}\!\underset{..}{O}\!-\!\underset{|}{\overset{..}{S}}\!-\!\underset{..}{\overset{..}{O}}\!: \right]^{2-}$$

$$\left[\ddot{:}\!\underset{..}{O}\!-\!\underset{|}{\overset{..}{P}}\!-\!\underset{..}{\overset{..}{O}}\!: \right]^{3-}$$

ClO_3^-, $7 + 3(6) + 1 = 26$ e⁻

$$\left[\ddot{:}\!\underset{..}{O}\!-\!\underset{|}{\overset{..}{Cl}}\!-\!\underset{..}{\overset{..}{O}}\!: \right]^{-}$$

Note: Species with the same number of atoms and valence electrons have similar Lewis structures.

c. ClO_2^- has $7 + 2(6) + 1 = 20$ valence electrons.

O – Cl – O

$$\left[\ddot{:}\!\underset{..}{O}\!-\!\underset{..}{\overset{..}{Cl}}\!-\!\underset{..}{\overset{..}{O}}\!: \right]^{-}$$

All atoms obey the octet rule and we have used all the valence electrons. This is a valid Lewis structure.

skeletal   complete
structure   octets

SCl_2, $6 + 2(7) = 20$ e⁻ PCl_2^-, $5 + 2(7) + 1 = 20$ e⁻

$$\ddot{:}\!\underset{..}{Cl}\!-\!\underset{..}{\overset{..}{S}}\!-\!\underset{..}{\overset{..}{Cl}}\!:$$

$$\left[\ddot{:}\!\underset{..}{Cl}\!-\!\underset{..}{\overset{..}{P}}\!-\!\underset{..}{\overset{..}{Cl}}\!: \right]^{-}$$

Note: Species with the same number of atoms and valence electrons have similar Lewis structures.

39.   Molecules/ions that have the same number of valence electrons and the same number of atoms will have similar Lewis structures.

40.   a. NO_2^- has $5 + 2(6) + 1 = 18$ valence electrons.

O—N—O $:\ddot{O}$—N—$\ddot{O}:$

skeletal structure   complete octet of more
                            electronegative oxygen
                            (16 e⁻ used)

We have 1 more pair of e⁻. Putting this pair on the nitrogen we get:

$$:\ddot{O}—\ddot{N}—\ddot{O}:$$ This accounts for all 18 electrons, but N does not obey the octet rule.

To get an octet about the nitrogen and not use any more electrons, one of the unshared pairs on an oxygen must be shared between N and O instead.

$$\left[\ddot{O}=\ddot{N}—\ddot{O}: \right]^{-} \longleftrightarrow \left[:\ddot{O}—\ddot{N}=\ddot{O} \right]^{-}$$
$$\quad\;\; 0 \quad\;\; 0 \quad\; -1 \qquad\qquad -1 \quad\;\; 0 \quad\;\; 0$$

Since there is no reason to have the double bond to either oxygen atom, we can draw two resonance structures. Formal charges are shown.

HNO₂ has 1 + 5 + 2(6) = 18 valence electrons.

$$:\ddot{O}=\ddot{N}—\ddot{O}—H$$
$$\;\; 0 \quad\;\; 0 \quad\;\; 0 \quad 0$$

NO₃⁻ has 5 + 3(6) + 1 = 24 valence electrons.

HNO₃ has 1 + 5 + 3(6) = 24 valence electrons.

b.   SO₄²⁻, See Exercise 13.38a.

HSO₄⁻ has 1 + 6 + 4(6) + 1 = 32 valence electrons

H₂SO₄ has 2 + 6 + 4(6) = 32 valence electrons

c. CN⁻, 4 + 5 + 1 = 10 e⁻   HCN, 1 + 4 + 5 = 10 e⁻

$$\left[\; :C\!\equiv\!N: \atop {\;\;-1 \quad\;\; 0} \right]^{-}$$ $$H\!-\!C\!\equiv\!N: \atop {\;0 \quad\;\; 0 \quad\;\; 0}$$

d. OCN⁻ has 6 + 4 + 5 + 1 = 16 valence electrons.

O—C—N   :Ö—C—N̈:   16 e⁻, but no octet about C.

skeletal   complete octets
structure   of O and N

We need to convert 2 nonbonding pairs to bonding pairs.

$$\left[\; :\ddot{O}\!-\!C\!\equiv\!N: \atop {\;\; -1 \quad\;\; 0 \quad\;\; 0} \right]^{-} \longleftrightarrow \left[\; :\ddot{O}\!=\!C\!=\!N\dot{\cdot} \atop {\;\; 0 \quad\;\; 0 \quad\;\; -1} \right]^{-} \longleftrightarrow \left[\; :O\!\equiv\!C\!-\!\ddot{N}: \atop {\;\; +1 \quad\;\; 0 \quad\;\; -2} \right]^{-}$$

Third structure is not likely because of the large formal charges.

SCN⁻ has 6 + 4 + 5 + 1 = 16 valence electrons.

$$\left[\; :\ddot{S}\!-\!C\!\equiv\!N: \atop {\;\; -1 \quad\;\; 0 \quad\;\; 0} \right]^{-} \longleftrightarrow \left[\; :\ddot{S}\!=\!C\!=\!N\dot{\cdot} \atop {\;\; 0 \quad\;\; 0 \quad\;\; -1} \right]^{-} \longleftrightarrow \left[\; :S\!\equiv\!C\!-\!\ddot{N}: \atop {\;\; +1 \quad\;\; 0 \quad\;\; -2} \right]^{-}$$

First two structures, and especially the second are best because of lowest formal charges.

N₃⁻ has 3(5) + 1 = 16 valence electrons.

$$\left[\; :\ddot{N}\!-\!N\!\equiv\!N: \atop {\;\; -2 \quad\;\; +1 \quad\;\; 0} \right]^{-} \longleftrightarrow \left[\; \dot{\cdot}\ddot{N}\!=\!N\!=\!N\dot{\cdot} \atop {\;\; -1 \quad\;\; +1 \quad\;\; -1} \right]^{-} \longleftrightarrow \left[\; :N\!\equiv\!N\!-\!\ddot{N}: \atop {\;\; 0 \quad\;\; +1 \quad\;\; -2} \right]^{-}$$

Second structure best because of formal charge.

e. NCCN has 2(5) + 2(4) = 18 valence electrons.

:N≡C—C≡N:
  0   0   0   0

41.   Ozone: O_3 has $3(6) = 18$ valence electrons.

Sulfur dioxide: SO_2 has $6 + 2(6) = 18$ valence electrons.

Sulfur trioxide: SO_3 has $6 + 3(6) = 24$ valence electrons.

42.   PAN ($H_3C_2NO_5$)) has $3 + 2(4) + 5 + 5(6) = 46$ valence electrons.

Skeletal structure with complete octets about oxygen atoms (46 electrons used).

This structure has used all 46 electrons, but there are only six electrons around one of the carbon atoms and the nitrogen atom. Two unshared pairs must become shared, that is we form two double bonds.

(last form not important)

43. CH_3NCO has $4 + 3(1) + 5 + 4 + 6 = 22$ valence electrons.

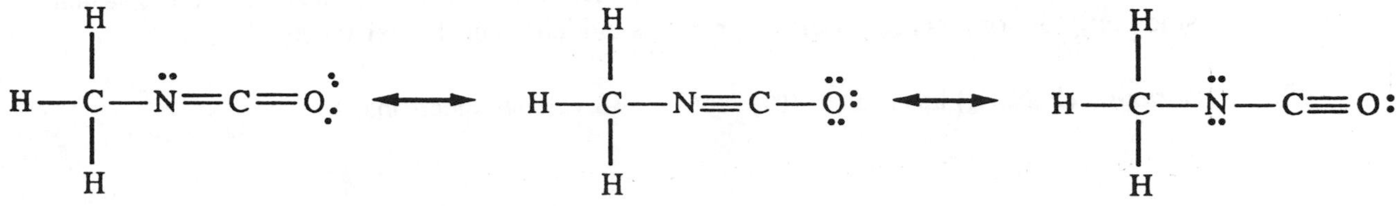

44. S_2Cl_2 has $2(6) + 2(7) = 26$ valence electrons.

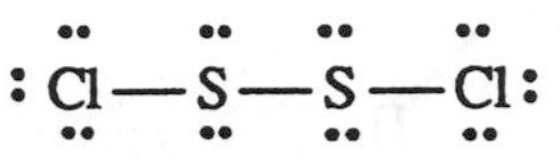

SCl$_2$ has $6 + 2(7) = 20$ valence electrons.

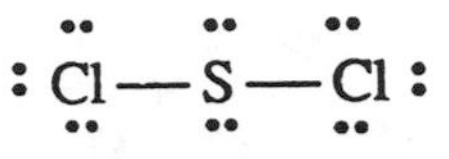

45.

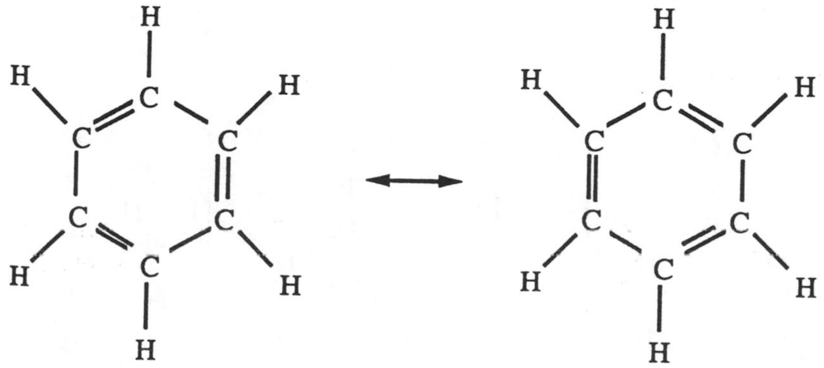

Benzene has $6(4) + 6(1) = 30$ valence electrons.

46.   If no resonance:

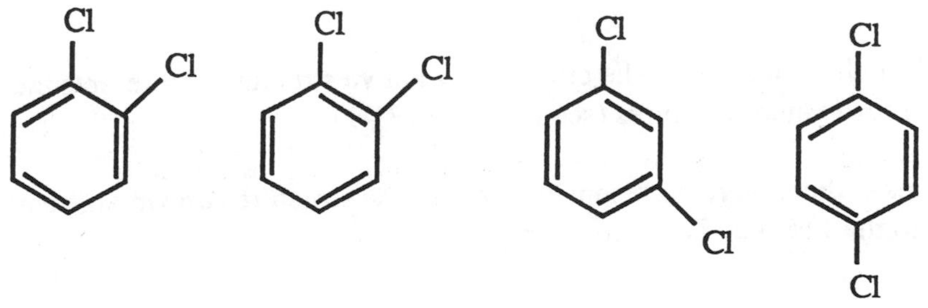

4 different molecules

With resonance:

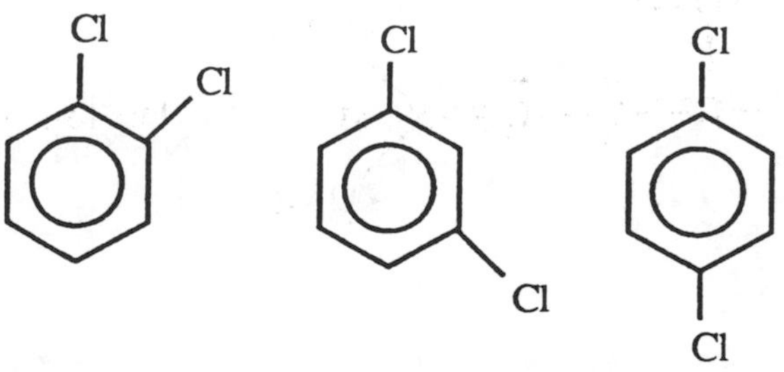

3 different molecules

If resonance is present, we can't distinguish between a single and double bond between adjacent carbons. All carbon-carbon bonds are equivalent (represented by the circle in the structures). Since only 3 isomers are observed provides evidence for the existence of resonance.

47.   Borazine ($B_3N_3H_6$) has $3(3) + 3(5) + 6(1) = 30$ valence electrons.

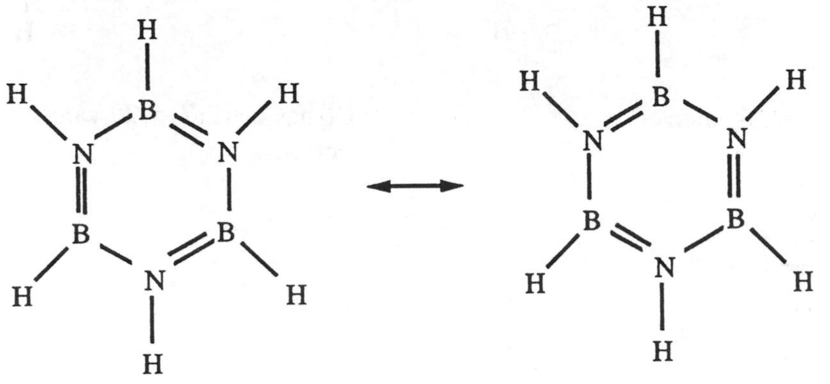

48.

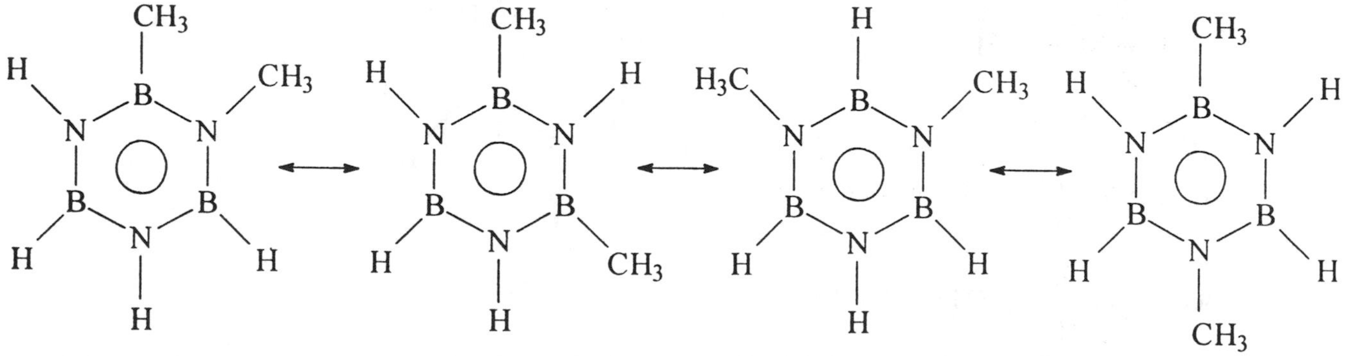

There are four different dimethylborazines. The circle in the above structures represents the ability of borazine to form resonance structures (see Exercise 13.47).

There would be 5 structures if there were no resonance. All of the structures drawn above plus an additional one related to the first Lewis structure above.

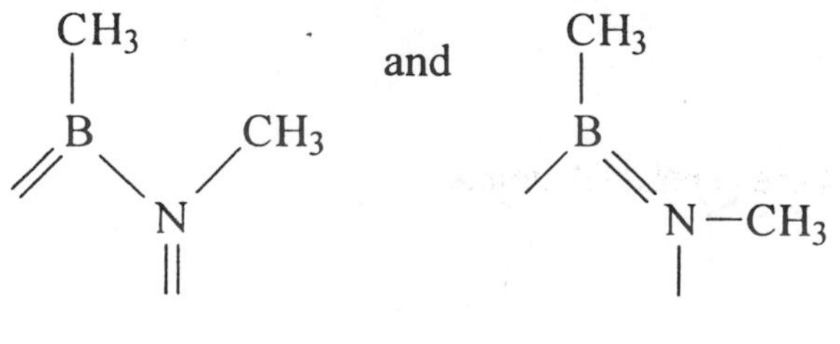

49. PF_5, 5 + 5(7) = 40 e⁻ BrF_3, 7 + 3(7) = 28 e⁻

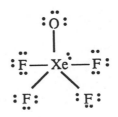

Be(CH₃)₂, 2 + 2[4 + 3(1)] = 16 e⁻ BCl_3, 3 + 3(7) = 24 e⁻

XeOF₄ 8 + 6 + 4(7) = 42 e⁻   XeF₆, 8 + 6(7) = 50 e⁻

SeF₄ has 6 + 4(7) = 34 valence electrons.

50.   ClF₃ has 7 + 3(7) = 28 valence   BrF₃ also has 28 valence
         electrons.   electrons.

We expand the octet of the central Cl atom.

51. CO_3^{2-} has $4 + 3(6) + 2 = 24$ valence electrons.

HCO_3^- has $1 + 4 + 3(6) + 1 = 24$ valence electrons.

H_2CO_3 has $2 + 4 + 3(6) = 24$ valence electrons.

Bonds broken: Bonds formed:

  2 C – O (358 kJ/mol)   1 C = O (799 kJ/mol)

$\Delta H = 2(358) - (799) = -83$ kJ; The carbon-oxygen double bond is stronger than two carbon-oxygen single bonds.

52.   The nitrogen-nitrogen bond length of 112 pm is between a double (120 pm) and a triple (110 pm) bond. The nitrogen-oxygen bond length of 119 pm is between a single (147 pm) and a double bond (115 pm). The last resonance structure doesn't appear to be as important as the other two and we can adequately describe the structure of N_2O using the resonance forms:

$$:\!N\!=\!\!=\!N\!=\!\!=\!O\!: \longleftrightarrow :N\!\equiv\!N\!-\!\ddot{O}\!:$$

There is no evidence from bond lengths for a nitrogen-oxygen triple bond as assigned in the third resonance form.

53.   Resonance is possible in CO_3^{2-} (see Exercise 13-51). We would expect all C – O bonds to be equal and intermediate in strength and length to a single and a double bond. The bond length of 136 pm is consistent with this view of the structure.

54.

CO: $:\!C\!\equiv\!O\!:$ Triple bond between C and O.

CO_2: $:\!\ddot{O}\!=\!C\!=\!\ddot{O}\!:$ Double bond between C and O.

CO_3^{2-}:

Average of 1 1/3 bond between C and O.

CH_3OH:

Single bond between C and O.

Longest → shortest: $CH_3OH > CO_3^{2-} > CO_2 > CO$

Weakest → strongest: $CH_3OH < CO_3^{2-} < CO_2 < CO$

55.

H$_2$NOH:

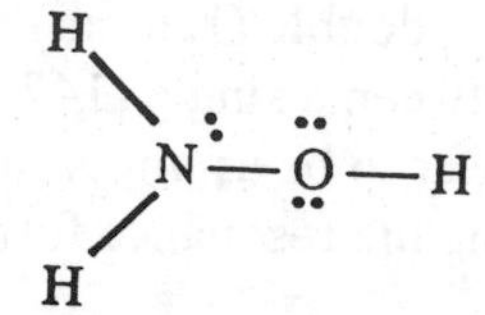

N$_2$O:

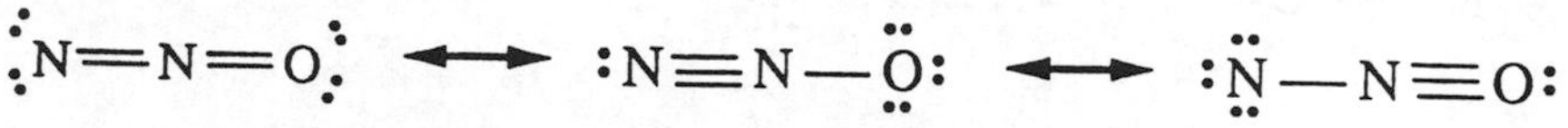

NO$^+$: $[:N\equiv O:]^+$

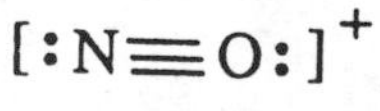

NO$_2^-$:

NO$_3^-$:

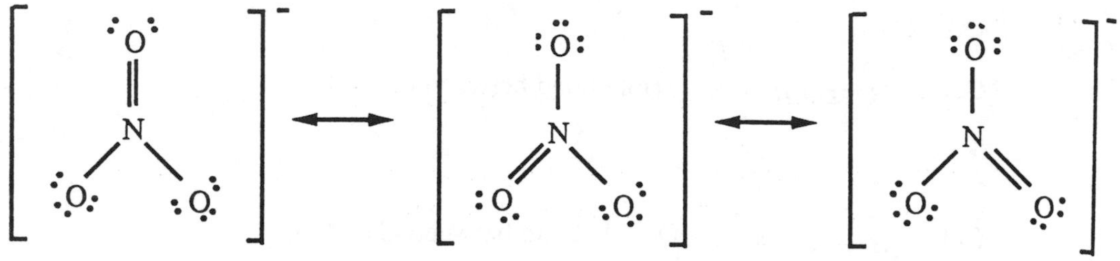

Shortest to Longest: NO$^+$ < N$_2$O < NO$_2^-$ < NO$_3^-$ < H$_2$NOH

Formal Charge

56.

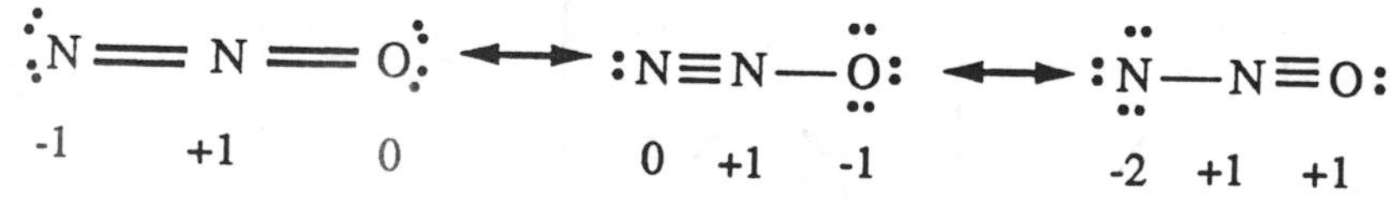

For: ($:\!N\!=\!=$)   FC = 5 - 4 - 1/2(4) = -1

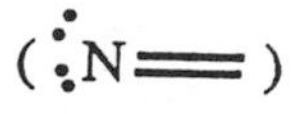

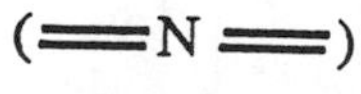

($=\!=\!N\!=\!=$)   FC = 5 - 1/2(8) = +1   Same for $\equiv$N— and —N$\equiv$

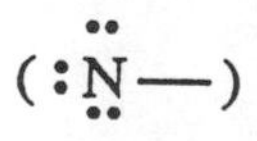

($:\!N\!—$)   FC = 5 - 6 - 1/2(2) = -2

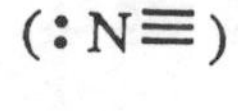

($:N\!\equiv$)   FC = 5 - 2 - 1/2(6) = 0

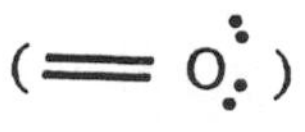

($=\!=\;O:$)   FC = 6 - 4 - 1/2(4) = 0

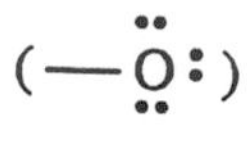

($—\ddot{O}:$)   FC = 6 - 6 - 1/2(2) = -1

($\equiv O:$)   FC = 6 - 2 - 1/2(6) = +1

We should eliminate $N - N \equiv O$ since it has a formal charge of +1 on the most electronegative element present (O). This is consistent with the observation that the N - N bond is between a double and triple bond and that the N – O bond is between a single and double bond.

57.   The Lewis structure

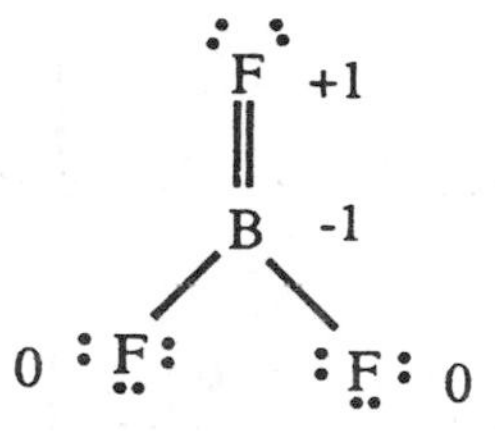

$$\begin{array}{c}
\text{:\ddot{F}: } +1 \\
\| \\
\text{B } -1 \\
0 \text{ :\ddot{F}: } \quad \text{:\ddot{F}: } 0
\end{array}$$

obeys the octet rule, but has a +1 formal charge on the most electronegative element there is, fluorine, and a negative formal charge on a much less electronegative element (boron). This is just the opposite of what we expect.

58. $:C \equiv O:$

Carbon: FC = 4 - 2 - 1/2(6) = -1; Oxygen: FC = 6 - 2 - 1/2(6) = +1

Electronegativity predicts the opposite polarization. The two opposing effects seem to cancel to give a much less polar molecule than expected.

59.   See Exercise 8.38a for Lewis structures of a, b, c and d.

a. $POCl_3$: P, FC = 5 - 1/2(8) = +1   b. SO_4^{2-}: S, FC = 6 - 1/2(8) = +2

c. ClO_4^-: Cl, FC = 7 - 1/2(8) = +3   d. PO_4^{3-}: P, FC = 5 - 1/2(8) = +1

e. SO_2Cl_2, 6 + 2(6) + 2(7) = 32 e⁻   f. XeO_4, 8 + 4(6) = 32 e⁻

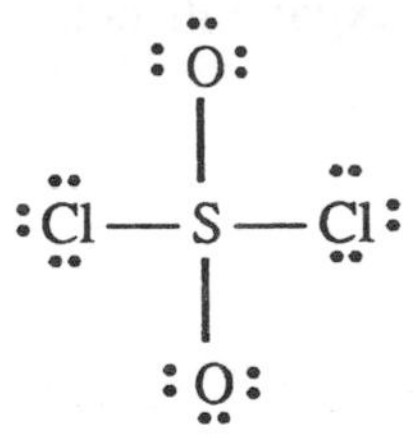

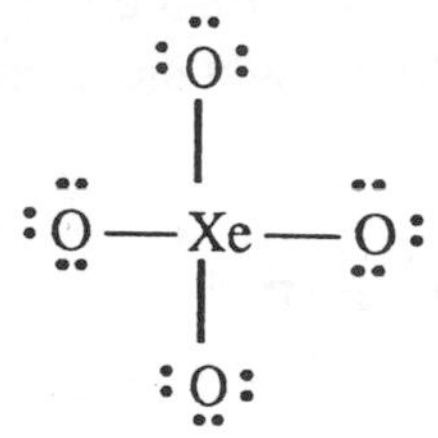

S, FC = 6 - 1/2(8) = +2   Xe, FC = 8 - 1/2(8) = +4

g. ClO_3^-, 7 + 3(6) + 1 = 26 e⁻   h. NO_4^{3-}, 5 + 4(6) + 3 = 32 e⁻

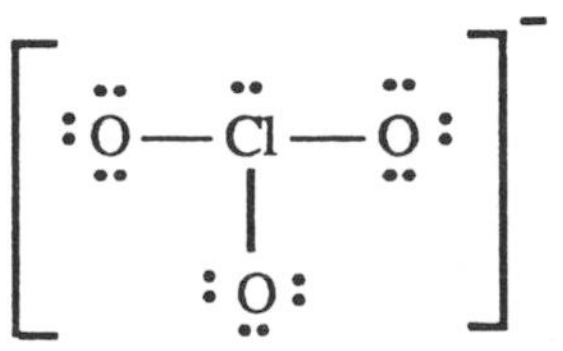

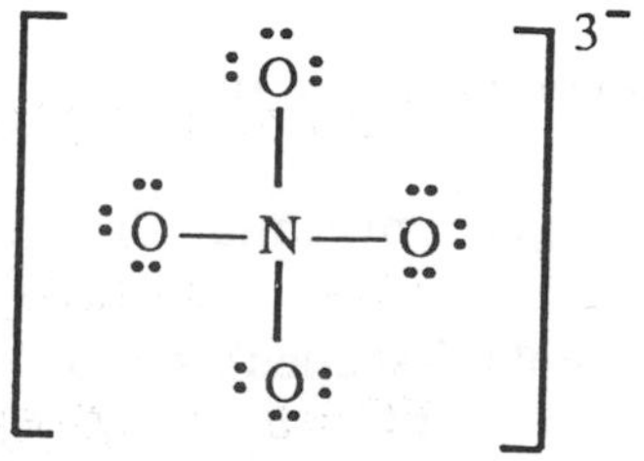

Cl, FC = 7 - 2 - 1/2(6) = +2   N, FC = 5 - 1/2(8) = +1

60.   a.

$$\begin{array}{c} :\!\ddot{O}\!: \\ \| \\ :\!\ddot{Cl}\!-\!P\!-\!\ddot{Cl}\!: \\ | \\ :\!\ddot{Cl}\!: \end{array}$$ P, FC = 0

b.

$$\left[\begin{array}{c} :\!\ddot{O}\!: \\ \| \\ :\!\ddot{O}\!-\!S\!-\!\ddot{O}\!: \\ | \\ :\!\ddot{O}\!: \end{array}\right]^{2-}$$ S, FC = 0

c.

$$\left[\begin{array}{c} :\!\ddot{O}\!: \\ \| \\ :\!\ddot{O}\!=\!Cl\!-\!\ddot{O}\!: \\ \| \\ :\!\ddot{O}\!: \end{array}\right]^{-}$$ Cl, FC = 0

d.

$$\left[\begin{array}{c} :\!\ddot{O}\!: \\ \| \\ :\!\ddot{O}\!-\!P\!-\!\ddot{O}\!: \\ | \\ :\!\ddot{O}\!: \end{array}\right]^{3-}$$ P, FC = 0

e.

$$\begin{array}{c} :\!\ddot{O}\!: \\ \| \\ :\!\ddot{Cl}\!-\!S\!-\!\ddot{Cl}\!: \\ \| \\ :\!\ddot{O}\!: \end{array}$$ S, FC = 0

f.

$$\begin{array}{c} :\!\ddot{O}\!: \\ \| \\ :\!\ddot{O}\!=\!Xe\!=\!\ddot{O}\!: \\ \| \\ :\!\ddot{O}\!: \end{array}$$ Xe, FC = 0

g.

$$\left[\begin{array}{c} :\!\ddot{O}\!=\!\ddot{Cl}\!=\!\ddot{O}\!: \\ | \\ :\!\ddot{O}\!: \end{array}\right]^{-}$$ Cl, FC = 0

h.   We can't. To minimze the formal charge on N would require N to expand its octet. Nitrogen will never have more than eight electrons around itself.

Molecular Geometry and Polarity

61.   13.37   a.   linear, 180°   b.   trigonal pyramid, < 109.5°

            c.   tetrahedral, ≈ 109.5°   d.   tetrahedral, 109.5°

            e.   tetrahedral, 109.5°   f.   V-shaped or bent, < 109.5°

        13.38   a.   all are tetrahedral, ≈ 109.5°

            b.   all are trigonal pyramid, < 109.5°

            c.   all are bent (V-shaped), < 109.5°

13.40   a. NO_2^-:   bent, $< 120°$

HNO$_2$:   bent about N, $< 120°$; bent about $N - O - H$, $< 109.5°$

NO_3^-:   trigonal planar, $120°$

HNO$_3$:   trigonal planar about N, $120°$; bent about $N - O - H$, $< 109.5°$

b.   tetrahedral about S in all three, $≈ 109.5°$; bent about $S - O - H$, $< 109.5°$

c.   HCN: linear, $180°$; CN$^-$ has no bond angle.

d.   all are linear; $180°$

e.   linear about both carbons, $180°$

62.   a. I_3^- has $3(7) + 1 = 22$ valence electrons.

:I—I—I: Complete octets, but only 20 electrons used.

Expand octet of central I:

$$\left[:I—I—I: \right]^-$$

In general we add extra pairs of electrons to the central atom.   We can see a more specific reason when we look at how I_3^- forms in solution:

$$I_2(aq) + I^-(aq) \rightarrow I_3^-(aq)$$

$$:I—I: + \left[:I: \right]^- \rightarrow \left[:I—I—I: \right]^-$$

There are 5 pairs of electrons about the central ion. The structure will be based on a trigonal bipyramid. The lone pair requires more room and will occupy the equatorial positions. The bonding pairs will occupy the axial position.

$$\left[\begin{array}{c} :I: \\ | \\ :I: \\ | \\ :I: \end{array} \right]^-$$

axial: apex of the trigonal pyramid

equatorial: around the middle of the pyramid, the corners of the trigonal plane.

Thus, I_3^- is linear with a $180°$ bond angle.

b. ClF_3 has $7 + 3(7) = 28$ valence electrons.

:F:
| ..
Cl—F:
|
:F:

T-shaped, FClF angles are ≈ 90°C. Since the lone pairs will take up more space, the FClF bond angles will probably be slightly less than 90°.

c. IF_4^+ has $7 + 4(7) - 1 = 34$ valence electrons.

< 120°

< 90°

See-saw or teeter-totter

Also called bisphenoid.

d. SF_5^+ has $6 + 5(7) - 1 = 40$ valence electrons.

120°

90°

trigonal bipyramid

Note: All of the species in this exercise have 5 pairs of electrons around the central atom. All of the structures are based on a trigonal bipyramid, but only in SF_5^+ are all of the pairs bonding pairs. Thus, SF_5^+ is the only one we describe the molecular structure as trigonal bipyramidal. Still, we had to begin with the trigonal bipyramid to get to the structures of the others.

63. SeO_3^{2-}, $6 + 3(6) + 2 = 26$ e⁻ SeH_2, $6 + 2(1) = 8$ e⁻

trigonal pyramid, < 109.5°

bent, < 109.5°

SeO_4^{2-} has $6 + 4(6) + 2 = 32$ valence electrons.

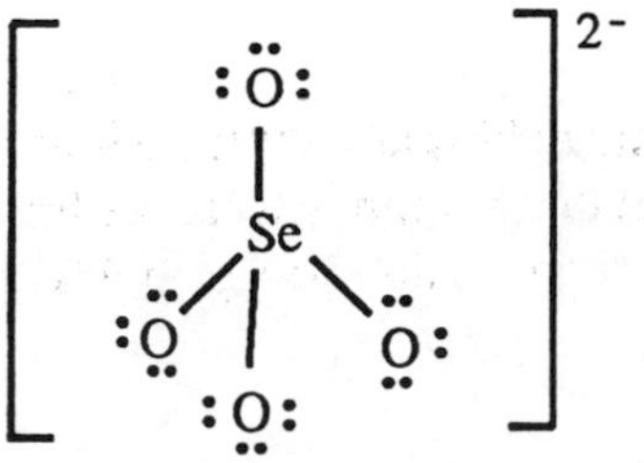

tetrahedral, 109.5°

Note: There are 4 pairs of electrons about the Se atom in each case in this exercise. All of the structures are based on a tetrahedral geometry, but only SeO_4^{2-} has a tetrahedral structure. We consider only the positions of the atoms when describing the molecular structure.

64. BrF_5, $7 + 5(7) = 42$ e⁻ KrF_4, $8 + 4(7) = 36$ e⁻

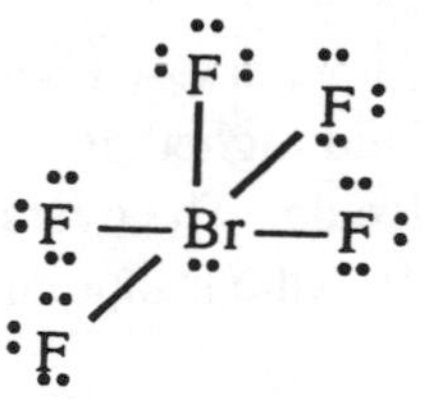

square pyramid, ≈ 90° 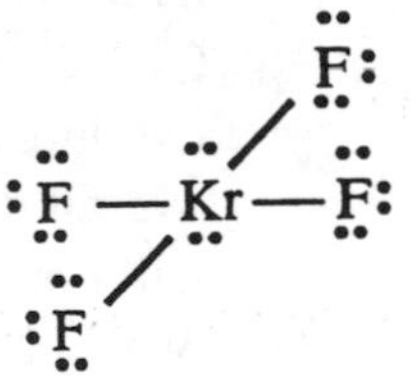   square planar, 90°

IF_6^+ has $7 + 6(7) - 1 = 48$ valence electrons.

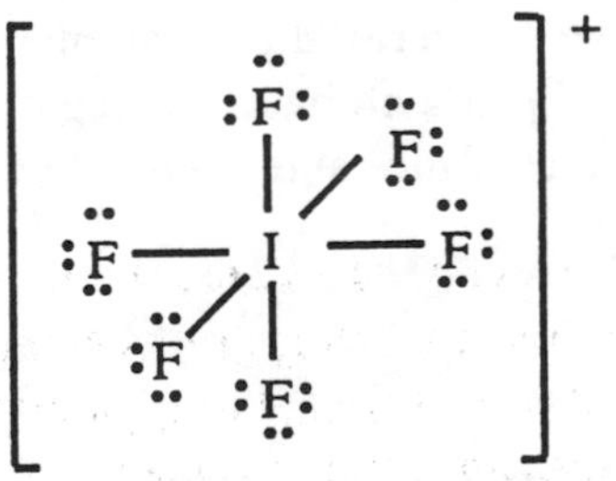

octahedral, 90° bond angles

Note: All these species have 6 pairs of electrons around the central atom. All three structures are based on the octahedron, but only IF_6^+ has octahedral molecular structure.

65. SbF_5, $5 + 5(7) = 40$ e⁻   HF, $1 + 7 = 8$ e⁻

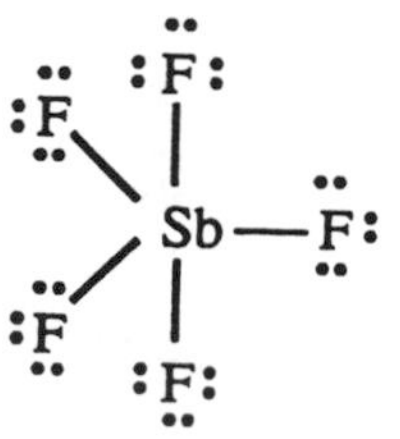

trigonal bipyramid 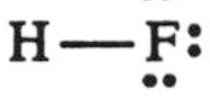   linear

SbF$_6^-$, 5 + 6(7) + 1 = 48 e$^-$ H$_2$F$^+$, 2(1) + 7 - 1 = 8 e$^-$

octahedral   bent

66.   a.

The C–H bonds are assumed nonpolar since the electronegativities of C and H are about equal.

δ+   δ-
C – Cl is the charge distribution for each C–Cl bond. The two individual C–Cl bond dipoles add togheter to give an overall dipole moment for the molecule. The overall dipole will point from the C (positive end) to the midpoint of the two Cl atoms (negative end).

The C–H bond is essentially nonpolar. The three C–Cl bond dipoles add together to give an overall dipole moment for the molecule. The overall dipole will have the negative end at the midpoint of the three chlorines and the positive end at the carbon.

CCl$_4$ is nonpolar. CCl$_4$ is a tetrahedral molecule. All four C–Cl bond dipoles cancel when added together. Let's consider just the C and two of the Cl atoms. There will be a net dipole pointing in the direction of the middle of the two Cl atoms.

There will be an equal and opposite dipole arising from the other two Cl atoms. Combining:

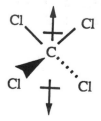

The two dipoles will cancel and CCl$_4$ is nonpolar.

b. CO_2 is nonpolar. CO_2 is a linear molecule with two equivalent bond dipoles that cancel.
N_2O is polar since the bond dipoles do not cancel.

$$\overset{\longrightarrow}{\underset{}{}}$$
$$\overset{\delta+}{} \quad \overset{\delta-}{}$$
$$:N = N = O:$$

c. NH_3 is polar. The 3 N–H bond dipoles add together to give a net dipole in the direction of
the lone pair. We would predict PH_3 to be nonpolar on the basis of electronegativity, i.e.,
P–H bonds are nonpolar. The presence of the lone pair makes the molecule slightly polar
and the net dipole points in the direction of the lone pair on P. The magnitude of the dipole
in PH_3 is about one third that of the NH_3 dipole.

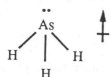

The As–H bonds are essentially nonpolar ($\Delta EN = 0.1$). The lone pair on AsH_3 makes the
molecule slightly polar with a dipole moment even less than the PH_3 dipole.

67.   a. CrO_4^{2-} has $6 + 4(6) + 2 = 32$ valence electrons.

$$\left[\begin{array}{c} :\overset{..}{O}: \\ | \\ :\overset{..}{O} - Cr - \overset{..}{O}: \\ | \\ :\overset{..}{O}: \end{array}\right]^{2-}$$

tetrahedral

Cr has an electron configuration of
$[Ar]4s^23d^4$. We count all six electrons
beyond the argon core.

b. Dichromate ion ($Cr_2O_7^{2-}$) has $2(6) + 7(6) + 2 = 56$ valence electrons.

$$\left[\begin{array}{c} :\overset{..}{O}: \quad\quad :\overset{..}{O}: \\ | \quad\quad\quad | \\ Cr \quad\quad\quad Cr \\ \end{array}\right]^{2-}$$

There is no simple way to describe the
structure. We would expect there to be a
roughly tetrahedral arrangement of O atoms
about each Cr and the Cr – O – Cr bond to
be bent with an angle < 109.5°.

c. $S_2O_3^{2-}$ has $2(6) + 3(6) + 2 = 32$ valence electrons.

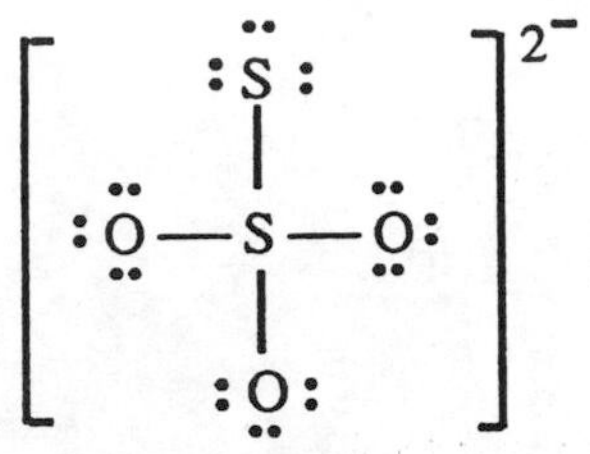

tetrahedral

d. $S_2O_8^{2-}$ has $2(6) + 8(6) + 2 = 62$ valence electrons.

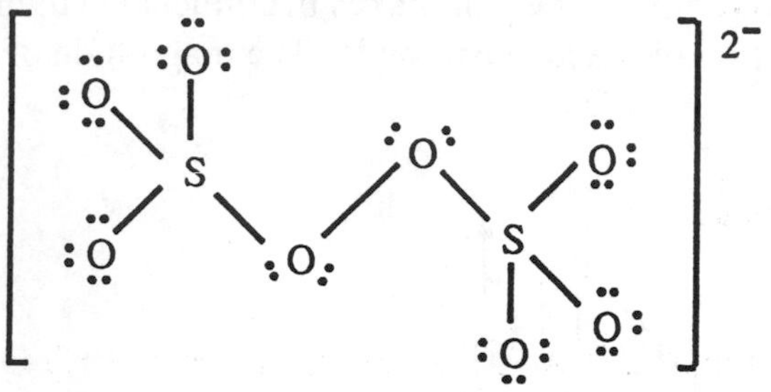

Again, there is no simple description for the entire structure. Oxygen atoms are roughly tetrahedrally arranged around each S and the S – O – O bonds are bent with angles slightly less than 109.5°.

68.   The two requirements for a polar molecule are:

1.   polar bonds

2.   a structure such that the polar bonds do not cancel.

In addition, some molecules that have no polar bonds, but contain unsymmetrical lone pairs are polar. For example, PH_3 and AsH_3 are slightly polar.

69.   a. OCl_2, $6 + 2(7) = 20$ e⁻ Br_3^-, $3(7) + 1 = 22$ e⁻

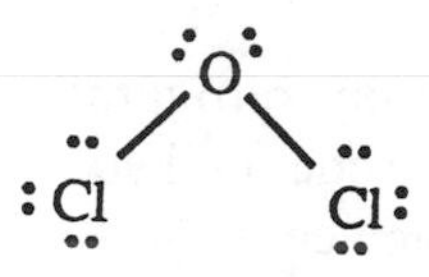

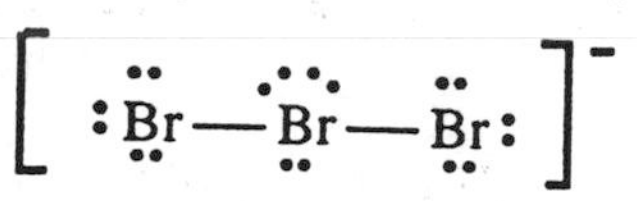

bent, polar   linear, nonpolar

BeH_2, $2 + 2(1) = 4$ e⁻ BH_2^-, $3 + 2(1) + 1 = 6$ e⁻

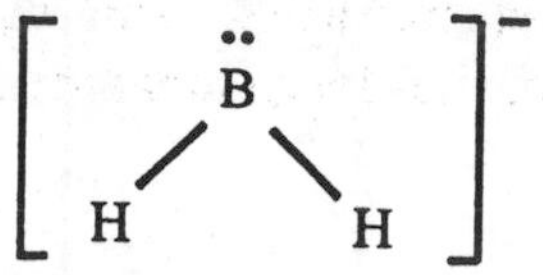

H — Be — H

linear, nonpolar   bent, polar

Note: All four species contain three atoms. They have different structures because the number of lone pairs of electrons around the central atom are different in each case.

b. BCl_3, $3 + 3(7) = 24$ e⁻ NF_3, $5 + 3(7) = 26$ e⁻

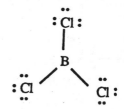

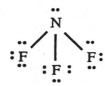

trigonal planar, nonpolar   Trigonal pyramid, polar

IF_3 has $7 + 3(7) = 28$ valence electrons.

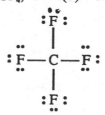   T-shaped, polar

Note: Each molecule has the same number of atoms, but the structures are different because of differing numbers of lone pairs around each central atom.

c. CF_4, $4 + 4(7) = 32$ e⁻ SeF_4, $6 + 4(7) = 34$ e⁻

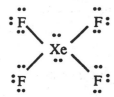

tetrahedral, nonpolar   See-saw, polar

XeF_4, $8 + 4(7) = 36$ valence electrons

square planar, nonpolar

Again, each molecule has the same number of atoms, but a different structure because of differing numbers of lone pairs around the central atom.

d. IF_5, 7 + 5(7) = 42 e⁻ AsF_5, 5 + 5(7) = 40 e⁻

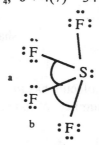

square pyramid, polar   trigonal bipyramid, nonpolar

Yet again, the molecules have the same number of atoms, but different structures because of the presence of differing numbers of lone pairs.

70. SF_2, 6 + 2(7) = 20 e⁻ SF_4, 6 + 4(7) = 34 e⁻

bent, polar   see-saw, polar
a)   109.5°   a) ≈ 120°
                                         b) ≈ 90°

SF_6, 6 + 6(7) = 48 e⁻

octahedral, nonpolar, 90°

S_2F_4, 2(6) + 4(7) = 40 e⁻

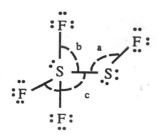

a)   < 109.5°
b)   ≈ 90°
c)   ≈ 120°

There is no easy description of the S_2F_4 structure. There is a trigonal bipyramidal arrangement (3 F, 1 S, 1 lone pair) about one sulfur and a tetrahedral (1 F, 1 S, 2 lone pairs) arrangement of e⁻ pairs about the other sulfur. With this picture we can predict values for all bond angles.   The molecule will be polar.

71.   The polar bonds are symmetrically arranged about the central atom and all of the individual bond dipoles cancel to give no net dipole moment for each molecule. All the molecules are nonpolar even though they all contain polar bonds.

72.   a.

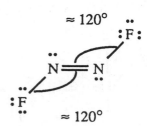

          ≈ 120°   < 120° < 120°

        :F: :F   F:

        N═══N   N═══N

        :F

          ≈ 120°

        b. nonpolar   polar

73.   a.   b.

            :O̤—C≡N: :F—Be—F:

        H

            polar   nonpolar

        c.   d.

            :F   F: :F   F:

                Kr   C

            :F   F: :Cl   Cl:

            nonpolar   polar

        e.   f.

            H—N̈—N̈—H   H

                | | C══O̤:
                H   H   H

            polar   polar

Additional Exercises

74.   The general structure of the trihalide ions is:

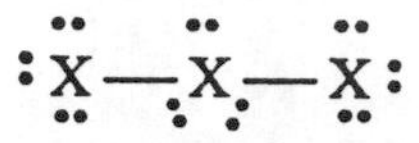

Bromine and iodine are large enough and have low energy, empty d-orbitals to accommodate the expanded octet. Fluorine is small and its valence shell contains only 2s and 2p orbitals (4 orbitals) and cannot expand its octet. The lowest energy d orbitals in F are the 3d orbitals. They are too high in energy, compared to the 2s and 2p orbitals, to be used in bonding.

75.

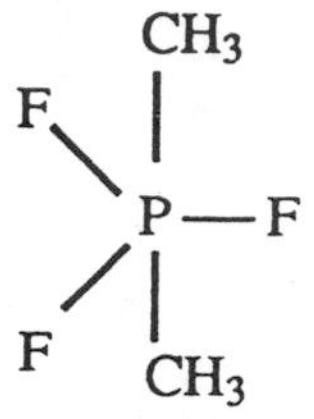

(i) is non-polar. (ii) and (iii) are polar with (ii) probably the more polar. Thus, dipole moments would help distinguish the three isomers; particularly in distinguishing (i) from the other two.

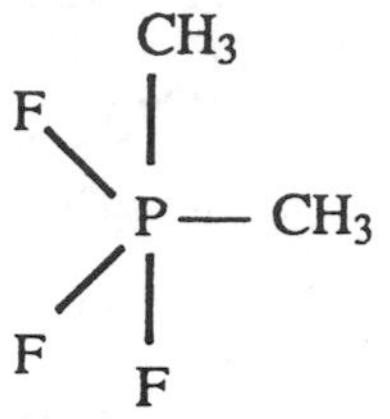

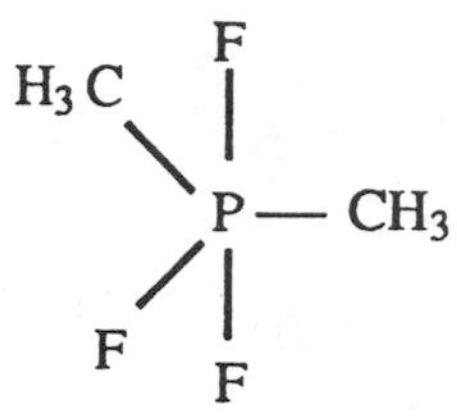

76. $C \equiv O$ (1072 kJ/mol) and $N \equiv N$ (941 kJ/mol); CO is polar while N_2 is nonpolar. This may lead to a greater reactivity.

77.   The stable species are:

a. SO_4^{2-}:   We can't draw a Lewis structure that obeys the octet rule for SO_4. The two extra electrons (from the -2 charge) complete octets.

b. PF_5:   N is too small and doesn't have low energy d-orbitals to expand its octet.

c. SF_6:   O is too small and doesn't have low energy d-orbitals to expand its octet.

d. BH_4^-: BH_3 doesn't obey octet rule.

e. MgO: In MgF we would have to have Mg^+F^- or Mg^{2+} F^{2-}. Neither Mg^+ nor F^{2-} are stable.

f. CsCl: Cs doesn't form Cs^{2+} ion.

g. KBr: Br doesn't form Br^{2-} ion.

78.   If we can draw resonance forms for the anion after loss of H^+, we can argue that the extra
       stability of the anion causes the proton to be more readily lost (is more acidic).

a.

b.

c.

In all 3 cases, extra resonance forms can be drawn for the anion that are not possible when the H^+
is present, which leads to enhanced stability.

79.   No, we would expect the more highly charged ions to have a greater attraction for an electron.
       Thus, the electronegativity of an element does depend on its oxidation state.

80.   a. SiF_4, $4 + 4(7) = 32$ e⁻   b. SeF_4, $6 + 4(7) = 34$ e⁻

       tetrahedral, nonpolar   see-saw, polar

c. KrF_4 has $8 + 4(7) = 36$ valence electrons.

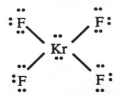

square planar, nonpolar

Although all three molecules have the same number of atoms, there is a different number of valence electrons in each one. This results in differing numbers of lone pairs on each central atom; hence, different structures.

81.   a. BF_3, $3 + 3(7) = 24$ e⁻   b. PF_3, $5 + 3(7) = 26$ e⁻

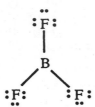

trigonal planar, nonpolar

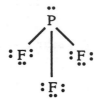

trigonal pyramid, polar

c. BrF_3 has $7 + 3(7) = 28$ valence electrons.

T-shape, polar

Again we see a series of molecules with the same number of atoms, but different numbers of lone pairs. They all have different structures.

82.   Nonmetals form covalent compounds. Nonmetals have valence electrons in the s and p orbitals. Since there are 4 total s and p orbitals, there is only room for eight electrons.

83.

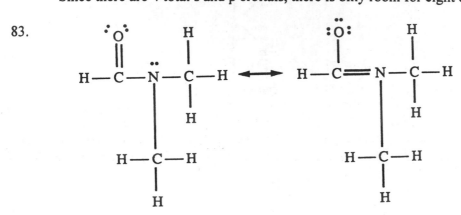

84. S_2N_2 has $2(6) + 2(5) = 22$ valence electrons.

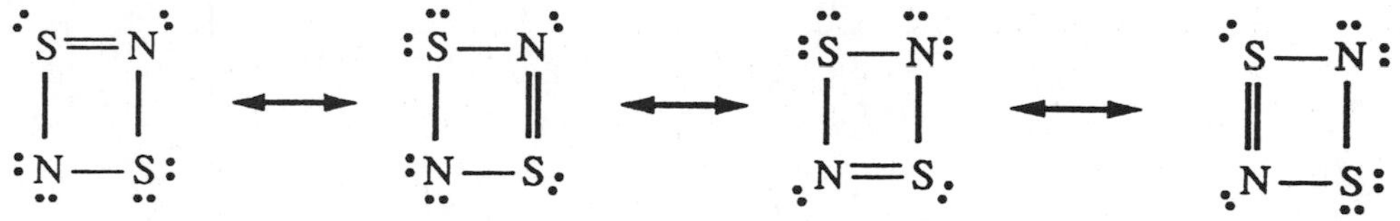

S—N
| |
N—S

:S̈=N̈: ⟷ :S̈—N̈: ⟷ :S̈—N̈: ⟷ S̈—N̈:
| | | ‖ | | ‖ |
:N̈—S̈: ⟷ :N̈—S̈: ⟷ :N̈=S̈: ⟷ :N̈—S̈:

85.   a. $Na^+(g) + Cl^-(g) \rightarrow NaCl(s)$ b. $NH_4^+(g) + Br^-(g) \rightarrow NH_4Br(s)$

        c. $Mg^{2+}(g) + S^{2-}(g) \rightarrow MgS(s)$ d. $1/2\ O_2(g) \rightarrow O(g)$

        e. $O_2(g) \rightarrow 2\ O(g)$

86.   a. Al_2Cl_6 has $2(3) + 6(7) = 48$ valence electrons.

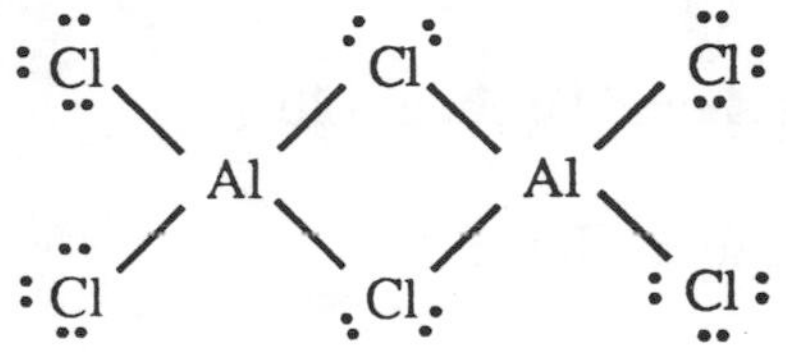

:C̈l   C̈l   C̈l:
   \\   / \\   /
     Al   Al
   /   \\ /   \\
:C̈l   C̈l   C̈l:

        b. There are 4 pairs of electrons about each Al, we would predict the bond angles to be close to a tetrahedral angle of 109.5°.

        c. nonpolar; The individual bond dipoles will cancel.

87.   Yes, each structure has the same number of effective pairs around the central atom giving the same molecular structure. (A multiple bond is counted as a single group of electrons.)

88.   This is a very weak bond. The structure

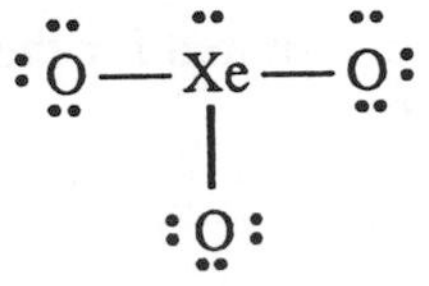

:Ö—Xe—Ö:
      |
     :Ö:

might be more accurate. It is difficult to think of a double bond as being so weak. This illustrates that no model is totally effective in describing the bonding in all molecules. We need to recognize that all models have limitations and that we can't apply all models to all molecules.

89.   a. $N(NO_2)_2^-$ contains $5 + 2(5) + 4(6) + 1 = 40$ valence electrons.

The most likely structures are:

There are other possible resonance structures, but these are most likely.

b.   The NNN and all ONN and ONO bond angles should be about 120°.

c. $NH_4N(NO_2)_2 \rightarrow 2\ N_2 + 2\ H_2O + O_2$; Break and form all bonds.

Bonds broken:	Bonds formed:
4 N – H (391 kJ/mol)	2 N ≡ N (941 kJ/mol)
1 N – N (160 kJ/mol)	4 H – O (467 kJ/mol)
1 N = N (418 kJ/mol)	1 O = O (495 kJ/mol)
3 N – O (201 kJ/mol)	
1 N = O (607 kJ/mol)	$\Sigma D_{formed} = 4245$ kJ
$\Sigma D_{broken} = 3352$ kJ	

$\Delta H = \Sigma D_{broken} - \Sigma D_{formed} = 3352\ kJ - 4245\ kJ = -893\ kJ$

The ΔH value that we calculated is too negative since we haven't accounted for the ionic bonding in $NH_4N(NO_2)_2$.

CHAPTER FOURTEEN

COVALENT BONDING: ORBITALS

Localized Electron Model and Hybrid Orbitals

1.   a.

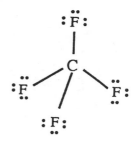

tetradedral   sp^3
109.5°   nonpolar

b.

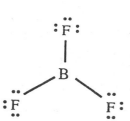

trigonal pyramid   sp^3
< 109.5°   polar

The angle should be slightly less than
109.5° because the lone pair requires more
room than the bonding pairs.

c.

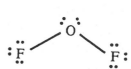

bent   sp^3
< 109.5°   polar

d.

trigonal planar   sp^2
120°   nonpolar

e.

H — Be — H

linear   sp
180°   nonpolar

f.

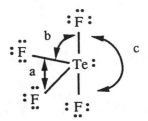

see-saw
a. ≈ 120°, b. ≈ 90°, c. ≈ 180°
dsp^3 polar

g.

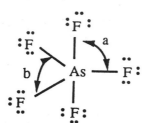

trigonal bipyramid   dsp^3
a. 90°, b. 120°   nonpolar

h.

:F — Kr — F:

linear   dsp^3
180°   nonpolar

i.

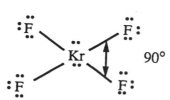

90°

square planar   d^{2}sp^3
90°   nonpolar

j.

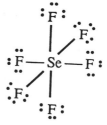

octahedral   d^{2}sp^3
90°   nonpolar

k.

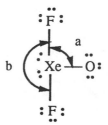

square pyramid   d^{2}sp^3
≈ 90°   polar

l.

T-shaped
a. ≈ 90°, b. ≈ 180°
dsp^3 polar

m.

:O:
|
Xe
/ / \
:O: :O:
:O:

tetrahedral
109.5°
sp³
nonpolar

2.   a.

S
/ ‖
:O: :O:

bent
sp²

Only one resonance form is shown. Resonance does not change the position of the atoms.
We can predict the geometry from only one of the resonance structures.

b.

:O:
‖
S
/ \
:O: :O:

trigonal planar
sp²

$\left[\begin{array}{c} :O: \\ | \\ :S—S—O: \\ | \\ :O: \end{array}\right]^{2-}$

tetradedral
sp³

Only one resonance form is shown.

d.

$\left[\begin{array}{c} :O: \quad :O: \\ | \quad\quad | \\ :O—S—O—O—S—O: \\ | \quad\quad | \\ :O: \quad :O: \end{array}\right]^{2-}$

Tetrahedral geometry about each S, sp³
hybrids; bent arrangement about peroxide
O's, sp³ hybrids

e.

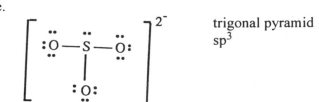

trigonal pyramid
sp³

f.

$$\left[\begin{array}{c} :\ddot{O}: \\ | \\ :\ddot{O} - S - \ddot{O}: \\ | \\ :\ddot{O}: \end{array}\right]^{2-}$$

tetrahedral
sp³

g.

bent
sp³

h.

< 90°
< 120°

see-saw
dsp³

i.

octahedral
d²sp³

See-saw about S atom with one lone pair (dsp³); bent about S atom with two lone pairs (sp³)

j.

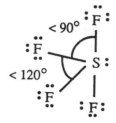

3.   a.   P₄, 4(5) = 20 valence electrons

Each P - sp³ hybridization.

P – P – P bond angle, 60° (Each P is at the corner of a tetrahedron. The faces of a tetrahedron are equilateral triangles.)

Note: We predict sp³ orbitals because there are 4 pairs of e⁻ about each P. There is considerable strain in this molecule.

b. P_4O_6, $4(5) + 6(6) = 56$ valence electrons

Each P - sp^3

POP angle, $< 109.5°$

OPO angle, $< 109.5°$

c. P_4O_{10}, $4(5) + 10(6) = 80$ valence electrons

Each P - sp^3

POP angle, $< 109.5°$

OPO angle, $109.5°$

4.

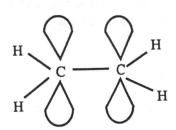

For the p-orbitals to properly align to form the π bond, all six atoms are forced into the same plane.

5.   No, the CH$_2$ planes are mutually perpendicular. The center C-atom is involved in two π-bonds. The p-orbitals used in each bond must be perpendicular to the other. This forces the two CH$_2$ planes to be perpendicular since different p orbitals are used to form the hybrid orbitals.

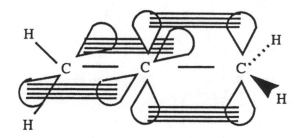

6.   Biacetyl ($C_4H_6O_2$) has $4(4) + 6(1) + 2(6) = 34$ valence electrons.

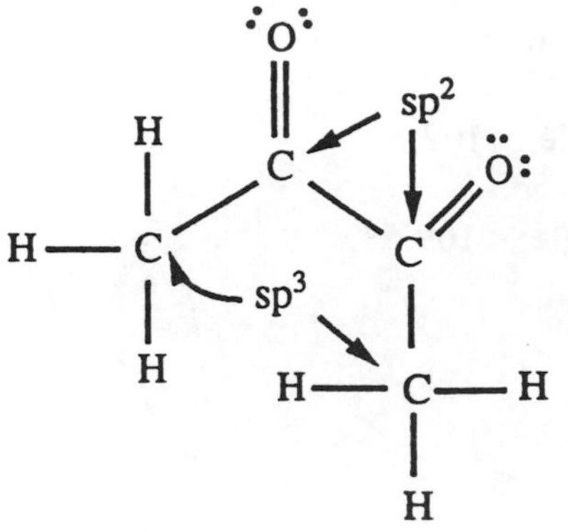

All CCO angles are 120° The six atoms are not in the same plane because of free rotation about the carbon-carbon single bonds.

Acetoin ($C_4H_8O_2$) has $4(4) + 8(1) + 2(6) = 36$ valence electrons.

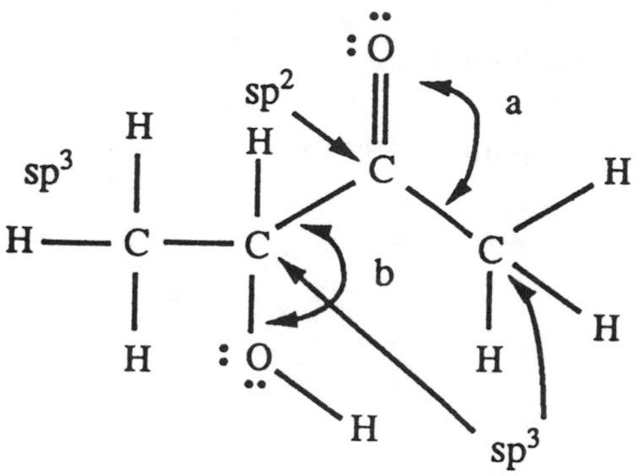

The carbon with the doubly bonded O is sp^2 hybridized. The other 3-C atoms are sp^3 hybridized.

a.   120°, b.   109.5°

7.   A single bond is a σ bond. Double and triple bonds are composed of a σ bond and 1 or 2 π bonds. In biacetyl there are 11 σ and 2 π bonds. In acetoin there are 13 σ and 1 π bonds. Note: CH_3 is shorthand for a carbon singly bonded to three hydrogen atoms.

8.   Acrylonitrile (C_3H_3N) has $3(4) + 3(1) + 5 = 20$ valence electrons.

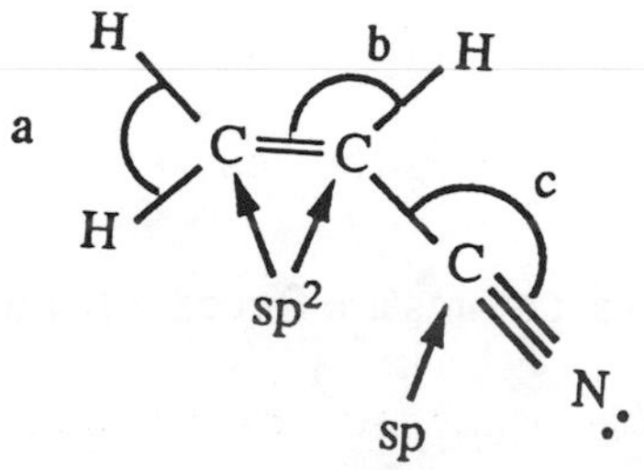

a.   120°

b.   120°

c.   180°

Methyl methacrylate ($C_5H_8O_2$) has $5(4) + 8(1) + 2(6) = 40$ valence electrons.

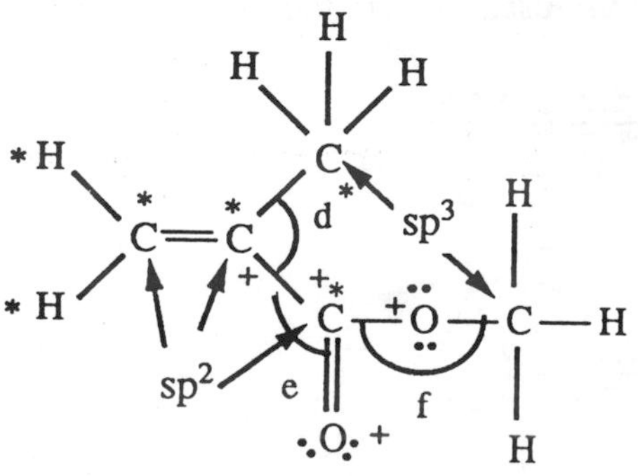

d.   120°

e.   120°

f.   < 109.5°

The atoms marked with a * and a + are all coplanar with each other. The two planes, however, do not have to coincide due to the rotations of sigma bonds.

9.   Methyl methacrylate: 14 σ and 2 π bonds

       Acrylonitrile: 6 σ and 3 π bonds

10.

a. 6   b. 4   c. The center N in $-N=N=N$ group

d. 33 σ   e. 5 π bonds   f. The six membered ring is planar.

g. 180°   h. < 109.5°   i. sp³

Note: For f, the six membered ring is planar due to other resonance structures that can be drawn for this ring.

11.   a. NCN²⁻ has 5 + 4 + 5 + 2 = 16 valence electrons.

H_2NCN has $2(1) + 5 + 4 + 5 = 16$ valence electrons.

favored by formal charge

$NCNC(NH_2)_2$ has $5 + 4 + 5 + 4 + 2(5) + 4(1) = 32$ valence electrons.

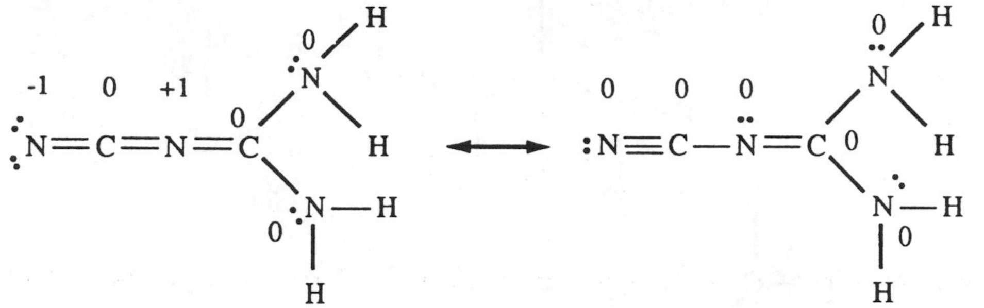

favored by formal charge

Melamine $(C_3N_6H_6)$ has $3(4) + 6(5) + 6(1) = 48$ valence electrons.

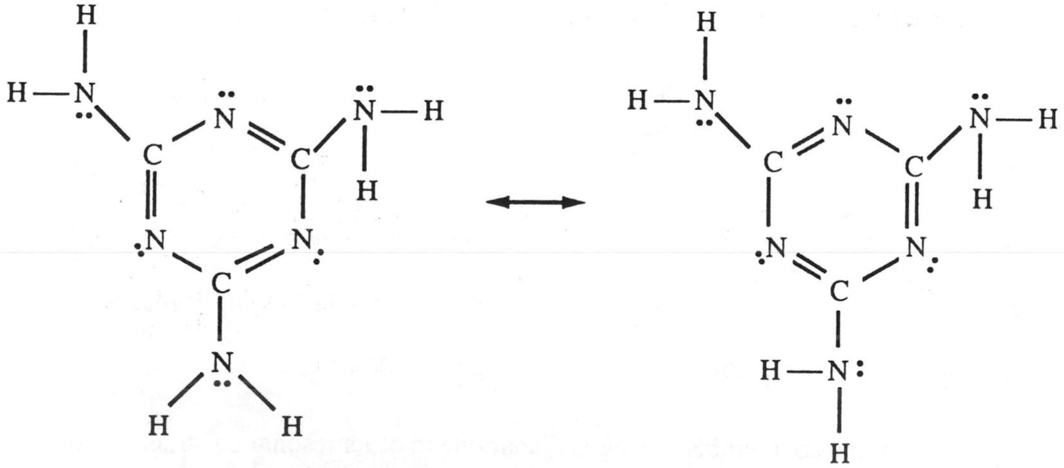

b. NCN^{2-}: C is sp hybridized. Depending on the resonance form, N can be sp, sp^2, or sp^3 hybridized. For the remaining compounds, we will predict hybrids for the favored resonance structures.

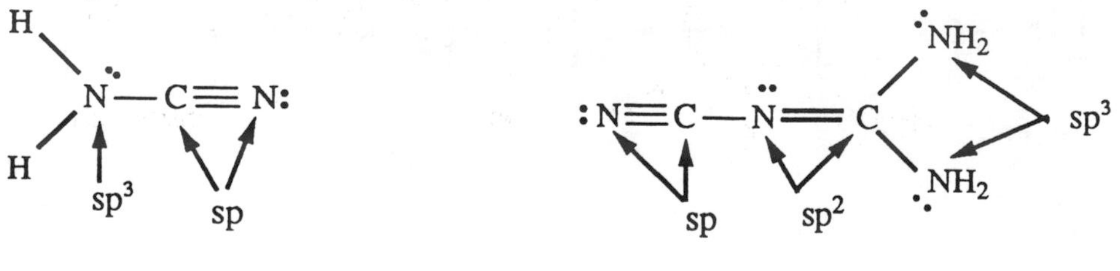

Melamine: N in NH_2 groups are all sp^3 hybridized. Atoms in ring are all sp^2 hybridized.

c. NCN^{2-}: 2 σ and 2 π bonds; H_2NCN: 4 σ and 2 π bonds; dicyandiamide: 9 σ and 3 π bonds; melamine: 15 σ and 3 π bonds

d. The π-system forces the ring to be planar just as the benzene ring is planar.

e. The structure:

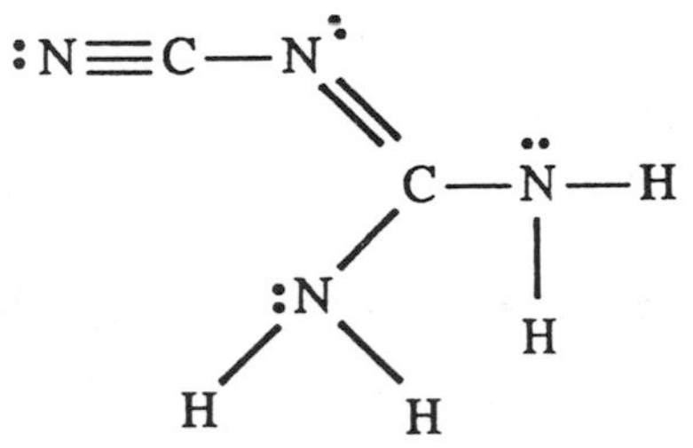

$$:N\equiv C-\overset{\cdot\cdot}{N}$$

is the most important since it has three different CN bonds. This structure is also favored on the basis of formal charge.

12. a. Piperine and capsaicin are molecules classified as organic compounds, i.e., compounds based on carbon. The majority of Lewis structures for organic compounds have all atoms with zero formal charge. Therefore, carbon atoms in organic compounds will usually form four bonds, nitrogen atoms will form three bonds and complete the octet with one lone pair of electrons, and oxygen atoms will form two bonds and complete the octet with two lone pairs of electrons. Using these guidelines, the Lewis structures are:

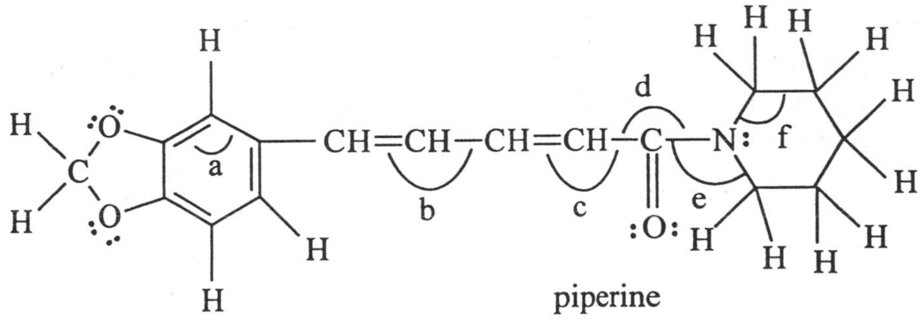

piperine

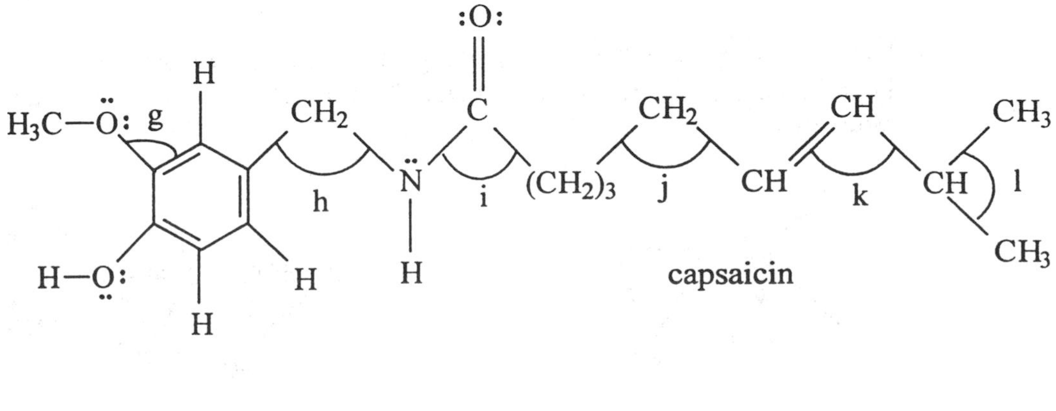

capsaicin

Note: The ring structures are all shorthand notation for rings of carbon atoms. In piperine, the first ring contains 6 carbon atoms and the second ring contains 5 carbon atoms (plus nitrogen).

b. piperine: 0 sp, 11 sp^2, and 6 sp^3 carbons; capsaicin: 0 sp, 9 sp^2, and 9 sp^3 carbons

c. The nitrogens are sp^3 hybridized in each molecule.

d.
a.	120°	e.	≈ 109.5°	i.	120°
b.	120°	f.	109.5°	j.	109.5°
c.	120°	g.	120°	k.	120°
d.	120°	h.	109.5°	l.	109.5°

The Molecular Orbital Model

13. Bonding molecular orbitals have electron density between the bonded atoms and are lower in energy than the atomic orbitals from which they are formed. Antibonding molecular orbitals have minimal electron density between the bonded atoms and are higher in energy than the atomic orbitals from which they are formed.

14. If we calculate a non-zero bond order for a molecule, we predict that it can exist (is stable).

a: H_2^+: $(\sigma_{1s})^1$ B.O. = (1-0)/2 = 1/2, stable
    H_2: $(\sigma_{1s})^2$ B.O. = (2-0)/2 = 1, stable
    H_2^-: $(\sigma_{1s})^2(\sigma_{1s}*)^1$ B.O. = (2-1)/2 = 1/2, stable
    H_2^{2-}: $(\sigma_{1s})^2(\sigma_{1s}*)^2$ B.O. = (2-2)/2 = 0, not stable

b. He_2^{2+}: $(\sigma_{1s})^2$ B.O. = (2-0)/2 = 1, stable
    He_2^+ $(\sigma_{1s})^2(\sigma_{1s}*)^1$ B.O. = (2-1)/2 = 1/2, stable
    He_2: $(\sigma_{1s})^2(\sigma_{1s}*)^2$ B.O. = (2-2)/2 = 0, not stable

c. Be_2: $(\sigma_{2s})^2(\sigma_{2s}*)^2$ B.O. = (2-2)/2 = 0, not stable
    B_2: $(\sigma_{2s})^2(\sigma_{2s}*)^2(\pi_{2p})^2$ B.O. = (4-2)/2 = 1, stable
    Li_2: $(\sigma_{2s})^2$ B.O. = (2-0)/2 = 1, stable

15. The electron configurations are:

O_2^+: $(\sigma_{2s})^2(\sigma_{2s}*)^2(\pi_{2p})^4(\sigma_{2p})^2(\pi_{2p}*)^1$

O_2: $(\sigma_{2s})^2(\sigma_{2s}*)^2(\pi_{2p})^4(\sigma_{2p})^2(\pi_{2p}*)^2$

O_2^-: $(\sigma_{2s})^2(\sigma_{2s}*)^2(\pi_{2p})^4(\sigma_{2p})^2(\pi_{2p}*)^3$

O_2^{2-}: $(\sigma_{2s})^2(\sigma_{2s}*)^2(\pi_{2p})^4(\sigma_{2p})^2(\pi_{2p}*)^4$

	O_2^+	O_2	O_2^-	O_2^{2-}
Bond order	2.5	2	1.5	1
# of unpaired electrons	1	2	1	0

Bond energy: $O_2^{2-} < O_2^- < O_2 < O_2^+$; Bond length: $O_2^+ < O_2 < O_2^- < O_2^{2-}$

Bond energy is directly proportional to bond order and bond length is inversely proportional to bond order.

16.   For NO: $(\sigma_{2s})^2(\sigma_{2s}*)^2(\pi_{2p})^4(\sigma_{2p})^2(\pi_{2p}*)^1$ B.O. $= \dfrac{8-3}{2} = 2.5$, 1 unpaired e$^-$

MO model does a better job. It predicts a bond order of 2.5. Measured bond energies and bond lengths are consistent with the bond in NO being stronger than a double bond.

	Bond Length	Bond Energy
$N - O$	147 pm	201 kJ/mol
$N = O$	115 pm	607 kJ/mol
NO (nitric oxide)	112 pm	678 kJ/mol

NO has 11 valence electrons. It is impossible to draw a Lewis structure satisfying the octet rule for odd electron species. The most reasonable attempt is to draw:

$$:\!N\!=\!\!O\!: \longleftrightarrow :\!N\!=\!\!O\!: \longleftrightarrow :\!N\!=\!\!O\!:$$

none of which makes the correct prediction for the bond strength and the number of unpaired electrons.

17.   Bond energy is directly proportional to bond order. Bond length is inversely proportional to bond order. Bond energy and length can be measured.

18.

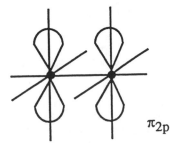

π_{2p}

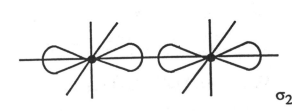

σ_{2p}

19.

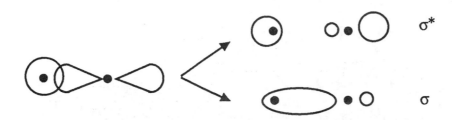

These molecular orbitals are sigma MOs since electron density is cylindrically symmetric about the internuclear axis.

20.   a. H_2: $(\sigma_{1s})^2$ B.O. = (2-0)/2 = 1, diamagetic
      b. B_2: $(\sigma_{2s})^2(\sigma_{2s}*)^2(\pi_{2p})^2$ B.O. = (4-2)/2 = 1, paramagnetic
      c. F_2: $(\sigma_{2s})^2(\sigma_{2s}*)^2(\pi_{2p})^4(\sigma_{2p})^2(\pi_{2p}*)^4$ B.O. = (8-6)/2 = 1, diamagnetic
      d. CN^+: $(\sigma_{2s})^2(\sigma_{2s}*)^2(\pi_{2p})^4$ B.O. = (6-2)/2 = 2, diamagnetic
      e. CN: $(\sigma_{2s})^2(\sigma_{2s}*)^2(\pi_{2p})^4(\sigma_{2p})^1$ B.O. = (7-2)/2 = 2.5, paramagnetic
      f. CN^-: $(\sigma_{2s})^2(\sigma_{2s}*)^2(\pi_{2p})^4(\sigma_{2p})^2$ B.O. = 3, diamagnetic
      g. N_2: $(\sigma_{2s})^2(\sigma_{2s}*)^2(\pi_{2p})^4(\sigma_{2p})^2$ B.O. = 3, diamagnetic
      h. N_2^+: $(\sigma_{2s})^2(\sigma_{2s}*)^2(\pi_{2p})^4(\sigma_{2p})^1$ B.O. = 2.5, paramagnetic
      i. N_2^-: $(\sigma_{2s})^2(\sigma_{2s}*)^2(\pi_{2p})^4(\sigma_{2p})^2(\pi_{2p}*)^1$ B.O. = 2.5, paramagnetic

21.   The bond orders are: CN^+, 2; CN, 2.5; CN^-, 3

       Shortest → longest bond length: $CN^- < CN < CN^+$

       Lowest → highest bond energy: $CN^+ < CN < CN^-$

22.   Bond orders: N_2, 3; N_2^+, 2.5; N_2^-, 2.5

       Bond length: $N_2 < N_2^+ \approx N_2^-$

       Bond energy: $N_2^- \approx N_2^+ < N_2$

       We would expect the strengths and lengths of the bonds in N_2^+ and N_2^- to be close to each other since both have bond orders of 2.5. Since there are more electrons in N_2^- than N_2^+, electron repulsions will be greater in N_2^- than N_2^+ and the bond in N_2^- probably is a little longer and a little weaker than the bond in N_2^+.

23. N_2: The π and $\pi*$ orbitals are symmetrical.

       CO: The π orbitals would place more electron density near the more electronegative oxygen and the $\pi*$ orbitals would place more electron density near the carbon.

24.   The π bonds between S atoms and between C and S atoms are not as strong. The atomic orbitals do not overlap with each other as well as the smaller atomic orbitals of C and O overlap.

25.

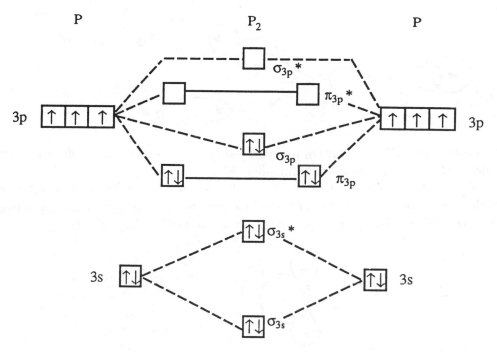

26. C_2^{2-} has 10 valence electrons. The Lewis structure predicts sp hybridization.

$$[:C\equiv C:]^{2-}$$

MO: $(\sigma_{2s})^2(\sigma_{2s}*)^2(\pi_{2p})^4(\sigma_{2p})^2$, B.O. = 3

Both give the same picture, a triple bond composed of a σ and two π-bonds. Both predict the ion
to be diamagnetic. Lewis structures deal well with diamagnetic (all electrons paired) species.
The Lewis model cannot really predict magnetic properties. C_2^{2-} is isoelectronic with the CO
molecule.

27.   a.   The electrons would be closer to F on the average. The F atom is more electronegative than
             the H atom and the 2p orbital of F is lower in energy than the 1s orbital of H.

        b.   The bonding MO would have more fluorine 2p character since it is closer in energy to the
             fluorine 2p orbital.

        c.   The antibonding MO would place more electron density closer to H and would have a
             greater contribution from the higher energy hydrogen 1s atomic orbital.

28.   a.   The electron removed from N_2 is in the σ_{2p} molecular orbital which is lower in energy than
             the 2p atomic orbital from which the electron in atomic nitrogen is removed. Since the
             electron removed from N_2 is lower in energy than the electron in N, the ionization energy of
             N_2 should be greater than for N.

        b. F_2 should have a lower first ionization energy than F. The electron removed from F_2 is in a
             $\pi_{2p}*$ antibonding molecular orbital which is higher in energy than the 2p atomic orbitals.
             Thus, it is easier to remove an electron from F_2 than from F.

29.   a.

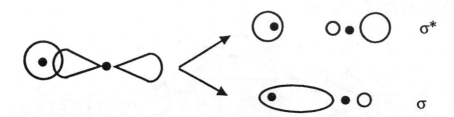

   b. The antibonding MO will have more hydrogen 1s character since the hydrogen 1s atomic
       orbital is closer in energy to the antibonding MO.

   c. No, the overall overlap is zero. The p_x orbital does not have proper symmetry to overlap
       with a 1s orbital. The $2p_x$ and $2p_y$ oribtals are called nonbonding orbitals.

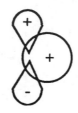

   d.

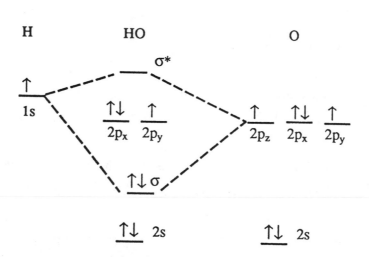

   e. Bond order = (2-0)/2 = 1;   Note: The $2p_x$ and $2p_y$ electrons have no effect on the bond order.

   f. To form OH$^+$ a nonbonding electron is removed from OH. Since the number of bonding
       electrons and antibonding electrons are unchanged, the bond order is still equal to one.

Additional Exercises

30.   a. $FClO$, $7 + 7 + 6 = 20$ e⁻

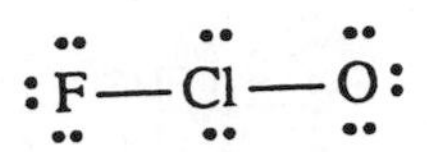

bent, sp^3 hybridization

b. $FClO_2$, $7 + 7 + 2(6) = 26$ e⁻

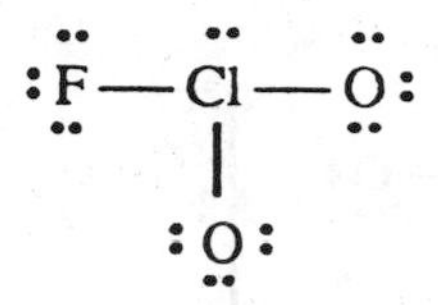

trigonal pyramid, sp^3

c. $FClO_3$, $7 + 7 + 3(6) = 32$ e⁻

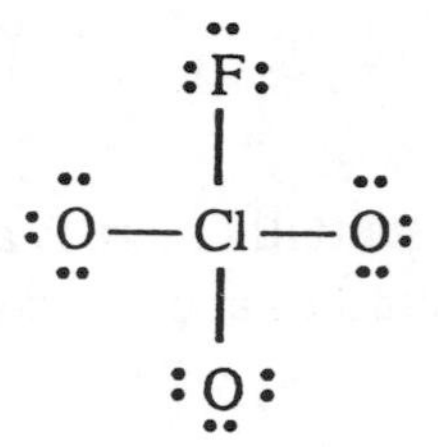

tetrahedral, sp^3

d. F_3ClO, $3(7) + 7 + 6 = 34$ e⁻

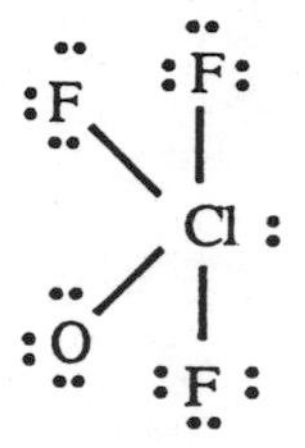

see-saw, dsp^3

e. F_3ClO_2, $3(7) + 7 + 2(6) = 40$ valence e⁻

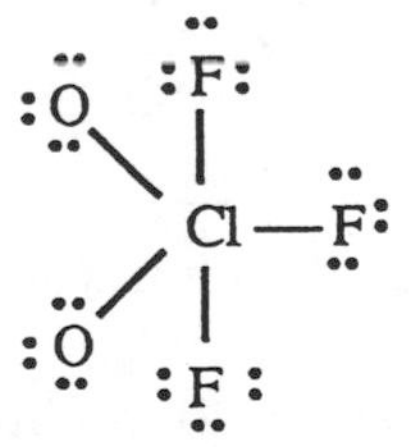

trigonal bipyramid, dsp^3

Note: Two additional Lewis structures are possible, depending on the positions of the oxygen atoms.

31. $FClO_2 + F^- \rightarrow F_2ClO_2^-$

$F_2ClO_2^-$, $2(7) + 7 + 2(6) + 1 = 34$ e⁻

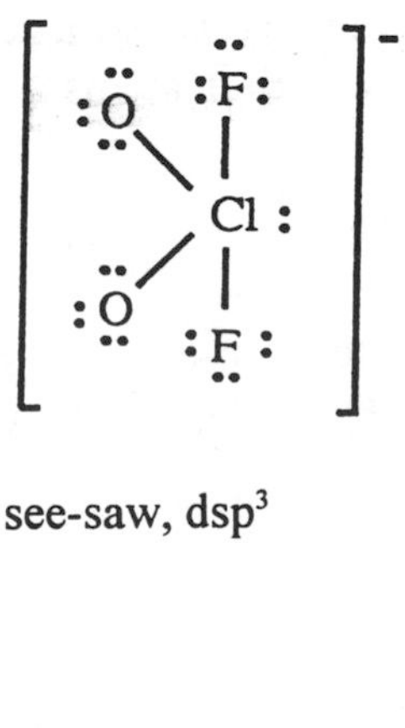

see-saw, dsp^3

$F_3ClO + F^- \rightarrow F_4ClO^-$

F_4ClO^-, $4(7) + 7 + 6 + 1 = 42$ e⁻

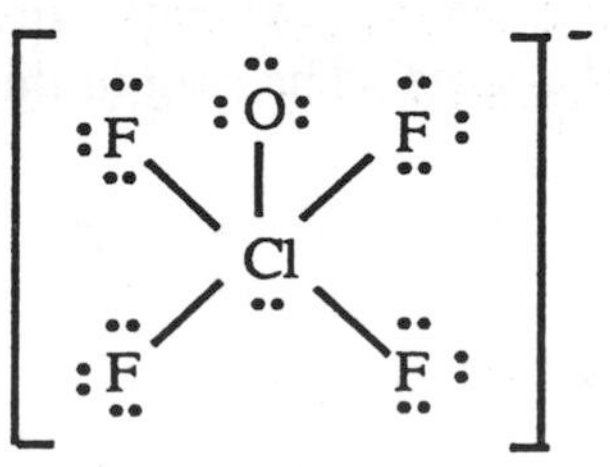

square pyramid, d^2sp^3

$F_3ClO \rightarrow F^- + F_2ClO^+$

F_2ClO^+, $2(7) + 7 + 6 - 1 = 26$ e$^-$

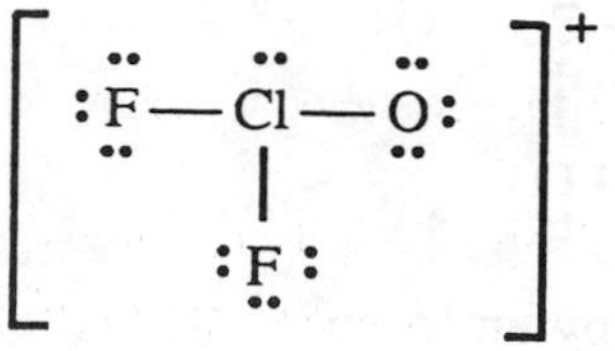

$$\left[\begin{array}{c} :\ddot{F}-\overset{\displaystyle ..}{Cl}-\ddot{O}: \\ | \\ :\ddot{F}: \end{array} \right]^+$$

trigonal pyramid, sp^3

$F_3ClO_2 \rightarrow F^- + F_2ClO_2^+$

$F_2ClO_2^+$, $2(7) + 7 + 2(6) - 1 = 32$ e$^-$

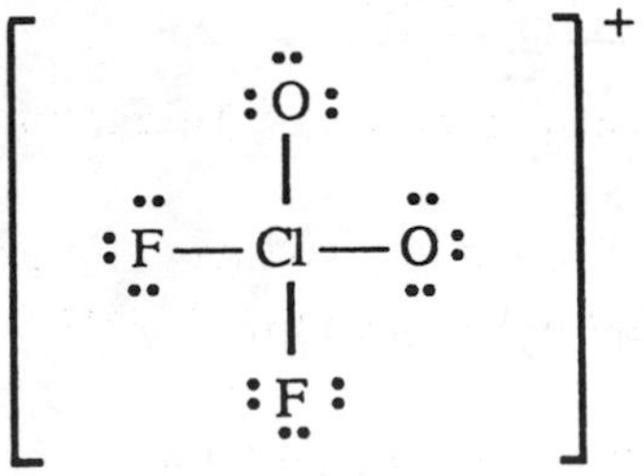

$$\left[\begin{array}{c} :\ddot{O}: \\ | \\ :\ddot{F}-\overset{\displaystyle}{Cl}-\ddot{O}: \\ | \\ :\ddot{F}: \end{array} \right]^+$$

tetrahedral, sp^3

32.   SbF$_5$ monomer:

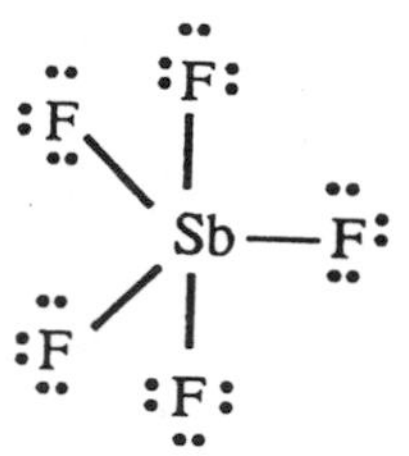

trigonal bipyramid, dsp^3

SbF$_5$ polymer: Essentially there is an octahdral arrangement of fluorines about each antimony atom (d^{2}sp^3):

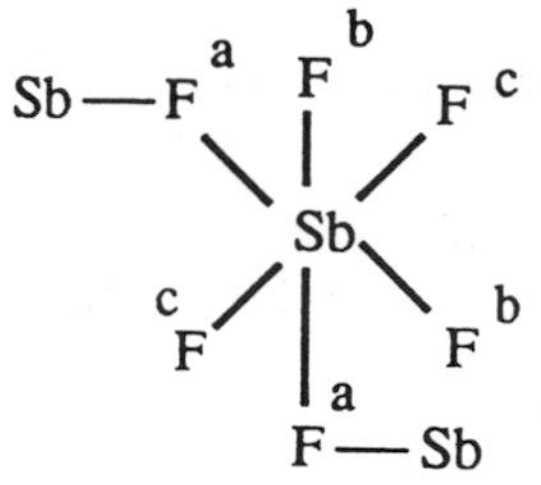

The three types of fluorines are:

a. bridging, bonded to two Sb atoms

b. 90° away from one type (a) and 180° away from the other type (a)

c. 90° away from both type (a)

33.   a. To complete the Lewis structure, add two lone pairs to each sulfur atom. This is an organic compound and Lewis structures for organic compounds usually have all atoms with zero formal charge. The ring in thiarubin-A is shorthand notation for four carbon atoms.

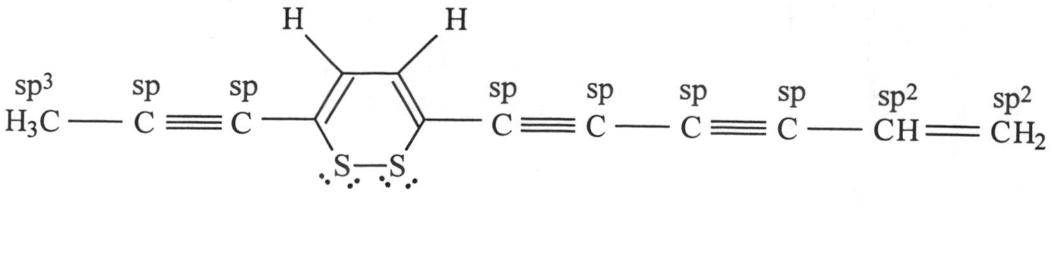

b. See Lewis structure. The four carbon atoms in the ring are all sp^2 hybridized. The two sulfur atoms are sp^3 hybridized.

c. 23σ and 9π bonds. Note: CH_3, CH_2 and CH are shorthand for carbon atoms singly bonded to hydrogen atoms.

34. a. No, some atoms are in different places. Thus, these are not resonance structures.

b. For the first Lewis structure, all nitrogens are sp^3 hybridized and all carbons are sp^2 hybridized. For the second Lewis structure, all nitrogens and carbons are sp^2 hybridized.

c. For the reaction:

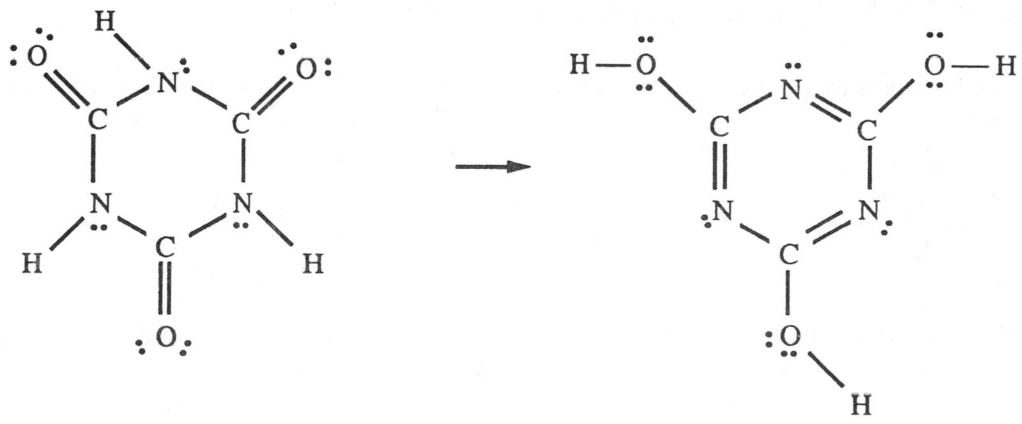

Bonds broken:

    3 C = O (799 kJ/mol)
    3 C – N (305 kJ/mol)
    3 N – H (391 kJ/mol)

Bonds formed:

    3 C = N (615 kJ/mol)
    3 C – O (358 kJ/mol)
    3 O – H (467 kJ/mol)

$\Delta H = 3(799) + 3(305) + 3(391) - [3(615) + 3(358) + 3(467)]$

$\Delta H = 4485 \text{ kJ} - 4320 \text{ kJ} = +165 \text{ kJ}$

Since ΔH is endothermic, the bonds are stronger in the first structure with the carbon-oxygen double bonds.

35.

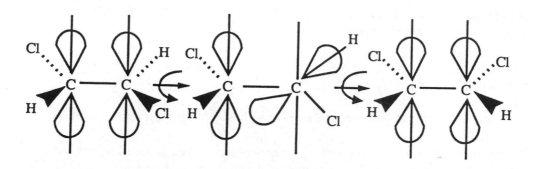

In order to rotate about the double bond, the molecule must go through an intermediate stage where the π bond is broken while the sigma bond remains intact. Bond energies are 347 kJ/mol for C – C and 614 kJ/mol for C = C. If we take the single bond as the strength of the σ bond, then the strength of the π bond is (614 - 347) = 267 kJ/mol. Thus, 267 kJ/mol must be supplied to rotate about a carbon-carbon double bond.

36. O_3 and NO_2^- are isoelectronic, so we only need consider one of them. The Lewis structures for O_3 are:

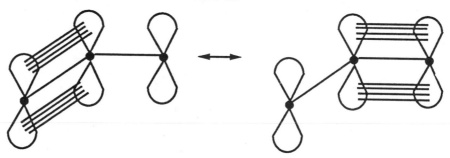

Two resonance forms: central atom is sp^2 hybridized with one unhybridized p atomic orbital. The resonance view of the two resonance forms is:

The MO picture of the π-bond would utilize all three p-orbitals to form three new MOs. The lowest energy MO would look like:

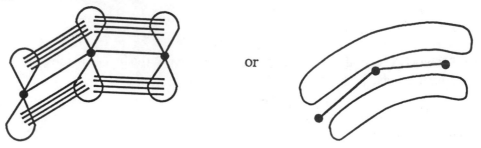

or

In the MO view, there would be two localized σ-bonds and a π-bond delocalized over all 3 atoms in the molecule.

37.   a. Add lone pairs to complete octets for each O and N.

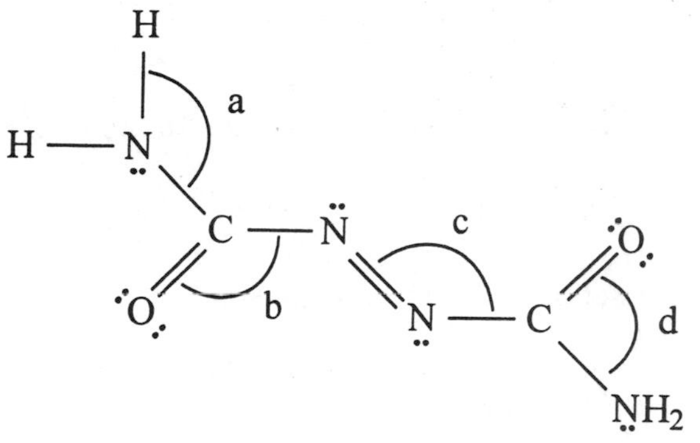

azidocarbonamide

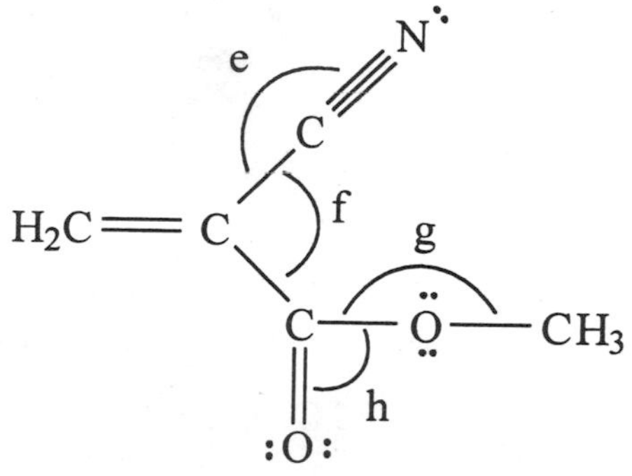

methyl cyanoacrylate

b. In azidocarbonamide, the two carbon atoms are sp^2 hybridized. The two nitrogen atoms with hydrogens attached are sp^3 hybridized and the other two nitrogens are sp^2 hybridized. In methyl cyanoacrylate, the CH_3 carbon is sp^3 hybridized, the carbon with the triple bond is sp hybridized, and the other three carbons are sp^2 hybridized.

c. Azidocarbonamide contains 3π bonds and methyl cyanoacrylate contains 4π bonds.

d. a. $< 109.5°$ b. $120°$ c. $< 120°$ d. $120°$ e. $180°$

f. $120°$ g. $< 109.5°$ h. $120°$

38.

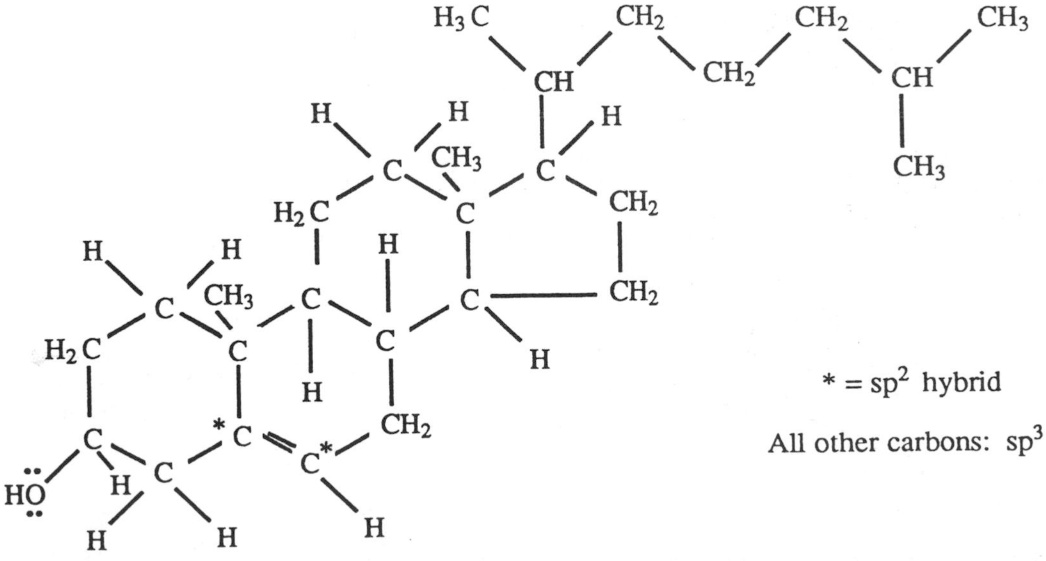

* = sp^2 hybrid

All other carbons: sp^3

No, not all carbon atoms are in the same plane since a majority of carbon atoms exhibit tetrahedral structure.

39. NO^+ has $5 + 6 - 1 = 10$ valence electrons. The Lewis structure shows a triple bond.

$$\left[:N \equiv O: \right]^+$$

NO has $5 + 6 = 11$ valence electrons. The Lewis structures can't predict a stable species.

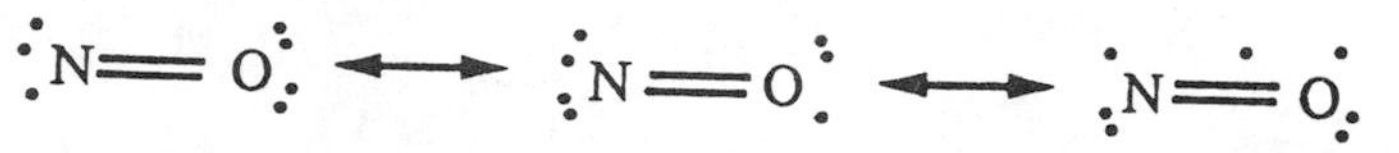

$$:\ddot{N} = \ddot{O}: \longleftrightarrow :\ddot{N} = \ddot{O}: \longleftrightarrow :\ddot{N} = \ddot{O}:$$

NO^- has 12 valence electrons. The Lewis structure predicts a double bond.

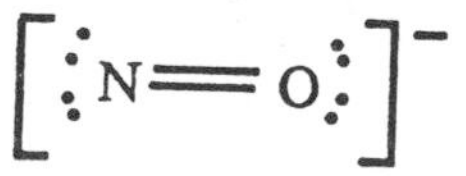

$$\left[:\ddot{N} = \ddot{O}: \right]^-$$

M.O. model:

   NO^+ $(\sigma_{2s})^2(\sigma_{2s}*)^2(\pi_{2p})^4(\sigma_{2p})^2$, B.O. = 3, 0 unpaired e^- (diamagnetic)

   NO $(\sigma_{2s})^2(\sigma_{2s}*)^2(\pi_{2p})^4(\sigma_{2p})^2(\pi_{2p}*)^1$, B.O. = 2.5, 1 unpaired e^- (paramagnetic)

   NO^- $(\sigma_{2s})^2(\sigma_{2s}*)^2(\pi_{2p})^4(\sigma_{2p})^2(\pi_{2p}*)^2$, B.O. = 2, 2 unpaired e^- (paramagnetic)

Bond Energies: $NO^- < NO < NO^+$; Bond Lengths: $NO^+ < NO < NO^-$

The two models only give the same results for NO^+. The M.O. view is the correct one for NO and NO^-. Lewis structures are inadequate for odd electron species and for predicting magnetic properties. NO^- is paramagnetic but the Lewis structure shows all electrons paired.

40. N_2 (ground state): $(\sigma_{2s})^2(\sigma_{2s}*)^2(\pi_{2p})^4(\sigma_{2p})^2$, B.O. = 3, diamagnetic

N_2 (1st excited state): $(\sigma_{2s})^2(\sigma_{2s}*)^2(\pi_{2p})^4(\sigma_{2p})^1(\pi_{2p}*)^1$

B.O. = (7-3)/2 = 2, paramagnetic (2 unpaired e^-)

The bond strengths are different as well as the magnetic properties.

41.   Be: $1s^2 2s^2$; Each Be has 2 valence electrons. We can draw a Lewis structure:

   Be = Be

This doesn't obey the octet rule, but is consistent with structures we've drawn earlier for Be compounds. There are two pairs of electrons around each Be atom just as in earlier drawings, e.g., BeH_2. From this Lewis structure we would predict Be_2 to be stable.

The MO model (see Exercise 14.14c) predicts a bond order of zero; Be_2 shoudn't exist. Be_2 has not yet been observed. Be is very small; electron-electron repulsions would be extremely large in the Lewis structure drawn.

42.   Paramagnetic: Unpaired electrons are present. Measure the mass of a substance in the presence and absence of a magnetic field. A substance with unpaired electrons will be attracted by the magnetic field, giving an apparent increase in mass in the presence of the field. Greater number of unpaired electrons will give greater attraction and greater observed mass increase.

43.   Considering only the twelve valence electrons in O_2, the MO models would be:

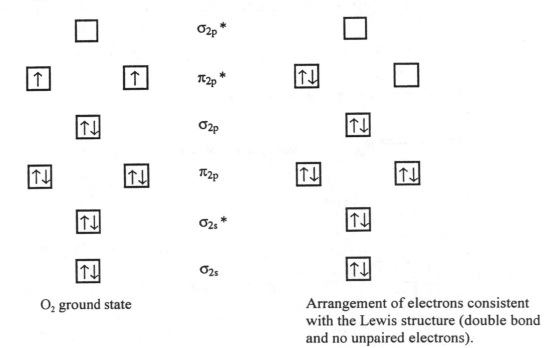

σ_{2p}^*

π_{2p}^*

σ_{2p}

π_{2p}

σ_{2s}^*

σ_{2s}

O_2 ground state

Arrangement of electrons consistent with the Lewis structure (double bond and no unpaired electrons).

It takes energy to pair electrons in the same orbital. Thus, the structure with no unpaired electrons is at a higher energy; it is an excited state.

44.   a. $COCl_2$ has $4 + 6 + 2(7) = 24$ valence electrons.

trigonal planar
polar
120°
sp^2

b. N_2F_2 has $2(5) + 2(7) = 24$ valence electrons.

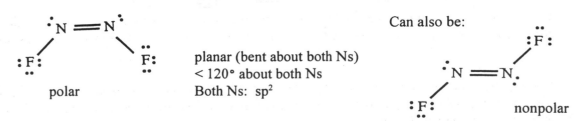

planar (bent about both Ns)
< 120° about both Ns
Both Ns: sp^2

Can also be:

nonpolar

These are distinctly different molecules.

c. N_2O_3 has $2(5) + 3(6) = 28$ valence electrons.

nonplanar (bent and trigonal planar about nitrogens)
polar
$\approx 120°$ about both Ns
Both Ns: sp^2

d.   COS has $4 + 6 + 6 = 16$ valence electrons.

linear, polar, 180°, sp

e. ICl_3 has $7 + 3(7) = 28$ valence electrons.

T-shaped
polar
a. $\approx 90°$, b. $\approx 180°$
dsp^3

45.   c.

d.

$$:O = C = S: \longleftrightarrow :O \equiv C - S: \longleftrightarrow :O - C \equiv S:$$
$$\quad 0 \quad\quad 0 \quad\quad 0 \quad\quad\quad\quad +1 \quad\quad 0 \quad -1 \quad\quad\quad\quad -1 \quad\quad 0 \quad +1$$

46.   We can determine molecular structures from experiment. Our view is that if electrons are in orbitals in atoms then they must be in orbitals in molecules. However, atomic orbitals don't point in the right direction for some molecules. We had to invent the concept of hybrid orbitals. Once we know the structure, we then determine the hybridization. Structures are facts. Hybrid orbitals come from a model to rationalize the facts.

47.   a.   Yes, both have 4 sets of electrons about the P. We would predict a tetrahedral structure for both. See below for Lewis structures.

b.   The hybridization is sp³ for each P since both structures are tetrahedral.

c.   P has to use one of its d orbitals to form the π bond.

d.

$$\begin{array}{ccc} :O:^{-1} & & \cdot O \cdot \;\; 0 \\ |^{+1} & & \| \;\; 0 \\ :Cl - P - Cl: & \longleftrightarrow & :Cl - P - Cl: \\ | & & | \\ :Cl: & & :Cl: \end{array}$$

The structure with the P = O bond is favored on the basis of formal charge.

48.   a.

$$:N = N = O: \longleftrightarrow :N \equiv N - O: \longleftrightarrow :N - N \equiv O:$$

$$:N = O = N: \longleftrightarrow :N \equiv O - N: \longleftrightarrow :N - O \equiv N:$$

The NNO structure is correct. From the Lewis structures we would predict both NNO and NON to be linear. However, we would predict NNO to be polar and NON to be nonpolar. Since experiments show N₂O to be polar, then NNO is the correct structure.

b.

$$:N = N = O: \longleftrightarrow :N \equiv N - O: \longleftrightarrow :N - N \equiv O:$$
$$\;\; -1 \quad +1 \quad 0 \quad\quad\quad\quad 0 \quad +1 \quad -1 \quad\quad\quad\quad -2 \quad +1 \quad +1$$

The central N is sp hybridized. We can probably ignore the 3rd resonance structure on the basis of formal charge.

c. The sp hybrid orbitals on the center N overlap with atomic orbitals (or hybrid orbitals) on the other two atoms to form two sigma bonds. The remaining p orbitals on the center N overlap with p orbitals on the peripheral N to form two π-bonds.

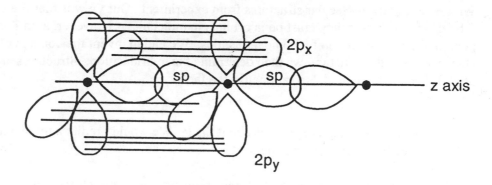

49.   O=N–Cl: The bond order of the NO bond in NOCl is 2 (a double bond).

NO: The bond order of this NO bond is 2.5 (see Exercise 14.16).

Both reactions apparently only involve the breaking of the N–Cl bond. However, in the reaction: ONCl → NO + Cl some energy is released in forming the stronger NO bond, lowering the value of ΔH. Therefore, the apparent N–Cl bond energy is artifically low for this reaction. The first reaction only involves the breaking of the N–Cl bond.

50.   a. The CO bond is polar with the negative end around the more electronegative oxygen atom. We would expect metal cations to be attracted to and to bond to the oxygen end of CO on the basis of electronegativity.

b.   FC (carbon) = 4 - 2 - 1/2(6) = -1

:C≡O:

FC (oxygen) = 6 - 2 - 1/2(6) = +1

From formal charge, we would expect metal cations to bond to the carbon (with the negative formal charge).

c. In molecular orbital theory, only orbitals with proper symmetry overlap to form bonding orbitals. The metals that form bonds to CO are usually transition metals, all of which have outer electrons in the d orbitals. The only molecular orbitals of CO that have proper symmetry to overlap with d orbitals are the π_{2p}* orbitals, whose shape is similar to the d orbitals (see Figure 14.36). Since the antibonding molecular orbitals have more carbon character, one would expect the bond to form through carbon.

CHAPTER FIFTEEN

CHEMICAL KINETICS

Reaction Rates

1. $\text{Rate} = \dfrac{-\Delta[S_2O_3^{2-}]}{\Delta t} = \dfrac{0.0080 \text{ mol}}{\text{L s}}$; I_2 is consumed at half this rate. $\dfrac{-\Delta[I_2]}{\Delta t} = 0.0040 \text{ mol L}^{-1} \text{ s}^{-1}$

   $\dfrac{\Delta[S_4O_6^{2-}]}{\Delta t} = 0.0040 \text{ mol L}^{-1} \text{ s}^{-1}$; $\dfrac{\Delta[I^-]}{\Delta t} = 0.0080 \text{ mol L}^{-1} \text{ s}^{-1}$

2. $\dfrac{d[H_2]}{dt} = 3\dfrac{d[N_2]}{dt}$ and $\dfrac{d[NH_3]}{dt} = -2\dfrac{d[N_2]}{dt}$; So, $\dfrac{1}{3}\dfrac{d[H_2]}{dt} = -\dfrac{1}{2}\dfrac{d[NH_3]}{dt}$ or $\dfrac{d[NH_3]}{dt} = -\dfrac{2}{3}\dfrac{d[H_2]}{dt}$

   Ammonia is produced at a rate equal to 2/3 of the rate of consumption of hydrogen.

3. a. $\text{mol L}^{-1} \text{ s}^{-1}$
    b. $\text{Rate} = k$; k has units of $\text{mol L}^{-1} \text{ s}^{-1}$

    c. $\text{Rate} = k[A]$, $\dfrac{\text{mol}}{\text{L s}} = k\left(\dfrac{\text{mol}}{\text{L}}\right)$
    d. $\text{Rate} = k[A]^2$, $\dfrac{\text{mol}}{\text{L s}} = k\left(\dfrac{\text{mol}}{\text{L}}\right)^2$

    k must have units of s^{-1}.
    k must have units $\text{L mol}^{-1} \text{ s}^{-1}$.

    e. $\text{L}^2 \text{ mol}^{-2} \text{ s}^{-1}$

4. $\text{Rate} = k[Cl]^{1/2}[CHCl_3]$, $\dfrac{\text{mol}}{\text{L s}} = k\left(\dfrac{\text{mol}}{\text{L}}\right)^{1/2}\left(\dfrac{\text{mol}}{\text{L}}\right)$; k must have units of $\text{L}^{1/2} \text{ mol}^{-1/2} \text{ s}^{-1}$.

5. a. molecules $\text{cm}^{-3} \text{ s}^{-1}$
    b. molecules $\text{cm}^{-3} \text{ s}^{-1}$
    c. s^{-1}

    d. cm^3 molecules$^{-1} \text{ s}^{-1}$
    e. cm^6 molecules$^{-2} \text{ s}^{-1}$

6. $\dfrac{1.24 \times 10^{-12} \text{ cm}^3}{\text{molecules s}} \times \dfrac{1 \text{ L}}{1000 \text{ cm}^3} \times \dfrac{6.022 \times 10^{23} \text{ molecules}}{\text{mol}} = 7.47 \times 10^8 \text{ L mol}^{-1} \text{ s}^{-1}$

7.   a. $25 \text{ min} \times \dfrac{60 \text{ s}}{\text{min}} \times \dfrac{1.20 \times 10^{-4} \text{ mol}}{\text{L s}} = \dfrac{0.18 \text{ mol}}{\text{L}}$, 0.18 mol/L HI has been consumed.

[HI] = 0.25 - 0.18 = 0.07 mol/L

b. $t \times \dfrac{1.20 \times 10^{-4} \text{ mol}}{\text{L s}} = \dfrac{0.250 \text{ mol}}{\text{L}}$, $t = 2.08 \times 10^{3}$ s

c. $\dfrac{1.20 \times 10^{-4} \text{ mol HI}}{\text{L s}} \times \dfrac{1 \text{ mol H}_2}{2 \text{ mol HI}} = 6.00 \times 10^{-5} \text{ mol L}^{-1} \text{ s}^{-1} = \dfrac{d[\text{H}_2]}{dt}$

Both H_2 and I_2 are formed at the rate of 6.00×10^{-5} mol L^{-1} s^{-1}.

Rate Laws from Experimental Data: Initial Rates Method

8.   a.   In the first two experiments, [NO] is held constant and [Cl$_2$] is doubled. The rate also doubled. Thus, the reaction is first order with respect to Cl$_2$. Mathematically: Rate = k[NO]^x[Cl$_2$]^y

$\dfrac{0.35}{0.18} = \dfrac{k(0.10)^x(0.20)^y}{k(0.10)^x(0.10)^y} = \dfrac{(0.20)^y}{(0.10)^y}$, $1.9 = 2.0^y$, $y \approx 1$

We get the dependence on NO from the second and third experiments.

$\dfrac{1.45}{0.35} = \dfrac{k(0.20)^x(0.20)}{k(0.10)^x(0.20)} = \dfrac{(0.20)^x}{(0.10)^x}$, $4.1 = 2.0^x$, $\log 4.1 = x \log 2.0$

$0.61 = x(0.30)$, $x \approx 2$; So, Rate = k[NO]^2[Cl$_2$]

b.   The rate constant k can be determined from the experiments. From experiment 1:

$\dfrac{0.18 \text{ mol}}{\text{L min}} = k \left(\dfrac{0.10 \text{ mol}}{\text{L}} \right)^2 \left(\dfrac{0.10 \text{ mol}}{\text{L}} \right)$, $k = 180 \text{ L}^2 \text{ mol}^{-2} \text{ min}^{-1}$

From the other experiments:

$k = 175 \text{ L}^2 \text{ mol}^{-2} \text{ min}^{-1}$ (2nd exp.); $k = 181 \text{ L}^2 \text{ mol}^{-2} \text{ min}^{-1}$ (3rd exp.)

To two significant figures, $k_{\text{mean}} = 1.8 \times 10^2 \text{ L}^2 \text{ mol}^{-2} \text{ min}^{-1}$.

9.   a.   Rate = k[I$^-$]^x[S$_2$O$_8^{2-}$]^y; $\dfrac{12.50 \times 10^{-6}}{6.250 \times 10^{-6}} = \dfrac{k(0.080)^x(0.040)^y}{k(0.040)^x(0.040)^y}$, $2.000 = 2.0^x$, $x = 1$

$\dfrac{12.50 \times 10^{-6}}{5.560 \times 10^{-6}} = \dfrac{k(0.080)^x(0.040)^y}{k(0.080)^x(0.020)^y}$, $2.248 = 2.0^y$

$\log 2.248 = y \log 2.0, \ 0.3518 = 0.30 \, y, \ y = 1.2 \approx 1$

$Rate = k[I^-][S_2O_8^{2-}]$

b. For the first experiment:

$$\frac{12.50 \times 10^{-6} \text{ mol}}{L \, s} = k\left(\frac{0.080 \text{ mol}}{L}\right)\left(\frac{0.040 \text{ mol}}{L}\right), \ k = 3.9 \times 10^{-3} \text{ L mol}^{-1} \text{ s}^{-1}$$

The other values are:

	Initial Rate (mol L⁻¹ s⁻¹)	k (L mol⁻¹ s⁻¹)
	12.5×10^{-6}	3.9×10^{-3}
	6.25×10^{-6}	3.9×10^{-3}
	5.56×10^{-6}	3.5×10^{-3}
	4.35×10^{-6}	3.4×10^{-3}
	6.41×10^{-6}	3.6×10^{-3}

$k_{mean} = 3.7 \times 10^{-3} \text{ L mol}^{-1} \text{ s}^{-1} \pm 0.3 \times 10^{-3} \text{ L mol}^{-1} \text{ s}^{-1}$

10. a. $Rate = k[I^-]^x[OCl^-]^y;$ $\dfrac{7.91 \times 10^{-2}}{3.95 \times 10^{-2}} = \dfrac{k(0.12)^x(0.18)^y}{k(0.060)^x(0.18)^y} = 2.0^x, \ 2.00 = 2.0^x, \ x = 1$

$\dfrac{3.95 \times 10^{-2}}{9.88 \times 10^{-3}} = \dfrac{k(0.060)(0.18)^y}{k(0.030)(0.090)^y}, \ x = 1; \ 4.00 = 2.0 \times 2.0^y, \ 2.0 = 2.0^y, \ y = 1$

$Rate = k[I^-][OCl^-]$

b. From the first experiment: $\dfrac{7.91 \times 10^{-2} \text{ mol}}{L \, s} = k\left(\dfrac{0.12 \text{ mol}}{L}\right)\left(\dfrac{0.18 \text{ mol}}{L}\right), \ k = 3.7 \text{ L mol}^{-1} \text{ s}^{-1}$

All four experiments give the same value of k to two significant figures.

11. a. $Rate = k[Hb]^x[CO]^y$

Comparing the first two experiments, [CO] is unchanged, [Hb] doubles, and the rate doubles. Therefore, $x = 1$ and the reaction is first order in Hb. Comparing the second and third experiments, [Hb] is unchanged, [CO] triples and the rate triples. Therefore, $y = 1$ and the reaction is first order in CO.

b. $Rate = k[Hb][CO]$

c. From the first experiment:

$0.619 \ \mu\text{mol L}^{-1} \text{ s}^{-1} = k \ (2.21 \ \mu\text{mol/L})(1.00 \ \mu\text{mol/L}), \ k = 0.280 \text{ L } \mu\text{mol}^{-1} \text{ s}^{-1}$

The second and third experiments give similar k values, so $k_{mean} = 0.280 \text{ L } \mu\text{mol}^{-1} \text{ s}^{-1}$.

d.   Rate = k[Hb][CO] = $\dfrac{0.280 \text{ L}}{\mu \text{mol s}} \times \dfrac{3.36\ \mu\text{mol}}{\text{L}} \times \dfrac{2.40\ \mu\text{mol}}{\text{L}} = 2.26\ \mu\text{mol L}^{-1}\text{ s}^{-1}$

12.   a.   Rate = $k[ClO_2]^x[OH^-]^y$; From the first two experiments:

$2.30 \times 10^{-1} = k(0.100)^x(0.100)^y$

$5.75 \times 10^{-2} = k(0.0500)^x(0.100)^y$

Dividing the two rate laws: $4.00 = \dfrac{(0.100)^x}{(0.0500)^x} = 2.00^x,\ x = 2$

Comparing the second and third experiments:

$2.30 \times 10^{-1} = k(0.100)(0.100)^y$

$1.15 \times 10^{-1} = k(0.100)(0.050)^y$

Dividing: $2.00 = \dfrac{(0.100)^y}{(0.050)^y} = 2.0^y,\ y = 1$

The rate law is: Rate = $k[ClO_2]^2[OH^-]$

2.30×10^{-1} mol L^{-1} s^{-1} = $k(0.100$ mol/L$)^2(0.100$ mol/L$)$, $k = 2.30 \times 10^2$ L^2 mol^{-2} s$^{-1} = k_{mean}$

b.   Rate = $\dfrac{2.30 \times 10^2 \text{ L}^2}{\text{mol}^2\text{ s}} \times \left(\dfrac{0.175 \text{ mol}}{\text{L}} \right)^2 \times \dfrac{0.0844 \text{ mol}}{\text{L}} = 0.594$ mol L^{-1} s^{-1}

Integrated Rate Laws from Experimental Data

13.   The first guess to make is that the reaction is first order. For a first order reaction a graph of ln [C$_4$H$_6$] vs. t should yield a straight line. If this isn't linear, then try the second order plot of 1/[C$_4$H$_6$] vs. t. The data for these two plots follow and the actual plots are on the next page.

Time	195	604	1246	2180	6210 s
[C$_4$H$_6$]	1.6×10^{-2}	1.5×10^{-2}	1.3×10^{-2}	1.1×10^{-2}	0.68×10^{-2} M
ln [C$_4$H$_6$]	-4.14	-4.20	-4.34	-4.51	-4.99
1/[C$_4$H$_6$]	62.5	66.7	76.9	90.9	147 M^{-1}

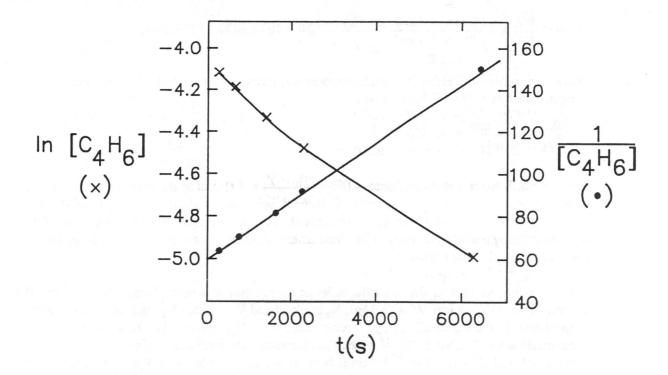

The natural log plot is not linear, so the reaction is not first order. The next guess is that the reaction is second order. For a second order reaction, a plot of $1/[C_4H_6]$ vs. t should yield a straight line. Since we get a straight line for this graph, we conclude the reaction is second order in butadiene. The differential rate law is:

$$\text{Rate} = \frac{-d[C_4H_6]}{dt} = k[C_4H_6]^2$$

For a second order reaction, the integrated rate law is: $\dfrac{1}{[C_4H_6]_t} = \dfrac{1}{[C_4H_6]_o} + kt$

The slope of the straight line equals the value of the rate constant. Using the points on the line at 1000. and 6000. s:

$$k = \text{slope} = \frac{144 \text{ L/mol} - 73 \text{ L/mol}}{6000.\text{ s} - 1000.\text{ s}} = 1.4 \times 10^{-2} \text{ L mol}^{-1} \text{ s}^{-1}$$

This value was obtained from reading the graph and "eyeballing" a straight line with a ruler.

14.   This problem differs in two ways from the previous problem:
        1.   a product is measured instead of a reactant and
        2.   only the volume of a gas is given and not the concentration.
We can find the initial concentration of $C_6H_5N_2Cl$ from the amount of N_2 evolved after infinite time when all the $C_6H_5N_2Cl$ has decomposed (assuming the reaction goes to completion).

$$n = \frac{PV}{RT} = \frac{1.00 \text{ atm} \times (58.3 \times 10^{-3} \text{ L})}{\frac{0.08206 \text{ L atm}}{\text{mol K}} \times 323 \text{ K}} = 2.20 \times 10^{-3} \text{ mol } N_2$$

Since each mole of $C_6H_5N_2Cl$ that decomposes produces one mole of N_2, then the initial concentration (t = 0) of $C_6H_5N_2Cl$ was:

$$\frac{2.20 \times 10^{-3} \text{ mol}}{40.0 \times 10^{-3} \text{ L}} = 0.0550 \ M$$

We can similarly calculate the moles of N_2 evolved at each point of the experiment, subtract that from 2.20×10^{-3} mol to get the moles of $C_6H_5N_2Cl$ remaining, and calculate $[C_6H_5N_2Cl]$ at each time. We would then use these results to make the appropriate graph to check the order of the reaction. Since the rate constant is the slope of a straight line, we would favor this approach to get a value for the rate constant.

There is a simpler way to check for the order of the reaction and saves doing a lot of math. The quantity $(V_\infty - V_t)$ where $V_\infty = 58.3$ mL N_2 evolved and $V_t = $ mL of N_2 evolved at time t will be proportional to the moles of $C_6H_5N_2Cl$ remaining; $(V_\infty - V_t)$ will also be proportional to the concentration of $C_6H_5N_2Cl$. Thus, we can get the same information by using $(V_\infty - V_t)$ as our measure of $[C_6H_5N_2Cl]$. If the reaction is first order a graph of $\ln (V_\infty - V_t)$ vs. t would be linear. The data for such a graph are:

t (s)	V_t (mL)	$(V_\infty - V_t)$	$\ln (V_\infty - V_t)$
0	0	58.3	4.066
6	19.3	39.0	3.664
9	26.0	32.3	3.475
14	36.0	22.3	3.105
22	45.0	13.3	2.588
30	50.4	7.9	2.07

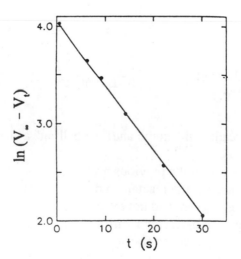

We can see from the graph that this plot is linear, so the reaction is first order. The differential rate law is: $-d[C_6H_5N_2Cl]/dt = k[C_6H_5N_2Cl]$ and the integrated rate law is: $\ln[C_6H_5N_2Cl] = -kt + \ln[C_6H_5N_2Cl]_0$. From separate data, k was determined to be 6.9×10^{-2} s^{-1}.

15.

Time (s)	[H₂O₂] (mol/L)	ln[H₂O₂]
0	1.0	0
120.	0.91	-0.094
300.	0.78	-0.25
600.	0.59	-0.53
1200.	0.37	-0.99
1800.	0.22	-1.51
2400.	0.13	-2.04
3000.	0.082	-2.50
3600.	0.050	-3.00

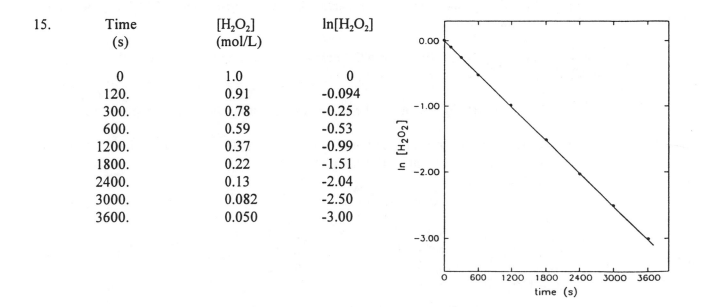

The plot of $\ln[H_2O_2]$ vs. t is linear. Thus, the reaction is first order. The integrated and differential rate laws are:

$$\ln[H_2O_2] = -kt + \ln[H_2O_2]_0 \text{ and Rate} = -d[H_2O_2]/dt = k[H_2O_2]$$

$$\text{Slope} = -k = \frac{0 - (3.00)}{0 - 3600.} = -8.3 \times 10^{-4} \text{ s}^{-1}, \quad k = 8.3 \times 10^{-4} \text{ s}^{-1}$$

16.

Time (s)	[NO₂] (M)	ln [NO₂]	1/[NO₂] (M⁻¹)
0	0.500	-0.693	2.00
1.20×10^3	0.444	-0.812	2.25
3.00×10^3	0.381	-0.965	2.62
4.50×10^3	0.340	-1.079	2.94
9.00×10^3	0.250	-1.386	4.00
1.80×10^4	0.174	-1.749	5.75

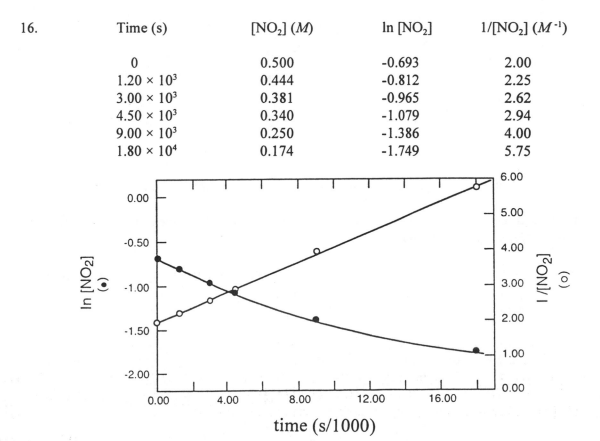

time (s/1000)

The plot of $1/[NO_2]$ vs. time is linear. The reaction is second order in NO_2. The integrated and differential rate laws are:

$$\frac{1}{[NO_2]} = kt + \frac{1}{[NO_2]_o} \text{ and Rate} = -d[NO_2]/dt = k[NO_2]^2$$

The slope of the plot $1/[NO_2]$ vs. t gives the value of k. Slope $= 2.08 \times 10^{-4}$ L mol^{-1} s^{-1} = k

17. a. First, we guess the reaction to be first order with respect to O. Hence, a graph of ln [O] vs. t should be linear if the reaction is first order.

t (s)	[O] (atoms/cm^3)	ln [O]
0	5.0×10^9	22.33
$10. \times 10^{-3}$	1.9×10^9	21.37
$20. \times 10^{-3}$	6.8×10^8	20.34
$30. \times 10^{-3}$	2.5×10^8	19.34

Since the graph is linear, we can conclude the reaction is first order with respect to O.

b. The overall rate law is: Rate = $k[NO_2][O]$; With an excess of NO_2, the rate law becomes: Rate = $k'[O]$ where $k' = k[NO_2]$.

For a first order reaction: ln $[A]_t = -kt + $ ln $[A]_o$; We can get k' from the slope of the graph:

$$k' = -\text{slope} = -\frac{19.34 - 22.33}{(30. \times 10^{-3} - 0) \text{ s}}, \ k' = 1.0 \times 10^2 \text{ s}^{-1}$$

$k' = k[NO_2]$, 1.0×10^2 s$^{-1} = k(1.0 \times 10^{13}$ molecules/cm^3), k $= 1.0 \times 10^{-11}$ cm^3 molecules^{-1} s^{-1}

18.   a.   We check for first order dependence by graphing ln C vs. t for each set of data. The rate
dependence on NO is determined from the first experiment.

time (ms)	[NO] (molecules/cm³)	ln [NO]
0	6.0×10^8	20.21
100.	5.0×10^8	20.03
500.	2.4×10^8	19.30
700.	1.7×10^8	18.95
1000.	9.9×10^7	18.41

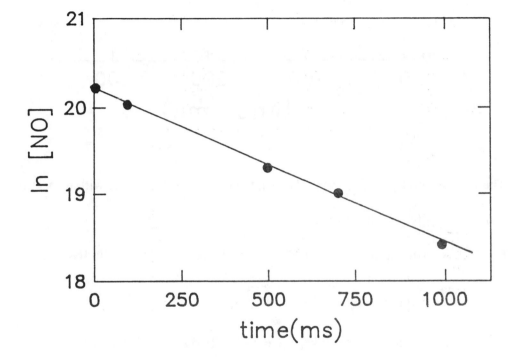

Since ln [NO] vs. t is linear, the reaction is first order with respect to nitric oxide.

We follow the same procedure for ozone. The data are:

time (ms)	[O₃] (molecules/cm³)	ln [O₃]
0	1.0×10^{10}	23.03
50.	8.4×10^9	22.85
100.	7.0×10^9	22.67
200.	4.9×10^9	22.31
300.	3.4×10^9	21.95

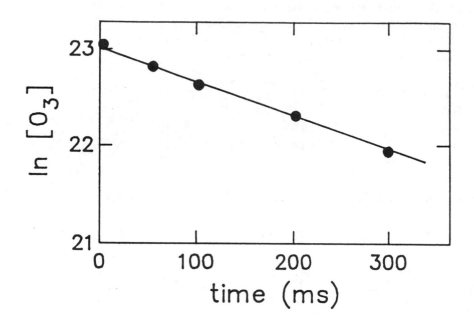

The plot of ln $[O_3]$ vs. t is linear. Hence, the reaction is first order with respect to ozone.

b.   Rate = $k[NO][O_3]$ is the overall rate law.

c.   For NO experiment, Rate = $k'[NO]$ and k' = -(slope from graph of ln $[NO]$ vs. t).

$$k' = \text{-slope} = -\frac{18.41 - 20.21}{(1000. - 0) \times 10^{-3}\text{ s}} = 1.8\text{ s}^{-1}$$

For ozone experiment, Rate = $k''[O_3]$ and k'' = -(slope from ln $[O_3]$ vs. t).

$$k'' = \text{-slope} = -\frac{(21.95 - 23.03)}{(300. - 0) \times 10^{-3}\text{ s}} = 3.6\text{ s}^{-1}$$

d.   From NO experiment, Rate = $k[NO][O_3]$ = $k'[NO]$ where k' = $k[O_3]$.

k' = 1.8 s^{-1} = k(1.0 × 10^{14} molecules/cm^3), k = 1.8 × 10^{-14} cm^3 molecules^{-1} s^{-1}

We can check this from the ozone data. Rate = $k''[O_3]$ = $k[NO][O_3]$ where k'' = $k[NO]$.

k'' = 3.6 s^{-1} = k(2.0 × 10^{14} molecules/cm^3), k = 1.8 × 10^{-14} cm^3 molecules^{-1} s^{-1}

Both values of k agree.

Half-Life

19.   a. $k = \dfrac{\ln 2}{t_{1/2}} = \dfrac{0.69315}{t_{1/2}}$; For ^{239}Pu, $k = \dfrac{0.69315}{24{,}360 \text{ yr}} = 2.845 \times 10^{-5} \text{ yr}^{-1}$

$\dfrac{2.845 \times 10^{-5}}{\text{yr}} \times \dfrac{1 \text{ yr}}{365 \text{ d}} \times \dfrac{1 \text{ d}}{24 \text{ hr}} \times \dfrac{1 \text{ hr}}{60 \text{ min}} \times \dfrac{1 \text{ min}}{60 \text{ s}} = 9.021 \times 10^{-13} \text{ s}^{-1}$

For ^{241}Pu, $k = \dfrac{0.693}{13 \text{ yr}} = 0.0533 \text{ yr}^{-1} = 1.7 \times 10^{-9} \text{ s}^{-1}$

   b.   The rate constant is larger for the decay of ^{241}Pu, hence, ^{241}Pu decays more rapidly.

   c.   For radioactive decay, Rate = kN where N is the number of radioactive nuclei.

Rate $= 1.7 \times 10^{-9} \text{ s}^{-1} \times \left(5.0 \text{ g} \times \dfrac{1 \text{ mol}}{241 \text{ g}} \times \dfrac{6.02 \times 10^{23} \text{ atoms}}{\text{mol}} \right) = 2.1 \times 10^{13}$ disintegrations/s

$\ln \dfrac{[A]_t}{[A]_o} = -kt$; $\ln\left(\dfrac{N}{5.0} \right) = -0.0533 \text{ yr}^{-1}(1.0 \text{ yr})$, $\ln N = -0.0533 + \ln 5.0$

Note: We will carry an extra significant figure for the quantity kt.

$\ln N = -0.0533 + \ln 5.0 = -0.0533 + 1.609 = 1.556$, $N = e^{1.556} = 4.7$ g left after 1.0 yr

$\ln N = -0.0533 \text{ yr}^{-1}(10. \text{ yr}) + \ln 5.0 = -0.533 + 1.609 = 1.076$, $N = e^{1.076} = 2.9$ g ^{241}Pu left after 10. years

$\ln N = -0.0533 \text{ yr}^{-1}(100. \text{ yr}) + \ln 5.0 = -5.33 + 1.609 = -3.72$, $N = e^{-3.72} = 0.024$ g left after 100. yr

20.   If $[A]_o = 100.0$, then after 65 s, 45.0% of A has reacted or $[A]_{65} = 55.0$.

$\ln\left(\dfrac{[A]_t}{[A]_o} \right) = -kt$, $\ln\left(\dfrac{55.0}{100.0} \right) = -k(65 \text{ s})$, $k = 9.2 \times 10^{-3} \text{ s}^{-1}$; $t_{1/2} = \dfrac{0.693}{k} = 75 \text{ s}$

21. $\ln\left(\dfrac{[A]_t}{[A]_o} \right) = -kt$; $k = \dfrac{0.6931}{t_{1/2}} = \dfrac{0.6931}{14.3 \text{ d}} = 4.85 \times 10^{-2} \text{ d}^{-1}$

If $[A]_o = 100.0$, then after 95.0% completion, $[A]_t = 5.0$.

$\ln (0.050) = -4.85 \times 10^{-2} \text{ d}^{-1} \times t$, $t = 62$ days

22.   For a second order reaction: $t_{1/2} = \dfrac{1}{k[A]_o}$ or $k = \dfrac{1}{t_{1/2}[A]_o}$

$k = \dfrac{1}{143\ s(0.060\ mol/L)} = 0.12\ L\ mol^{-1}\ s^{-1}$

23.   a.   If the reaction is 38.5% complete, then 38.5% of the original concentration is consumed, leaving 61.5%.

$[A]_t = 61.5\%$ of $[A]_o$ or $[A]_t = 0.615[A]_o$; $\ln\left(\dfrac{[A]_t}{[A]_o}\right) = -kt$

$\ln(0.615) = -k(480.\ s),\ \ -0.486 = -k(480.\ s),\ \ k = 1.01 \times 10^{-3}\ s^{-1}$

b. $t_{1/2} = (\ln 2)/k = 0.6931/1.01 \times 10^{-3}\ s^{-1} = 686\ s$

c.   25% complete: $[A]_t = 0.75[A]_o$; $\ln(0.75) = -1.01 \times 10^{-3}\ (t),\ \ t = 280\ s$

75% complete: $[A]_t = 0.25[A]_o$; $\ln(0.25) = -1.01 \times 10^{-3}\ (t),\ \ t = 1.4 \times 10^3\ s$

Or, we know it takes 2 $t_{1/2}$ for reaction to be 75% complete.

$t = 2(686\ s) = 1372\ s \approx 1.4 \times 10^3\ s$

95% complete: $[A]_t = 0.05[A]_o$; $\ln(0.05) = -1.01 \times 10^{-3}\ (t),\ \ t = 3 \times 10^3\ s$

24.   a. $k = \dfrac{0.6931}{t_{1/2}} = \dfrac{0.6931}{12.8\ day} \times \dfrac{1\ day}{24\ hour} \times \dfrac{1\ hour}{3600\ s} = 6.27 \times 10^{-7}\ s^{-1}$

b.   Rate $= kN = 6.27 \times 10^{-7}\ s^{-1}\left(28.0 \times 10^{-3}\ g \times \dfrac{6.022 \times 10^{23}\ atoms}{64.0\ g}\right) = 1.65 \times 10^{14}$ decays/s

c. $\ln(N_t/N_o) = -kt,\ \ \ln(0.03) = -\left(\dfrac{0.6931}{12.8\ day}\right)(t),\ \ t = 64.8\ days \approx 60\ days$

Reaction Mechanisms

25.   a.   An elementary step (reaction) is one in which the rate law can be written from the molecularity, i.e., from the coefficients in the balanced equation.

b.   The mechanism of a reaction is the series of elementary reactions that occur to give the overall reaction. The sum of all the steps in the mechanism gives the balanced chemical reaction.

c.   The rate determining step is the slowest elementary reaction in any given mechanism.

26.   a.   Rate = $k[CH_3NC]$ b.   Rate = $k[O_3][NO]$

       c.   Rate = $k[O_3]$ d.   Rate = $k[O_3][O]$

       e.   Rate = $k[{}^{14}_{6}C]$ or Rate = kN where N = the number of ${}^{14}_{6}C$ atoms

27.   The rate law is: Rate = $k[NO]^2[Cl_2]$. If we assume the first step is rate determining, we would
       expect the rate law to be: Rate = $k_1[NO][Cl_2]$. This isn't correct. However, if we assume the
       second step to be rate determining: Rate = $k_2[NOCl_2][NO]$. To see if this agrees with
       experiment, we must substitute for the intermediate $NOCl_2$ concentration. Assuming a fast
       equilibrium first step (rate reverse = rate forward):

$$k_{-1}[NOCl_2] = k_1[NO][Cl_2], \quad [NOCl_2] = \frac{k_1}{k_{-1}}[NO][Cl_2]; \text{ Substituting into the rate equation:}$$

$$\text{Rate} = \frac{k_2 k_1}{k_{-1}}[NO]^2[Cl_2] = k[NO]^2[Cl_2]$$

       This is a possible mechanism with the second step the rate determining step since the derived
       rate law agrees with experiment.

28.   We know from Exercise 15.15 that Rate = $k[H_2O_2]$. The mechanism agrees with the experi-
       mental rate law if the first step, $H_2O_2 \rightarrow 2\,OH$, is the slow step. Assuming this is the slow step,
       then Rate = $k[H_2O_2]$ which agrees with the experiment.

29.   Let's determine the rate law for each mechanism. If this rate law is the same as the experimental
       rate law, then the mechanism is possible. When deriving rate laws from a mechanism, we must
       substitute for all intermediate concentrations.

       a.   Rate = $k[NO][O_2]$ <u>not possible</u>

       b.   Rate = $k[NO_3][NO]$ and $\dfrac{[NO_3]}{[NO][O_2]} = K_{eq} = k_1/k_{-1}$ or $[NO_3] = K_{eq}[NO][O_2]$

            Rate = $kK_{eq}[NO]^2[O_2]$ <u>possible</u>

       c.   Rate = $k[NO]^2$ <u>not possible</u>

       d.   Rate = $k[N_2O_2]$ and $[N_2O_2] = K_{eq}[NO]^2$; Rate = $kK_{eq}[NO]^2$ <u>not possible</u>

30.   If the first step is slow, Rate = $k_1[NO_2Cl]$ which has the same form as the experimental rate law.
       Thus, the first step is rate determining.

31.   Rate = $k_3[Br^-][H_2BrO_3^+]$; We must substitute for the intermediate concentrations. Since steps 1
       and 2 are fast equilibrium steps, then rate forward reaction = rate reverse reaction.

       $k_2[HBrO_3][H^+] = k_{-2}[H_2BrO_3^+]$; $k_1[BrO_3^-][H^+] = k_{-1}[HBrO_3]$

$$[HBrO_3] = \frac{k_1}{k_{-1}}[BrO_3^-][H^+]; \quad [H_2BrO_3^+] = \frac{k_2}{k_{-2}}[HBrO_3][H^+] = \frac{k_2k_1}{k_{-2}k_{-1}}[BrO_3^-][H^+]^2$$

$$\text{Rate} = \frac{k_3k_2k_1}{k_{-2}k_{-1}}[Br^-][BrO_3^-][H^+]^2 = k[Br^-][BrO_3^-][H^+]^2$$

32.   a.   This rate law occurs when the first step is rate determining.

$$\text{Rate} = k_1[Br_2] = k'[Br_2]$$

        b.   This rate law occurs when the second step is rate determining and the first step is a fast equilibrium step.

$$\text{Rate} = k_2[Br][H_2]; \quad \text{From the fast equilibrium first step:}$$

$$k_1[Br_2] = k_{-1}[Br]^2, \quad [Br] = (k_1/k_{-1})^{1/2}[Br_2]^{1/2}$$

Substituting into the rate equation:

$$\text{Rate} = k_2(k_1/k_{-1})^{1/2}[Br_2]^{1/2}[H_2] = k''[H_2][Br_2]^{1/2}$$

        c.   From a, $k' = k_1$; From b, $k'' = k_2(k_1/k_{-1})^{1/2}$

Temperature Dependence of Rate Constants and the Collision Model

33.   a.   The greater the frequency of collisions, the greater the opportunities for molecules to react, and, hence, the greater the rate.

        b.   Chemical reactions involve the making and breaking of chemical bonds. The kinetic energy of the collision can be used to break bonds.

        c.   For a reaction to occur, it is the reactive portion of each molecule that must be involved in a collision. Thus, only some collisions have the correct orientation.

34.   In a unimolecular reaction, a single reactant molecule decomposes to products. In a bimolecular reaction, two molecules collide to give products.

35.   The probability of the simultaneous collision of three molecules with the correct energy and orientation is exceedingly small.

36. $H_3O^+(aq) + OH^-(aq) \rightarrow 2\ H_2O(l)$ should have the faster rate. H_3O^+ and OH^- will be electrostatically attracted to each other; Ce^{4+} and Hg_2^{2+} will repel each other.

37. $k = A \exp(-E_a/RT)$ or $\ln k = \dfrac{-E_a}{RT} + \ln A$

For two conditions: $\ln\left(\dfrac{k_2}{k_1}\right) = \dfrac{E_a}{R}\left(\dfrac{1}{T_1} - \dfrac{1}{T_2}\right)$ (Assuming A is temperature independent.)

Let $k_1 = 2.0 \times 10^3$ s^{-1}, $T_1 = 298$ K; $k_2 = ?$, $T_2 = 348$ K; $E_a = 15.0 \times 10^3$ J/mol

$\ln\left(\dfrac{k_2}{2.0 \times 10^3 \text{ s}^{-1}}\right) = \dfrac{15.0 \times 10^3 \text{ J/mol}}{8.3145 \text{ J mol}^{-1}\text{ K}^{-1}}\left(\dfrac{1}{298 \text{ K}} - \dfrac{1}{348 \text{ K}}\right) = 0.87$

$\ln\left(\dfrac{k_2}{2.0 \times 10^3}\right) = 0.87$, $\dfrac{k_2}{2.0 \times 10^3} = e^{0.87} = 2.4$, $k_2 = 2.4(2.0 \times 10^3) = 4.8 \times 10^3$ s^{-1}

38. $k_1 = A \exp(-E_a/RT_1)$, $\ln k_1 = -E_a/RT_1 + \ln A$; $k_2 = A \exp(-E_a/RT_2)$, $\ln k_2 = -E_a/RT_2 + \ln A$

$\ln k_1 - \ln k_2 = -E_a/RT_1 - (-E_a/RT_2)$, $\ln\left(\dfrac{k_2}{k_1}\right) = \dfrac{E_a}{R}\left(\dfrac{1}{T_1} - \dfrac{1}{T_2}\right)$

$\ln\left(\dfrac{8.1 \times 10^{-2} \text{ s}^{-1}}{4.6 \times 10^{-2} \text{ s}^{-1}}\right) = \dfrac{E_a}{8.3145 \text{ J K}^{-1}\text{ mol}^{-1}}\left(\dfrac{1}{273 \text{ K}} - \dfrac{1}{293 \text{ K}}\right)$

$0.566 = \dfrac{E_a}{8.3145}(2.5 \times 10^{-4})$, $E_a = 1.9 \times 10^4$ J/mol $= 19$ kJ/mol

39. a. $\ln\left(\dfrac{k_2}{k_1}\right) = \dfrac{E_a}{R}\left(\dfrac{1}{T_1} - \dfrac{1}{T_2}\right)$; $\ln\left(\dfrac{0.950}{2.45 \times 10^{-4}}\right) = \dfrac{E_a}{8.3145}\left(\dfrac{1}{575} - \dfrac{1}{781}\right)$

$8.263 = \dfrac{E_a}{8.3145}(4.6 \times 10^{-4})$, $E_a = 1.5 \times 10^5$ J/mol $= 150$ kJ/mol

$\ln k_1 = -E_a/RT + \ln A$, $\ln(2.45 \times 10^{-4}) = \left(\dfrac{-1.5 \times 10^5 \text{ J/mol}}{8.3145 \text{ J mol}^{-1}\text{ K}^{-1}}\right)\left(\dfrac{1}{575 \text{ K}}\right) + \ln A$

$-8.314 = -31.38 + \ln A$, $\ln A = 23.07$, $A = e^{23.07} = 1.0 \times 10^{10}$ L mol^{-1} s^{-1}

b. $k = A \exp(-E_a/RT) = 1.0 \times 10^{10} \exp\left(\dfrac{-1.5 \times 10^5}{8.3145 \times 648}\right) = 8.1 \times 10^{-3}$ L mol^{-1} s^{-1}

40. From the Arrhenius equation, $k = A \exp(-E_a/RT)$ or in logarithmic form, $\ln k = -E_a/RT + \ln A$. Hence, a graph of $\ln k$ vs. $1/T$ should yield a straight line with a slope equal to $-E_a/R$.

T (K)	1/T (K⁻¹)	k (L mol⁻¹ s⁻¹)	ln k
195	5.13×10^{-3}	1.08×10^{9}	20.80
230	4.35×10^{-3}	2.95×10^{9}	21.81
260	3.85×10^{-3}	5.42×10^{9}	22.41
298	3.36×10^{-3}	12.0×10^{9}	23.21
369	2.71×10^{-3}	35.5×10^{9}	24.29

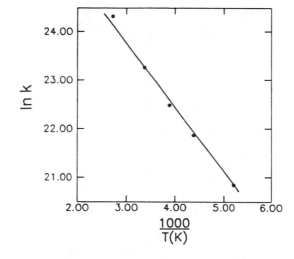

From the "eyeball" line on the graph:

$$\text{slope} = \frac{20.95 - 23.65}{5.00 \times 10^{-3} - 3.00 \times 10^{-3}} = \frac{-2.70}{2.00 \times 10^{-3}} = -1.35 \times 10^{3} \text{ K} = \frac{-E_a}{R}$$

$$E_a = 1.35 \times 10^{3} \text{ K} \times \frac{8.3145 \text{ J}}{\text{K mol}} = 1.12 \times 10^{4} \text{ J/mol} = 11.2 \text{ kJ/mol}$$

From the best straight line (by computer): slope $= -1.43 \times 10^{3}$ K and $E_a = 11.9$ kJ/mol

41.   If we double the rate, k doubles and $\dfrac{k_{35}}{k_{25}} = 2.00 = \dfrac{A \exp\left[-E_a/R(308)\right]}{A \exp\left[-E_a/R(298)\right]}$

$$\ln 2.00 = \frac{-E_a}{308 \, R} + \frac{E_a}{298 \, R} = E_a\left(\frac{1}{298 \, R} - \frac{1}{308 \, R}\right), \quad 0.693 = E_a \, (4.04 \times 10^{-4} - 3.91 \times 10^{-4})$$

$$\frac{0.693}{0.13 \times 10^{-4}} = E_a = 5.3 \times 10^{4} \text{ J/mol} = 53 \text{ kJ/mol}$$

42.   A graph of ln k vs. 1/T should be linear with slope = $-E_a/R$.

T (K)	1/T (K^{-1})	k (s^{-1})	ln k
338	2.96×10^{-3}	4.9×10^{-3}	-5.32
318	3.14×10^{-3}	5.0×10^{-4}	-7.60
298	3.36×10^{-3}	3.5×10^{-5}	-10.26

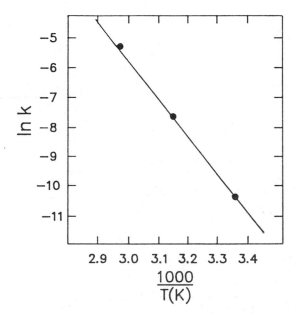

$$\text{Slope} = \frac{-10.76 - (-5.85)}{3.40 \times 10^{-3} - 3.00 \times 10^{-3}} = -1.2 \times 10^4 \text{ K} = -E_a/R$$

$$E_a = -\text{slope} \times R = 1.2 \times 10^4 \text{ K} \times \frac{8.3145 \text{ J}}{\text{K mol}}$$

$$E_a = 1.0 \times 10^5 \text{ J/mol} = 1.0 \times 10^2 \text{ kJ/mol}$$

43.

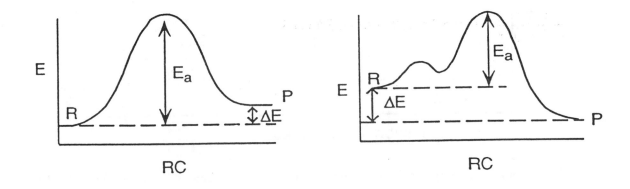

44.

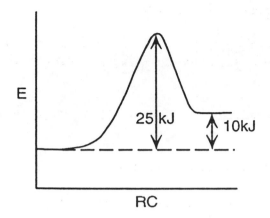

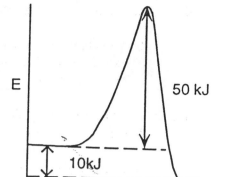

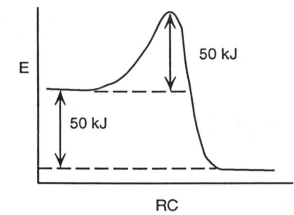

The reaction in a will have the greatest rate since it has the smallest activation energy.

45.

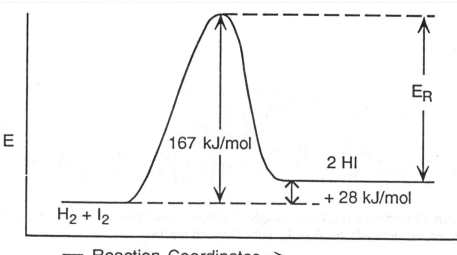

The activation energy for the reverse reaction is E_R in the diagram. $E_R = 167 - 28 = 139$ kJ/mol

Catalysis

46.   a.   A homogeneous catalyst is in the same phase as the reactants.

b.   A heterogeneous catalyst is in a different phase than the reactants. The catalyst is usually a solid, although a catalyst in a liquid phase can act as a heterogeneous catalyst for some gas phase reactions.

47.   A catalyst increases the rate of a reaction by providing an alternate pathway (mechanism) with a lower activation energy.

48.   Yes, the catalyst takes part in the mechanism and will appear in the rate law.

49.   No, the catalyzed reaction has a different mechanism and hence, a different rate law.

50.   a.   NO is the catalyst.

b. NO_2 is an intermediate

c. $k = A \exp(-E_a/RT)$

$$\frac{k_{cat}}{k_{un}} = \frac{A \exp\left[-E_a(cat)/RT\right]}{A \exp\left[-E_a(un)/RT\right]} = \exp\left(\frac{E_a(un) - E_a(cat)}{RT}\right)$$

$$\frac{k_{cat}}{k_{un}} = \exp\left(\frac{2100 \text{ J/mol}}{8.3145 \text{ J mol}^{-1} \text{ K}^{-1} \times 298 \text{ K}}\right) = e^{0.85} = 2.3$$

The catalyzed reaction is 2.3 times faster than the uncatalyzed reaction at 25°C.

51. $O_3 + Cl \rightarrow O_2 + ClO$ (slow)
$ClO + O \rightarrow O_2 + Cl$ (fast)

$O_3 + O \rightarrow 2 O_2$

Since the chlorine atom catalyzed reaction has the lower activation energy, then the Cl catalyzed rate is faster. Hence, Cl is a more effective catalyst.

At 25°C: $\dfrac{k_{Cl}}{k_{NO}} = \exp\left(\dfrac{-E_a(Cl)}{RT} + \dfrac{E_a(NO)}{RT}\right) = \exp\left(\dfrac{(-2100 + 11,900) \text{ J/mol}}{8.3145 \times 298 \text{ J/mol}}\right) = e^{3.96} = 52$

At 25°C, the Cl catalyzed reaction is roughly 52 times faster than the NO catalyzed reaction, assuming the frequency factor A is the same for each reaction.

52.   The surface of the catalyst should be:

metal surface

Thus, CH$_2$D–CH$_2$D should be the product.

53.   At high [S], the enzyme is completely saturated with substrate. Once the enzyme is completely saturated, the rate of decomposition of ES can no longer increase and the overall rate remains constant.

Additional Exercises

54.   Rate = k[H$_2$SeO$_3$]^x[H$^+$]^y[I$^-$]^z; Comparing the first and second experiments:

$$\frac{3.33 \times 10^{-7}}{1.66 \times 10^{-7}} = \frac{k(2.0 \times 10^{-4})^x\,(2.0 \times 10^{-2})^y\,(2.0 \times 10^{-2})^z}{k(1.0 \times 10^{-4})^x\,(2.0 \times 10^{-2})^y\,(2.0 \times 10^{-2})^z},\ 2.01 = 2.0^x,\ x = 1$$

Comparing the first and fourth experiments:

$$\frac{6.66 \times 10^{-7}}{1.66 \times 10^{-7}} = \frac{k(1.0 \times 10^{-4})\,(4.0 \times 10^{-2})^y\,(2.0 \times 10^{-2})^z}{k(1.0 \times 10^{-4})\,(2.0 \times 10^{-2})^y\,(2.0 \times 10^{-2})^z},\ 4.01 = 2.0^y,\ y = 2$$

Comparing the first and sixth experiments:

$$\frac{13.2 \times 10^{-7}}{1.66 \times 10^{-7}} = \frac{k(1.0 \times 10^{-4})\,(2.0 \times 10^{-2})^2\,(4.0 \times 10^{-2})^z}{k(1.0 \times 10^{-4})\,(2.0 \times 10^{-2})^2\,(2.0 \times 10^{-2})^z}$$

$$7.95 = 2.0^z,\ z = \frac{\log 7.95}{\log 2.0} = 2.99 \approx 3$$

Rate = k[H$_2$SeO$_3$][H$^+$]^2[I$^-$]^3

Experiment #1:

$$\frac{1.66 \times 10^{-7}\,\text{mol}}{\text{L s}} = k\left(\frac{1.0 \times 10^{-4}\,\text{mol}}{\text{L}}\right)\left(\frac{2.0 \times 10^{-2}\,\text{mol}}{\text{L}}\right)^2\left(\frac{2.0 \times 10^{-2}\,\text{mol}}{\text{L}}\right)^3$$

k = 5.19 × 10^5 L^5 mol^{-5} s^{-1} = 5.2 × 10^5 L^5 mol^{-5} s^{-1}

For all experiments:

Exp #	k (L^5 mol^{-5} s^{-1})
1	5.19×10^5
2	5.20×10^5
3	5.20×10^5
4	5.20×10^5
5	5.25×10^5
6	5.16×10^5
7	5.25×10^5

$k_{mean} = 5.2 \times 10^5$ L^5 mol^{-5} s^{-1}

55.   Rate = $k[I^-]^x[OCl^-]^y[OH^-]^z$; Comparing the first and second experiments:

$$\frac{18.7 \times 10^{-3}}{9.4 \times 10^{-3}} = \frac{k(0.0026)^x (0.012)^y (0.10)^z}{k(0.0013)^x (0.012)^y (0.10)^z}, \quad 2.0 = 2.0^x, \quad x = 1$$

Comparing the first and third experiments:

$$\frac{9.4 \times 10^{-3}}{4.7 \times 10^{-3}} = \frac{k(0.0013) (0.012)^y (0.10)^z}{k(0.0013) (0.006)^y (0.10)^z}, \quad 2.0 = 2^y, \quad y = 1$$

Comparing the first and sixth experiments:

$$\frac{4.7 \times 10^{-3}}{9.4 \times 10^{-3}} = \frac{k(0.0013) (0.012) (0.20)^z}{k(0.0013) (0.012) (0.10)^z}, \quad 1/2 = 2.0^z, \quad z = -1$$

Rate = $\dfrac{k[I^-][OCl^-]}{[OH^-]}$

For the first experiment:

$$\frac{9.4 \times 10^{-3} \text{ mol}}{L \text{ s}} = k \frac{(0.0013 \text{ mol/L}) (0.012 \text{ mol/L})}{(0.10 \text{ mol/L})} = 60.3 \text{ s}^{-1} = 60. \text{ s}^{-1}$$

For all experiments:

Exp #	1	2	3	4	5	6	7	
k (s^{-1})	60.3	59.9	60.3	59.8	59.9	60.3	59.8	$k_{mean} = 6.0 \times 10^1$ s^{-1}

56.   Rate = $k[N_2O_5]^x$; The rate laws for the first two experiments are:

$2.26 \times 10^{-3} = k(0.190)^x$ and $8.90 \times 10^{-4} = k(0.075)^x$

Dividing: $2.54 = \dfrac{(0.190)^x}{(0.075)^x} = (2.5)^x, \quad x = 1; \quad$ Rate = $k[N_2O_5]$

$$k = \frac{\text{Rate}}{[\text{N}_2\text{O}_5]} = \frac{8.90 \times 10^{-4} \text{ mol L}^{-1} \text{ s}^{-1}}{0.075 \text{ mol/L}} = 1.2 \times 10^{-2} \text{ s}^{-1}; \quad k_{\text{mean}} = 1.2 \times 10^{-2} \text{ s}^{-1}$$

57.

Heating Time	Untreated		Deacidifying		Antioxidant	
(days)	s	ln s	s	ln s	s	ln s
0.00	100.0	4.605	100.1	4.606	114.6	4.741
1.00	67.9	4.218	60.8	4.108	65.2	4.177
2.00	38.9	3.661	26.8	3.288	28.1	3.336
3.00	16.1	2.779	-	-	11.3	2.425
6.00	6.8	1.92	-	-	-	-

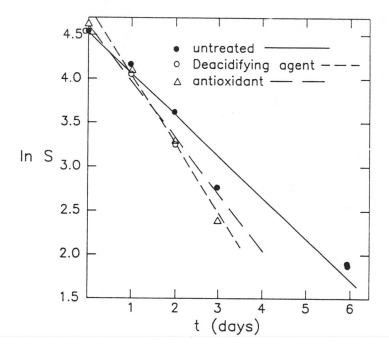

We used a calculator to fit the data by least squares. The results follow.

Untreated: ln s = -0.465 t + 4.55, k = -0.465 day^{-1}

Deacidifying agent: ln s = -0.659 t + 4.66, k = 0.659 day^{-1}

Antioxidant: ln s = -0.779 t + 4.84, k = 0.779 day^{-1}

b. No, the silk degrades more rapidly with the additives.

58. $t_{1/2} = (\ln 2)/k$

Untreated: $t_{1/2} = 1.49$ day; Deacidifying agent: $t_{1/2} = 1.05$ day; Antioxidant: $t_{1/2} = 0.890$ day

59.   The pressure of a gas is proportional to concentration. Therefore, we will use the pressure data to solve the problem.

$SO_2Cl_2(g) \rightarrow SO_2(g) + Cl_2(g)$; Let P_0 = initial partial pressure of SO_2Cl_2

If $x = P_{SO_2}$ at some time, then $x = P_{SO_2} = P_{Cl_2}$ and $P_{SO_2Cl_2} = P_0 - x$

$P_{tot} = P_{SO_2Cl_2} + P_{SO_2} + P_{Cl_2} = P_0 - x + x + x$, $P_{tot} = P_0 + x$, $P_{tot} - P_0 = x$

At time = 0 hour, $P_{tot} = P_0 = 4.93$ atm. The data for other times are:

Time (hour)	0.00	1.00	2.00	4.00	8.00	16.00
P_{tot} (atm)	4.93	5.60	6.34	7.33	8.56	9.52
$P_{SO_2Cl_2}$ (atm)	4.93	4.26	3.52	2.53	1.30	0.34
$\ln P_{SO_2Cl_2}$	1.595	1.449	1.258	0.928	0.262	-1.08

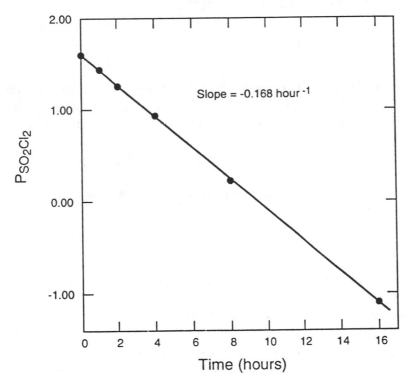

Since pressure of a gas is proportional to concentration and since the $\ln P_{SO_2Cl_2}$ vs. time plot is linear, then the reaction is first order in SO_2Cl_2.

a.   Slope of $\ln(P)$ vs. t plot is -0.168 hour^{-1} = -k, k = 0.168 hour^{-1} = 4.67 × 10^{-5} s^{-1}

b. $t_{1/2} = \dfrac{\ln 2}{k} = \dfrac{0.6931}{k} = \dfrac{0.6931}{0.168\ h^{-1}} = 4.13$ hour

c. $\ln P_{SO_2Cl_2} = -kt + \ln P_o = -(0.168\ h^{-1})(0.500\ h) + \ln 4.93$

$\ln P_{SO_2Cl_2} = -0.0840 + 1.595 = 1.511,\ P_{SO_2Cl_2} = e^{1.511} = 4.53$ atm

$P_{Cl_2} = P_{SO_2} = 4.93$ atm - 4.53 atm = 0.40 atm

$P_{tot} = P_{SO_2Cl_2} + P_{Cl_2} + P_{SO_2} = 4.53 + 0.40 + 0.40 = 5.33$ atm

After 12.0 hours: from graph, $\ln P_{SO_2Cl_2} = -0.403$

$P_{SO_2Cl_2} = e^{-0.403} = 0.668$ atm, $P_{SO_2} = 4.93 - 0.668 = 4.26$ atm, $P_{Cl_2} = 4.26$ atm

$P_{tot} = 0.668 + 4.26 + 4.26 = 9.19$ atm

d. $\ln\left(\dfrac{P_{SO_2Cl_2}}{P_o}\right) = -0.168\ h^{-1}\ (20.0\ hr) = -3.36,\quad \left(\dfrac{P_{SO_2Cl_2}}{P_o}\right) = e^{-3.36} = 3.47 \times 10^{-2}$

Fraction left = 0.0347 = 3.47%

60.   The rate depends on the number of reactant molecules adsorbed on the surface of the catalyst. This quantity is proportional to the concentration of reactant. However, when all of the surface sites are occupied, the rate becomes independent of the concentration of reactant.

61.   a.   W since it has a lower activation energy than the Os catalyst.

b. $k_w = A_w \exp[-E_a(W)/RT];\ k_{uncat} = A_{uncat} \exp[-E_a(uncat)/RT];$ Assume $A_w = A_{uncat}$

$\dfrac{k_w}{k_{uncat}} = \exp\left(\dfrac{-E_a(W)}{RT} + \dfrac{E_a(uncat)}{RT}\right)$

$\dfrac{k_w}{k_{uncat}} = \exp\left(\dfrac{-163{,}000\ J/mol + 335{,}000\ J/mol}{(8.3145\ J\ mol^{-1}\ K^{-1})(298\ K)}\right) = 1.41 \times 10^{30}$

The W catalyzed reaction is approximately 10^{30} times faster than the uncatalyzed reaction.

c. H_2 decreases the rate of the reaction. For the decomposition to occur, NH_3 molecules must be adsorbed on the surface of the catalyst. If H_2 is also adsorbed on the catalyst, then there are fewer sites for NH_3 molecules to be adsorbed and the rate decreases.

62.   Rate = $k_2[I^-][HOCl]$; From the fast equilibrium first step:

$$k_1[OCl^-] = k_{-1}[HOCl][OH^-], \quad [HOCl] = \frac{k_1[OCl^-]}{k_{-1}[OH^-]}; \quad \text{Substituting into the rate equation:}$$

$$\text{Rate} = \frac{k_2k_1[I^-][OCl^-]}{k_{-1}[OH^-]} = \frac{k[I^-][OCl^-]}{[OH^-]}$$

63.   Rate = $k[BrO_3^-][SO_3^{2-}][H^+]$; First step: $SO_3^{2-} + H^+ \rightleftharpoons HSO_3^-$ (fast)

From the rate law we can say that the rate determining step contains one of each ion. Thus, a possible second step could be:

$$BrO_3^- + HSO_3^- \rightarrow \text{Products}   \text{(slow)}$$

A likely choice is: $BrO_3^- + HSO_3^- \rightarrow SO_4^{2-} + HBrO_2$ (slow)

Followed by: $HBrO_2 + SO_3^{2-} \rightarrow HBrO + SO_4^{2-}$ (fast);
$HBrO + SO_3^{2-} \rightarrow SO_4^{2-} + H^+ + Br^-$ (fast)

All species in this mechanism (HSO_3^-, $HBrO_2$, $HBrO$) are known substances. This mechanism also gives the correct experimentally determined rate law.

Rate = $k_2[BrO_3^-] [HSO_3^-]$; Assuming reaction one is a fast equilibrium step:

$$k_1[SO_3^{2-}][H^+] = k_{-1}[HSO_3^-], \quad [HSO_3^-] = \frac{k_1}{k_{-1}}[SO_2^{2-}][H^+]$$

$$\text{Rate} = \frac{k_2k_1}{k_{-1}}[BrO_3^-][SO_3^{2-}][H^+] = k[BrO_3^-][SO_3^{2-}][H^+]$$

64.   a.   Rate = $k_3[COCl][Cl_2]$; From fast equilibrium reactions 1 and 2:

$$\frac{[COCl]}{[Cl][CO]} = \frac{k_2}{k_{-2}}, \quad [COCl] = \frac{k_2}{k_{-2}}[CO][Cl]$$

$$\frac{[Cl]^2}{[Cl_2]} = \frac{k_1}{k_{-1}}, \quad [Cl] = \left(\frac{k_1}{k_{-1}}[Cl_2]\right)^{1/2}$$

$$\text{Thus, } [COCl] = \frac{k_2}{k_{-2}}\left(\frac{k_1}{k_{-1}}\right)^{1/2}[CO][Cl_2]^{1/2}; \quad \text{Substituting into rate law:}$$

$$\text{Rate} = k_3\frac{k_2}{k_{-2}}\left(\frac{k_1}{k_{-1}}\right)^{1/2}[CO][Cl_2]^{3/2} = k[CO][Cl_2]^{3/2}$$

b.   Cl and COCl are intermediates.

65.   The integrated rate law for each reaction is:

$$\ln[A] = -4.50 \times 10^{-4} \text{ s}^{-1}(t) + \ln[A]_o \text{ and } \ln[B] = -3.70 \times 10^{-3} \text{ s}^{-1}(t) + \ln[B]_o$$

Subtracting the second equation from the first equation ($\ln[A]_o = \ln[B]_o$):

$$\ln[A] - \ln[B] = -4.50 \times 10^{-4} (t) + 3.70 \times 10^{-3} (t), \ \ln\left(\frac{[A]}{[B]}\right) = 3.25 \times 10^{-3} \text{ s}^{-1} (t)$$

When $[A] = 4.00[B]$, $\ln 4.00 = 3.25 \times 10^{-3}$ (t), $t = 427$ s

66.   Comparing the last two experiments, doubling the [B] (keeping [A] constant) doubles the rate. The order with respect to B is 1. We can't get the order with respect to A by inspection. Comparing the 2nd and 3rd experiments:

$$11.8 \text{ mol L}^{-1} \text{ s}^{-1} = (0.5 \ M)^x (1.0 \ M), \ 33.5 \text{ mol L}^{-1} \text{ s}^{-1} = (1.0 \ M)^x (1.0 \ M)$$

Dividing: $\dfrac{33.5}{11.8} = \dfrac{(1.0)^x}{(0.5)^x} = 2.84 = 2^x, \ \ln 2.84 = x \ln 2, \ x = \dfrac{1.044}{0.693} = 1.5$

The order with respect to A is 1.5.

67. $\ln\left(\dfrac{k_2}{k_1}\right) = \dfrac{E_a}{R}\left(\dfrac{1}{T_1} - \dfrac{1}{T_2}\right); \ \dfrac{k_2}{k_1} = 7.00; \ T_1 = 275 \text{ K}; \ T_2 = 300. \text{ K}$

$$\ln 7.00 = \frac{E_a}{8.3145 \text{ J mol}^{-1} \text{K}^{-1}}\left(\frac{1}{275 \text{ K}} - \frac{1}{300. \text{ K}}\right)$$

$$E_a = \frac{(8.3145 \text{ J mol}^{-1} \text{K}^{-1})(\ln 7.00)}{3.0 \times 10^{-4} \text{ K}^{-1}} = 5.4 \times 10^4 \text{ J/mol} = 54 \text{ kJ/mol}$$

68.   a. $[A] = -(5.0 \times 10^{-2} \text{ mol L}^{-1} \text{ s}^{-1}) t + [A]_o$; If $[A]_o = 1.0 \times 10^{-3} \ M$, then:

$$[A] = -(5.0 \times 10^{-2} \text{ mol L}^{-1} \text{ s}^{-1}) t + 1.0 \times 10^{-3} \text{ mol/L}$$

b. $\dfrac{[A]_o}{2} = (-5.0 \times 10^{-2}) t_{1/2} + [A]_o$

$-0.50[A]_o = (-5.0 \times 10^{-2}) t_{1/2}, \ t_{1/2} = \dfrac{0.50(1.0 \times 10^{-3})}{5.0 \times 10^{-2}} = 1.0 \times 10^{-2}$ s

c.   Rate $= \dfrac{d[B]}{dt} = \dfrac{-d[A]}{dt} = k; \ [B] - [B]_o = kt$ where $[B]_o = 0$

$$[B] = kt = (5.0 \times 10^{-2} \text{ mol L}^{-1} \text{ s}^{-1})(5.0 \times 10^{-3} \text{ s}) = 2.5 \times 10^{-4} \ M$$

69.   a.   Let P_o = initial partial pressure of C_2H_5OH = 250. torr. If x torr of C_2H_5OH reacts, then at any time:

$$P_{C_2H_5OH} = 250. - x, \quad P_{CH_3CHO} = P_{H_2} = x; \quad P_{tot} = 250. - x + x + x = 250. + x$$

Therefore, $P_{C_2H_5OH}$ at any time can be calculated from the data by determining x (= P_{tot} - 250.) and then subtracting from 250. torr. Using the $P_{C_2H_5OH}$ data, a plot of $P_{C_2H_5OH}$ vs. t is linear. The reaction is zero order in $P_{C_2H_5OH}$. One could also use the P_{tot} vs. t data since P_{tot} increases at the same rate that $P_{C_2H_5OH}$ decreases. Note: The ln P vs. t plot is also linear. The reaction hasn't been followed for enough time for curvature to be seen. However, since $P_{C_2H_5OH}$ decreases at steady increments, we can conclude that the reaction is zero order in C_2H_5OH.

From the data, the integrated rate law involving P_{tot} is:

$$P_{tot} = \frac{13}{100.}t + 250.; \quad \text{At } t = 900. \text{ s}, \; P_{tot} = \frac{13}{100.}(900.) + 250. = 370 \text{ torr}$$

b.   From the $P_{C_2H_5OH}$ vs. t plot:

$$\text{slope} = -k = \frac{-13 \text{ torr}}{100. \text{ s}} \times \frac{1 \text{ atm}}{760 \text{ torr}}, \quad k = 1.7 \times 10^{-4} \text{ atm/s}$$

Note: P_{tot} increases at the same rate that $P_{C_2H_5OH}$ decreases. Therefore, the rate constant k could have been determined from a P_{tot} vs. t plot.

c.   Zero order in C_2H_5OH since $P_{C_2H_5OH}$ vs. t is linear.

70.   a.   A plot of $P_{C_2H_5OH}$ vs. t is linear. The reaction is zero order in C_2H_5OH. This problem is similar to Exercise 15.69. See discussion in 15.69 for more details.

From the data, the integrated rate equation involving P_{tot} is: $P_{tot} = \frac{15}{10.}t + 250.$

At $t = 80.$ s, $P_{tot} = \frac{15}{10.}(80.) + 250. = 370$ torr

b.   slope = -k = -1.5 torr/s (from $P_{C_2H_5OH}$ vs. t plot), k = 1.5 torr/s; The P_{tot} vs. t plot would give the same rate constant since P_{tot} increases at the same rate that $P_{C_2H_5OH}$ decreases.

c.   zero order

d. $P_{tot} = \frac{15}{10.}(300.) + 250. = 7.0 \times 10^2$ torr

This is an impossible answer. Since only 250. torr of C_2H_5OH are present initially, the maximum pressure can only be 500. torr when all of the C_2H_5OH is consumed, i.e., $P_{tot} = P_{C_2H_4} + P_{H_2O} = 250. + 250. = 500.$ torr. Therefore, at 300. s, $P_{tot} = 500.$ torr.

71.   a.   t (s) $[C_4H_6]$ (M) $\ln[C_4H_6]$ $1/[C_4H_6]$ (M^{-1})

            0   0.01000   -4.6052 1.000×10^2
            1000.   0.00629   -5.069 1.59×10^2
            2000.   0.00459   -5.384 2.18×10^2
            3000.   0.00361   -5.624 2.77×10^2

The plot of $1/[C_4H_6]$ vs. t is linear, thus the reaction is second order in butadiene. From the plot, the integrated rate law is:

$$\frac{1}{[C_4H_6]} = 5.90 \times 10^{-2}\, t + 100.0$$

b.   When dimerization is 1.0% complete, 99.0% of C_4H_6 is left.

$[C_4H_6] = 0.990(0.01000) = 0.00990\ M;\ \ \dfrac{1}{0.00990} = 5.90 \times 10^{-2}\, t + 100.0,\ \ t = 17.1\ s \approx 20\ s$

c.   2.0% complete, $[C_4H_6] = 0.00980\ M;\ \ \dfrac{1}{0.00980} = 5.90 \times 10^{-2}\, t + 100.0,\ \ t = 34.6\ s \approx 30\ s$

d. $\dfrac{1}{[C_4H_6]} = kt + \dfrac{1}{[C_4H_6]_o};\ \ [C_4H_6]_o = 0.020\ M;\ \ $ At $t = t_{1/2},\ [C_4H_6] = 0.010\ M$

$\dfrac{1}{0.010} = (5.90 \times 10^{-2})t_{1/2} + \dfrac{1}{0.020};\ \ t = 847\ s = 850\ s \approx 800\ s$

Or, $t_{1/2} = \dfrac{1}{k[A]_o} = \dfrac{1}{(5.90 \times 10^{-2})(2.0 \times 10^{-2})} = 850\ s$

e.   From Exercise 15.13, $k = 1.4 \times 10^{-2}\ L\ mol^{-1}\ s^{-1}$ at 500. K. From this problem, $k = 5.90 \times 10^{-2}\ L\ mol^{-1}\ s^{-1}$ at 620. K.

$$\ln\left(\frac{k_2}{k_1}\right) = \frac{E_a}{R}\left(\frac{1}{T_1} - \frac{1}{T_2}\right);\ \ \ln\left(\frac{5.90 \times 10^{-2}}{1.4 \times 10^{-2}}\right) = \frac{E_a}{8.3145\ J\ mol^{-1}\ K^{-1}}\left(\frac{1}{500.\ K} - \frac{1}{620.\ K}\right)$$

$12 = E_a\,(3.9 \times 10^{-4}),\ E_a = 3.1 \times 10^4\ J/mol = 31\ kJ/mol$

72.   For second order kinetics: $\dfrac{1}{[A]} - \dfrac{1}{[A]_o} = kt$

a. $\dfrac{1}{[A]} = (0.250\ L\ mol^{-1}\ s^{-1})t + \dfrac{1}{[A]_o},\ \ \dfrac{1}{[A]} = (0.250)\,(180.\ s) + \dfrac{1}{1.00 \times 10^{-2}\ M}$

$\dfrac{1}{[A]} = 145\ M^{-1},\ \ [A] = 6.90 \times 10^{-3}\ M$

Amount of A that reacted $= 0.0100 - 0.00690 = 0.0031\ M$

$$[A_2] = \frac{1}{2}(3.1 \times 10^{-3} \, M) = 1.6 \times 10^{-3} \, M$$

b. After 3 minutes (180. s): $[A] = 3.00 \, [B]$, $6.90 \times 10^{-3} \, M = 3.00 \, [B]$, $[B] = 2.30 \times 10^{-3} \, M$

$$\frac{1}{[B]} = k_2 t + \frac{1}{[B]_o}; \quad \frac{1}{2.30 \times 10^{-3} \, M} = k_2(180. \, s) + \frac{1}{2.50 \times 10^{-2} \, M}, \quad k_2 = 2.19 \text{ L mol}^{-1} \text{ s}^{-1}$$

c. $[A]_o = 1.00 \times 10^{-2} \, M$; At $t = t_{1/2}$, $[A] = 5.00 \times 10^{-3} \, M$

$$\frac{1}{5.00 \times 10^{-3}} = 0.250 \, t + \frac{1}{1.00 \times 10^{-2}}, \quad t_{1/2} = 4.00 \times 10^2 \, s$$

73. $100\% \rightarrow 50\% \rightarrow 25\% \rightarrow 12.5\%$; This process is 3 half-lives = 3(14 h) = 42 hours.

74. Sucessive half-lives increase in time for a second order reaction. Therefore, assume reaction is second order in A.

$$t_{1/2} = \frac{1}{k[A]_o}, \quad k = \frac{1}{t_{1/2}[A]_o} = \frac{1}{(10.0 \text{ min})(0.10 \, M)} = 1.0 \text{ L mol}^{-1} \text{ min}^{-1}$$

a. $\dfrac{1}{[A]} = 1.0(80.0 \text{ min}) + \dfrac{1}{0.10 \, M} = 90. \, M^{-1}$, $[A] = 1.1 \times 10^{-2} \, M$

b. 30.0 min = 2 half-lives, so 25% of original A is remaining.

$[A] = 0.25(0.10 \, M) = 0.025 \, M$

75. Rate $= k[A]^x[B]^y[C]^z$; During the course of experiment 1, [A] and [C] are essentially constant, and Rate $= k'[B]^y$ where $k' = k[A]_o^x [C]_o^z$.

[B] (M)	time (s)	ln[B]	1/[B] (M^{-1})
1.0×10^{-3}	0	-6.91	1.0×10^3
2.7×10^{-4}	1.0×10^5	-8.22	3.7×10^3
1.6×10^{-4}	2.0×10^5	-8.74	6.3×10^3
1.1×10^{-4}	3.0×10^5	-9.12	9.1×10^3
8.5×10^{-5}	4.0×10^5	-9.37	12×10^3
6.9×10^{-5}	5.0×10^5	-9.58	14×10^3
5.8×10^{-5}	6.0×10^5	-9.76	17×10^3

The plot of 1/[B] vs. t is linear. The reaction is second order in B and the integrated rate equation is:

$$1/[B] = (2.7 \times 10^{-2} \, M^{-1} \, s^{-1}) \, t + 1.0 \times 10^3; \quad k' = 2.7 \times 10^{-2} \, M^{-1} \, s^{-1}$$

For experiment 2, [B] and [C] are essentially constant and Rate = $k''[A]^x$ where k'' = $k[B]_o^y [C]_o^z = k[B]_o^2 [C]_o^z$.

[A] (M)	time (s)	ln[A]	1/[A] (M^{-1})
1.0×10^{-2}	0	-4.61	1.0×10^2
8.9×10^{-3}	1.0	-4.72	110
7.1×10^{-3}	3.0	-4.95	140
5.5×10^{-3}	5.0	-5.20	180
3.8×10^{-3}	8.0	-5.57	260
2.9×10^{-3}	10.0	-5.84	340
2.0×10^{-3}	13.0	-6.21	5.0×10^2

The plot of ln[A] vs. t is linear. The reaction is first order in A and the integrated raw law is:

$$\text{ln}[A] = -(0.123 \text{ s}^{-1}) t - 4.61; \quad k'' = 0.123 \text{ s}^{-1}$$

Note: We will carry an extra significant figure in k''.

Experiment 3:   [A] and [B] are constant; Rate = $k'''[C]^z$

The plot of [C] vs. t is linear. Thus, z = 0.

The overall rate law is: Rate = $k[A][B]^2$

From Experiment 1 (to determine k):

$$k' = 2.7 \times 10^{-2} M^{-1} \text{ s}^{-1} = k[A]_o^x [C]_o^z = k[A]_o = k(2.0 M), \quad k = 1.4 \times 10^{-2} L^2 \text{ mol}^{-2} \text{ s}^{-1}$$

From Experiment 2: $k'' = 0.123 \text{ s}^{-1} = k[B]_o^2$, $k = \dfrac{0.123 \text{ s}^{-1}}{(3.0 M)^2} = 1.4 \times 10^{-2} L^2 \text{ mol}^{-2} \text{ s}^{-1}$

Thus, rate = $k[A][B]^2$ and $k = 1.4 \times 10^{-2} L^2 \text{ mol}^{-2} \text{ s}^{-1}$.

76.   From 338 K data, a plot of $\text{ln}[N_2O_5]$ vs. t is linear. The integrated rate law is:

$$\text{ln}[N_2O_5] = -4.86 \times 10^{-3} t - 2.30; \quad k = 4.86 \times 10^{-3} \text{ s}^{-1} \text{ at } 338 \text{ K}$$

From 318 K data: $\text{ln}[N_2O_5] = -4.98 \times 10^{-4} t - 2.30; \quad k = 4.98 \times 10^{-4} \text{ s}^{-1}$ at 318 K

$$\text{ln}\left(\frac{k_2}{k_1}\right) = \frac{E_a}{R}\left(\frac{1}{T_1} - \frac{1}{T_2}\right); \quad \text{ln}\left(\frac{4.86 \times 10^{-3}}{4.98 \times 10^{-4}}\right) = \frac{E_a}{8.3145 \text{ J K}^{-1} \text{mol}^{-1}}\left(\frac{1}{318 \text{ K}} - \frac{1}{338 \text{ K}}\right)$$

$E_a = 1.0 \times 10^5$ J/mol = 1.0×10^2 kJ/mol

77.   a.   Rate $= k[CH_3X]^x[Y]^y$; For experiment 1, [Y] is constant so Rate $= k'[CH_3X]^x$ where $k' =$ $k(3.0\ M)^y$.

A plot of $\ln[CH_3X]$ vs t is linear (x = 1). The integrated rate law is:

$\ln[CH_3X] = -0.93\ t - 3.99$; $k' = 0.93\ h^{-1}$

For Exeriment 2, [Y] is again constant with Rate $= k''[CH_3X]^x$ where $k'' = k(4.5\ M)^y$. The ln plot is linear again with an integrated rate law:

$\ln[CH_3X] = -0.93\ t - 5.40$; $k'' = 0.93\ h^{-1}$

Dividing the rate constant values: $\dfrac{k'}{k''} = \dfrac{0.93}{0.93} = \dfrac{k(3.0)^y}{k(4.5)^y}$, $1.0 = (0.67)^y$, $y = 0$

Reaction is first order in CH_3X and zero order in Y. The overall rate law is:

Rate $= k[CH_3X]$ where $k = 0.93\ h^{-1}$ at 25°C.

b. $t_{1/2} = (\ln 2)/k = 0.6931/(7.88 \times 10^8\ h^{-1}) = 8.80 \times 10^{-10}$ hour

c. $\ln(k_2/k_1) = (E_a/R)(1/T_1 - 1/T_2) = \ln\left(\dfrac{7.88 \times 10^8}{0.93}\right) = \dfrac{E_a}{8.3145\ J\ K^{-1}\ mol^{-1}}\left(\dfrac{1}{298\ K} - \dfrac{1}{358\ K}\right)$

$E_a = 3.0 \times 10^5\ J/mol = 3.0 \times 10^2\ kJ/mol$

d.   The activation energy is close to the C-X bond energy. A plausible mechanism is:

$CH_3X \rightarrow CH_3 + X$ (slow)

$CH_3 + Y \rightarrow CH_3Y$ (fast)

78.   a.

T (K)	1/T (K^{-1})	k (min^{-1})	ln k
298.2	3.353×10^{-3}	178	5.182
293.5	3.407×10^{-3}	126	4.836
290.5	3.442×10^{-3}	100.	4.605

The plot of ln k vs. 1/T gives a straight line. The equation for the straight line is:

$\ln k = -6.48 \times 10^3\ (1/T) + 26.9$

For a ln k vs. 1/T plot, the slope $= -E_a/R$.

$-6.48 \times 10^3\ K = -E_a/8.3145\ J\ mol^{-1}\ K^{-1}$, $E_a = 5.39 \times 10^4\ J/mol = 53.9\ kJ/mol$

b. $\ln k = -6.48 \times 10^3(1/288.2) + 26.9 = 4.42$, $k = e^{4.42} = 83\ min^{-1}$

About 83 chirps per minute per insect.

c.   k gives the number of chirps per minute. The number or chirps in 15 sec is k/4.

T (C°)	T (°F)	k (min⁻¹)	42 + 0.80 (k/4)
25.0	77.0	178	78° F
20.3	68.5	126	67°F
17.3	63.1	100.	62°F
15.0	59.0	83	59°F

The rule of thumb appears to be fairly accurate, almost ± 1°F.

79.   a.   If the interval between flashes is 16.3 sec, then the rate is:

1 flash/16.3 s = 6.13×10^{-2} s⁻¹ = k

Interval	k	T
16.3 s	6.13×10^{-2} s⁻¹	21.0°C (294.2 K)
13.0 s	7.69×10^{-2} s⁻¹	27.8°C (301.0 K)

$$\ln\left(\frac{k_2}{k_1}\right) = \frac{E_a}{R}\left(\frac{1}{T_1} - \frac{1}{T_2}\right); \text{ Solving: } E_a = 2.5 \times 10^4 \text{ J/mol} = 25 \text{ kJ/mol}$$

b. $$\ln\left(\frac{k}{6.13 \times 10^{-2}}\right) = \frac{2.4 \times 10^4 \text{ J/mol}}{8.3145 \text{ J K}^{-1} \text{ mol}^{-1}}\left(\frac{1}{294.2 \text{ K}} - \frac{1}{303.2 \text{ K}}\right)$$

ln k = 0.29 + ln(6.13 × 10⁻²) = -2.50

k = $e^{-2.50}$ = 8.2×10^{-2} s⁻¹; Interval = 1/k = 12 seconds

c.

T	Interval	54-2(Intervals)
21.0 °C	16.3 s	21 °C
27.8 °C	13.0 s	28 °C
30.0 °C	12 s	30. °C

This rule of thumb gives excellent agreement to two significant figures.

80.   a. $MoCl_5^-$

b.   Rate = $\dfrac{d[NO_2^-]}{dt}$ = $k_2[NO_3^-][MoCl_5^-]$; Apply the steady-state approximation to $MoCl_5^-$.

$$\frac{d[MoCl_5^-]}{dt} = 0, \text{ so } k_1[MoCl_6^{2-}] = k_{-1}[MoCl_5^-][Cl^-] + k_2[NO_3^-][MoCl_5^-]$$

$$[MoCl_5^-] = \frac{k_1[MoCl_6^{2-}]}{k_{-1}[Cl^-] + k_2[NO_3^-]}; \quad Rate = \frac{d[NO_2^-]}{dt} = \frac{k_1k_2[NO_3^-][MoCl_6^{2-}]}{k_{-1}[Cl^-] + k_2[NO_3^-]}$$

81. $$Rate = \frac{-d[O_3]}{dt} = k_1[M][O_3] + k_2[O][O_3] - k_{-1}[M][O_2][O]; \quad \text{Apply steady-state approx. to O:}$$

$$\frac{d[O]}{dt} = 0, \text{ so } k_1[M][O_3] = k_{-1}[M][O_2][O] + k_2[O][O_3], \quad k_1[M][O_3] - k_{-1}[M][O_2][O] = k_2[O][O_3]$$

Substitute this expression into the rate law: $$Rate = \frac{-d[O_3]}{dt} = 2\,k_2[O][O_3]$$

Rearranging the steady-state approx. for [O]: $$[O] = \frac{k_1[O_3][M]}{k_{-1}[M][O_2] + k_2[O_3]}$$

Substituting into the rate law: $$Rate = \frac{-d[O_3]}{dt} = \frac{2\,k_2k_1[O_3]^2[M]}{k_{-1}[M][O_2] + k_2[O_3]}$$

82. $$Rate = \frac{-d[N_2O_5]}{dt} = k_1[M][N_2O_5] - k_{-1}[NO_3][NO_2][M]$$

Assume $d[NO_3]/dt = 0$, so $k_1[N_2O_5][M] = k_{-1}[NO_3][NO_2][M] + k_2[NO_3][NO_2] + k_3[NO_3][NO]$

$$[NO_3] = \frac{k_1[N_2O_5][M]}{k_{-1}[NO_2][M] + k_2[NO_2] + k_3[NO]}$$

Assume $\dfrac{d[NO]}{dt} = 0$, so $k_2[NO_3][NO_2] = k_3[NO_3][NO]$, $[NO] = \dfrac{k_2}{k_3}[NO_2]$

Substituting: $$[NO_3] = \frac{k_1[N_2O_5][M]}{k_{-1}[NO_2][M] + k_2[NO_2] + \dfrac{k_3k_2}{k_3}[NO_2]} = \frac{k_1[N_2O_5][M]}{[NO_2]\,(k_{-1}[M] + 2\,k_2)}$$

Solving for the rate law:

$$Rate = \frac{-d[N_2O_5]}{dt} = k_1[N_2O_5][M] - \frac{k_{-1}k_1[NO_2][N_2O_5][M]^2}{[NO_2]\,(k_{-1}[M] + 2\,k_2)} = k_1[N_2O_5][M] - \frac{k_{-1}k_1[M]^2[N_2O_5]}{k_{-1}[M] + 2\,k_2}$$

$$Rate = \frac{-d[N_2O_5]}{dt} = \left(k_1 - \frac{k_{-1}k_1[M]}{k_{-1}[M] + 2\,k_2} \right)[N_2O_5][M]; \quad \text{Simplifying:}$$

$$Rate = \frac{-d[N_2O_5]}{dt} = \frac{2\,k_1k_2[M][N_2O_5]}{k_{-1}[M] + 2\,k_2}$$

83.   Rate = $\dfrac{d[Cl_2]}{dt}$ = k$_2$[NO$_2$Cl][Cl]; Assume $\dfrac{d[Cl]}{dt}$ = 0, then:

k$_1$[NO$_2$Cl] = k$_{-1}$[NO$_2$][Cl] + k$_2$[NO$_2$Cl][Cl], [Cl] = $\dfrac{k_1[NO_2Cl]}{k_{-1}[NO_2] + k_2[NO_2Cl]}$

Rate = $\dfrac{d[Cl_2]}{dt}$ = $\dfrac{k_1k_2[NO_2Cl]^2}{k_{-1}[NO_2] + k_2\,[NO_2Cl]}$

84.   Assume A is the same for the catalyzed and the uncatalyzed reaction.

$\dfrac{k_{un}}{k_{cat}} = \dfrac{A\,\exp[-E_a(un)/RT]}{A\,\exp[-E_a(cat)/RT]} = \exp\left(\dfrac{E_a(cat) - E_a(un)}{RT}\right) = \exp\left(\dfrac{-2.50 \times 10^4\ J/mol}{RT}\right)$

$\dfrac{k_{un}}{k_{cat}} = \exp\left(\dfrac{-2.50 \times 10^4\ J/mol}{8.3145\ J\ K^{-1}\ mol^{-1} \times 298\ K}\right) = 4.15 \times 10^{-5}$ at 25°C

$\dfrac{k_{un}}{k_{cat}} = \exp\left(\dfrac{-2.50 \times 10^4}{8.3145 \times 523}\right) = 3.19 \times 10^{-3}$ at 250.°C

85.   a.   rate = $\dfrac{d[E]}{dt}$ = k$_2$[B*]; Assume $\dfrac{d[B^*]}{dt}$ = 0, then k$_1$[B]^2 = k$_{-1}$[B][B*] + k$_2$[B*]

[B*] = $\dfrac{k_1[B]^2}{k_{-1}[B] + k_2}$; The rate law is: Rate = $\dfrac{d[E]}{dt}$ = $\dfrac{k_1k_2[B]^2}{k_{-1}[B] + k_2}$

b.   When k$_2$ << k$_{-1}$[B], then Rate = $\dfrac{d[E]}{dt}$ = $\dfrac{k_1k_2[B]^2}{k_{-1}[B]}$ = $\dfrac{k_1k_2}{k_{-1}}$[B]; Reaction is first order.

c.   Collisions between B molecules only transfer energy from one B to another. This occurs at a much faster rate than the decomposition of an energetic B molecule (B*).

86.   a.   Experiments 1 and 2: [H$_2$O$_2$] doubles, [I$^-$] is constant, and the rate doubles; First order in [H$_2$O$_2$]

Experiments 1 and 3: [I$^-$] doubles, [H$_2$O$_2$] is constant, and the rate doubles; First order in [I$^-$]

Rate = k[H$_2$O$_2$][I$^-$]

b.   7.0 × 10^{-4} mol L^{-1} min^{-1} = k(0.10 mol/L) (0.10 mol/L), k = 7.0 × 10^{-2} L mol^{-1} min^{-1} = k$_{mean}$

c.   Rate = $\dfrac{7.0 \times 10^{-2}\ L}{mol\ min} \times \dfrac{0.50\ mol}{L} \times \dfrac{0.25\ mol}{L}$ = 8.8 × 10^{-3} mol L^{-1} min^{-1} = $\dfrac{d[I_3^-]}{dt}$

$$0.50 \text{ L} \times \frac{8.8 \times 10^{-3} \text{ mol}}{\text{L min}} = \frac{4.4 \times 10^{-3} \text{ mol I}_3^-}{\text{min}}$$

87.   a.   Rate = $(k_1 + k_2[H^+])[I^-]^m[H_2O_2]^n$

In all the experiments, the concentration of H_2O_2 is small compared to the concentrations of I^- and H^+. Therefore, the concentrations of I^- and H^+ are effectively constant and the rate law reduces to:

Rate = $k_{obs}[H_2O_2]^n$ where $k_{obs} = (k_1 + k_2[H^+])[I^-]^m$

Since all plots of $\ln[H_2O_2]$ vs. time are linear, the reaction is first order with respect to H_2O_2 (n = 1). The slopes of the $\ln[H_2O_2]$ vs. time plots equal $-k_{obs}$ which equals $(k_1 + k_2[H^+])[I^-]^m$. To determine the order of I^-, compare the slopes of two experiments where I^- changes and H^+ is constant. Comparing the first two experiments:

$$\frac{\text{slope (exp. 2)}}{\text{slope (exp. 1)}} = \frac{-0.360}{-0.120} = \frac{-[k_1 + k_2\,(0.0400\ M)]\,(0.3000\ M)^m}{-[k_1 + k_2\,(0.0400\ M)]\,(0.1000\ M)^m},$$

$$3.00 = \left(\frac{0.3000}{0.1000}\right)^m = (3.000)^m, \ m = 1$$

The reaction is also first order with respect to I^-.

b.   The slope equation has two unknowns, k_1 and k_2. To solve for k_1 and k_2, we must have two equations. We need to take one of the first set of three experiments and one of the 2nd set of three experiments to generate the two equations in k_1 and k_2.

Experiment 1: slope = $-(k_1 + k_2[H^+])[I^-]$

-0.120 min^{-1} = $-[k_1 + k_2\,(0.0400\ M)]\,0.1000\ M$ or $1.20 = k_1 + k_2\,(0.0400)$

Experiment 4:

-0.0760 min^{-1} = $-[k_1 + k_2\,(0.0200\ M)]\,0.0750\ M$ or $1.01 = k_1 + k_2\,(0.0200)$

Subtracting 4 from 1:

$$1.20 = k_1 + k_2\,(0.0400)$$
$$-1.01 = -k_1 - k_2\,(0.0200)$$
$$\overline{}$$
$$0.19 = k_2\,(0.0200), \ k_2 = 9.5 \text{ L}^2 \text{ mol}^{-2} \text{ min}^{-1}$$

$1.20 = k_1 + 9.5(0.0400), \ k_1 = 0.82 \text{ L mol}^{-1} \text{ min}^{-1}$

c.   There are two pathways, one involving H^+ with rate = $k_2[H^+][I^-][H_2O_2]$ and another pathway not involving H^+ with rate = $k_1[I^-][H_2O_2]$. The overall rate of reaction depends on which pathway dominates which depends on the H^+ concentration.

88.   Rate = $\dfrac{d[P]}{dt}$ = $k_2[ES]$; Apply the steady state approximation to ES, $\dfrac{d[ES]}{dt}$ = 0.

$k_1[E][S] = k_{-1}[ES] + k_2[ES]$; $[E]_T = [E] + [ES]$, so $[E] = [E]_T - [ES]$

Substituting: $k_1[S]([E]_T - [ES]) = (k_{-1} + k_2)[ES]$, $k_1[E]_T[S] = (k_{-1} + k_2 + k_1[S])[ES]$

$[ES] = \dfrac{k_1[E]_T[S]}{k_{-1} + k_2 + k_1[S]}$; Substituting into the rate equation, Rate = $k_2[ES]$:

$$\text{Rate} = \dfrac{d[P]}{dt} = \dfrac{k_1k_2[E]_T[S]}{k_{-1} + k_2 + k_1[S]}$$

CHAPTER SIXTEEN

LIQUIDS AND SOLIDS

Intermolecular Forces and Physical Properties

1. There is an electrostatic attraction between the permanent dipoles of the polar molecules. The greater the polarity, the greater the attraction among molecules.

2. London dispersion (LDF) < dipole-dipole < H-bonding < metallic bonding, covalent network, ionic.

   Yes, there is considerable overlap. Consider some of the examples in Exercise 16.12. Benzene (only LDF) has a higher boiling point than acetone (dipole-dipole). Also, there is even more overlap of the stronger forces (metallic, covalent, and ionic).

3. As the size of the molecule increases, the strength of the London dispersion forces also increases. As the electron cloud gets larger it is easier for the electrons to be drawn away from the nucleus (more polarizable).

4. Yes, there are some substances in which only dispersion forces are present such as naphthalene, $C_{10}H_{18}$, and polyethylene that are solids at room temperature. Since these substances are solids at room temperature, then their interparticle forces are stronger than those that are liquids at room temperature, such as water in which there are hydrogen bonds.

5. As the strengths of interparticle forces increase: surface tension, viscosity, melting point, and boiling point increase, while the vapor pressure decreases.

6. The nature of the forces stays the same. As the temperature increases and the phase changes, solid → liquid → gas, occur, a greater fraction of the forces are overcome by the increased thermal (kinetic) energy of the particles.

7. a. ionic   b. LDF, dipole   c. LDF only

   d. LDF mostly; For all practical purposes, we consider a C – H bond to be nonpolar even though there is a small difference in electronegativity.

e.   metallic   f.   metallic   g.   LDF only

h.   H-bonding, LDF   i.   ionic   j.   metallic

k.   LDF mostly; C – F bonds are polar, but polymers like teflon are so large the LDF are the predominant interparticle forces.

l.   LDF   m. LDF, dipole   n.   covalent network

o.   LDF, dipole   p.   LDF

8.   a. OCS: OCS is polar and has dipole-dipole forces in addition to LDF. All polar molecules have dipole forces. CO_2 is nonpolar and only has LD forces. In all of the following (b-d), only one molecule is polar and, in turn, has dipole-dipole forces. To predict polarity, draw the Lewis structure and deduce if the individual bond dipoles cancel.

     b. PF_3 is polar (PF_5 is nonpolar).   c. SF_2 is polar (SF_6 is nonpolar).

     d. SO_2 is polar (SO_3 is nonpolar).

9.   Dipole forces are generally weaker than hydrogen bonding. They are similar in that they arise from an unequal sharing of electrons. We can look at hydrogen bonding as a particularly strong dipole force.

10. $H_2NCH_2CH_2NH_2$; More extensive hydrogen bonding is possible.

     b. $B(OH)_3$; No hydrogen bonding in BH_3. For hydrogen bonding to occur, an H-atom must be covalently bonded to O, N, or F.

     c. CH_3OH d.   HF

11. a. Neopentane is more compact than n-pentane. There is less surface area contact between neopentane molecules. This leads to weaker LD forces and a lower boiling point.

     b. Ethanol is capable of H-bonding, dimethylether is not.

     c. HF is capable of H-bonding.

     d. LiCl is ionic. Ionic forces are much stronger than the intermolecular forces present in molecular solids. $TiCl_4$ is a nonpolar molecular substance with relatively weak London dispersion forces.

     e. LiCl is ionic and HCl is a molecular solid with only dipole forces and LD forces between HCl molecules. Ionic forces are much stronger than molecular forces.

     f. Both LiCl and CsCl are ionic. The lattice energy of LiCl is greater than the lattice energy of CsCl because the Li^+ ion is smaller than the Cs^+ ion. Thus, the forces are stronger in LiCl.

12.

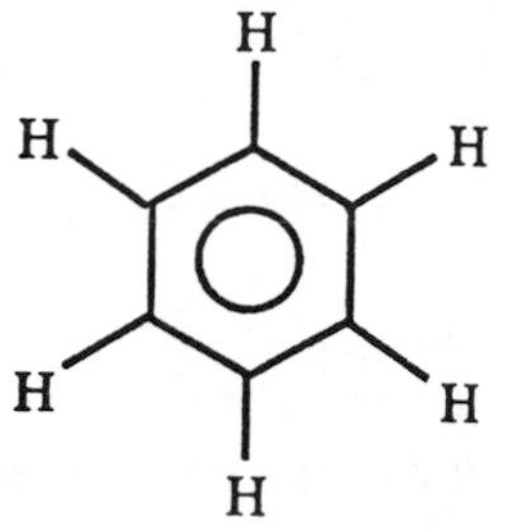

LDF

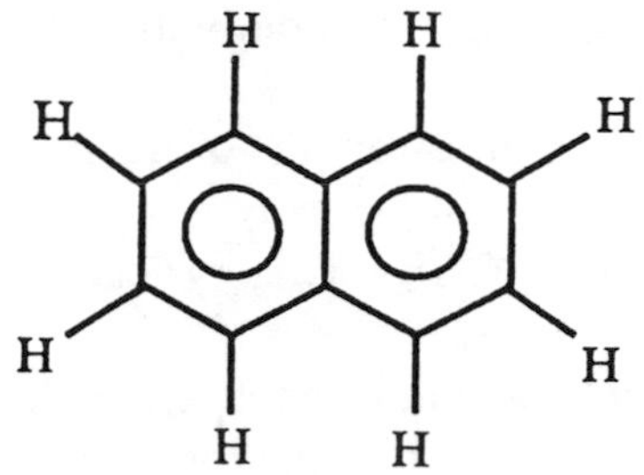

LDF

Note: LDF in molecules like benzene and naphthalene are fairly large. The molecules are flat and there is efficient surface area contact between molecules. Large surface area contact leads to stronger London dispersion forces.

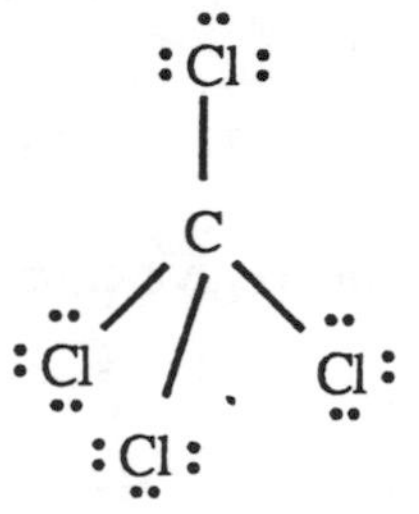

LDF (polar bonds but a nonpolar molecule)

In terms of size and shape: $CCl_4 < C_6H_6 < C_{10}H_8$

The strengths of the LDF are proportional to size and are related to shape. Although the size of CCl_4 is fairly large, the overall spherical shape gives rise to relatively weak LD forces as compared to flat molecules like benzene and and naphthalene. The physical properties are consistent with the order listed above. Each of the physical properties will increase with an increase in interparticle forces.

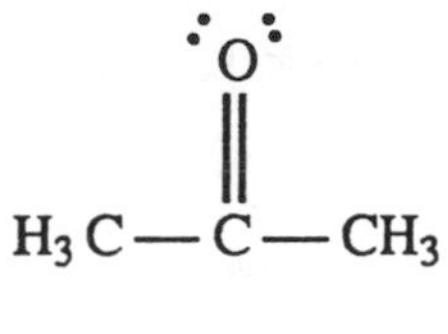

LDF, dipole

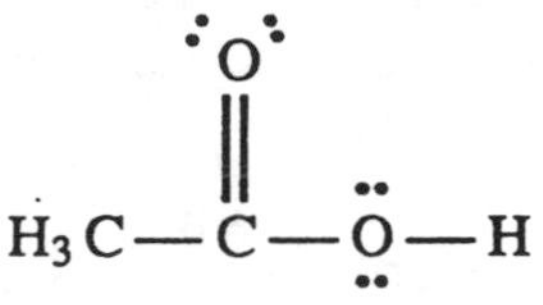

LDF, dipole, H-bonding

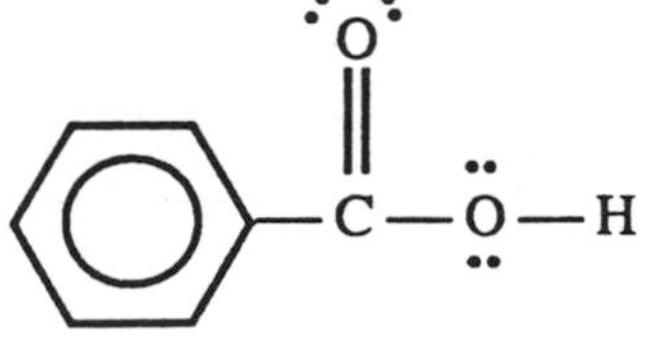

LDF, dipole, H-bonding

We would predict the strength of interparticle forces of the last three molecules to be:

acetone < acetic acid < benzoic acid

polar   H-bonding   H-bonding, but large LDF because of greater size and shape.

This ordering is consistent with the values given for bp, mp, and ΔH_{vap}.

The order of the strengths of interparticle forces based on physical properties are:

$$acetone < CCl_4 < C_6H_6 < acetic\ acid < naphthalene < benzoic\ acid$$

The order seems reasonable except for acetone and napthalene. Since acetone is polar, we would not expect it to boil at the lowest temperature. However, in terms of size and shape, acetone is the smallest molecule and the LDF in acetone must be very small compared to the other molecules. Napthalene must have very strong LDF because of its size and flat shape.

13.   As the electronegativity of the atoms bonded to H in a hydrogen bond increases, the strength of the hydrogen bond increases.

$$N\ \text{-}\text{-}\text{-}\ H\text{-}N\ <\ N\ \text{-}\text{-}\text{-}\ H\text{-}O\ <\ O\ \text{-}\text{-}\text{-}\ H\text{-}O\ <\ O\ \text{-}\text{-}\text{-}\ H\text{-}F\ <\ F\ \text{-}\text{-}\text{-}\ H\text{-}F$$

    weakest   strongest

14. FHF^- can be thought of as forming from a HF molecule and a F^- ion to give equal $F-H$ distances. The structure will be linear as predicted by VSEPR.

                                            linear

15.   Only a fraction of the hydrogen bonds are broken in going from the solid phase to the liquid phase. Most of the hydrogen bonds are still present in the liquid phase and must be broken during the liquid to gas phase transition.

16.   A single hydrogen bond in H_2O has a strength of 21 kJ/mol. Each H_2O molecule forms two H-bonds. Thus, it should take 42 kJ/mol of energy to break all of the H-bonds in water. For the phase transitions:

$$\text{solid}\ \xrightarrow{\ 6.0\ kJ\ }\ \text{liquid}\ \xrightarrow{\ 40.7\ kJ\ }\ \text{vapor}\qquad \Delta H_{sub} = \Delta H_{fus} + \Delta H_{vap}$$

it takes a total of 46.7 kJ/mol (ΔH_{sub}). This would be the amount of energy necessary to disrupt all of the interparticle forces in ice. Thus, $(42 \div 46.7) \times 100 = 90\%$ of the attraction in ice can be attributed to H-bonding.

17.   Both molecules are capable of H-bonding. However, in oil of wintergreen the hydrogen bonding is <u>intra</u>molecular.

In methyl-4-hydroxybenzoate, the H-bonding is <u>inter</u>molecular, resulting in stronger intermolecular forces and a higher melting point.

18. $NaCl$, $MgCl_2$, NaF, MgF_2 AlF_3 all have very high melting points indicative of strong interparticle forces. They are all ionic solids. $SiCl_4$, SiF_4, F_2, Cl_2, PF_5 and SF_6 are nonpolar covalent molecules. Only LDF are present. PCl_3 and SCl_2 are polar molecules. LDF and dipole forces are present. In these 8 molecular substances, the interparticle forces are weak and the melting points low. $AlCl_3$ doesn't seem to fit in as well. From the melting point, there are much stronger forces present than in the nonmetal halides, but they aren't as strong as we would expect for an ionic solid. $AlCl_3$ illustrates a gradual transition from ionic to covalent bonding; from an ionic solid to discrete molecules.

19.   a. $NaCl$; strong ionic bonding in lattice

     b. H_2; nonpolar like CH_4, smaller than CH_4, weaker LDF

     c. SiO_2; covalent network solid vs. a gas and a liquid

     d. $HOCH_2CH_2OH$ (ethylene glycol); Greatest amount of H-bonding since two –OH groups are present.

     e. NH_3: Need N – H, H – F, or O – H for hydrogen bonding. NH_3 can form H-bonds.

     f. H_2O; Water forms H-bonding interactions, others do not.

     g. CO_2; nonpolar molecular substance, weakest intermolecular forces

20. $C_{25}H_{52}$ has the stronger intermolecular forces because it has the higher boiling point. Even though $C_{25}H_{52}$ is nonpolar, it is so large that its London dispersion forces are much stronger than the sum of the London dispersion and hydrogen bonding interactions found in H_2O.

21.   If TiO_2 conducts electricity as a liquid it would be ionic. The electronegativity difference is 3.5 - 1.5 = 2.0. From this difference in electronegativities we might predict TiO_2 to be ionic. It is actually a covalent network solid. 1.5 is the electronegativity value for Ti^{2+}. Ti^{4+} should have a higher electronegativity. See Exercise 13.79.

22.   a.   Polarizability of an atom refers to the ease of distorting the electron cloud. It can also refer to distorting the electron clouds in molecules or ions. Polarity refers to the presence of a permanent dipole moment in a molecule.

     b.   London dispersion forces are present in all substances. LDF can be referred to as accidental dipole - induced dipole forces. Dipole - dipole forces involve the attraction of molecules with permanent dipoles for each other.

     c.   inter: between; intra: within

For example, in H_2 the covalent bond is an intramolecular force, holding the two H-atoms together in the molecule. The much weaker London dispersion forces are the intermolecular forces of attraction.

23.   1.   We can draw a reasonable Lewis structure:

$$:\!\overset{\displaystyle ..}{\underset{\displaystyle ..}{Cl}}\!:$$

$$:\!\overset{\displaystyle ..}{\underset{\displaystyle ..}{Cl}}\!-Ti-\overset{\displaystyle ..}{\underset{\displaystyle ..}{Cl}}\!:$$

$$:\!\overset{\displaystyle ..}{\underset{\displaystyle ..}{Cl}}\!:$$

2.   If we start with Ti^{4+} ions and Cl^- anions, the Ti^{4+} will have a strong attraction for the electrons on Cl^- and will want to share those electrons.

3.   Consider the cases:

    $H^+ + Cl^- \rightarrow H - Cl$ $Ti^{4+} + 4\,Cl^- \rightarrow TiCl_4$

    empty   covalent bond   empty 4s, 3d,   covalent bonds
    1s orbital   4p orbitals

    They really aren't so very different in how the covalent bonds form. A pair of electrons on a chloride ion form a bond using empty orbitals on the cation.

Properties of Liquids

24.   Liquids and solids both have characteristic volume and are not very compressible. Liquids and gases flow and assume the shape of their container.

25.   Critical temperature: The temperature above which a liquid cannot exist, i.e., the gas cannot be liquified by increased pressure.

    Critical pressure: The pressure that must be applied to a substance at its critical temperature to produce a liquid.

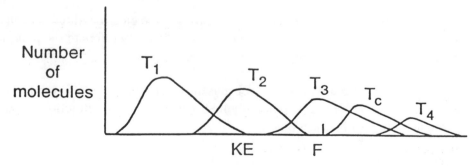

The kinetic energy distribution changes as one raises the temperature ($T_4 > T_c > T_3 > T_2 > T_1$). At the critical temperature, T_c, all molecules have kinetic energies greater than the interparticle forces, F, and a liquid can't form.

26.   As the interparticle forces increase the critical temperature increases.

27.   When a liquid evaporates, the molecules that escape have higher kinetic energies. The average kinetic energy of the remaining molecules is lower, thus, the temperature of the liquid is lower.

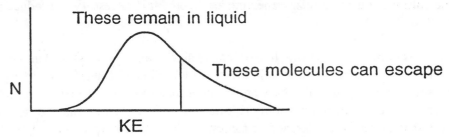

28.   The attraction of H_2O for glass is stronger than the $H_2O - H_2O$ attraction. The miniscus is concave to increase the area of contact between glass and H_2O. The $Hg - Hg$ attraction is greater than the $Hg -$ glass attraction. The miniscus is convex to minimize the $Hg -$ glass contact. Polyethylene is a nonpolar substance. The $H_2O - H_2O$ attraction is stronger than the $H_2O -$ polyethylene attraction. Thus, the miniscus will have a convex shape.

29. H_2O will rise higher in a glass tube because of a greater attraction for glass than polyethylene.

30.   The interparticle attractions are slightly stronger in D_2O than H_2O.

31.   The structure of H_2O_2 is $H - O - O - H$, which produces greater hydrogen bonding than water. Long chains of hydrogen bonded H_2O_2 molecules then get tangled together.

32. CO_2 is a gas at room temperature. Strengths of forces: $CO_2 < CS_2 < CSe_2$; As mp and bp increase, the strength of the interparticle forces increases. The above order is the same as the order of mp and bp. From a structural standpoint this is expected. All three are linear, nonpolar molecules. Thus, only London dispersion forces are present. Since the molecules increase in size from $CO_2 < CS_2 < CSe_2$, the interparticle forces will be in the same order.

Structures and Properties of Solids

33.   a. Crystalline solid: Regular, repeating structure

          Amorphous solid: Irregular arrangement of atoms or molecules

      b. Ionic solid: Made up of ions held together by ionic bonding.

          Molecular solid: Made up of discrete covalently bonded molecules held together in the solid by weaker forces (LDF, dipole, or hydrogen bonds).

      c. Molecular solid: Discrete molecules

          Covalent network solid: No discrete molecules; A covalent network solid is one large molecule. The interparticle forces are the covalent bonds between atoms.

d. Metallic solid: Completely delocalized electrons, conductor of electricity (ions in a sea of electrons)

Covalent network: Localized electrons, insulator or semi-conductor

34. A crystalline solid because a regular, repeating arrangement is necessary to produce planes of atoms that will diffract the x-rays.

35. No, an example is common glass which is primarily amorphous SiO_2 (a covalent network solid) as compared to ice (a crystalline solid held together by weaker H-bonds). The interparticle forces in the amorphous solid in this case are stronger than those in the crystalline solid. Whether or not a solid is amorphous or crystalline depends on the long range order in the solid and not on the strengths of the interparticle forces.

36. a. CO_2: molecular   b. SiO_2: covalent network   c. Si: atomic, covalent network

d. CH_4: molecular   e. Ru: atomic, metallic   f. I_2: molecular

g. KBr: ionic   h. H_2O: molecular   i. NaOH: ionic

j. U: atomic, metallic   k. $CaCO_3$: ionic   l. PH_3: molecular

m. GaAs: covalent network   n. BaO: ionic   o. NO: molecular

p. GeO_2: ionic

37. $n\lambda = 2d \sin\theta$, $\lambda = \dfrac{2d \sin\theta}{n} = \dfrac{2 \times (201 \times 10^{-12} \text{ m}) \sin 34.68°}{1}$, $\lambda = 2.29 \times 10^{-10}$ m = 229 pm

38. $n\lambda = 2d \sin\theta$, $\sin\theta = \dfrac{n\lambda}{2d} = \dfrac{1 \times 0.712 \text{ Å}}{19.93 \text{ Å}} = 0.0357$, $\theta = \sin^{-1} 0.0357 = 2.05°$

39. $n\lambda = 2d \sin\theta$; $1.54 \text{ Å} = 1.54 \times 10^{-10}$ m = 154 pm

$d = \dfrac{n\lambda}{2 \sin\theta} = \dfrac{1 \times 154 \text{ pm}}{2 \times \sin 14.22°} = 313$ pm = 3.13×10^{-10} m = 3.13 Å

40.

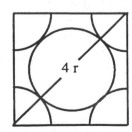

3.92×10^{-8} cm = length of cube edge = l

4r = length of diagonal

From the Pythagorean theorem:

$$(4r)^2 = l^2 + l^2 = (3.92 \times 10^{-8})^2 + (3.92 \times 10^{-8})^2, \quad r = 1.39 \times 10^{-8} \text{ cm} = 139 \text{ pm} = 1.39 \text{ Å}$$

Note: For all face-centered unit cells, the radius and the cube edge length are related by $4r = \sqrt{2} \, l$

In a face centered cubic unit cell: $8 \text{ corners} \times \dfrac{1/8 \text{ atom}}{\text{corner}} + 6 \text{ faces} \times \dfrac{1/2 \text{ atom}}{\text{face}} = 4 \text{ atoms}$

$\text{density} = \dfrac{\text{mass}}{\text{volume}}$; For a unit cell where AM = atomic mass:

$$\text{mass} = 4 \text{ atoms} \times \dfrac{1 \text{ mol}}{6.022 \times 10^{23} \text{ atoms}} \times \dfrac{\text{AM g}}{\text{mol}}; \quad \text{volume} = (\text{edge})^3 = (3.92 \times 10^{-8} \text{ cm})^3$$

$$\text{density} = 21.45 \text{ g/cm}^3 = \dfrac{4 \times \text{AM} \times \dfrac{1}{6.022 \times 10^{23}}}{(3.92 \times 10^{-8} \text{ cm})^3}; \quad \text{Solving:}$$

AM = 194.5 g/mol ≈ 195 g/mol; The metal is platinum.

41.   A cubic closest packed structure has a face centered cubic unit cell. A face centered cubic unit cell will contain 4 Co atoms. The length of the face diagonal is related to the radius of Co by $\sqrt{2} \, l = 4r$, where l is the cube edge length. See solution to Exercise 16.40 for details.

$$l = 4r/\sqrt{2} - 4(1.25 \times 10^{-8} \text{ cm})/\sqrt{2} = 3.54 \times 10^{-8} \text{ cm}$$

For a unit cell of the β form:

$$d = \dfrac{\text{mass}}{\text{volume}} = \dfrac{4 \text{ atoms} \times \dfrac{1 \text{ mol}}{6.022 \times 10^{23} \text{ atoms}} \times \dfrac{58.93 \text{ g}}{\text{mol}}}{(3.54 \times 10^{-8} \text{ cm})^3} = \dfrac{3.91 \times 10^{-22} \text{ g}}{4.44 \times 10^{-23} \text{ cm}^3} = 8.81 \text{ g/cm}^3$$

There appears to be a slight difference in density between the two forms.

42.   There are 4 Ir atoms in the unit cell. (See Exercise 16.40.)

$$22.61 \text{ g/cm}^3 = \dfrac{4 \text{ atoms} \times \dfrac{1 \text{ mol}}{N_o \text{ atoms}} \times \dfrac{192.2 \text{ g}}{\text{mol}}}{(3.833 \times 10^{-8} \text{ cm})^3}, \quad N_o = \dfrac{4 \times 192.2}{22.61 \times (3.833 \times 10^{-8})^3} = 6.038 \times 10^{23}$$

43.   Body centered unit cell: $8 \text{ corners} \times \dfrac{1/8 \text{ Ti}}{\text{corner}} + \text{Ti at body center} = 2 \text{ Ti atoms}$

All body centered unit cells have 2 atoms per unit cell. For a unit cell:

$$d = 4.50 \text{ g/cm}^3 = \dfrac{2 \text{ atoms} \times \dfrac{1 \text{ mol}}{6.022 \times 10^{23} \text{ atoms}} \times \dfrac{47.88 \text{ g}}{\text{mol}}}{l^3}, \quad l = \text{cube edge length}$$

Solving: l = edge length of unit cell = 3.28×10^{-8} cm = 328 pm

Assume Ti atoms just touch along the body diagonal of the cube, so body diagonal = 4 r.

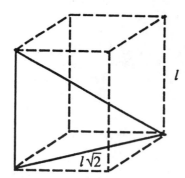

The triangle we need to solve is:

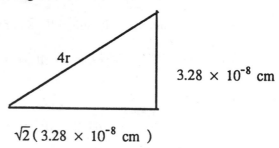

$(4r)^2 = (3.28 \times 10^{-8}$ cm$)^2 + [\sqrt{2}(3.28 \times 10^{-8}$ cm$)]^2$, $r = 1.42 \times 10^{-8}$ cm = 142 pm = 1.42 Å

For a body centered unit cell, the radius of the atom is related to the cube edge length by $4r = \sqrt{3}\, l$.

44.   From Exercise 16.43:

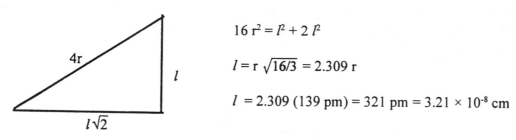

$16\, r^2 = l^2 + 2\, l^2$

$l = r\, \sqrt{16/3} = 2.309\, r$

$l = 2.309\, (139$ pm$) = 321$ pm = 3.21×10^{-8} cm

In bcc, there are 2 atoms/unit cell. For a unit cell:

$$d = \frac{\text{mass}}{\text{volume}} = \frac{2 \text{ atoms} \times \dfrac{1 \text{ mol}}{6.022 \times 10^{23} \text{ atoms}} \times \dfrac{183.9 \text{ g}}{\text{mol}}}{(3.21 \times 10^{-8} \text{ cm})^3} = \frac{18.5 \text{ g}}{\text{cm}^3}$$

45.   There are 4 Ni atoms in each unit cell: For a unit cell:

$$\text{density} = \frac{\text{mass}}{\text{volume}} = 6.84 \text{ g/cm}^3 = \frac{4 \text{ atoms} \times \dfrac{1 \text{ mol}}{6.022 \times 10^{23} \text{ atoms}} \times \dfrac{58.69 \text{ g}}{\text{mol}}}{l^3}$$

$l = 3.85 \times 10^{-8}$ cm = cube edge length

For a face centered cube:

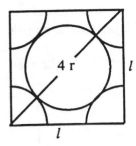

$$(4r)^2 = l^2 + l^2 = 2\ l^2$$

$$16\ r^2 = 2(3.85 \times 10^{-8}\ \text{cm})^2$$

$$r = 1.36 \times 10^{-8}\ \text{cm} = 1.36\ \text{Å} = 136\ \text{pm}$$

46.   If fcc structure: 4 atoms/unit cell

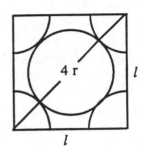

$$2\ l^2 = 16\ r^2$$

$$l = r\sqrt{8} = 144\ \text{pm}\ \sqrt{8} = 407\ \text{pm}$$

$$l = 407 \times 10^{-12}\ \text{m} = 407 \times 10^{-10}\ \text{cm}$$

$$d = \frac{4\ \text{atoms} \times \dfrac{1\ \text{mol}}{6.022 \times 10^{23}\ \text{atoms}} \times \dfrac{197.0\ \text{g}}{\text{mol}}}{(407 \times 10^{-10}\ \text{cm})^3} = \frac{19.4\ \text{g}}{\text{cm}^3}$$

If bcc: 2 atoms/unit cell

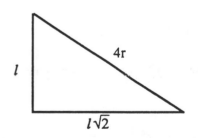

$$16\ r^2 = l^2 + 2\ l^2$$

$$l = r\sqrt{16/3} = 333\ \text{pm} = 333 \times 10^{-10}\ \text{cm}$$

$$d = \frac{2\ \text{atoms} \times \dfrac{1\ \text{mol}}{6.022 \times 10^{23}\ \text{atoms}} \times \dfrac{197.0\ \text{g}}{\text{mol}}}{(333 \times 10^{-10}\ \text{cm})^3} = \frac{17.7\ \text{g}}{\text{cm}^3}$$

The measured density is consistent with a face centered cubic unit cell.

47.

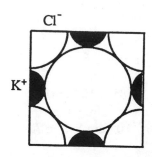

Assuming K^+ and Cl^- just touch along the edge:

$$l = 2(314 \text{ pm}) = 628 \text{ pm} = 6.28 \times 10^{-8} \text{ cm}$$

The unit cell contains 4 K^+ and 4 Cl^- ions. For a unit cell:

$$d = \frac{4 \text{ KCl molecules} \times \dfrac{1 \text{ mol KCl}}{6.022 \times 10^{23} \text{ molecules}} \times \dfrac{74.55 \text{ g KCl}}{\text{mol KCl}}}{(6.28 \times 10^{-8} \text{ cm})^3} = \frac{2.00 \text{ g}}{\text{cm}^3}$$

48.   Since magnesium oxide has the same structure as NaCl, each unit cell contains 4 Mg^{2+} ions and 4 O^{2-} ions. The mass of a unit cell is:

$$4 \text{ MgO molecules} \left(\frac{1 \text{ mol MgO}}{6.022 \times 10^{23} \text{ molecules}} \right) \left(\frac{40.31 \text{ g MgO}}{1 \text{ mol MgO}} \right) = 2.678 \times 10^{-22} \text{ g MgO}$$

$$\text{Volume of unit cell} = 2.678 \times 10^{-22} \text{ g MgO} \left(\frac{1 \text{ cm}^3}{3.58 \text{ g}} \right) = 7.48 \times 10^{-23} \text{ cm}^3$$

Volume of unit cell $= l^3$, l = cube edge length; $l = (7.48 \times 10^{-23} \text{ cm}^3)^{1/3} = 4.21 \times 10^{-8} \text{ cm} = 421 \text{ pm}$

From the NaCl structure:

$$l = 2 \, r_{Mg^{2+}} + 2 \, r_{O^{2-}} = 2 \, (65 \text{ pm}) + 2 \, (140. \text{ pm}) = 410. \text{ pm}$$

The two values agree within 3%. In the actual crystals the Mg^{2+} and O^{2-} ions may not touch which is assumed in calculating the 410. pm value.

49.   a. $8 \text{ corners} \times \dfrac{1/8 \text{ Xe}}{\text{corner}} + 1 \text{ Xe inside cell} = 2 \text{ Xe}; \quad 8 \text{ edges} \times \dfrac{1/4 \text{ F}}{\text{edges}} + 2 \text{ F inside cell} = 4 \text{ F}$

Empirical formula is XeF_2. This is also the molecular formula.

b.   For a unit cell:

$$\text{mass} = 2 \text{ XeF}_2 \text{ molecules} \times \frac{1 \text{ mol XeF}_2}{6.022 \times 10^{23} \text{ molecules}} \times \frac{169.3 \text{ g XeF}_2}{\text{mol XeF}_2} = 5.62 \times 10^{-22} \text{ g}$$

$$\text{volume} = (702 \times 10^{-12} \text{ m}) \, (432 \times 10^{-12} \text{ m}) \, (432 \times 10^{-12} \text{ m}) = 1.31 \times 10^{-28} \text{ m}^3$$

$$\text{volume} = 1.31 \times 10^{-28} \text{ m}^3 \times \left(\frac{100 \text{ cm}}{\text{m}}\right)^3 = 1.31 \times 10^{-22} \text{ cm}^3$$

$$\text{density} = d = \frac{\text{mass}}{\text{volume}} = \frac{5.62 \times 10^{-22} \text{ g}}{1.31 \times 10^{-22} \text{ cm}^3} = 4.29 \text{ g/cm}^3$$

50.   Al: 8 corners $\times \dfrac{1/8 \text{ Al}}{\text{corner}} = 1$ Al; Ni: 6 face centers $\times \dfrac{1/2 \text{ Ni}}{\text{face center}} = 3$ Ni

Composition: $AlNi_3$

51.   a.   The unit cell consists of Ni at the cube corners and Ti at the body center, or Ti at the cube corners and Ni at the body center.

       b. $8 \times 1/8 = 1$ atom from corners + 1 atom at body center; Empirical formula = NiTi

       c.   Both have coordination numbers of 8.

52.   a.   1 Ti at body center; 8 corners $\times \dfrac{1/8 \text{ Ca}}{\text{corner}} = 1$ Ca atom

            6 face centers $\times \dfrac{1/2 \text{ oxygen}}{\text{face center}} = 3$ O atoms; Formula = $CaTiO_3$

       b.   The Ti atoms are at the corners of each unit cell and the oxygen atoms are at the center of each edge in the unit cell.

            Since each of the 12 cube edges is shared by 4 unit cells:

            $12 \times 1/4 = 3$ O atoms; $8 \times 1/8 = 1$ Ti atom; 1 Ca at center; Formula = $CaTiO_3$

       c.   Six oxygen atoms surround each Ti atom.

53.   a.   Y: 1 Y in center; Ba: 2 Ba in center

            Cu: 8 corners $\times \dfrac{1/8 \text{ Cu}}{\text{corner}} = 1$ Cu, 8 edges $\times \dfrac{1/4 \text{ Cu}}{\text{edge}} = 2$ Cu, total = 3 Cu atoms

            O: 20 edges $\times \dfrac{1/4 \text{ O}}{\text{edge}} = 5$ oxygen, 8 faces $\times \dfrac{1/2 \text{ O}}{\text{face}} = 4$ oxygen, total = 9 O atoms

            Formula: $YBa_2Cu_3O_9$

       b.   The structure of this superconductor material follows the alternate perovskite structure described in Exercise 52.b. The $YBa_2Cu_3O_9$ structure is three of these cubic perovskite unit cells stacked on top of each other. The oxygen atoms are in the same places, Cu takes the place of Ti, and two Ca are replaced by Ba and one Ca is replaced by Y.

c.   Y, Ba, and Cu are the same. Some oxygen atoms are missing.

$$12 \text{ edges} \times \frac{1/4 \text{ O}}{\text{edge}} = 3 \text{ O}, \quad 8 \text{ faces} \times \frac{1/2 \text{ O}}{\text{face}} = 4 \text{ O}, \quad \text{total} = 7 \text{ O atoms}$$

Superconductor is $YBa_2Cu_3O_7$

54.   a.   Structure i:

Ba: 2 Ba inside unit cell; Tl: $8 \text{ corners} \times \frac{1/8 \text{ Tl}}{\text{corner}} = 1 \text{ Tl}$; Cu: $4 \text{ edges} \times \frac{1/4 \text{ Cu}}{\text{edge}} = 1 \text{ Cu}$

O: $6 \text{ faces} \times \frac{1/2 \text{ O}}{\text{face}} + 8 \text{ edges} \times \frac{1/4 \text{ O}}{\text{edge}} = 5 \text{ O}$

Formula = $TlBa_2CuO_5$

Structure ii:

Tl and Ba are the same as in structure i.

Ca: 1 Ca inside unit cell; Cu: $8 \text{ edges} \times \frac{1/4 \text{ Cu}}{\text{edge}} = 2 \text{ Cu}$

O: $10 \text{ faces} \times \frac{1/2 \text{ O}}{\text{face}} + 8 \text{ edges} \times \frac{1/4 \text{ O}}{\text{edge}} = 7 \text{ O}$

Formula = $TlBa_2CaCu_2O_7$

Structure iii:

Tl and Ba are the same and two Ca are located inside the unit cell.

Cu: $12 \text{ edges} \times \frac{1/4 \text{ Cu}}{\text{edge}} = 3 \text{ Cu}$; O: $14 \text{ faces} \times \frac{1/2 \text{ O}}{\text{face}} + 8 \text{ edges} \times \frac{1/4 \text{ O}}{\text{edge}} = 9 \text{ O}$

Formula: $TlBa_2Ca_2Cu_3O_9$

Structure iv: Following similar calculations, formula = $TlBa_2Ca_3Cu_4O_{11}$

b.   Structure i has one planar sheet of Cu and O atoms and the number increases by one for each of the remaining structures. The order of superconductivity temperature from lowest to highest temperature is: i < ii < iii < iv.

c. $TlBa_2CuO_5$: $3 + 2(2) + x + 5(-2) = 0$, $x = +3$
Only Cu^{3+} is present in each formula unit.

$TlBa_2CaCu_2O_7$: $3 + 2(2) + 2 + 2(x) + 7(-2) = 0$, $x = +5/2$
Each formula unit contains 1 Cu^{2+} and 1 Cu^{3+}.

$TlBa_2Ca_2Cu_3O_9$: $3 + 2(2) + 2(2) + 3(x) + 9(-2) = 0$, $x = +7/3$
Each formula unit contains 2 Cu^{2+} and 1 Cu^{3+}.

$TlBa_2Ca_3Cu_4O_{11}$: $3 + 2(2) + 3(2) + 4(x) + 11(-2) = 0$, $x = +9/4$
Each formula unit contains 3 Cu^{2+} and 1 Cu^{3+}.

d. This superconductor material achieves variable copper oxidation states by varying the numbers of Ca, Cu and O in each unit cell. The mixtures of copper oxidation states is discussed in 54c. The superconductor material in Exercise 53 achieves variable copper oxidation states by omitting oxygen at various sites in the lattice.

55. Conductor: Partially filled valence band; Electrons can move through the valence band (from filled to unfilled orbitals).

Insulator: Filled valence band, large band gap; Electrons cannot move in the valence band and cannot jump to the conduction band.

Semiconductor: Filled valence band, small band gap; Electrons cannot move in the valence band but can jump to the conduction band.

56. a. As the temperature is increased, more electrons in the valence band have sufficient kinetic energy to jump from the valence band to the conduction band.

b. A photon of light is absorbed by an electron which then has sufficient energy to jump from the valence band to the conduction band.

c. An impurity either adds electrons at an energy near that of the conduction band (n-type) or creates holes (empty energy levels) at energies in the valence band (p-type).

57. In has fewer valence electrons than Se, thus, Se doped with In would be a p-type semiconductor.

58. To make a p-type semiconductor we need to dope the material with atoms with fewer valence electrons. The average number of valence electrons is four. We could dope with more of the Group 3A elements or with atoms of Zn or Cd. Cadmium is the most common impurity used to produce p-type GaAs semiconductors. To make a n-type GaAs semiconductor, dope with an excess group 5 element or dope with a Group 6 element such as sulfur.

59. $E = 2.5$ eV $\times (1.6 \times 10^{-19}$ J/eV$) = 4.0 \times 10^{-19}$ J

$$E = \frac{hc}{\lambda}, \ \lambda = \frac{hc}{E} = \frac{(6.63 \times 10^{-34} \text{ J s}) (3.00 \times 10^8 \text{ m/s})}{4.0 \times 10^{-19} \text{ J}} = 5.0 \times 10^{-7} \text{ m} = 5.0 \times 10^2 \text{ nm}$$

60. $$E = \frac{hc}{\lambda} = \frac{(6.626 \times 10^{-34} \text{ J s}) (2.998 \times 10^8 \text{ m/s})}{730. \times 10^{-9} \text{ m}} = 2.72 \times 10^{-19} \text{ J} = \text{energy of band gap}$$

Phase Changes and Phase Diagrams

61.   a.   Condensation: vapor → liquid   b.   Evaporation: liquid → vapor

       c.   Sublimation: solid → vapor

       d.   A supercooled liquid is a liquid which is at a temperature below its freezing point.

62.   Equilibrium: There is no change in composition; the vapor pressure is constant.

       Dynamic: Two processes, vapor → liquid and liquid → vapor, are both occurring with equal
                       rates.

63.   At 100.°C (373 K), the vapor pressure of H_2O is 1.00 atm. For water, $\Delta H_{vap} = 40.7$ kJ/mol.

$$\ln\left(\frac{P_1}{P_2}\right) = \frac{\Delta H_{vap}}{R}\left(\frac{1}{T_2} - \frac{1}{T_1}\right) \text{ or } \ln\left(\frac{P_2}{P_1}\right) = \frac{\Delta H_{vap}}{R}\left(\frac{1}{T_1} - \frac{1}{T_2}\right)$$

$$\ln\left(\frac{P_2}{1.00 \text{ atm}}\right) = \frac{40.7 \times 10^3 \text{ J/mol}}{8.3145 \text{ J K}^{-1}\text{ mol}^{-1}}\left(\frac{1}{373 \text{ K}} - \frac{1}{388 \text{ K}}\right), \ln P_2 = 0.51, \ P_2 = e^{0.51} = 1.7 \text{ atm}$$

$$\ln\left(\frac{3.50}{1.00}\right) = \frac{40.7 \times 10^3 \text{ J/mol}}{8.3145 \text{ J K}^{-1}\text{ mol}^{-1}}\left(\frac{1}{373 \text{ K}} - \frac{1}{T_2}\right) = 1.253, \ 2.56 \times 10^{-4} = \left(\frac{1}{373} - \frac{1}{T_2}\right)$$

$$2.56 \times 10^{-4} = 2.68 \times 10^{-3} - \frac{1}{T_2}, \ \frac{1}{T_2} = 2.42 \times 10^{-3}, \ T_2 = \frac{1}{2.42 \times 10^{-3}} = 413 \text{ K or } 140.°C$$

64. $$\ln\left(\frac{P_2}{1.00}\right) = \frac{40.7 \times 10^3 \text{ J/mol}}{8.3145 \text{ J K}^{-1}\text{ mol}^{-1}}\left(\frac{1}{373 \text{ K}} - \frac{1}{623 \text{ K}}\right), \ \ln P_2 = 5.27, \ P_2 = e^{5.27} = 194 \text{ atm}$$

65. $H_2O(s, -20.°C) \rightarrow H_2O(s, 0°C)$

$$q_1 = s_{ice} \times m \times \Delta T = 2.1 \frac{J}{g\,°C} \times 5.00 \times 10^2 \text{ g} \times 20.°C = 2.1 \times 10^4 \text{ J} = 21 \text{ kJ}$$

$H_2O(s, 0°C) \rightarrow H_2O(l, 0°C)$, $\ q_2 = 5.00 \times 10^2 \text{ g } H_2O \times \frac{1 \text{ mol}}{18.02 \text{ g}} \times \frac{6.01 \text{ kJ}}{\text{mol}} = 167 \text{ kJ}$

$H_2O(l, 0°C) \rightarrow H_2O(l, 100.°C)$, $\ q_3 = 4.2 \frac{J}{g\,°C} \times 5.00 \times 10^2 \text{ g} \times 100.°C = 2.1 \times 10^5 \text{J} = 210 \text{ kJ}$

$H_2O(l, 100.°C) \rightarrow H_2O(g, 100.°C)$, $\ q_4 = 5.00 \times 10^2 \text{ g} \times \frac{1 \text{ mol}}{18.02 \text{ g}} \times \frac{40.7 \text{ kJ}}{\text{mol}} = 1130 \text{ kJ}$

$H_2O(g, 100.°C) \rightarrow H_2O(g, 250.°C)$, $\ q_5 = 2.0 \frac{J}{g\,°C} \times 5.00 \times 10^2 \text{ g} \times 150.°C = 1.5 \times 10^5 \text{ J} = 150 \text{ kJ}$

$q_{total} = q_1 + q_2 + q_3 + q_4 + q_5 = 21 + 167 + 210 + 1130 + 150 = 1680 \text{ kJ}$

66.   To melt the 10.0 g of ice at 0°C: $10.0 \text{ g} \times \dfrac{1 \text{ mol}}{18.02 \text{ g}} \times \dfrac{6.01 \text{ kJ}}{\text{mol}} = 3.34 \text{ kJ} = 3340 \text{ J}$

Only 850 J of heat are added. Some, but not all of the ice will melt and the temperature will still be 0°C.

67. $1.00 \text{ lb} \times \dfrac{454 \text{ g}}{\text{lb}} = 454 \text{ g } H_2O$; A change of 1.00°F is equal to a change of 5/9°C.

The amount of heat in J in 1 Btu is: $\dfrac{4.18 \text{ J}}{\text{g °C}} \times 454 \text{ g} \times \dfrac{5}{9} \text{ °C} = 1.05 \times 10^3 \text{ J} = 1.05 \text{ kJ}$

It takes 40.7 kJ to vaporize 1 mol H_2O (ΔH_{vap}). Combining these:

$$\dfrac{1.00 \times 10^4 \text{ Btu}}{\text{hr}} \times \dfrac{1.05 \text{ kJ}}{\text{Btu}} \times \dfrac{1 \text{ mol } H_2O}{40.7 \text{ kJ}} = 258 \text{ mol/hr}$$

or $\dfrac{258 \text{ mol}}{\text{hr}} \times \dfrac{18.02 \text{ g}}{\text{mol}} = 4650 \text{ g/hr} = 4.65 \text{ kg/hr}$

68.   We want to graph ln P vs 1/T. The slope of the resulting straight line will be -ΔH/R.

P	ln P	T (Li)	1/T	T (Mg)	1/T
1 torr	0	1023 K	$9.775 \times 10^{-4} \text{ K}^{-1}$	893 K	$11.2 \times 10^{-4} \text{ K}^{-1}$
10.	2.30	1163	8.598×10^{-4}	1013	9.872×10^{-4}
100.	4.605	1353	7.391×10^{-4}	1173	8.525×10^{-4}
400.	5.991	1513	6.609×10^{-4}	1313	7.616×10^{-4}
760.	6.633	1583	6.317×10^{-4}	1383	7.231×10^{-4}

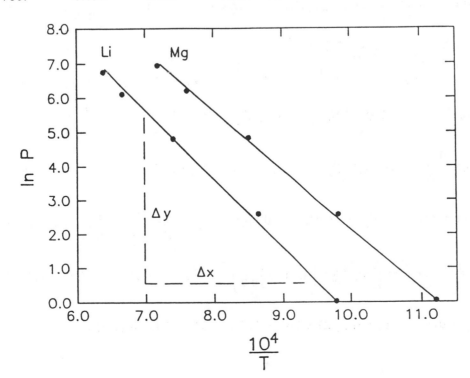

For Li:

We get the slope by taking two points (x, y) that are on the line we draw, not data points. There may be experimental error and individual data points may not fall directly on the best fit line. For a line:

$$\text{slope} = \frac{\Delta y}{\Delta x} = \frac{y_2 - y_1}{x_2 - x_1}$$

or we can fit the straight line using a computer or calculator.

The equation of this line is: $\ln P = -1.90 \times 10^4(1/T) + 18.6$, slope $= -1.90 \times 10^4$ K

Slope $= -\Delta H/R$, $\Delta H = -$slope $\times R = 1.90 \times 10^4$ K $\times 8.3145$ J K^{-1} mol^{-1}

$\Delta H = 1.58 \times 10^5$ J/mol $= 158$ kJ/mol

For Mg:

The equation of the line is: $\ln P = -1.67 \times 10^4(1/T) + 18.7$, slope $= -1.67 \times 10^4$ K

$\Delta H = -$slope $\times R = 1.67 \times 10^4$ K $\times 8.3145$ J K^{-1} mol^{-1}, $\Delta H = 1.39 \times 10^5$ J/mol $= 139$ kJ/mol

The bonding is stronger in Li since ΔH_{vap} is larger for Li.

69. $$\ln\left(\frac{P_1}{P_2}\right) = \frac{\Delta H_{vap}}{R}\left(\frac{1}{T_2} - \frac{1}{T_1}\right)$$

At normal boiling point, $P_1 = 760.$ torr, $T_1 = 56.5°C = 329.7$ K; $T_2 = 25.0°C = 298.2$ K, $P_2 = ?$

$$\ln\left(\frac{760.}{P_2}\right) = \frac{32.0 \times 10^3 \text{ J/mol}}{8.3145 \text{ J K}^{-1}\text{ mol}^{-1}}\left(\frac{1}{298.2} - \frac{1}{329.7}\right), \; 6.633 - \ln P_2 = 1.23$$

$\ln P_2 = 5.40$, $P_2 = e^{5.40} = 221$ torr

70. $$\ln\left(\frac{P_1}{P_2}\right) = \frac{\Delta H_{vap}}{R}\left(\frac{1}{T_2} - \frac{1}{T_1}\right)$$

$P_1 = 760.$ torr, $T_1 = 56.5°C$ (the normal bp from previous exercise); $P_2 = 630.$ torr, $T_2 = ?$

$$\ln\left(\frac{760.}{630.}\right) = \frac{32.0 \times 10^3 \text{ J/mol}}{8.3145 \text{ J K}^{-1}\text{ mol}^{-1}}\left(\frac{1}{T_2} - \frac{1}{329.7}\right), \; 0.188 = 3.85 \times 10^3\left(\frac{1}{T_2} - 3.033 \times 10^{-3}\right)$$

$$\frac{1}{T_2} - 3.033 \times 10^{-3} = 4.88 \times 10^{-5}, \quad \frac{1}{T_2} = 3.082 \times 10^{-3}, \quad T_2 = 324.5 \text{ K} = 51.3 \degree\text{C}$$

$$\ln\left(\frac{630.}{P_2}\right) = \frac{32.0 \times 10^3}{8.3145}\left(\frac{1}{298} - \frac{1}{324.5}\right), \quad 6.446 - \ln P_2 = 1.05 \quad \text{(Carry extra significant figure.)}$$

$$\ln P_2 = 5.40, \quad P_2 = e^{5.40} = 221 \text{ torr} = 220 \text{ torr}$$

71. $\ln\left(\dfrac{P_1}{P_2}\right) = \dfrac{\Delta H_{vap}}{R}\left(\dfrac{1}{T_2} - \dfrac{1}{T_1}\right); \quad P_1 = 760. \text{ torr}, \quad T_1 = 630. \text{ K}; \quad P_2 = ?, \quad T_2 = 298 \text{ K}$

$$\ln 760. - \ln P_2 = \frac{59.1 \times 10^3 \text{ J/mol}}{8.3145 \text{ J K}^{-1}\text{mol}^{-1}}\left(\frac{1}{298 \text{ K}} - \frac{1}{630. \text{ K}}\right), \quad 6.633 - \ln P_2 = 12.6$$

$$\ln P_2 = -6.0, \quad P_2 = e^{-6.0} = 2.5 \times 10^{-3} \text{ torr}$$

72.   a.   As the intermolecular forces increase, the rate of evaporation decreases.

        b.   Increase T, increase rate

        c.   Increase surface area, increase rate

73.

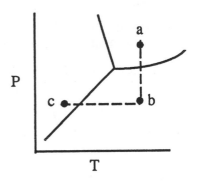

As P is lowered, we go from a to b on the phase diagram. The water boils. The evaporation of the water is endothermic and the water is cooled (b → c), leaving some ice. If the pump is left on, the ice will sublime until none is left. This is the basis of freeze drying.

74.   A: solid;   B: liquid;   C: vapor

        D: solid + vapor;   E: solid + liquid + vapor

        F: liquid + vapor;   G: liquid + vapor;   H: vapor

        triple point: E;   critical point: G

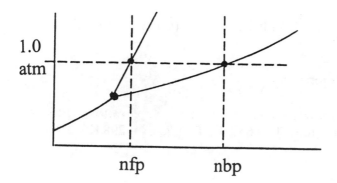

Since the solid-liquid equilibrium line has a positive slope, the solid phase is denser than the liquid phase.

75.   A typical heating curve looks like:

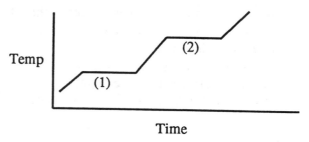

Plateau (1) gives a temperature at which the solid and liquid are in equilibrium. This temperature along with the pressure gives a point on the phase diagram. Plateau (2) will give a temperature and pressure point on the liquid-vapor line.

76.

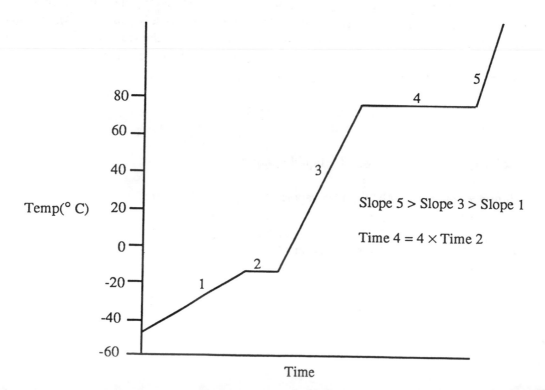

77.   The phase change: $H_2O(g) \rightarrow H_2O(l)$ releases heat that can cause additional damage. Also steam can be at a temperature greater than 100°C.

78.   Heat released = 0.250 g Na $\times \dfrac{1 \text{ mol}}{22.99 \text{ g}} \times \dfrac{368 \text{ kJ}}{2 \text{ mol}} = 2.00$ kJ

To melt 50.0 g of ice requires: 50.0 g ice $\times \dfrac{1 \text{ mol } H_2O}{18.02 \text{ g}} \times \dfrac{6.01 \text{ kJ}}{\text{mol}} = 16.7$ kJ

The reaction doesn't release enough heat to melt all of the ice. The temperature will remain at 0°C.

79.   a.   3

       b.   Triple point at 95.31°C: rhombic, monoclinic, gas
            Triple point at 115.18°C: monoclinic, liquid, gas
            Triple point at 153°C: rhombic, monoclinic, liquid

       c.   Rhombic is stable at T = 20°C and P = 1 atm.

       d.   Yes, monoclinic sulfur and vapor (gas) share a common boundary line in the phase diagram.

       e.   444.6°C; The normal boiling point occurs at P = 1 atm.

       f.   Rhombic, since the rhombic-monoclinic equilibrium line has a positive slope.

Additional Exercises

80.   One B atom and one N atom together have the same number of electrons as two C atoms. The description of physical properties sound a lot like the properties of graphite and diamond, the two solid forms of carbon. The two forms of BN have structures similar to graphite and diamond.

81.   Again we graph ln P vs 1/T. The slope of the line equals $-\Delta H_{vap}/R$.

T (K)	$10^3/T$ (K^{-1})	P (torr)	ln P
273	3.66	14.4	2.67
283	3.53	26.6	3.28
293	3.41	47.9	3.87
303	3.30	81.3	4.40
313	3.19	133	4.89
323	3.10	208	5.34
353	2.83	670	6.51

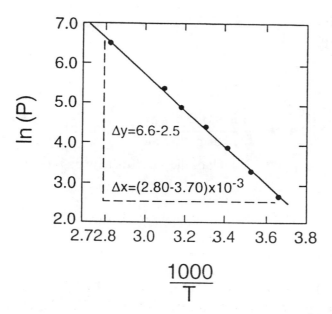

The slope of the line is -4600 K.

$$-4600 \text{ K} = \frac{-\Delta H_{vap}}{R} = \frac{-\Delta H_{vap}}{8.3145 \text{ J K}^{-1} \text{mol}^{-1}}, \quad \Delta H_{vap} = 38 \text{ kJ/mol}$$

$$\ln\left(\frac{P_1}{P_2}\right) = \frac{\Delta H_{vap}}{R}\left(\frac{1}{T_2} - \frac{1}{T_1}\right)$$

At the normal boiling point, the vapor pressure equals 1.00 atm or 760. torr. At 273 K, the vapor pressure is 14.4 torr.

$$\ln\left(\frac{14.4}{760.}\right) = \frac{38,000 \text{ J/mol}}{8.3145 \text{ J K}^{-1} \text{mol}^{-1}}\left(\frac{1}{T_2} - 3.66 \times 10^{-3}\right), \quad -3.966 = 4.6 \times 10^{3} (1/T_2 - 3.66 \times 10^{-3})$$

$$-8.6 \times 10^{-4} + 3.66 \times 10^{-3} = 1/T_2 = 2.80 \times 10^{-3}, \quad T_2 = 357 \text{ K}$$

82.   Although liquid water cannot exist, sublimation may still occur.

83.   There is one atom (radius = R) in a simple cubic unit cell.

$V_{occ} = 4/3\pi R^3$ = volume occupied by atom

The atom touches along the cube edge, so the cube edge length (l) of the unit cell is 2R.

$V_{tot} = l^3 = (2R)^3 = 8R^3$

$$\frac{V_{occ}}{V_{tot}} = \frac{4/3\pi R^3}{8R^3} = \frac{4\pi}{24} = 0.524 \text{ or } 52.4 \text{ \% of the volume is occupied.}$$

84.

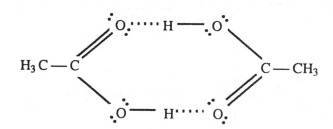

85. CH_3CO_2H: H-bonding

        CH_2ClCO_2H: H-bonding + larger electronegative atom replacing H (greater dipole)

        $CH_3CO_2CH_3$:   polar, no H-bonding

        From the intermolecular forces listed above, we predict $CH_3CO_2CH_3$ to have the weakest
        intermolecular forces and CH_2ClCO_2H to have the strongest. The boiling points are consistent
        with this view.

86.   The unit cell is a parallelpiped. There are three parallelpiped unit cells in a hexagon. In each
        parallelpiped, there are atoms at each of the 8 corners (shared by eight unit cells) plus one atom
        inside the unit cell: 2/3 of one atom plus 1/6 from each of two others. Thus, there are two atoms
        in the unit cell.

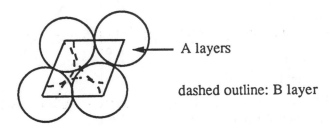

        — A layers

        dashed outline: B layer

87.   Energy added to ice cubes = $5.00 \text{ min} \times \dfrac{60 \text{ s}}{\text{min}} \times \dfrac{750. \text{ J}}{\text{s}} = 2.25 \times 10^5 \text{ J} = 225 \text{ kJ}$

        Energy to melt ice = $475 \text{ g} \times \dfrac{1 \text{ mol}}{18.02 \text{ g}} \times \dfrac{6.01 \text{ kJ}}{\text{mol}} = 158 \text{ kJ}$

        Energy to heat water = 225 kJ - 158 kJ = 67 kJ; $67{,}000 \text{ J} = 475 \text{ g} \times \dfrac{4.18 \text{ J}}{\text{g} \,^\circ\text{C}} \times \Delta T$, $\Delta T = 34^\circ\text{C}$

        The final temperature is 34°C. The water doesn't boil.

88.   Total mass H_2O = 18 cubes $\times \dfrac{30.0 \text{ g}}{\text{cube}}$ = 540. g;   540. g $H_2O \times \dfrac{1 \text{ mol } H_2O}{18.02 \text{ g}}$ = 30.0 mol H_2O

Heat needed to produce ice at -5.0°C:

$$\dfrac{4.18 \text{ J}}{\text{g °C}} \times 540. \text{ g} \times 22.0 \text{ °C} + \dfrac{6.01 \times 10^3 \text{ J}}{\text{mol}} \times 30.0 \text{ mol} + \dfrac{2.08 \text{ J}}{\text{g °C}} \times 540. \text{ g} \times 5.0 \text{ °C}$$

$$= 4.97 \times 10^4 \text{ J} + 1.80 \times 10^5 \text{ J} + 5.6 \times 10^3 \text{ J} = 2.35 \times 10^5 \text{ J}$$

$2.35 \times 10^5 \text{ J} \times \dfrac{1 \text{ g } CF_2Cl_2}{158 \text{ J}}$ = 1.49×10^3 g CF_2Cl_2 must be vaporized.

89.

T (°C)	T (K)	1/T (K^{-1})	P (torr)	ln P
-6.0	267.2	3.743×10^{-3}	20.0	2.996
5.0	278.2	3.595×10^{-3}	40.0	3.689
12.1	285.3	3.505×10^{-3}	60.0	4.094
21.2	294.4	3.397×10^{-3}	100.0	4.605
49.9	323.1	3.095×10^{-3}	400.0	5.991

From graph: ln P = -4620/T + 20.29

Slope = $\dfrac{-\Delta H_{vap}}{R}$ = -4.62×10^3 K

$\Delta H_{vap} = 4.62 \times 10^3 \text{ K} \left(\dfrac{8.3145 \text{ J}}{\text{mol K}} \right)$

ΔH_{vap} = 38,400 J/mol = 38.4 kJ/mol

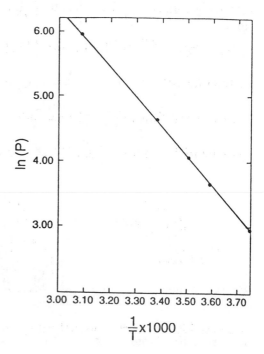

At the normal boiling point, P = 760. torr. From the straight line equation:

ln 760. = -4620/T_b + 20.29,   T = $\dfrac{-4620}{6.63 - 20.29}$ = 338 K = 65°C

90.   a.   two

        b.   higher pressure triple point: graphite, diamond, and liquid

             lower pressure triple point: graphite, liquid, and vapor

        c.   It is converted to diamond (the more dense solid form).

        d.   diamond

91.   CsCl is a simple cubic array of Cl^- ions with Cs^+ in the middle of each unit cell. There is one Cs^+ and one Cl^- ion in each unit cell. Cs^+ and Cl^- touch along the body diagonal.

        body diagonal $= 2r_{Cs^+} + 2r_{Cl^-} = \sqrt{3}\, l$, l = length of cube edge

        In each unit cell:

            mass = 1 CsCl molecule $(1\ mol/6.022 \times 10^{23}\ molecules)\ (168.4\ g/mol) = 2.796 \times 10^{-22}\ g$

            volume $= l^3 = $ mass/density $= 2.796 \times 10^{-22}\ g/3.97\ g\ cm^{-3} = 7.04 \times 10^{-23}\ cm^3$

        $l^3 = 7.04 \times 10^{-23}\ cm^3$, $l = 4.13 \times 10^{-8}\ cm = 413\ pm = $ length of cube edge

        $2r_{Cs^+} + 2r_{Cl^-} = \sqrt{3}\, l = \sqrt{3}(413\ pm) = 715\ pm$

        The distance between ion centers $= r_{Cs^+} + r_{Cl^-} = 715\ pm/2 = 358\ pm$

        From ionic radius: $r_{Cs^+} = 169\ pm$ and $r_{Cl^-} = 181\ pm$; $r_{Cs^+} + r_{Cl^-} = 169 + 181 = 350.\ pm$

        The actual distance is 8 pm (2.3%) greater than that calculated from tables of ionic radii.

92.

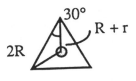   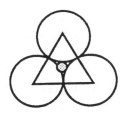   R = radius of sphere

                                                     r = radius of trigonal hole

        For the right angle triangle in the diagram, $\cos 30° = \dfrac{adjacent}{hypotenuse} = \dfrac{R}{R+r}$, $0.866 = \dfrac{R}{R+r}$

        $0.866\ r = R - 0.866\ R$, $r = \left(\dfrac{0.134}{0.866}\right) \times R = 0.155\ R$

        The cation must have a radius which is 0.155 times the radius of the spheres to just fit into the trigonal hole.

93.   There are 2 tetrahedral holes per packing atom. Let f = fraction of tetrahedral holes filled

K_2O: cation to anion ratio = $\dfrac{2}{1} = \dfrac{2f}{1}$, f = 1; All tetrahedral holes are filled.

CuI: cation to anion ratio = $\dfrac{1}{1} = \dfrac{2f}{1}$, f = $\dfrac{1}{2}$; $\dfrac{1}{2}$ of the tetrahedral holes are filled.

ZrI_4:   cation to anion ratio = $\dfrac{1}{4} = \dfrac{2f}{1}$, f = $\dfrac{1}{8}$; $\dfrac{1}{8}$ of the tetrahedral holes are filled.

94.   For an octahedral hole; the geometry is:

R = Cl$^-$ radius

r = Li$^+$ radius

From the diagram: $\cos 45° = \dfrac{\text{adjacent}}{\text{hypotenuse}} = \dfrac{2R}{2R + 2r} = \dfrac{R}{R + r}$, $0.707 = \dfrac{R}{R + r}$

R = 0.707 (R + r), r = 0.414 R

LiCl unit cell:

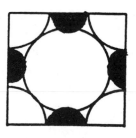

length of cube edge = 2R + 2r = 514 pm

2R + 2(0.414 R) = 514 pm, R = 182 pm = Cl$^-$ radius

2(182 pm) + 2r = 514 pm, r = 75 pm = Li$^+$ radius

From Figure 13.7, the Li$^+$ radius is 60 pm and the Cl$^-$ radius is 181 pm. The Li$^+$ ion is much smaller than calculated. This probably means that the ions are not actually in contact with each other. The octahedral holes are larger than the Li$^+$ ion.

95.   For a cubic hole to be filled, the cation to anion radius ratio is between 0.732 < r_+/r_- < 1.00.

CsBr: Cs$^+$ radius = 169 pm, Br$^-$ radius = 195 pm; $r_+/r_- = 169/195 = 0.867$

From the radius ratio, Cs$^+$ should occupy cubic holes. The structure should be the CsCl structure. The actual structure is the CsCl structure.

KF: K^+ radius = 133 pm, F^- radius = 136 pm; $r_+/r_- = 133/136 = 0.978$

Again, we would predict a structure similar to CsCl, i.e., cations in the middle of a simple cubic array of anions. The actual structure is the NaCl structure.

The rule fails for KF. Exceptions are common for crystal structures.

96.   Ar is cubic closest packed. There are 4 Ar atoms per unit cell and with a face centered unit cell, the atoms touch along the face diagonal.

face diagonal = $4r = \sqrt{2}\, l$, l = length of cube edge; $l = 4(190. \text{ pm})/\sqrt{2} = 537$ pm $= 5.37 \times 10^{-8}$ cm

$$d = \frac{\text{mass}}{\text{volume}} = \frac{4 \text{ atoms} \times \dfrac{1 \text{ mol}}{6.022 \times 10^{23} \text{ atoms}} \times \dfrac{39.95 \text{ g}}{\text{mol}}}{(5.37 \times 10^{-8} \text{ cm})^3} = 1.71 \text{ g/cm}^3$$

97.   The ionic radius is 148 pm for Rb^+ and 181 pm for Cl^-. Using these values, lets calculate the density of the two structures.

Normal pressure: Rb^+Cl^- touch along cube edge (forms NaCl structure).

cube edge = $l = 2(148 + 181) = 658$ pm $= 6.58 \times 10^{-8}$ cm; There are 4 RbCl molecules/unit cell.

$$\text{density} = \frac{4(85.47) + 4(35.45)}{6.022 \times 10^{23} \, (6.58 \times 10^{-8})^3} = 2.82 \text{ g/cm}^3$$

High pressure: Rb^+Cl^- touch along body diagonal (forms CsCl structure).

$2r_- + 2r_+ = 658$ pm = body diagonal = $\sqrt{3}\, l$, $l = 658$ pm$/\sqrt{3} = 380.$ pm

Since each unit cell contains 1 RbCl molecule: $d = \dfrac{85.47 + 35.45}{(6.022 \times 10^{23})\,(3.80 \times 10^{-8})^3} = 3.66 \text{ g/cm}^3$

The high pressure form has the higher density. The density ratio is 3.66/2.82 = 1.30. We would expect this since the effect of pressure is to push things closer together and thus, increase density.

98.   Re at 8 corners: 8(1/8) = 1 Re; O at 12 edges: 12(1/4) = 3 O

Formula is ReO_3. If O has 2- charge, then charge on Re is +6.

99.   First we need to get the empirical formula of spinel. Assume 100.0 g of spinel.

$$37.9 \text{ g Al} \times \frac{1 \text{ mol Al}}{26.98 \text{ g Al}} = 1.40 \text{ mol Al}$$

The mole ratios are 2:1:4.

$$17.1 \text{ g Mg} \times \frac{1 \text{ mol Mg}}{24.31 \text{ g Mg}} = 0.703 \text{ mol Mg}$$

Empirical Formula = Al_2MgO_4

$$45.0 \text{ g O} \times \frac{1 \text{ mol O}}{16.00 \text{ g O}} = 2.81 \text{ mol O}$$

Each unit cell will contain an integral value (n) of empirical formula units. Each formula unit has a mass of: 24.31 + 2(26.98) + 4(16.00) = 142.27 g/mol

$$3.57 \text{ g/cm}^3 = \frac{n(142.27 \text{ g/mol})}{(6.022 \times 10^{23}/\text{mol})(8.09 \times 10^{-8} \text{ cm})^3}, \; n = 8.00$$

Each unit cell has 16 Al, 8 Mg, and 32 O.

100.   Total charge of all Fe is +2 to balance the -2 charge from the one O atom. Sum of Fe is 0.950.

Setting up two equations: x + y = 0.950 and 2x + 3y = 2.000

Solving: 2x + 3(0.950 - x) = 2.000, x = 0.85 and y = 0.10

$$\frac{0.10}{0.95} = 0.11 = \text{fraction of Fe as } Fe^{3+}$$

If all Fe^{2+}, then 1.000 Fe^{2+}/O^{2-}; 1.000 - 0.950 = 0.050 = vacant sites

5.0% of Fe sites are vacant.

101.   Out of 100.00 g: $28.31 \text{ g O} \times \dfrac{1 \text{ mol}}{15.999 \text{ g}} = 1.769 \text{ mol O}$; $71.69 \text{ g Ti} \times \dfrac{1 \text{ mol}}{47.88 \text{ g}} = 1.497 \text{ mol Ti}$

Formula is $TiO_{1.182}$ or $Ti_{0.8462}O$.

For $Ti_{0.8462}O$, let x = Ti^{2+} and y = Ti^{3+}. Setting up two equations and solving:

x + y = 0.8462 and 2x + 3y = 2; 2x + 3(0.8462 - x) = 2

x = 0.539 mol Ti^{2+}/mol O^{2-} and y = 0.307 mol Ti^{3+}/mol O^{2-}

$$\frac{0.539}{0.8462} \times 100 = 63.7\% \text{ of the titanium is } Ti^{2+} \text{ and } 36.3\% \text{ is } Ti^{3+}.$$

102.   For water vapor at 30.0°C and 31.824 torr:

$$\text{density} = \frac{P(\text{molar mass})}{RT} = \frac{\left(\dfrac{31.824 \text{ atm}}{760}\right)\left(\dfrac{18.015 \text{ g}}{\text{mol}}\right)}{\dfrac{0.08206 \text{ L atm}}{\text{mol K}} \times 303.2 \text{ K}} = 0.03032 \text{ g/L}$$

The volume of one molecule is proportional to d^3 where d is the average distance between molecules. For a large sample of molecules, the volume is still proportional to d^3. So:

$$\frac{V_{gas}}{V_{liq}} = \frac{d_{gas}^3}{d_{liq}^3}$$

If we have 0.99567 g H_2O then $V_{liq} = 1.0000 \text{ cm}^3 = 1.0000 \times 10^{-3}$ L.

$V_{gas} = 0.99567 \text{ g} \times 1 \text{ L}/0.03032 \text{ g} = 32.84$ L

$$\frac{d_{gas}^3}{d_{liq}^3} = \frac{32.84 \text{ L}}{1.0000 \times 10^{-3} \text{ L}} = 3.284 \times 10^4, \quad \frac{d_{gas}}{d_{liq}} = (3.284 \times 10^4)^{1/3} = 32.02, \quad \frac{d_{liq}}{d_{gas}} = 0.03123$$

103. $24.7 \text{ g C}_6\text{H}_6 \times \dfrac{1 \text{ mol}}{78.11 \text{ g}} = 0.316 \text{ mol C}_6\text{H}_6$

$$P = \frac{nRT}{V} = \frac{0.316 \text{ mol} \times \dfrac{0.08206 \text{ L atm}}{\text{mol K}} \times 293.2 \text{ K}}{100.0 \text{ L}} = 0.0760 \text{ atm or } 57.8 \text{ torr}$$

104. $100.0 \text{ g N}_2 \times \dfrac{1 \text{ mol N}_2}{28.014 \text{ g N}_2} = 3.570 \text{ mol N}_2; \quad P_{H_2O} = \chi_{H_2O} P_{tot}, \quad \chi_{H_2O} = \dfrac{23.8 \text{ torr}}{700. \text{ torr}} = 0.0340$

$$\chi_{H_2O} = 0.0340 = \frac{n_{H_2O}}{n_{N_2} + n_{H_2O}} = \frac{n_{H_2O}}{3.570 + n_{H_2O}}, \quad n_{H_2O} = 0.121 + 0.0340 \, n_{H_2O}$$

$n_{H_2O} = 0.125 \text{ mol}; \quad 0.125 \text{ mol} \times 18.02 \text{ g/mol} = 2.25 \text{ g H}_2\text{O}$

105. $w = -P\Delta V;$ Assume const P of 1 atm.

$$V_{373} = \frac{nRT}{P} = \frac{1.00 \, (0.08206) \, (373)}{1.00} = 30.6 \text{ L for one mol of water vapor}$$

Since the density of $H_2O(l)$ is 1.00 g/cm³, 1.00 mol of $H_2O(l)$ occupies 18.0 cm³ or 0.0180 L.

$w = -1.00 \text{ atm} \, (30.6 \text{ L} - 0.0180 \text{ L}) = -30.6 \text{ L atm}$

$w = -30.6 \text{ L atm} \times 101.3 \text{ J L}^{-1} \text{ atm}^{-1} = -3.10 \times 10^3 \text{ J} = -3.10 \text{ kJ}$

$\Delta E = q + w = 41.16$ kJ $- 3.10$ kJ $= 38.06$ kJ

$\dfrac{38.06}{41.16} \times 100 = 92.47\%$ of heat goes to increase the internal energy of the water.

The remainder of the energy (7.53%) goes to do work against the atmosphere.

106.   If we extend the liquid-vapor line of the water phase diagram to below the freezing point, we find that supercooled water will have a higher vapor pressure than ice at -10°C. To achieve equilibrium there must be a constant vapor pressure. Over time supercooled water will be transformed through the vapor into ice in an attempt to equilibrate the vapor pressure. Eventually there will only be ice at -10°C and its vapor at the vapor pressure given by the solid-vapor line in the phase diagram.

107.   Mn at 8 corners × (1/8) + 1 Mn at body center = 2 Mn; There are 2 Mn atoms in each unit cell. The coordination number of Mn is 8.

For Mn at center: coordination # is from 8 Mn at corners.

For Mn at corner: coordination # is from 8 Mn at the center of 8 different cubes that touch each corner Mn.

108.   The three sheets of B's form a cube. In this cubic unit cell there are B atoms at every face, B atoms at every edge, B atoms at every corner and one B atom in the center. A representation of this cube is:

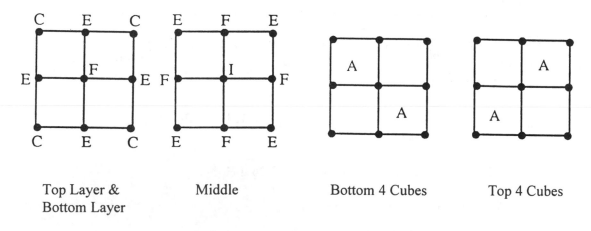

Top Layer &   Middle   Bottom 4 Cubes   Top 4 Cubes
Bottom Layer

C = corner;   E = edge;   F = face;   I = inside

Each unit cell of B atoms contains 4 A atoms. The number of B atoms in each unit cell is:

8 corners × 1/8 + 6 faces × 1/2 + 12 edges × 1/4 + 1 middle B = 8 B atoms

The empirical formula is AB_2.

Each A atom is in a cubic hole of B atoms so 8 B atoms surround each A atom. This will also be true in the extended lattice. The structure of B atoms in the unit cell is a cubic arrangement with B atoms at every face, edge, corner and center of the cube.

109.   A face centered cubic unit cell contains 4 atoms. For a unit cell:

$$\text{mass of X} = \text{volume} \times \text{density} = (4.09 \times 10^{-8} \text{ cm})^3 \times 10.5 \text{ g/cm}^3 = 7.18 \times 10^{-22} \text{ g}$$

$$\text{mol X} = 4 \text{ atoms X} \times \frac{1 \text{ mol X}}{6.022 \times 10^{23} \text{ atoms}} = 6.642 \times 10^{-24} \text{ mol X}$$

$$\text{Atomic mass} = \frac{7.18 \times 10^{-22} \text{ g X}}{6.642 \times 10^{-24} \text{ mol X}} = 108 \text{ g/mol}; \quad \text{The metal is silver (Ag)}.$$

110.   a.   The arrangement of the layers are:

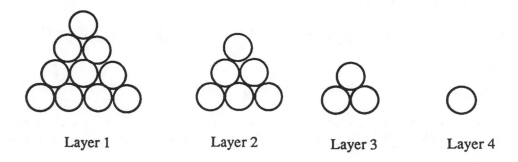

Layer 1   Layer 2   Layer 3   Layer 4

A total of 20 cannon balls will be needed.

b.   The layering alternates abcabc which is cubic closest packing.

c.   tetrahedron

CHAPTER SEVENTEEN

PROPERTIES OF SOLUTIONS

1. $125 \text{ g sucrose} \times \dfrac{1 \text{ mol}}{342.3 \text{ g}} = 0.365 \text{ mol}; \quad M = \dfrac{0.365 \text{ mol}}{1.00 \text{ L}} = \dfrac{0.365 \text{ mol sucrose}}{\text{L}}$

2. $0.250 \text{ L} \times \dfrac{0.100 \text{ mol}}{\text{L}} \times \dfrac{134.0 \text{ g}}{\text{mol}} = 3.35 \text{ g Na}_2\text{C}_2\text{O}_4$

3. $25.00 \times 10^{-3} \text{ L} \times \dfrac{0.308 \text{ mol}}{\text{L}} = 7.70 \times 10^{-3} \text{ mol}; \quad \dfrac{7.70 \times 10^{-3} \text{ mol}}{0.500 \text{ L}} = \dfrac{1.54 \times 10^{-2} \text{ mol NiCl}_2}{\text{L}}$

   $\text{NiCl}_2(s) \rightarrow \text{Ni}^{2+}(aq) + 2 \text{ Cl}^-(aq); \quad M_{\text{Ni}^{2+}} = \dfrac{1.54 \times 10^{-2} \text{ mol}}{\text{L}}; \quad M_{\text{Cl}^-} = \dfrac{3.08 \times 10^{-2} \text{ mol}}{\text{L}}$

4. a. $158.5 \times 10^{-3} \text{ g Cu} \times \dfrac{1 \text{ mol}}{63.55 \text{ g}} = 2.494 \times 10^{-3} \text{ mol Cu}$

   $M_{\text{Cu}^{2+}} = \dfrac{2.494 \times 10^{-3} \text{ mol}}{1.00 \text{ L}} = \dfrac{2.49 \times 10^{-3} \text{ mol}}{\text{L}}$

   b. $\text{ppm} = \dfrac{\text{g Cu}}{\text{g solution}} \times 10^6 = \dfrac{0.1585 \text{ g}}{1.0 \times 10^3 \text{ g}} \times 10^6 = 158.5 \text{ ppm} \approx 160 \text{ ppm Cu}$

5. a. $\dfrac{2.8 \times 10^{-3} \text{ g}}{1.0 \text{ g}} \times 10^6 = 2800 \text{ ppm Cd}^{2+} \text{ (assuming density} = 1.0 \text{ g/mL)}$

   $2800 \text{ ppm} = 2.8 \times 10^6 \text{ ppb Cd}^{2+}$

   b. $\dfrac{2.8 \times 10^{-3} \text{ g}}{1.0 \times 10^{-3} \text{ L}} \times \dfrac{1 \text{ mol}}{112.4 \text{ g}} = \dfrac{2.5 \times 10^{-2} \text{ mol Cd}^{2+}}{\text{L}}$

6. a. $\text{Ca(NO}_3)_2(s) \rightarrow \text{Ca}^{2+}(aq) + 2 \text{ NO}_3^-(aq); \quad M_{\text{Ca}^{2+}} = \dfrac{1.06 \times 10^{-3} \text{ mol}}{\text{L}}; \quad M_{\text{NO}_3^-} = \dfrac{2.12 \times 10^{-3} \text{ mol}}{\text{L}}$

   b. $1.0 \times 10^{-3} \text{ L} \times \dfrac{1.06 \times 10^{-3} \text{ mol Ca}^{2+}}{\text{L}} \times \dfrac{40.08 \text{ g Ca}^{2+}}{\text{mol}} = 4.2 \times 10^{-5} \text{ g Ca}^{2+}$

c. $\quad 1.0 \times 10^{-6}\,L \times \dfrac{2.12 \times 10^{-3}\,\text{mol NO}_3^-}{L} \times \dfrac{6.02 \times 10^{23}\,\text{NO}_3^-\,\text{ions}}{\text{mol NO}_3^-} = 1.3 \times 10^{15}\,\text{NO}_3^-\,\text{ions}$

7.$\quad 1.00\,L \times \dfrac{0.040\,\text{mol HCl}}{L} = 0.040\,\text{mol HCl};\ \ 0.040\,\text{mol HCl} \times \dfrac{1\,L}{0.25\,\text{mol HCl}} = 0.16\,L = 160\,mL$

8.

a. $\quad HNO_3(l) \rightarrow H^+(aq) + NO_3^-(aq)$ b. $\quad Na_2SO_4(s) \rightarrow 2\,Na^+(aq) + SO_4^{2-}(aq)$

c. $\quad AlCl_3(s) \rightarrow Al^{3+}(aq) + 3\,Cl^-(aq)$ d. $\quad SrBr_2(s) \rightarrow Sr^{2+}(aq) + 2\,Br^-(aq)$

e. $\quad KClO_4(s) \rightarrow K^+(aq) + ClO_4^-(aq)$ f. $\quad NH_4Br(s) \rightarrow NH_4^+(aq) + Br^-(aq)$

g. $\quad NH_4NO_3(s) \rightarrow NH_4^+(aq) + NO_3^-(aq)$ h. $\quad CuSO_4(s) \rightarrow Cu^{2+}(aq) + SO_4^{2-}(aq)$

i. $\quad NaOH(s) \rightarrow Na^+(aq) + OH^-(aq)$

Concentration of Solutions

9.$\quad$ molarity $= \dfrac{3.0\,\text{g H}_2\text{O}_2}{100.0\,\text{g soln}} \times \dfrac{1.0\,\text{g soln}}{\text{cm}^3\,\text{soln}} \times \dfrac{1000\,\text{cm}^3}{L} \times \dfrac{1\,\text{mol H}_2\text{O}_2}{34.0\,\text{g H}_2\text{O}_2} = 0.88\,\text{mol/L}$

molality $= \dfrac{\text{mol H}_2\text{O}_2}{\text{kg H}_2\text{O}} = \dfrac{3.0\,\text{g H}_2\text{O}_2}{97.0\,\text{g H}_2\text{O}} \times \dfrac{1000\,\text{g}}{\text{kg}} \times \dfrac{1\,\text{mol H}_2\text{O}_2}{34.0\,\text{g H}_2\text{O}_2} = 0.91\,\text{mol/kg}$

$3.0\,\text{g H}_2\text{O}_2 \times \dfrac{1\,\text{mol}}{34.0\,\text{g}} = 8.8 \times 10^{-2}\,\text{mol H}_2\text{O}_2;\ \ 97.0\,\text{g H}_2\text{O} \times \dfrac{1\,\text{mol}}{18.02\,\text{g}} = 5.38\,\text{mol H}_2\text{O}$

mole fraction of $H_2O_2 = \chi_{H_2O_2} = \dfrac{\text{mol H}_2\text{O}_2}{\text{total mol}} = \dfrac{8.8 \times 10^{-2}}{5.38 + 0.088} = 1.6 \times 10^{-2}$

10.$\quad$ Hydrochloric acid:

molarity $= \dfrac{38\,\text{g HCl}}{100.\,\text{g soln}} \times \dfrac{1.19\,\text{g soln}}{\text{cm}^3\,\text{soln}} \times \dfrac{1000\,\text{cm}^3}{L} \times \dfrac{1\,\text{mol HCl}}{36.5\,\text{g}} = 12\,\text{mol/L}$

molality $= \dfrac{38\,\text{g HCl}}{62\,\text{g solvent}} \times \dfrac{1000\,\text{g}}{\text{kg}} \times \dfrac{1\,\text{mol HCl}}{36.5\,\text{g}} = 17\,\text{mol/kg}$

$38\,\text{g HCl} \times \dfrac{1\,\text{mol}}{36.5\,\text{g}} = 1.0\,\text{mol HCl};\ \ 62\,\text{g H}_2\text{O} \times \dfrac{1\,\text{mol}}{18.0\,\text{g}} = 3.4\,\text{mol H}_2\text{O}$

mole fraction of $HCl = \chi_{HCl} = \dfrac{1.0}{3.4 + 1.0} = 0.23$

Nitric acid:

$$\frac{70.\ g\ HNO_3}{100.\ g\ soln} \times \frac{1.42\ g\ soln}{cm^3\ soln} \times \frac{1000\ cm^3}{L} \times \frac{1\ mol\ HNO_3}{63.0\ g} = 16\ mol/L$$

$$\frac{70.\ g\ HNO_3}{30.\ g\ solvent} \times \frac{1000\ g}{kg} \times \frac{1\ mol\ HNO_3}{63.0\ g} = 37\ mol/kg$$

$$70.\ g\ HNO_3 \times \frac{1\ mol}{63.0\ g} = 1.1\ mol\ HNO_3; \quad 30.\ g\ H_2O \times \frac{1\ mol}{18.0\ g} = 1.7\ mol\ H_2O$$

$$\chi_{HNO_3} = \frac{1.1}{1.7 + 1.1} = 0.39$$

Sulfuric acid:

$$\frac{95\ g\ H_2SO_4}{100.\ g\ soln} \times \frac{1.84\ g\ soln}{cm^3\ soln} \times \frac{1000\ cm^3}{L} \times \frac{1\ mol\ H_2SO_4}{98.1\ g\ H_2SO_4} = 18\ mol/L$$

$$\frac{95\ g\ H_2SO_4}{5\ g\ H_2O} \times \frac{1000\ g}{kg} \times \frac{1\ mol}{98.1\ g} = 194\ mol/kg \approx 200\ mol/kg$$

$$95\ g\ H_2SO_4 \times \frac{1\ mol}{98.1\ g} = 0.97\ mol\ H_2SO_4; \quad 5\ g\ H_2O \times \frac{1\ mol}{18.0\ g} = 0.3\ mol\ H_2O$$

$$\chi_{H_2SO_4} = \frac{0.97}{0.97 + 0.3} = 0.76$$

Acetic Acid:

$$\frac{99\ g\ CH_3CO_2H}{100.\ g\ soln} \times \frac{1.05\ g\ soln}{cm^3\ soln} \times \frac{1000\ cm^3}{L} \times \frac{1\ mol}{60.05\ g} = 17\ mol/L$$

$$\frac{99\ g\ CH_3CO_2H}{1\ g\ H_2O} \times \frac{1000\ g}{kg} \times \frac{1\ mol}{60.05\ g} = 1600\ mol/kg \approx 2000\ mol/kg$$

$$99\ g\ HOAc \times \frac{1\ mol}{60.05\ g} = 1.6\ mol\ HOAc; \quad 1\ g\ H_2O \times \frac{1\ mol}{18.0\ g} = 0.06\ mol\ H_2O$$

$$\chi_{HOAc} = \frac{1.6}{1.6 + 0.06} = 0.96$$

Ammonia:

$$\frac{28 \text{ g NH}_3}{100. \text{ g soln}} \times \frac{0.90 \text{ g}}{\text{cm}^3} \times \frac{1000 \text{ cm}^3}{\text{L}} \times \frac{1 \text{ mol}}{17.0 \text{ g}} = 15 \text{ mol/L}$$

$$\frac{28 \text{ g NH}_3}{72 \text{ g H}_2\text{O}} \times \frac{1000 \text{ g}}{\text{kg}} \times \frac{1 \text{ mol}}{17.0 \text{ g}} = 23 \text{ mol/kg}$$

$$28 \text{ g NH}_3 \times \frac{1 \text{ mol}}{17.0 \text{ g}} = 1.6 \text{ mol NH}_3; \quad 72 \text{ g H}_2\text{O} \times \frac{1 \text{ mol}}{18.0 \text{ g}} = 4.0 \text{ mol H}_2\text{O}$$

$$\chi_{\text{NH}_3} = \frac{1.6}{4.0 + 1.6} = 0.29$$

11. $50.0 \text{ mL toluene} \times \dfrac{0.867 \text{ g}}{\text{mL}} = 43.4 \text{ g toluene}; \quad 125 \text{ mL benzene} \times \dfrac{0.874 \text{ g}}{\text{mL}} = 109 \text{ g benzene}$

$$\text{mass \% toluene} = \frac{\text{mass of toluene}}{\text{total mass}} \times 100 = \frac{43.4}{43.4 + 109} \times 100 = 28.5\%$$

$$\frac{43.4 \text{ g toluene}}{175 \text{ mL soln}} \times \frac{1000 \text{ mL}}{\text{L}} \times \frac{1 \text{ mol toluene}}{92.13 \text{ g toluene}} = 2.69 \text{ mol/L}$$

$$\frac{43.4 \text{ g toluene}}{109 \text{ g benzene}} \times \frac{1000 \text{ g}}{\text{kg}} \times \frac{1 \text{ mol toluene}}{92.13 \text{ g toluene}} = 4.32 \text{ mol/kg}$$

$$43.4 \text{ g toluene} \times \frac{1 \text{ mol}}{92.13 \text{ g}} = 0.471 \text{ mol toluene}$$

$$109 \text{ g benzene} \times \frac{1 \text{ mol benzene}}{78.11 \text{ g benzene}} = 1.40 \text{ mol benzene}$$

$$\chi_{\text{toluene}} = \frac{0.471}{0.471 + 1.40} = 0.252$$

12. $\dfrac{40.0 \text{ g EG}}{60.0 \text{ g H}_2\text{O}} \times \dfrac{1000 \text{ g}}{\text{kg}} \times \dfrac{1 \text{ mol EG}}{62.07 \text{ g}} = 10.7 \text{ mol/kg}$

$$\frac{40.0 \text{ g EG}}{100.0 \text{ g solution}} \times \frac{1.05 \text{ g}}{\text{cm}^3} \times \frac{1000 \text{ cm}^3}{\text{L}} \times \frac{1 \text{ mol}}{62.07 \text{ g}} = 6.77 \text{ mol/L}$$

$$40.0 \text{ g EG} \times \frac{1 \text{ mol}}{62.07 \text{ g}} = 0.644 \text{ mol EG}; \quad 60.0 \text{ g H}_2\text{O} \times \frac{1 \text{ mol}}{18.02 \text{ g}} = 3.33 \text{ mol H}_2\text{O}$$

$$\chi_{\text{EG}} = \frac{0.644}{3.33 + 0.644} = 0.162$$

13.   If we have 1.00 L of solution:

$$1.37 \text{ mol citric acid} \times \frac{192.1 \text{ g}}{\text{mol}} = 263 \text{ g citric acid}$$

$$1.00 \times 10^3 \text{ mL solution} \times \frac{1.10 \text{ g}}{\text{mL}} = 1.10 \times 10^3 \text{ g solution}$$

$$\text{mass \% of citric acid} = \frac{263}{1.10 \times 10^3} \times 100 = 23.9\%$$

In 1.00 L of solution, we have 263 g citric acid and $(1.10 \times 10^3 - 263) = 840$ g of H_2O.

$$\frac{1.37 \text{ mol citric acid}}{0.84 \text{ kg } H_2O} = 1.6 \text{ mol/kg}$$

$$840 \text{ g } H_2O \times \frac{1 \text{ mol}}{18.0 \text{ g}} = 47 \text{ mol } H_2O; \quad \chi_{\text{citric acid}} = \frac{1.37}{47 + 1.37} = 0.028$$

14. $\text{Molarity} = \dfrac{\text{moles solute}}{\text{L solution}}; \quad \text{Molality} = \dfrac{\text{moles solute}}{\text{kg solvent}}$

Since volume is temperature dependent and mass isn't, then molarity is temperature dependent and molality is temperature independent. When determining ΔT_f and ΔT_b, we are interested in how temperature depends on composition. Thus, we don't want our expression of composition to also be dependent on temperature.

15. $\dfrac{1.00 \text{ mol acetone}}{1.00 \text{ kg ethanol}} = 1.00 \text{ molal}; \quad 1.00 \times 10^3 \text{ g } C_2H_5OH \times \dfrac{1 \text{ mol}}{46.07 \text{ g}} = 21.7 \text{ mol } C_2H_5OH$

$$\chi_{\text{acetone}} = \frac{1.00}{1.00 + 21.7} = 0.0441$$

$$1.00 \text{ mol } CH_3COCH_3 \times \frac{58.08 \text{ g } CH_3COCH_3}{\text{mol } CH_3COCH_3} \times \frac{1 \text{ mL}}{0.788 \text{ g}} = 73.7 \text{ mL}$$

$$1.00 \times 10^3 \text{ g ethanol} \times \frac{1 \text{ mL}}{0.789 \text{ g}} = 1270 \text{ mL}; \quad \text{Total V} = 1270 + 73.7 = 1340 \text{ mL}$$

$$\text{molarity} = \frac{1.00 \text{ mol}}{1.34 \text{ L}} = 0.746 \, M$$

16. $0.40 \text{ mol } CH_3COCH_3 \times \dfrac{58.1 \text{ g}}{\text{mol}} = 23 \text{ g acetone}; \quad 0.60 \text{ mol } C_2H_5OH \times \dfrac{46.1 \text{ g}}{\text{mol}} = 28 \text{ g ethanol}$

$$\text{mass \% acetone} = \frac{23 \text{ g}}{23 \text{ g} + 28 \text{ g}} \times 100 = 45\%$$

17. $1.0 \text{ L} \times \dfrac{1000 \text{ mL}}{\text{L}} \times \dfrac{1.0 \text{ g}}{\text{mL}} = 1.0 \times 10^3 \text{ g solution}$

mass % NaCl $= \dfrac{25 \text{ g NaCl}}{1.0 \times 10^3 \text{ g solution}} \times 100 = 2.5\%$

$\dfrac{25 \text{ g NaCl}}{\text{L}} \times \dfrac{1 \text{ mol NaCl}}{58.4 \text{ g NaCl}} = 0.43 \ M$

1.0×10^3 g solution contains 25 g NaCl and 975 g $H_2O \approx 980$ g H_2O.

molality $= \dfrac{0.43 \text{ mol NaCl}}{0.98 \text{ kg solvent}} = 0.44 \text{ molal}$

0.43 mol NaCl; $980 \text{ g} \times \dfrac{1 \text{ mol}}{18.0 \text{ g}} = 54 \text{ mol } H_2O; \ \ \chi_{NaCl} = \dfrac{0.43}{0.43 + 54} = 7.9 \times 10^{-3}$

18. $\dfrac{50.0 \text{ g CsCl}}{100.0 \text{ g solution}} \times 100 = 50.0\% \text{ CsCl by mass}$

$100.0 \text{ g solution} \times \dfrac{1 \text{ mL}}{1.58 \text{ g}} = 63.3 \text{mL}; \ \ M = \dfrac{50.0 \text{ g}}{63.3 \text{ mL}} \times \dfrac{1000 \text{ mL}}{\text{L}} \times \dfrac{1 \text{ mol}}{168.4 \text{ g}} = 4.69 \text{ mol/L}$

$m = \dfrac{50.0 \text{ g CsCl}}{50.0 \text{ g solvent}} \times \dfrac{1000 \text{ g}}{\text{kg}} \times \dfrac{1 \text{ mol CsCl}}{168.4 \text{ g}} = 5.94 \text{ mol/kg}$

$50.0 \text{ g CsCl} \times \dfrac{1 \text{ mol}}{168.4 \text{ g}} = 0.297 \text{ mol CsCl}; \ \ 50.0 \text{ g } H_2O \times \dfrac{1 \text{ mol}}{18.02 \text{ g}} = 2.77 \text{ mol } H_2O$

$\chi_{CsCl} = \dfrac{0.297}{0.297 + 2.77} = 9.68 \times 10^{-2}$

Thermodynamics of Solutions and Solubility

19.   The nature of the interparticle forces. Polar solutes and ionic solutes dissolve in polar solvents and nonpolar solutes dissolve in nonpolar solvents.

20.   As the length of the hydrocarbon chain increases, the solubility decreases. The –OH end of the alcohols can hydrogen bond with water. The hydrocarbon chain, however, is relatively nonpolar and interacts poorly with water. As the chain gets longer, a greater portion of the molecule cannot interact with the water molecules and the solubility decreases, i.e., the effect of the –OH group decreases as the compounds get larger.

21.   a. Mg^{2+}; smaller, higher charge   b. Be^{2+}; smaller

c. Fe^{3+}; smaller size, higher charge   d. F^-; smaller

e. Cl^-; smaller   f. SO_4^{2-}; higher charge

22.

$$KCl(s) \rightarrow K^+(g) + Cl^-(g) \qquad \Delta H = -\Delta H_{LE} = -(-715 \text{ kJ/mol})$$
$$K^+(g) + Cl^-(g) \rightarrow K^+(aq) + Cl^-(aq) \qquad \Delta H = \Delta H_{hyd} = -684 \text{ kJ/mol}$$

_____

$$KCl(s) \rightarrow K^+(aq) + Cl^-(aq) \qquad \Delta H_{soln} = 31 \text{ kJ/mol}$$

23.

$$CsI(s) \rightarrow Cs^+(g) + I^-(g) \qquad \Delta H = -\Delta H_{LE} = 604 \text{ kJ}$$
$$Cs^+(g) + I^-(g) \rightarrow Cs^+(aq) + I^-(aq) \qquad \Delta H = \Delta H_{hyd}$$

_____

$$CsI(s) \rightarrow Cs^+(aq) + I^-(aq) \qquad \Delta H_{soln} = 33 \text{ kJ}$$

$$33 \text{ kJ} = 604 \text{ kJ} + \Delta H_{hyd}, \quad \Delta H_{hyd} = -571 \text{ kJ}$$

$$CsOH(s) \rightarrow Cs^+(g) + OH^-(g) \qquad \Delta H = -\Delta H_{LE} = 724 \text{ kJ}$$
$$Cs^+(g) + OH^-(g) \rightarrow Cs^+(aq) + OH^-(aq) \qquad \Delta H = \Delta H_{hyd}$$

_____

$$CsOH(s) \rightarrow Cs^+(aq) + OH^-(aq) \qquad \Delta H_{soln} = -72 \text{ kJ}$$

$$-72 \text{ kJ} = 724 \text{ kJ} + \Delta H_{hyd}, \quad \Delta H_{hyd} = -796 \text{ kJ}$$

24.   The enthalpy of hydration of CsOH is more exothermic than for CsI. Any differences must be because of differences in hydrations between OH^- and I^-. Thus, hydroxide ion is more strongly hydrated.

25.   Water is a polar molecule capable of hydrogen bonding. Polar molecules, molecules capable of hydrogen bonding, and ions can be hydrated. For covalent compounds, as polarity increases, hydration increases. For ionic compounds, as the size of an ion decreases and/or the charge increases, hydration increases.

   a. CH_3CH_2OH; CH_3CH_2OH is polar while $CH_3CH_2CH_3$ is nonpolar.

   b. $CHCl_3$; $CHCl_3$ is polar while CCl_4 is nonpolar.

   c. CH_3CO_2H; CH_3CO_2H is much more polar than $CH_3(CH_2)_{14}CO_2H$.

   d. Na_2S;   Most sulfides are insoluble. However, Na_2S is an exception.

   e. $AlCl_3$; Al_2O_3 is an insoluble covalent network solid.

   f. CO_2; CO_2 is nonpolar and not very soluble, but SiO_2 is a covalent network solid and is virtually insoluble in water.

26.   a.   Water; $Cu(NO_3)_2$ is an ionic solid.

   b. CCl_4; CS_2 is a nonpolar molecule.   c.   Water; CH_3CO_2H is polar.

   d. CCl_4; The long nonpolar hydrocarbon chain favors a nonpolar solvent.

   e.   Water; HCl is polar.   f. CCl_4; C_6H_6 is nonpolar.

27.   Both $Al(OH)_3$ and NaOH are ionic. Since the lattice energy is proportional to charge, the lattice energy of aluminum hydroxide is greater than that of sodium hydroxide. The attraction of water molecules for Al^{3+} and OH^- cannot overcome the larger lattice energy and $Al(OH)_3$ is insoluble. For NaOH, the hydration energy is large enough to overcome the smaller lattice energy and NaOH is soluble.

28.   The interparticle forces are:

          hexane: LDF; chloroform ($CHCl_3$): dipole-dipole, LDF

          methanol: H-bonding, LDF: H_2O: H-bonding (dominate)

        There is a gradual change in the nature of the interparticle forces (weaker to stronger). Each preceding solvent is miscible in its predecessor because there is not a great change in the strengths of interparticle forces from one solvent to the next.

29.   The micelles form so the ionic end, $-OSO_3^-$, is exposed to water on the outside and the nonpolar hydrocarbon chain is hidden from the water by pointing toward the inside of the micelle.

30.   hydrophobic: water hating; hydrophilic: water loving

31. $CO_2 + OH^- \rightarrow HCO_3^-$; No, the reaction of CO_2 with OH^- greatly increases the solubility of CO_2 in basic solution by forming the soluble bicarbonate ion.

32. $P_{gas} = kC$, $120 \text{ torr} \times \dfrac{1 \text{ atm}}{760 \text{ torr}} = \dfrac{780 \text{ atm}}{M} \times C$, $C = 2.0 \times 10^{-4}$ mol/L

33.   As the temperature increases the gas molecules will have a greater average kinetic energy. A greater fraction of the gas molecules in solution will have a kinetic energy greater than the attractive forces between the gas molecules and the solvent molecules. More gas molecules are able to escape to the vapor phase and the solubility of the gas decreases.

Vapor Pressures of Solutions

34. $P_{H_2O} = \chi_{H_2O}^L P_{H_2O}^\circ$; $\chi_{H_2O}^L = \dfrac{\text{mol } H_2O \text{ in solution}}{\text{total mol in solution}}$

$50.0 \text{ g } C_6H_{12}O_6 \times \dfrac{1 \text{ mol } C_2H_{12}O_6}{180.16 \text{ g } C_6H_{12}O_6} = 0.278 \text{ mol glucose}$

$600.0 \text{ g } H_2O \times \dfrac{1 \text{ mol}}{18.015 \text{ g}} = 33.31 \text{ mol } H_2O$

$\chi_{H_2O}^L = \dfrac{33.31}{33.59} = 0.9917$; $P_{H_2O} = \chi_{H_2O}^L P_{H_2O}^\circ = 0.9917 \times 23.8 \text{ torr} = 23.6 \text{ torr}$

35. $50.0 \text{ g } CH_3COCH_3 \times \dfrac{1 \text{ mol}}{58.08 \text{ g}} = 0.861 \text{ mol acetone}$

$50.0 \text{ g } CH_3OH \times \dfrac{1 \text{ mol}}{32.04 \text{ g}} = 1.56 \text{ mol methanol}$

$\chi_{acetone}^L = \dfrac{0.861}{0.861 + 1.56} = 0.356$; $\chi_{methanol}^L = 1 - \chi_{acetone}^L = 0.644$

$P_{total} = P_{methanol} + P_{acetone} = 0.644(143 \text{ torr}) + 0.356(271 \text{ torr}) = 92.1 \text{ torr} + 96.5 \text{ torr} = 188.6 \text{ torr}$

Since partial pressures are proportional to the number of moles, then in the vapor phase:

$\chi_{acetone}^V = \dfrac{P_{acetone}}{P_{total}} = \dfrac{96.5 \text{ torr}}{188.6 \text{ torr}} = 0.512$; $\chi_{methanol}^V = 1.000 - 0.512 = 0.488$

It is probably not a good assumption that this solution is ideal because of methanol-acetone hydrogen bonding, resulting in stronger forces in solution as compared to the separate phases.

36.   Compared to H_2O, solution d (methanol/water) will have the highest vapor pressure since methanol is more volatile than water. Both solution b (glucose/water) and solution c (NaCl/water) will have a lower vapor pressure than water by Raoult's law. NaCl dissolves to give Na^+ ions and Cl^- ions; glucose is a nonelectrolyte. Since there are more particles in solution c, the vapor pressure of solution c will be the lowest.

37.   If solute-solvent attractions > solvent-solvent and solute-solute attractions, then there is a negative deviation from Raoult's law. If solute-solvent attractions < solvent-solvent and solute-solute attractions, then there is a positive deviation from Raoult's law.

38.   A positive deviation from Raoult's law means the vapor pressure of the solution is greater than if the solution were ideal. At the boiling point, the vapor pressure equals atmospheric pressure. For a solution with positive deviations, it will take a lower temperature to achieve a vapor pressure of one atmosphere. Therefore, the boiling point is lower than if the solution were ideal.

39.   a. CF_3CF_3 and $CF_3CF_2CF_3$; Solutions will be ideal when the intermolecular forces between the two substances are about equal. This usually occurs for two nonpolar substances of similar size since the strength of the London dispersion forces will be similar. CF_3CF_3 and $CF_3CF_2CF_3$ are both relatively nonpolar with similar size. Hydrogen bonding will be important in water/acetone solutions and will likely form solutions with a negative deviation from Raoult's law.

       b. C_7H_{16} and C_6H_{14}; Both are nonpolar with similar size. The LD forces are approximately equal. Dipole forces are present in both CHF_3 and CH_3OCH_3 and it is unlikely that the dipole forces in solution are the same as in the two pure substances.

       c. CCl_4 and CF_4 since both are nonpolar substances. There is hydrogen bonding in aqueous solutions of phosphoric acid and it is unlikely that the forces in solution are the same as in the pure substances.

40. $25.8 \text{ g CH}_4\text{N}_2\text{O} \times \dfrac{1 \text{ mol}}{60.06 \text{ g}} = 0.430 \text{ mol}; \quad 275 \text{ g H}_2\text{O} \times \dfrac{1 \text{ mol}}{18.02 \text{ g}} = 15.3 \text{ mol}$

       $\chi^L_{H_2O} = \dfrac{15.3}{15.3 + 0.430} = 0.973; \quad P_{H_2O} = \chi^L_{H_2O} P^\circ_{H_2O} = 0.973 \,(23.8 \text{ torr}) = 23.2 \text{ torr at } 25°C$

       $P_{H_2O} = 0.973 \,(71.9 \text{ torr}) = 70.0 \text{ torr at } 45°C$

41. $P_{tol} = \chi^L_{tol} P^\circ_{tol}; \; P_{ben} = \chi^L_{ben} P^\circ_{ben}$; For the vapor, $\chi^V_A = P_A/P_{total}$. Since the mole fractions of benzene and toluene are equal in the vapor phase, then $P_{tol} = P_{ben}$.

       $\chi^L_{tol} P^\circ_{tol} = \chi^L_{ben} P^\circ_{ben} = (1.00 - \chi^L_{tol})P^\circ_{ben}, \quad \chi^L_{tol}(28 \text{ torr}) = (1.00 - \chi^L_{tol})95 \text{ torr}$

       $123 \, \chi^L_{tol} = 95, \; \chi^L_{tol} = 0.77; \; \chi^L_{ben} = 1.00 - 0.77 = 0.23$

42.   a. $25 \text{ mL C}_5\text{H}_{12} \times \dfrac{0.63 \text{ g}}{\text{mL}} \times \dfrac{1 \text{ mol}}{72.1 \text{ g}} = 0.22 \text{ mol C}_5\text{H}_{12}$

            $45 \text{ mL C}_6\text{H}_{14} \times \dfrac{0.66 \text{ g}}{\text{mL}} \times \dfrac{1 \text{ mol}}{86.2 \text{ g}} = 0.34 \text{ mol C}_6\text{H}_{14}$

            $\chi^L_{pen} = \dfrac{\text{mol pentane in solution}}{\text{total mol in solution}} = \dfrac{0.22}{0.56} = 0.39, \; \chi^L_{hex} = 1.00 - 0.39 = 0.61$

            $P_{pen} = \chi^L_{pen} P^\circ_{pen} = 0.39(511 \text{ torr}) = 2.0 \times 10^2 \text{ torr}; \; P_{hex} = 0.61(150. \text{ torr}) = 92 \text{ torr}$

            $P_{total} = P_{pen} + P_{hex} = 2.0 \times 10^2 + 92 = 292 \text{ torr} \approx 290 \text{ torr}$

       b.   In the vapor phase:

            $\chi^V_{pen} = \dfrac{P_{pen}}{P_{total}} = \dfrac{2.0 \times 10^2 \text{ torr}}{290 \text{ torr}} = 0.69$

43. $P_{total} = P_{pen} + P_{hex}$,   350. torr $= \chi^L_{pen}(511 \text{ torr}) + \chi^L_{hex}(150. \text{ torr})$

350. $= 511 \chi^L_{pen} + (1 - \chi^L_{pen})150.$, $\dfrac{200.}{361} = \chi^L_{pen} = 0.554$; $\chi^L_{hex} = 1.000 - 0.554 = 0.446$

For the vapor composition:

$P_{pen} = 0.554 \, (511 \text{ torr}) = 283 \text{ torr}$; $P_{hex} = 0.446 \, (150. \text{ torr}) = 66.9 \text{ torr}$

$\chi^V_{pen} = \dfrac{P_{pen}}{P_{tot}} = \dfrac{283 \text{ torr}}{350. \text{ torr}} = 0.809$; $\chi^V_{hex} = 1 - 0.809 = 0.191$

Colligative Properties

44. $m = \dfrac{4.9 \text{ g sucrose}}{175 \text{ g solvent}} \times \dfrac{1000 \text{ g}}{\text{kg}} \times \dfrac{1 \text{ mol } C_{12}H_{22}O_{11}}{342.3 \text{ g } C_{12}H_{22}O_{11}} = 0.082 \text{ molal}$

$\Delta T_b = K_b m = \dfrac{0.51°C}{\text{molal}} \times 0.082 \text{ molal} = 0.042°C$; $T_b = 100.042°C$ at 1 atm barometric pressure

$\Delta T_f = K_f m = \dfrac{1.86°C}{\text{molal}} \times 0.082 \text{ molal} = 0.15°C$; $T_f = -0.15°C$

$\pi = MRT$;   In dilute aqueous solutions the density of the solution is about 1.0 g/mL, thus we can assume that molarity $= M =$ molality $= 0.082$.

$\pi = \dfrac{0.082 \text{ mol}}{\text{L}} \times \dfrac{0.08206 \text{ L atm}}{\text{mol K}} \times 298 \text{ K} = 2.0 \text{ atm}$

45.   Assume molarity and molality are equal since solution is dilute.

$M = m = \dfrac{1.0 \text{ g}}{\text{L}} \times \dfrac{1 \text{ mol}}{9.0 \times 10^4 \text{ g}} = 1.1 \times 10^{-5} \text{ mol/L}$; $\pi = MRT$

At 298 K: $\pi = \dfrac{1.1 \times 10^{-5} \text{ mol}}{\text{L}} \times \dfrac{0.08206 \text{ L atm}}{\text{mol K}} \times 298 \text{ K} \times \dfrac{760 \text{ mm Hg}}{\text{atm}}$, $\pi = 0.20 \text{ mm Hg}$

$\Delta T_f = K_f m = \dfrac{1.86°C}{\text{molal}} \times 1.1 \times 10^{-5} \text{ molal} = 2.0 \times 10^{-5}°C$

46.   Osmotic pressure is better for determining the molar mass of large molecules. A temperature change of $10^{-5}°C$ is very difficult to measure. A change in height of a column of mercury by 0.2 mm is not as hard to measure precisely.

47. $\pi = MRT$, $M = \dfrac{\pi}{RT} = \dfrac{0.56 \text{ torr}}{\dfrac{0.08206 \text{ L atm}}{\text{mol K}} \times 298 \text{ K}} \times \dfrac{1 \text{ atm}}{760 \text{ torr}}$, $M = 3.0 \times 10^{-5} \text{ mol/L}$

$$3.0 \times 10^{-5} \text{ mol/L} = \frac{20.0 \times 10^{-3} \text{ g}}{25.0 \times 10^{-3} \text{ L}} \times \frac{1}{\text{molar mass}}, \text{ molar mass} = 27,000 \text{ g/mol}$$

48. $\Delta T_f = K_f m, \quad m = \dfrac{\Delta T_f}{K_f} = \dfrac{2.63 \,°C}{40.°C \text{ kg mol}^{-1}} = \dfrac{6.6 \times 10^{-2} \text{ mol reserpine}}{\text{kg solvent}}$

$$\frac{6.6 \times 10^{-2} \text{ mol reserpine}}{\text{kg solvent}} = \frac{1.00 \text{ g reserpine}}{25.0 \text{ g solvent}} \times \frac{1}{\text{MM}} \times \frac{1000 \text{ g solvent}}{\text{kg solvent}}$$

$$\text{MM} = \text{molar mass of reserpine} = \frac{1000}{25.0 \times (6.6 \times 10^{-2})} = 606 \text{ g/mol} \approx 610 \text{ g/mol}$$

49.   Cellophane has several –OH groups that can hydrogen bond to water molecules. Water should be "soluble" in cellophane.

50. $\Delta T_f = 5.51 - 2.81 = 2.70°C; \quad m = \dfrac{\Delta T_f}{K_f} = \dfrac{2.70°C}{5.12°C/\text{molal}} = 0.527 \text{ molal}$

Let x = mass of naphthalene (molar mass = 128.2 g/mol). Then $1.60 - x$ = mass of anthracene (molar mass = 178.2 g/mol).

$$\frac{x}{128.2} = \text{moles naphthalene and } \frac{1.60 - x}{178.2} = \text{moles anthracene}$$

$$\frac{0.527 \text{ moles solute}}{\text{kg solvent}} = \frac{\dfrac{x}{128.2} + \dfrac{1.60 - x}{178.2}}{0.0200 \text{ kg solvent}}, \quad 1.05 \times 10^{-2} = \frac{178.2\,x + 1.60(128.2) - 128.2\,x}{128.2(178.2)}$$

$50.0\,x + 205 = 240., \quad 50.0\,x = 240. - 205, \quad 50.0\,x = 35, \quad x = 0.70 \text{ g naphthalene}$

So the mixture is:

$$\frac{0.70 \text{ g}}{1.60 \text{ g}} \times 100 = 44\% \text{ naphthalene by mass and } 56\% \text{ anthracene by mass}$$

51.   Assume $T = 25°C; \quad \pi = MRT = \dfrac{0.1 \text{ mol}}{\text{L}} \times \dfrac{0.08206 \text{ L atm}}{\text{mol K}} \times 298 \text{ K} = 2.45 \text{ atm} \approx 2 \text{ atm}$

$$\pi = 2 \text{ atm} \times \frac{760 \text{ mm Hg}}{\text{atm}} \approx 2000 \text{ mm} \approx 2 \text{ m}$$

The osmotic pressure would support a mercury column of ≈ 2 m. The height of a fluid column in a tree will be higher because Hg is more dense than the fluid in a tree. If we assume the fluid in a tree is mostly H_2O, then the fluid has a density of 1.0 g/cm^3. The density of Hg is 13.6 g/cm^3.

Height of fluid ≈ 2 m $\times$ 13.6 ≈ 30 m

52.   With addition of salt or sugar, the osmotic pressure inside the fruit cells (and bacteria) is less than outside the cell. Water will leave the cells which will dehydrate any bacteria present, causing them to die.

53. $$m = \frac{\Delta T_f}{K_f} = \frac{30.0°C}{1.86°C \text{ kg mol}^{-1}} = 16.1 \text{ mol } C_2H_6O_2/kg$$

Since the density of water is 1.00 g/cm^3, the moles of $C_2H_6O_2$ needed are:

$$15.0 \text{ L H}_2O \times \frac{1.00 \text{ kg H}_2O}{\text{L H}_2O} \times \frac{16.1 \text{ mol } C_2H_6O_2}{\text{kg H}_2O} = 242 \text{ mol } C_2H_6O_2$$

$$\text{Volume } C_2H_6O_2 = 242 \text{ mol } C_2H_6O_2 \times \frac{62.07 \text{ g}}{\text{mol}} \times \frac{1 \text{ cm}^3}{1.11 \text{ g}} = 13,500 \text{ cm}^3 = 13.5 \text{ L}$$

$$\Delta T_b = K_b m = \frac{0.51°C}{\text{molal}} \times 16.1 \text{ molal} = 8.2°C; \quad T_b = 100.0°C + 8.2°C = 108.2°C$$

Properties of Electrolyte Solutions

54. $NaCl(s) \rightarrow Na^+(aq) + Cl^-(aq)$; The total concentration of particles is $2(0.6 \text{ } M) = 1.2 \text{ } M$.

$$\pi = MRT = \frac{1.2 \text{ mol}}{L} \times \frac{0.08206 \text{ L atm}}{\text{mol K}} \times 298 \text{ K} = 29.3 \text{ atm} = 29 \text{ atm}$$

A pressure greater than 30. atm should be applied to insure purification by reverse osmosis.

55.   a.   Water would migrate from right to left. The level of liquid in the right arm would go down and the level in the left arm would go up.

        b.   The levels would be equal. The concentration of NaCl would be equal in both chambers.

56. $MgCl_2$ and NaCl are strong electrolytes, HOCl is a weak electrolyte and glucose is a nonelectrolyte. The effective particle concentrations are ~3.0 m $MgCl_2$, ~2.0 m NaCl, 2.0 < m HOCl < 1.0, and 1.0 m glucose. The order of freezing point depressions ($\Delta T_f = K_f m$) from lowest to highest are: glucose < HOCl < NaCl < $MgCl_2$.

57.   The solutions of glucose, NaCl and $CaCl_2$ will all have lower freezing points, higher boiling points and higher osmotic pressures than pure water. The solution with the largest particle concentration will have the lowest freezing point, the highest boiling point and the highest osmotic pressure. The $CaCl_2$ solution will have the largest effective particle concentration since it produces three ions per 1 mol of compound.

        a.   pure water   b. $CaCl_2$ solution   c. $CaCl_2$ solution

        d.   pure water   e. $CaCl_2$ solution

58. $Ca(NO_3)_2(s) \rightarrow Ca^{2+}(aq) + 2\ NO_3^-(aq);\ i = 3$ mol particles/mol $Ca(NO_3)_2$

$\Delta T_f = iK_f m = 3 \times \dfrac{1.86°C}{molal} \times 0.5$ molal $= 2.8°C \approx 3°C;\ T_f = -3°C$

59.   The measured freezing point should be higher. Due to ion pairing, the depression of the freezing point will be less than expected since the effective particle concentration is less than 3(0.5) molal. This results in a higher freezing point.

60.   a.   Assuming $MgCO_3(s)$ does not dissociate, the solute concentration in water is:

$$\frac{560\ \mu g\ MgCO_3(s)}{mL} = \frac{560\ mg}{L} = \frac{560 \times 10^{-3}\ g}{L} \times \frac{1\ mol\ MgCO_3}{84.32\ g} = 6.6 \times 10^{-3}\ mol\ MgCO_3/L$$

An applied pressure of 8.0 atm will purify water up to a solute concentration of:

$$M = \frac{\pi}{RT} = \frac{8.0\ atm}{0.08206\ L\ atm\ mol^{-1}\ K^{-1}\ (300.\ K)} = \frac{0.32\ mol}{L}$$

When the concentration of $MgCO_3(s)$ reaches 0.32 mol/L, the reverse osmosis unit can no longer purify the water. Let V = volume (L) of water remaining after purifying 45 L of H_2O. When V + 45 L of water has been processed, the moles of solute particles will equal:

6.6×10^{-3} mol/L (45 L + V) = 0.32 mol/L (V)

Solving: 0.30 = (0.32 - 0.0066) V, V − 0.96 L

The minimum total volume of water that must be processed is 45 L + 0.96 L = 46 L.

Note: If $MgCO_3$ does dissociate into Mg^{2+} and CO_3^{2-} ions, then the solute concentration increases to $1.3 \times 10^{-2}\ M$ and at least 47 L of water must be processed.

b.   No; A reverse osmosis system that applies 8.0 atm can only purify water with a solute concentration less than 0.32 mol/L. Salt water has a solute concentration of 2(0.60 *m*) = 1.20 *m* ions. The solute concentration of salt water is much too high for this reverse osmosis unit to work.

61.   Ion pairing can occur, resulting in fewer particles than expected. This results in smaller freezing point depressions and smaller boiling point elevations ($\Delta T = Km$). Ion pairing will increase as the concentration of electrolyte increases.

62. $\Delta T_f = iK_f m,\ i = \dfrac{\Delta T_f}{K_f m} = \dfrac{0.440°C}{1.86°C/molal \times 0.091\ molal} = 2.6$ for 0.091 *m* $CaCl_2$

$i = \dfrac{1.330}{1.86 \times 0.279} = 2.56$ for 0.279 *m* $CaCl_2$; $i = \dfrac{2.345}{1.86 \times 0.475} = 2.65$ for 0.475 *m* $CaCl_2$

63.   For $CaCl_2$, i = 2.6 (from Exercise 17.62); % $CaCl_2$ ionized = $\dfrac{2.6 - 1.0}{3.0 - 1.0} \times 100 = 80.\%$

For CsCl: i = $\dfrac{\Delta T_f}{K_f m}$ = $\dfrac{0.302°C}{1.86°C/molal \times 0.091\ molal}$ = 1.8

% CsCl ionized = $\dfrac{1.8 - 1.0}{2.0 - 1.0} \times 100 = 80.\%$

It appears that the i value for $CaCl_2$ is further from the ideal value. However, since both compounds are equally ionized, we can conclude that the extent of ion association is about the same in each solution.

64.   a.   °C = 5(°F - 32)/9 = 5(-29 -32)/9 = -34°C

Assuming the solubility of $CaCl_2$ is temperature independent, the molality of a saturated $CaCl_2$ solution is:

$$\dfrac{74.5\ g\ CaCl_2}{100.0\ mL\ H_2O} \times \dfrac{1.00\ mL\ H_2O}{1.00\ g\ H_2O} \times \dfrac{1000\ g}{kg} \times \dfrac{1\ mol\ CaCl_2}{111.0\ g\ CaCl_2} = \dfrac{6.71\ mol\ CaCl_2}{kg\ H_2O}$$

ΔT_f = $i K_f m$ = 3(1.86°C kg mol^{-1}) (6.71 mol kg^{-1}) = 37.4°C

Assuming i = 3, a saturated solution of $CaCl_2$ can lower the freezing point of water to -37.4°C. Assuming these conditions, a saturated $CaCl_2$ solution should melt ice at -34°C (-29°F).

b.   From Exercise 62, i ≈ 2.6; ΔT_f = $i K_f m$ = 2.6(1.86) (6.71) = 32°C; T_f = -32°C

Assuming i = 2.6, a saturated $CaCl_2$ solution will not melt ice at -34°C(-29°F).

Additional Exercises

65.

Benzoic acid is capable of hydrogen bonding. However, it is more soluble in nonpolar benzene than in water. In benzene, a nonpolar hydrogen bonded dimer forms.

The dimer is relatively nonpolar and thus more soluble in benzene than in water. Since benzoic acid forms dimers in benzene, the effective solute particle concentration will be less than 1.0 molal. Therefore, the freezing point depression would be less than 5.12°C ($\Delta T_f = K_f m$).

66.   Benzoic acid would be more soluble in a basic solution because of the reaction:

$$C_6H_5CO_2H + OH^- \rightarrow C_6H_5CO_2^- + H_2O$$

67.   a.   The average values for each ion are:

   300. mg Na^+; 15.7 mg K^+; 5.45 mg Ca^{2+}; 388 mg Cl^-; 246 mg lactate, $C_3H_5O_3^-$

   Note: Since we can precisely weigh to ± 0.1 mg on an analytical balance, we'll carry extra significant figures and calculate results to ± 0.1 mg.

The only source of lactate is $NaC_3H_5O_3$.

$$246 \text{ mg lactate} \times \frac{112.06 \text{ mg } NaC_3H_5O_3}{89.07 \text{ mg } C_3H_5O_3^-} = 309.5 \text{ mg sodium lactate}$$

The only source of Ca^{2+} is $CaCl_2\cdot 2H_2O$.

$$5.45 \text{ mg } Ca^{2+} \times \frac{147.0 \text{ mg } CaCl_2\cdot 2H_2O}{40.08 \text{ mg } Ca^{2+}} = 19.99 \text{ or } 20.0 \text{ mg } CaCl\cdot 2H_2O$$

The only source of K^+ is KCl.

$$15.7 \text{ mg } K^+ \times \frac{74.55 \text{ mg KCl}}{39.10 \text{ mg } K^+} = 29.9 \text{ mg KCl}$$

From what we have used already, let's calculate the mass of Na^+ and Cl^- added.

   309.5 mg sodium lactate = 246.0 mg lactate + 63.5 mg Na^+

Thus, we need to add an additional 236.5 mg Na^+ to get the desired 300. mg.

$$236.5 \text{ mg } Na^+ \times \frac{58.44 \text{ mg NaCl}}{22.99 \text{ mg } Na^+} = 601.2 \text{ mg NaCl}$$

Let's check the mass of Cl^- added:

$$20.0 \text{ mg } CaCl_2\cdot 2H_2O \times \frac{70.90 \text{ mg } Cl^-}{147.0 \text{ mg } CaCl_2\cdot 2H_2O} = 9.6 \text{ mg } Cl^-$$

$$20.0 \text{ mg CaCl}_2 \cdot 2\text{H}_2\text{O} = 9.6 \text{ mg Cl}^-$$

$$29.9 \text{ mg KCl} - 15.7 \text{ mg K}^+ = 14.2 \text{ mg Cl}^-$$

$$601.2 \text{ mg NaCl} - 236.5 \text{ mg Na}^+ = 364.7 \text{ mg Cl}^-$$

$$\text{Total Cl}^- = 388.5 \text{ mg Cl}^-$$

This is the quantity of Cl$^-$ we want (the average amount of Cl$^-$).

An analytical balance can weigh to the nearest 0.1 mg. We would use 309.5 mg sodium lactate, 20.0 mg CaCl$_2 \cdot$2H$_2$O, 29.9 mg KCl and 601.2 mg NaCl.

b. To get the range of osmotic pressure, we need to calculate the molar concentration of each ion at its minimum and maximum values. At minimum concentrations, we have:

$$\frac{285 \text{ mg Na}^+}{100. \text{ mL}} \times \frac{1 \text{ mmol}}{22.99 \text{ mg}} = 0.124 \ M; \quad \frac{14.1 \text{ mg K}^+}{100. \text{ mL}} \times \frac{1 \text{ mmol}}{39.10 \text{ mg}} = 0.00361 \ M$$

$$\frac{4.9 \text{ mg Ca}^{2+}}{100. \text{ mL}} \times \frac{1 \text{ mmol}}{40.08 \text{ mg}} = 0.0012 \ M; \quad \frac{368 \text{ mg Cl}^-}{100. \text{ mL}} \times \frac{1 \text{ mmol}}{35.45 \text{ mg}} = 0.104 \ M$$

$$\frac{231 \text{ mg lactate}}{100. \text{ mL}} \times \frac{1 \text{ mmol}}{89.07 \text{ mg}} = 0.0259 \ M$$

Total = 0.124 + 0.00361 + 0.0012 + 0.104 + 0.0259 = 0.259 M

$$\pi = MRT = \frac{0.259 \text{ mol}}{L} \times \frac{0.08206 \text{ L atm}}{\text{mol K}} \times 310. \text{ K} = 6.59 \text{ atm}$$

Similarly at maximum concentrations, the concentration of each ion is:

Na$^+$: 0.137 M; K$^+$: 0.00442 M; Ca^{2+}: 0.0015 M; Cl$^-$: 0.115 M; C$_3$H$_5$O$_3^-$: 0.0293 M

The total concentration of all ions is the sum, 0.287 M.

$$\pi = \frac{0.287 \text{ mol}}{L} \times \frac{0.08206 \text{ L atm}}{\text{mol K}} \times 310. \text{ K} = 7.30 \text{ atm}$$

Osmotic pressure ranges from 6.59 atm to 7.30 atm.

68.   No, the solution is not ideal. For an ideal solution, the strength of interparticle forces in the solution are the same as in the pure solute and pure solvent. This results in $\Delta H_{soln} = 0$ for an ideal solution. ΔH_{soln} for methanol/water is not zero. Since $\Delta H_{soln} < 0$, this solution shows negative deviation from Raoult's law.

69. $\chi_{pen}^{V} = 0.15 = \dfrac{P_{pen}}{P_{tot}}$, $P_{pen} = 0.15\, P_{tot} = \chi_{pen}^{L}(511\text{ torr})$

$P_{tot} = P_{pen} + P_{hex} = \chi_{pen}^{L}(511) + \chi_{hex}^{L}(150.)$, $P_{tot} = \chi_{pen}^{L}(511) + (1 - \chi_{pen}^{L})(150.) = 150. + 361\,\chi_{pen}^{L}$

From above: $P_{pen} = 0.15\, P_{tot} = \chi_{pen}^{L}(511)$ and $P_{tot} = 150. + 361\,\chi_{pen}^{L}$; Solving using simultaneous equations:

$$0.15\,P_{tot} = 0.15(361)\,\chi_{pen}^{L} + 0.15(150.)$$
$$-0.15\,P_{tot} = \qquad\quad -511\,\chi_{pen}^{L}$$

$$\overline{\qquad\qquad 0 = (54 - 511)\,\chi_{pen}^{L} + 23,\;457\,\chi_{pen}^{L} = 23,\;\chi_{pen}^{L} = 0.050 \qquad}$$

70. $P = \chi^{L}P°$, 710.0 torr $= \chi^{L}(760.0$ torr$)$, $\chi^{L} = 0.9342 =$ mole fraction of methanol

71. $\Delta T_f = K_f m$, $m = \dfrac{\Delta T}{K_f} = \dfrac{5.40°C}{1.86°C/molal} = 2.90$ molal

$\dfrac{2.90 \text{ mol solute}}{\text{kg solvent}} = \dfrac{n}{0.0500 \text{ kg}}$, n = 0.145 mol of ions in solution

Since $NaNO_3$ and $Mg(NO_3)_2$ are strong electrolytes:

n = 2(x mol of $NaNO_3$) + 3[y mol $Mg(NO_3)_2$] = 0.145 mol ions

In addition: 6.50 g = x mol $NaNO_3\left(\dfrac{85.00 \text{ g}}{\text{mol}}\right) + y$ mol $Mg(NO_3)_2\left(\dfrac{148.3 \text{ g}}{\text{mol}}\right)$

We have two equations: 2 x + 3 y = 0.145 and 85.00 x + 148.3 y = 6.50

Solving by simultaneous equations:

$$-85.00\,x - 127.5\,y = -6.16$$
$$85.00\,x + 148.3\,y = 6.50$$

$$\overline{\qquad\quad 20.8\,y = 0.34,\; y = 0.016 \text{ mol } Mg(NO_3)_2 \qquad}$$

mass of $Mg(NO_3)_2$ = 0.016 mol (148.3 g/mol) = 2.4 g $Mg(NO_3)_2$

mass of $NaNO_3$ = 6.50 g - 2.4 g = 4.1 g $NaNO_3$

72.   14.22 mg $CO_2 \times \dfrac{12.011 \text{ mg C}}{44.009 \text{ mg } CO_2} = 3.881$ mg C; % C $= \dfrac{3.881}{4.80} \times 100 = 80.9\%$ C

1.66 mg $H_2O \times \dfrac{2.016 \text{ mg H}}{18.02 \text{ mg } H_2O} = 0.186$ mg H; % H $= \dfrac{0.186}{4.80} \times 100 = 3.88\%$ H

% O = 100.00 - (80.9 + 3.88) = 15.2% O

Out of 100.00 g:

$$80.9 \text{ g C} \times \frac{1 \text{ mol}}{12.01 \text{ g}} = 6.74 \text{ mol C}; \quad \frac{6.74}{0.950} = 7.09 \approx 7$$

$$3.88 \text{ g H} \times \frac{1 \text{ mol}}{1.008 \text{ g}} = 3.85 \text{ mol H}; \quad \frac{3.85}{0.950} = 4.05 \approx 4$$

$$15.2 \text{ g O} \times \frac{1 \text{ mol}}{16.00 \text{ g}} = 0.950 \text{ mol O}; \quad \frac{0.950}{0.950} = 1$$

Therefore, the empirical formula is C_7H_4O.

$$\Delta T_f = K_f m, \quad m = \frac{\Delta T_f}{K_f} = \frac{22.3°C}{40.°C/molal} = 0.56 \text{ molal}$$

$$0.56 \text{ molal} = \frac{1.32 \text{ g anthraquinone}}{0.0114 \text{ kg camphor}} \times \frac{1}{MM}, \quad MM = \text{molar mass anthraquinone}$$

$$MM = \frac{1.32}{0.0114 \times 0.56} = 210 \text{ g/mol}$$

The empirical mass of C_7H_4O is: $7(12) + 4(1) + 16 \approx 104$ g/mol. Since the molar mass is twice the empirical mass, then the molecular formula is $C_{14}H_8O_2$.

73.   a. $m = \dfrac{\Delta T_f}{K_f} = \dfrac{1.32°C}{5.12°C \text{ kg mol}^{-1}} = 0.258 \text{ mol/kg}$

$$\frac{0.258 \text{ mol X}}{\text{kg solvent}} = \frac{1.22 \text{ g X}}{0.01560 \text{ kg}} \times \frac{1}{MM}, \quad MM = \text{molar mass of unknown, X}$$

Solving for the molar mass: MM = 303 g/mol for the unknown compound.

Uncertainty in temperature $= \dfrac{0.04}{1.32} \times 100 = 3\%$; A 3% uncertainty in 303 g/mol = 9 g/mol.

So, molar mass = 303 ± 9 g/mol.

b. No, codeine could not be eliminated since its molar mass is in the possible range including the uncertainty.

c. We would really like the uncertainty to be ± 1 g/mol. We need the freezing point depression to be about 10 times what it was in this problem. Two possibilities are:

    1. make the solution ten times more concentrated (may be solubility problem) or
    2. use camphor ($K_f = 40.$) as the solvent.

74. $\pi = MRT$, $\pi = 18.6 \text{ torr} \times \dfrac{1 \text{ atm}}{760 \text{ torr}} = M \times \dfrac{0.08206 \text{ L atm}}{\text{mol K}} \times 298 \text{ K}$

$M = 1.00 \times 10^{-3} \text{ mol/L}$; $\dfrac{1.00 \times 10^{-3} \text{ mol}}{L} = \dfrac{0.15 \text{ g}}{2.0 \times 10^{-3} \text{ L}} \times \dfrac{1}{MM}$

MM = 7.5×10^{4} g/mol = molar mass of protein

75.   Out of 100.00 g, there are:

$31.57 \text{ g C} \times \dfrac{1 \text{ mol}}{12.011 \text{ g}} = 2.628 \text{ mol C}$; $\dfrac{2.628}{2.628} = 1.000$

$5.30 \text{ g H} \times \dfrac{1 \text{ mol}}{1.008 \text{ g}} = 5.26 \text{ mol H}$; $\dfrac{5.26}{2.628} = 2.00$

$63.13 \text{ g O} \times \dfrac{1 \text{ mol}}{15.999 \text{ g}} = 3.946 \text{ mol O}$; $\dfrac{3.946}{2.628} = 1.502$

Empirical formula: $C_2H_4O_3$

$m = \dfrac{\Delta T_f}{K_f} = \dfrac{5.20°C}{1.86°C/\text{molal}} = 2.80 \text{ molal}$; $\dfrac{2.80 \text{ mol}}{\text{kg}} = \dfrac{10.56 \text{ g}}{0.0250 \text{ kg H}_2\text{O}} \times \dfrac{1}{MM}$

MM = 151 g/mol = experimental molar mass of compound

The empirical mass of $C_2H_4O_3$ = 76.051 g/mol. Since the molar mass is about twice the empirical mass, then the molecular formula is $C_4H_8O_6$ which has a molar mass of 152.101 g/mol.

Note: We use the experimental molar mass to get the molecular formula. Knowing this, we calculate the molar mass precisely from the molecular formula.

76. $50 \text{ g NaOH} \times \dfrac{1 \text{ mol}}{40.0 \text{ g}} = 1.3 \text{ mol NaOH}$ (carry extra significant figure)

$50 \text{ g H}_2\text{O} \times \dfrac{1 \text{ mol}}{18.0 \text{ g}} = 2.8 \text{ mol H}_2\text{O}$

$\chi_{\text{NaOH}} = \dfrac{1.3}{1.3 + 2.8} = 0.3$; $m = \dfrac{1.3 \text{ mol NaOH}}{50 \text{ g H}_2\text{O}} \times \dfrac{1000 \text{ g}}{\text{kg}} = 26 \text{ mol/kg} \approx 30 \text{ mol/kg}$

77.   If there are 100.0 mL of wine:

$12.5 \text{ mL C}_2\text{H}_5\text{OH} \times \dfrac{0.79 \text{ g}}{\text{mL}} = 9.9 \text{ g C}_2\text{H}_5\text{OH}$ and $87.5 \text{ mL H}_2\text{O} \times \dfrac{1.0 \text{ g}}{\text{mL}} = 87.5 \text{ g H}_2\text{O}$

mass % ethanol = $\dfrac{9.9}{87.5 + 9.9} \times 100 = 10.\% \text{ by mass}$

$$m = \frac{9.9 \text{ g C}_2\text{H}_5\text{OH}}{0.0875 \text{ kg H}_2\text{O}} \times \frac{1 \text{ mol}}{46.07 \text{ g}} = 2.5 \text{ molal}$$

78.   a.   If we use 100. mL (100. g) of H_2O, we need:

$$0.100 \text{ kg} \times \frac{2.0 \text{ mol KCl}}{\text{kg}} \times \frac{74.55 \text{ g}}{\text{mol}} = 14.9 \text{ g} \approx 15 \text{ g KCl}$$

Dissolve 15 g KCl in 100. mL H_2O. This will give us slightly more than 100 mL, but this will be the easiest way to make the solution. Since we don't know the density, we can't calculate the molarity and use a volumetric flask to make exactly 100 mL of solution.

b.   If we took 15 g NaOH and 85 g H_2O, the volume would be less than 100 mL. To make sure we have enough solution, use 100. mL H_2O (100. g).

$$\text{mass \%} = 15 = \frac{x}{100. + x}(100), \quad 1500 + 15 x = 100 x, \quad x = 17.6 \text{ g} \approx 18 \text{ g}$$

Dissolve 18 g NaOH in 100. mL H_2O.

c.   In a fashion similar to 17.78b, use 100. mL CH_3OH, which is equal to 100. mL × 0.79 g/mL = 79 g CH_3OH.

$$\text{mass \%} = 25 = \frac{x}{79 + x}(100), \quad 25(79) + 25 x = 100 x, \quad x = 26.3 \text{ g} \approx 26 \text{ g}$$

Dissolve 26 g NaOH in 100. mL CH_3OH.

d.   Dissolve 30. ml C_2H_5OH in 70. mL H_2O. The total volume will be close to 100. mL.

79. $$m = \frac{40.0 \text{ g C}_2\text{H}_6\text{O}_2}{60.0 \text{ g H}_2\text{O}} \times \frac{1000 \text{ g}}{\text{kg}} \times \frac{1 \text{ mol}}{62.07 \text{ g}} = 10.7 \text{ mol/kg}$$

$$\Delta T_f = K_f m = 1.86°\text{C/molal} \times 10.7 \text{ molal} = 19.9°\text{C}; \quad T_f = -19.9°\text{C}$$

$$\Delta T_b = K_b m = 0.51°\text{C/molal} \times 10.7 \text{ molal} = 5.5°\text{C}; \quad T_b = 105.5°\text{C}$$

80. $$\pi = MRT, \quad M = \frac{\pi}{RT} = \frac{15}{(0.08206)(298)} = 0.61 \ M; \quad \frac{0.61 \text{ mol}}{L} \times \frac{342.3 \text{ g}}{\text{mol}} = 209 \text{ g/L} \approx 210 \text{ g/L}$$

Dissolve 209 g sucrose in water and dilute to 1.0 L in a volumetric flask. To get 0.61 ± 0.01 mol/L, we need 209 ± 3 g sucrose.

81. $\Delta T = 25.50°C - 24.59°C = 0.91°C = K_f m$, $m = \dfrac{0.91°C}{9.1°C/molal} = 0.10$ molal

mass H_2O = 0.0100 kg t-butanol $\left(\dfrac{0.10 \text{ mol } H_2O}{\text{kg t-butanol}} \right) \left(\dfrac{18.0 \text{ g } H_2O}{\text{mol } H_2O} \right) = 0.018$ g H_2O

82.   a. $m = \dfrac{5.0 \text{ g } C_6H_{12}O_6}{0.025 \text{ kg}} \times \dfrac{1 \text{ mol}}{180.2 \text{ g}} = 1.1$ molal

$\Delta T_f = K_f m = \dfrac{1.86°C}{\text{molal}} \times 1.1$ molal $= 2.0°C$; $T_f = -2.0°C$

b. $m = \dfrac{5.0 \text{ g NaCl}}{0.025 \text{ kg}} \times \dfrac{1 \text{ mol}}{58.44 \text{ g}} = 3.4$ molal;   NaCl is a strong electrolyte with i = 2.

$\Delta T_f = iK_f m = 2 \times 1.86°C/molal \times 3.4$ molal $= 13°C$; $T_f = -13°C$

c.   aluminum nitrate, $Al(NO_3)_3$,   i = 4; $m = \dfrac{2.0 \text{ g}}{0.015 \text{ kg}} \times \dfrac{1 \text{ mol}}{213.0 \text{ g}} = 0.63$ mol/kg

$\Delta T_f = iK_f m = 4 \times 1.86°C/molal \times 0.63$ molal $= 4.7°C$; $T_f = -4.7°C$

d. $m = \dfrac{1.0 \text{ g } C_6H_5CO_2H}{0.010 \text{ kg}} \times \dfrac{1 \text{ mol}}{122.1 \text{ g}} = 0.82$ mol/kg

$\Delta T_f = K_f m = 5.12°C/molal \times 0.82$ molal $= 4.2°C$; $T_f = 5.5°C - 4.2°C = 1.3°C$

Note: This ignores dimer formation from benzoic acid (see Exercise 17.65).

83.   750. mL grape juice $\times \dfrac{12 \text{ mL } C_2H_5OH}{100. \text{ mL juice}} \times \dfrac{0.79 \text{ g } C_2H_5OH}{\text{mL}} \times \dfrac{1 \text{ mol } C_2H_5OH}{46.07 \text{ g}} \times \dfrac{2 \text{ mol } CO_2}{2 \text{ mol } C_2H_5OH}$

$= 1.54$ mol CO_2 (carry extra significant figure)

1.54 mol CO_2 = total mol CO_2 = mol CO_2(g) + mol CO_2(aq) = $n_g + n_{aq}$

$P_{CO_2} = \dfrac{n_g RT}{V} = \dfrac{n_g \left(\dfrac{0.08206 \text{ L atm}}{\text{mol K}} \right) (298 \text{ K})}{75 \times 10^{-3} \text{ L}} = 326 \, n_g$; $P_{CO_2} = kC = \dfrac{32 \text{ L atm}}{\text{mol}} \times \dfrac{n_{aq}}{0.750 \text{ L}} = 42.7 \, n_{aq}$

$P_{CO_2} = 326 \, n_g = 42.7 \, n_{aq}$ and from above $n_{aq} = 1.54 - n_g$;   Solving:

$326 \, n_g = 42.7(1.54 - n_g)$, $369 \, n_g = 65.8$, $n_g = 0.18$ mol

$P_{CO_2} = 326(0.18) = 59$ atm in gas phase;   59 atm $= \dfrac{32 \text{ L atm}}{\text{mol}} \times C$, $C = 1.8$ mol CO_2/L in wine

84. $m = \dfrac{24.0 \text{ g} \times \dfrac{1 \text{ mol}}{58.0 \text{ g}}}{0.600 \text{ kg}} = 0.690 \text{ mol/kg}; \quad \Delta T_b = K_b m = (0.51 °C \text{ kg mol}^{-1})(0.690 \text{ mol/kg}) = 0.35 °C$

$T_b = 99.725 °C + 0.35 °C = 100.08 °C$

85. $m = \dfrac{\Delta T_b}{K_b} = \dfrac{0.55 °C}{1.71 \text{ °C kg mol}^{-1}} = 0.32 \text{ mol/kg}$

$0.32 \text{ mol/kg} = \dfrac{3.75 \text{ g}}{0.095 \text{ kg}} \times \dfrac{1}{\text{MM}}, \quad \text{MM} = 120 \text{ g/mol} = \text{molar mass of hydrocarbon}$

86. $P_{H_2O} = \chi_{H_2O} P^\circ_{H_2O}; \quad \chi_{H_2O} = \dfrac{\text{mol H}_2\text{O}}{\text{total mol}} = \dfrac{1000. \text{ g H}_2\text{O}(1 \text{ mol}/18.015 \text{ g})}{68.0 \text{ g sucrose }(1 \text{ mol}/342.30 \text{ g}) + 1000.(1/18.015)} = 0.9964$

$P_{H_2O} = 0.9964 (28.35 \text{ torr}) = 28.25 \text{ torr}$

87. $M = \dfrac{1.75 \text{ g} \times \dfrac{1 \text{ mol}}{342.3 \text{ g}}}{0.150 \text{ L}} = 0.0341 \text{ mol/L}$

$\pi = MRT = \dfrac{0.0341 \text{ mol}}{L} \times \dfrac{0.08206 \text{ L atm}}{\text{mol K}} \times 290.2 \text{ K} = 0.812 \text{ atm} = 617 \text{ torr}$

88. $m = \dfrac{\Delta T_f}{K_f} = \dfrac{0.52 °C}{1.86 °C \text{ kg mol}^{-1}} = 0.28 \text{ mol/kg} \approx M \approx 0.28 \text{ mol/L}$

$\pi = MRT = \dfrac{0.28 \text{ mol}}{L} \times \dfrac{0.08206 \text{ L atm}}{\text{mol K}} \times 310. \text{ K} = 7.1 \text{ atm}$

89. a. Water boils when the vapor pressure equals the pressure above the water. In an open pan $P_{atm} \approx 1$ atm. In a pressure cooker, $P_{inside} > 1$ atm and water boils at a higher temperature. The higher the cooking temperature, the faster the cooking time.

b. Salt dissolves in water forming a solution with a melting point lower than pure water $(\Delta T_f = K_f m)$. This happens in water on the surface of ice. If it is not too cold, the ice melts. This won't work if the ambient temperature is lower than the depressed freezing point of the salt solution.

c. When water freezes from a solution, it freezes as pure water, leaving behind a more concentrated salt solution.

d. On the CO_2 phase diagram, the triple point is above 1 atm and $CO_2(g)$ is the stable phase at 1 atm and room temperature. $CO_2(l)$ can't exist at normal atmospheric pressures. Therefore, dry ice sublimes instead of boils. In a fire extinguisher, $P > 1$ atm and $CO_2(l)$ can exist. When CO_2 is released from the fire extinguisher, $CO_2(g)$ forms as predicted from the phase diagram.

90. $$m = \frac{0.100 \text{ g} \times \dfrac{1 \text{ mol}}{100.0 \text{ g}}}{0.5000 \text{ kg}} = 2.00 \times 10^{-3} \text{ mol/kg} \approx 2.00 \times 10^{-3} \text{ mol/L} \quad \text{(dilute solution)}$$

$\Delta T_f = iK_f m$, $0.0056°C = i(1.86°C/\text{molal}) (2.00 \times 10^{-3} \text{ molal})$, $i = 1.5$

If $i = 1.0$, % dissociation = 0% and if $i = 2.0$, % dissociation = 100%. Since $i = 1.5$, then the weak acid is 50.% dissociated.

$$HA \rightleftharpoons H^+ + A^- \qquad K_a = \frac{[H^+][A^-]}{[HA]}$$

Since the weak acid is 50.% dissociated, then:

$$[H^+] = [A^-] = [HA]_o \times 0.50 = 2.00 \times 10^{-3} \, M \times 0.50 = 1.0 \times 10^{-3} \, M$$

$$[HA] = 2.00 \times 10^{-3} \, M - 1.0 \times 10^{-3} \, M = 1.0 \times 10^{-3} \, M$$

$$K_a = \frac{[H^+][A^-]}{[HA]} = \frac{(1.0 \times 10^{-3})(1.0 \times 10^{-3})}{1.0 \times 10^{-3}} = 1.0 \times 10^{-3}$$

91.   If ideal, NaCl dissociates completely and $i = 2$. $\Delta T_f = iK_f m$

$1.28°C = 2 \times 1.86°C \text{ kg/mol} \times m$, $m = 0.344 \text{ mol NaCl/kg H}_2\text{O}$

$0.344 \text{ mol NaCl} \times 58.44 \text{ g/mol} = 20.1 \text{ g NaCl}$; mass % NaCl $= \dfrac{20.1 \text{ g}}{1.00 \times 10^3 \text{ g} + 20.1 \text{ g}} \times 100 = 1.97\%$

92.   a.   Assuming no ion association betwee SO_4^{2-}(aq) and Fe^{3+}(aq), then $i = 5$ for $Fe_2(SO_4)_3$.

$\pi = iMRT = 5(0.0500 \text{ mol/L})(0.08206 \text{ L atm mol}^{-1} \text{ K}^{-1})(298 \text{ K}) = 6.11 \text{ atm}$

b. $Fe_2(SO_4)_3$(aq) $\rightarrow$ 2 Fe^{3+}(aq) + 3 SO_4^{2-}(aq)

Under ideal circumstances, 2/5 of π calculated above results from Fe^{3+} and 3/5 results from SO_4^{2-}. The contribution to π from SO_4^{2-} is 3/5 $\times$ 6.11 atm = 3.67 atm. Since SO_4^{2-} is assumed unchanged in solution, then the SO_4^{2-} contribution in the actual solution will also be 3.67 atm. The contribution to the actual π from the $Fe(H_2O)_6^{3+}$ dissociation reaction is 6.73 - 3.67 = 3.06 atm.

The initial concentration of $Fe(H_2O)_6^{2+}$ is 2(0.0500) = 0.100 M. The set-up for the weak acid problem is:

$$Fe(H_2O)_6^{3+} \rightleftharpoons H^+ + Fe(OH)(H_2O)_5^{2+} \qquad K_a = \frac{[H^+][Fe(OH)(H_2O)_5^{2+}]}{[Fe(H_2O)_6^{3+}]}$$

Initial   0.100 M 0   0
            x mol/L of $Fe(H_2O)_6^{3+}$ reacts to reach equilibrium
Equil.   0.100 - x x x

$\pi = iMRT;$ Total ion concentration $= iM = \dfrac{\pi}{RT} = \dfrac{3.06 \text{ atm}}{0.08206 \text{ L atm K}^{-1} \text{mol}^{-1} (298 \text{ K})} = 0.125 \ M$

$0.125 \ M = 0.100 - x + x + x = 0.100 + x, \ x = 0.025 \ M$

$K_a = \dfrac{[\text{H}^+][\text{Fe(OH)(H}_2\text{O)}_5^{2+}]}{[\text{Fe(H}_2\text{O)}_6^{3+}]} = \dfrac{x^2}{0.100 - x} = \dfrac{(0.025)^2}{(0.100 - 0.025)} = \dfrac{(0.025)^2}{0.075}, \ K_a = 8.3 \times 10^{-3}$

93. $P_{CS_2} = \chi_{CS_2}^{V} P_{tot} = 0.855 \ (263 \text{ torr}) = 225 \text{ torr}$

$P_{CS_2} = \chi_{CS_2}^{L} P_{CS_2}^{\circ}, \ \chi_{CS_2}^{L} = \dfrac{P_{CS_2}}{P_{CS_2}^{\circ}} = \dfrac{225 \text{ torr}}{375 \text{ torr}} = 0.600$

94. $iM = \dfrac{\pi}{RT} = \dfrac{0.3950 \text{ atm}}{0.08206 \text{ L atm mol}^{-1} \text{K}^{-1} (298.2 \text{ K})} = 0.01614 \text{ mol/L} = $ total ion concentration

$0.01614 \ M = C_{Mg^{2+}} + C_{Na^+} + C_{Cl^-}; \ C_{Cl^-} = 2 \, C_{Mg^{2+}} + C_{Na^+}$ (charge balance)

Combining: $0.01614 = 3 \, C_{Mg^{2+}} + 2 \, C_{Na^+}$

Let $x = $ mass MgCl$_2$ and $y = $ mass NaCl, then $x + y = 0.5000$ g.

$C_{Mg^{2+}} = \dfrac{x}{95.218}$ and $C_{Na^+} = \dfrac{y}{58.443}$ (Since V = 1.000 L)

Total ion concentration $= \dfrac{3 \, x}{95.218} + \dfrac{2 \, y}{58.443} = 0.01614 \ M;$ Rearranging: $3 \, x + 3.2585 \, y = 1.537$

Solving by simultaneous equations:

$$\begin{array}{rcl} 3 \, x & + \ 3.2585 \, y = & 1.537 \\ -3 \, (x & + \quad\quad y) = & -3(0.5000) \\ \hline & 0.2585 \, y = & 0.037, \ y = 0.14 \text{ g NaCl} \end{array}$$

mass MgCl$_2$ = 0.5000 g - 0.14 g = 0.36 g;   mass % MgCl$_2$ = $\dfrac{0.36 \text{ g}}{0.5000 \text{ g}} \times 100 = 72\%$

95.   Initial moles VCl$_4$ = 6.6834 g VCl$_4$ $\times$ 1 mol VCl$_4$/192.74 g VCl$_4$ = 3.4676×10^{-2} mol VCl$_4$

Total molality of solute particles $= im = \dfrac{\Delta T_f}{K_f} = \dfrac{5.97°\text{C}}{29.8°\text{C kg mol}^{-1}} = 0.200 \text{ mol/kg}$

Since we have 0.1000 kg CCl_4, the total moles of solute particles present is:

0.200 mol/kg (0.1000 kg) = 0.0200 mol

$$2\ VCl_4 \quad \rightleftharpoons \quad V_2Cl_8 \qquad K = \frac{[V_2Cl_8]}{[VCl_4]^2}$$

Initial 3.4676×10^{-2} mol   0

         $2x$ mol VCl_4 reacts to reach equilibrium

Equil. 3.4676×10^{-2} - $2x$ x

Total moles solute particles = 0.0200 mol = mol VCl_4 + mol V_2Cl_8 = 3.4676×10^{-2} -$2x$ + x

$0.0200 = 3.4676 \times 10^{-2}$ - x, $x = 0.0147$ mol

At equilibrium we have 0.0147 mol V_2Cl_8 and 0.0200 - 0.0147 = 0.0053 mol VCl_4. To determine the equilibrium constant, we need the total volume of solution in order to calculate equilibrium concentrations. The total mass of solution is 100.0 g + 6.6834 g = 106.7 g.

Total volume = 106.7 g $\times$ 1 cm^3/1.696 g = 62.91 cm^3 = 0.06291 L

The equilibrium concentrations are:

$$[V_2Cl_8] = \frac{0.0147\ \text{mol}}{0.06291\ \text{L}} = 0.234\ \text{mol/L}; \quad [VCl_4] = \frac{0.0053\ \text{mol}}{0.06291\ \text{L}} = 0.084\ \text{mol/L}$$

$$K = \frac{[V_2Cl_8]}{[VCl_4]^2} = \frac{0.234}{(0.084)^2} = 33$$

CHAPTER EIGHTEEN

THE REPRESENTATIVE ELEMENTS: GROUPS 1A THROUGH 4A

Group 1A Elements

1.   The gravity of the earth is not strong enough to keep H_2 in the atmosphere.

2.   a. $\Delta H° = -110.5 - [-242 - 75] = 207$ kJ; $\Delta S° = 3(131) + 198 - [186 + 189] = 216$ J/K

      b. $\Delta G° = \Delta H° - T\Delta S°$; $\Delta G° = 0$ when $T = \dfrac{\Delta H°}{\Delta S°} = \dfrac{207 \times 10^3 \text{ J}}{216 \text{ J/K}} = 958$ K

      At T > 958 K and standard pressures, the favorable $\Delta S°$ term dominates and the reaction is spontaneous ($\Delta G° < 0$).

3.   For $3\ Fe(s) + 4\ H_2O(g) \rightarrow Fe_3O_4(s) + 4\ H_2(g)$

      a. $\Delta H° = -1117 - [4(-242)] = -149$ kJ; $\Delta S° = 146 + 4(131) - [3(27) + 4(189)] = -167$ J/K

      b. $\Delta G° = 0$ when $T = \dfrac{\Delta H°}{\Delta S°} = \dfrac{-149 \times 10^3 \text{ J}}{-167 \text{ J/K}} = 892$ K

      At T < 892 K and standard pressures, the favorable $\Delta H°$ term dominates and the reaction is spontaneous ($\Delta G° < 0$).

      For $C(s) + H_2O(g) \rightarrow CO(g) + H_2(g)$

      a. $\Delta H° = -110.5 - (-242) = 132$ kJ; $\Delta S° = 198 + 131 - [6 + 189] = 134$ J/K

      $T = \dfrac{\Delta H°}{\Delta S°} = \dfrac{132 \times 10^3 \text{ J}}{134 \text{ J/K}} = 985$ K

      b.   This reaction is spontaneous when the favorable $\Delta S°$ term dominates, which occurs at T > 985 K (assuming standard pressures).

4.   1. Ammonia production and 2. Hydrogenation of vegetable oils

5.   Ionic, covalent, and metallic (or interstitial); The ionic and covalent hydrides are true compounds obeying the law of definite proportions and differ from each other in the type of bonding. The interstitial hydrides are more like solid solutions of hydrogen with a transition metal, and do not obey the law of definite proportions.

6.   The small size of the Li^+ cation means that there is a much greater attraction to water. The attraction to water is not so great for other alkali metal ions. Thus, lithium salts tend to absorb water.

7.   a. $Li_3N(s) + 3\ HCl(aq) \rightarrow 3\ LiCl(aq) + NH_3(aq)$ b. $Rb_2O(s) + H_2O(l) \rightarrow 2\ RbOH(aq)$

      c. $Cs_2O_2(s) + 2\ H_2O(l) \rightarrow 2\ CsOH(aq) + H_2O_2(aq)$

      d. $NaH(s) + H_2O(l) \rightarrow NaOH(aq) + H_2(g)$

8. $K^+(out) \rightarrow K^+(in);\ \ E = E° - \dfrac{0.0592}{1} \log\left(\dfrac{[K^+]_{in}}{[K^+]_{out}}\right);\ \ E° = 0$

      $E = -0.0592 \log\left(\dfrac{0.15}{5.0 \times 10^{-3}}\right) = -0.087\ V$

      $\Delta G = work = -nFE = -(1\ mol\ e^-)(96{,}485\ C/mol\ e^-)(-0.087\ J/C) = 8400\ J = 8.4\ kJ = work$

9.

	Li	Na	K	Rb	Cs	Fr
Atomic number	3	11	19	37	55	87
MP (°C)	180	98	63	39	29	≈22

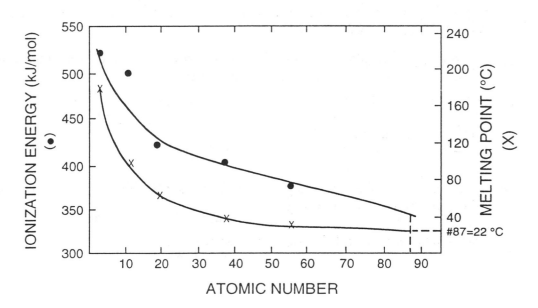

Using the above plot from a previous exercise, we would estimate the melting point of Fr to be around 20°C. Thus, it would be a liquid at room temperature.

10. $2 Li(s) + 2 C_2H_2(g) \rightarrow 2 LiC_2H(s) + H_2(g)$; This is an oxidation-reduction reaction.

11.   Hydrogen forms many compounds in which the oxidation state is +1, as do the Group 1A elements. For example H_2SO_4 and HCl compared to Na_2SO_4 and $NaCl$. On the other hand, hydrogen forms diatomic H_2 molecules and is a nonmetal, while the Group 1A elements are metals. Hydrogen also forms compounds with a -1 oxidation state, which is not characteristic of Group 1A metals, e.g., NaH.

12. $NaH(s) + H_2O(l) \rightarrow Na^+(aq) + OH^-(aq) + H_2(g)$

        Acid-base: A proton is transferred from an acid, H_2O, to a base, H^-, forming the conjugate base of water, OH^-, and the conjugate acid of H^-, H_2.

        Oxidation-reduction: The oxidation number of H is -1 in NaH, +1 in H_2O, and zero in H_2. Thus, an electron is transferred from the hydride ion to a hydrogen in water when forming $H_2(g)$.

13.   a. $K(s) + O_2(g) \rightarrow KO_2(s)$ b. $16 K(s) + S_8(s) \rightarrow 8 K_2S(s)$

        c. $12 K(s) + P_4(s) \rightarrow 4 K_3P(s)$ d. $2 K(s) + H_2(g) \rightarrow 2 KH(s)$

        e. $2 K(s) + 2 H_2O(l) \rightarrow H_2(g) + 2 KOH(aq)$

14.   a.   sodium oxide: Na_2O; b.   sodium superoxide: NaO_2; c.   sodium peroxide: Na_2O_2

15. $4 KO_2(s) + 2 CO_2(g) \rightarrow 2 K_2CO_3(s) + 3 O_2(g)$; Potassium superoxide can react with exhaled CO_2 to produce O_2 which then can be breathed.

Group 2A Elements

16.   Group IA and IIA metals are all easily oxidized. They must be produced in the absence of materials (H_2O, O_2) that are capable of oxidizing them.

17.

        Geometry is trigonal planar.

        Be uses sp^2 hybrid orbitals.

        N uses sp^3 hybrid orbitals.

        $BeCl_2$ is a Lewis acid.

18.   The acidity decreases. Solutions of Be^{2+} are acidic, while solutions of the other M^{2+} ions are neutral.

19. $BeCl_2(NH_3)_2$ would form in excess ammonia. $BeCl_2(NH_3)_2$ has $2 + 2(7) + 2(5) + 6(1) = 32$ valence electrons. A structure for this molecule can be drawn that obeys the octet rule.

$$
\begin{array}{ccc}
H & :\ddot{Cl}: & H \\
| & | & | \\
H-N-&Be-&N-H \\
| & | & | \\
H & :\ddot{Cl}: & H
\end{array}
$$

20.   In the gas phase, linear molecules would exist.

$$:\ddot{F}-Be-\ddot{F}:$$

In the solid state, BeF_2 would exist as a polymeric solid with a structure:

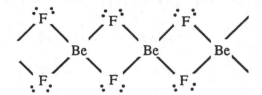

21.   a. $2\ Ca(s) + O_2(g) \rightarrow 2\ CaO(s)$ b. $8\ Ca(s) + S_8(s) \rightarrow 8\ CaS(s)$

      c. $3\ Ca(s) + N_2(g) \rightarrow Ca_3N_2(s)$ d. $6\ Ca(s) + P_4(s) \rightarrow 2\ Ca_3P_2(s)$

      e. $Ca(s) + H_2(g) \rightarrow CaH_2(s)$ f. $Ca(s) + 2\ H_2O(l) \rightarrow H_2(g) + Ca(OH)_2(aq)$

22.   barium oxide: BaO; barium peroxide: BaO_2

23. $Mg_3N_2(s) + 6\ H_2O(l) \rightarrow 2\ NH_3(g) + 3\ Mg^{2+}(aq) + 6\ OH^-(aq)$

      $Mg_3P_2(s) + 6\ H_2O(l) \rightarrow 2\ PH_3(g) + 3\ Mg^{2+}(aq) + 6\ OH^-(aq)$

24. $1.00 \times 10^3\ kg \times \dfrac{1000\ g}{kg} \times \dfrac{1\ mol\ Ca}{40.08\ g} \times \dfrac{2\ mol\ e^-}{mol\ Ca} \times \dfrac{96,485\ C}{mol\ e^-} = 4.81 \times 10^9\ C$

      $current = \dfrac{4.81 \times 10^9\ C}{8.00\ hr} \times \dfrac{1\ hr}{60\ min} \times \dfrac{1\ min}{60\ s} = \dfrac{1.67 \times 10^5\ C}{s} = 1.67 \times 10^5\ A$

      $1.00 \times 10^3\ kg\ Ca \times \dfrac{70.90\ g\ Cl_2}{40.08\ g\ Ca} = 1.77 \times 10^3\ kg\ of\ Cl_2$

25. $Mg(OH)_2(s) \rightleftharpoons Mg^{2+}(aq) + 2\ OH^-(aq)$ $K_{sp} = 8.9 \times 10^{-12} = [Mg^{2+}]\ [OH^-]^2$

      Since pH = 8.00, $[OH^-] = 1.0 \times 10^{-6}\ M$; $[Mg^{2+}](1.0 \times 10^{-6})^2 = 8.9 \times 10^{-12}$, $[Mg^{2+}] = 8.9\ M$

      Considering only the solubility product equilibria, the solubility is 8.9 mol $Mg(OH)_2$/L. In fact, the solubility is less than this and is controlled by other factors.

Group 3A Elements

26. $B_2H_6 + 3\ O_2 \rightarrow 2\ B(OH)_3$

27.   a. Thallium(I) hydroxide   b. Indium(III) sulfide   c. Gallium(III) oxide

28.   A single $AlCl_3$ has the Lewis structure:

$$:\ddot{C}l \!-\! Al \!-\! \ddot{C}l:$$
$$|$$
$$:\ddot{C}l:$$

The Al has room for a pair of electrons. It can act as a Lewis acid. The only lone pairs are on Cl, so the structure is:

$$:\ddot{C}l \qquad \dot{C}\dot{l} \qquad \ddot{C}l:$$
$$\searrow \qquad \swarrow \searrow \qquad \swarrow$$
$$Al \qquad Al$$
$$\swarrow \qquad \searrow \qquad \searrow$$
$$:\ddot{C}l \qquad :\ddot{C}l: \qquad \ddot{C}l:$$

29.   Element 113 would fall below Tl in the periodic table.

Element	AN	r of M^{3+} (Å)
B	5	0.2
Al	13	0.51
Ga	31	0.62
In	49	0.81
Tl	81	0.95
	113	~1.0

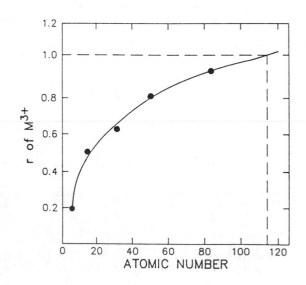

From the graph, we would predict the radius of the 3^+ ion of element 113 to be about 1.0 Å (100 pm). From the data in Table 18.9, we would expect the ionization energy of element 113 to be close to that of Tl. The trend is too erratic to make a more definitive prediction.

30. $LiAlH_4$: H, -1 (metal hydride); Li, +1; Al, +3; $0 = +1 + x + 4(-1)$, $x = +3$

31. $In_2O_3(s) + 6 H^+(aq) \rightarrow 2 In^{3+}(aq) + 3 H_2O(l)$

$In_2O_3(s) + OH^-(aq) \rightarrow$ no reaction

$Ga_2O_3(s) + 6 H^+(aq) \rightarrow 2 Ga^{3+}(aq) + 3 H_2O(l)$

$Ga_2O_3(s) + 2 OH^-(aq) + 3 H_2O(l) \rightarrow 2 Ga(OH)_4^-(aq)$

32.   Group 3A elements have one fewer valence electron than Si or Ge. A p-type semiconductor would form.

33.   a.   Out of 100.0 g of compound there are:

$$44.4 \text{ g Ca} \times \frac{1 \text{ mol}}{40.08 \text{ g}} = 1.11 \text{ mol Ca}; \quad 20.0 \text{ g Al} \times \frac{1 \text{ mol}}{26.98 \text{ g}} = 0.741 \text{ mol Al}$$

$$35.6 \text{ g O} \times \frac{1 \text{ mol}}{16.00 \text{ g}} = 2.23 \text{ mol O}$$

$$\frac{1.11}{0.741} = 1.5, \quad \frac{0.741}{0.741} = 1, \quad \frac{2.23}{0.741} = 3; \quad \text{Empirical formula is } Ca_3Al_2O_6.$$

b. $Ca_9Al_6O_{18}$

c.   There are covalent bonds between Al and O atoms in the $[Al_6O_{18}]^{18-}$ anion; sp^3 hybrid orbitals on aluminum overlap with sp^3 hybrid orbitals on oxygen to form the sigma bonds.

34.   a. $2 In(s) + 3 F_2(g) \rightarrow 2 InF_3(s)$ b. $2 In(s) + 3 Cl_2(g) \rightarrow 2 InCl_3(s)$

c. $4 In(s) + 3 O_2(g) \rightarrow 2 In_2O_3(s)$

d. $2 In(s) + 6 HCl(aq) \rightarrow 3 H_2(g) + 2 InCl_3(aq)$ or $2 In(s) + 6 HCl(g) \rightarrow 3 H_2(g) + 2 InCl_3(s)$

35. $2 Al(s) + 2 NaOH(aq) + 6 H_2O(l) \rightarrow 2 Al(OH)_4^-(aq) + 2 Na^+(aq) + 3 H_2(g)$

36. Al_2O_3 is an amphoteric oxide.

As a base: $Al_2O_3(s) + 6 H^+(aq) \rightarrow 2 Al^{3+}(aq) + 3 H_2O(l)$

As a Lewis acid: $Al_2O_3(s) + 3 H_2O(l) + 2 OH^-(aq) \rightarrow 2 Al(OH)_4^-(aq)$

Group 4A Elements

37.   Compounds containing Si – Si single and multiple bonds are rare, unlike compounds of carbon. The bond strengths of the Si – Si and C – C single bonds are similar. The difference in bonding properties must be for other reasons. One reason is that silicon does not form strong π bonds, unlike carbon. Another reason is that silicon forms particularly strong sigma bonds to oxygen, resulting in compounds with Si – O bonds instead of Si – Si bonds.

38.   Si – H: 393 kJ/mol; C – H: 413 kJ/mol; Since the Si – H bond is slightly weaker, we would expect it to be more reactive than a C – H bond.

39.   a. Linear about all carbons;   b. sp

40.

Bonds broken: 2 C – O (358 kJ/mol);   Bonds formed: 1 C = O (799 kJ/mol)

$\Delta H = 2(358) - 799 = -83$ kJ; ΔH is favorable for the decomposition of H_2CO_3 to CO_2 and H_2O. ΔS is also favorable for the decomposition as there is an increase in disorder. Hence, H_2CO_3 will ⌐, ⌐ntaneously decompose to CO_2 and H_2O.

41. CS_2 has $4 + 2(6) = 16$ valence electrons. C_3S_2 has $3(4) + 2(6) = 24$ valence electrons.

42.   The bonds in SnX_4 compounds have a large covalent character. SnX_4 acts as discrete molecules held together by weak dispersion forces. SnX_2 compounds are ionic and are held in the solid state by strong ionic forces. Since the intermolecular forces are weaker for SnX_4 compounds, they are more volatile.

43.   White tin is stable at normal temperatures. Gray tin is stable at temperatures below 13.2°C. Thus for the phase change: Sn(gray) → Sn(white), ΔG is (-) at T > 13.2°C and ΔG is (+) at T < 13.2°C. This is only possible if ΔH is (+) and ΔS is (+). Thus, gray tin has the more ordered structure.

44. $SiCl_4(l) + 2 H_2O(l) \rightarrow SiO_2(s) + 4 H^+(aq) + 4 Cl^-(aq)$

$\Delta H° = 4(-167) + (-911) - [-687 + 2(-286)] = -320.$ kJ

$\Delta S° = 4(57) + (42) - [240 + 2(70)] = -110.$ J/K

$T = \Delta H°/\Delta S° = -320. \times 10^3$ J/-110. J K^{-1} = 2910 K

The reaction is spontaneous at temperatures below 2910 K, due to the favorable $\Delta H°$ term. There are, overall, stronger bonds in SiO_2 and $HCl(aq)$ than in $SiCl_4$ and H_2O.

The corresponding reaction for CCl_4 is:

$$CCl_4(l) + 2\ H_2O(l) \rightarrow CO_2(g) + 4\ H^+(aq) + 4\ Cl^-(aq)$$

$$\Delta H° = 4(-167) + (-393.5) - [-135 + 2(-286)] = -355\ kJ$$

$$\Delta S° = 4(57) + 214 - [216 + 2(70)] = 86\ J/K$$

Thermodynamics predicts that this reaction would be spontaneous at any temperature.

The answer must lie with kinetics. $SiCl_4$ reacts because an activated complex can form by a water molecule attaching to silicon in $SiCl_4$. The activated complex requires silicon to form a fifth bond. Silicon has low energy 3d orbitals available to expand the octet. Carbon will not break the octet rule, therefore, CCl_4 cannot form this activated complex. CCl_4 and H_2O require a different pathway to get to products. The different pathway has a higher activation energy and, in turn, the reaction is much slower. (See Exercise 18.65.)

45.   Tin(II) fluoride

46.   SiC would have a covalent network structure similar to diamond.

47. $SiO_2(s) + 4\ HF(aq) \rightarrow SiF_4(g) + 2\ H_2O(l)$

48. Pb_3O_4: We assign -2 for the oxidation number of O. The sum of the oxidation numbers of Pb must be +8. We get this if two of the lead atoms are Pb(II) and one is Pb(IV). So 2/3 of the lead is Pb(II).

49.
$$
\begin{array}{ll}
(Pb + 2\ OH^- \rightarrow Pb(OH)_2 + 2\ e^-) \times 2 & -E° = +0.57\ V \\
4\ e^- + 2\ H_2O + O_2 \rightarrow 4\ OH^- & E° = +0.40\ V \\
\hline
2\ Pb + 2\ H_2O + O_2 \rightarrow 2\ Pb(OH)_2 & E°_{cell} = +0.97\ V
\end{array}
$$

Comparing cell potentials, Fe pipes corrode more easily than Pb pipes. However, the corrosion of Pb pipes is still spontaneous and Pb(II) is very toxic. Pb pipes were extensively used by the Romans and it has been proposed that chronic lead poisoning is one of the factors leading to the decline and fall of the Roman Empire.

50.
$$Pb(OH)_2(s) \rightleftharpoons Pb^{2+} + 2\ OH^- \quad K_{sp} = 1.2 \times 10^{-15} = [Pb^{2+}]\,[OH^-]^2$$

s = solubility
in mol/L $\longrightarrow$ s   2s   (Ignore OH^- from H_2O)

$K_{sp} = (s)(2s)^2 = 1.2 \times 10^{-15}$, $4s^3 = 1.2 \times 10^{-15}$, $s = 6.7 \times 10^{-6}$ mol/L; Assumption good.

51.   Sn and Pb can reduce H^+ to H_2.

$$Sn(s) + 2\,H^+(aq) \rightarrow Sn^{2+}(aq) + H_2(g); \quad Pb(s) + 2\,H^+(aq) \rightarrow Pb^{2+}(aq) + H_2(g)$$

Additional Exercises

52.   a. Na^+ can oxidize Na^- to Na. The purpose of the cryptand is to encapsulate the Na^+ ion so that it does not come in contact with the Na^- ion and oxidize it to sodium metal.

53.   a. $2\,Na + 2\,NH_3 \rightarrow 2\,NaNH_2 + H_2$

b. $\dfrac{251.4\text{ g}}{1000.\text{ g} + 251.4\text{ g}} \times 100 = 20.09\ \%$ by mass

mol Na $= 251.4 \times \dfrac{1\text{ mol}}{22.99\text{ g}} = 10.94$ mol Na; mol $NH_3 = 1000.\text{ g} \times \dfrac{1\text{ mol}}{17.031\text{ g}} = 58.72$ mol NH_3

$\chi_{Na} = \dfrac{10.94}{10.94 + 58.72} = 0.1570$

molality $= \dfrac{251.4\text{ g Na}}{kg} \times \dfrac{1\text{ mol Na}}{22.99\text{ g Na}} = 10.94$ mol/kg

54. $\,$ crown ether surrounds the cation. The ion pair:

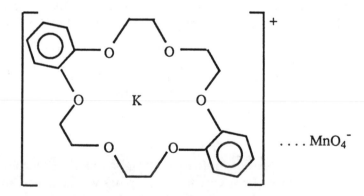

is soluble in nonpolar solvents because the solvent is attracted to the crown ether.

55. $Be + 4\,OH^- \rightarrow Be(OH)_4^{2-} + 2\,e^-$
                            $2\,H_2O + 2\,e^- \rightarrow H_2 + 2\,OH^-$

$$Be(s) + 2\,H_2O(l) + 2\,OH^-(aq) \rightarrow Be(OH)_4^{2-}(aq) + H_2(g)$$

Be is the reducing agent. H_2O is the oxidizing agent.

56.   a. Be_3N_2 b. SrO_2

57.   The Be^{2+} ion is a Lewis acid and has a strong affinity for the lone pairs of electrons on oxygen in water. The ion in solution is $Be(H_2O)_4^{2+}$. The acidic solution results from the reaction:
$Be(H_2O)_4^{2+}(aq) \rightleftharpoons Be(H_2O)_3(OH)^+(aq) + H^+(aq)$

58. $B_2O_3(s) + 3\ Mg(s) \rightarrow 3\ MgO(s) + 2\ B(s)$

59. $Tl^{3+} + 2\ e^- \rightarrow Tl^+$ $E° = +1.25$ V
                $3\ I^- \rightarrow I_3^- + 2\ e^-$ $-E° = -0.55$ V

       _____

       $Tl^{3+} + 3\ I^- \rightarrow Tl^+ + I_3^-$ $E_{cell}^· = +0.70$ V

       In solution, Tl^{3+} can oxidize I^- to I_3^-. Thus, we expect TlI_3 to be thallium(I) triiodide.

60.   Ga(I): $[Ar]3d^{10}4s^2$, no unpaired e^-; Ga(III): $[Ar]3d^{10}$, no unpaired e^-

       Ga(II): $[Ar]3d^{10}4s^1$, 1 unpaired e^-

       If the compound contained Ga(II) it would be paramagnetic and if the compound contained Ga(I) and Ga(III) it would be diamagnetic. This can easily be determined by measuring the mass of a sample in the presence and in the absence of a magnetic field. Paramagnetic compounds will have an apparent greater mass in a magnetic field.

61.   The π electrons are free to move in graphite, thus giving it a greater conductivity (lower resistance). The electrons have the greatest mobility within sheets of carbon atoms, resulting in a lower resistance in the basal plane. Electrons in diamond are not mobile (high resistance). The structure of diamond is uniform in all directions; thus, there is no directional dependence of the resistivity.

62.   Planes of carbon atoms slide easily along each other. Graphite is not volatile. The lubricant will not be lost when used in a high vacuum environment.

63.   a. $K_2SiF_6(s) + 4\ K(l) \rightarrow 6\ KF(s) + Si(s)$

       b. K_2SiF_6 is an ionic compound, composed of K^+ cations and SiF_6^{2-} anions. The SiF_6^{2-} anion is held together by covalent bonds. The structure is:

The anion is octahedral.

64.

$$CH_3(CH_2)_6CH_2$$

$$:\overset{\cdot\cdot}{\underset{\cdot\cdot}{Cl}} - Sn$$

$$:\overset{\cdot\cdot}{\underset{\cdot\cdot}{Cl}} \qquad CH_2(CH_2)_6CH_3$$

The compound is held together by covalent bonds. The structure is tetrahedral about the central Sn.

65.   Carbon cannot form the fifth bond necessary for the transition state since carbon doesn't have low energy d orbitals available to expand the octet.

66.   Quartz: Crystalline, long range order; The structure is an ordered arrangement of 12 membered rings, each containing 6 – Si and 6 – O atoms.

Amorphous SiO_2:   No long range order; Irregular arrangement that contains many different ring sizes. See Section 16.5.

67.   Size decreases from left to right and increases going down the periodic table. So going one element right and one element down would result in a similar size for the two elements diagonal to each other. The ionization energies will be similar for the diagonal elements since the periodic trends also oppose each other. Electron affinities are harder to predict, but atoms with similar size and ionization energy will also have similar electron affinities.

68.   The "inert pair effect" refers to the difficulty of removing the pair of 6s electrons from some of the elements in the sixth period of the periodic table. Tl^+, Tl^{3+}, Pb^{2+}, and Pb^{4+} ions are all important in the chemistry of Tl and Pb.

CHAPTER NINETEEN

THE REPRESENTATIVE ELEMENTS: GROUPS 5A THROUGH 8A

Group 5A Elements

1. NO_4^{3-}

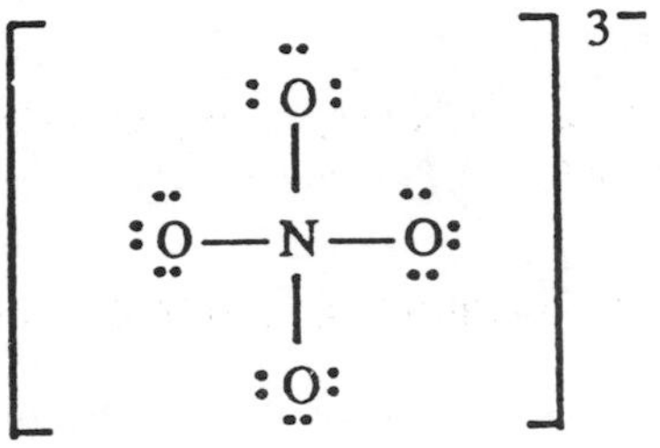

N is small. There is probably not enough room for all 4 oxygen atoms around N. P is larger, thus, PO_4^{3-} is stable.

      PO_3^-

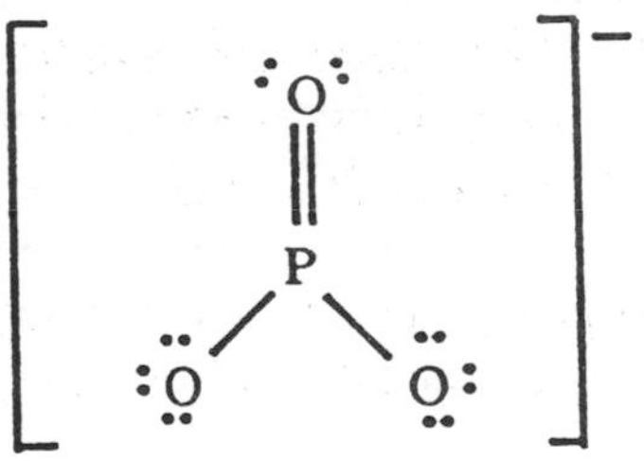

$P = O$ bonds are not particularly stable, while $N = O$ bonds are stable. Thus, NO_3^- is stable.

2. $\Delta H° = 2(90 \text{ kJ}) - [0] = 180. \text{ kJ}$; $\Delta S° = 2(211 \text{ J/K}) - [192 + 205] = 25 \text{ J/K}$

      $\Delta G° = 2(87 \text{ kJ}) - [0] = 174 \text{ kJ}$

At high temperature the reaction $N_2 + O_2 \rightarrow 2 \text{ NO}$ becomes spontaneous. In the atmosphere, even though $2 \text{ NO} \rightarrow N_2 + O_2$ is spontaneous, it doesn't occur because the rate is slow.

3. NH_3: sp^3; N_2H_4: sp^3; NH_2OH: sp^3; N_2: sp; N_2O: central N, sp

      NO: sp^2; N_2O_3: both Ns are sp^2; NO_2: sp^2; HNO_3: sp^2

4.   Resonance is possible for N_2O, NO, N_2O_3, NO_2, and HNO_3.

N_2O dinitrogen monoxide (nitrous oxide)

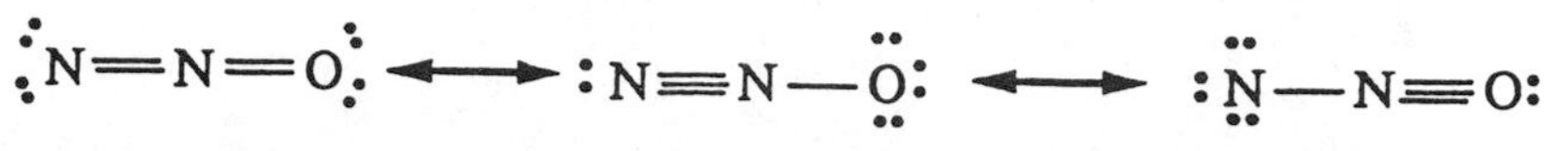

last form
not important

NO nitrogen monoxide (nitric oxide)

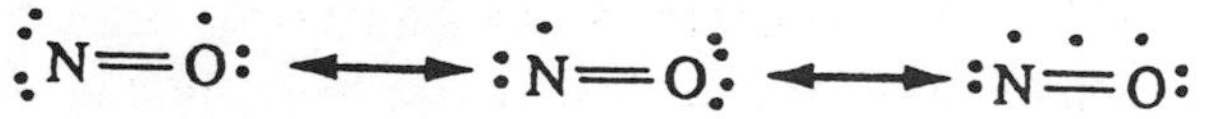

N_2O_3 dinitrogen trioxide

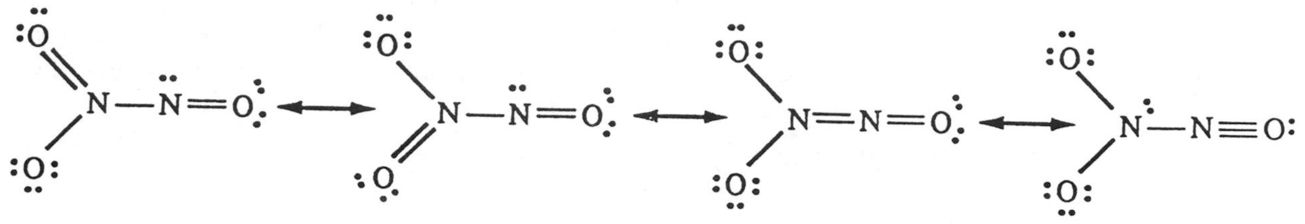

last 2 not important

NO_2 nitrogen dioxide

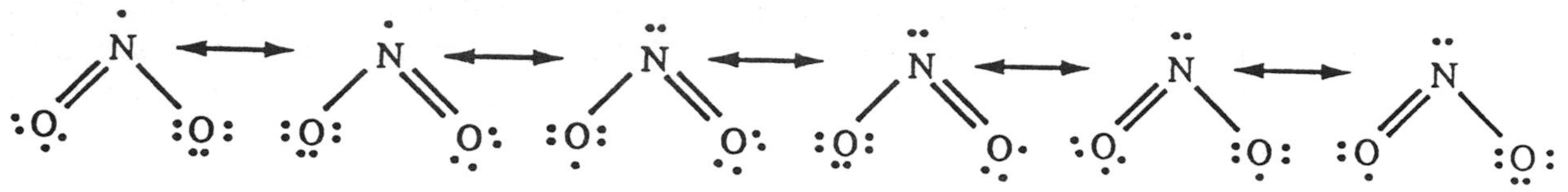

HNO_3 nitric acid

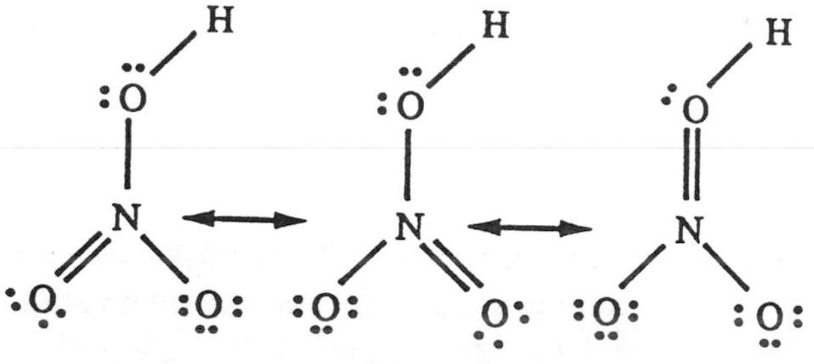

last one not important

5.   For the reaction:

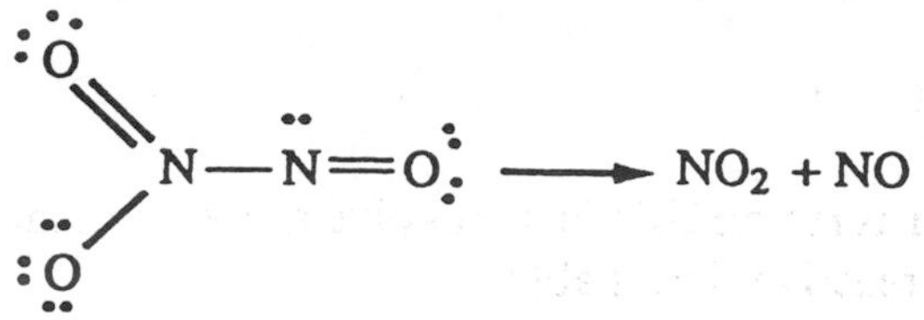

$\longrightarrow$ $NO_2 + NO$

the activation energy must in some way involve the breaking of a nitrogen-nitrogen single bond.

For the reaction:

$$\rightarrow \quad O_2 + N_2O$$

at some point nitrogen-oxygen bonds must be broken. N – N single bonds (160 kJ/mol) are weaker than N – O single bonds (201 kJ/mol). In addition, resonance structures indicate that there is more double bond character in the N – O bonds than in the N— N bond. Thus, NO_2 and NO are preferred by kinetics because of the lower activation energy.

6.   In the solid state: NO_2^+ and NO_3^-; In the gas phase: molecular N_2O_5

7.   The reaction $N_2(g) + 3 H_2(g) \rightarrow 2 NH_3(g)$ is exothermic. Thus, K_p decreases as the temperature increases. Lower temperatures are favored for maximum yield of ammonia. However, at lower temperatures the rate is slow; without a catalyst the rate is too slow for the process to be feasible. The discovery of a catalyst increased the rate of reaction at a lower temperature favored by thermodynamics.

8. OCN^- has $6 + 4 + 5 + 1 = 16$ valence electrons.

| Formal Charge | 0 | 0 | -1 | | -1 | 0 | 0 | | +1 | 0 | -2 |

Only the first two resonance structures should be important. The third places a positive formal charge on the most electronegative atom in the ion and a -2 formal charge on N.

CNO^-:

| Formal Charge | -2 | +1 | 0 | | -1 | +1 | -1 | | -3 | +1 | +1 |

All of the resonance structures for fulminate involve greater formal charges than in cyanate, making fulminate more reactive (less stable).

9. $N_2H_4 + 2 F_2 \rightarrow 4 HF + N_2$

    Bonds broken: Bonds formed:

        1 N – N (160 kJ/mol)   4 H – F (565 kJ/mol)
        4 N – H (391 kJ/mol)   1 N≡ N (941 kJ/mol)
        2 F – F (154 kJ/mol)

    $\Delta H = 160 + 4(391) + 2(154) - [4(565) + 941] = 2032 \text{ kJ} - 3201 \text{ kJ} = -1169 \text{ kJ}$

10. a. $8 H^+(aq) + 2 NO_3^-(aq) + 3 Cu(s) \rightarrow 3 Cu^{2+}(aq) + 4 H_2O(l) + 2 NO(g)$

    b. $NH_4NO_3(s) \xrightarrow{Heat} N_2O(g) + 2 H_2O(g)$

    c. $NO(g) + NO_2(g) + 2 KOH(aq) \rightarrow 2 KNO_2(aq) + H_2O(l)$

11. a. PF_5;   N is too small and doesn't have low energy d-orbitals to expand its octet to form NF_5.

    b. AsF_5;   I is too large to fit 5 atoms around As.

    c. NF_3;   N is too small for three large bromine atoms to fit around it.

12. Production of antimony:

        $2 Sb_2S_3(s) + 9 O_2(g) \rightarrow 2 Sb_2O_3(s) + 6 SO_2(g)$

        $2 Sb_2O_3(s) + 3 C(s) \rightarrow 4 Sb(s) + 3 CO_2(g)$

    Production of bismuth:

        $2 Bi_2S_3(s) + 9 O_2(g) \rightarrow 2 Bi_2O_3(s) + 6 SO_2(g)$

        $2 Bi_2O_3(s) + 3 C(s) \rightarrow 4 Bi(s) + 3 CO_2(g)$

13. a. $H_3PO_4 > H_3PO_3$;   The strongest acid has the most oxygen atoms.

    b. $H_3PO_4 > H_2PO_4^- > HPO_4^{2-}$;   Acid strength decreases as protons are removed.

14. The acidic protons are attached to oxygen.

    $H_4P_2O_6$: $H_4P_2O_5$:

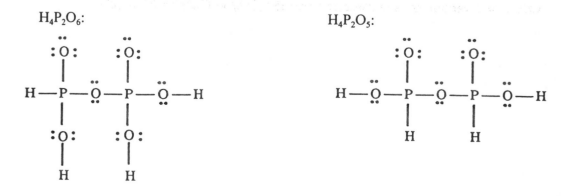

15.   White phosphorus consists of P_4 tetradedra that react readily with oxygen. In red phosphorus the P_4 tetrahedra are bonded to each other in chains and are less reactive. They need a source of energy to react with oxygen, such as when one strikes a match. Black phosphorus is crystalline with the P atoms more tightly bonded in the crystal and are fairly unreactive towards oxygen.

16.   Hypochlorite can act as an oxidizing agent. For example it is capable of oxidizing I^- to I_2. If a solution containing I^- turns brown when BiOCl is added, then BiOCl is bismuth(I) hypochlorite. The brown color indicates production of I_2. If the solution doesn't change color, then it is bismuthylchloride.

17. $4 As(s) + 3 O_2(g) \rightarrow As_4O_6(s)$; $4 As(s) + 5 O_2(g) \rightarrow As_4O_{10}(s)$

$As_4O_6(s) + 6 H_2O(l) \rightarrow 4 H_3AsO_3(aq)$; $As_4O_{10}(s) + 6 H_2O(l) \rightarrow 4 H_3AsO_4(aq)$

18.

		Bond order	# unpaired e^-
M.O model:	NO:	2.5	1
	NO^+:	3	0
	NO^-:	2	2

Lewis structures: NO^+: $[:N\equiv O:]^+$

NO:

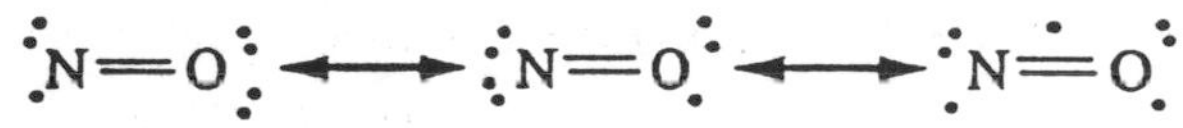

NO^-:

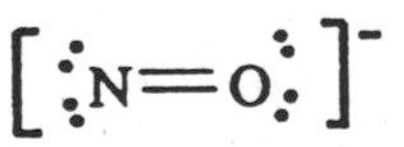

Lewis structure are not adequate for NO and NO^-. M.O. model gives correct results for all three species. For NO, Lewis structures are poor for odd electron species. For NO^-, Lewis structures fail to predict that NO^- is paramagnetic.

19. $Mg^{2+} + P_3O_{10}^{5-} \rightleftharpoons MgP_3O_{10}^{3-}$; $[Mg^{2+}]_o = \dfrac{50. \times 10^{-3} \text{ g}}{L} \times \dfrac{1 \text{ mol}}{24.3 \text{ g}} = 2.1 \times 10^{-3} M$

$[P_3O_{10}^{5-}]_o = \dfrac{40. \text{ g Na}_5\text{P}_3\text{O}_{10}}{L} \times \dfrac{1 \text{ mol}}{367.9 \text{ g}} = 0.11 M$

Assume the reaction goes to completion since K is large ($10^{8.60} = 4.0 \times 10^8$).

$$Mg^{2+} \quad + \quad P_3O_{10}^{5-} \quad \rightleftharpoons \quad MgP_3O_{10}^{3-}$$

Before	$2.1 \times 10^{-3}\ M$	$0.11\ M$	0	
Change	-2.1×10^{-3}	-2.1×10^{-3} $\rightarrow$	$+2.1 \times 10^{-3}$	React completely
After	0	0.11	2.1×10^{-3}	New initial

x mol/L $MgP_3O_{10}^{3-}$ dissociates to reach equilibrium

Change	$+x$	$+x$ $\leftarrow$	$-x$	
Equil.	x	$0.11 + x$	$2.1 \times 10^{-3} - x$	

$$K = 4.0 \times 10^8 = \frac{[MgP_3O_{10}^{3-}]}{[Mg^{2+}][P_3O_{10}^{5-}]} = \frac{2.1 \times 10^{-3} - x}{x(0.11 + x)}$$

$$4.0 \times 10^8 \approx \frac{2.1 \times 10^{-3}}{x(0.11)}, \quad x = [Mg^{2+}] = 4.8 \times 10^{-11}\ M; \quad \text{Assumptions good.}$$

20. a.

$$(4\ H_2O + Mn^{2+} \rightarrow MnO_4^- + 8\ H^+ + 5e^-) \times 2$$
$$(2e^- + NaBiO_3 \rightarrow BiO_3^{3-} + Na^+) \times 5$$

$$8\ H_2O(l) + 2\ Mn^{2+}(aq) + 5\ NaBiO_3(s) \rightarrow 2\ MnO_4^-(aq) + 16\ H^+(aq) + 5\ BiO_3^{3-}(aq) + 5\ Na^+(aq)$$

b. Bismuthate exists as a covalent network solid: $(BiO_3^-)_x$.

21. a. SbF_5 HSO_3F $H_2SO_3F^+$

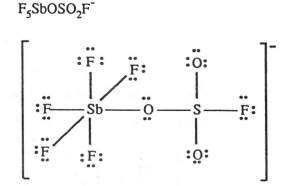

F₅SbOSO₂FH   F₅SbOSO₂F⁻

b. The active protonating species is $H_2SO_3F^+$.

22. $AsCl_4^+$, $5 + 4(7) - 1 = 32$ e⁻ $AsCl_6^-$, $5 + 6(7) + 1 = 48$ e⁻

The reaction is a Lewis acid-base reaction. Chloride ion acts as a Lewis base when it is transferred from one $AsCl_5$ to another. Arsenic is the Lewis acid (electron pair acceptor).

23.   a. $P_4O_6(s) + 2\ O_2(g) \rightarrow P_4O_{10}(s)$

        b. $P_4O_{10}(s) + 6\ H_2O(l) \rightarrow 4\ H_3PO_4(aq)$

        c. $PCl_5(l) + 4\ H_2O(l) \rightarrow H_3PO_4(aq) + 5\ HCl(aq)$

Group 6A Elements

24. $O = O - O \rightarrow O = O + O$

Break O – O bond: $\Delta H = 146\ kJ/mol \times \dfrac{1\ mol}{6.022 \times 10^{23}} = 2.42 \times 10^{-22}\ kJ = 2.42 \times 10^{-19}\ J$

A photon of light must contain at least 2.42×10^{-19} J to break one O – O bond.

$E_{photon} = \dfrac{hc}{\lambda}$, $\lambda = \dfrac{hc}{E} = \dfrac{(6.626 \times 10^{-34}\ J\ s)\ (2.998 \times 10^8\ m/s)}{2.42 \times 10^{-19}\ J} = 8.21 \times 10^{-7}\ m = 821\ nm$

25.   In the upper atmosphere, O_3 acts as a filter for UV radiation:

$O_3 \overset{h\nu}{\rightarrow} O_2 + O$ (See Exercise 19.24.)

O_3 is also a powerful oxidizing agent. It irritates the lungs and eyes, and at high concentration it is toxic. The smell of a "fresh spring day" is O_3 formed during lightning discharges. Toxic materials don't necessarily smell bad. For example, HCN smells like almonds.

26. $OTeF_5^-$ has $6 + 6 + 5(7) + 1 = 48$ valence electrons.

$F_5TeO - \overset{\cdot\cdot}{\underset{\underset{OTeF_5}{|}}{P}} - OTeF_5$

27.   a.   As we go down the family, K_a increases. This is consistent with the bond to hydrogen getting weaker.

        b.   Po is below Te, so K_a should be larger. The K_a for H_2Po should be on the order of 10^{-2}.

28. TeF_5^- has $6 + 5(7) + 1 = 42$ valence electrons.

The lone pair of electrons around Te exerts a stronger repulsion than the bonding pairs, pushing the four equatorial F's away from the lone pair.

29.   Sulfur forms polysulfide ions, S_n^{2-}, which are soluble, e.g., $S_8 + S^{2-} \rightleftharpoons S_9^{2-}$. Nitric acid oxidizes S^{2-} to S, which then precipitates out of solution.

30.   a.   oxidation - reduction reaction   b. NO

        c.

$$(S^{2-} \rightarrow S + 2e^-) \times 3$$
$$(3e^- + 3 H^+ + HNO_3 \rightarrow NO + 2 H_2O) \times 2$$

$$\overline{6 H^+(aq) + 2 HNO_3(aq) + 3 S^{2-}(aq) \rightarrow 3 S(s) + 2 NO(g) + 4 H_2O(l)}$$

31. $H_2SeO_4(aq) + 3 SO_2(g) \rightarrow Se(s) + 3 SO_3(g) + H_2O(l)$

32.   Light from violet to green will work.

33. $2.42 \text{ eV} \times \dfrac{96.5 \text{ kJ mol}^{-1}}{\text{eV}} \times \dfrac{1 \text{ mol photons}}{6.022 \times 10^{23} \text{ photons}} \times \dfrac{1000 \text{ J}}{\text{kJ}} = \dfrac{3.88 \times 10^{-19} \text{ J}}{\text{photon}}$

$E = \dfrac{hc}{\lambda}$, $\lambda = \dfrac{hc}{E} = \dfrac{(6.626 \times 10^{-34} \text{ J s})(2.998 \times 10^8 \text{ m/s})}{3.88 \times 10^{-19} \text{ J}} = 5.12 \times 10^{-7} \text{ m} = 512 \text{ nm};$ Green light

34. SF_5^- has $6 + 5(7) + 1 = 42$ valence electrons.

square pyramid

Group 7A Elements

35.   a. ClF_5, $7 + 5(7) = 42$ e⁻   b. IF_3, $7 + 3(7) = 28$ e⁻

Square pyramid   T-shaped

c. Cl_2O_7, $2(7) + 7(6) = 56$ e-   d. $FBrO_2$, $7 + 7 + 2(6) = 26$ e⁻

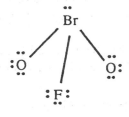

The four O atoms are tetra-   Trigonal pyramidal
hedrally arranged about each
Cl. The Cl – O – Cl bond angle
is close to the tetrahedral angle.

36.   a. $F_2 + H_2O \rightarrow HOF + HF$; $2 HOF \rightarrow 2 HF + O_2$; $HOF + H_2O \rightarrow HF + H_2O_2$ (acid)

In dilute base, HOF exists as OF⁻ and HF exists as F⁻.

$(2e⁻ + H_2O + OF⁻ \rightarrow F⁻ + 2 OH⁻) \times 2$
$4 OH⁻ \rightarrow O_2 + 2 H_2O + 4e⁻$

―――――――――――――――――――――――

$2 OF⁻ \rightarrow O_2 + 2 F⁻$

b.   HOF: Assign +1 to H and -1 to F. The oxidation number of oxygen is then zero. Oxygen is
very electronegative. A zero oxidation state is not very stable since oxygen is a very good
oxidizing agent.

37.

$$\ddot{\text{:F}}\text{---}\ddot{\text{O}}\text{---}\ddot{\text{O}}\text{---}\ddot{\text{F:}}$$

| Formal Charge | 0 | 0 | 0 | 0 |
| Oxid. Number | | -1 | +1 | +1 | -1 |

Oxidation numbers are more useful. We are forced to assign +1 as the oxidation number for oxygen. Oxygen is very electronegative and +1 is not a stable oxidation state for this element.

38.   Fluorine is the most reactive of the halogens because it is the most electronegative atom and the bond in the F_2 molecule is very weak.

39.   a.

$$2e^- + Cl_2 \rightarrow 2\ Cl^- \qquad\qquad E° = +1.36\ V$$
$$2\ H_2O + Cl_2 \rightarrow 2\ OCl^- + 4\ H^+ + 2e^- \qquad -E° = -1.63\ V$$

_____

$$2\ H_2O + 2\ Cl_2 \rightarrow 2\ Cl^- + 2\ OCl^- + 4\ H^+ \qquad E°_{cell} = -0.27\ V$$

b.   Since $E°_{cell} < 0$, then the reaction is not spontaneous at standard conditions.

c.   The forward reaction is favored as the solution becomes more basic (as H^+ is removed).

40.   A disproportionation reaction is an oxidation-reduction reaction in which one species will act as both an oxidizing and reducing agent. The species reacts with itself forming products with higher and lower oxidation states. For example, $2\ Cu^+ \rightarrow Cu + Cu^{2+}$ is a disproportionation reaction.

$HClO_2$ will disproportionate since $E°_{cell} > 0$:

$$HClO_2 + 2\ H^+ + 2\ e^- \rightarrow HClO + H_2O \qquad\qquad E° = +1.65\ V$$
$$HClO_2 + H_2O \rightarrow ClO_3^- + 3\ H^+ + 2\ e^- \qquad -E° = -1.21\ V$$

_____

$$2\ HClO_2 \rightarrow HClO + ClO_3^- + H^+ \qquad\qquad E°_{cell} = +0.44\ V$$

41.   a.

$$ClO_3^- + H_2O \rightarrow ClO_4^- + 2\ H^+ + 2e^- \qquad -E° = -1.19\ V$$
$$2\ H^+ + 2\ e^- \rightarrow H_2 \qquad\qquad E° = \ \ 0.0\ V$$

_____

$$ClO_3^- + H_2O \rightarrow ClO_4^- + H_2 \qquad\qquad E°_{cell} = -1.19\ V$$

A minimum potential of 1.19 V must be applied assuming standard conditions.

b. $3\ Al(s) + 3\ NH_4ClO_4(s) \rightarrow Al_2O_3(s) + AlCl_3(s) + 3\ NO(g) + 6\ H_2O(g)$

$\Delta H° = 3(90) + (-704) + (-1676) + 6(-242) - [3(-295) + 3(0)] = -2677\ kJ$

$$7 \times 10^8\ g\ NH_4ClO_4 \times \frac{1\ mol}{117.5\ g} \times \frac{-2677\ kJ}{3\ mol\ NH_4ClO_4} = -5 \times 10^9\ kJ\ of\ heat\ released$$

42.   a. $AgCl(s) \xrightarrow{h\nu} Ag(s) + Cl$; The reactive chlorine atom is trapped in the crystal. When light is removed, Cl reacts with silver atoms to reform AgCl, i.e., the reverse reaction occurs. In pure AgCl, the Cl atoms escape, making the reverse reaction impossible.

       b.   Over time chlorine is lost and the dark silver metal is permanent.

Group 8A Elements

43.   In Mendeleev's time none of the noble gases were known. Since an entire family was missing, no gaps seemed to appear in the periodic arrangement. Mendeleev had no evidence to predict the existence of such a family.

44.   Helium is unreactive and doesn't combine with any other elements. It is a very light gas and would easily escape the earth's gravitational pull as the planet was formed.

45.   The heavier members are not really inert. Xe and Kr have been shown to react and form compounds with other elements.

46. $10.0\ m \times 5.0\ m \times 3.0\ m = 1.5 \times 10^2\ m^3$; From Table 19.12, volume % Xe $= 9 \times 10^{-6}$.

$$1.5 \times 10^2\ m^3 \times \left(\frac{10\ dm}{m}\right)^3 \times \frac{1\ L}{dm^3} \times \frac{9 \times 10^{-6}\ L\ Xe}{100\ L\ air} = 1 \times 10^{-2}\ L\ of\ Xe\ in\ the\ room$$

$$PV = nRT,\ \ n = \frac{PV}{RT} = \frac{(1.0\ atm)\ (1 \times 10^{-2}\ L)}{(0.08206\ L\ atm\ mol^{-1}\ K^{-1})\ (298\ K)} = 4 \times 10^{-4}\ mol\ Xe$$

$$4 \times 10^{-4}\ mol\ Xe \times \frac{131.3\ g}{mol} = 5 \times 10^{-2}\ g\ Xe\ in\ the\ room$$

$$4 \times 10^{-4}\ mol\ Xe \times \frac{6.022 \times 10^{23}\ atoms}{mol} = 2 \times 10^{20}\ atoms\ Xe\ in\ the\ room$$

A 2 L breath contains: $2\ L\ air \times \dfrac{9 \times 10^{-6}\ L\ Xe}{100\ L\ air} = 2 \times 10^{-7}\ L\ Xe$

$$n = \frac{PV}{RT} = \frac{(1.0\ atm)\ (2 \times 10^{-7}\ L)}{(0.08206\ L\ atm\ mol^{-1}\ K^{-1})\ (298\ K)} = 8 \times 10^{-9}\ mol\ Xe$$

$$8 \times 10^{-9}\ mol\ Xe \times \frac{6.022 \times 10^{23}\ atoms}{mol} = 5 \times 10^{15}\ atoms\ of\ Xe\ in\ a\ 2\ L\ breath$$

47.   a. XeO_3, $8 + 3(6) = 26$ e⁻

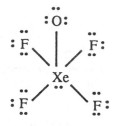

trigonal pyramid

b. XeO_4, $8 + 4(6) = 32$ e⁻

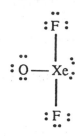

tetrahedral

c. $XeOF_4$, $8 + 6 + 4(7) = 42$ e⁻

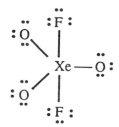

square pyramid

d. $XeOF_2$, $8 + 6 + 2(7) = 28$ e⁻

T-shaped

e. XeO_3F_2 has $8 + 3(6) + 2(7) = 40$ valence electrons.

trigonal bipyramid

48. XeF_2 can react with oxygen to produce explosive xenon oxides and oxyfluorides.

Additional Exercises

49.   As the halogen atoms get larger, it becomes more difficult to fit three halogen atoms around the small N, and the NX_3 molecule becomes less stable.

50.

51.   Xe has one more valence electron than I. Thus, the isoelectric species will have I plus one extra electron substituted for Xe, giving a species with a net minus one charge.

    a. IO_4^- b. IO_3^- c. IF_2^- d. IF_4^- e. IF_6^- f. IOF_3^-

52.   For $NCl_3 \rightarrow NCl_2 + Cl$, only the $N - Cl$ bond is broken. For $O = N - Cl \rightarrow NO + Cl$, the NO bond gets stronger (bond order increases from 2.0 to 2.5) when the $N - Cl$ bond is broken. This makes ΔH for the reaction smaller than just the energy necessary to break the $N - Cl$ bond.

53. $\Delta H° = 82 + 34 - [3(90)] = -154$ kJ; $\Delta S° = 220 + 240 - [3(211)] = -173$ J/K

    $\Delta G° = \Delta H° - T\Delta S° = -154$ kJ $- 298$ K$(-0.173$ kJ/K$) = -102$ kJ

    $\Delta G° = 0$ when $T = \dfrac{\Delta H°}{\Delta S°} = \dfrac{-154{,}000 \text{ J}}{-173 \text{ J/K}} = 890.$ K

    At $T < 890°$ K and standard pressures, the reaction is spontaneous since the favorable $\Delta H°$ term dominates.

54.   Plastic sulfur consists of long S_n chains of sulfur atoms. As plastic sulfur becomes brittle, the long chains break down into S_8 rings.

55.   The pollution provides nitrogen and phosphorous nutrients so the algae can grow. The algae consume oxygen, causing fish to die.

56.
$$H_2O + BrO_3^- \rightarrow BrO_4^- + 2\ H^+ + 2\ e^-$$
$$2\ e^- + 2\ H^+ + XeF_2 \rightarrow Xe + 2\ HF$$

$$H_2O(l) + BrO_3^-(aq) + XeF_2(aq) \rightarrow BrO_4^-(aq) + Xe(g) + 2\ HF(aq)$$

CHAPTER TWENTY

TRANSITION METALS AND COORDINATION CHEMISTRY

Transition Metals

1. a. Co: $[Ar]4s^23d^7$

     Co^{2+}: $[Ar]3d^7$

     Co^{3+}: $[Ar]3d^6$

     b. Pt: $[Xe]6s^14f^{14}5d^9$

      Pt^{2+}: $[Xe]4f^{14}5d^8$

      Pt^{4+}: $[Xe]4f^{14}5d^6$

     c. Fe: $[Ar]4s^23d^6$

      Fe^{2+}: $[Ar]3d^6$

      Fe^{3+}: $[Ar]3d^5$

    d. Au: $[Xe]6s^14f^{14}5d^{10}$

     Au$^+$: $[Xe]4f^{14}5d^{10}$

     Au^{3+}: $[Xe]4f^{14}5d^8$

    e. Cu: $[Ar]4s^13d^{10}$

     Cu$^+$: $[Ar]3d^{10}$

     Cu^{2+}: $[Ar]3d^9$

2.

$$Fe^{2+} \rightarrow Fe^{3+} + e^- \qquad \Delta H = 2.957 \times 10^6 \text{ J}$$
$$e^- + Ti^{4+} \rightarrow Ti^{3+} \qquad \Delta H = -4.175 \times 10^6 \text{ J}$$

$$\overline{Fe^{2+} + Ti^{4+} \rightarrow Fe^{3+} + Ti^{3+} \qquad \Delta H = -1.218 \times 10^6 \text{ J}}$$

From this, we would predict that Fe^{3+} and Ti^{3+} is more stable than Fe^{2+} and Ti^{4+}.

$$Fe^{3+} + e^- \rightarrow Fe^{2+} \qquad E° = 0.77 \text{ V}$$
$$Ti^{3+} + H_2O \rightarrow TiO^{2+} + 2 H^+ + e^- \qquad -E° = -0.099 \text{ V}$$

$$\overline{Fe^{3+} + Ti^{3+} + H_2O \rightarrow Fe^{2+} + TiO^{2+} + 2 H^+ \qquad E°_{cell} = 0.67 \text{ V}}$$

From E° values, we would predict Fe(II) and Ti(IV) to be more stable. The electrochemical data are consistent with the information in the text. These data are for solutions while ionization energies are for gas phase reactions. Solution data are more representative of the conditions from which ilmenite was initially formed.

3. a. molybdenum(IV) sulfide and molybdenum(VI) oxide

    b. MoS_2, +4; MoO_3, +6; $(NH_4)_2Mo_2O_7$, +6; $(NH_4)_6Mo_7O_{24} \cdot 4H_2O$, +6

c. $2\ MoS_2(s) + 7\ O_2(g) \rightarrow 2\ MoO_3(s) + 4\ SO_2(g)$

$2\ NH_3(aq) + 2\ MoO_3(s) + H_2O(l) \rightarrow (NH_4)_2Mo_2O_7(s)$

$6\ NH_3(aq) + 7\ MoO_3(s) + 7\ H_2O(l) \rightarrow (NH_4)_6Mo_7O_{24}\cdot4H_2O(s)$

4.   a. 4 O atoms on faces × 1/2 O/face = 2 O atoms, 2 O atoms inside body, Total: 4 O atoms

8 Ti atoms on corners × 1/8 Ti/corner + 1 Ti atom/body center = 2 Ti atoms

Formula of the unit cell is Ti_2O_4. The empirical formula is TiO_2.

b.
$$\overset{+4\ -2}{2\ TiO_2} + \overset{0}{3\ C} + \overset{0}{4\ Cl_2} \rightarrow \overset{+4\ -1}{2\ TiCl_4} + \overset{+4\ -2}{CO_2} + \overset{+2\ -2}{2\ CO}$$

Cl is reduced and C is oxidized. Cl_2 is the oxidizing agent and C is the reducing agent.

$$\overset{+4\ -1}{TiCl_4} + \overset{0}{O_2} \rightarrow \overset{+4\ -2}{TiO_2} + \overset{0}{2\ Cl_2}$$

O is reduced and Cl is oxidized. O_2 is the oxidizing agent and $TiCl_4$ is the reducing agent.

5. TiF_4: Ionic compound containing Ti^{4+} ions and F^- ions. $TiCl_4$, $TiBr_4$, and TiI_4: Covalent compounds containing discrete, tetrahedral TiX_4 molecules. As these molecules get larger, the bp and mp increase because the London dispersion forces increase.

6.   pyrolusite, manganese(IV)oxide; rhodochrosite, manganese(II)carbonate

7.   a. $2\ CoAs_2(s) + 4\ O_2(g) \rightarrow 2\ CoO(s) + As_4O_6(s)$

As_4O_6, tetraarsenic hexoxide; As_4O_6 has a cage structure similar to P_4O_6.

b.
$$(Co^{2+} + 3\ OH^- \rightarrow Co(OH)_3 + e^-) \times 2$$
$$2\ e^- + H_2O + OCl^- \rightarrow Cl^- + 2\ OH^-$$

$$2\ Co^{2+}(aq) + 4\ OH^-(aq) + H_2O(l) + OCl^-(aq) \rightarrow 2\ Co(OH)_3(s) + Cl^-(aq)$$

c. $\qquad$ $Co(OH)_3(s)$ $\rightleftharpoons$ $Co^{3+}(aq)$ + $3\ OH^-(aq)$

Initial	0	0	$1.0 \times 10^{-7}\ M$

s mol/L $Co(OH)_3$ dissolves to reach equilibrium

Change	-s	$\rightarrow$ +s	+3s
Equil.		s	$1.0 \times 10^{-7} + 3s$

$2.5 \times 10^{-43} = [Co^{3+}][OH^-]^3 = s(1.0 \times 10^{-7} + 3s)^3 \approx s(1.0 \times 10^{-21})$

$s = [Co^{3+}] = 2.5 \times 10^{-22}$ mol/L;   Assumptions good.

d. $Co(OH)_3 \rightleftharpoons Co^{3+} + 3\ OH^-$; pH = 10.00, pOH = 4.00, $[OH^-] = 1.0 \times 10^{-4}\ M$

$2.5 \times 10^{-43} = [Co^{3+}][OH^-]^3 = [Co^{3+}](1.0 \times 10^{-4})^3$, $[Co^{3+}] = 2.5 \times 10^{-31}\ M$

8.

$$\begin{array}{ll} (2\ OH^- + Zn \rightarrow Zn(OH)_2 + 2\ e^-) \times 2 & -E^\circ = 1.24\ V \\ 4\ e^- + O_2 + 2\ H_2O \rightarrow 4\ OH^- & E^\circ = 0.40\ V \end{array}$$

$\overline{\qquad\qquad\qquad\qquad\qquad\qquad\qquad\qquad\qquad\qquad\qquad\qquad\qquad\qquad}$

$2\ Zn(s) + O_2(g) + 2\ H_2O(l) \rightarrow 2\ Zn(OH)_2(s)$ $\qquad\qquad$ $E^\circ_{cell} = 1.64\ V$

$\Delta G^\circ = -nFE^\circ = -(4\ mol\ e^-)(96{,}485\ C/mol\ e^-)(1.64\ J/C) = -6.33 \times 10^5\ J = -633\ kJ$

$\log K = \dfrac{nE^\circ}{0.0592} = \dfrac{4(1.64)}{0.0592} = 111,\ K \approx 10^{111}$

Coordination Compounds

9.   a. ligand: $\qquad$ Species that donates a pair of electrons to form a covalent bond to a metal ion (a Lewis base).

   b. chelate: $\qquad$ Ligand that can form more than one bond.

   c. bidentate: $\qquad$ Ligand that can form two bonds.

   d. complex ion: $\quad$ Metal ion plus ligands.

10.   a. hexaamminecobalt(II) chloride $\qquad\qquad$ b. hexaaquacobalt(III) iodide

   c. potassium tetrachloroplatinate(II) $\qquad$ d. potassium hexachloroplatinate(II)

   e. pentaamminechlorocobalt(III) chloride $\quad$ f. triamminetrinitrocobalt(III)

11.   a. pentaamminechlororuthenium(III) ion $\qquad$ b. hexacyanoferrate(II) ion

   c. tris(ethylenediamine)manganese(II) ion $\quad$ d. pentaamminenitrocobalt(III) ion

12.   a.   sodium tris(oxalato)nickelate(II)   b.   potassium tetrachlorocobaltate(II)

       c.   tetraamminecopper(II) sulfate

       d.   chlorobis(ethylenediamine)thiocyanatocobalt(III) chloride

13.   a. $[Co(C_5H_5N)_6]Cl_3$ b. $[Cr(NH_3)_5I]I_2$ c. $[Ni(NH_2CH_2CH_2NH_2)_3]Br_2$

       d. $K_2[Ni(CN)_4]$ e. $[Pt(NH_3)_4Cl_2]PtCl_4$

14.   a. $FeCl_4^-$ b. $[Ru(NH_3)_5H_2O]^{3+}$

       c. $[Pt(C_5H_5N)_5I]^{3+}$ d. $[Pt(NH_3)Cl_3]^-$

15.   a.   2; Forms bonds through the lone pairs on the two oxygen atoms.

       b.   3; Forms bonds through the lone pairs on the three nitrogen atoms.

       c.   4; Forms bonds through the two nitrogen atoms and the two oxygen atoms.

       d.   4; Forms bonds through the four nitrogen atoms.

16.   a.   isomers: Species with the same formulas but different properties. See text for examples
                        of the following types of isomers.

       b.   structural isomers: Isomers that have one or more bonds that are different.

       c.   stereoisomers: Isomers that contain the same bonds but differ in how the atoms are arranged
                        in space.

       d.   coordination isomers: Structural isomers that differ in the atoms that make up the complex
                        ion.

       e.   linkage isomers: Structural isomers that differ in how one or more ligands are attached to
                        the transition metal.

       f.   geometric isomers: (cis - trans isomerism) Stereoisomers that differ in the positions of
                        atoms with respect to a rigid ring, bond, or each other.

       g.   optical isomers: Stereoisomers that are nonsuperimposable mirror images of each
                        other; that is, they are different in the same way that our left and right
                        hands are different.

17. **a.**

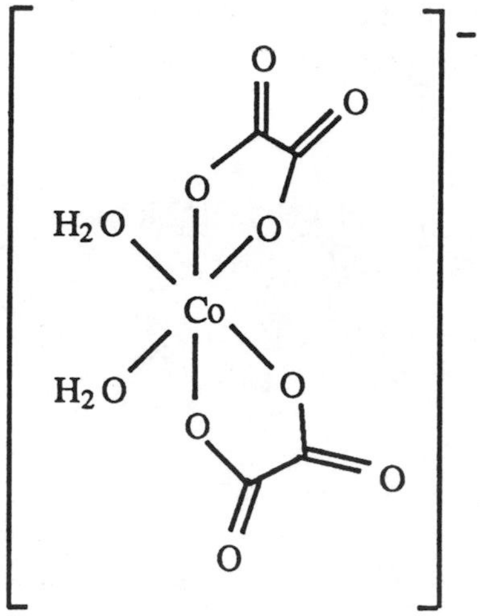

cis 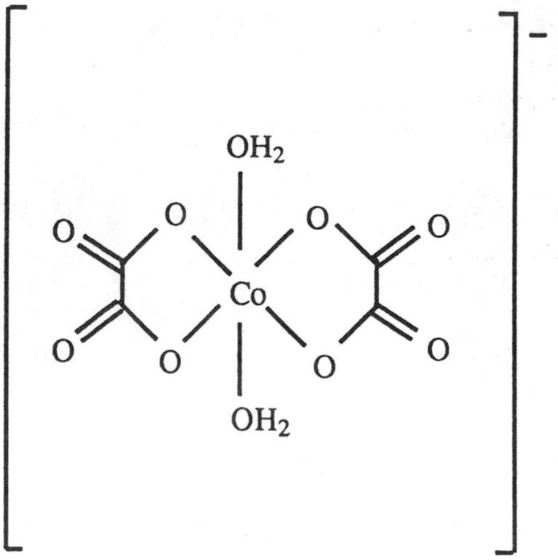   trans

b.

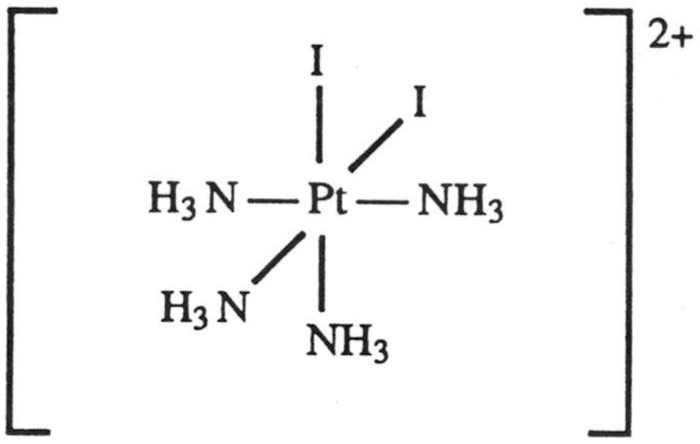

cis 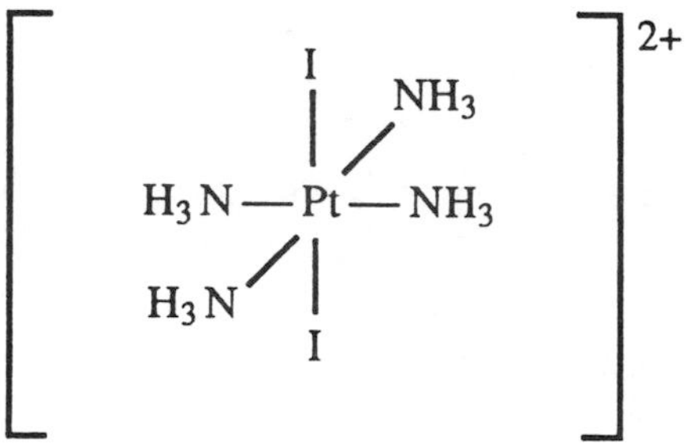   trans

c.

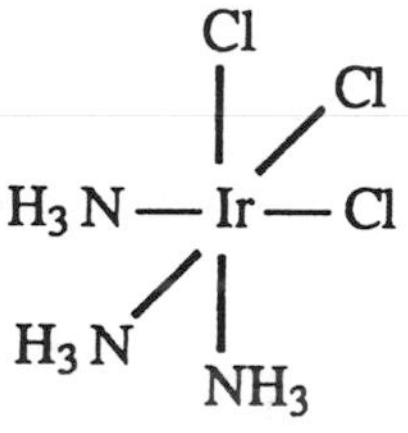

cis 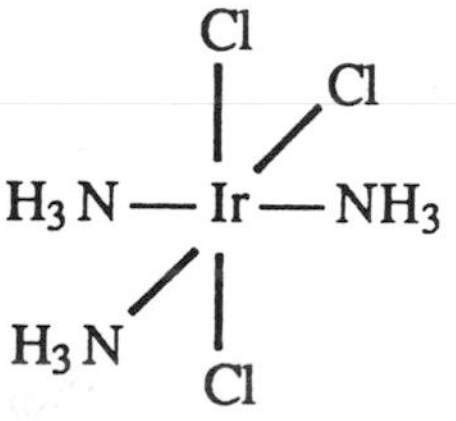   trans

d.

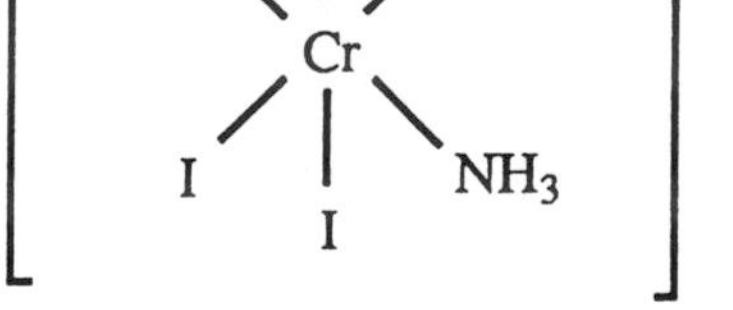

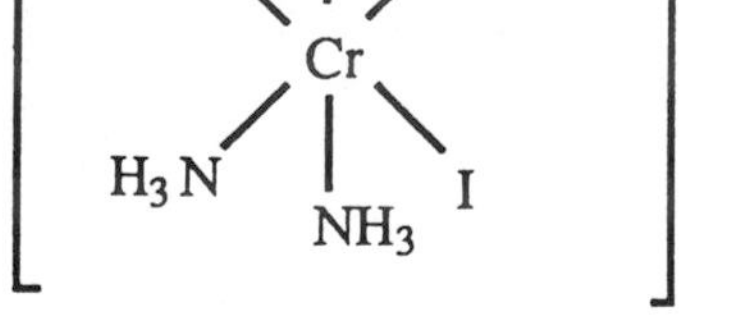

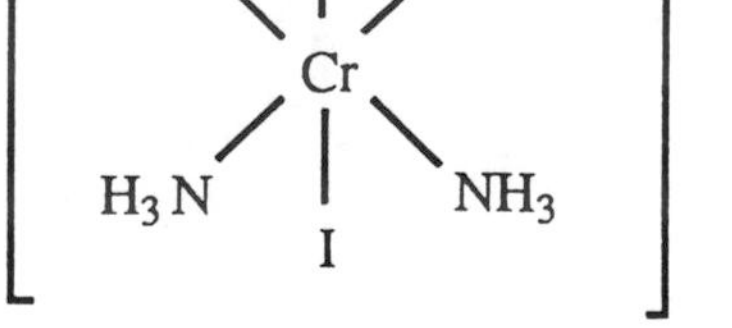

18.   Linkage isomers differ in the way the ligand bonds to the metal. SCN⁻ can bond through the sulfur or through the nitrogen atom. NO₂⁻ can bond through the nitrogen or through the oxygen atom. OCN⁻ can bond through the oxygen or through the nitrogen atom. N₃⁻, en, and I⁻ are not capable of linkage isomerism.

19.

```
        SCN                         SCN                         NCS
         /                           /                           /
H₃N —— Pt —— NH₃            H₃N —— Pt —— NH₃            H₃N —— Pt —— NH₃
     /                           /                           /
   NCS                         SCN                         SCN
```

```
        SCN                         NCS                         NCS
         /                           /                           /
H₃N —— Pt —— SCN           H₃N —— Pt —— SCN           H₃N —— Pt —— NCS
     /                           /                           /
   H₃N                         H₃N                         H₃N
```

20.

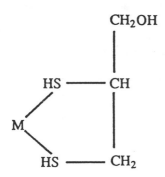

```
        CH₂SH                               HS ———— CH₂
         |                                 /            \
HS ———— CH                             M                CH—SH
 \       |                              \              /
  M.     |                               HO ———— CH₂
   \     |
   HO ——— CH₂
```

21.   a.   b.

```
           NH₂      Cl                        Cl
          /        /                    CH₂—H₂N │ NH₂—CH₂
  H₂C —               \                   |      │      |
   |          Pt                          |      Co     |
  H₂C —      /        \                   |      │      |
          \ NH₂      Cl                 CH₂—H₂N │ NH₂—CH₂
                                            Cl
```

c.

$$\left[\begin{array}{c} \text{H}_3\text{N} \quad \overset{\text{NH}_3}{\underset{\text{NH}_3}{\overset{|}{\text{Co}}}} \quad \text{Cl} \\ \text{H}_3\text{N} \qquad \qquad \text{NO}_2 \end{array} \right]^{+}$$

d.

$$\left[\begin{array}{c} \text{H}_3\text{N} \quad \overset{\text{Cl}}{\underset{\text{ONO}}{\overset{|}{\text{Co}}}} \quad \text{NH}_3 \\ \text{H}_3\text{N} \qquad \qquad \text{NH}_3 \end{array} \right]^{+}$$

e.

$$\left[\begin{array}{c} \text{CH}_2 - \text{H}_2\text{N} \quad \overset{\text{H}_2\text{O}}{\underset{\text{H}_2\text{O}}{\overset{|}{\text{Cu}}}} \quad \text{NH}_2 - \text{CH}_2 \\ \text{CH}_2 - \text{H}_2\text{N} \qquad \qquad \text{NH}_2 - \text{CH}_2 \end{array} \right]^{2+}$$

22.

and

23. $BaCl_2$ gives no precipitate so SO_4^{2-} must be in the coordination sphere. A precipitate with $AgNO_3$ means that the Cl^- is not in the coordination sphere. Since there are only four ammonia molecules in the coordination sphere, then the SO_4^{2-} must be acting as a bidentate ligand. The structure is:

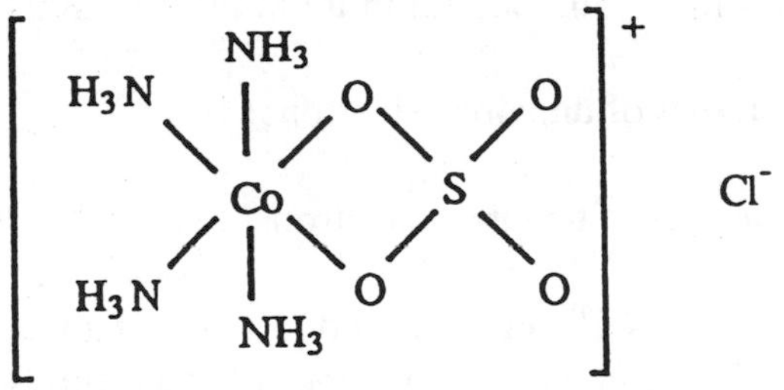

24.

monodentate

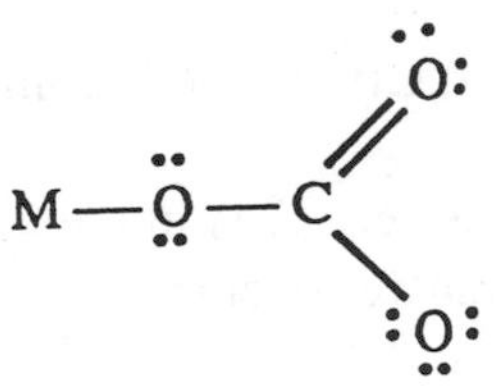

bidentate

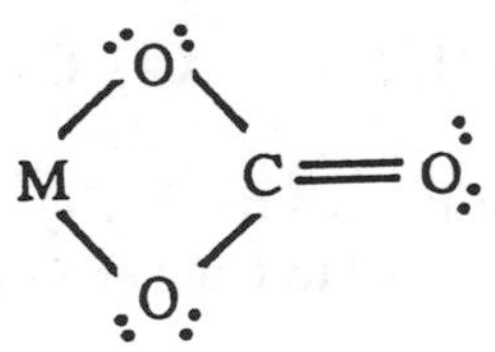

bridging

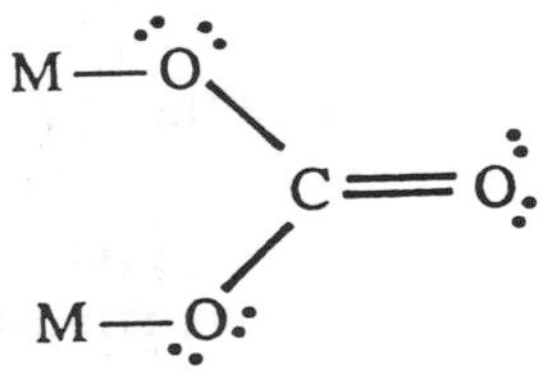

25.

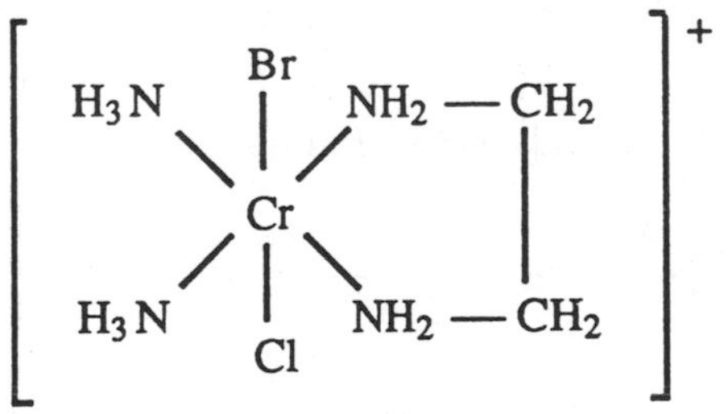

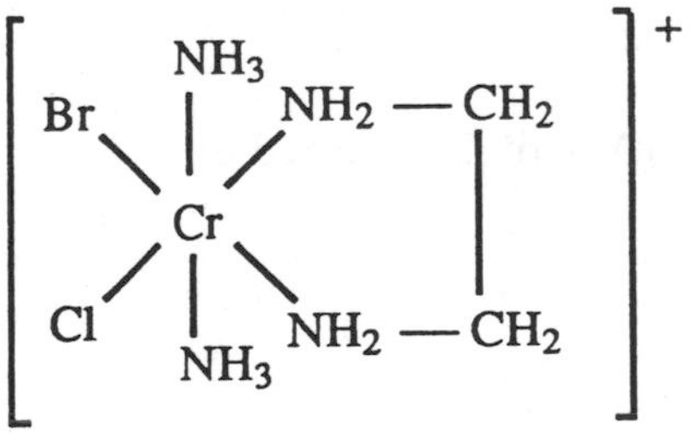

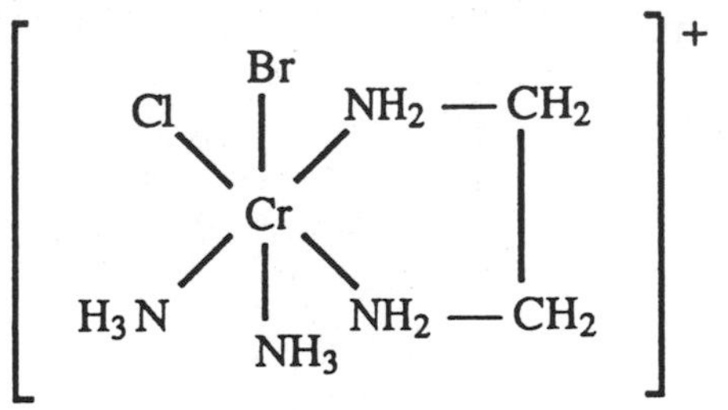

optically active
(mirror image not shown)

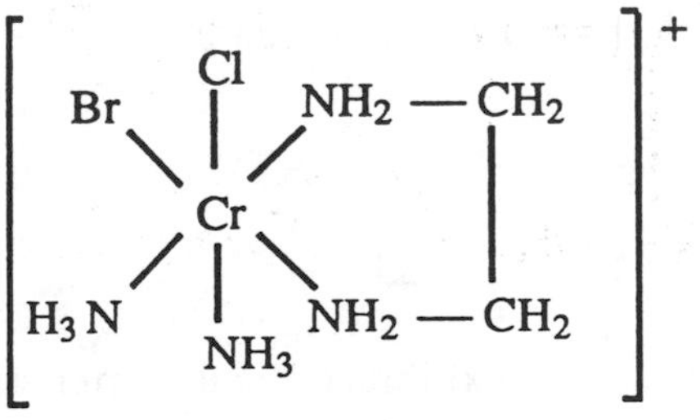

optically active
(mirror image not shown)

Bonding, Color, and Magnetism in Coordination Compounds

26.   a.   Ligand that will give complex ions with the maximum number of unpaired electrons.

b.   Ligand that will give complex ions with the minimum number of unpaired electrons.

c.   Complex with a minimum number of unpaired electrons.

d.   Complex with a maximum number of unparied electrons.

27. Cu^{2+}: $[Ar]3d^9$; Cu^+: $[Ar]3d^{10}$; Cu(II) is d^9 and Cu(I) is d^{10}. Color is a result of the electron transfer between split d orbitals. This cannot occur for the filled d orbitals in Cu(I).

28.   No, the d-orbitals are completely filled since Cd^{2+} is d^{10}. Electron transfer cannot occur when the d-orbitals are completely filled. See Exercise 20.27.

29.
$$\underset{II}{(H_2O)_5Cr} - Cl - \underset{III}{Co(NH_3)_5} \rightarrow \underset{III}{(H_2O)_5Cr} - Cl - \underset{II}{Co(NH_3)_5} \rightarrow Cr(H_2O)_5Cl^{2+} + Co(II) \text{ complex}$$

Yes; After the oxidation, the ligands on Cr(III) won't exchange. Since Cl^- is in the coordination sphere it must have formed a bond to Cr(II) before the electron transfer occurred.

30.   a. Ru^{2+}: $[Kr]4d^6$, no unpaired e^- b. Fe^{3+}: $[Ar]3d^5$, 1 unpaired e^-

c. Ni^{2+}: $[Ar]3d^8$, 2 unpaired e^- d. V^{3+}: $[Ar]3d^2$, 2 unpaired e^-

e. Co^{2+}: $[Ar]3d^7$, 3 unpaired e^- (tetrahedral splitting)

31.   a. $Ru(phen)_3^{2+}$ exhibits optical isomerism.

b. Ru^{2+}: $[Kr]4d^6$; Since there are no unpaired electrons, then Ru^{2+} is a strong-field case.

32. $NiCl_4^{2-}$ is tetrahedral and $Ni(CN)_4^{2-}$ is square planar. The corresponding d-orbital splitting diagrams for the d^8 Ni^{2+} are:

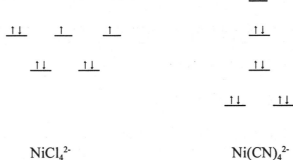

          $NiCl_4^{2-}$ $Ni(CN)_4^{2-}$

33. Co^{2+}: $[Ar]3d^7$; The corresponding d-orbital splitting diagram for tetrahedral Co^{2+} complexes is:

    ↑   ↑   ↑

    ↑↓   ↑↓

All tetrahedral complexes are high spin since the d-orbital splitting is small. Ions with 2 or 7 d-electrons should give the most stable tetrahedral complexes since they have the greatest number of electrons in the lower energy orbitals as compared to the number of electrons in the higher energy orbitals.

34.   a. Fe^{2+} (d^6):

    ↑   ↑   —   —

  ↑↓   ↑   ↑   ↑↓ ↑↓ ↑↓

    High spin   Low spin

    b. Fe^{3+} (d^5):   c. Ni^{2+} (d^8):

    ↑   ↑   ↑   ↑

  ↑   ↑   ↑   ↑↓ ↑↓ ↑↓

    High spin

d. Zn^{2+} (d^{10}):

e. Co^{2+}(d^7):

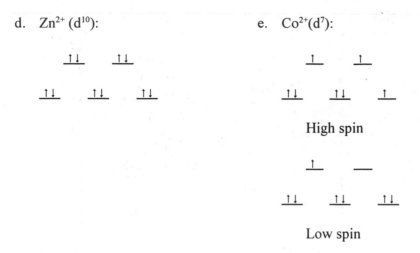

35.   Transition compounds exhibit the color compementary to that absorbed. Using Table 20.16, $Ni(H_2O)_6Cl_2$ absorbs red light and $Ni(NH_3)Cl_2$ absorbs yellow-green light. $Ni(NH_3)_6Cl_2$ absorbs the shorter wavelength light which is the higher energy light. Therefore, Δ is larger for $Ni(NH_3)_6Cl_2$. NH_3 is a stronger field ligand than H_2O, consistent with the spectrochemical series.

36. Fe^{3+} complexes have one unpaired electron when a strong-field case and five unpaired electrons when a weak-field case. $Fe(CN)_6^{2-}$ is a strong-field case and $Fe(SCN)_6^{3-}$ is a weak-field case. Therefore, cyanide, CN^-, is a stronger field ligand than thiocyanate, SCN^-.

37.   Octahedral Cr^{2+} complexes should be used. Cr^{2+}: $[Ar]3d^4$; High-spin Cr^{2+} complexes have 4 unpaired electrons and low-spin Cr^{2+} complexes have 2 unpaired electrons. Ni^{2+}: $[Ar]3d^8$; Octahedral Ni^{2+} complexes will always have 2 unpaired electrons, whether high or low-spin. Therefore, Ni^{2+} complexes cannot be used to distinguish weak from strong-field ligands by examining magnetic properties. Alternatively, the ligand field strengths can be measured using visible spectra. Either Cr^{2+} or Ni^{2+} complexes can be used for this method.

Additional Exercises

38. $Ni(CO)_4$ is composed of 4 CO molecules and Ni. Thus, nickel has an oxidation state of zero.

39.   Since transition metals form bonds to species with lone pairs of electrons, they are Lewis acids (electron pair acceptors).

40. CN^- is a weak base. The two reactions are:

$$Ni^{2+}(aq) + 2\ OH^-(aq) \rightarrow Ni(OH)_2(s)$$

$$Ni(OH)_2(s) + 4\ CN^-(aq) \rightarrow Ni(CN)_4^{2-}(aq) + 2\ OH^-\ (aq)$$

41.   a.

mirror

NH$_3$ molecules are not shown.

b.   All are Co(III).

c.   Co^{3+} is d^6. There are zero unpaired electrons if a low-spin case.

42.   i. $0.0203 \text{ g CrO}_3 \times \dfrac{52.00 \text{ g Cr}}{100.0 \text{ g CrO}_3} = 0.0106 \text{ g Cr};$ % Cr $= \dfrac{0.0106}{0.105} \times 100 = 10.1\%$ Cr

ii. $32.93 \text{ mL HCl} \times \dfrac{0.100 \text{ mmol HCl}}{\text{mL}} \times \dfrac{1 \text{ mmol NH}_3}{\text{mmol HCl}} \times \dfrac{17.03 \text{ mg NH}_3}{\text{mmol}} = 56.1 \text{ mg NH}_3$

% NH$_3$ = $\dfrac{56.1 \text{ mg}}{341 \text{ mg}} \times 100 = 16.5\%$ NH$_3$

iii. 75.53 + 16.5 + 10.1 = 100.1; The compound is composed of only Cr, NH$_3$, and I.

Out of 100.00 of compound:

$10.1 \text{ g Cr} \times \dfrac{1 \text{ mol}}{52.00 \text{ g}} = 0.194$ $\dfrac{0.194}{0.194} = 1.00$

$16.5 \text{ g NH}_3 \times \dfrac{1 \text{ mol}}{17.03 \text{ g}} = 0.969$ $\dfrac{0.969}{0.194} = 4.99$

$73.53 \text{ g I} \times \dfrac{1 \text{ mol}}{126.9 \text{ g}} = 0.5794$ $\dfrac{0.5794}{0.194} = 2.99$

Cr(NH$_3$)$_5$I$_3$ is the empirical formula. Cr(III) forms octahedral complexes. So compound A is made of [Cr(NH$_3$)$_5$I]$^{2+}$ and two I$^-$ ions or [Cr(NH$_3$)$_5$I]I$_2$.

iv. $\Delta T_f = iK_f m$; For $[Cr(NH_3)_5I]I_2$, $i = 3$ ions.

$$m = \frac{0.601 \text{ g complex}}{10.0 \text{ g H}_2\text{O}} \times \frac{1 \text{ mol complex}}{517.9 \text{ g complex}} \times \frac{1000 \text{ g H}_2\text{O}}{\text{kg}} = 0.116 \text{ molal}$$

$\Delta T_f = 3 \times 1.86°\text{C/molal} \times 0.116 \text{ molal} = 0.65°\text{C}$

Since ΔT_f is close to the measured value, then this is consistent with the formula $[Cr(NH_3)_5I]I_2$.

43. CN^- and CO form much stronger complexes with Fe(II) than O_2. Thus, O_2 cannot be transported by hemoglobin in the presence of CN^- or CO.

44. $Fe_2O_3(s) + 6 H_2C_2O_4(aq) \rightarrow 2 Fe(C_2O_4)_3{}^{3-}(aq) + 3 H_2O(l) + 6 H^+(aq)$

The oxalate anion forms a soluble complex ion with iron in rust.

45. Most transition metals have unfilled d orbitals which creates a large number of valence electrons that can be removed.

46. Sc^{3+} has no electrons in d orbitals. V^{3+} and Ti^{3+} have d electrons present. Color of transition metal complexes results from electron transfer between split d orbitals. If no d electrons are present, no electron transfer can occur and the compounds are not colored.

47. No; In all three cases, six bonds are formed between Ni^{2+} and nitrogen, so ΔH values should be similar. $\Delta S°$ for formation of the complex ion is most negative for 6 NH_3 molecules reacting with a metal ion (7 independent species become 1). For penten reacting with a metal ion, 2 independent species become 1, so $\Delta S°$ is less negative. Thus, the chelate effect occurs because the more bonds a chelating agent can form to the metal, the more favorable $\Delta S°$ is for the formation of the complex ion and the larger the formation constant.

48. a. Consider the following electrochemical cell:

$$Co^{3+} + e^- \rightarrow Co^{2+} \qquad E° = 1.82 \text{ V}$$
$$Co(en)_3{}^{2+} \rightarrow Co(en)_3{}^{3+} + e^- \qquad -E° = ?$$

$$Co^{3+} + Co(en)_3{}^{2+} \rightarrow Co^{2+} + Co(en)_3{}^{3+} \qquad E°_{cell} = 1.82 - E°$$

The equilibrium constant for this overall reaction is:

$$Co^{3+} + 3 \text{ en} \rightarrow Co(en)_3{}^{3+} \qquad K_1 = K_f = 2.0 \times 10^{47}$$
$$Co(en)_3{}^{2+} \rightarrow Co^{2+} + 3 \text{ en} \qquad K_2 = 1/K_f = 1/1.5 \times 10^{12}$$

$$Co^{3+} + Co(en)_3{}^{2+} \rightarrow Co(en)_3{}^{3+} + Co^{2+} \qquad K = K_1K_2 = \frac{2.0 \times 10^{47}}{1.5 \times 10^{12}} = 1.3 \times 10^{35}$$

From the Nernst equation for the overall raction:

$$E_{cell}^{\bullet} = \frac{0.0592}{n} \log K = \frac{0.0592}{1} \log (1.3 \times 10^{35}), \quad E_{cell}^{\bullet} = 2.08 \text{ V}$$

$$E_{cell}^{\bullet} = 1.82 - E^{\circ} = 2.08 \text{ V}, \quad E^{\circ} = 1.82 \text{ V} - 2.08 \text{ V} = -0.26 \text{ V}$$

b. The strongest oxidizing agent is the species most easily reduced. From the reduction potentials, Co^{3+} ($E^{\circ} = 1.82$ V) is a much stronger oxidizing agent than $Co(en)_3^{3+}$ ($E^{\circ} = -0.26$ V).

c. In aqueous solution, Co^{3+} forms the hydrated transition metal complex, $Co(H_2O)_6^{3+}$. In both complexes, $Co(H_2O)_6^{3+}$ and $Co(en)_3^{3+}$, cobalt exists as Co^{3+} which has 6 d electrons. Assuming a strong-field case, the d-orbital splitting diagram for each is:

When each complex gains an electron, the electron enters the higher energy e_g orbitals. Since en is a stronger field ligand than H_2O, then the d-orbital splitting is larger for $Co(en)^{3+}$ and it takes more energy to add an electron to $Co(en)^{3+}$ than to $Co(H_2O)_6^{3+}$. Therefore, it is more favorable for $Co(H_2O)_6^{3+}$ to gain an electron than for $Co(en)^{3+}$ to gain an electron.

49. a.

b. cis-$Cr(acac)_2(H_2O)_2$ and $Cr(acac)_3$ are optically active. There is a plane of symmetry in trans-$Cr(acac)_2(H_2O)_2$, so it is not optically active.

c. $0.112 \text{ g Eu}_2O_3 \times \dfrac{304.0 \text{ g Eu}}{352.0 \text{ g Eu}_2O_3} = 0.0967 \text{ g Eu};$ % Eu $= \dfrac{0.0967 \text{ g}}{0.286 \text{ g}} \times 100 = 33.8\%$ Eu

% O = 100.00 - (33.8 + 40.1 + 4.71) = 21.4% O

Out of 100.00 g of compound:

$$33.8 \text{ g Eu} \times \frac{1 \text{ mol}}{152.0 \text{ g}} = 0.222 \text{ mol Eu}; \quad 40.1 \text{ g C} \times \frac{1 \text{ mol}}{12.01 \text{ g}} = 3.34 \text{ mol C}$$

$$4.71 \text{ g H} \times \frac{1 \text{ mol}}{1.008 \text{ g}} = 4.67 \text{ mol H}; \ 21.4 \text{ g O} \times \frac{1 \text{ mol}}{16.00 \text{ g}} = 1.34 \text{ mol O}$$

$$\frac{3.34}{0.222} = 15.0, \ \frac{4.67}{0.222} = 21.0, \ \frac{1.34}{0.222} = 6.04$$

The empirical and molecular formula is $EuC_{15}H_{21}O_6$. Since each acac⁻ is $C_5H_7O_2^-$, then an abbreviated molecular formula is $Eu(acac)_3$.

50.   a. $Be(tfa)_2$ exhibits optical isomerism. A representation for the tetrahedral optical isomers are:

mirror

b. Square planar $Cu(tfa)_2$ molecules exhibit geometric isomerism. In one geometric isomer, the CF_3 groups are cis to each other and in the other isomer, the CF_3 groups are trans.

cis   trans

51.   There is no plane symmetry in the complex, thus it is optically active.

52.

The $d_{x^2-y^2}$ and d_{xy} are in the plane of the three ligands and should be destabilized the most. The d_{z^2} has some electron density in the xy plane (the doughnut) and should be destabilized a lesser amount. The d_{xz} and d_{yz} have no electron density in the plane and should be lowest in energy.

53.

The d_{z^2} will be destabilized much more than in the trigonal planar case (see Exercise 20.52).

54.   There is a steady decrease in the atomic radii of the lanthanide elements, going from left to right. As a result of the lanthanide contraction, the properties of the 4d and 5d elements in each group are very similar because of the similar sizes of atoms and ions (see Exercise 12.60).

CHAPTER TWENTY-ONE

THE NUCLEUS: A CHEMIST'S VIEW

Radioactive Decay and Nuclear Transformations

1. a. $^{3}_{1}H \rightarrow {^{0}_{-1}e} + {^{3}_{2}He}$ c. $^{7}_{4}Be + {^{0}_{-1}e} \rightarrow {^{7}_{3}Li}$

   b. $^{8}_{3}Li \rightarrow {^{8}_{4}Be} + {^{0}_{-1}e}$ d. $^{8}_{5}B \rightarrow {^{8}_{4}Be} + {^{0}_{+1}e}$

   $$\frac{^{8}_{4}Be \rightarrow 2\,{^{4}_{2}He}}{}$$ e. $^{32}_{15}P \rightarrow {^{32}_{16}S} + {^{0}_{-1}e}$

   $^{8}_{3}Li \rightarrow 2\,{^{4}_{2}He} + {^{0}_{-1}e}$

2. a. $^{60}_{27}Co \rightarrow {^{60}_{28}Ni} + {^{0}_{-1}e}$ b. $^{97}_{43}Tc + {^{0}_{-1}e} \rightarrow {^{97}_{42}Mo}$

   c. $^{99}_{43}Tc \rightarrow {^{99}_{44}Ru} + {^{0}_{-1}e}$ d. $^{239}_{94}Pu \rightarrow {^{235}_{92}U} + {^{4}_{2}He}$

3. ^{8}B and ^{9}B contain too many protons or too few neutrons. Electron capture or positron production are both possible decay mechanisms that increase the neutron to proton ratio. ^{12}B and ^{13}B contain too many neutrons or too few protons. Beta production lowers the neutron to proton ratio, so we expect ^{12}B and ^{13}B to be β-emitters.

4. a. $^{1}_{1}H + {^{14}_{7}N} \rightarrow {^{11}_{6}C} + {^{4}_{2}He}$ b. $2\,{^{3}_{2}He} \rightarrow {^{4}_{2}He} + 2\,{^{1}_{1}H}$

   c. $^{1}_{1}H + {^{1}_{1}H} \rightarrow {^{2}_{1}H} + {^{0}_{+1}e}$ (positron)   d. $^{1}_{1}H + {^{12}_{6}C} \rightarrow {^{13}_{7}N}$

5. a. $^{240}_{95}Am + {^{4}_{2}He} \rightarrow {^{243}_{97}Bk} + {^{1}_{0}n}$ b. $^{238}_{92}U + {^{12}_{6}C} \rightarrow {^{244}_{98}Cf} + 6\,{^{1}_{0}n}$

   c. $^{249}_{98}Cf + {^{18}_{8}O} \rightarrow {^{263}_{106}Unh} + 4\,{^{1}_{0}n}$ d. $^{249}_{98}Cf + {^{10}_{5}B} \rightarrow {^{257}_{103}Lr} + 2\,{^{1}_{0}n}$

6.   Fission: Splitting of a heavy nucleus into two (or more) lighter nuclei.

Fusion: Combining two light nuclei to form a heavier nucleus.

Fusion is more likely for elements lighter than Fe; fission is more likely for elements heavier than Fe.

7.   Characteristic frequencies of energies emitted in a nuclear reaction suggests that discrete energy levels exist in the nucleus. Extra stability of certain numbers of nucleons and the predominance of nuclei with even numbers of nucleons suggests that the nuclear structure might be described by using quantum numbers.

8. $^{53}_{26}$Fe has too many protons. It will undergo either positron production, electron capture and/or alpha particle production. $^{59}_{26}$Fe has too many neutrons and will undergo beta particle production.

9.   a. $^{207}_{82}$Pb; Complete decay is 7α and 4β: $^{235}_{92}U \rightarrow\ ^{207}_{82}Pb + 7\ ^{4}_{2}He + 4\ ^{0}_{-1}e$

b. $^{235}_{92}U \rightarrow\ ^{231}_{90}Th +\ ^{4}_{2}He \rightarrow\ ^{231}_{91}Pa +\ ^{0}_{-1}e \rightarrow\ ^{227}_{89}Ac +\ ^{4}_{2}He$

$^{215}_{84}Po +\ ^{4}_{2}He \leftarrow\ ^{219}_{86}Rn +\ ^{4}_{2}He \leftarrow\ ^{223}_{88}Ra +\ ^{4}_{2}He \leftarrow\ ^{227}_{90}Th +\ ^{0}_{-1}e$

$^{4}_{2}He +\ ^{211}_{82}Pb \rightarrow\ ^{211}_{83}Bi +\ ^{0}_{-1}e \rightarrow\ ^{207}_{81}Tl +\ ^{4}_{2}He \rightarrow\ ^{207}_{82}Pb +\ ^{0}_{-1}e$

10.   a. $^{241}_{95}Am \rightarrow\ ^{4}_{2}He +\ ^{237}_{93}Np$

b. $^{241}_{95}Am \rightarrow 8\ ^{4}_{2}He + 4\ ^{0}_{-1}e +\ ^{209}_{83}Bi$; The final product is $^{209}_{83}Bi$.

c. $^{241}_{95}Am \rightarrow\ ^{237}_{93}Np + \alpha \rightarrow\ ^{233}_{91}Pa + \alpha \rightarrow\ ^{233}_{92}U + \beta \rightarrow\ ^{229}_{92}Th + \alpha \rightarrow\ ^{225}_{88}Ra + \alpha$

$^{213}_{84}Po + \beta \leftarrow\ ^{213}_{83}Bi + \alpha \leftarrow\ ^{217}_{85}At + \alpha \leftarrow\ ^{221}_{87}Fr + \alpha \leftarrow\ ^{225}_{89}Ac + \beta$

$^{209}_{82}Pb + \alpha \rightarrow\ ^{209}_{83}Bi + \beta$

The intermediate radionuclides are:

$^{237}_{93}Np,\ ^{233}_{91}Pa,\ ^{233}_{92}U,\ ^{229}_{90}Th,\ ^{225}_{88}Ra,\ ^{225}_{89}Ac,\ ^{221}_{87}Fr,\ ^{217}_{85}At,\ ^{213}_{83}Bi,\ ^{213}_{84}Po,$ and $^{209}_{82}Pb$.

11.   The most abundant isotope is generally the most stable isotope. The periodic table predicts that the most stable isotopes for exercises a - d are ^{39}K, ^{56}Fe, ^{23}Na and ^{204}Tl.

   a.   Unstable; ^{45}K has too many neutrons and will undergo beta particle production.

   b.   Stable

   c.   Unstable; ^{20}Na has too few neutrons and will most likely undergo electron capture or positron production.

   d.   Unstable; ^{194}Tl has too few neutrons and will undergo electron capture, positron production and/or alpha particle production.

12.   a. $^{249}_{98}Cf + ^{18}_{8}O \rightarrow ^{263}_{106}Sg + 4\,^{1}_{0}n$

   b. $^{259}_{104}Unq;\ ^{263}_{106}Sg \rightarrow ^{4}_{2}He + ^{259}_{104}Unq$

Kinetics of Radioactive Decay

13.   For $t_{1/2} = 12{,}000$ yr:

$$k = \frac{\ln 2}{t_{1/2}} = \frac{0.693}{t_{1/2}} = \frac{0.693}{12{,}000\ \text{yr}} \times \frac{1\ \text{yr}}{365\ \text{d}} \times \frac{1\ \text{d}}{24\ \text{h}} \times \frac{1\ \text{h}}{3600\ \text{s}} = 1.8 \times 10^{-12}\ \text{s}^{-1}$$

$$\text{Rate} = kN = 1.8 \times 10^{-12}\ \text{s}^{-1} \times 6.02 \times 10^{23}\ \text{nuclei} = \frac{1.1 \times 10^{12}\ \text{disintegrations}}{\text{s}}$$

For $t_{1/2} = 12$ h:

$$k = \frac{0.693}{t_{1/2}} = \frac{0.693}{12\ \text{h}} \times \frac{1\ \text{h}}{3600\ \text{s}} = 1.6 \times 10^{-5}\ \text{s}^{-1}$$

$$\text{Rate} = 1.6 \times 10^{-5}\ \text{s}^{-1} \times 6.02 \times 10^{23}\ \text{nuclei} = \frac{9.6 \times 10^{18}\ \text{disintegrations}}{\text{s}}$$

For $t_{1/2} = 12$ min:

$$\text{Rate} = \frac{0.693}{12\ \text{min}} \times \frac{1\ \text{min}}{60\ \text{s}} \times 6.02 \times 10^{23}\ \text{nuclei} = \frac{5.8 \times 10^{20}\ \text{disintegrations}}{\text{s}}$$

For $t_{1/2} = 12$ s:

$$\text{Rate} = \frac{0.693}{12\ \text{s}} \times 6.02 \times 10^{23}\ \text{nuclei} = \frac{3.5 \times 10^{22}\ \text{disintegrations}}{\text{s}}$$

14. $t_{1/2} = 5730$ yr; $k = (\ln 2)/t_{1/2}$; $\ln (N/N_o) = -kt$

$$\ln\left(\frac{N}{N_o}\right) = \frac{-0.6931\,t}{t_{1/2}} = \frac{-0.6931\,(2200\text{ yr})}{5730\text{ yr}} = -0.27, \quad \frac{N}{N_o} = e^{-0.27} = 0.76 = 76\% \text{ of } {}^{14}\text{C remains}$$

15.   The assumptions are that the ^{14}C levels in the atmosphere are constant or the ^{14}C level at the time the plant died can be calculated. Constant ^{14}C level is a poor assumption and accounting for variation is complicated.

16. $k = (\ln 2)/t_{1/2}$; $\ln\left(\dfrac{N}{N_o}\right) = -kt = \dfrac{-0.693\,t}{t_{1/2}}$, $\ln\left(\dfrac{5.0\ \mu\text{g Ca}}{m}\right) = \dfrac{-0.693\,(2.0\text{ d})}{4.5\text{ d}} = -0.31$

$\ln 5.0 - \ln m = -0.31$, $1.61 + 0.31 = \ln m$, $m = e^{1.92} = 6.8\ \mu$g of Ca

$$6.8\ \mu\text{g }{}^{47}\text{Ca} \times \frac{107.0\ \mu\text{g }{}^{47}\text{CaCO}_3}{47.0\ \mu\text{g }{}^{47}\text{Ca}} = 15\ \mu\text{g }{}^{47}\text{CaCO}_3$$

Note: Units for N and N_o are usually number of nuclei but can also be grams if the units are the same for both N and N_o.

17. $k = (\ln 2)/t_{1/2}$; $\ln\left(\dfrac{N}{N_o}\right) = -kt = \dfrac{-0.6931\,t}{t_{1/2}}$, If 10.0% decays, then 90.0% is left.

$$\ln\left(\frac{90.0}{100.0}\right) = \frac{-0.6931\,t}{5.26\text{ yr}}, \quad t = 0.800 \text{ years}$$

18. $k = (\ln 2)/t_{1/2}$; $\ln\left(\dfrac{N}{N_o}\right) = -kt = \dfrac{-0.6931\,t}{t_{1/2}}$, $\ln\left(\dfrac{N}{15.3}\right) = \dfrac{-0.6931\,(15{,}000\text{ yr})}{5730\text{ yr}}$

$\ln N = -1.81 + \ln 15.3 = -1.81 + 2.728 = 0.92$, $N = 2.5$ disintegrations per minute per g of C

If we had 10. mg C, we would see:

$$10.\text{ mg} \times \frac{1\text{ g}}{1000\text{ mg}} \times \frac{2.5\text{ disintegrations}}{\text{min g}} = \frac{0.025\text{ disintegrations}}{\text{min}}$$

It would take roughly 40 min to see a single disintegration. This is too long to wait and the background radiation would probably be much greater than the ^{14}C activity. Thus, ^{14}C dating is not practical for small samples.

19.   Since 4.5×10^9 years is equal to the half-life, one-half of the ^{238}U atoms will have been converted to ^{206}Pb. The numbers of atoms of ^{206}Pb and ^{238}U will be equal. Thus, the mass ratio is equal to the molar mass ratio:

$$\frac{206}{238} = 0.866$$

20.   a.   The decay of ^{40}K is not the sole source of ^{40}Ca.

       b.   Decay of ^{40}K is the sole source of ^{40}Ar and that no ^{40}Ar is lost over the years.

       c. $\dfrac{0.95 \text{ g } ^{40}\text{Ar}}{1.00 \text{ g } ^{40}\text{K}}$ = current mass ratio

             0.95 g of ^{40}K decayed to ^{40}Ar. 0.95 g of ^{40}K is only 10.7% of the total ^{40}K that decayed, or:

             0.107 (m) = 0.95 g, m = 8.9 g = total mass of ^{40}K that decayed

             Mass of ^{40}K when the rock was formed was 1.00 g + 8.9 g = 9.9 g.

$$\ln\left(\frac{1.00 \text{ g } ^{40}\text{K}}{9.9 \text{ g } ^{40}\text{K}}\right) = -kt = \frac{-(\ln 2)\, t}{t_{1/2}} = \frac{-0.6931\, t}{1.27 \times 10^9 \text{ yr}},\ \ t = 4.2 \times 10^9 \text{ years old}$$

       d.   If some ^{40}Ar escaped then the measured ratio of ^{40}Ar/^{40}K is less than it should be. We would calculate the age of the rocks to be less than it actually is.

21. $175 \text{ mg Na}_3^{32}\text{PO}_4 \times \dfrac{32.0 \text{ mg } ^{32}\text{P}}{165.0 \text{ mg Na}_3^{32}\text{PO}_4} = 33.9 \text{ mg } ^{32}\text{P};\ \ k = \dfrac{\ln 2}{t_{1/2}}$

$$\ln\left(\frac{N}{N_o}\right) = -kt = \frac{-0.6931\, t}{t_{1/2}},\ \ \ln\left(\frac{m}{33.9 \text{ mg}}\right) = \frac{-0.6931\,(35.0 \text{ d})}{14.3 \text{ d}}$$

       $\ln(m) = -1.696 + 3.523 = 1.827,\ \ m = e^{1.827} = 6.22 \text{ mg } ^{32}\text{P}$ remains

22.   a.   In 175 mg Na$_3^{32}$PO$_4$ there are 33.9 mg ^{32}P (see Exercise 21.21).

$$33.9 \times 10^{-3} \text{ g} \times \frac{1 \text{ mol}}{32.0 \text{ g}} \times \frac{6.022 \times 10^{23} \text{ atoms}}{\text{mol}} = 6.38 \times 10^{20} \text{ atoms } ^{32}\text{P};\ \ k = (\ln 2)/t_{1/2}$$

$$\text{Rate} = kN = \frac{0.6931}{14.3 \text{ d}} \times \frac{1 \text{ d}}{24 \text{ h}} \times \frac{1 \text{ h}}{3600 \text{ s}} \times 6.38 \times 10^{20} \text{ atoms} = 3.58 \times 10^{14} \text{ disintegrations/s}$$

$$\text{Rate} = \frac{3.58 \times 10^{14} \text{ disint.}}{\text{s}} \times \frac{1 \text{ Ci}}{\dfrac{3.7 \times 10^{10} \text{ disint.}}{\text{s}}} \times \frac{1000 \text{ mCi}}{\text{Ci}} = 9.7 \times 10^6 \text{ mCi}$$

b.   Rate = kN = $\dfrac{0.693}{24,000 \text{ yr}} \times \dfrac{1 \text{ yr}}{365 \text{ d}} \times \dfrac{1 \text{ d}}{24 \text{ h}} \times \dfrac{1 \text{ h}}{3600 \text{ s}} \times 6.0 \times 10^{23}$ nuclei = 5.5×10^{11} disint./s

Rate = $\dfrac{5.5 \times 10^{11} \text{ disint.}}{\text{s}} \times \dfrac{1 \text{ Ci}}{\dfrac{3.7 \times 10^{10} \text{ disint.}}{\text{s}}} \times \dfrac{1000 \text{ mCi}}{\text{Ci}} = 1.5 \times 10^{4}$ mCi

23.   a. $10.0 \text{ mCi} \times \dfrac{\dfrac{3.7 \times 10^{7} \text{ disint.}}{\text{s}}}{\text{mCi}} = 3.7 \times 10^{8}$ dis./s;   k = (ln 2)/$t_{1/2}$

Rate = kN, $\dfrac{3.7 \times 10^{8} \text{ dis.}}{\text{s}} = \dfrac{0.6931}{2.87 \text{ h}} \times \dfrac{1 \text{ h}}{3600 \text{ s}} \times N$,   N = 5.5×10^{12} atoms of ^{38}S

5.5×10^{12} atoms ^{38}S $\times \dfrac{1 \text{ mol } ^{38}\text{S}}{6.02 \times 10^{23} \text{ atoms}} \times \dfrac{1 \text{ mol Na}_2^{38}\text{SO}_4}{\text{mol } ^{38}\text{S}} = 9.1 \times 10^{-12}$ mol $Na_2^{38}SO_4$

9.1×10^{-12} mol $Na_2^{38}SO_4 \times \dfrac{148.0 \text{ g Na}_2^{38}\text{SO}_4}{\text{mol Na}_2^{38}\text{SO}_4} = 1.3 \times 10^{-9}$ g = 1.3 ng $Na_2^{38}SO_4$

b.   99.99% decays, 0.01% left; $\ln\left(\dfrac{0.01}{100}\right) = \dfrac{-0.6931 \, t}{2.87 \text{ h}}$,   t = 38.1 hours ≈ 40 hours

24.   k = (ln 2)/$t_{1/2}$; $\ln\left(\dfrac{N}{N_0}\right) = -kt = \dfrac{-0.6931 \, (50.0 \text{ yr})}{28.8 \text{ yr}} = -1.203$, $\left(\dfrac{N}{N_0}\right) = e^{-1.203} = 0.300$

30.0% of the ^{90}Sr remains (as of July 16, 1995).

25. $\ln(N/N_0) = -kt$;   N = 0.020 N_0; $t_{1/2}$ = (ln 2)/k, k = 0.6931/12.3 yr = 0.0563 yr^{-1}

$\ln(0.020) = -(0.0563 \text{ yr}^{-1})t$, t = 69 yr

Energy Changes in Nuclear Reactions

26. $\Delta E = \Delta mc^2$, $\Delta m = \dfrac{\Delta E}{c^2} = \dfrac{3.9 \times 10^{23} \text{ kg m}^2/\text{s}^2}{(3.00 \times 10^{8} \text{ m/s})^2} = 4.3 \times 10^{6}$ kg

The sun loses 4.3×10^{6} kg of mass each second.

27. $\dfrac{1.8 \times 10^{14} \text{ kJ}}{\text{s}} \times \dfrac{1000 \text{ J}}{\text{kJ}} \times \dfrac{3600 \text{ s}}{\text{h}} \times \dfrac{24 \text{ h}}{\text{day}} = 1.6 \times 10^{22}$ J

$\Delta E = \Delta mc^2$, $\Delta m = \dfrac{\Delta E}{c^2} = \dfrac{1.6 \times 10^{22} \text{ J}}{(3.00 \times 10^{8} \text{ m/s})^2} = 1.8 \times 10^{5}$ kg of solar material provides 1 day of solar energy to the earth.

$$1.6 \times 10^{22} \text{ J} \times \frac{1 \text{ kJ}}{1000 \text{ J}} \times \frac{1 \text{ g}}{32 \text{ kJ}} \times \frac{1 \text{ kg}}{1000 \text{ g}} = 5.0 \times 10^{14} \text{ kg of coal is needed to provide the same amount of energy.}$$

28. $12 \, {}^{1}_{1}\text{H} + 12 \, {}^{1}_{0}\text{n} + 12 \, {}^{0}_{-1}\text{e} \rightarrow {}^{24}_{12}\text{Mg}$; mass of proton = 1.00728 amu

$\Delta m = 23.9850$ amu $-[12(1.00728) + 12(1.00866) + 12(5.49 \times 10^{-4})]$amu $= -0.2129$ amu

$\Delta E = \Delta mc^2 = -0.2129 \text{ amu} \times \dfrac{1 \text{ g}}{6.0221 \times 10^{23} \text{ amu}} \times \dfrac{1 \text{ kg}}{1000 \text{ g}} \times (2.9979 \times 10^8 \text{ m/s})^2 = -3.177 \times 10^{-11} \text{ J}$

$$\frac{\text{BE}}{\text{nucleon}} = \frac{-3.177 \times 10^{-11} \text{ J}}{24} = \frac{-1.324 \times 10^{-12} \text{ J}}{\text{nucleon}}$$

For ${}^{27}\text{Mg}$: $12 \, {}^{1}_{1}\text{H} + 15 \, {}^{1}_{0}\text{n} + 12 \, {}^{0}_{-1}\text{e} \rightarrow {}^{27}_{12}\text{Mg}$

$\Delta m = 26.9843$ amu $-[12(1.00728) + 15(1.00866) + 12(5.49 \times 10^{-4})]$ amu $= -0.2395$ amu

$\Delta E = \Delta mc^2 = -0.2395 \text{ amu} \times \dfrac{1 \text{ g}}{6.0221 \times 10^{23} \text{ amu}} \times \dfrac{1 \text{ kg}}{1000 \text{ g}} \times (2.9979 \times 10^8 \text{ m/s})^2 = -3.574 \times 10^{-11} \text{ J}$

$$\frac{\text{BE}}{\text{nucleon}} = \frac{-3.574 \times 10^{-11} \text{ J}}{27 \text{ nucleons}} = \frac{-1.324 \times 10^{-12} \text{ J}}{\text{nucleon}}$$

29.   Mass of nucleus = mass of atom - mass of electrons = 6.015126 - 3(0.0005486) = 6.013480 amu

$3 \, {}^{1}_{1}\text{H} + 3 \, {}^{1}_{0}\text{n} \rightarrow {}^{6}_{3}\text{Li}$; $\Delta m = 6.013480 - [3(1.00728) + 3(1.00866)] = -0.03434$ amu

For 1 mol of ${}^{6}\text{Li}$, the mass defect is -0.03434 g.

$\Delta E = \Delta mc^2 = -3.434 \times 10^{-2} \text{ g} \times \dfrac{1 \text{ kg}}{1000 \text{ g}} \times (2.9979 \times 10^8 \text{ m/s})^2 = -3.086 \times 10^{12} \text{ J/mol}$

30. ${}^{1}_{1}\text{H} + {}^{1}_{0}\text{n} \rightarrow {}^{2}_{1}\text{H}$; Atomic mass - mass of electrons = mass of nucleus

$\Delta m = 2.01410$ amu $- 0.000549$ amu $- [1.00728$ amu $+ 1.00866$ amu$] = -2.39 \times 10^{-3}$ amu

$\Delta E = \Delta mc^2 = -2.39 \times 10^{-3} \text{ amu} \times \dfrac{1 \times 10^{-3} \text{ kg}}{6.022 \times 10^{23} \text{ amu}} \times (2.998 \times 10^8 \text{ m/s})^2 = -3.57 \times 10^{-13} \text{ J}$

$$\frac{\text{BE}}{\text{nucleon}} = \frac{-3.57 \times 10^{-13} \text{ J}}{2 \text{ nucleons}} = -1.79 \times 10^{-13} \text{ J/nucleon}$$

${}^{1}_{1}\text{H} + 2 \, {}^{1}_{0}\text{n} \rightarrow {}^{3}_{1}\text{H}$; $\Delta m = 3.01605 - 0.000549 - [1.00728 + 2(1.00866)] = -9.10 \times 10^{-3}$ amu

$\Delta E = -9.10 \times 10^{-3} \text{ amu} \times \dfrac{1 \times 10^{-3} \text{ kg}}{6.022 \times 10^{23} \text{ amu}} \times (2.998 \times 10^8 \text{ m/s})^2 = -1.36 \times 10^{-12} \text{ J}$

$$\frac{BE}{\text{nucleon}} = \frac{-1.36 \times 10^{-12} \text{ J}}{3 \text{ nucleons}} = -4.53 \times 10^{-13} \text{ J/nucleon}$$

31. $^{1}_{1}H + ^{1}_{0}n \rightarrow 2\ ^{1}_{1}H + ^{1}_{0}n + ^{1}_{-1}H$; mass $^{1}_{-1}H$ = mass $^{1}_{1}H$ = 1.00728 amu = mass of proton

$\Delta m = +2(1.00728) = 2.01456$ amu

$$\Delta E = \Delta mc^2 = 2.01456 \text{ amu} \times \frac{1 \times 10^{-3} \text{ kg}}{6.02214 \times 10^{23} \text{ amu}} \times (2.997925 \times 10^8 \text{ m/s})^2$$

$\Delta E = 3.00657 \times 10^{-10}$ J of energy is absorbed per nuclei or 1.81060×10^{14} J/mol.

The source of energy is the kinetic energy of the proton and the neutron in the particle accelerator.

32. $\Delta m = -2(5.486 \times 10^{-4} \text{ amu}) = -10.97 \times 10^{-4}$ amu

$$\Delta E = \Delta mc^2 = -10.97 \times 10^{-4} \text{ amu} \times \frac{1 \times 10^{-3} \text{ kg}}{6.0221 \times 10^{23} \text{ amu}} \times (2.9979 \times 10^8 \text{ m/s})^2 = -1.637 \times 10^{-13} \text{ J}$$

$E_{photon} = 1/2(1.637 \times 10^{-13} \text{ J}) = 8.185 \times 10^{-14} \text{ J} = hc/\lambda$

$$\lambda = \frac{hc}{E} = \frac{6.6261 \times 10^{-34} \text{ J s} \times (2.9979 \times 10^8 \text{ m/s})}{8.185 \times 10^{-14} \text{ J}} = 2.427 \times 10^{-12} \text{ m} = 2.427 \times 10^{-3} \text{ nm}$$

33. $20{,}000 \text{ ton TNT} \times \dfrac{4 \times 10^9 \text{ J}}{\text{ton TNT}} \times \dfrac{1 \text{ mol } ^{235}U}{2 \times 10^{13} \text{ J}} \times \dfrac{235 \text{ g } ^{235}U}{\text{mol } ^{235}U} = 940 \text{ g } ^{235}U \approx 900 \text{ g } ^{235}U$

This assumes all of the ^{325}U undergoes fission.

34.   Mass of nucleus = atomic mass - mass of electron = 2.01410 amu - 0.000549 amu = 2.01355 amu

$$u_{rms} = \left(\frac{3\,RT}{M}\right)^{1/2} = \left(\frac{3(8.3145 \text{ J K}^{-1} \text{ mol}^{-1})(4 \times 10^7 \text{ K})}{2.01355 \text{ g (1 kg/1000 g)}}\right)^{1/2} = 7 \times 10^5 \text{ m/s}$$

$$E_K = \frac{1}{2}mu^2 = \frac{1}{2}\left(2.01355 \text{ amu} \times \frac{1 \times 10^{-3} \text{ kg}}{6.02214 \times 10^{23} \text{ amu}}\right)(7 \times 10^5 \text{ m/s})^2 = 8 \times 10^{-16} \text{ J}$$

35. $^{1}_{1}H + ^{1}_{1}H \rightarrow ^{2}_{1}H + ^{1}_{+1}e$; $\Delta m = (2.01410 \text{ amu} - m_e + m_e) - 2(1.00782 \text{ amu} - m_e)$

$\Delta m = 2.01410 - 2(1.00782) + 2(0.000549) = -4.4 \times 10^{-4}$ amu for two protons reacting

When two mol of protons undergo fusion, $\Delta m = -4.4 \times 10^{-4}$ g.

$$\Delta E = \Delta mc^2 = -4.4 \times 10^{-7} \text{ kg} \times (3.00 \times 10^8 \text{ m/s})^2 = -4.0 \times 10^{10} \text{ J}$$

$$\frac{-4.0 \times 10^{10} \text{ J}}{2 \text{ mol protons}} \times \frac{1 \text{ mol}}{1.01 \text{ g}} = -2.0 \times 10^{10} \text{ J/g of hydrogen nuclei}$$

36. $^2_1\text{H} + {}^3_1\text{H} \rightarrow {}^4_2\text{He} + {}^1_0\text{n}$; Mass of electrons cancel when determining Δm for the nuclear reaction.

$$\Delta m = [4.00260 + 1.00866 - (2.01410 + 3.01605)] \text{ amu} = -1.889 \times 10^{-2} \text{ amu}$$

For production of 1.0 mol of ^4_2He: $\Delta m = -1.889 \times 10^{-2} \text{ g} = -1.889 \times 10^{-5} \text{ kg}$

$$\Delta E = \Delta mc^2 = -1.889 \times 10^{-5} \text{ kg} \times (2.9979 \times 10^8 \text{ m/s})^2 = -1.698 \times 10^{12} \text{ J/mol}$$

For 1 nuclei of ^4_2He:

$$-1.698 \times 10^{12} \text{ J/mol} \times \frac{1 \text{ mol}}{6.0221 \times 10^{23} \text{ nuclei}} = -2.820 \times 10^{-12} \text{ J/nuclei}$$

Detection, Uses, and Health Effects of Radiation

37.   The chemical properties determine where a radioactive material may be concentrated in the body or how easily it may be excreted. The length of time of exposure and what is exposed to radiation significantly affects the health hazard. (See exercise 21.46 for a specific example.)

38.   Not all of the emitted radiation enters the Geiger-Müller tube. The fraction of radiation entering the tube must be constant.

39.   The Geiger-Müller tube has a certain response time. After the gas in the tube ionizes to produce a "count," some time must elapse for the gas to return to an electrically neutral state. The response of the tube levels because at high activities radioactive particles are entering the tube faster than the tube can respond to them.

40.   All evolved O_2 comes from water.

41.   A nonradioactive substance can be put in equilibrium with a radioactive substance. The two materials can then be checked to see if all the radioactivity remains in the original material or if it has been scrambled by the equilibrium.

42.   No, coal fired power plants also pose risks. A partial list of risks:

Coal	Nuclear
Air pollution	Radiation exposure to workers
Coal mine accidents	Disposal of wastes
Health risks to miners	Meltdown
(black lung disease)	Terrorists
	Public fear

43.   Assuming that the radionuclide is long lived enough such that no significant decay occurs during the time of the experiment, the total counts of radioactivity injected are:

$$0.10 \text{ mL} \times \frac{5.0 \times 10^3 \text{ cpm}}{\text{mL}} = 5.0 \times 10^2 \text{ cpm}$$

Assuming that the total activity is uniformly distributed only in the rats blood, the blood volume is:

$$\frac{48 \text{ cpm}}{\text{mL}} \times V = 5.0 \times 10^2 \text{ cpm}, \quad V = 10.4 \text{ mL} = 10. \text{ mL}$$

44.   The reaction mechanism is:

$$N^{16}O + {}^{18}O_2 \rightleftharpoons N^{16}O^{18}O^{18}O$$

$$N^{16}O^{18}O^{18}O + N^{16}O \rightarrow 2 \text{ NO}_2$$

Since any one of the three oxygen atoms in NO_3 can be transferred to NO in the last step, then 2/3 of NO_2 will be $N^{16}O^{18}O$, 1/6 of NO_2 will be $N^{16}O_2$, and 1/6 of NO_2 will be $N^{18}O_2$.

Additional Exercises

45. $\Delta x \bullet \Delta(mv) \geq h/4\pi$; $\Delta(mv) = v\Delta m$; Assume $\Delta x(v\Delta m) = h/4\pi$

$$\Delta m = \frac{h}{4\pi(\Delta x)(v)} = \frac{6.63 \times 10^{-34} \text{ J s}}{4(3.14)(1 \times 10^{-35} \text{ m})(3.0 \times 10^7 \text{ m/s})} = 2 \times 10^{-7} \text{ kg}$$

mass of electron, $m_e = 9 \times 10^{-31}$ kg; mass of proton, $m_p = 1.7 \times 10^{-27}$ kg

$$\frac{\Delta m}{m_e} = \frac{2 \times 10^{-7}}{9 \times 10^{-31}} = 2 \times 10^{23}; \quad \frac{\Delta m}{m_p} = \frac{2 \times 10^{-7}}{1.7 \times 10^{-27}} = 1 \times 10^{20}$$

The uncertainty in the superstring mass is 2×10^{23} times the electron mass and 1×10^{20} times the proton mass.

46.   i) and ii) mean that Pu is not a significant threat outside the body. Our skin is sufficient to keep out the α particles. If Pu gets inside the body, it is easily oxidized to Pu^{4+} (iv), which is chemically similar to Fe^{3+} (iii). Thus, Pu^{4+} will concentrate in tissues where Fe^{3+} is found. One of these is the bone marrow where red blood cells are produced. Once inside the body, α particles cause considerable damage.

47.   The maximum binding energy per nucleon occurs at about Fe. Smaller nuclei become more stable by fusing to form heavier nuclei. Larger nuclei form more stable nuclei by splitting to form lighter nuclei.

48.   For fusion reactions, a collision of sufficient energy must occur between two positively charged particles to initiate the reaction. This requires high temperatures. In fission, an electrically neutral neutron collides with the positively charged nucleus. This has a much lower activation energy.

49.   The temperatures of fusion reactions are so high that all physical containers would be destroyed. At these high temperatures most of the electrons are stripped from the atoms. A plasma of gaseous ions is formed which can be controlled by magnetic fields.

50.   Magnetic fields are needed to contain the fusion reaction. Superconductors allow the production of very strong magnetic fields by their ability to carry large electric currents. Stronger magnetic fields should be more capable of containing the fusion reaction.

51.   Moderator: Slows the neutrons.

        Control rods: Absorbs neutrons to slow or halt the fission reaction.

52.   The radiation may cause nuclei in the metal to undergo nuclear reaction. This changes the identity of the element and as the electrons rearrange themselves, the new atom may not fit in the crystal lattice. One result is the crystal becomes brittle. One needs to look for materials that do not undergo nuclear reactions when subjected to radiation (particularly neutrons) or ones that produce products of a similar atomic size as the original atoms.

53.   a. $2 H_2O + 2 e^- \rightarrow H_2 + 2 OH^-$, $E° = -0.83$ V; $E_{cell}° = E_{H_2O}° - E_{Zr}° = -0.83$ V $+ 2.36$ V $= 1.53$ V;

            Yes, the reduction of H_2O to H_2 by Zr is spontaneous at standard conditions since $E_{cell}° > 0$.

        b. $4 H_2O + 4 e^- \rightarrow 2 H_2 + 4 OH^-$
                $Zr + 4 OH^- \rightarrow ZrO_2 \cdot H_2O + H_2O + 4 e^-$
                _____
                $3 H_2O(l) + Zr(s) \rightarrow 2 H_2(g) + ZrO_2 \cdot H_2O(s)$

        c. $\Delta G° = -nFE° = -(4$ mol $e^-)$ $(96,485$ C/mol $e^-)$ $(1.53$ J/C$) = -5.90 \times 10^5$ J $= -590.$ kJ

            $E = E° - \dfrac{0.0592}{n} \log Q$; At equilibrium, $E = 0$ and $Q = K$.

$$E° = \frac{0.0592}{n} \log K, \quad \log K = \frac{4(1.53)}{0.0592} = 103, \quad K \approx 10^{103}$$

d. $1.00 \times 10^3 \text{ kg Zr} \times \dfrac{1000 \text{ g}}{\text{kg}} \times \dfrac{1 \text{ mol Zr}}{91.22 \text{ g Zr}} \times \dfrac{2 \text{ mol H}_2}{\text{mol Zr}} = 2.19 \times 10^4 \text{ mol H}_2$

$2.19 \times 10^4 \text{ mol H}_2 \times \dfrac{2.016 \text{ g H}_2}{\text{mol H}_2} = 4.42 \times 10^4 \text{ g H}_2$

$$V = \frac{nRT}{P} = \frac{(2.19 \times 10^4 \text{ mol}) (0.08206 \text{ L atm mol}^{-1} \text{K}^{-1}) (1273 \text{ K})}{1 \text{ atm}} = 2 \times 10^6 \text{ L H}_2$$

e. Probably yes; Less radioactivity overall was released by venting the H_2 than what would have been released if the H_2 exploded inside the reactor (as happened at Chernoboyl). Neither alternative is pleasant, but venting the radioactive hydrogen is the less unpleasant of the two alternatives.

54. a. ^{12}C; It takes part in the first step of the reaction but is regenerated in the last step. ^{12}C is not consumed.

b. ^{13}N, ^{13}C, ^{14}N, ^{15}O, and ^{15}N are intermediates.

c. Since protons are reacting in this fusion process, we will calculate the energy released per mole of hydrogen nuclei (^1H).

$4\,^1_1H \rightarrow\, ^4_2He + 2\,^0_{+1}e; \quad \Delta m = 4.00260 \text{ amu} - 2\, m_e + 2\, m_e - 4(1.00782 \text{ amu} - m_e)$

$\Delta m = 4.00260 - 4(1.00782) + 4(0.000549) = -0.02648 \text{ amu for 4 protons reacting}$

For 4 mol of protons, $\Delta m = -0.02648$ g.

$\Delta E = \Delta mc^2 = -2.648 \times 10^{-5} \text{ kg} (2.9979 \times 10^8 \text{ m/s})^2 = -2.380 \times 10^{12} \text{ J}$

$\dfrac{-2.380 \times 10^{12} \text{ J}}{4 \text{ mol } ^1H} = \dfrac{-5.950 \times 10^{11} \text{ J}}{\text{mol } ^1H}$

55. Release of Sr is probably more harmful. Xe is chemically unreactive. Strontium is in the same family as calcium and could be absorbed and concentrated in the body in a fashion similar to Ca. This puts the radioactive Sr in the bones: red blood cells are produced in bone marrow. Xe would not be readily incorporated in the body.

CHAPTER TWENTY-TWO

ORGANIC CHEMISTRY

Hydrocarbons

1. $CH_3-CH_2-CH_2-CH_2-CH_2-CH_3$ hexane or n-hexane (highest b.p., least branched)

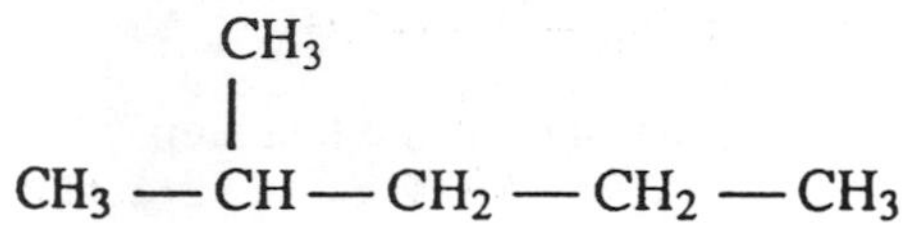

$$
\begin{array}{c}
CH_3 \\
| \\
CH_3 - CH - CH_2 - CH_2 - CH_3
\end{array}
\qquad \text{2-methylpentane}
$$

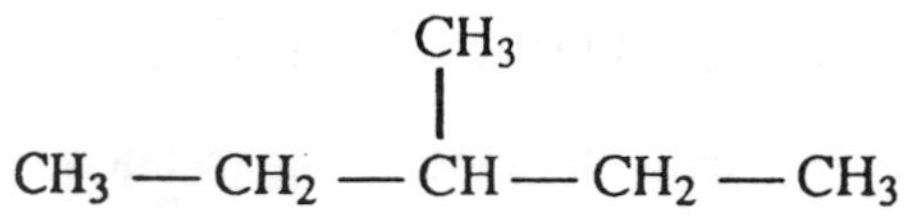

$$
\begin{array}{c}
CH_3 \\
| \\
CH_3 - CH_2 - CH - CH_2 - CH_3
\end{array}
\qquad \text{3-methylpentane}
$$

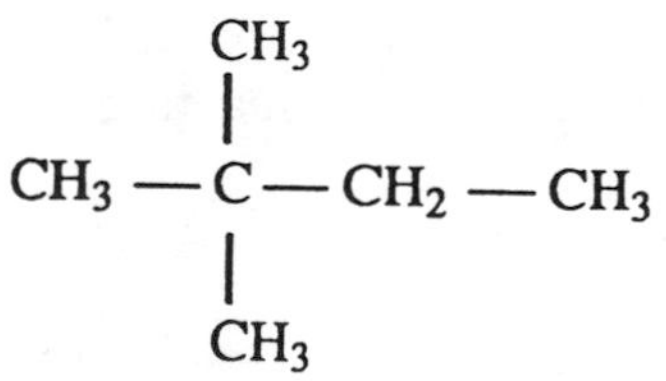

$$
\begin{array}{c}
CH_3 \\
| \\
CH_3 - C - CH_2 - CH_3 \\
| \\
CH_3
\end{array}
\qquad \text{2,2-dimethylbutane}
$$

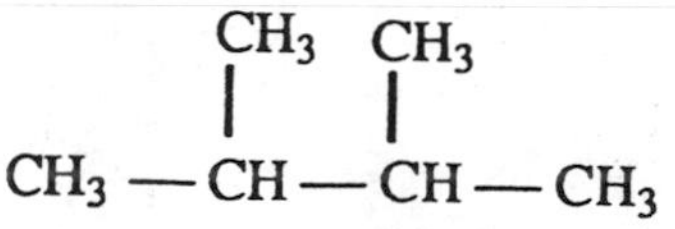

$$
\begin{array}{c}
CH_3 \ \ CH_3 \\
| \ \ \ \ \ | \\
CH_3 - CH - CH - CH_3
\end{array}
\qquad \text{2,3-dimethylbutane}
$$

n-Hexane would have the highest boiling point. It is the least branched and, therefore, will have the strongest London dispersion forces between molecules.

2.   a. 2,3,3-trimethylhexane   b. 8-ethyl-2,5,5-trimethyldecane

   c. 3-methylhexane

3.   There is only one consecutive chain of C-atoms. They are not all in a true straight line since the bond angle at each carbon is a tetrahedral angle of 109.5°.

4.   a. 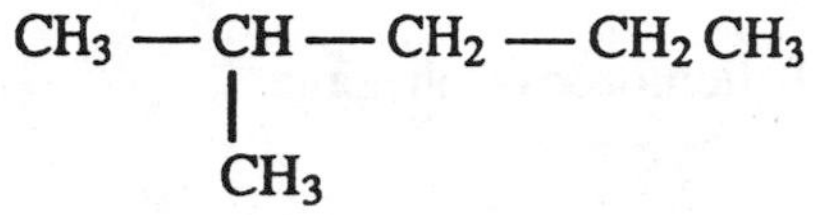 $CH_3 - CH - CH_2 - CH_2 CH_3$
                      |
                    CH_3

b. 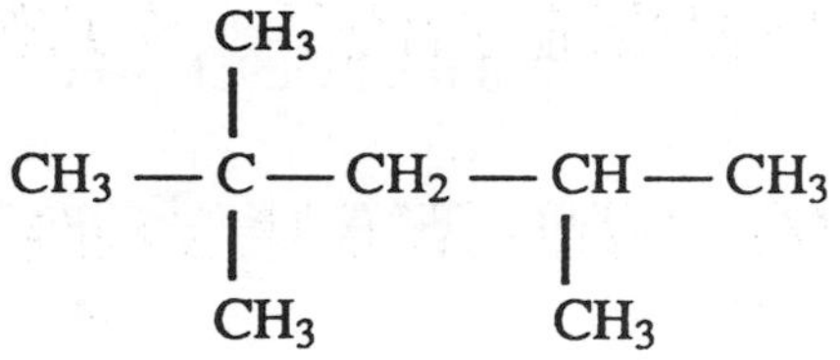
                    CH_3
                      |
      $CH_3 - C - CH_2 - CH - CH_3$
                      | |
                    CH_3 CH_3

c. $CH_3 - CH - CH_2 CH_2 CH_3$
                      |
      $CH_3 - C - CH_3$
                      |
                    CH_3

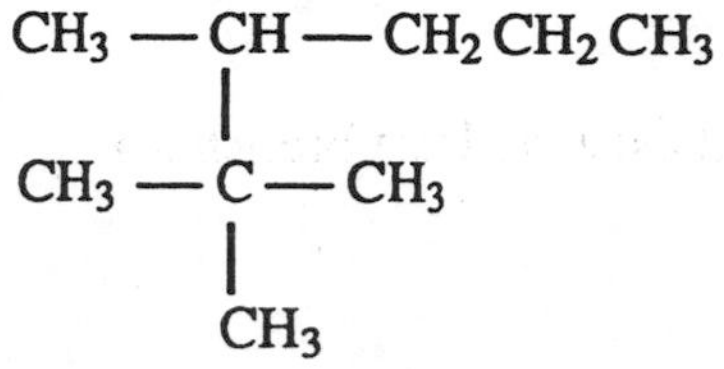

d.   The longest chain is 6 carbons long.

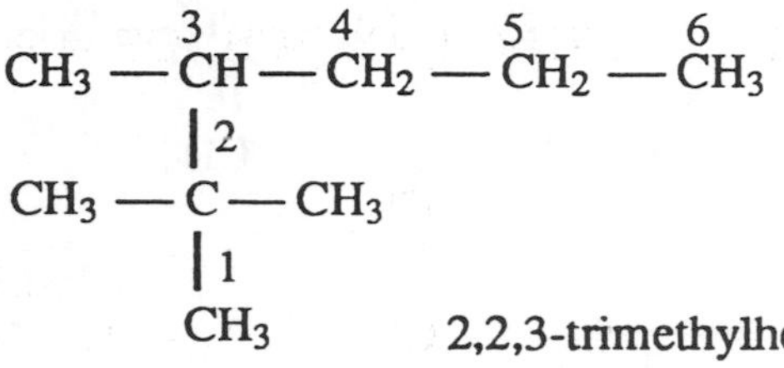

            3   4   5   6
      $CH_3 - CH - CH_2 - CH_2 - CH_3$
                    | 2
      $CH_3 - C - CH_3$
                    | 1
                CH_3 2,2,3-trimethylhexane

5.   a.   1-butene   b.   2-methyl-2-butene   c.   2,5-dimethyl-3-heptene

6.   a. $CH_3 - CH_2 - CH = CH - CH_2 - CH_3$ b. $CH_3CH = CHCH = CHCH_2CH_3$

c. 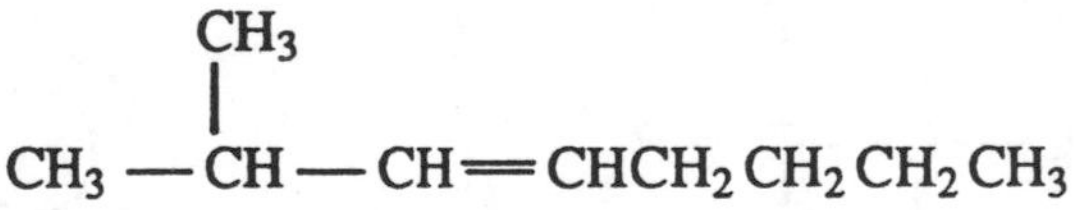
            CH_3
              |
      $CH_3 - CH - CH = CHCH_2 CH_2 CH_2 CH_3$

7.   a. 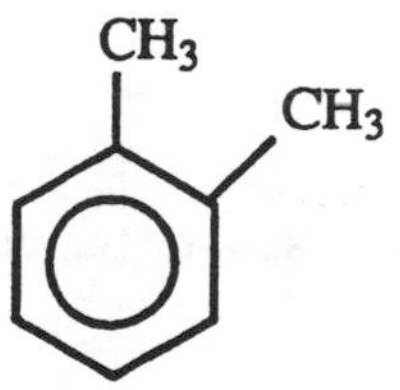
            CH_3
              CH_3

b. 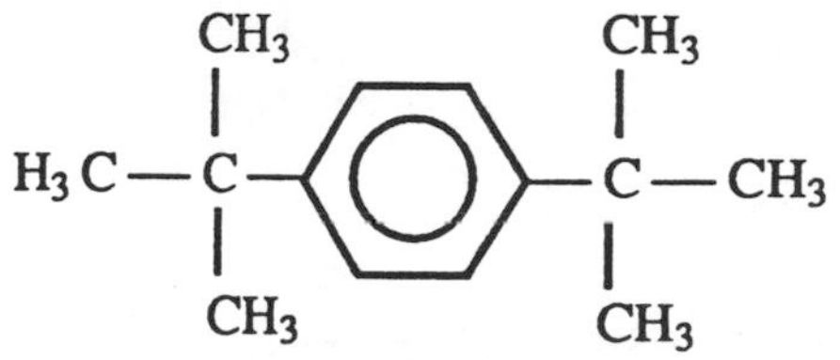 $H_3C - \overset{\displaystyle CH_3}{\underset{\displaystyle CH_3}{C}} - \bigcirc - \overset{\displaystyle CH_3}{\underset{\displaystyle CH_3}{C}} - CH_3$

c. $CH_2 CH_3$

            $CH_2 CH_3$

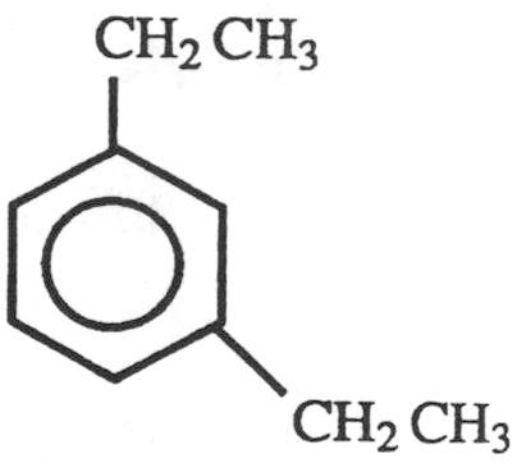

8.   a.   1,3-dichlorobutane   b.   1,1,1-trichlorobutane

c.   2,3-dichloro-2,4-dimethylhexane   d.   1,2-difluoroethane

e.   chlorobenzene   f.   chlorocyclohexane

g.   3-chlorocyclohexene (double bond assumed between C_1 and C_2)

9.   a.   methylcyclopropane   b.   t-butylcyclohexane

         c.   3,4-dimethylcyclopentene   d.   chloroethene (vinyl chloride)

         e.   1,2-dimethylcyclopentene   f.   1,1-dichlorocyclohexane

10.   isopropylbenzene or 2-phenylpropane

11.

$$
\begin{array}{c}
\overset{7}{CH_2} - \overset{8}{CH_3} \\
| \\
CH_3 - \overset{6}{CH} - \overset{5}{CH} - \overset{4}{CH_2} - \overset{3}{CH} - CH_3 \\
\quad\quad | \quad\quad\quad\quad\quad | \\
\quad\quad CH_3 \quad\quad\quad \overset{2}{CH} \\
\quad\quad\quad\quad\quad\quad / \quad \backslash \\
\quad\quad\quad\quad H_3C \quad\quad \underset{1}{CH_3}
\end{array}
$$

2,3,5,6-tetramethyloctane

12.   a.   b.   c.

$$
\begin{array}{c}
F \\
| \\
H - C - F \\
| \\
H
\end{array}
$$

$$
\begin{array}{c}
Br \quad H \quad H \\
| \quad\quad | \quad\quad | \\
H - C - C - C - H \\
| \quad\quad | \quad\quad | \\
Cl \quad Cl \quad H
\end{array}
$$

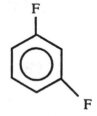

Isomerism

13.   Structural isomers: Differ in bonding, either the kinds of bonds present or the way in which
                                  the bonds connect atoms to each other.

         Geometrical isomers: Same bonds but differ in arrangement in space about a rigid bond or
                                  ring.

14.   Resonance: All atoms are in the same position. Only the position of π electrons are different.

         Isomerism: Atoms are in different locations in space.

         Isomers are distinctly different substances. Resonance is the use of more than one Lewis
         structure to describe the bonding in a single compound. Resonance structures are <u>not</u> isomers.

15.

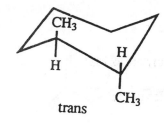

trans

cis

16.

$$\underset{Cl}{\overset{Cl}{C}} = CH - CH_3 \qquad CH_2 = CCl - CH_2Cl \qquad CH_2 = CH - CHCl_2$$

$$\underset{Cl}{\overset{H}{C}} = \underset{CH_3}{\overset{Cl}{C}} \qquad \underset{H}{\overset{Cl}{C}} = \underset{CH_3}{\overset{Cl}{C}} \qquad \underset{H}{\overset{Cl}{C}} = \underset{CH_2Cl}{\overset{H}{C}}$$

$$\underset{Cl}{\overset{H}{C}} = \underset{CH_2Cl}{\overset{H}{C}}$$

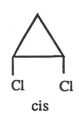

cis

trans

17.

CH₃

CH₃

CH₃

CH₃

CH₃   CH₃

CH₃

CH₃

CH₃

18.   HCBrCl — CH = CH$_2$

Note: 1-bromo-1-chlorocyclopropane, cis-1-bromo-2-chlorocyclopropane and trans-1-bromo-2-chlorocyclopropane are the ring structures that are isomers of bromochloropropene. We did not include the ring structures in the answer since their base name is not bromochloropropene.

19.

polar   polar   nonpolar

20. $CH_2Cl–CH_2Cl$, 1-2-dichloroethane; There is free rotation about the C–C single bond which doesn't constitute different compounds. $CHCl=CHCl$, 1,2-dichloroethene; There is no rotation about the C=C double bond. This creates the cis and trans isomers which are different compounds.

21.   a.

Bonds broken:

 1 O – H (467 kJ/mol)
 1 C – O (358 kJ/mol)
 1 C = C (614 kJ/mol)

Bonds formed:

 1 C – H (413 kJ/mol)
 1 C = O (799 kJ/mol)
 1 C – C (347 kJ/mol)

$\Delta H = 467 + 358 + 614 - (413 + 799 + 347) = 1439 - 1559 = -120.$ kJ

Since ΔH is negative, the ketone is more stable because it contains the stronger bonds.

b.

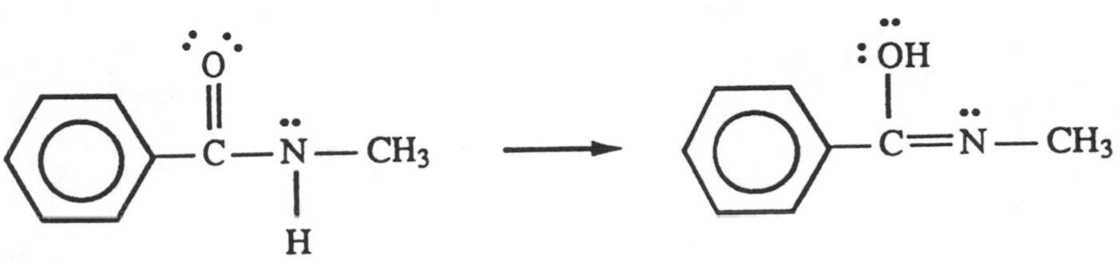

Bonds broken:

 1 C = O (799 kJ/mol)
 1 C – N (305 kJ/mol)
 1 N – H (391 kJ/mol)

Bonds formed:

 1 C – O (358 kJ/mol)
 1 C = N (615 kJ/mol)
 1 O – H (467 kJ/mol)

$\Delta H = 799 + 305 + 391 - (358 + 615 + 467) = 55$ kJ

Since ΔH is positive, the amide with C = O is more stable because it contains the stronger bonds.

c.

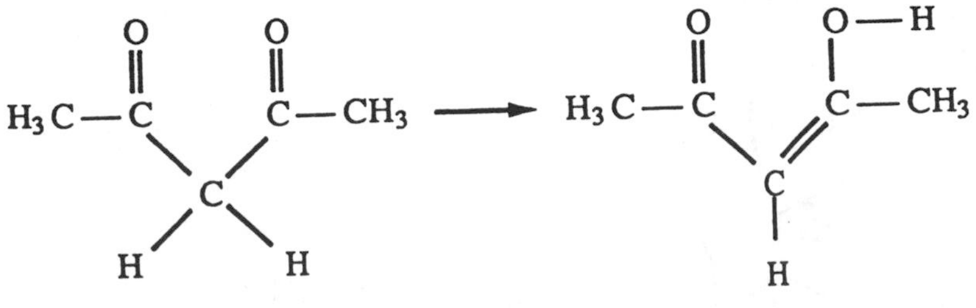

Bonds broken:　　　　　　　　Bonds formed:

$$1\ C=O\ (799\ kJ/mol)\qquad 1\ O-H\ \ (467\ kJ/mol)$$
$$1\ C-C\ \ (347\ kJ/mol)\qquad 1\ C-O\ \ (358\ kJ/mol)$$
$$1\ C-H\ \ (413\ kJ/mol)\qquad 1\ C=C\ (614\ kJ/mol)$$

$$\Delta H = 799 + 347 + 413 - (467 + 358 + 614) = 1559 - 1439 = 120.\ kJ$$

Since ΔH is positive, the form with two carbonyl groups is more stable because it contains the stronger bonds.

22.

There are many possibilities for isomers. Any structure with four chlorines in any four of the numbered positions would be an isomer, i.e., 1,2,3,4-tetrachloro-dibenzo-p-dioxin is a possible isomer.

Functional Groups

23.　a.

b.

c.

24. $HC\equiv C - C\equiv C - CH=C=CH - CH=CH - CH=CH - CH_2 - \overset{\overset{\displaystyle O}{\|}}{C} - OH$
     13   12   11   10   9   8   7   6   5   4   3   2   1

25.   The C with the –OH group is number 1.

26.   a.

       b. 5 carbons in ring and in –CO$_2$H: sp^2; the other two carbons: sp^3

       c. 24 sigma bonds; 4 pi bonds

27.

a. Minoxidil would be more soluble in acidic solution. The nitrogens with lone pairs can be protonated, forming a water soluble ion.

b. The two nitrogens in the ring with double bonds are sp^2 hybridized. The other three N's are sp^3 hybridized.

c. See structure. The five carbon atoms in the ring with one nitrogen are all sp^3 hybridized. The four carbon atoms in the other ring with double bonds are all sp^2 hybridized.

d. Angle a, b, and e $\approx$ 109.5°; Angles c, d, and f $\approx$ 120°

e. Hydrogen atoms are often omitted from ring structures. In organic compounds, carbon atoms usually contain four bonds. With this in mind, each of the five sp^3 hybridized carbon atoms has two C – H bonds and one of the sp^2 hybridized carbon atoms has one C – H bond. Including the eleven C – H bonds, there are 31 sigma bonds in minoxidil.

f. 3 pi bonds

28. The amine group is protonated. Morphine hydrochloride is ionic and, hence, more soluble in water.

$$\left[\begin{array}{c} \text{NH} \\ \text{CH}_3 \end{array}\right]^{+} \quad \text{Cl}^-$$

29. Out of 100.00 g:

$$71.89 \text{ g C} \times \frac{1 \text{ mol C}}{12.011 \text{ g C}} = 5.985 \text{ mol} \approx 6 \text{ mol C}$$

$$12.13 \text{ g H} \times \frac{1 \text{ mol H}}{1.0079 \text{ g H}} = 12.03 \text{ mol} \approx 12 \text{ mol H} \qquad \text{The empirical formula is } C_6H_{12}O.$$

$$15.98 \text{ g O} \times \frac{1 \text{ mol O}}{15.999 \text{ g O}} = 0.9988 \text{ mol} \approx 1 \text{ mol O}$$

$$R_1 - \overset{\overset{\text{O}}{\|}}{C} - O - R_2 + H_2O \longrightarrow R_1 \overset{\overset{\text{O}}{\|}}{C} - OH + HOCH_2CH_3$$

R_2 must be CH_3CH_2 since CH_3CH_2OH is one of the products. The molar mass of $-CO_2H$ is $\approx$ 45 g/mol, so the mass of R_1 is 172 - 45 = 127. If R_1 is $CH_3-(CH_2)_n-$, then 15 + n(14) = 127 and n = 112/14 = 8. Ethyl caprate is:

$$CH_3(CH_2)_8 \overset{\overset{\text{O}}{\|}}{C} - OCH_2CH_3$$

The molecular formula of $C_{12}H_{24}O_2$ agrees with the empirical formula.

30.   a.   **acetone:**

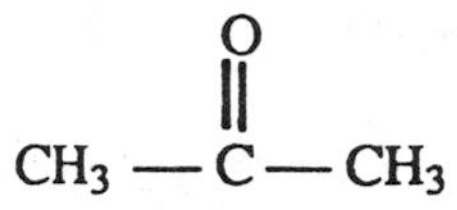

$$CH_3 - \overset{\overset{\displaystyle O}{\|}}{C} - CH_3$$

aldehyde that is an isomer of acetone:

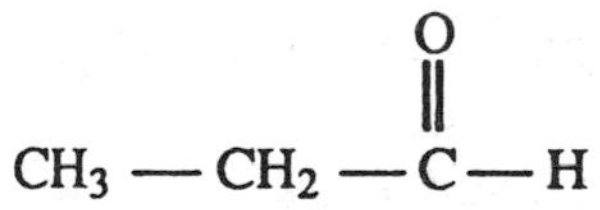

$$CH_3 - CH_2 - \overset{\overset{\displaystyle O}{\|}}{C} - H \qquad \text{propanal}$$

b.   **2-propanol:**

$$CH_3 - \underset{\underset{\displaystyle OH}{|}}{CH} - CH_3$$

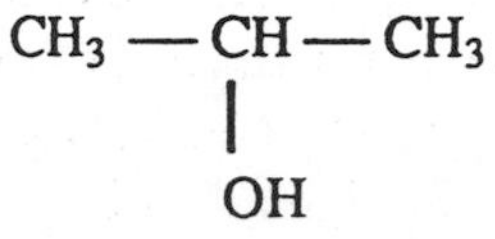

ether: $CH_3 - O - CH_2 - CH_3$ ethylmethyl ether

c.   **cis-2-butene:**

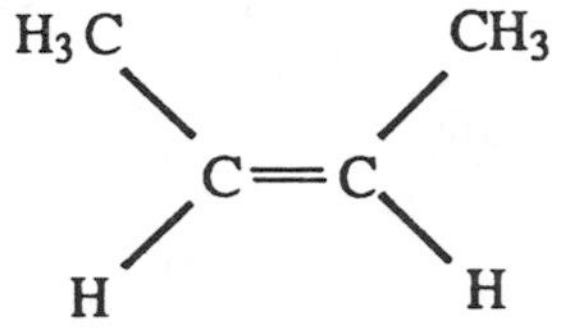

$$\begin{array}{ccc} H_3C & & CH_3 \\ & \diagdown \quad \diagup & \\ & C=C & \\ & \diagup \quad \diagdown & \\ H & & H \end{array}$$

geometrical isomer:

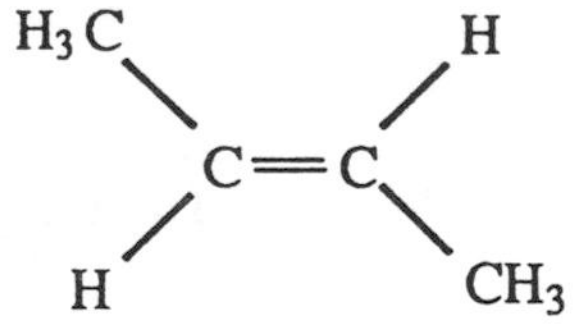

$$\begin{array}{ccc} H_3C & & H \\ & \diagdown \quad \diagup & \\ & C=C & \qquad \text{trans-2-butene} \\ & \diagup \quad \diagdown & \\ H & & CH_3 \end{array}$$

d.   **trimethylamine:**

$$H_3C - \underset{\underset{\displaystyle CH_3}{|}}{N} - CH_3$$

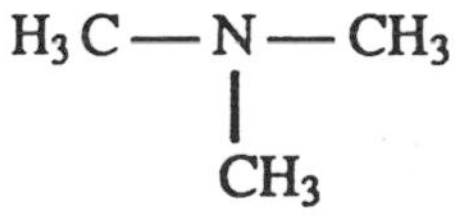

primary amine: $CH_3CH_2CH_2NH_2$ propylamine or 1-aminopropane

e.   **ethylmethylamine (secondary amine):**

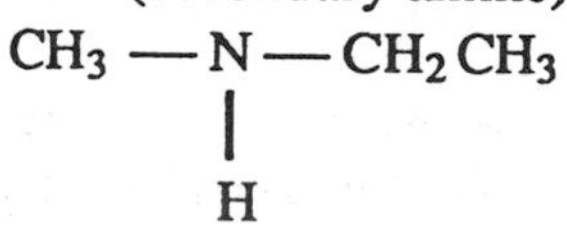

$$CH_3 - \underset{\underset{\displaystyle H}{|}}{N} - CH_2CH_3$$

f. **1-propanol (primary alcohol):** $CH_3CH_2CH_2OH$

31.   a.   ketone   b.   aldehyde   c.   ketone   d.   amine

32.

Reactions of Organic Compounds

33.   Substitution: An atom or group is replaced by another atom or group.

e.g., H in benzene is replaced by Cl. $C_6H_6 + Cl_2 \xrightarrow{\text{catalyst}} C_6H_5Cl + HCl$

Addition: Atoms or groups are added to a molecule.

e.g., Cl_2 adds to ethylene. $CH_2 = CH_2 + Cl_2 \rightarrow CH_2Cl - CH_2Cl$

34.   a.   Two monochloro products are formed: $CH_2ClCH_2CH_3$ and $CH_3CHClCH_3$

b.   Four dichloro products are formed:

$CHCl_2CH_2CH_3$, $CH_3CCl_2CH_3$, $CH_2ClCHClCH_3$ and $CH_2ClCH_2CH_2Cl$

35.   a. $CH_2 = CH_2 + Br_2 \rightarrow CH_2Br - CH_2Br$

b. $C_6H_6 + Br_2 \xrightarrow{\text{Fe}} C_6H_5Br + HBr$

c. $CH_3CO_2H + CH_3OH \rightarrow CH_3CO_2CH_3 + H_2O$

36.   a. $CH_3CH = CH_2 + HCl \rightarrow CH_3CHClCH_3$

b. $CH_3CH = CH_2 + H_2O \rightarrow CH_3CHCH_3$
                                                        |
                                                       OH

c.

d. $H_2C{=}C\begin{smallmatrix}CH_3\\[2pt]\\[2pt]CH_3\end{smallmatrix}$ $+ H_2O \longrightarrow$ $H_3C{-}\underset{\underset{CH_3}{|}}{\overset{\overset{CH_3}{|}}{C}}{-}O{-}H$

e. $+ H_2O \longrightarrow$

f. $CH_3C{\equiv}CH \xrightarrow{\;HCl\;} CH_3{-}\underset{\underset{Cl}{|}}{C}{=}CH_2 \xrightarrow{\;HCl\;} CH_3 CCl_2 CH_3$

37.   a.   b.

$CH_3{-}\overset{\overset{O}{\|}}{C}{-}H$ $CH_3{-}\overset{\overset{O}{\|}}{C}{-}CH_3$

c.   d.

No reaction occurs

e.

$H_3C{-}\underset{\underset{CH_3}{|}}{\overset{\overset{CH_3}{|}}{C}}{-}CH_2OH \xrightarrow{[Ox]} H_3C{-}\underset{\underset{CH_3}{|}}{\overset{\overset{CH_3}{|}}{C}}{-}\overset{\overset{O}{\|}}{CH}$

38.   a.

i. $CH_3{-}\overset{\overset{O}{\|}}{C}{-}OH$ ii.   iii. $(CH_3)_2CHC{-}OH$ with $\overset{O}{\|}$ above the C

b.

$$CH_3CH_2OH \xrightarrow{[Ox]} CH_3\overset{\overset{\displaystyle O}{\|}}{C}H + CH_3\overset{\overset{\displaystyle O}{\|}}{C}OH$$

$$(CH_3)_3C-CH_2OH \xrightarrow{[Ox]} (CH_3)_3C-\overset{\overset{\displaystyle O}{\|}}{C}H + (CH_3)_3C-\overset{\overset{\displaystyle O}{\|}}{C}-OH$$

39.   a. $CH_3CH=CH_2 + Br_2 \rightarrow CH_3CHBrCH_2Br$

b. $CH_3C\equiv CH + H_2 \xrightarrow{catalyst} CH_3CH=CH_2 + Br_2 \rightarrow CH_3CHBrCH_2Br$

c. $CH_3CO_2H + HOCH_2CH_2CH_2CH_3 \longrightarrow CH_3\overset{\overset{\displaystyle O}{\|}}{C}-O-CH_2CH_2CH_2CH_3 + H_2O$

ethanoic acid   butanol   butyl acetate or butylethanoate

d.

$$CH_3CH_2CH_2\overset{\overset{\displaystyle O}{\|}}{C}-OH + HO-CH_2CH_3 \longrightarrow CH_3CH_2CH_2\overset{\overset{\displaystyle O}{\|}}{C}-OCH_2CH_3 + H_2O$$

butanoic acid   ethanol   ethyl butyrate or ethylbutanoate

e.

$$CH_3-\overset{\overset{\displaystyle OH}{|}}{C}H-CH_3 \xrightarrow{[Ox]} CH_3-\overset{\overset{\displaystyle O}{\|}}{C}-CH_3$$

2-propanol

Polymers

40.   a.   Addition polymer: Polymer formed by adding monomer units to a double bond. Teflon, polyvinyl chloride and polyethylene are examples of addition polymers.

b.   Condensation polymer: Polymer that forms when two monomers combine, eliminating a small molecule. Nylon and dacron are examples of condensation polymers.

c.   Copolymer: Polymer formed from more than one type of monomer. Nylon and dacron are also copolymers.

41.   a.

$$H_2N-\hspace{-0.5em}\bigcirc\hspace{-0.5em}-NH_2 \quad \text{and} \quad HO_2C-\hspace{-0.5em}\bigcirc\hspace{-0.5em}-CO_2H$$

b. Repeating unit:

$$\left(\text{HN}-\!\!\bigcirc\!\!-\text{NHC}(=\!O)-\!\!\bigcirc\!\!-\text{C}(=\!O)\right)_n$$

The two polymers differ in the substitution pattern on the benzene rings. The Kevlar chain is straighter and there is more efficient hydrogen bonding between Kevlar chains than between Nomex chains.

42.

$$\left(-O-\overset{\overset{\textstyle CH_3}{|}}{CH}-\overset{\overset{\textstyle O}{\|}}{C}-O-\overset{\overset{\textstyle CH_3}{|}}{CH}-\overset{\overset{\textstyle O}{\|}}{C}-O-\overset{\overset{\textstyle CH_3}{|}}{C}-\overset{\overset{\textstyle O}{\|}}{C}-\right)_n$$

43.

$$H_2N-\!\!\bigcirc\!\!-NH_2 \quad \text{and} \quad \begin{matrix} HO_2C & & CO_2H \\ & \bigcirc & \\ HO_2C & & CO_2H \end{matrix}$$

44.

$$\left(\overset{\overset{\textstyle CN}{|}}{\underset{\underset{\textstyle C-OCH_3}{|}}{\underset{\textstyle \|}{\underset{\textstyle O}{}}}}C - CH_2 - \overset{\overset{\textstyle CN}{|}}{\underset{\underset{\textstyle C-OCH_3}{|}}{\underset{\textstyle \|}{\underset{\textstyle O}{}}}}C - CH_2\right)_n$$

45. a. The bond angles in the ring are about 60°. VSEPR predicts bond angles close to 109°. The bonding electrons are closer together than they prefer resulting is strong electron-electron repulsions. Thus, ethylene oxide is unstable (reactive).

b. $\left(O-CH_2CH_2-O-CH_2CH_2-O-CH_2CH_2\right)_n$

46.

a.

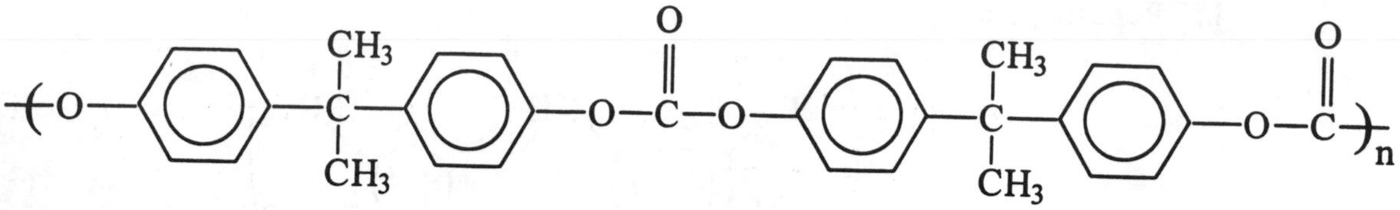

b. Condensation; HCl is eliminated when the polymer bonds form.

47. Divinylbenzene crosslinks different chains to each other. The chains cannot move past each other because of the crosslinks making the polymer more rigid.

48. The stronger interparticle forces would be found in polyvinyl chloride since there are also dipole-dipole forces in PVC that are not present in polyethylene.

49.   a.   repeating unit: $-(CHF-CH_2)_n$ monomer: $CHF=CH_2$

b.   repeating unit: monomer: $HO-CH_2CH_2-CO_2H$

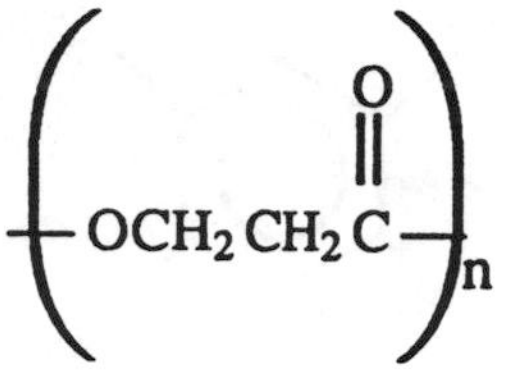

c.   repeating unit:

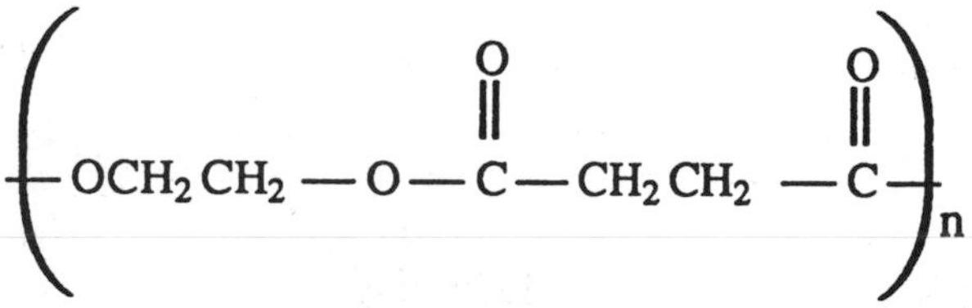

copolymer of: $HOCH_2CH_2OH$ and $HO_2CCH_2CH_2CO_2H$

d.   monomer:

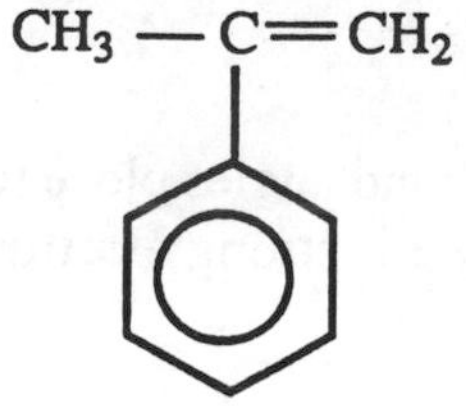

$CH_3-C=CH_2$

e.   monomer:

$CH=CHCH_3$

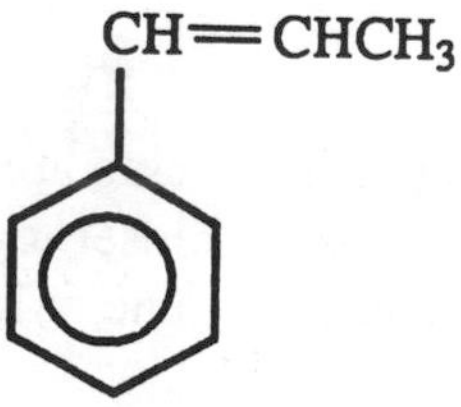

f.   monomer: $CClF=CF_2$

g. copolymer of:

$$HOCH_2 - \underset{}{\bigcirc} - CH_2OH \quad and \quad HO_2C - \underset{}{\bigcirc} - CO_2H$$

50.   Addition polymers: a, d, e, and f; Condensation polymers: b, c, and g; Copolymer: c and g

Additional Exercises

51.   a. 2-methyl-1,3-butadiene

        b.

cis-polyisoprene (natural rubber)

trans-polyisoprene (gutta percha)

52.   a.

b.

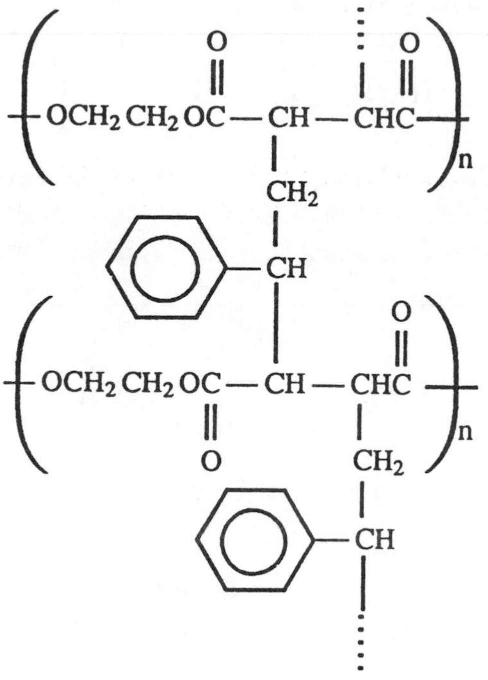

53.

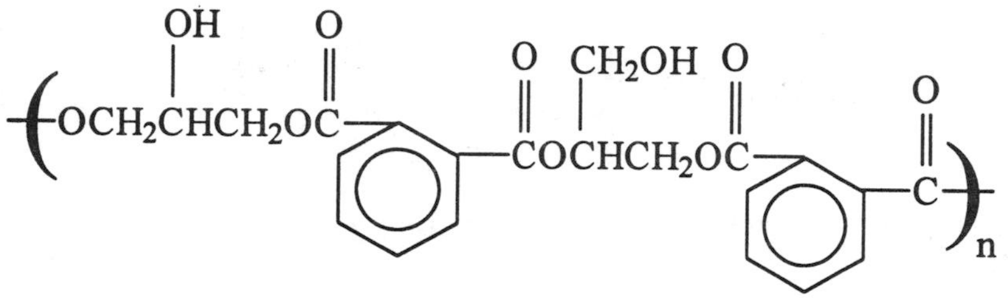

Two linkages a possible with glycerol. A possible repeating unit with both types of linkages is shown above. With either linkage, there are free OH groups on the polymer chains. These can react with the acid groups of phthalic acid to form crosslinks between various polymer chains.

54.   For the reaction: $3\ CH_2 = CH_2(g) + 3\ H - H(g) \rightarrow 3\ CH_3 - CH_3(g)$

Bonds broken: Bonds formed:

$3\ C = C$ (614 kJ/mol) $3\ C - C$ (347 kJ/mol)
$3\ H - H$ (432 kJ/mol) $6\ C - H$ (413 kJ/mol)

$\Delta H = 3(614) + 3(432) - [3(347) + 6(413)] = -381$ kJ

From enthalpies of formation: $\Delta H° = 3\ \Delta H_f^{\cdot}(C_2H_6) - 3\ \Delta H_f^{\cdot}(C_2H_4)$

$\Delta H° = 3$ mol(-84.7 kJ/mol) - 3 mol(52 kJ/mol) = -410. kJ

The two values agree fairly well.

For $C_6H_6(g) + 3 H_2(g) \rightarrow C_6H_{12}(g)$, we would get the same ΔH from bond energies as the first reaction since the same number and type of bonds are broken and formed. $\Delta H = -381$ kJ.

From enthalpies of formation: $\Delta H° = -90.3$ kJ $- (82.9$ kJ$) = -173.2$ kJ

There is about a 208 kJ discrepancy. Benzene is more stable by about 208 kJ/mol (lower in energy) than we expect from bond energies. This extra stability is evidence for resonance stabilization.

55.

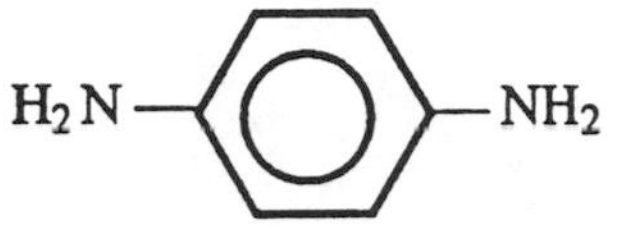

and

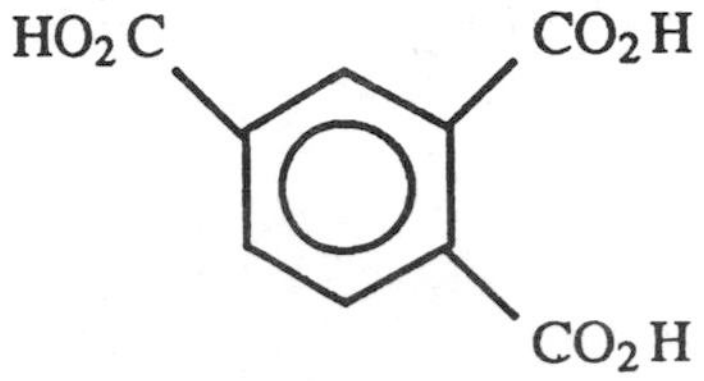

56.

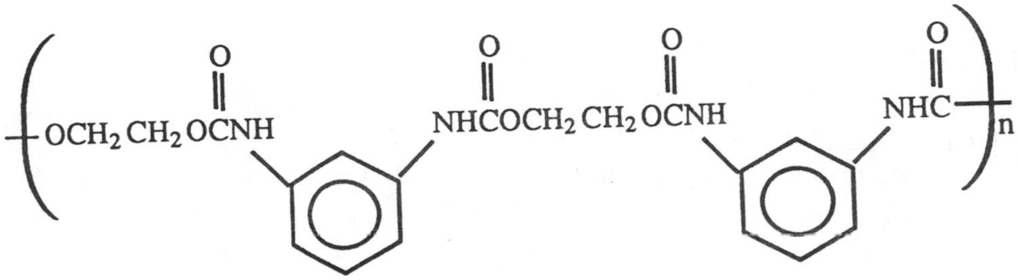

57.

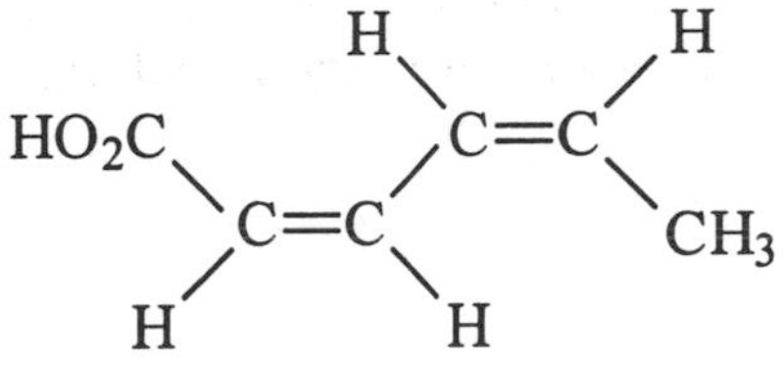

cis-2-cis-4-hexadienoic acid

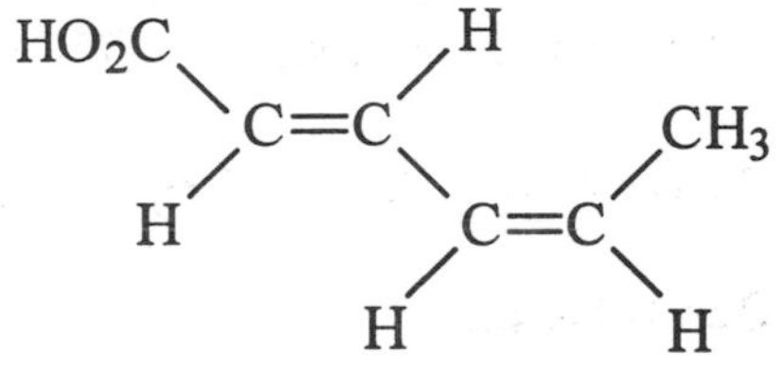

trans-2-cis-4-hexadienoic acid

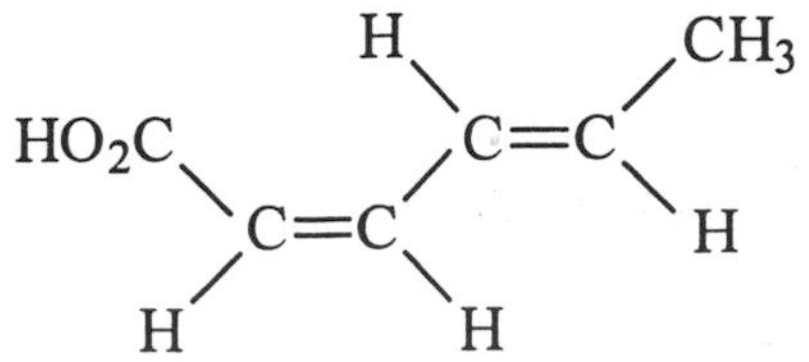

cis-2-trans-4-hexadienoic acid

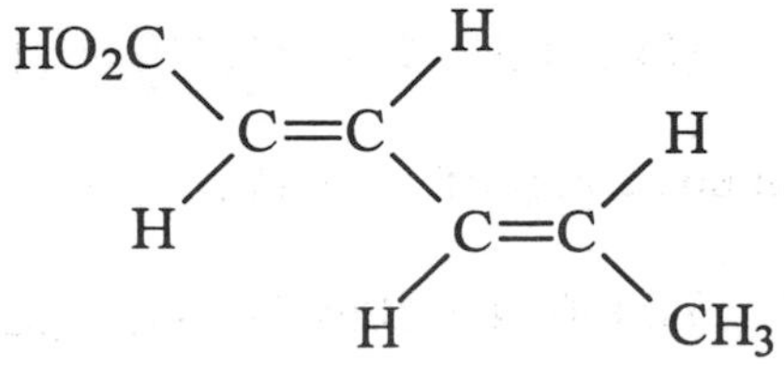

trans-2-trans-4-hexadienoic acid

58.   At low temperatures, the polymer is coiled into balls. The forces between poly(lauryl methacrylate) and oil molecules will be minimal and the effect on viscosity will be minimal. At higher temperatures, the chains of the polymer will unwind and become tangled with the oil molecules, increasing the viscosity. Thus, the presence of the polymer counteracts the temperature effect and the viscosity of the oil remains relatively constant.

CHAPTER TWENTY THREE

BIOCHEMISTRY

Proteins and Amino Acids

1.   a. $H_2NCH_2CO_2H \rightleftharpoons H^+ + H_2NCH_2CO_2^-$ $K_a = 4.3 \times 10^{-3}$

           $H_2NCH_2CO_2^- + H^+ \rightleftharpoons {}^+H_3NCH_2CO_2^-$ $K_2 = 1/K_a(amino) = K_b/K_w$

                                               $K_2 = 6.0 \times 10^{-3}/10^{-14} = 6.0 \times 10^{11}$

           $H_2NCH_2CO_2H \rightleftharpoons {}^+H_3NCH_2CO_2^-$ $K = K_aK_2 = 2.6 \times 10^9$

        Equilibrium lies far to the right since K >> 1.

   b. ${}^+H_3NCH_2CO_2H$, 1.0 M H$^+$; $H_2NCH_2CO_2^-$, 1.0 M OH$^-$

2.   a.   Aspartic acid and phenylalanine

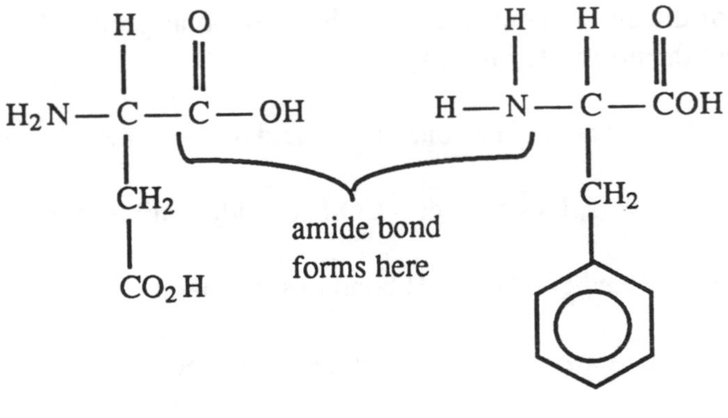

        amide bond forms here

   b.   Aspartame contains the methyl ester of phenylalanine. This ester can hydrolyze to form methanol:

        $R\text{–}CO_2CH_3 + H_2O \rightleftharpoons RCO_2H + CH_3OH$

3.   Crystalline amino acids exist as zwitterions, ${}^+H_3N\text{–}CH\text{–}CO_2^-$ with R substituent. The ionic interparticle forces are strong. Before the temperature gets high enough to break the ionic bonds, the amino acid decomposes.

4.   They are both hydrophilic amino acids because both contain highly polar R groups.

5.   Glutamic acid:

$$H_2N \text{—} CH \text{—} CO_2H$$
$$|$$
$$CH_2\,CH_2\,CO_2H$$

Monosodium glutamate: One of the acidic protons is lost.

$$H_2N \text{—} CH \text{—} CO_2H$$
$$|$$
$$CH_2\,CH_2\,CO_2^-\,Na^+$$

6.   Primary: The amino acid sequence in the protein.

Secondary: Includes structural features known as α-helix or pleated sheet. Both are maintained mostly through hydrogen bonding interactions.

Tertiary: The overall shape of a protein, long and narrow or globular. Maintained by hydrophobic and hydrophillic interactions, such as salt linkages, hydrogen bonds, disulfide linkages, and dispersion forces.

7.   Glutamic acid: R = -CH₂CH₂CO₂H; Valine: R = -CH-(CH₃)₂

A polar side chain of the amino acid is replaced by a nonpolar group. This could affect the tertiary structure and the ability to bind oxygen.

8.   a.   Ionic: Need - NH₂ on side chain of one amino acid and - CO₂H on side chain of the other.

- NH₂ on side chain = His, Lys or Arg; - CO₂H on side chain = Asp or Glu

b.   Hydrogen bonding: Need N–H or O–H bond in side chain.

- X - H · · · · · · · O = C (carbonyl group from peptide bond)

Ser	Asn	Any amino acid
Glu	Thr	
Tyr	Asp	
His	Gln	
Arg	Lys	

c.   Covalent: cys – cys

d.   London dispersion: all nonpolar amino acids

e.   dipole-dipole: Tyr, Thr, and Ser

9.   phenylalanine - isoleucine: London disperson forces; aspartic acid - lysine: ionic

10.

          ser - ala   ala - ser

11. Writing peptides from amino to carboxyl ends: ala-ala-gly, ala-gly-ala, gly-ala-ala
     Three peptides are possible.

12. For a dipeptide there are $5 \times 5 = 25$ different cases. For a tripeptide there are $5 \times 5 \times 5 = 125$
     different cases. So for 25 amino acids in the polypeptide, there are $5^{25} = 2.98 \times 10^{17}$ different
     polypeptides.

13. Both denaturation and inhibition reduce the catalytic activity of an enzyme. Denaturation
     changes the structure of an enzyme. Inhibition involves the attachment of an incorrect molecule
     at the active site, preventing the substrate from interacting with the enzyme.

14. The initial increase in rate is a result of the effect of temperature on the rate constant. At higher
     temperatures the enzyme begins to denature, losing its activity and the rate decreases.

15. All amino acids can act as both a weak acid and a weak base; this is the requirement for a buffer.

16. The secondary and tertiary structures are changed by denaturation.

17. a.   The new amino acid is most similar to methionine due to its $-CH_2CH_2SCH_3$ R group.

     b.

c.   The new amino acid replaces methionine. The structure is:

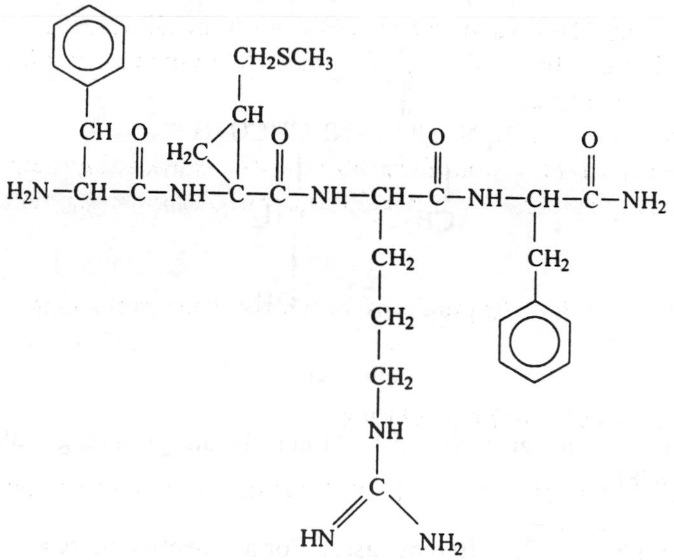

Carbohydrates

18.

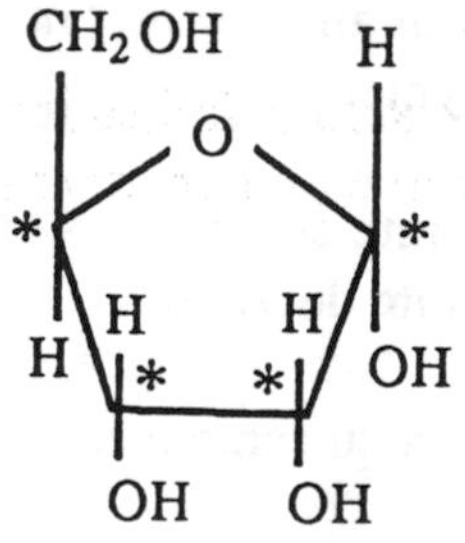

D-ribose

D-mannose

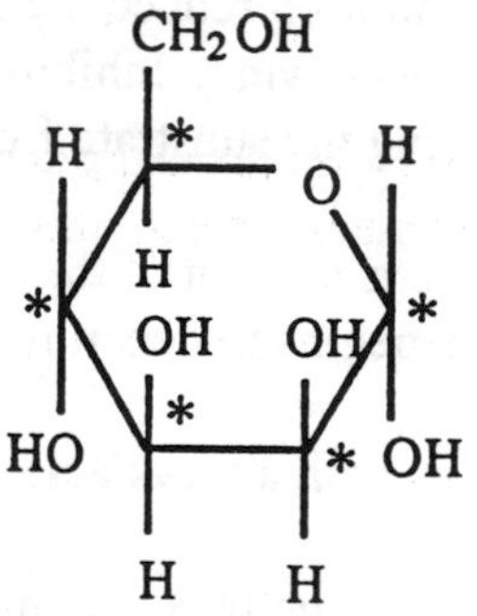

19.   Chiral carbons are marked with an asterisk in Exercise 23.18.

20.   a.

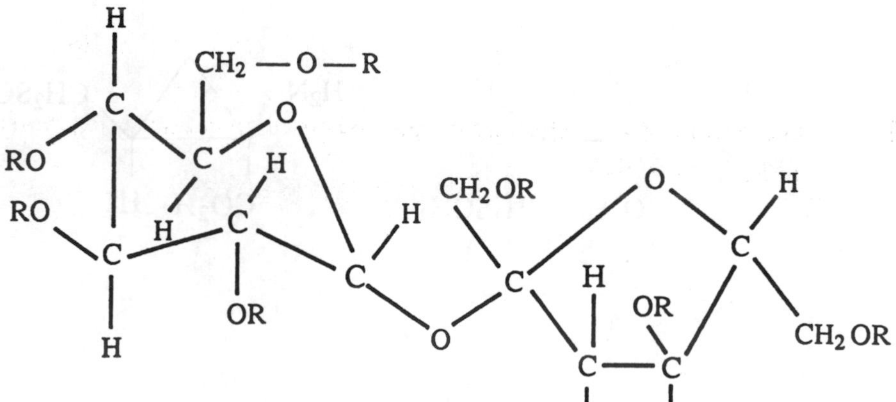

$$R = \!-\!C\!-\!(CH_2)_7\!-\!CH\!=\!CH\!-\!(CH_2)_7\!-\!CH_3$$

b.   Eight isomers; Any one of the eight -OH groups in sucrose is left unreacted.

c.   8 × 7/2 = 28 isomers; The first unreacted -OH has 8 possibilities. The second unreacted -OH has 7 possibilities. We divide by two because it doesn't matter which order we leave the two hydroxyls unreacted.

d.   The long chains prevent the sucrose from fitting in the active sites of enzymes that metabolize carbohydrates, and the fatty acid chains don't fit in enzymes that metabolize triglycerides.

21.   Hydrogen bonding between the -OH groups of starch and water molecules.

Optical Isomerism and Chiral Carbon Atoms

22.   Structural isomers: Same formula, different functional groups or chain lengths (different bonds).

Geometrical isomers: Same functional groups (same bonds), but different arrangement of some groups in space.

Optical isomers: Compounds that are nonsuperimposable mirror images of each other.

23.   A chiral carbon has four different groups attached to it. A compound with a chiral carbon is optically active. Isoleucine and threonine contain more than the one chiral carbon atom (see asterisks).

isoleucine   threonine

$$H_3C - \overset{H}{\underset{H}{\overset{|}{\underset{|}{C^*}}}} - CH_2CH_3$$
$$H_2N - C^* - CO_2H$$

$$H_3C - \overset{H}{\underset{H}{\overset{|}{\underset{|}{C^*}}}} - OH$$
$$H_2N - C^* - CO_2H$$

24.   There is no chiral carbon atom in glycine since it contains no carbon atoms with four different groups bonded to it.

25.

Each chiral carbon atoms (marked with *).

26.

27.

is optically active. The chiral carbon is marked with an asterisk.

28.   A chiral carbon has four different groups attached to it. The two chiral carbon atoms in α-pinene are marked with a *. Since it has chiral carbons, α-pinene is optically active.

29.   The chiral carbons are indicated with an asterisk.

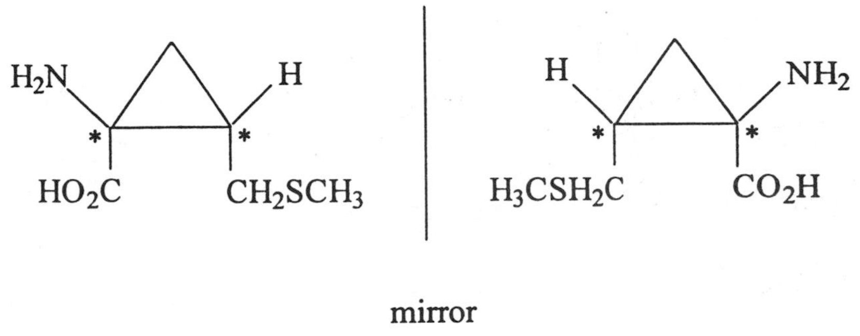

mirror

Nucleic Acids

30.   Nitrogen atoms with lone pairs of electrons.

31. 5×10^9 pairs $\times \dfrac{340 \times 10^{-12} \text{ m}}{\text{pair}} = 1.7$ m ≈ 2m

      1.7 m corresponds to 5' 7".

32.   DNA: Deoxyribose sugar; double stranded; A, T, G, and C are the major bases.

      RNA: Ribose sugar; single stranded; A, G, C, and U are the bases.

33.   T-A-C-G-C-C-G-T-A

34.   For each letter, there are 4 choices; A, T, G, or C. Hence, the total number of codons is
      $4 \times 4 \times 4 = 64$.

35.   Uracil will H-bond to adenine.

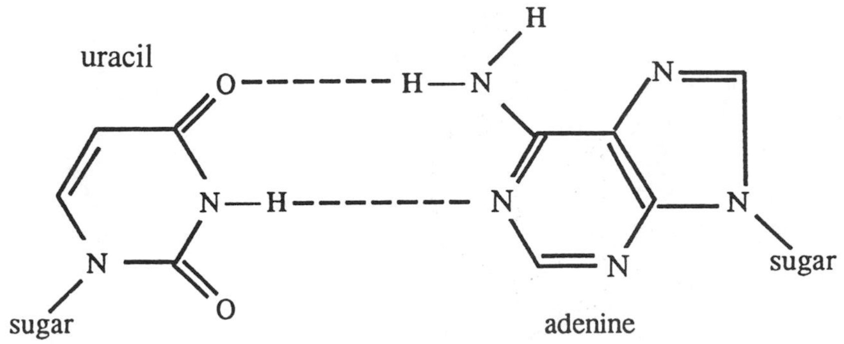

36.   Base pair: a.   Glu: CTT, CTC

       RNA   DNA   Val: CAA, CAG, CAT, CAC

       A . . . . T   Met: TAC

       G C   Trp: ACC

       C . . . . G   Phe: AAA, AAG

       U . . . . A   Asp: CTA, CTG

b.   DNA sequence for Met - Met - Phe - Asp - Trp:

        TAC - TAC - AAA - CTA - ACC
                 or   or
                AAG CTG

c.   Due to phe and asp, there are four possible different DNA sequences.

d.   C - T - T - A - C - C - A - A - A
        Glu -   Trp   -   Phe

e.   C - T - C - A - C - C - A - A - A

     C - T - T - A - C - C - A - A - G

     C - T - C - A - C - C - A - A - G

37.   In sickle cell anemia, glutamic acid is replaced by valine.

DNA codons: Glu: CTT, CTC; Val: CAA, CAG, CAT, CAC

Replacing a T with an A in the code for Glu will code for Val.

    CTC → CAC or CTT → CAT
    Glu   Val   Glu   Val

38.   A deletion may change the entire code for a protein. A substitution will change only one single amino acid in a protein.

39.   The Cl⁻ ions are lost upon binding to DNA. The dimension is just right for cisplatin to bond to two adjacent bases in one strand of the helix, which inhibits DNA synthesis.

40.   a.

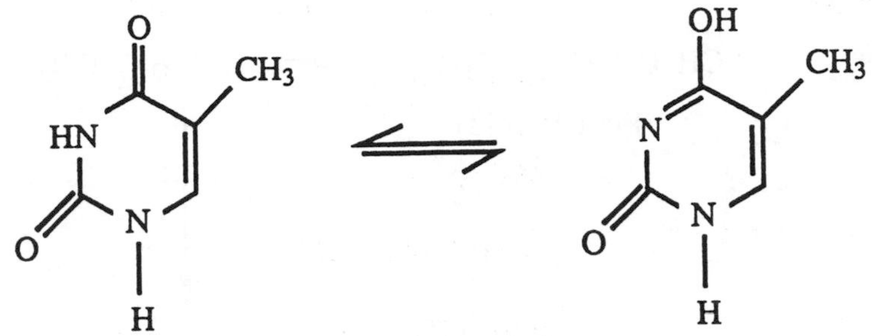

Bonds broken: Bonds formed:

   1 C$=$O (799 kJ/mol)   1 C$-$O (358 kJ/mol)
   1 N$-$C (305 kJ/mol)   1 N$=$C (615 kJ/mol)
   1 N$-$H (391 kJ/mol)   1 O$-$H (467 kJ/mol)

$\Delta H = 799 + 305 + 391 - (358 + 615 + 467) = +55$ kJ

Since ΔH is positive, the structure with two C$=$O bonds is more stable.

b.   The tautomer could hydrogen bond to guanine, forming a G–T base pair instead of A–T.

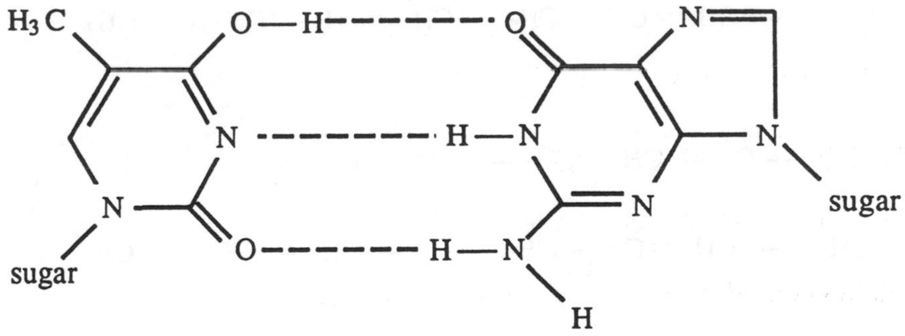

Lipids and Steroids

41.   Organic solvents are generally nonpolar solvents. Lipids are nonpolar and will be soluble in organic solvents. Carbohydrates contain several -OH groups capable of hydrogen bonding and are soluble in polar solvents (water).

42. CH₂ — OH
     |
     CH — OH   +   3 CH₃ CH₂ CH₂ CH₂ (CH₂ CH=CH)₂ (CH₂)₇ CO₂ H
     | (linoleic acid)
     CH₂ — OH

     (glycerol)

                                              O
                                              ‖
                          CH₂ — OC(CH₂)₇ (CH=CHCH₂)₂ CH₂ CH₂ CH₂ CH₃
                          |
             (triglyceride) CH — OC(CH₂)₇ (CH=CHCH₂)₂ CH₂ CH₂ CH₂ CH₃
                          | ‖
                          | O
                          CH₂ — OC(CH₂)₇ (CH=CHCH₂)₂ CH₂ CH₂ CH₂ CH₃
                                         ‖
                                         O

43.   O
                          ‖
     CH₂ — OC(CH₂)₇ — CH=CH — CH₂ — CH=CH — (CH₂)₄ — CH₃
     | O
     | ‖
     CH — OC(CH₂)₇ — CH=CH — CH₂ — CH=CH — (CH₂)₄ — CH₃
     |
     CH₂ — OC(CH₂)₇ — CH=CH — CH₂ — CH=CH — (CH₂)₄ — CH₃
                 ‖
                 O

It will take 6 mol of H₂ to completely hydrogenate this triglyceride.

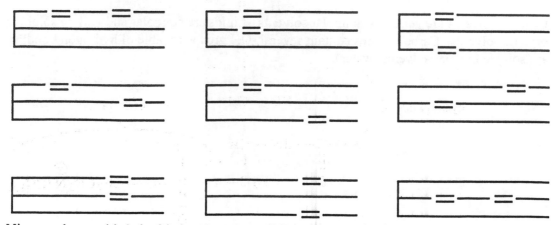

Nine products with 2 double bonds are possible as illustrated above.

44.   A polyunsaturated fat contains carbon-carbon double bonds.

45.

$$\underset{\overset{\|}{O}}{HOC} - (CH_2)_{11} - CH = CH - CH_2 - CH_3 \qquad \text{16 carbon omega-3 fatty acid}$$

$$\underset{\overset{\|}{O}}{HOC} - (CH_2)_{13} - CH = CH - CH_2 - CH_3 \qquad \text{18 carbon omega-3 fatty acid}$$

46.   Triglycerides (fats) are hydrolyzed to glycerol and fatty acids (soap) in the presence of base. This is how soap is produced. The products of the reaction (soap) feel slippery.

47.   The R groups are all hydrophobic, alkyl chains from fatty acids.

Lecithin:

Sphingomyelin:

48.   Cetyl palmitate: $CH_3(CH_2)_{14}COOCH_2(CH_2)_{14}CH_3$; Surfactants have the ability to suspend nonpolar materials in water by forming micelles (see Exercise 17.29). Cetyl palmitate is an ester and forms micelles in aqueous solution. The polar carbonyl groups point to the outside of the micelle and the nonpolar R groups point toward the inside of the micelle. Nonpolar dirt or grease molecules are entrapped in the micelle and are washed away.

49.

a)

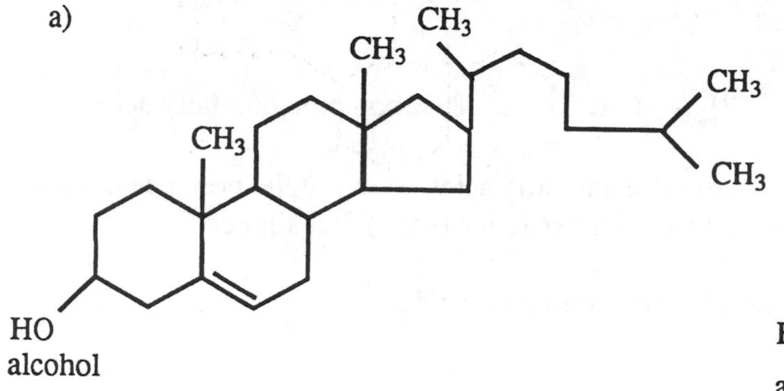

alcohol

b)

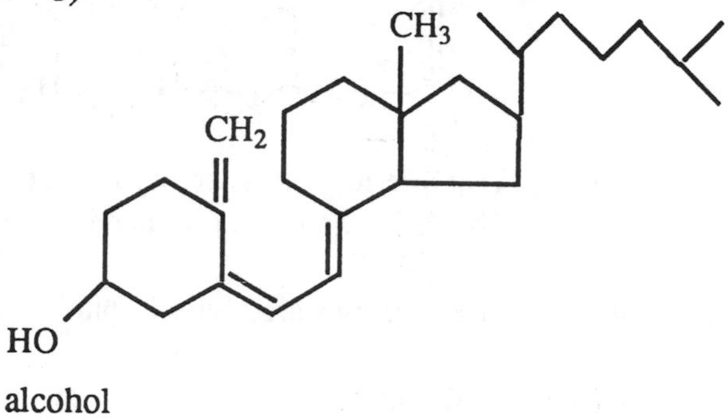

alcohol

c)

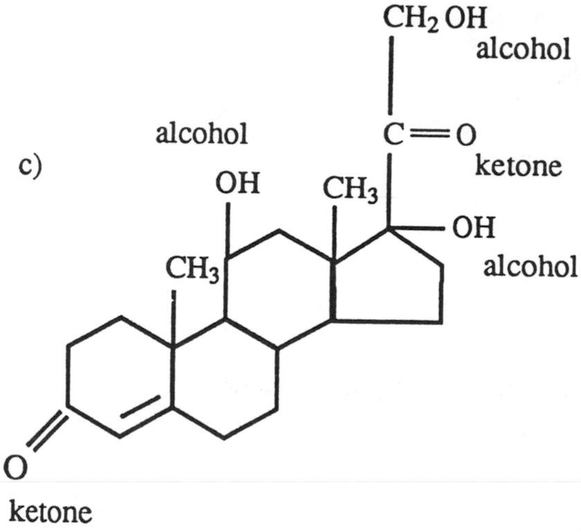

ketone

d)

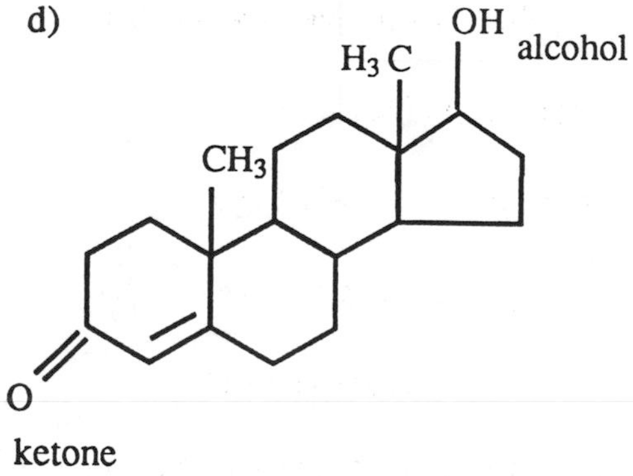

ketone

e)

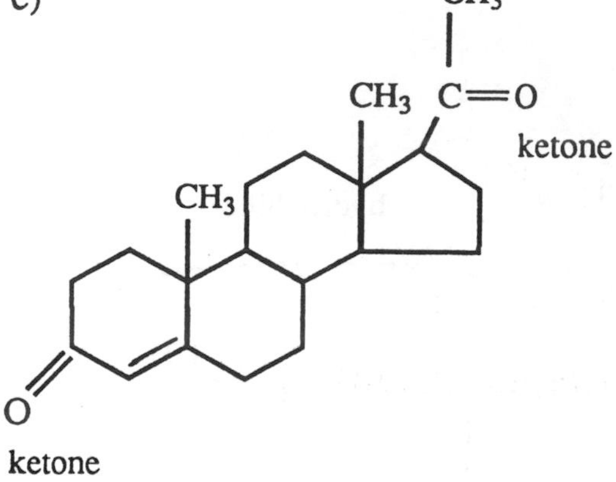

ketone

f)

alcohol

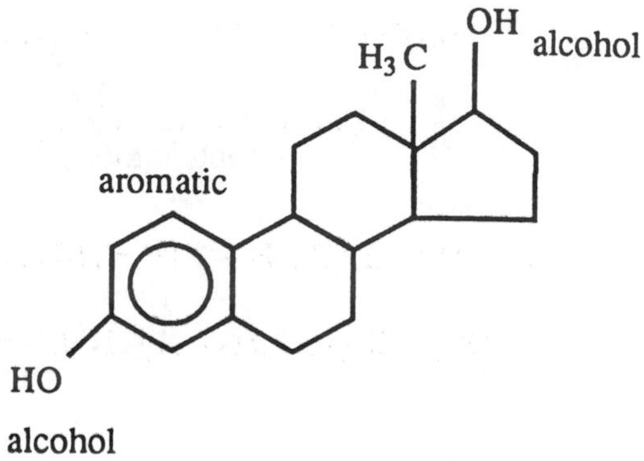

g)

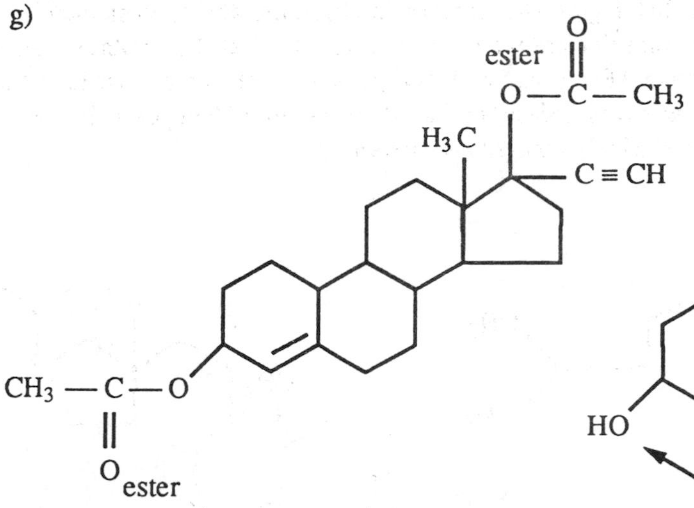

h)

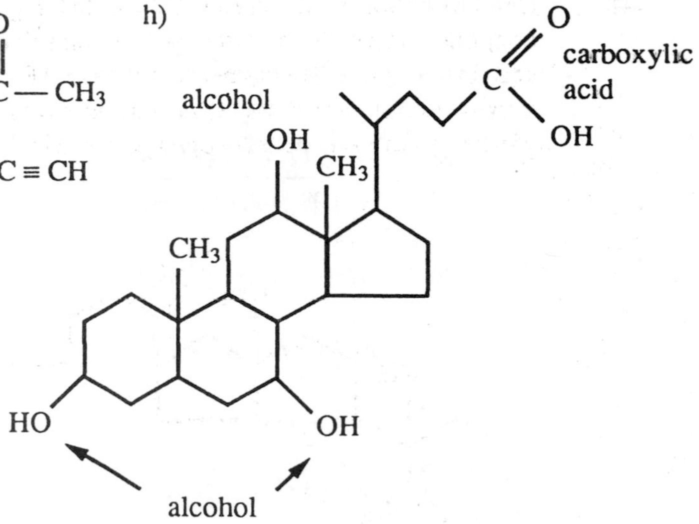

Additional Exercises

50.   From $-NH_2$ to $-CO_2H$ end:

phe-phe-gly-gly, gly-gly-phe-phe, gly-phe-phe-gly,

phe-gly-gly-phe, phe-gly-phe-gly, gly-phe-gly-phe

Six tetrapeptides are possible.

51.   From $-NH_2$ to $-CO_2H$ end:

gly-phe-ala, gly-ala-phe, phe-gly-ala, phe-ala-gly, ala-phe-gly, ala-gly-phe

Six tripeptides are possible.

52.   a. $^+H_3N-CH_2-CO_2H + H_2O \rightleftharpoons H_2CH_2CO_2H + H_3O^+$

$$K_{eq} = K_a\,(-NH_3^+) = \frac{K_w}{K_b\,(NH_2)} = \frac{1.0 \times 10^{-14}}{6.0 \times 10^{-3}} = 1.7 \times 10^{-12}$$

b. $H_2NCH_2CO_2^- + H_2O \rightleftharpoons H_2NCH_2CO_2H + OH^-$

$$K_{eq} = K_b\,(-CO_2^-) = \frac{K_w}{K_a\,(CO_2H)} = \frac{1.0 \times 10^{-14}}{4.3 \times 10^{-3}} = 2.3 \times 10^{-12}$$

c. $^+H_3NCH_2CO_2H \rightleftharpoons 2\,H^+ + H_2NCH_2CO_2^-$

$$K_{eq} = K_a(-CO_2H) \times K_a(-NH_3^+) = (4.3 \times 10^{-3})(1.7 \times 10^{-12}) = 7.3 \times 10^{-15}$$

53.   For the reaction:

$$^+H_3NCH_2CO_2H \rightleftharpoons 2\ H^+ + H_2NCH_2CO_2^- \qquad K_{eq} = 7.3 \times 10^{-15}$$

$$7.3 \times 10^{-15} = \frac{[H^+]^2[H_2NCH_2CO_2^-]}{[^+H_3NCH_2CO_2H]} = [H^+]^2, \ [H^+] = (7.3 \times 10^{-15})^{1/2}$$

$[H^+] = 8.5 \times 10^{-8}; \ \ pH = -\log [H^+] = 7.07 = $ isoelectric point

54.   From structures in *The Merck Index*:

a.   A - fat soluble

b.   E - fat soluble

c. K_5 - water soluble

d. K_6 - water soluble

55.   a.

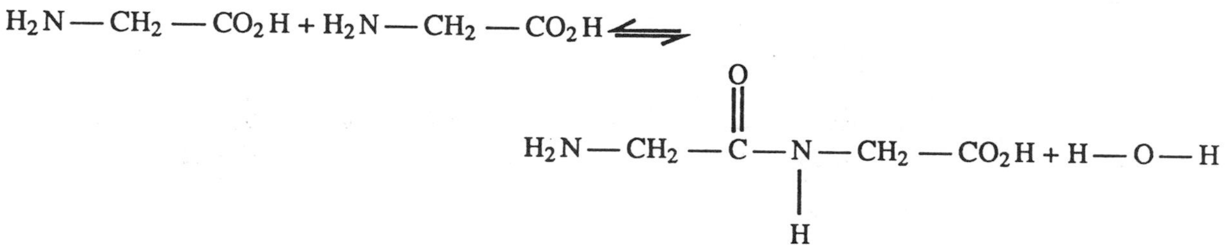

$$H_2N-CH_2-CO_2H + H_2N-CH_2-CO_2H \rightleftharpoons$$

$$H_2N-CH_2-\overset{\overset{\displaystyle O}{\|}}{C}-\underset{\underset{\displaystyle H}{|}}{N}-CH_2-CO_2H + H-O-H$$

Bonds broken: Bonds formed:

$\ \ \ \ $ 1 C – O (358 kJ/mol) $\ \ \ \ $ 1 C – N (305 kJ/mol)
$\ \ \ \ $ 1 H – N (391 kJ/mol) $\ \ \ \ $ 1 H – O (467 kJ/mol)

$\Delta H = 358 + 391 - (305 + 467) = -23$ kJ

b. ΔS for this process is negative (unfavorable) since order increases.

c. $\Delta G = \Delta H - T\Delta S$; ΔG is positive because of the unfavorable entropy change. The reaction is not spontaneous.

56. $\Delta G = \Delta H - T\Delta S$; For the reaction, we break a P–O and O–H bond and form a P–O and O–H bond. Thus, $\Delta H \approx 0$. $\Delta S < 0$, since 2 molecules are going to form one molecule (order increases). Thus, $\Delta G > 0$ and the reaction is not spontaneous.

57.   Both proteins and nucleic acids must form for life to exist. From the simple analysis, it looks as if life can't exist, an obviously incorrect assumption. A cell is not an isolated system. There is an external source of energy to drive the reaction. A photosynthetic plant uses sunlight and animals use the carbohydrates produced by plants as sources of energy. For a cell, $\Delta S_{sys} < 0$, but $\Delta S_{surr} > 0$ and $\Delta S_{surr} > |\Delta S_{sys}|$. Therefore, ΔS_{univ} increases (the second law of thermodynamics).

58.   a.   No; Since there is a plane of symmetry in the molecule, it is not optically active.

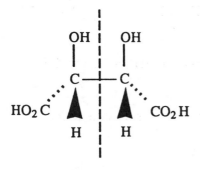

symmetry plane

b.   The optically active forms of tartaric acid have no plane of symmetry. The structures are:

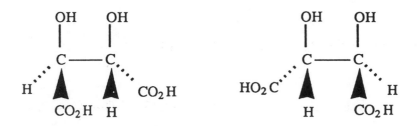

59.   The index numbers for the compounds are from *The Merck Index*, 11th Ed.

Ceclor, 1920;   Xanax, 310;   Claritin, 5455;

Elocon, 6151;   Wellbutrin, 1488;   Femstat, 1525;

Bonefos, 2369;   Aclovate, 213;   Fareston, 9474;

Halfan, 4508;   Sporanox, 5131;   Selepam, 8039;

Melox, 6091

All are pharmaceuticals with some use for human disease and all contain chlorine. It would probably not be wise to institute a total ban on all chlorine containing compounds since they have many important uses.